Textbook of
Engineering Physics

Textbook of Engineering Physics

As per Anna University Syllabus, Chennai

S.L. Kakani
M.Sc. (Physics), Ph.D.

Formerly
Principal, Govt. SMM PG College, Bhilwara, Rajasthan
Convenor, Board of Studies and Research Board,
MDS University, Ajmer, Rajasthan
Director, MGIMT, Chittorgarh, Rajasthan

Presently
Executive Director, Institute of Technology & Management,
Bhilwara, Rajasthan

Shubhra Kakani
M.Sc.
Physics Department, Sukhadia University, Udaipur,
Rajasthan

CBS PUBLISHERS & DISTRIBUTORS
NEW DELHI • BANGLORE • PUNE (INDIA)

ISBN : 978-81-239-1655-2

First Edition : 2008

Published by Satish Kumar Jain and produced by Vinod K. Jain for
CBS Publishers & Distributors,
4819/XI Prahlad Street, 24 Ansari Road, Daryaganj, New Delhi - 110 002, India.
e-mail: cbspubs@vsnl.com, cbspubs@airtelmail.in
Website: www.cbspd.com

Branches:

- Seema House 2975, 17th Cross, K.R. Road,
  Bansankari 2nd Stage, Bangalore - 560 070
  Fax: 080-26771680 • e-mail: cbsbng@dataone.in
- Shaan Bramha Complex, Basement, Appa Balwant Chowk,
  Budhwar Peth, Next to Ratan Talkies, Pune - 411 002
  Fax: 020-24464059 • e-mail: pune@cbspd.com

Printed at :
India Binding House, Noida (U.P.)

Preface

Physics has a key role to play in Engineering and Technology. Keeping this in view, this textbook is written and designed to fulfill the needs of the first year B.E/B.Tech/B.Architect students of various Univeristies. Its coverage is comprehensive and has been planned each to cover the prescribed syllabus of **Anna** and Roma after other Universities.

The present text is an attempt to present the subject matter in simple and systematic manner, covering latest syllabi of technical Universities/Boards. Subject matter has been arranged in such a manner that a student will have no difficulty in understanding the topics. Each chapter is followed by good number of typical worked out and exercise problems together with review questions, short question answers, and objective type questions. University question papers and suggested readings are provided at the end of the book. The book will also be useful for AMIE students.

The authors thank CBS publishers for their full cooperation to bring out the book in its present form. Suggestions for improvement will be gratefully acknowledged.

Bhilwara

S. L. Kakani
Shubhra Kakani

Contents

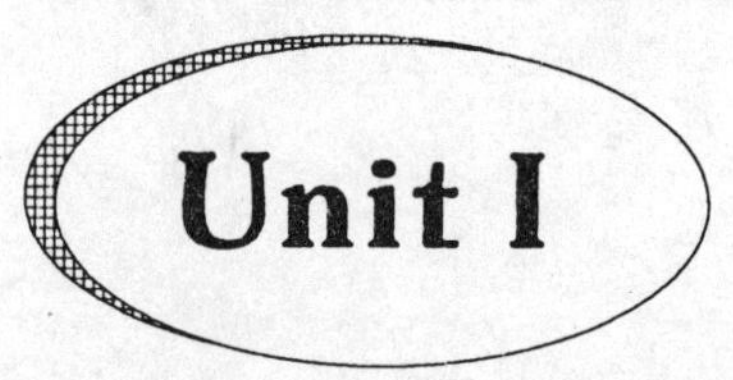

Electromagnetic Phenomena and Applications

Scalar and Vector Fields

1.1 FIELDS

A physical quantity can be expressed as a continuous function of the position of a region of space. If the cartesian coordinates of a point P are x, y, z and the physical quantity at this is A, then if

$$A = A(x, y, z,) \qquad (1)$$

This function is called a point function and the region in which it specifies the physical quantity is known as a '*field*'.

Fields are of two kinds: (*i*) *Scalar Field* and (*ii*) *Vector Field*, depending upon the nature of the physical quantity concerned.

1.2 SCALAR FIELDS

When a 'scalar' physical quantity is expressed from point to point in a region of space by a continuous point function.

$$\phi = \phi(x, y, z)$$

which gives the value of the quantity at each point, then the region is a '*scalar field*' and the function ϕ is a 'scalar point function'. Distribution of density, temperature, electric potential, gravitational pötential in space etc. are few familiar examples of scalar fields.

One can represent scalar quantities graphically by drawing such surfaces at which the value of the quantity remains a constant. We must note that one point in space corresponds to only one value of the quantity, *i.e.* scalar fields are single valued function at each point, *i.e.* two level surfaces cannot intersect (other wise ϕ should have to values at the point of intersection which contradicts our definition) (Fig. 1.1). These surfaces are called *equiscalar surfaces.* If we put a point charge at any place, then the electric potential around the charge will depend on the position of the point. Electric potential is a scalar quantity and hence the field around the charge is known as *scalar potential field*. If we join all such points at which the potential is constant, then such a surface is called a *equipotential surface*. Fig. 1.1 shows S_1 and S_2 as equipotential surfaces, with constant potential ϕ_1 and ϕ_2 respectively.

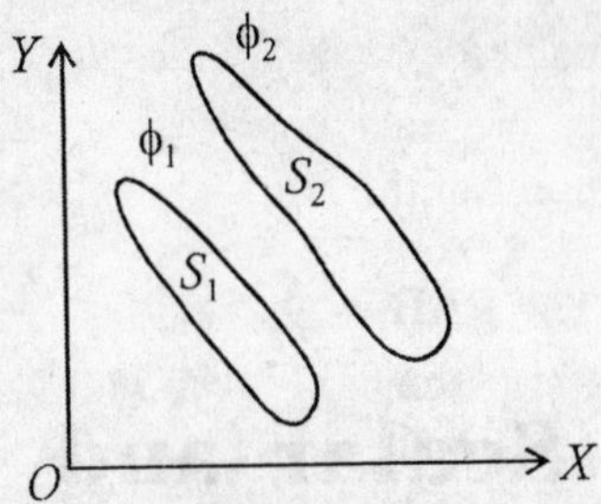

Fig. 1.1: Graphical representation of scalar fields

1.3 VECTOR FIELDS

When a vector physical quantity is expressed from point to point in a region of space by a continuous vector function $A(x, y, z)$ then the region is said to be a 'vector field' and the function A is called a 'vector point function'. At any given point in the field the function A is represented by a vector of definite magnitude and distribution of gravitational or magnetic field intensity in space are few examples of vector fields.

Vector fields are graphically represented by lines and these lines are called *field lines* or *flux lines*. These lines are drawn in the field in such a way that the tangent at any point of the line represent the direction of the vector field at that point. The magnitude of A at any point on a flux line is given by the number of flux lines crossing unit area perpendicular to their direction drawn at that point. Two flux lines cannot intersect (otherwise the directions of the vector would become indefinite at the point of intersection).

Let us consider that the number of field lines, or flux going out of an area element ds perpendicular to field lines at any point is $d\phi$ (Fig. 1.2).

Obviously, the magnitude of the intensity of the vector field will be

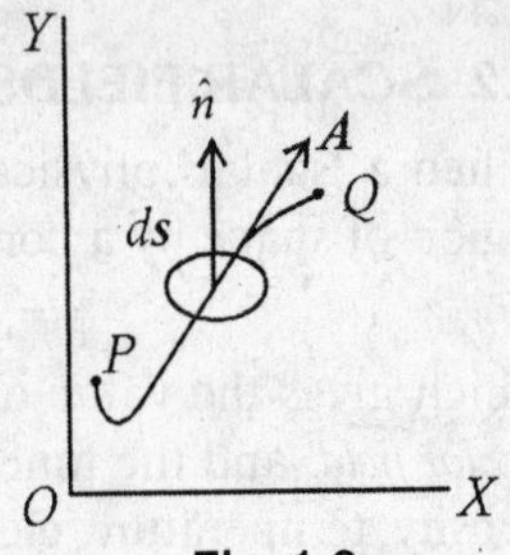

Fig. 1.2

$$|A| = \frac{d\phi}{ds} \tag{2}$$

$$\therefore \quad A = \left(\frac{d\phi}{ds}\right)\hat{n} \tag{3}$$

where $\hat{n}$ is unit vector perpendicular to the area ds. Thus

$$\hat{n}\, ds = ds$$

Now,

$$A \cdot \hat{n} = \frac{d\phi}{ds} \cdot \hat{n} \cdot \hat{n} = \frac{d\phi}{ds}$$

or $$d\phi = (\boldsymbol{A} \cdot \hat{n})ds = \boldsymbol{A} \cdot d\boldsymbol{s} \tag{4}$$

We must remember that point function $\phi(x, y, z)$ or $\boldsymbol{A}(x, y, z)$ at every point of the field is single valued.

1.4 PARTIAL DERIVATIVE

Let us consider that a physical quantity f is a positional function of the coordinates x, y and z of the point of observation, where x, y and z are independent variables, then

$$f = f(x, y, z) \tag{5}$$

Keeping y and z constant, let us change the value of x by an infinitesimal amount Δx and the value of f changes by an amount Δf, then for constant y and z the rate of change f with respect to x, $\left(\frac{\Delta f}{\Delta x}\right)$ is known as partial derivative of f with respect to x. Usually it is expressed as $\left(\frac{\partial f}{\partial x}\right)_{y,z}$ or simply $\frac{\partial f}{\partial x}$. Obviously, Δx is considered infinitely small, *i.e.* $\Delta x \to 0$ and hence

$$\frac{\partial f}{\partial x} = Lt_{\Delta x \to 0} \frac{f(x + \Delta x, y, z) - f(x, y, z)}{\Delta x} \tag{6}$$

Similarly if one keeps x and z constant and the value of y is changed by an infinitesimal amount Δy, then

$$\frac{\partial f}{\partial x} = Lt_{\Delta y \to 0} \frac{f(x, y, + \Delta y, z) - f(x, y, z)}{\Delta y} \tag{7}$$

and if one keeps x and y constant and the value of z is changed by an infinitesimal amount Δz, then

$$\frac{\partial f}{\partial z} = Lt_{\Delta z \to 0} \frac{f(x, y, z + \Delta z) - f(x, y, z)}{\Delta z} \tag{8}$$

Now, consider that all the three independent variables x, y, z are simultaneously changed by infinitesimal amounts, say δx, δy and δz respectively and the value of f is changed by an amount δf. When $\delta x \to 0$, $\delta y \to 0$ and $\delta z \to 0$, *i.e.* all the three vanish simultaneously $\delta f \to 0$, *i.e.* δf will also vanish. Obviously, in this state, the total change in the value of f, *i.e.* δf is known as total derivative. Thus

$$\delta f = \left(\frac{\partial f}{\partial x}\right)\delta x + \left(\frac{\partial f}{\partial y}\right)\delta y + \left(\frac{\partial f}{\partial z}\right)\delta z$$

One can write the total derivative for the limiting change as

$$df = \left(\frac{\partial f}{\partial x}\right)dx + \left(\frac{\partial f}{\partial y}\right)dy + \left(\frac{\partial f}{\partial z}\right)dz \tag{9}$$

The partial derivatives $\frac{\partial f}{\partial x}$, $\frac{\partial f}{\partial y}$ and $\frac{\partial f}{\partial z}$ are called *first partial derivatives* or partial derivatives of the first order. By differentiating these derivatives once more, one can obtain the *second partial derivatives* (or partial derivatives of the second order) e.g., $\frac{\partial^2 f}{\partial x^2}$, $\frac{\partial^2 f}{\partial y \partial x}$, $\frac{\partial^2 f}{\partial x \partial y}$, $\frac{\partial^2 f}{\partial y^2}$, $\frac{\partial^2 f}{\partial z^2}$ etc. Obviously, $\frac{\partial^2 f}{\partial x^2}$ is the partial derivative of $\frac{\partial f}{\partial x}$ with respect to *x*, $\frac{\partial^2 f}{\partial x \partial y}$ is the partial derivative of $\frac{\partial f}{\partial x}$ with respect to *y* etc. If all the derivatives concerned are continuous, then $\frac{\partial^2 f}{\partial x \partial y} = \frac{\partial^2 f}{\partial x \partial y}$, *i.e.*, then the order of differentiation is immaterial. By differentiating the second partial derivatives again with respect to *x*, *y* and *z* respectively, one obtains the third partial derivatives and so on.

If A is vector depending on more than one scalar variables such as cartesian coordinates *x*, *y*, *z* of a point in space, then $\frac{\partial A}{\partial x}$, $\frac{\partial A}{\partial y}$ and $\frac{\partial A}{\partial z}$ are partial derivatives of A with respect to *x* (when *y* and *z* remain constant), *y* (when *x* and *z* remain constant) and *z* (when *x* and *y* remain constant) respectively, if now *x*, *y* and *z* change simultaneously by small increments *dx*, *dy* and *dz*, then the total change or total derivative of A will be

$$dA = \frac{\partial A}{\partial x} dx + \frac{\partial A}{\partial y} dy + \frac{\partial A}{\partial z} dz$$

$$= \left(\frac{\partial}{\partial x} dx + \frac{\partial}{\partial y} dy + \frac{\partial}{\partial z} dz \right) A$$

$$= \left(\frac{\partial}{\partial x} \hat{i} + \frac{\partial}{\partial y} \hat{j} + \frac{\partial}{\partial z} \hat{k} \right) \cdot \left(dx\hat{i} + dy\hat{j} + dz\hat{k} \right) A \tag{10}$$

where $\hat{i}$, $\hat{j}$ and $\hat{k}$ are unit vectors in the directions of *x*, *y* and *z* axes respectively. $\frac{\partial}{\partial x} \hat{i} + \frac{\partial}{\partial y} \hat{j} + \frac{\partial}{\partial z} \hat{k}$ is defined as the *vector differential operator*, written as ∇ and read as '*del*'. Thus

$$\nabla = \frac{-\partial}{\partial x} \hat{i} + \frac{\partial}{\partial y} \hat{j} + \frac{\partial}{\partial z} + \hat{k} \tag{11}$$

Further if r is the position vector of the point (*x*, *y*, *z*) with respect to origin, then

$$r = \hat{i}x + \hat{j}y + \hat{k}z \tag{12}$$

so that

$$d\boldsymbol{r} = \hat{i}dx + \hat{j}dy + \hat{k}dz \tag{13}$$

The Eq. (10) can be expressed as

$$dA = (\nabla \cdot d\boldsymbol{r})A \tag{14}$$

Here ∇ is a vector quantity possessing all the properties of an ordinary vector. Since it is an '*operator*', its magnitude has no physical significance.

The differential operator $\nabla \cdot \nabla = \nabla^2$ is defined as

$$\nabla^2 = \frac{\partial^2}{\partial x^2} + \frac{\partial^2}{\partial y^2} + \frac{\partial}{\partial z^2} \tag{15}$$

∇^2 is a scalar and is known as '*Laplacian Operator.*'

1.5 GRADIENT OF A SCALAR FUNCTION

A scalar field can be mapped out by a series of level surfaces. Let us consider that in a scalar field two surfaces S_1 and S_2 represent surfaces with constant scalar quantities ϕ and $\phi + \delta\phi$ respectively (Fig. 1.3). $\boldsymbol{r}$ is the position vector of any point P on S, relative to the origin O.

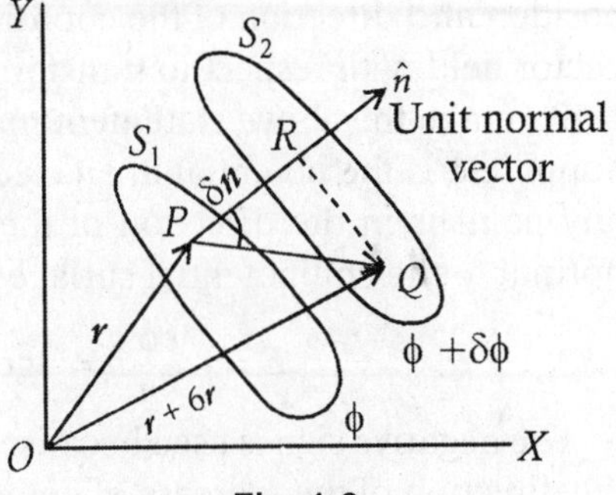

Fig. 1.3

The position vector of another point Q on S_2 is $\boldsymbol{r} + \delta\boldsymbol{r}$. Obviously, the displacement of Q with respect to P is $\boldsymbol{PQ} = \delta\boldsymbol{r}$. The rate of change of scalar function in the direction of $\boldsymbol{PQ}$ is $\frac{\delta\phi}{\delta r}$. Now, draw a perpendicular PR from P on surface S_2.

The minimum distance between the surfaces S_1 and S_2, *i.e.* $PR = \delta n$ will be in the direction of their normal. Therefore, the rate of change of scalar function ϕ in the direction of $PR = \frac{\delta\phi}{\delta n}$. If the distance δr is infinitesimal then one can consider ΔPQR, a right angled triangle. If Q is the angle between $\delta\boldsymbol{r}$ and $\delta\boldsymbol{n}$, then

$$PR = PQ\cos\theta \text{ or } \qquad \delta n = \delta r\cos\theta$$

$$\therefore \qquad PQ = \frac{\delta\phi}{\delta n}\cos\theta \tag{16}$$

obviously, Eq. (16) represents the rate of change of ϕ in the direction of PQ. Since the maximum value of cos θ is 1 and hence the maximum rate of change of

$$\phi = \left(\frac{\delta\phi}{\delta r}\right)_{\max} = \left(\frac{\delta\phi}{\delta n}\right)$$

$$\therefore \qquad Lt\delta r \to 0\left(\frac{\delta\phi}{\delta r}\right)_{\max} = \frac{d\phi}{dn}$$

obviously, the rate of change of a scalar function is maximum in a direction perpendicular to the equiscalar surface.

Thus in a *scalar field, the maximum rate of change of the scalar function is known as the gradient of the scalar field.*

Gradient is a vector quantity and its direction is perpendicular to the equiscalar surface, *i.e.* the direction in which rate of change is maximum. If $\hat{n}$ is the unit vector normal to the equiscalar surface, then

$$\text{grad}\phi = \left(\frac{d\phi}{dn}\right)\hat{n} = \left(\frac{d\phi}{dr}\right)_{\max}\hat{n} \tag{17}$$

obviously, the gradient of a scalar field ϕ depends only (*i*) on the way the values of ϕ are distributed in space and (*ii*) entirely independent of the position and direction of the coordinate axes. Thus grad ϕ is an invariant vector field with respect to transformation of coordinates.

To make the above statement more clear, let us consider an example. Suppose ϕ is the potential in an electric field. The intensity of the field at any point is in the direction of the greatest rate of fall of potential, *i.e.* normal to the equipotential surface and is equal to the rate, *i.e.*

$$\boldsymbol{E} = -\frac{\partial\phi}{\partial n}\hat{n} = -\text{grad}\phi \tag{18}$$

The negative sign is used because the direction of the field is opposite to the direction of the increase of potential.

The Gradient of a Scalar Field in Rectangular Cartesian Coordinates

Suppose the scalar field ϕ is a function of (x, y, z), then with the use of partial differentiation, one can write

$$d\phi = \frac{\partial\phi}{\partial x}\cdot dx + \frac{\partial\phi}{\partial y}\cdot dy + \frac{\partial\phi}{\partial z}\cdot dz \tag{19}$$

Now, we have the gradient of ϕ is a vector given by

$$\nabla\phi = \frac{\partial\phi}{\partial n}\hat{n}$$

where $\hat{n}$ is a unit vector. Let us take its scalar product with an element of radius vector dr, one obtains

$$(\text{grad}\phi)\cdot \boldsymbol{dr} = \frac{\partial\phi}{\partial n}\hat{n}\cdot \boldsymbol{dr}$$

$$= \frac{\partial\phi}{\partial n}dr\cos\theta \qquad (\because \hat{n}\cdot \boldsymbol{dr} = dr\cos\theta)$$

$$= \frac{\partial\phi}{\partial n} \cdot dn$$

$$\left(\because \frac{\partial\phi}{\partial n} \text{ is the rate of change of } \theta \text{ in the direction } \hat{n}\right)$$

$$= d\phi$$

$$= \left(\frac{\partial\phi}{\partial x}\right)dx + \left(\frac{\partial\phi}{\partial y}\right)dy + \left(\frac{\partial\phi}{\partial z}\right)dz$$

$$= \left(\frac{\partial\phi}{\partial x}\hat{i} + \frac{\partial\phi}{\partial y}\hat{j} + \frac{\partial\phi}{\partial z}\hat{k}\right) \cdot (dx\hat{i} + dy\hat{j} + dz\hat{k})$$

$$= \left(\frac{\partial\phi}{\partial x}\hat{i} + \frac{\partial\phi}{\partial y}\hat{j} + \frac{\partial\phi}{\partial z}\hat{k}\right) \cdot dr$$

$$\therefore \qquad \nabla\phi = \frac{\partial\phi}{\partial x}\hat{i} + \frac{\partial\phi}{\partial y}\hat{j} + \frac{\partial\phi}{\partial z}\hat{k} = \text{grad}\phi \qquad (20)$$

Obviously, $\nabla\phi$ is such a vector which represents the maximum rate of change of scalar function ϕ.

It is worthwhile to note that the gradient of a scalar field is always a vector field, however, its converse is always not true. All vector fields cannot be expressed as gradient of scalar fields.

A vector field which can be expressed as gradient of a scalar field is called a *lamellar* or *non-curl field*.

A pure electrostatic field is lamellar vector field, because it can be expressed as (negative) gradient of a scalar potential field. If V is the electric potential, then one can write the electric field as

$$\boldsymbol{E} = -\nabla V \qquad (21)$$

A scalar field $\phi(x, y, z)$ evaluated at a particular point is independent of the coordinates of that point. For example, the temperature at a point is not dependent on whether co-ordinates (x, y, z) or (x', y', z') are used. This means a scalar field is invariant with respect to the transformation of coordinates. Its gradient which is a vector field, must therefore be also invariant with respect to the transformation.

1.6 LINE INTEGRAL OF VECTOR FIELD

Let us consider two points P and Q in a vector field. One can divide the path between P and Q into displacement elements of lengths dr_1, dr_2, dr_3, ..., etc. (not shown in Fig. 1.4) respectively. Since these are very small and one can consider them as straight lines, dr is one of such element shown in Fig. 1.4. Let A be

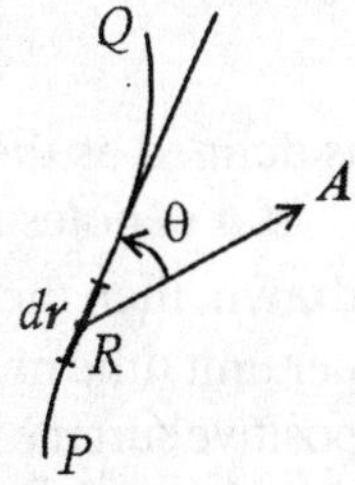

Fig. 1.4

the vector at R in a direction making an angle ϕ with dr. If A varies in magnitude and direction from point to point along the curve PQ, then the integral

$$\int_B^A \boldsymbol{A} \cdot d\boldsymbol{r} = \int_P^Q (A\cos\theta)dr = \int_P^Q \left(A_x dx + A_y dy + A_z dz\right) \quad (22)$$

is defined as the line integral of vector $\boldsymbol{A}$ along the curve AB. Few familiar examples of line integral are:

(*i*) If $\boldsymbol{A}$ denotes a force acting on a particle which moves along the curve PQ then the integral $\int_P^Q \boldsymbol{A} \cdot d\boldsymbol{r}$ represents the work done by the force in moving from P to Q.

(*ii*) If $\boldsymbol{A}$ represents the electric field intensity at any point, then the line integral represents the potential difference between P and Q.

(*iii*) If $\boldsymbol{A}$ represents the velocity at any point in a fluid then the line integral of $\boldsymbol{A}$ round a closed path is known as circulation of the fluid.

If the value of the line integral depends only upon the coordinates of the two points in the vector field and not upon the actual path taken between them, then the vector field $\boldsymbol{A}$ (x, y, z) is called a 'conservative' or 'noncurl' or 'lamellar field.'

1.7 SURFACE INTEGRAL

Let S be a surface drawn in a vector field, and ds an infinite small element of the surface (Fig. 1.5). Let $\hat{n}$ be unit positive vector normal to ds. The surface element of area ds is represented by a vector $\boldsymbol{ds}$ whose magnitude is $|ds|$ and direction is that of $\hat{n}$. Thus

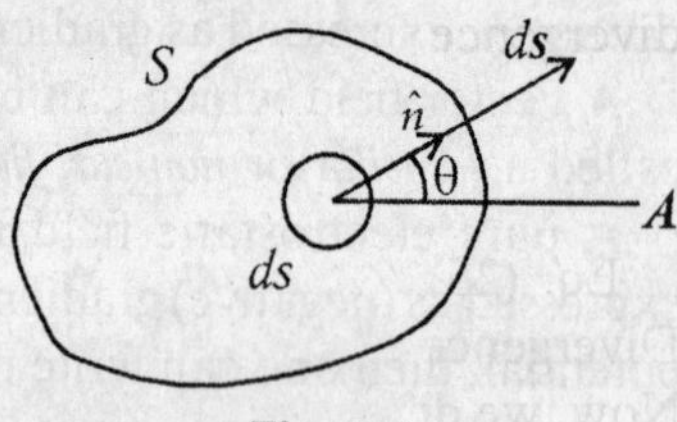

Fig. 1.5

$$\boldsymbol{ds} = \hat{n}\, ds$$

$\boldsymbol{ds}$ is called the **area vector**.

Now, if $\boldsymbol{A}$ be a vector at the middle of the element ds in a direction making an angle θ with $\hat{n}$, then the integral

$$\iint \boldsymbol{A} \cdot \boldsymbol{ds} \text{ or } \int_S \boldsymbol{A} \cdot \hat{n} ds = \int_S (A\cos\theta) ds \quad (23)$$

is defined as the 'surface integral' or 'Flux' of $\boldsymbol{A}$ across the surface S.

If $\boldsymbol{A}$ denotes the velocity of a moving fluid in which a fixed surface S is drawn, then the surface integral $\int_S \boldsymbol{A} \cdot \boldsymbol{ds}$ gives the amount of fluid flowing per unit time normally through the surface S. In case of a closed surface, a positive surface integral means that the fluid is flowing outwards, if negative then inwards.

Other examples of surface integrals are $\int_S A \times ds$ and $\int_S \phi ds$ where ϕ is a scalar function.

1.8 VOLUME INTEGRAL

Let $dV = dxdydz$ denote the element of volume V. Volume element dV is a scalar. If A be a vector function inside it then the integral $\int_V AdV$ is called the volume integral $\iiint_V AdV$ (we will write it as $\int_V AdV$) of A over the volume V, *i.e.*

$$\int_V AdV = \int_V \left(A_x\hat{i} + A_y\hat{j} + A_z\hat{k}\right)dxdydz \tag{24}$$

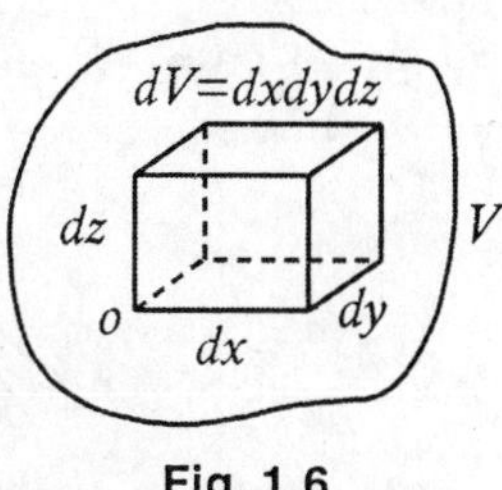

Fig. 1.6

In case of scalar function ϕ, the integral $\int_V \phi dV$ represents the volume integral over V.

1.9 DIVERGENCE OF A VECTOR FIELD

The divergence of a vector field at any point is defined as the amount of flux per unit volume diverging from that point. Since the divergence is the amount of flux, it is essentially a scalar. One can express the divergence of a vector field $A(x, y, x)$ as

$$\nabla \cdot A = \frac{\partial A_x}{\partial x} + \frac{\partial A_y}{\partial y} + \frac{\partial A_z}{\partial z} = \text{div}A \tag{25}$$

Eq. (25) is the expression for divergence in cartesian coordinates. Divergence of a vector field has important applications in hydrodynamics. Now, we derive the expression for divergence in cartesian coordinates.

Let us consider a small rectangular parallelopiped whose centre is $C(x, y, z)$ and whose sides have lengths dx, dy and dz parallel to the coordinate axes x, y and z respectively as shown in Fig. 1.7.

Suppose the vector A has components A_x, A_y and A_z along x, y and z axes respectively.

The value of the x component of A at the centre of the face $EFGH$ is

$$A_x - \frac{1}{2}\frac{\partial A_x}{\partial x}dx \tag{26}$$

and that at the centre of the face $PQRS$ is

$$A_x + \frac{1}{2}\frac{\partial A_x}{\partial x}dx \tag{27}$$

Now, the volume of the fluid flowing per unit time through a face is given by the product of the area of the face and the normal component of the velocity upon it. We call it the 'flux' through the face. The area of each of faces $EFGH$ and $PQRS$ is $dydz$, one finds.

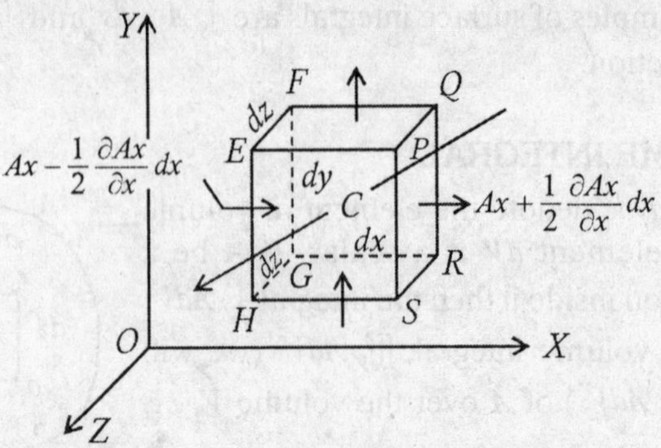

Fig. 1.7

$$\text{Flux entering the face } EFGH = \left(A_x - \frac{1}{2}\frac{\partial A_x}{\partial x}dx\right)dydz \tag{28}$$

and

$$\text{Flux leaving the face } PQRS = \left(A_x + \frac{1}{2}\frac{\partial A_x}{\partial x}dx\right)dydz \tag{29}$$

obviously, the excess of the flux leaving the parallelopiped over that entering it in the X-direction is

$$\left(A_x + \frac{1}{2}\frac{\partial A_x}{\partial x}dx\right)dydz - \left(A_x - \frac{1}{2}\frac{\partial A_x}{\partial x}dx\right)dydz = \frac{\partial A_x}{\partial x}dxdydz \tag{30}$$

Similarly, one obtains the net flux leaving the parallelopiped in the Y and Z directions are $\frac{\partial A_y}{\partial y}dxdydz$ and $\frac{\partial A_z}{\partial z}dxdydz$ respectively. Therefore to total net flux diverging from (*i.e.* leaving the) parallelopiped is

$$\left(\frac{\partial A_x}{\partial x} + \frac{\partial A_y}{\partial y} + \frac{\partial A_z}{\partial z}\right)dxdydz \tag{31}$$

Since $dx\,dy\,dz$ is the volume of the elementary parallelopiped and hence the amount of flux diverging per unit volume, which is defined as divergence of A is:

$$\nabla \cdot A = \text{div}A = \frac{\partial A_x}{\partial x} + \frac{\partial A_y}{\partial y} + \frac{\partial A_z}{\partial z} \tag{32}$$

1.10 PHYSICAL MEANING OF DIVERGENCE OF A VECTOR ($\nabla \cdot \mathbf{A}$)

$\nabla \cdot A$ at a point gives the amount of flux diverging per unit volume from that point. If A, denotes the velocity of the moving fluid, then $\nabla \cdot A$ tells the rate at which the fluid is diverging from the point per unit volume.

If the divergence of a vector is positive at a point is a fluid, then either the fluid is expanding and its density at that point is falling with time or the point as a source of fluid. In case, if the divergence of a vector is negative, then the fluid is contracting and its density is rising or the point is a negative source *i.e. sink.*

If the flux, entering any element of field space is exactly balanced by that leaving it, the quantity $\nabla \cdot \boldsymbol{A} = 0$. Obviously, there is no source or sink, nor its density is changing, *i.e.* the fluid is incompressible. If the flux entering and leaving an element are equal, the lines of flow of vector $\boldsymbol{A}$ should form either closed curves (e.g, in case of magnetic field due to a current) or extend to infinity. A vector which satisfies this condition is called *Solenoidal.*

In the case of non-material fluxes (e.g. magnetic or electric flux) the existence of divergence means the presence of source or sink of flux at that point.

1.11 GAUSS'S DIVERGENCE THEOREM

This theorem states that the volume integral of the divergence of a vector field $\boldsymbol{A}$ taken over any volume V is equal to the surface integral of A taken over the closed surface surrounding the volume, *i.e.*

$$\iiint_v (\nabla \cdot \boldsymbol{A}) dV = \iint_s \boldsymbol{A} \cdot ds \tag{33}$$

Proof. In Cartesian coordinates

$$\nabla \cdot \boldsymbol{A} = \left(\hat{i}\frac{\partial}{\partial A} + \hat{j}\frac{\partial}{\partial y} + \hat{k}\frac{\partial}{\partial z}\right) \cdot \left(A_x\hat{i} + A_y\hat{j} + A_z\hat{k}\right)$$

$$= \frac{\partial A_x}{\partial x} + \frac{\partial A_y}{\partial y} + \frac{\partial A_z}{\partial z}$$

and $\qquad dv = dx\,dy\,dz$

Also

$$\boldsymbol{A} \cdot ds = \left(A_x\hat{i} + A_y\hat{j} + A_z\hat{k}\right) \cdot \left(dS_x\hat{i} + dS_y\hat{j} + dS_z\hat{k}\right)$$

$$= A_x dS_x + A_y dS_y + A_z dS_z$$

$$= A_x dydz + A_y dxdz + A_z dxdy$$

Using the above results, one can express Gauss's theorem in cartesian coordinates as

$$\iiint_v \left(\frac{\partial A_x}{\partial x} + \frac{\partial A_y}{\partial y} + \frac{\partial A_z}{\partial z}\right) dxdydz = \iint_s \left(A_x dS_x + A_y dS_y + A_z dS_z\right) \tag{34}$$

Let us consider the first part of the L.H.S. integral of Eq. (34). As shown in Fig. 1.8, integrate it with respect to x along a stripe of cross section $dydz$ from $P_1(x_1, y, z)$ to $P_2(x_2, y, z)$, *i.e.*

$$\iiint_v \frac{\partial A_x}{\partial x} dxdydz = \iint_s \left| A_x \right|_{P_1}^{P_2} dydz$$

$$= \iint_s \left[A_x(P_2) - A_x(P_1) \right] dydz$$

$$= \iint_s \left[A_x(x_2, y, z) - A_x(x_1, y, z) \right] dydz$$

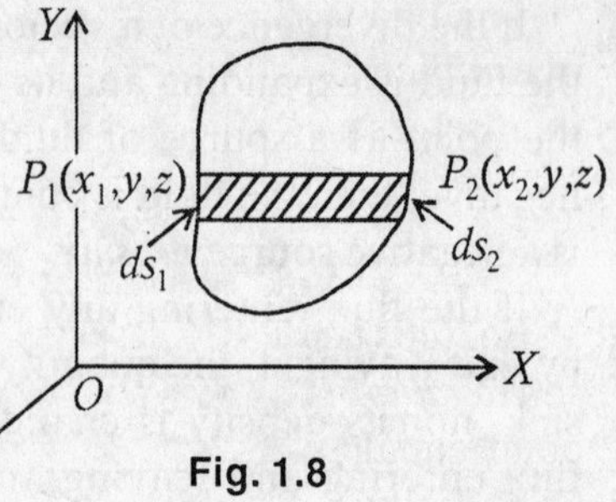

Evidently, at P_1, $A_x dydz = -A_x dS_x$ and at P_2,

$$A_x dydz = A_x dS_x$$

Fig. 1.8

($\because$ at P_1 and P_2 the x components of area have opposite directions)

$$\therefore \quad \iiint_v \frac{\partial A_x}{\partial x} dxdydz = \iint_s A_x dS_x \tag{35}$$

We should not confuse that we have written for

$$\iiint_v A_x(P_2) - A_x(P_1) dydz = \iint_s (A_x dS_x + A_x dS_x)$$

only $\iint_s A_x dS_x$ because if we take all the stripes then each integral $\iint_s A_x dS_x$ account for the half integral. Hence $\iint_s A_x dS_x$ accounts for the total surface integral for x component. Similarly, one obtains

$$\iiint_v \frac{\partial A_y}{\partial y} dxdydz = \iint_s A_y dS_y \tag{36}$$

and $$\iiint_v \frac{\partial A_z}{\partial z} dxdydz = \iint_s A_z dS_z \tag{37}$$

Adding Eq. (35) to Eq. (37), one obtains

$$\iiint_v \left(\frac{\partial A_x}{\partial x} + \frac{\partial A_y}{\partial y} + \frac{\partial A_z}{\partial z} \right) dxdydz = \iint_s \left(A_x dS_x + A_y dS_y + A_z dS_z \right)$$

or $$\iiint_v (\nabla \cdot \boldsymbol{A}) dv = \iint_s \boldsymbol{A} \cdot ds$$

or $$\iiint_v (div \boldsymbol{A}) dv = \iint_s \boldsymbol{A} \cdot ds \tag{38}$$

This proves the *Gauss's divergence theorem.*

Now we apply this theorem to an electric field $\boldsymbol{E}$. Let us consider that there exist throughout the volume V enclosed by the surface S, a charge distribution of varying density ρ, then

$$\frac{1}{\varepsilon_0} \int_v \rho dv = \int_s \boldsymbol{E} \cdot ds = \int_v (\nabla \cdot \boldsymbol{E}) dv \tag{39}$$

The two volume integrals in Eq. (39) are equal, whatever the volume over which the integration takes place. It therefore follows that integrands themselves must be equal, *i.e.*

$$\nabla \cdot \boldsymbol{E} = \frac{\rho}{\varepsilon_0} \tag{40}$$

Eq. (40) is sometimes referred to as the *differential form of Gauss's law in electrostatics.* According to this, the divergence of the electric field at a point equals ρ/ε_0 in the immediate neighbourhood of that point. If the point is situated in a material of dielectric constant K, then R.H.S. of Eq. (40) must be divided by K.

The differential or point form Eq. (40) is valid only where the derivatives exist. The integral form Eq. (39) is valid everywhere.

1.12 THE LAPLACIAN OPERATOR

Let ϕ be a scalar potential function. If ϕ represents an electric field $\boldsymbol{E}$, then

$$\boldsymbol{E} = -\nabla\phi = -\text{grad}\phi \tag{41}$$

According to Gauss's law in electrostatics, the divergence of an electric field emerging from a closed surface is equal to $\frac{1}{\varepsilon}$ times the charge density enclosed in the surface , *i.e.*

$$\nabla \cdot \boldsymbol{E} = \frac{\rho}{\varepsilon_0} \tag{42}$$

Here ε is the permittivity of the vacuum. Substituting Eq. (41) in Eq. (42), one obtains

$$-\nabla \cdot \nabla\phi = \frac{\rho}{\varepsilon_0} \tag{43}$$

But

$$\begin{aligned} \nabla \cdot \nabla\phi &= \text{div}(\text{grad}\phi) \\ &= \left(\hat{i}\frac{\partial}{\partial x} + \hat{j}\frac{\partial}{\partial y} + \hat{k}\frac{\partial}{\partial z}\right)\cdot\left(\hat{i}\frac{\partial}{\partial x} + \hat{j}\frac{\partial}{\partial y} + \hat{k}\frac{\partial}{\partial z}\right)\phi \\ &= \left(\frac{\partial^2}{\partial x^2} + \frac{\partial^2}{\partial y^2} + \frac{\partial^2}{\partial z^2}\right)\phi \\ &= \nabla^2\phi \end{aligned} \tag{44}$$

Thus $\nabla^2 = \frac{\partial^2}{\partial x^2} + \frac{\partial^2}{\partial y^2} + \frac{\partial^2}{\partial z^2}$ is known is *Laplacian operator* or simply, the *Laplacian.* $\nabla^2 = \nabla \cdot \nabla$ is usually pronounced as del squared.

1.13 POISSON'S AND LAPLACE'S EQUATIONS

Substituting Eq. (44) in Eq. (43), one obtains

$$\nabla^2\phi = -\frac{\rho}{\varepsilon_0} \tag{45}$$

Eq. (45) is known as *Poisson's equation.* One can calculate the electric potential at a point due to a charge distribution with the help of this equation. In a medium of dielectric constant K, Eq. (45) takes the form

$$\nabla^2\phi = -\frac{\rho}{K\varepsilon_0} \tag{46}$$

In the charge free region, volume charge density $\rho = 0$, so the Eq. (45) or (46) become

$$\nabla^2\phi = 0 \tag{47}$$

Eq. (47) is known as *Laplace's equation.* By solving this equation, one can calculate the potential in an electric field, which is devoid of the source charge.

In *spherical polar co-ordinates.* Poison's equation becomes

$$\frac{1}{r^2}\frac{\partial}{\partial r}\left(r^2\frac{\partial\phi}{\partial r}\right)+\frac{1}{r^2}\frac{1}{\sin\theta}\frac{\partial}{\partial\theta}\left(\sin\theta\frac{\partial\phi}{\partial\theta}\right)+\frac{1}{r^2\sin^2\theta}\frac{\partial^2\phi}{\partial\theta^2}=\frac{\rho}{k\varepsilon_0} \tag{48}$$

In *cylindrical polar co-ordinates,* Poisson's equation becomes

$$\frac{1}{r}\frac{\partial}{\partial r}\left(r\frac{\partial\phi}{\partial r}\right)+\frac{1}{r^2}\frac{\partial^2\phi}{\partial\theta^2}+\frac{\partial^2\phi}{\partial z^2}=\frac{\rho}{k\varepsilon_0} \tag{49}$$

1.14 THE CURL OF A VECTOR FUNCTION

We have seen that divergence of a vector field is a scalar. Now, we define a vector called curl A. Also written as $\nabla \times A$, associated with a vector field A.

If a vector field is derived as the gradient of a scalar field then the line integral of the vector around any closed path in the field is zero. Such a field is called '*lamellar*' or *non curl field.* There are, however, such vector fields for which the line-integral round any closed path is non-zero and have a finite value. Such vector fields cannot be derived as gradient of any scalar field and they exhibit the property of curl.

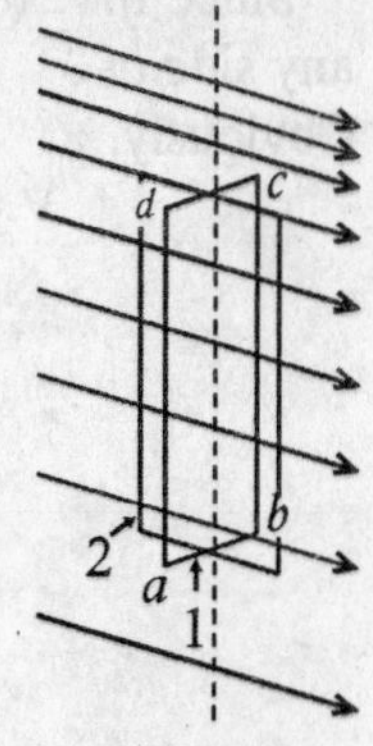

Fig. 1.9

Suppose a non-lamellar vector field is represented by several lines of flow as shown in Fig. (1.9). Let us consider a plane rectangular area in this field. When the plane *abcd* is perpendicular to the field (position 1) none line of the flow lies along its boundary so that the line integral round it is zero. When the plane, *abcd* is parallel to the field (position 2) lines of flow lie along the upper and lower edges of the plane. Since the value of the vector at the upper edge is different from that at the lower edge, the line integral round the boundary has a finite value. This value will be different for different orientations of the plane, and shall be maximum for a particular orientation. This maximum line integral computed per unit area along the boundary is called the curl of the vector field. Thus one can define the curl of a vector field as:

The curl of a (non-lameller) vector field at any point is defined as the maximum line integral of the vector component per unit area along the boundary of an infinitesimal area at that point. Obviously, a curl of a vector field is a vector.

Curl in Cartesian Coordinates

We now calculate curl *A*, where *A* is function of cartesian coordinates (x, y, z). Let us consider a rectangular shaped infinitely small element of area $dxdy$ in the (x, y) plane as shown in Fig. 1.10.

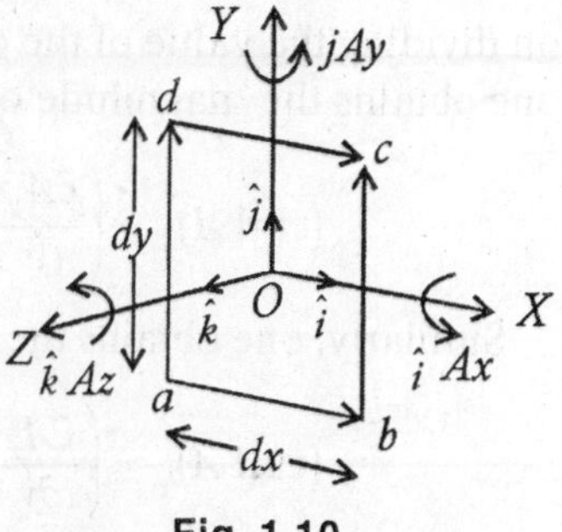

Fig. 1.10

Let the magnitude of the vector *A* at a point *O* is $|A|$ and A_x, A_y and A_z be its components along x, y and z axes respectively. Let the sides dx, dy of the rectangle be parallel to the x and y axes respectively. So the normal to the area is along the z-axis. The arrow heads on the sides of the rectangle show the directions in which the component of *A* act.

Since the rectangle is infinitely small, the average value of *A* along any side of it may be taken as the value at the middle point of that side. Obviously, the average values along the four sides will be as follows:

$$\text{along } ab;\ A_x - \frac{1}{2}\frac{\partial A_x}{\partial y}dy$$

$$\text{along } bc;\ A_y + \frac{1}{2}\frac{\partial A_y}{\partial x}dx$$

$$\text{along } dc;\ A_x + \frac{1}{2}\frac{\partial A_x}{\partial x}dy$$

$$\text{and along } ad;\ A_y - \frac{1}{2}\frac{\partial A_y}{\partial x}dx$$

Here we have considered only the first order of small quantities dx, dy in the Taylor's expansion.

Therefore, the closed path line integral around the boundary $abcd$ of the rectangle is given by

$$= \left[\left(A_x - \frac{1}{2}\frac{\partial A_x}{\partial y}dy\right) - \left(A_x + \frac{1}{2}\frac{\partial A_x}{\partial y}dy\right)\right]dx$$

$$+ \left[\left(A_y + \frac{1}{2}\frac{\partial A_y}{\partial x}dx\right) - \left(A_y - \frac{1}{2}\frac{\partial A_y}{\partial x}dx\right)\right]dy$$

$$= \left(\frac{\partial A_y}{\partial x} - \frac{\partial A_x}{\partial y}\right)dxdy$$

on dividing the value of the circulation by the area $dxdy$ of the rectangle, one obtains the magnitude of the component of curl A along Z-axis as

$$(\text{curl } A)_z = \left(\frac{\partial A_y}{\partial x} - \frac{\partial A_x}{\partial y}\right)\hat{k}$$

Similarly, one obtains the components of curl A along X and Y-axes as

$$(\text{curl } A)_x = \left(\frac{\partial A_z}{\partial y} - \frac{\partial A_y}{\partial z}\right)\hat{i}$$

and $$(\text{curl } A)_y = \left(\frac{\partial A_x}{\partial z} - \frac{\partial A_z}{\partial x}\right)\hat{j}$$

Adding the three components, one obtains

$$\text{curl } A = \left(\frac{\partial A_z}{\partial y} - \frac{\partial A_y}{\partial z}\right)\hat{i} + \left(\frac{\partial A_x}{\partial z} - \frac{\partial A_z}{\partial x}\right)\hat{j} + \left(\frac{\partial A_y}{\partial x} - \frac{\partial A_x}{\partial y}\right)\hat{k} \quad (50)$$

One can express $\nabla \times A$ in the form of a determinant as

$$\text{curl } A = \begin{vmatrix} \hat{i} & \hat{j} & \hat{k} \\ \frac{\partial}{\partial x} & \frac{\partial}{\partial y} & \frac{\partial}{\partial z} \\ A_x & A_y & A_z \end{vmatrix} \quad (51)$$

We must note that Eq. (51) is independent of axes. The name curl indicates that a vector field at a point in space has circulation. The determinate in Eq. (51) is equal to the vector product of del operator (∇) and vector A, *i.e.*

$$\text{Curl}A = \nabla \times A \quad (52)$$

1.15 STOKE'S CURL THEOREM

This states that *the flux of the curl of a vector field **A** over any surface of any shape is equal to the line integral of the vector field **A** over the boundary of that surface.* We must note that the sense of line integral is related with the positive normal to the surface.

Mathematically one can express it as

$$\iint_s \text{curl } \boldsymbol{A} \cdot \boldsymbol{ds} = \oint \boldsymbol{A} \cdot \boldsymbol{dl} \tag{53}$$

Here the line integration is being carried out for the boundary of the surface.

The path C bounds the surface S as shown in Fig. 1.11. Let the surface S be divided into large number of small areas $\boldsymbol{ds_1}$, $\boldsymbol{ds_2}$, $\boldsymbol{ds_3}$, ..., etc. Since all the areas are to be traversed by the curves bounding them in the same sense. The sum of the line integrals around these areas is equal to the line integral around the closed path C, *i.e.*

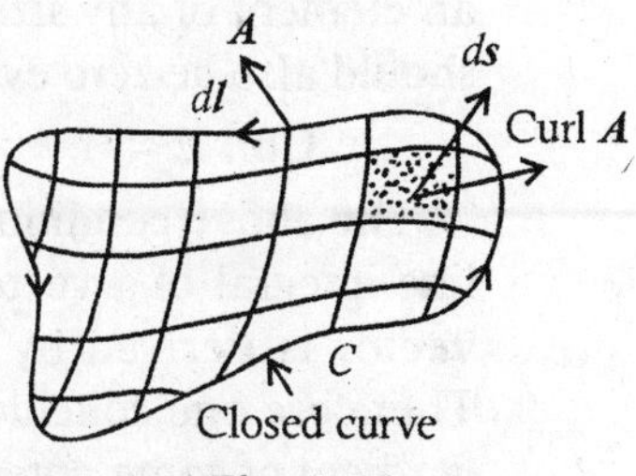

Fig. 1.11

$$\oint \boldsymbol{A} \cdot \boldsymbol{dl} = \Sigma_i \oint_c \boldsymbol{A} \cdot \boldsymbol{dl}$$

From the definition of curl $\boldsymbol{A}$, at a point inside ds_i, we have

$$\text{Curl}\boldsymbol{A} \cdot \boldsymbol{ds_i} = \oint_{c_i} \boldsymbol{A} \cdot \boldsymbol{dl}$$

Since the number of these areas is very large and hence in the limiting case

$$\oint \boldsymbol{A} \cdot \boldsymbol{dl} = \left[\lim \partial s \to 0 \, \Sigma_i \frac{\partial}{\partial s_i} \oint \boldsymbol{A} \cdot \boldsymbol{dl} \right] \partial s_i \tag{54}$$

Since the curl is the line integral per unit area in the limiting case, when area tends to zero, we have

$$\oint \boldsymbol{A} \cdot \boldsymbol{dl} = \iint_s \text{curl}\boldsymbol{A} \cdot \boldsymbol{ds} \tag{55}$$

Eq. (55) represents the *Stoke's curl theorem. Obviously, the line integral of a vector field along the limiting boundary of the area enclosed by a closed path is equal to the surface integral of the curl of the vector field.* This theorem is used in converting line integral into surface integral and *vice versa.*

1.16 PHYSICAL MEANING OF CURL OF A VECTOR FIELD

The existence of the curl of a vector field is always due to its rotation or circulation. A vector field having non-zero curl has circulation and has curly or rotational field lines about the point of non-zero curl. In German literature curl is written as 'rot' which is derived from rotation. The following examples illustrates clearly the physical meaning of curl.

(*i*) In an electrostatic field ***E***, the curl ***E*** is zero everywhere. The line integral of the field ***E*** between two points *A* and *B* is independent of the path. Therefore line integral along *A* to *B* is negative of the line integral along *B* to *A* or in other words.

$$\int_A^B \boldsymbol{E} \cdot dl = -\int_B^A \boldsymbol{E} \cdot dl$$

Hence $$\int_A^B \boldsymbol{E} \cdot dl + \int_B^A \boldsymbol{E} \cdot dl = 0$$

Obviously, the line integral of the field ***E*** around any closed path is zero. Also the circulation is zero around any closed path, then from the Stoke's theorem, the surface integral of curl ***E*** is over an element of any size and shape. This clearly reveals that curl ***E*** should also be zero everywhere, *i.e.*

$$\text{Curl } \boldsymbol{E} = 0 \tag{56}$$

The above condition is for a field to be conservative. Since the line integral of a vector around the closed path is zero when the vector is derived by taking the gradient of a scalar function. Therefore one conclude that electric field can be described as the gradient of some potential function.

(*ii*) Let us consider an example of a vector field having non-zero curl. We can show that the curl of the magnetic induction vector is proportional to the current density vector at the given point, *i.e.*

$$\nabla \times \boldsymbol{B} = \mu_0 \boldsymbol{J} \tag{57}$$

where μ_0 is the permeability of free space ($\mu_0 = 4\pi \times 10^{-7}$ NA^{-2}). However, we must note that Eq. (57) hold good only for the magnetic field in a vacuum in the absence of time-varying electric fields.

A comparison of Eq. (56) and Eq. (57) shows that electrostatic and magnetic fields are of an appreciably different nature. The curl of an electrostatic field is zero, *i.e.* an electrostatic field is conservative and can be characterized by the scalar potential ϕ. The curl of a magnetic field at points where a current differs from zero. Accordingly, the circulation of ***B*** is proportional to the current enclosed by a loop. This is why we cannot ascribe to a magnetic field a scalar potential that would be related to ***B*** by an equation similar to Eq. (41). This potential would not be unique-upon each circumvention of the look and return to the initial point it would receive an increment equal to $\mu_0 I$. A field whose curl differs from zero is called a *Vortex* or a *Solenoidal* one.

Since $\nabla \cdot \boldsymbol{B}$ is zero everywhere and hence ***B*** can be represented as the curl of another vector ***A***. Thus

$$\boldsymbol{B} = \nabla \times \boldsymbol{A} \tag{58}$$

The divergence of a curl is always zero. ***A*** is called the *vector potential.*

1.17 GREEN'S THEOREM

Green's theorem is a corollary of the divergence theorem. The theorem states that if ϕ and Ψ are two scalar point functions such that these functions and their first derivatives are continuously differentiable, then

$$\iiint_v (\phi\nabla^2\psi + \nabla\phi\cdot\nabla\psi)dV = \iint_s (\phi\nabla\psi)\cdot ds \tag{59}$$

and $$\iiint_v (\phi\nabla^2\psi - \psi\nabla^2\phi)dV = \iint_s (\phi\nabla\psi - \psi\nabla\phi)\cdot ds \tag{60}$$

Eq. (59) and Eq. (60) are respectively known as first and second form of Green's theorem. One can derive it as follows:

We start with Gauss's divergence theorem which states

$$\iiint_v \nabla\cdot \boldsymbol{A} dV = \iint_s \boldsymbol{A}\cdot ds \tag{61}$$

Let us take $\boldsymbol{A} = \phi\nabla\psi$ or

$$\hat{i}A_x + \hat{j}A_y + \hat{k}A_z = \phi\left(\hat{i}\frac{\partial\psi}{\partial x} + \hat{j}\frac{\partial\psi}{\partial y} + \hat{k}\frac{\partial\psi}{\partial z}\right)$$

Obviously,

$$A_x = \phi\frac{\partial\psi}{\partial x},\ A_y = \phi\frac{\partial\psi}{\partial y} \text{ and } A_z = \phi\frac{\partial\psi}{\partial z}$$

Now,

$$\nabla\cdot\boldsymbol{A} = \frac{\partial A_x}{\partial x} + \frac{\partial A_y}{\partial y} + \frac{\partial A_z}{\partial z}$$

$$= \frac{\partial}{\partial x}\left(\phi\frac{\partial\psi}{\partial x}\right) + \frac{\partial}{\partial y}\left(\phi\frac{\partial\psi}{\partial y}\right) + \frac{\partial}{\partial z}\left(\phi\frac{\partial\psi}{\partial z}\right)$$

$$= \left(\phi\frac{\partial^2\psi}{\partial x^2} + \frac{\partial\phi}{\partial x}\frac{\partial\psi}{\partial x}\right) + \left(\phi\frac{\partial^2\psi}{\partial y^2} + \frac{\partial\phi}{\partial y}\frac{\partial\psi}{\partial y}\right) + \left(\phi\frac{\partial^2\psi}{\partial z^2} + \frac{\partial\phi}{\partial z}\frac{\partial\psi}{\partial z}\right)$$

$$= \phi\left(\frac{\partial^2\psi}{\partial x^2} + \frac{\partial^2\psi}{\partial y^2} + \frac{\partial^2\psi}{\partial z^2}\right) + \frac{\partial\phi}{\partial x}\frac{\partial\psi}{\partial x} + \frac{\partial\phi}{\partial y}\frac{\partial\psi}{\partial y} + \frac{\partial\phi}{\partial z}\frac{\partial\psi}{\partial z}$$

$$= \phi\nabla^2\psi + \nabla\phi\cdot\nabla\psi$$

Making use of these results in Eq. (61), one obtains

$$\iiint_v (\phi\nabla^2\psi + \nabla\phi\cdot\nabla\psi)dV = \iint_s (\phi\nabla\psi)\cdot ds \tag{62}$$

This is Green's first equation.

Interchanging ϕ and ψ in Eq. (62), one obtains

$$\iiint_v (\psi\nabla^2\phi + \nabla\psi\cdot\nabla\psi)dV = \iint_s (\psi\nabla\phi)\cdot ds \tag{63}$$

Subtracting Eq. (63) from Eq. (62), one obtains

$$\iiint_v (\phi\nabla^2\psi - \psi\nabla^2\phi)dV = \iint_s (\phi\nabla\psi - \psi\nabla\phi)\cdot ds \tag{64}$$

This is Green's second equation. This is also referred as *symmetrical theorem.*

In the R.H.S. of Eq. (64), we have

$$\nabla\psi \cdot ds = \frac{\partial\psi}{\partial n} ds$$

and
$$\nabla\phi \cdot ds = \frac{\partial\phi}{\partial n} ds$$

where $\frac{\partial\psi}{\partial n}$ and $\frac{\partial\phi}{\partial n}$ are the directional derivatives of ψ and ϕ respectively along the outwards normal to *ds*. In the light of these results Eq. (64) can also be expressed as

$$\iiint_v (\phi\nabla^2\psi - \psi\nabla^2\phi)dV = \iint_s \left(\phi\frac{\partial\psi}{\partial n} - \psi\frac{\partial\phi}{\partial n}\right)ds \tag{65}$$

Eq. (65) is an alternative from of Green's Symmetrical theorem.

One can easily derive Gauss's divergence theorem from Green's theorem. We have Green's symmetrical theorem (Eq. (64)) if ϕ is constant, then we have

$$\phi\iiint_V \nabla^2\psi dV = \phi\iint_s \nabla\psi \cdot ds$$

or
$$\iiint_V (\nabla\cdot\nabla\psi)dV = \iint_s \nabla\psi\cdot ds$$

Putting $\nabla\psi = \boldsymbol{A}$, one obtains

$$\iiint_V (\nabla\cdot \boldsymbol{A})dV = \iint_s \boldsymbol{A}\cdot ds \tag{66}$$

Which is *Gauss's divergence theorem.*

ILLUSTRATIVE EXAMPLES

Example 1

A field is expressed by the potential function $V = 3x^2z - xy^3 + z$. Calculate the potential at the point (1, –2, +2).

Solution

We have

$$V = 3x^2z - xy^3 + z$$

Potential at the point (1, –2, +2) is obtained as

$$V_{(1,\,-2,\,+2)} = 3(1)^2 \times 2 - (1)\,(-2)^3 + 2 = 16 \text{ Volts.}$$

Example 2

A vector field $\boldsymbol{A}$ is represented by the function $\boldsymbol{A} = \hat{i}(2x^2y - x^4) + \hat{j}(yz^2) - \hat{k}(xy^2)$, then obtains the values of $\frac{\partial \boldsymbol{A}}{\partial x}$, $\frac{\partial^2 \boldsymbol{A}}{\partial x^2}$ and $\frac{\partial^2 \boldsymbol{A}}{\partial y \partial z}$.

Solution

We have

$$\boldsymbol{A} = \hat{i}(2x^2y - x^4) + \hat{j}yz^2 - \hat{k}xy^2$$

$$\therefore \quad \frac{\partial \boldsymbol{A}}{\partial x} = \frac{\partial}{\partial x}\left[\hat{i}(2x^2y - x^4) + \hat{j}yz^2 - \hat{k}xy^2\right]$$

$$= \hat{i}(4xy - 4x^3) + \hat{j}0 - \hat{k}y^2$$

Now
$$\frac{\partial^2 \boldsymbol{A}}{\partial x^2} = \frac{\partial}{\partial x}\left[\hat{i}(4xy - 4x^3) - \hat{k}y^2\right]$$

$$= \hat{i}(4y - 12x^2) - \hat{k}0$$

and
$$\frac{\partial^2 \boldsymbol{A}}{\partial y \partial z} = \frac{\partial}{\partial z}\left(\frac{\partial \boldsymbol{A}}{\partial y}\right)$$

$$= \frac{\partial}{\partial z}\left[\frac{\partial}{\partial y}\left\{\hat{i}(2x^2y - x^4) - \hat{j}yz^2 - \hat{k}xy^2\right\}\right]$$

$$= \frac{\partial}{\partial z}\left[\hat{i}(2x^2) + \hat{j}z^2 - \hat{k}2xy\right] = 2\hat{j}z$$

Example 3

Find the value of div (r^n $\hat{r}$), where $\hat{r}$ is the unit vector.

Solution

We have
$$\boldsymbol{r} = \hat{i}x + \hat{j}y + \hat{k}z$$

$$\therefore \quad |\boldsymbol{r}| = \sqrt{x^2 + y^2 + z^2}$$

$$\therefore \quad \text{div}\,(r^n\hat{r}) = \nabla \cdot (r^n\hat{r}) = \left(\hat{i}\frac{\partial}{\partial x} + \hat{j}\frac{\partial}{\partial y} + \hat{k}\frac{\partial}{\partial z}\right) \cdot (r^{n-1}r\,\hat{r})$$

$$= \left(\hat{i}\frac{\partial}{\partial x} + \hat{j}\frac{\partial}{\partial y} + \hat{k}\frac{\partial}{\partial z}\right) \cdot (r^{n-1}\boldsymbol{r})$$

$$= \left(\hat{i}\frac{\partial}{\partial x} + \hat{j}\frac{\partial}{\partial y} + \hat{k}\frac{\partial}{\partial z}\right) \cdot r^{n-1}\left(\hat{i}x + \hat{j}y + \hat{k}z\right)$$

$$= \frac{\partial}{\partial x}\left(xr^{n-1}\right) + \frac{\partial}{\partial y}\left(yr^{n-1}\right) + \frac{\partial}{\partial z}\left(zr^{n-1}\right)$$

$$= \left[r^{n-1} + x\frac{\partial}{\partial x}\left(x^2 + y^2 + z^2\right)^{n-\frac{1}{2}}\right]$$

$$+\left[r^{n-1} + y\frac{\partial}{\partial y}\left(x^2 + y^2 + z^2\right)^{n-\frac{1}{2}}\right]$$

$$+\left[r^{n-1} + z\frac{\partial}{\partial z}\left(x^2 + y^2 + z^2\right)^{n-\frac{1}{2}}\right]$$

$$= r^{n-1} + x\left(\frac{n-1}{2}\right)\left(x^2 + y^2 + z^2\right)^{n-3/2} 2x + \ldots$$

$$= 3r^{n-1} + (n-1)\left[x^2 + y^2 + z^2\right]^{n-\frac{1}{2}}$$

$$= 3r^{n-1} + (n-1)r^{n-1} = (n+2)r^{n-1}$$

Example 4

Find (*a*) $\nabla \cdot \boldsymbol{r}$, where $\boldsymbol{r}$ is the position vector and

(*b*) $\nabla. \left[\hat{i}(x + y) + \hat{j}(x - y) + \hat{k}(4z)\right]$

Solution: (*a*)

$$\nabla \cdot \boldsymbol{r} = \left(\hat{i}\frac{\partial}{\partial x} + \hat{j}\frac{\partial}{\partial y} + \hat{k}\frac{\partial}{\partial z}\right) \cdot \left(\hat{i}x + \hat{j}y + \hat{k}z\right)$$

$$= \frac{\partial x}{\partial x} + \frac{\partial y}{\partial y} + \frac{\partial z}{\partial z} = 3$$

(*b*)

$$\nabla \cdot \left[\hat{i}(x + y) + \hat{j}(x - y) + \hat{k}(4z)\right]$$

$$= \left(\hat{i}\frac{\partial}{\partial x} + \hat{j}\frac{\partial}{\partial y} + \hat{k}\frac{\partial}{\partial z}\right) \cdot [\hat{i}(x + y) + \hat{j}(x - y) + \hat{k}(4z)]$$

$$= \frac{\partial}{\partial x}(x + y) + \frac{\partial}{\partial y}(x - y) + \frac{\partial}{\partial z}(4z)$$

$$= 1 - 1 + 4 = 4$$

Example 5

If $\boldsymbol{r}$ is the position vector of a point, calculate the gradient of $\frac{1}{r}$.

Solution: $\boldsymbol{r} = \hat{i}x + \hat{j}y + \hat{k}z$

$$\therefore \quad |\boldsymbol{r}| = \sqrt{x^2 + y^2 + z^2}$$

Now $$\nabla\left(\frac{1}{r}\right) = \nabla\left(x^2 + y^2 + z^2\right)^{-1/2}$$

or $$\nabla\left(\frac{1}{r}\right) = \left(\hat{i}\frac{\partial}{\partial x} + \hat{j}\frac{\partial}{\partial y} + \hat{k}\frac{\partial}{\partial z}\right)\left(x^2 + y^2 + z^2\right)^{-1/2}$$

$$= \hat{i}\frac{\partial}{\partial x}\left(x^2 + y^2 + z^2\right)^{-1/2} + \hat{j}\frac{\partial}{\partial y}\left(x^2 + y^2 + z^2\right)^{-1/2} + \hat{k}\frac{\partial}{\partial z}\left(x^2 + y^2 + z^2\right)^{-1/2}$$

$$\therefore \quad \frac{\partial}{\partial x}\left(x^2 + y^2 + z^2\right)^{-1/2} = -\frac{1}{2}\left(x^2 + y^2 + z^2\right)^{-3/2}(2x) = \frac{-x}{r^3}$$

Similarly,

$$\frac{\partial}{\partial y}\left(x^2 + y^2 + z^2\right)^{-1/2} = \frac{-y}{r^3} \text{ and } \frac{\partial}{\partial z}\left(x^2 + y^2 + z^2\right)^{-1/2} = \frac{-z}{r^3}$$

$$\therefore \quad \nabla\left(\frac{1}{r}\right) = -\frac{\hat{i}x + \hat{j}y + \hat{k}z}{r^3} = \frac{-\boldsymbol{r}}{r^3}$$

Example 6

The electric field due to a point charge is expressed as $\boldsymbol{E} = \frac{Q}{r^2}\hat{r}$. Show that the divergence of electric field due to that point charge is zero.

Solution

$$\nabla \cdot \boldsymbol{E} = \nabla \cdot \left(\frac{Q}{r^2}\hat{r}\right) = \nabla \cdot \left(\frac{Q}{r^3}\hat{r}r\right) = \nabla \cdot \left(\frac{Q}{r^3}\boldsymbol{r}\right)$$

$$= \left(\hat{i}\frac{\partial}{\partial x} + \hat{j}\frac{\partial}{\partial y} + \hat{k}\frac{\partial}{\partial z}\right) \cdot Q\left(\frac{\hat{i}x + \hat{j}y + \hat{k}z}{\left(x^2 + y^2 + z^2\right)^{3/2}}\right)$$

$$\left(\because \boldsymbol{r} = \hat{i}x + \hat{j}y + \hat{k}z \;\therefore |\boldsymbol{r}| = \sqrt{x^2 + y^2 + z^2}\right)$$

$$= Q\left[\frac{\partial}{\partial x}\left\{\frac{x}{\left(x^2+y^2+z^2\right)^{3/2}}\right\}\right] + Q\left[\frac{\partial}{\partial y}\left(\frac{y}{\left(x^2+y^2+z^2\right)^{3/2}}\right)\right]$$

$$+Q\left[\frac{\partial}{\partial z}\left\{\frac{z}{\left(x^2+y^2+z^2\right)^{3/2}}\right\}\right]$$

$$= Q\left[\frac{1}{\left(x^2+y^2+z^2\right)^{3/2}} - \frac{3x^2}{\left(x^2+y^2+z^2\right)^{5/2}} + \frac{1}{\left(x^2+y^2+z^2\right)^{3/2}}\right.$$

$$\left. - \frac{3y^2}{\left(x^2+y^2+z^2\right)^{5/2}} + \frac{1}{\left(x^2+y^2+z^2\right)^{3/2}} - \frac{3z^2}{\left(x^2+y^2+z^2\right)^{5/2}}\right]$$

$$= 0$$

Example 7

Show that the potential function $V = q(x^2 + y^2 + z^2)^{1/2}$ does not satisfy the Laplace's equation.

Solution

The Laplace's equation is $\nabla^2 V = 0$, *i.e.*

$$\frac{\partial^2 V}{\partial x^2} + \frac{\partial^2 V}{\partial y^2} + \frac{\partial^2 V}{\partial z^2} = 0 \qquad (1)$$

Using $V = q(x^2 + y^2 + z^2)^{1/2}$, L.H.S. of Eq. (1) becomes

$$\text{L.H.S.} = q\frac{\partial}{\partial x}\left[\frac{\partial}{\partial x}\left(x^2+y^2+z^2\right)^{1/2}\right] + q\frac{\partial}{\partial y}\left[\frac{\partial}{\partial y}\left(x^2+y^2+z^2\right)^{1/2}\right]$$

$$+q\frac{\partial}{\partial z}\left[\frac{\partial}{\partial z}\left(x^2+y^2+z^2\right)^{1/2}\right]$$

$$= q\frac{\partial}{\partial x}\left[x\left(x^2+y^2+z^2\right)^{1/2}\right] + q\frac{\partial}{\partial y}\left[y\left(x^2+y^2+z^2\right)^{1/2}\right]$$

$$+q\frac{\partial}{\partial z}\left[z\left(x^2+y^2+z^2\right)^{1/2}\right]$$

$$= q\left\{-\frac{1}{4}\left(x^2+y^2+z^2\right)^{-3/2} 4x + \left(x^2+y^2+z^2\right)^{-1/2}\right.$$

$$-\frac{1}{4}\left(x^2+y^2+z^2\right)^{-3/2}4y+\left(x^2+y^2+z^2\right)^{-1/2}$$

$$-\frac{1}{4}\left(x^2+y^2+z^2\right)^{-3/2}4z+\left(x^2+y^2+z^2\right)^{-1/2}\Bigg\}$$

$$=q\left\{\left(x^2+y^2+z^2\right)^{-3/2}\left(x^2+y^2+z^2\right)+3\left(x^2+y^2+z^2\right)^{-1/2}\right\}$$

$$V=2q(x^2+y^2+z^2)^{-1/2}\neq 0$$

Obviously, the given potential function does not satisfy the Laplace's equation.

Example 8

Show that curl $\boldsymbol{r} = 0$, where r is position vector.

Solution

$$\nabla\times\boldsymbol{r}=\left(\hat{i}\frac{\partial}{\partial x}+\hat{j}\frac{\partial}{\partial y}+\hat{k}\frac{\partial}{\partial z}\right)\times\left(\hat{i}x+\hat{j}y+\hat{k}z\right)$$

$$=\begin{vmatrix}\hat{i} & \hat{j} & \hat{k}\\ \frac{\partial}{\partial x} & \frac{\partial}{\partial y} & \frac{\partial}{\partial z}\\ x & y & z\end{vmatrix}=0$$

Example 9

A potential function in an electric field is given by, $\phi = (2x^2 - y^2 + 3z^2)$ volts. Show that the intensity of electric field at point (1, –1, 0) m is $-\left(4\hat{i}-2\hat{j}\right)$ V/m.

Solution

$$\boldsymbol{E}=-\nabla\phi=-\left(\hat{i}\frac{\partial\phi}{\partial x}+\hat{j}\frac{\partial\phi}{\partial y}+\hat{k}\frac{\partial\phi}{\partial z}\right)$$

Now $\quad \dfrac{\partial\phi}{\partial x}=4x,\ \dfrac{\partial\phi}{\partial y}=-2y$ and $\dfrac{\partial\phi}{\partial z}=6z$

$$\therefore \quad \boldsymbol{E}=-\left(\hat{i}4x-\hat{j}2y+\hat{k}6z\right)$$

$\therefore$ $\boldsymbol{E}$ at the point (1, –1, 0)

$$=-\left(4\hat{i}-2\hat{j}\right)=\left(4\hat{i}+2\hat{j}\right)\ \text{V/m}.$$

Example 10

Show that curl gradϕ = 0, where ϕ is any scalar function.

Solution

$$\nabla \times \nabla\phi = \left(\hat{i}\frac{\partial}{\partial x} + \hat{j}\frac{\partial}{\partial y} + \hat{k}\frac{\partial}{\partial z}\right) \times \left(\hat{i}\frac{\partial\phi}{\partial x} + \hat{j}\frac{\partial\phi}{\partial y} + \hat{k}\frac{\partial\phi}{\partial z}\right)$$

$$= \begin{vmatrix} \hat{i} & \hat{j} & \hat{k} \\ \frac{\partial}{\partial x} & \frac{\partial}{\partial y} & \frac{\partial}{\partial z} \\ \frac{\partial\phi}{\partial x} & \frac{\partial\phi}{\partial y} & \frac{\partial\phi}{\partial z} \end{vmatrix}$$

$$= \hat{i}\left(\frac{\partial^2\phi}{\partial y\partial z} - \frac{\partial^2\phi}{\partial z\partial y}\right) + \hat{j}\left(\frac{\partial^2\phi}{\partial z\partial x} - \frac{\partial^2\phi}{\partial x\partial z}\right) + \hat{k}\left(\frac{\partial^2\phi}{\partial x\partial y} - \frac{\partial^2\phi}{\partial y\partial x}\right)$$

$$= 0\left(\because \frac{\partial^2\phi}{\partial y\partial z} = \frac{\partial^2\phi}{\partial z\partial y} \text{ and so on}\right)$$

Example 11

Show that div grad (R^n) = $n(n + 1)r^{n-2}$, where R is position vector.

Solution

$$\text{div grad } (R^n) = \nabla \cdot \nabla\,(R^n)$$

Now $$\nabla(R^n) = \left(\hat{i}\frac{\partial}{\partial x} + \hat{j}\frac{\partial}{\partial y} + \hat{k}\frac{\partial}{\partial z}\right)R^n$$

$$= \hat{i}nR^{n-1}\frac{\partial R}{\partial x} + \hat{j}nR^{n-1}\frac{\partial R}{\partial y} + \hat{k}nR^{n-1}\frac{\partial R}{\partial z}$$

$$= nR^{n-1}\left(\hat{i}\frac{\partial R}{\partial x} + \hat{j}\frac{\partial R}{\partial y} + \hat{k}\frac{\partial R}{\partial z}\right)$$

$\because$ $$\mathbf{R} = \hat{i}x + \hat{j}y + \hat{k}z$$

$\therefore$ $$R^2 = \mathbf{R} \cdot \mathbf{R} = x^2 + y^2 + z^2$$

$\therefore$ $$2R\frac{\partial R}{\partial x} = 2x$$

$$\therefore \quad \frac{\partial R}{\partial x} = \frac{x}{R}$$

Similarly $\dfrac{\partial R}{\partial y} = \dfrac{y}{R}$ and $\dfrac{\partial R}{\partial z} = \dfrac{z}{R}$

$$\therefore \quad \nabla R^n = nR^{n-1}\left(\hat{i}\frac{x}{R} + \hat{j}\frac{y}{R} + \hat{k}\frac{z}{R}\right)$$

$$\nabla \cdot \nabla R^n = \left(\hat{i}\frac{\partial}{\partial x} + \hat{j}\frac{\partial}{\partial y} + \hat{k}\frac{\partial}{\partial z}\right) \cdot nR^{n-2}\left(\hat{i}x + \hat{j}y + \hat{k}z\right)$$

$$= \frac{\partial}{\partial x}\left(nR^{n-2}x\right) + \frac{\partial}{\partial y}\left(nR^{n-2}y\right) + \frac{\partial}{\partial z}\left(nR^{n-2}z\right)$$

$$= n\left\{R^{n-2} + x(n-2)R^{n-3}\frac{\partial R}{\partial x} + R^{n-2}\right.$$

$$\left. + y(n-2)R^{n-3}\frac{\partial R}{\partial y} + R^{n-2} + z(n-2)R^{n-3}\frac{\partial R}{\partial z}\right\}$$

$$= 3nR^{n-2} + n(n-2)R^{n-3}\left\{x\frac{\partial R}{\partial x} + y\frac{\partial R}{\partial y} + z\frac{\partial R}{\partial z}\right\}$$

$$= 3nR^{n-2} + n(n-2)R^{n-3}\left\{x\frac{x}{R} + y\frac{y}{R} + z\frac{z}{R}\right\}$$

$$= 3nR^{n-2} + n(n-2)R^{n-3}\left(\frac{x^2 + y^2 + z^2}{R}\right)$$

$$= 3nR^{n-2} + n(n-2)R^{n-3}R$$

$$= 3R^{n-2}(3 + n - 2) = n(n+1)R^{n-2}$$

REVIEW QUESTIONS

1. Describe scalar and vector field. How can they be represented graphically?
2. What do you mean by the field of any physical quantity? What are scalar and vector fields? Give examples.
3. Define gradient of a scalar field. Show that the gradient of a scalar field is a vector. Give an example.
4. Prove that the gradient of a scalar field is normal to the surface, ϕ = constant.
5. Express the gradient of a scalar field $\phi(x, y, z)$ in terms of cartesian coordinates.
6. Show that when a vector field can be derived as the gradient as a scalar field the line integral of the vector taken round any closed path in the vector field is zero.

7. Explain the meaning of line integral of a vector field using at least one example.
8. Explain the physical meaning of the divergence of a vector field A. Express divergence A in orthogonal coordinates and show that div $A = \nabla \cdot A$.
9. Define curl of a vector field and give its physical significance. Derive expression for a vector field and show that curl $A = \nabla \times A$
10. What is a Laplacian operator? Express div grad ϕ in cartesian coordinates, where ϕ is a scalar quantity.
11. Derive Laplace' and Poisson's equations starting from the differential form of Gauss' law.
12. State and Prove Green's theorem.

PROBLEMS

1. For a position vector $\boldsymbol{R} = \hat{i}x + \hat{j}y + \hat{k}z$ and the values of (*i*) grad $\frac{1}{R}$ (*ii*) div $\left(\frac{\boldsymbol{R}}{R^3}\right)$ and (*iii*) curl $\left(\frac{\boldsymbol{R}}{R^3}\right)$. [**Ans.** (*i*) $-\frac{\boldsymbol{R}}{R^3}$ (*ii*) 0 (*iii*) 0]
2. A potential field is represented by the equation $\phi = 4yz^2 + 3xyz - z^2 + 2$. Calculate the potential at the points (1, –1, –2). [**Ans.** –36 V]
3. A vector field is represented by the equation, $A = \hat{i}x^2yz - \hat{j}2xz^3 + \hat{k}xz^2$. Calculate the values of the partial derivatives (*i*) $\frac{\partial A}{\partial x}$, (*ii*) $\frac{\partial A}{\partial z}$.
   [**Ans.** (*i*) $\hat{i}2xyz - \hat{j}2z^3 + \hat{k}z^2$ (*ii*) $\hat{i}x^2z - \hat{j}6xz^2 + \hat{k}2xz$]
4. The temperature of a body at any point changes according to the relation, $T = 4x^2 + 3y^2 - 2z^2$. Show that the grad T at the point (1, 2, 3) is equal to $-8\hat{i} - 12\hat{j} - 12\hat{k}$.
5. Show that the function $\phi = x^2 - y^2$ satisfies the Laplace's equation.
6. Show that the potential functions
   (*i*) $V = x^2 - y^2 + z$; (*ii*) $V = x^2 + y^2 - 2z$;
   (*iii*) $V = ax + by + cz$ and; (*iv*) $V = ax^2 - ay^2 + 4z$
   satisfy Laplace' equation.
7. Calculate the Curl of the following fields.
   (*a*) $\boldsymbol{E} = a\left\{3x^2yz\hat{i} + \left[x^3z + yz^2(yz)^{1/2}\right]\hat{j} + \left[x^3y + y^2z(yz)^{1/2}\right]\hat{k}\right\}$
   (*b*) $\boldsymbol{B} = y\hat{i} - x\hat{j}$ and
   (*c*) $\boldsymbol{B} = a\left\{\frac{y\hat{i}}{\left(x^2 + y^2\right)^{3/2}} - \frac{x\hat{k}}{\left(x^2 + y^2\right)^{3/2}}\right\}$
   [**Ans.** (*a*) 0, (*b*) $-2\hat{k}$ (*c*) $\frac{a\hat{k}}{\left(x^2 + y^2\right)^{3/2}}$]

8. Calculate the divergence at a point (x, y, z) for the field in problem 7,

[**Ans.** (*a*) $6axyz + \frac{3}{4}a(yz)^{1/2}$ (*b*) zero (*c*) zero]

9. If $A = xz^3\hat{i} - 2x^2yz\hat{j} - 2yz^4\hat{k}$, show that Curl A at the point (1, –1, 1) is $(3\hat{i} + 4\hat{k})$.

10. Evaluate $\iint_s F \cdot ds$, where $F = \hat{i}4xz - \hat{j}y^2 + \hat{k}yz$ and S is the surface of the cube bounded by $x = 0, x = 1, y = 0, y = 1, z = 0$ and $z = 1$. [**Ans.** 3]

11. Evaluate $\iint_s r \cdot \hat{n}ds$ where, S is a closed surface and V is the volume enclosed by S. [**Ans.** $3V$]

12. Show that $\iint_s \hat{n}ds = 0$ for any closed surface S.

13. Show that $\iint_s \text{curl}F \cdot dS = 0$ for any closed surface S.

14. A fluid of density $\rho(x, y, z, t)$ moves with velocity $v(x, y, z, t)$. If there are no sources or sinks, prove that

$$\nabla \cdot J + \frac{\partial \rho}{\partial t} = 0, \text{ where } J = eV$$

15. Show that the necessary and sufficient condition for $\oint_c A \cdot dr = 0$ to be true for every curve C is that curl $A = 0$.

16. Prove Stoke's theorem for the vector $A(x + y, 2x - z, y + z)$ taken over the triangle cut from the plane $3x + 2y + z = 6$ by the coordinate planes.

SHORT ANSWER QUESTIONS

1. How many kinds of fields are there? [**Ans.** Two: (*i*) Scalar (*ii*) vector]

2. Give two examples of vector field.

   [**Ans.** (*i*) Distribution of velocity in a fluid (*ii*) Distribution of gravitational, magnetic or electric field in space]

3. If $\phi(x, y, z)$ is a scalar field, is $\hat{i}\frac{\partial \phi}{\partial x} + 2\hat{j}\frac{\partial \phi}{\partial y} + 3\hat{k}\frac{\partial \phi}{\partial z}$ a vector field?

   [**Ans.** We know that a scalar field $\phi(x, y, z)$ evaluated at a particular point is independent of the coordinates of that point. Obviously, scalar field, is invariant with respect to the transformation of coordinates. Its gradient, which is a vector field, must also be invariant with respect to the transformation. Now, on applying coordinate transformation, the vector $\hat{i}\frac{\partial \phi}{\partial x} + 2\hat{j}\frac{\partial \phi}{\partial y} + 3\hat{k}\frac{\partial \phi}{\partial z}$ does not retain its form on applying a coordinate transformation. Thus it cannot be a vector field corresponding to given scalar field $\phi(x, y, z)$]

4. What are lamellar fields?

   [**Ans.** Certain vector fields can be expressed as gradient of a scalar field ϕ and such fields are known as lamellar fields]

5. What is name of the theorem which provides a method of reducing volume integrals to surface integrals?

   [**Ans.** Gauss's theorem of divergence]

OBJECTIVE QUESTIONS

1. In two dimension, the divergence transform as a __________. (scalar)
2. The curl of a vector function A is __________. (vector)
3. The divergence of a curl is always __________. (zero)
4. The curl of a gradient is always __________. (zero)
5. $\nabla^2 = \frac{\partial^2}{\partial x^2} + \frac{\partial^2}{\partial y^2} + \frac{\partial^2}{\partial z^2}$ is known as __________ operator. (Laplacian)

2

Electrostatics

2.1 INTRODUCTION

Physics is based on the concept that all matter is made of atoms and that these atoms in turn are made of electric charged particles whose behaviour is governed by electrodynamics.

There are two kinds of electric charges: *positive and negative.* Like charges repel each other while unlike charges attract each other. The algebraic sum of electric charges remains constant in a closed system. This is known as the *law of conservation of electric charges.*

The electric charge of any body consists of a whole number of elementary charges each of which equals 1.6×10^{-19} C charge. The smallest stable subatomic particle having a negative elementary charge is called the *electron.* The mass of an electron is 9.1×10^{-31} kg. The smallest stable subatomic particle having a positive charge is the proton. The mass of the proton is 1.67×10^{-27} kg. Electrons and protons are found in the atoms of any substance. A neutral body contains charges of opposite sign which are equal in absolute value. Electric charges are said to be point charges if the linear dimensions of the bodies on which the charges are concentrated are very much smaller than any other lengths pertitent to the problem under consideration. A body exhibiting positive electrification has a positive electric charge, and one with negative electrification has a negative electric charge. The net charge of a body is the algebraic sum of its positive and negative charges. A particle having a nonzero net charge is often called an *ion.* Since matter in bulk does not exhibit gross electrical forces, one may assume that it is composed of equal amounts of positive and negative charges. *The net or total charge does not change for any process occurring within an isolated system.* No exception has been found to this rule, *i.e. principle of conservation of charge.*

2.2 COULOMB'S LAW

The results of the electric interaction between two charged particles at rest in the observer's inertial frame of reference or, at most, moving with a very small velocity constitute what is called electrostatics. The *electrostatic* interaction for two charged particles is given by *Coulomb's law*.

Coulomb from experimental observations concluded that the *electrostatic interaction between two charged particles is proportional to the square of the distance between them, and its direction is along the lines joining the two charges.*

Let '1' and '2' are two particles carrying charges q_1 and q_2 respectively and separated by a distance r_{12} (Fig. 2.1). According to Coulomb's law, the electric force exerted by the particle '1' on the particle '2' is

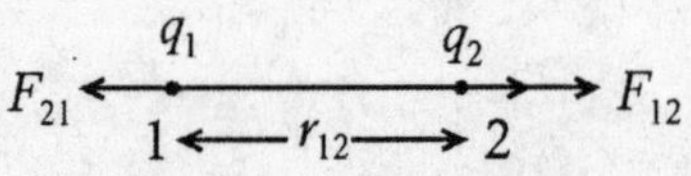

Fig. 2.1: Coulomb's Law

$$F_{12} \propto \frac{q_1 q_2}{r_{12}^2} \; i.e., \quad F_{12} = k\frac{q_1 q_2}{r_{12}^2} = -F_{12} \tag{1}$$

where k is the constant of proportionality and F_{21} is the force exerted by the particle '2' on particle '1'. Expressing Eq. (1) in vector notation, one obtains

$$\boldsymbol{F}_{12} = k\frac{q_1 q_2}{|\boldsymbol{r}_{12}|^2}\hat{r}_{12} = k\frac{q_1 q_2}{|\boldsymbol{r}_{12}|^3}\boldsymbol{r}_{12} \tag{2}$$

where $\hat{r}_{12}$ is the unit vector along $\boldsymbol{r}_{12}$.

Coulomb's law is very similar to the law of gravitational interaction. The sign of charges in Eq. (2) decides whether the force is *attractive* or *repulsive*. When both the charges are similar (*i.e.*, positive or negative, F_{12} is positive and represents the force of repulsion; while if one is positive and other is negative, $\boldsymbol{F}_{12}$ is negative and is the force of attraction.

The numerical value of the constant k in the MKSC system is equal to $10^{-7}c^2 = 8.9874 \times 10^9$, where c is the velocity of light. For practical purposes, we may say that $k = 9 \times 10^9$. Then, when the distance r_{12} is measured in metres and the force in Newtons, Eq. (1) becomes

$$F_{12} = 9 \times 10^9 \frac{q_1 q_2}{r_{12}^2} = -F_{21} \tag{3}$$

once we have decided on the value of k, the unit charge is fixed. This unit is called as Coulomb and is designated by C. *The Coulomb is that charge which, when placed one metre from an equal charge in vacuum, repels it with a force of* 8.987 × 10^9 *Newtons.* Formula in Eq. (3) hold only for two charged particles in vacuum; *i.e.*, for two charged particles in the absence of any other charge or matter. We must note that according to Eq. (2), the unit of k is Nm^2C^{-2} or $m^3kgs^{-2}C^{-2}$. The constant k is then equal to

$$k = \frac{1}{4\pi\varepsilon_0}$$

where ε_0 is the vacuum permeability or *permittivity of free space.* It is also some times referred to as *capacitivity* in vacuum, or the dielectric constant of vacuum. Thus

$$\varepsilon_0 = \frac{10^7}{4\pi c^2} = 8.854 \times 10^{12} N^{-1} m^{-2} C^2 \quad (\text{or } m^{-3} kg^{-1} s^2)$$

Now, the Coulomb's law is written as

$$\boldsymbol{F}_{12} = \frac{1}{4\pi\varepsilon_0} \frac{q_1 q_2}{|\boldsymbol{r}_{12}|^3} \boldsymbol{r}_{12} \quad (\text{Newtons}) \tag{4}$$

The forces of electrostatic interaction are central and long range. One can obtain an idea about the relative magnitudes of electrostatic and gravitational forces by comparing the force of attraction between two electrons due to gravitation and the force of repulsion due to electrostatic repulsion.

$$\frac{\text{Electrostatic repulsion between two electrons}}{\text{Gravitational attraction between two electrons}}$$

$$= \frac{1}{4\pi\varepsilon_0} \frac{e^2}{r^2} \frac{r^2}{Gm^2} = \frac{e^2}{4\pi\varepsilon_0 Gm^2}$$

$$= 9 \times 10^9 \times \frac{(1.6 \times 10^{-19})^2}{(6.67 \times 10^{-11})(9.1 \times 10^{-31})^2}$$

$$= 4.17 \times 10^{42}$$

Obviously, the electrical interaction is of the order of magnitude required to produce the binding between atoms to form molecules, or the binding between the electrons and protons to form atoms.

This reveals that the chemical processes (and in general the behaviour of matter in bulk) are due to electrical interactions between atoms and molecules.

Coulomb's law is based primarily on experiments. At this stage, one can raise a question whether the law is exactly that of inverse square, *i.e.*, of the force is proportional to $\frac{1}{r^n}$, is n exactly equal to 2?

Cavendish found that $n = 2 \pm 0.02$. Plimpton and Lawton in 1936 reported that n differs from 2 by not more than one part in 10^9. Lamb and Rutherford in 1947 found from their measurements of energy levels of the hydrogen atom reveal that the exponent in Coulomb's law is correct top one part in 10^9 at distances of the order of 10^{-10} m. Evidence from nuclear experimental observation has shown that the electrostatic forces vary approximately according to inverse square law even at distances of the order of 10^{-15} m.

2.3 ELECTRIC FIELD

Any region where an electric charge experiences a force is called an *electric field*. The force is due to the presence of other charges in that region. For example, a charge q placed in a region where there are other charges q_1, q_2, q_3, etc. (Fig. 2.2) experiences a force $\boldsymbol{F} = \boldsymbol{F}_1 + \boldsymbol{F}_2 + \boldsymbol{F}_3 + \ldots$, and we say that it is in an electric field produced by the charge q_1, q_2, q_3, ... We must note that charges q of course also exerts forces on q_1, q_2, q_3, ... but we are not concerned with them now. Since the force that each charge q_1, q_2, q_3, ... produces on the charge q is proportional to q, the resultant force $\boldsymbol{F}$ is also proportional to q. Obviously, the force on a particle placed in an electric field is proportional to the charge of the particle.

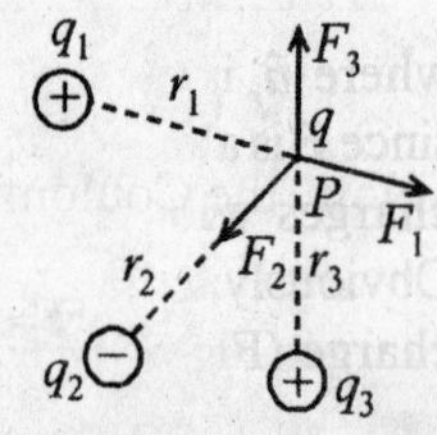

Fig. 2.2: Resultant electric field at P due to the presence of several charges in the region

The intensity of the electric field at a point is equal to the force per unit charge placed at that point. Thus

$$\boldsymbol{E} = \frac{\boldsymbol{F}}{q} \quad \text{or} \quad \boldsymbol{F} = q\boldsymbol{E} \tag{5}$$

The unit of electric field intensity is Newton/Coulomb or NC^{-1}.

It q is positive, the force $\boldsymbol{F}$ acting on the charge has the same direction as the field $\boldsymbol{E}$, but if q is negative, the force $\boldsymbol{F}$ has the direction opposite to $\boldsymbol{E}$ (Fig. 2.3).

Obviously, if we apply an electric field to a region where positive and negative particles or ions are present, the field will tend to move the positively and negatively charged bodies in opposite directions, resulting in a charge separation, an effect sometimes referred to as *polarization*.

Electric field $\boldsymbol{E}$

Positive charge $\boldsymbol{F} = q\boldsymbol{E}$

Negative charge $\boldsymbol{F} = q\boldsymbol{E}$

Fig. 2.3: Direction of the force produced by an electric field on a positive and a negative charge

Let us write Eq. (3) in the form $\boldsymbol{F} = q_2\left(\frac{q_1}{4\pi\varepsilon_0 r_{12}^2}\right)$. This gives the force produced by the charge q_1 on the charge q_2 placed at a distance r_{12} from q_1. Using Eq. (5), we may say that the electric field E at the point where q_2 is placed is such that $F = q_2 E$. Obviously, by comparing both expressions of F, we conclude that the electric field at a distance r_{12} from the charge q_1, is $E = \frac{q_1}{4\pi\varepsilon_0 r_{12}^2}$, or in the vector form,

$$\boldsymbol{E} = \frac{q_1}{4\pi\varepsilon_0 r_{12}^2}\,\hat{a}_r \tag{6}$$

where $\hat{a}_r$ is the unit vector in the radial direction away from the charge q, since $\boldsymbol{F}$ is along this direction. Eq. (6) is valid for both positibve and negative charges with the direction of $\boldsymbol{E}$ relative to $\hat{a}_r$ given by the sign of q_1. Obviously, $\boldsymbol{E}$ is directed away from a positive charge and toward a negative charge (Fig. 2.4).

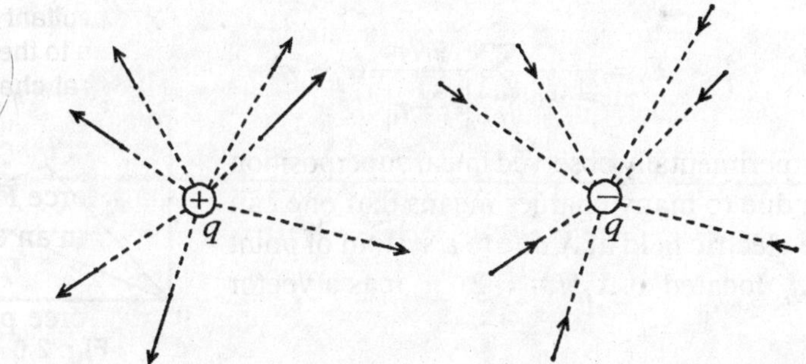

Fig. 2.4: Electric field produced by (*a*) positive and (*b*) a negative charge

Let $q_1 = q_0$ is sufficiently small so that it does not disort the field whose intensity is to be measured. The resultant force $\boldsymbol{F}$ exerted on a test charge q by a field set up by a systems of fixed charges $q_1, q_2, q_3, \ldots q_n$ is equal to the vector sum of the forces $\boldsymbol{F}_i$ exerted on the test charge by each of the fields of charges q_1, *i.e.*

$$\boldsymbol{F} = \sum_{i=1}^{n} \boldsymbol{F}_i \tag{7}$$

This leads to the principle of superposition of electric fields:

$$\boldsymbol{E} = \sum_{i}^{n} \boldsymbol{E}_i \tag{8}$$

2.4 PRINCIPLE OF SUPERPOSITION

If there are more than two particles present with charges, say $q_1, q_2, q_3, \ldots$, the total force on any one particle is the vector sum of forces it experiences due to all other particles present there separately. This is called the principle of superposition. Let there are three charges q_1, q_2 and q_3 as shown in Fig. 2.5. The force on q_3 is

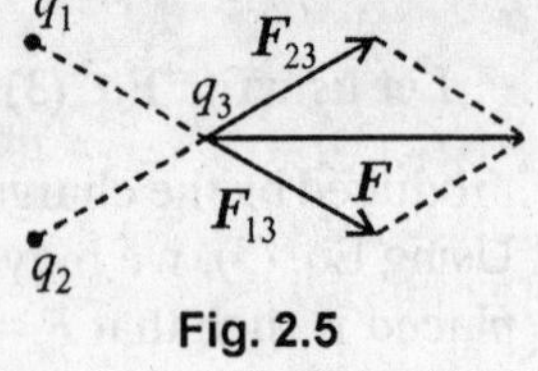

Fig. 2.5

$$F = F_{13} + F_{23}$$

$$= \frac{1}{4\pi\varepsilon_0}\frac{q_1q_3}{|\boldsymbol{r}_{13}|^3}\boldsymbol{r}_{13} + \frac{1}{4\pi\varepsilon_0}\frac{q_2q_3}{|\boldsymbol{r}_{23}|^3}\boldsymbol{r}_{23}$$

Generalising, one finds the force acting on a charge q_j due to a number of other charges present in the region is

$$\boldsymbol{F}_j = \frac{1}{4\pi\varepsilon_0}\sum_{i \neq j}\frac{q_iq_j}{|r_{13}^3|}\boldsymbol{r}_{ij} \tag{9}$$

Let r_i and r_j, are the vectors representing the location of charges q_i and q_j respectively (Fig. 2.6). One can write the Eq. (9) in terms of these position vectors as

$$\boldsymbol{F}_j = \frac{1}{4\pi\varepsilon_0}\sum_{i \neq j}\frac{q_iq_j}{|\boldsymbol{r}_j - \boldsymbol{r}_i|^3}(\boldsymbol{r}_j - \boldsymbol{r}_i) \tag{10}$$

The experimentally observed linear superposition of forces due to many charges means that one can write the electric field at $\boldsymbol{X}$ due to a system of point charges q_i, located at $\boldsymbol{X}_i$, $i = 1, 2, \ldots, n$ as a vector sum

Fig. 2.6

$$\boldsymbol{E}(X) = \frac{1}{4\pi\varepsilon_0}\sum_{i=1}^{n} q_i \frac{\boldsymbol{X} - \boldsymbol{X}_i}{|\boldsymbol{X} - \boldsymbol{X}_i|^3} \tag{11}$$

If the charges are so small and so numerous that they can be described by a charge density $\rho(\boldsymbol{x}')$ [if Δq is the charge in a small volume $\Delta x \Delta y \Delta z$ at the point $\boldsymbol{X}'$, then $\Delta q = \rho(X')\Delta x \Delta y \Delta z$], the sum in Eq. (11) is replaced by the integral

$$E(X) = \frac{1}{4\pi\varepsilon_0}\int \rho(X')\frac{X - X'}{|X - X'|^3}\, d^3x' \tag{12}$$

where $d^3x' = dx'dy'dz'$ is a three-dimensional volume element at $\boldsymbol{X}'$.

We have seen that the principle of superposition has facilitated considerably the mathematical handling of electromagnetic theory. However, this principle fails in nuclear interactions and this is one of the reasons why the nuclear theory is somewhat more complicated and troublesome than the theory of atomic interactions.

2.5 DIRAC DELTA FUNCTION

It is sometimes convenient to regard a point charge as a 'fictitious' continues charge distribution. One can achieve this with the help of Dirac delta function, written $\delta(x-a)$, is a mathematically improper function having the following properties:

(*i*) $\delta(x-a) = 0$ for $x \neq a$, and

(*ii*) $\int \delta(x-a)dx = 1$ if the region of integration includes $x = a$, and is zero otherwise.

(*iii*) $\int f(x)\delta(x-a)dx = f(a)$

(*iv*) $\int f(x)\delta'(x-a)dx = -f'(a)$ where prime denotes differentiation with respect to the argument.

(*v*) $\delta(f(x)) = \sum_i \frac{1}{\left|\frac{df}{dx}(x_i)\right|}\delta(x-x_i)$ where $f(x)$ is assumed to have only simple zeros, located at $x = x_i$. In more than one dimension, one merely take products of delta functions in each dimension. In three dimensions, for e.g., with cartesian coordinates.

(*vi*) $\delta(x-X) = \delta(x_1 - X_1)\delta(x_2 - X_2)\delta(x_3 - X_3)$ is a delta function that vanishes everywhere except at $x = X$, and is such that

(*vii*) $\int_{\Delta V} \delta(x-X)d^3x = \begin{cases} 1 \text{ if } \Delta V \text{ contains } x = X \\ 0 \text{ if } \Delta V \text{ does not contains } x = X \end{cases}$

We must remember that a delta function has the dimensions of an inverse volume in whatever number of dimensions the space has.

A discrete set of point charges can be described with a charge density by means of delta functions. For example,

$$\rho(x) = \sum_{i=1}^{n} q_i(x - x_i) \tag{13}$$

represents a distribution of n point charges q_i, located at the points x_1. Using Eq. (13) in Eq. (12) and integration, using the properties of the delta function, yields the discrete sum (Eq. 11).

2.6 LINES AND TUBES OF FORCE

Just as in the case of a gravitational field, an electric field may be represented by lines of force, which are lines that, at each point, are tangent to the direction of the electric field at that point, *i.e., lines of force* or *field lines* are parallel to the direction of the field every where. Since there is a single direction for the electric field at every point of the field and hence there is usually just one field line through any given point, *i.e.* two field lines can never intersect. The lines of force in Fig. 2.7 (*a*) depict the electric field of a positive charge, and those in Fig. 2.7 (*b*) show the electric field of a negative charge. We can see that they are straight lines passing through the charges.

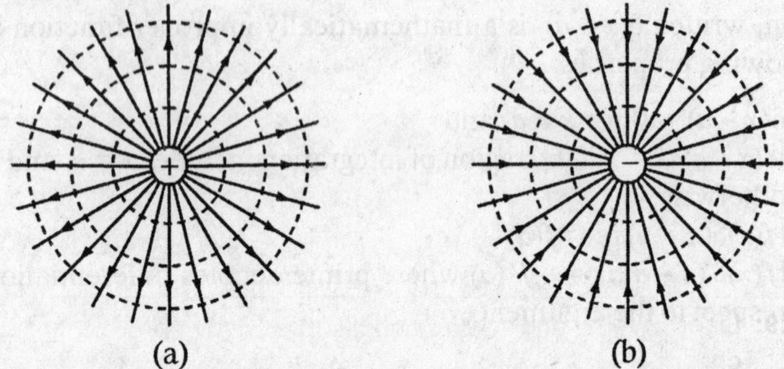

Fig. 2.7: Lines of force and equipotential surfaces of the electric field of a positive and negative charge

Figures 2.8 and 2.9 shows the lines of force near a pair of charges of equal magnitude one positive and one negative and for two equal positive charges respectively.

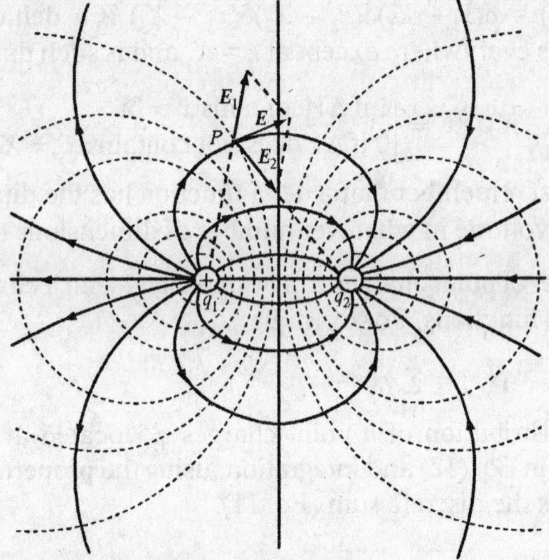

Fig. 2.8: Lines of force and equipotential surfaces of the electric field of two equal but opposite charges

In both figures the lines of force of the resultant electric field produced by the two charges have also been represented.

Let us consider an element of area *ds* in the field (small enough). Suppose that the electric intensity to be the same in magnitude and direction at every point of *ds*. So that the field of lines or lines of force at all points on the boundary will be approximately all parallel.

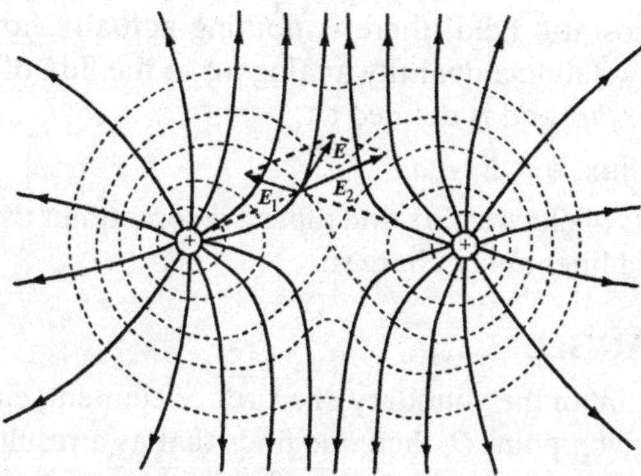

Fig. 2.9: Lines of force and equipotential surfaces of the two identical positive charges

One obtains the "tubular" surface as Fig. 2.10 shown in and it is called a '*tube of force*' we must note that the normal cross-section of the tube of force forms a part of an *equipotential surface*.

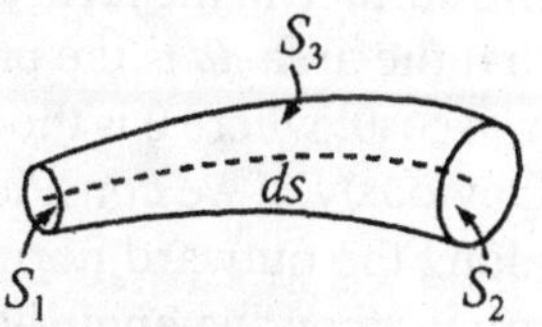

Fig. 2.10: Tube of flux

2.7 ELECTRIC FLUX

The concept of electric flux is of great usefulness in many physical problems. To understand the concept of electric flux let us imagine that a fluid is flowing with a speed V through a small flat surface ds in a direction normal to the surface as shown in Fig. 2.11 (*a*). The rate of flow of the fluid, *i.e.* the volume of the fluid crossing the area per unit time is termed as the flux of fluid and it is equal to vds. Now, if the normal to the surface is not parallel to the direction of flow of the fluid and makes an angle with the surface as shown in Fig. 2.11 (*b*), then one finds the projected area in a plane perpendicular to v is $ds\cos\theta$. Thus the flux F is

$$F = vds\cos\theta \tag{14}$$

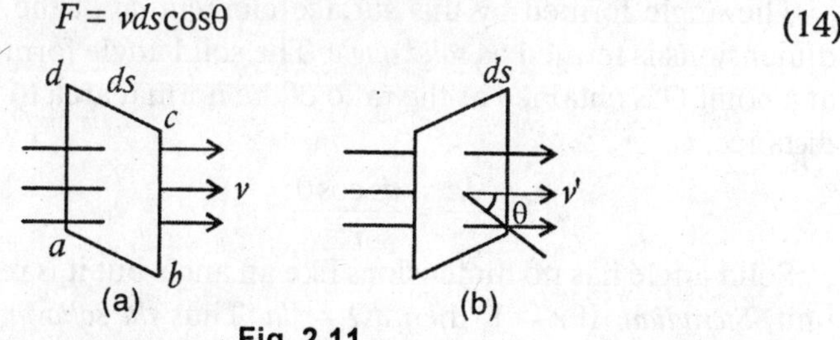

Fig. 2.11

Let us represent the surfaces by vector $ds\,\hat{a}_n$ having magnitude ds and direction along the normal to the surface. $\hat{a}_n$ being the unit vector in this direction. One can express the flux as

$$\boldsymbol{F} = \boldsymbol{v} \cdot \hat{a}_n ds \tag{15}$$

In an electrostatic field, there is nothing actually flowing, one can mathematically define a quantity analogous to the flux of a fluid. This is called as *electric flux* and is defined as

$$\text{Electric flux, } \phi = \boldsymbol{E} \cdot \hat{a}_n ds \qquad (16)$$

The concepts of electric flux and tubes are sometimes useful in drawing the electric field lines about charges.

2.8 SOLID ANGLE

When every point of the boundary of a surface element, e.g. *CD* of area *ds* is joined to a given point *O*, then one finds that as a result of it, a cone is formed. Let us imagine a spherical surface with centre at *O* and of radius *r* (=distance of surface element from *O*), then the above cone will intersect this surface in the form of another surface *AB* of area *da*. We can easily see that the area *da* is the projection of the area *ds* perpendicular to *OP i.e.* $da = ds\cos\theta$, where θ is the angle between the surfaces *AB* and *CD* (Fig. 2.12). Obviously, if we consider the surface element as vector whose direction is along the outward normal to the surface element and *O* is considered as origin, then the angle between **ds** and **r** will be θ. We have

$$da = \hat{r} \cdot \boldsymbol{ds} = ds\cos\theta$$

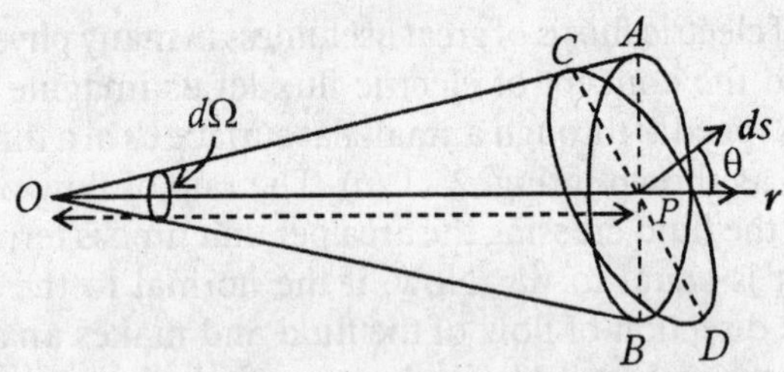

Fig. 2.12: Concept of a solid angle

The angle formed by this surface element *da* at the point *O* in three dimensions is termed as *solid angle.* The solid angle formed by the area *ds* at a point *O* is obtained as the ratio of the normal area to the square of the distance, *i.e.*

$$d\Omega = \frac{da}{r^2} = \frac{ds\cos\theta}{r^2} \qquad (16a)$$

Solid angle has no dimensions like an angle but it is represented by the unit, *Steradian.* If $r = 1$, then $d\Omega = da$. Thus *the solid angle formed by any surface at point O is equal in magnitude to the area intersected by the cone or pyramid formed by joining every point of that surface to the point O, on a spherical surface of unit radius and having centre at O.*

Let us now determine the solid angle formed by the element of the area *PQRS* situated on the spherical surface of radius *r*, at the (centre *O* (Fig.

2.13)). Let us draw planes *TOM* and *TON* making angles ϕ and $\phi + d\phi$ respectively from the plane *ZOX*, *i.e.* the angle formed by the side *SP* of area element *PQRS* on the Z-axis is $d\phi$. From Fig. 2.13 it is evident that lines *OP* and *OQ* make angles θ and $\theta + d\theta$ from the Z-axis and hence the angle formed at the centre *O* by the side *PQ* of the area element is $d\theta$ as shown in Fig. 2.13.

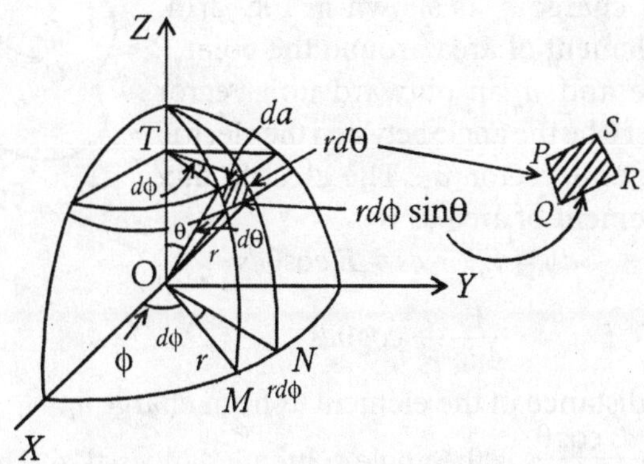

Fig. 2.13: Concept of a solid angle

The radius of the spherical surface *PQRS* is *r*, and hence

$$PQ = rd\theta$$

From ΔOPT, we have

$$\frac{TP}{OP} = \sin\theta$$

$\therefore$ $$TP = OP\sin\theta = r\sin\theta \tag{16b}$$

or $$SP = (r\sin\theta)d\phi$$

$\therefore$ Area of element $PQRS = PQ \times PS = (rd\theta)r\sin\theta d\phi$

or $$da = r^2\sin\theta d\theta d\phi \tag{16c}$$

$\therefore$ Soild angle $d\Omega = \dfrac{da}{r^2} = \dfrac{r^2 \sin\theta d\theta d\phi}{r^2}$

$$= \sin\theta d\theta d\phi \tag{16d}$$

Now, the solid angle formed at the centre by the entire spherical surface is equal to the sum of solid angles formed at the centre by all the area elements situated on the surface. Since the values of θ and ϕ very from 0 to π and 0 to 2π respectively for the entire spherical surface and hence solid angle subtended by spherical surface at its centre is

$$\Omega = \int_0^{2\pi}\int_0^{\pi}\sin\theta d\theta d\phi = 2\pi\int_0^{\pi}\sin\theta d\theta$$

$$-2\pi(-\cos\theta)_0^{\pi} = 4\pi \tag{16e}$$

2.9 GAUSS'S LAW FOR THE ELECTRIC FIELD (INTEGRAL FORM)

One can use relation (16) to obtain an important law in electrostatics, *i.e.* Gauss's law. This law relates the electric flux (ϕ) through any closed surface to the net amount of charge within the surface.

Let us now imagine a closed surface S surrounding a charge q_i as shown in Fig. 2.14. Let ds be an element of area around the point P on the surface and $\hat{a}_n$ an outward unit vector normal to it. Let θ be the angle between the electric field at P and unit vector $\hat{a}_n$. The electric flux through the element of area ds is

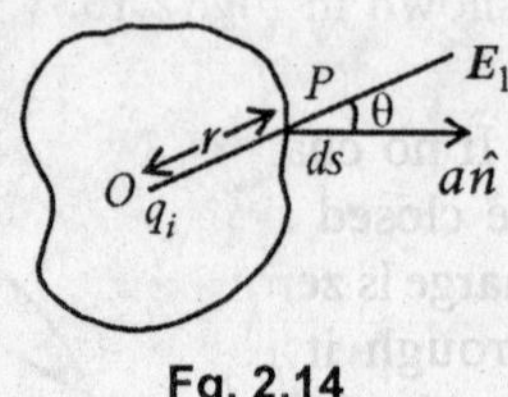

Fg. 2.14

$$d\phi = \boldsymbol{E}_i \cdot \hat{a}_n ds = \boldsymbol{E}_i \cos\theta ds$$

$$= \frac{1}{4\pi\varepsilon} \frac{q_i}{r^2} \cos\theta ds$$

where r is the distance of the element ds from charge q_i.

Now $d\Omega = \dfrac{ds \cos\theta}{r^2}$ is the angle subtended by ds at q_i. Thus

$$d\phi = \boldsymbol{E}_i \cdot \hat{a}_n ds = \frac{1}{4\pi\varepsilon_0} q_i d\Omega \tag{17}$$

Thus the total flux through the entire surface S is

$$\phi = \int_S \boldsymbol{E}_i \cdot \hat{a}_n ds = \frac{1}{4\pi\varepsilon_0} q_i \int d\Omega = \frac{q_i}{\varepsilon_0} \tag{18}$$

Obviously, the electric flux through the surface S, then is proportional to the charge and independent of the distance r (in case of a spherical surface r be radius). Therefore, if we draw several concentric spherical surfaces S_1, S_2, S_3, ... (Fig. 2.15) around the charge q_i, the electric flux through all of them is the same and equal to $\dfrac{q}{\varepsilon_0}$. The result is due to $\dfrac{1}{r^2}$ dependence of the field. The is also true in the case of gravitational field.

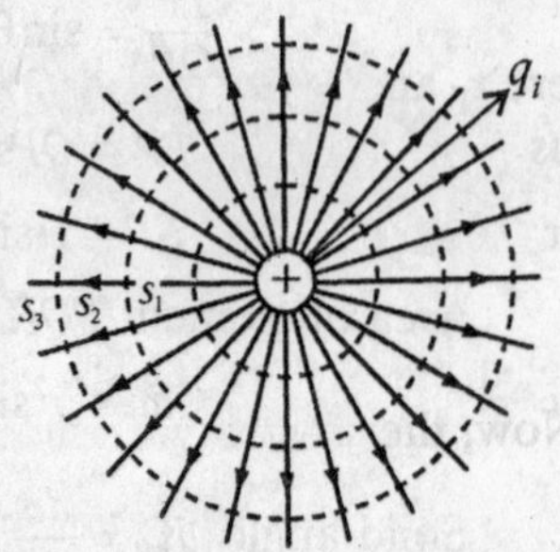

Fig. 2.15: The electric flux through concentric spheres surrounding the same charge is same.

If there is any arbitrary distribution of charges within the surface S, by the principle of superposition

$$\boldsymbol{E} = \sum \boldsymbol{E}_i \tag{19}$$

Eq. (19) leads to the relation

$$\oint_S \boldsymbol{E}_i \cdot \hat{a}_n ds = \frac{\sum q_i}{\varepsilon_0} = \frac{Q}{\varepsilon_0} \tag{20}$$

where Q is the total charge ($= \Sigma q_i$) within the closed surface S. This result is valid for any closed surface, irrespective of the position of the charge within the surface.

In the case of a continuous distribution of charge within the surface S, the relation (20) changes to

$$\oint_s \boldsymbol{E}_i \cdot \hat{a}_n ds = \frac{1}{\varepsilon_0} \int \rho dv \tag{21}$$

If no charges are present inside the closed surface S, or if the net charge is zero, the total electric flux through it is zero. The charges outside the closed surface do not contribute to the total flux. One can easily verify it. Let us suppose that charge q_i is outside the closed surface S (Fig. 2.16). The electric field vector $\boldsymbol{E}$ passes through two elements of area ds_1 and ds_2 cut out by a cone with its apex at charge q_i as shown in Fig. 2.16. $\hat{a}_n$ and $\hat{a}_{n_2}$ are the outward drawn unit vectors normal to ds_1, and ds_2 making angles θ_1 and θ_2 respectively with the field. The net outward flux through ds_1 and ds_1 area is

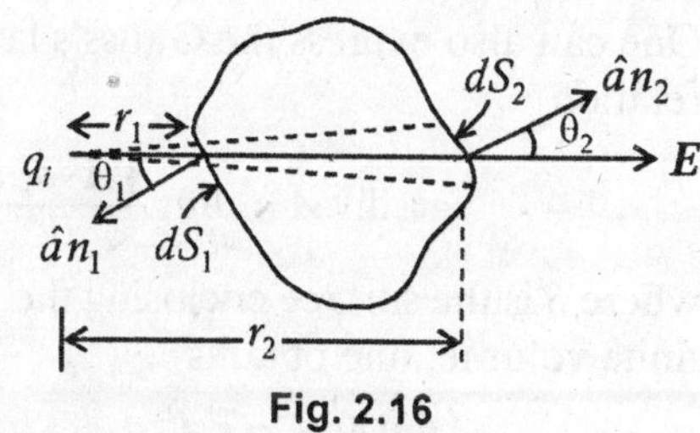

Fig. 2.16

$$d\phi = \frac{1}{4\pi\varepsilon_0}\left[\frac{q_i}{r_1^2}\cos(180°+\theta_1)ds_1 + \frac{q_i}{r_2^2}\cos\theta_2 ds_2\right]$$

where r_1 and r_2 are the distances of the elements ds_1, ds_2 from charge q_i. Thus

$$d\phi = \frac{1}{4\pi\varepsilon_0}\left[-\frac{\cos\theta_1}{r_1^2}ds_1 + \frac{\cos\theta_2}{r_2^2}ds_2\right]$$

Now, the solid angle subtended by ds_1 and ds_2 at q_i are the same, *i.e.*

$$\frac{\cos\theta_1}{r_1^2}ds_1 = \frac{\cos\theta_2}{r_2^2}ds_2 = d\Omega$$

Hence $\quad d\theta = 0$

One can see that this holds good for all the cones drawn from q_i through the surface and, hence, the total flux $\phi = 0$.

The results constitute what is known as Gauss's law for the electric field. One can state it as

$$\left.\begin{aligned}\phi_s \boldsymbol{E}\cdot\hat{a}_n ds &= \frac{Q}{\varepsilon_0} \text{ for charge } Q \text{ inside } S \\ &= 0 \text{ for charge } Q \text{ outside } S\end{aligned}\right\} \tag{22}$$

Relation (22) is the *integral form* of Gauss's law for the electric field.

Gauss's law is particularly useful when we wish to compute the electric field produced by charge distribution having certain geometrical symmetries.

2.10 GAUSS'S LAW FOR THE ELECTRIC FIELD (DIFFERENTIAL FORM)

One can also express the Gauss's law in a differential form. We have the relation

$$\text{div}\,\boldsymbol{A} = \lim_{dV \to 0} \frac{\oint \boldsymbol{A} \cdot \hat{a}_n ds}{dV} \tag{23}$$

where S is the surface enclosing the volume element. Integrating over the finite volume, one obtains

$$\int_v \text{div}\boldsymbol{A} dV = \oint \boldsymbol{A} \cdot \hat{a}_n ds \tag{24}$$

We are familiar with Eq. (24), which is the well known *Divergence theorem* usually used for transforming a volume integral into a surface integral and *vice versa*. Using Eq. (24), the Gauss's law can be expressed as

$$\oint \boldsymbol{E} \cdot \hat{a}_n ds = \oint_v \text{div}\boldsymbol{E} dV = \frac{1}{\varepsilon_0} \int_v \rho dV$$

i.e. $$\int_v \left(div\boldsymbol{E} - \frac{\rho}{\varepsilon_0} \right) dV = 0$$

The above is true for any arbitrary volume V. Thus

$$\text{div}\boldsymbol{E} - \frac{\rho}{\varepsilon_0} = 0$$

or $$\nabla \cdot \boldsymbol{E} = \frac{\rho}{\varepsilon_0} \tag{25}$$

Relation (25) is the differential form of Gauss's law for the electric field. In this form, the law shows that the sources of electric displacement are free electric charges (since $\rho \neq 0$), *i.e.* the electric flux lines either leave or enter a closed surface, therefore, there must be either a source or sink of flux lines. We know from convention that flux lines start from positive charge as sources and terminate in negative charges as sinks. This shows that positive and negative charges exist independently. Obviously, Gauss's law in differential form expresses a local relation between $\boldsymbol{E}$ and ρ. Thus we may say that electric charges are the sources of the electric field, and that their distribution and magnitude determine the electric field at each point of space.

We have adopted Coulomb's law in electrostatics as the fundamental law. We have also seen that the Gauss's law in electrostatics is a consequence of the fact that the electric force between charged particles is inversely proportional to the square of the distance between them, *i.e.* Coulomb's inverse square law. One can easily see that any other law, say, $\frac{1}{r_n}$ with $n \neq 2$ would not give Gauss's law. Obviously, we can take Gauss's law as also the fundamental law of electrostatics.

2.11 APPLICATIONS OF GAUSS'S LAW

This law provides a powerful method for evaluating the electric field intensity E and potential variation due to simple charge distribution whose *symmetry* is such that the field is constant and normal over the entire Gaussion surface. We shall discuss some useful applications of Gauss's law in this section. We must remember that *symmetry* is crucial to the application of Gauss's law. There are only three kinds of symmetry which are sufficient:

(*i*) *Spherical symmetry*: One can make Gaussian surface a concentric sphere

(*ii*) *Cylindrical Symmetry*: One can make Gaussian surface a coaxial cylinder (Fig. 2.17).

(*iii*) *Plane Symmetry*: One can make use of Gaussian 'pill box'. Which straddles the surface (Fig. 2.18).

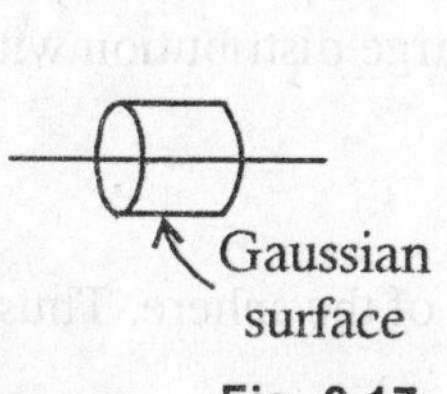

Fig. 2.17

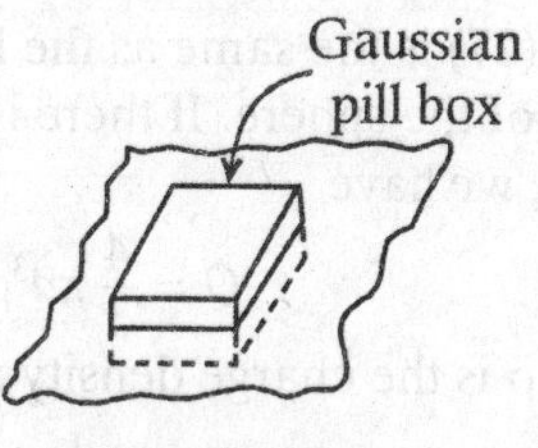

Fig. 2.18

Field due to an Infinite Layer of Positive Charge with Uniform Surface Density (σ)

Let us consider that *ABCD* be a plane as shown in Fig. 2.19. Symmetry leads that the field lines are perpendicular to the plane. Now, we consider a right circular cylinder *PQR* with cross section *ds*.

One obtains from Gauss's law

$$(E_1 - E_2)ds = \frac{\sigma ds}{\varepsilon_0}$$

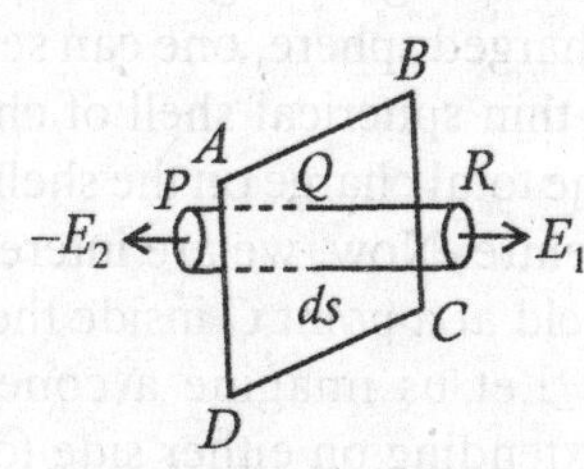

Fig. 2.19

$$\therefore \qquad E_1 - E_2 = \frac{\sigma}{\varepsilon_0} \tag{26}$$

Obviously, the field changes across a charge layer and the change is equal to $\frac{\sigma}{\varepsilon_0}$.

Field Outside an Isolated Charged Sphere

Let us suppose that a sphere A is filled with a uniform charge distribution (Fig. 2.20). We are interested in finding the electric field at a point P outside the sphere.

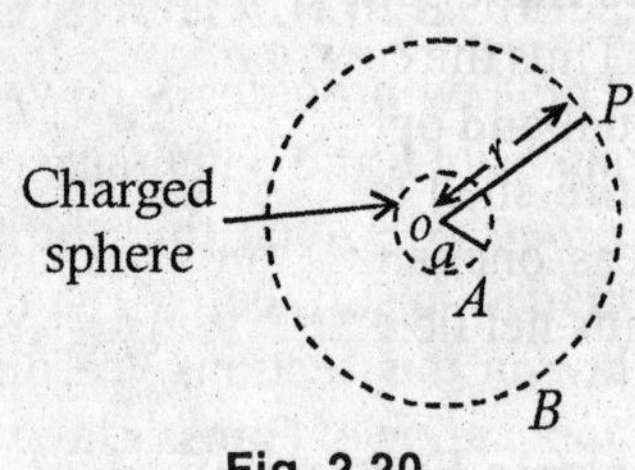

Fig. 2.20

Construct a spherical surface B concentric with the charged sphere A and passing through the point P. Let r is the radius of the imaginary sphere B and its surface area is $4\pi r^2$. By symmetry, we can say that the electric intensity E is the same at every point of the surface. Obviously, the flux outwards through the surface is

$$\oint_s \mathbf{E} \cdot \hat{a}_n ds = 4\pi r^2 E = \frac{Q}{\varepsilon_0} \quad \text{(Guass's law)}$$

where Q is the total charge within the sphere. Thus

$$E = \frac{1}{4\pi\varepsilon_0} \frac{Q}{r^2} \tag{27}$$

Eq. (27) is the same as the field produced by the point charge Q at the centre of the sphere. If there is continuous charge distribution within the sphere, we have

$$Q = \frac{4}{3}\pi a^3 \rho$$

where ρ is the charge density and a the radius of the sphere. Thus

$$E = \frac{1}{4\pi\varepsilon_0} \frac{4\pi a^3 \rho}{3r^2} = \frac{a^3 \rho}{3\varepsilon_0 r^2} \tag{28}$$

Electric Field Due to a Spherical Shell of Charge

Following the arguments applied in the case of a field outside an isolated charged sphere, one can see that the field outside a thin spherical shell of charge is the same as if the total charge on the shell is concentrated at the centre. Now, we are interested in finding out the field at a point O inside the shell.

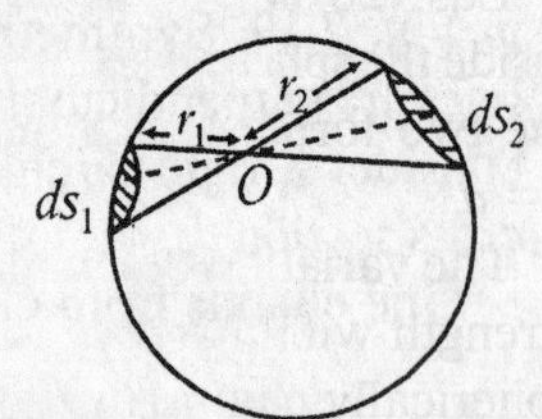

Fig. 2.21

Let us imagine a cone with apex at O and extending on either side to cut surface elements ds_1 and ds_2 as shown in Fig. 2.21. Let r_1 and r_2 be the distances of ds_1 and ds_2 from O respectively. Let σ is the surface density of charge, the fields at O due to elements are $\frac{\sigma s_1}{r_1^2}$ and $\frac{\sigma ds_2}{r_2^2}$ and these will act in opposite direction. However,

$$\frac{ds_1}{r_1^2} = \frac{ds_2}{r_2^2} = d\Omega$$

($d\Omega$ is the solid angle subtended at O by ds_1, and ds_2).

Thus the contribution to the electric field due to the two elements being equal and opposite and hence cancel exactly. Extending this to cover the entire shell by balancing the field contributions of opposite differential areas, one can see that each pair of differential areas gives a zero contribution. Thus net field at O is zero.

Electric Field Strength at an Internal Point

Let P be an internal point at a distance r from the centre of the charge distribution O as shown in Fig. 2.22. Imagine a sphere of radius [OP =] r concentric with the spherical charge. Let ρ be the volume charge density (*i.e.*, charge per unit volume), *i.e.*

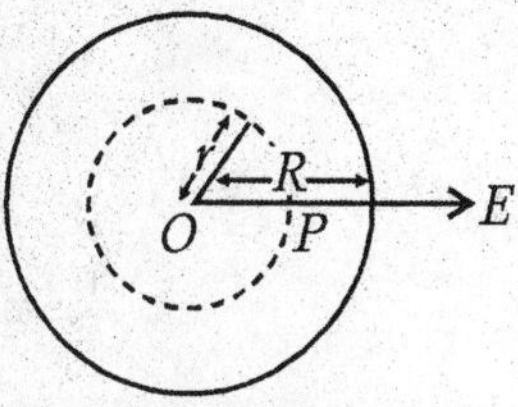

Fig. 2.22

$$\rho = \frac{\text{Charge}}{\text{Volume}} = \frac{Q}{\frac{4}{3}R^3}$$

We must note that the charge in the shell of thickness $R - r$ does not contribute to the field at P since the point P lies inside the shell. If E is the field at P, the flux through the imaginary or Gaussian surface is

$$4\pi r^2 E = \frac{1}{\varepsilon_0}\frac{4}{3}\pi r^3 \rho \qquad \text{(Guass's law)}$$

or
$$E = \frac{r\rho}{3\varepsilon_0} = \frac{1}{4\pi\varepsilon_0}\frac{Qr}{R^3}\frac{\sigma ds_2}{r_2^2} \qquad (29)$$

Thus the electric field strength at point P inside a spherical symmetric charge distribution is directly proportional to the distance of the point P from the centre of the spherical charge.

Eqs. (28) and (29) give the field outside and inside the sphere respectively. One can see that the two forms match, as they should, when $r = a$.

The variation of magnitude of electric field strength with distance from the centre of the spherically charge distribution is represented by the curve as shown in Fig. 2.23.

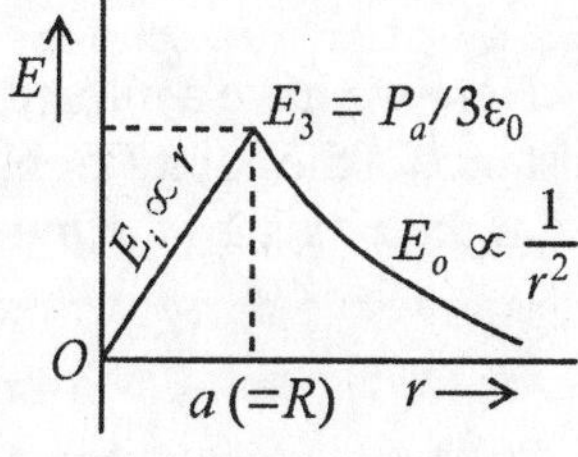

Fig. 2.23

Electric Field Strength Due to Uniformly Charged Infinite Cylinder

Let us consider a charged cylinder of infinite length of radius r as shown in Fig. 2.24. We are interested in finding the field at a point P at a distance 'd' from the axis. For this, let us imagine a closed cylindrical surface coaxial with the cylinder and passing through P, its two end faces being perpendicular to the axis of the cylinder. One can easily see by symmetry, that the field will be normal to the axis directed away from the axis and is the same at equal distances from the axis. If 'l' is the height of the cylindrical Gaussian surface, the flux through this surface is $= 4\pi lE$. We must remember that the end faces do not contribute to the field since E is tangential to these faces.

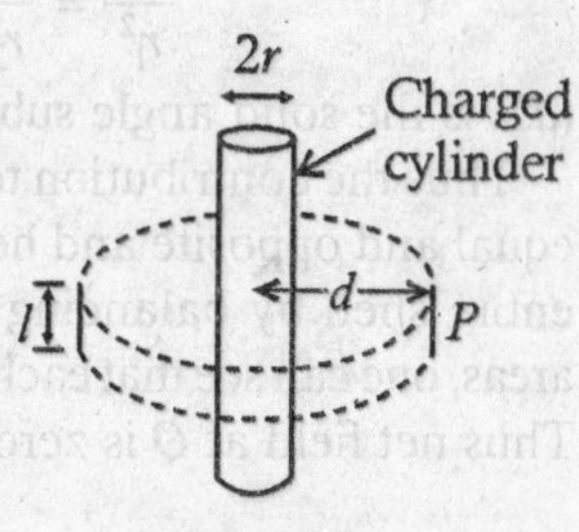

Fig. 2.24

Applying Gauss's law we have

$$2\pi lE = \frac{\lambda l}{\varepsilon_0}$$

where λ is the charge per unit length. Thus

$$E = \frac{\lambda}{2\pi\varepsilon_d} = \frac{1}{4\pi\varepsilon_0}\frac{2\lambda}{d} \tag{30}$$

Obviously, the electric field strength at a distance d from an infinite cylinder uniformly charged with linear charge density λ has magnitude $\left(\frac{\lambda}{2\pi\varepsilon_0}\right)$ which is inversely proportional to d. We can further see that the result is independent of the radius of charged cylinder and holds also for the rectilinear distribution of charges.

One could obtain the above result by direct integration. Let us consider a filament of finite length of constant line density of charge λ Coulomb/m as shown in Fig. 2.25. Now, the field at a point P due to a line element dz is

$$dE_d = \frac{\lambda dz \cos\theta}{4\pi\varepsilon_0 r^2}$$

Fig. 2.25

Here we have assumed that the filament coincide with the z-axis. The distance of the point P from the element is dz and the perpendicular distance from the z-axis is 'd'. One obtains the total field as

$$E_d = \frac{\lambda}{4\pi\varepsilon_0}\int_{-\infty}^{\infty}\frac{\cos\theta}{r^2}dz$$

We have $z = d\tan\theta$ and $r = d\sec\theta$

$$\therefore \quad E_d = \frac{\lambda}{4\pi\varepsilon_0}\int_{-\pi/2}^{\pi/2}\frac{\cos\theta}{d}d\theta = \frac{\lambda}{2\pi\varepsilon_0 d}$$

Electrostatic Field between two Concentric Spheres with Equal and Opposite Charge

Fig. 2.26 exhibit two concentric spheres 1 and 2 which have equal and opposite charges. Imagine a Gaussian surface drawn through *P*. Obviously, the field at *P* due to the outer sphere is zero and that due to the inner sphere is $E = \frac{1}{4\pi\varepsilon_0}\frac{q}{r^2}$.

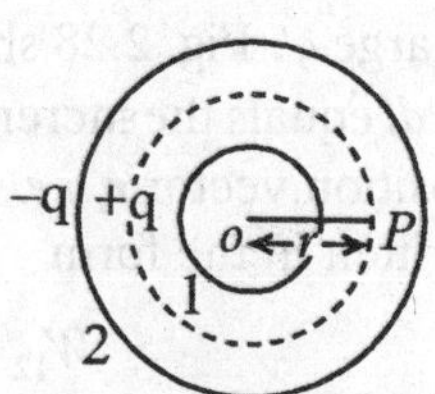

Fig. 2.26

Let us now consider the case of two nonconcentric charged spheres as shown in Fig. 2.27. Figure shows the redistribution of charge and field lines. Draw a cone with *P* as the apex. We can see that the two charge densities on ds_1, and ds_2 are not equal $dE_2 > dE_1$, and hence the two different areas yield a net *dE* in the negative direction. One can find a net E_{-} due to the outer sphere in the negative *x*-direction by covering the entire sphere. The field due to the inner sphere at *P* is E_+. Thus the effective field at *P*, $E = E_+ + E_-$.

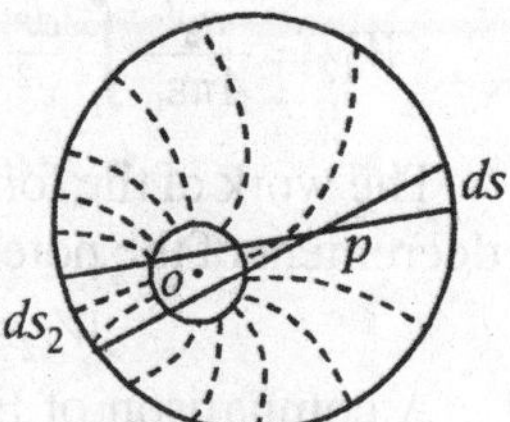

Fig. 2.27

2.12 ELECTROSTATIC POTENTIAL

A charged particle placed in an electric field has potential energy because of its interaction with the field. One can define the electric potential at a point as the potential energy per unit charge placed at that point. The electric potential is usually designated by *V* and the potential energy of a charge *q* by E_p, we have

$$V = \frac{E_p}{q} \quad \text{or} \quad E_p = qV \tag{31}$$

The electric potential is measured in Joules/Coulomb or JC^{-1}, a unit called a volt, abbreviated V. In terms of international units, $V = m^2kgs^{-2}C^{-1}$.

Let us consider the field produced by a stationary point charge *q*. At any point of this field, the point charge *q'* experiences the force

$$\boldsymbol{F} = \frac{1}{4\pi\varepsilon_0}\frac{q'q}{r^2}\hat{a}_r = F(r)\hat{a}_r \tag{32}$$

Here *F*(*r*) is the magnitude of the force ***F***, and $\hat{a}_r$ is the unit vector of the position vector ***r*** determining the position of the charge *q'* relative to the charge *q*.

The force (Eq. 32) is a central one. A central field of forces is conservative. Consequently, the work done by the forces of the field on the charge q' when it is moved from one point to another does not depends on the path. This work is

$$W_{12} = \int_1^2 F(r)\hat{a}_r dl \tag{33}$$

where dl is the elementary displacement of the charge q'. Fig. 2.28 shows that the scaler product $\hat{a}_r\, dl$ equals the increment of the magnitude of the position vector r, *i.e.* dr. Eq (33) can therefore be written in the form

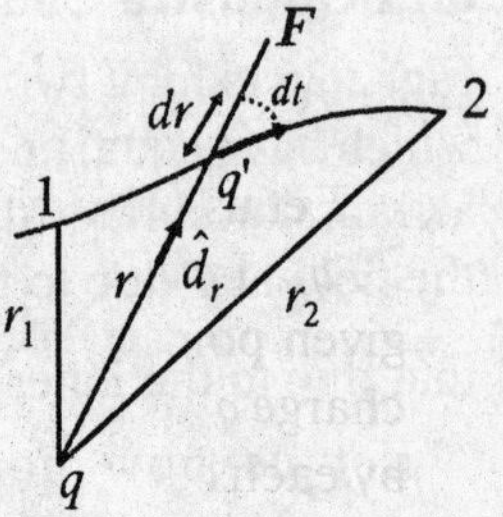

$$W_{12} = \int_1^2 F(R)dr$$

Using Eq. (32), one obtains

Fig. 2.28

$$W_{12} = \frac{qq'}{4\pi\varepsilon_0}\int_1^2 \frac{dr}{r^2} = \frac{1}{4\pi\varepsilon_0}\left(\frac{qq'}{r_1} - \frac{qq'}{r_2}\right) \tag{34}$$

The work of the forces of the conservative field can be represented as a decrement of the potential energy

$$W_{12} = V_{p,1} - V_{p,2} \tag{35}$$

A comparison of Eqs. (34) and (35) leads to the following expression for the potential energy of the charge q' in the field of charge q.

$$V_p = \frac{1}{4\pi\varepsilon_0}\frac{qq'}{r} + \text{Const}$$

The value of the constant in the expression for the potential energy is usually chosen so that when the charge moves away to infinity (*i.e.*, when $r = \infty$), the potential energy vanishes. When this condition is observed, one obtains

$$V_p = \frac{1}{4\pi\varepsilon_0}\frac{qq'}{r} \tag{36}$$

Let us use the charge q' as a test charge for studying the field. By Eq. (36), the potential energy which the test charge has dependence not only on its magnitude q', but also on the quantities q and r determining the field. Obviously one can use this energy to describe the field just like we used the force acting on the charge for this purpose.

Different test charges q_t', q_t'', etc. will have different energies V_p', V_p'' etc. at the same point of the field. But the ratio $\frac{V_p}{q_t}$ will be the same for all the charges [Eq. (36)]. The quantity

$$V \text{ or } \phi = \frac{V_p}{q_t} \tag{37}$$

is called the *field potential* at a given point and is used together with the field strength E to describe electric fields.

It can be seen from Eq. (37) that the potential numerically equals the potential energy which a unit positive charge would have at the given point of the field. Substituting for the potential energy in Eq. (37) its value from Eq. (36), one obtains the expression for the potential of a point charge

$$V \text{ or } \phi = \frac{1}{4\pi\varepsilon_0}\frac{q}{r} \tag{38}$$

Let us consider the field produced by a system of N point charges q_1, q_2 ... q_N. Let $r_1, r_2, \ldots, r_N$ be the distances from each of the charges to the given point of the field. The work done by the forces of this field on the charge q' will equal the algebraic sum of the work done by the forces set up by each of the charges separately.

$$W_{12} = \sum_{i=1}^{N} W_i$$

By Eq. (34), each work W_i equal

$$W_i = \frac{1}{4\pi\varepsilon_0}\left(\frac{q_i q'}{r_{i,1}} - \frac{q_i q'}{r_{i,2}}\right)$$

where $r_{i,1}$, is the distance from the charge q_i to the initial position of the charge q' and $r_{i,2}$ is the distance from q_i to the final position of the charge q'. Thus

$$W_{12} = \frac{1}{4\pi\varepsilon_0}\sum_{i=1}^{N}\frac{q_i q'}{r_{i,1}} - \frac{1}{4\pi\varepsilon_0}\sum_{i=1}^{N}\frac{q_i q'}{r_{i,2}}$$

Comparing this equation with Eq. (35), one obtains the following expression for the potential energy of the charge q' in the field of a system of charges.

$$V_p = \frac{1}{4\pi\varepsilon_0}\sum_{i=1}^{N}\frac{q_i q'}{r_i}$$

from which it can be seen that

$$V \text{ or } \phi = \frac{1}{4\pi\varepsilon_0}\sum_{i=1}^{N}\frac{q_i}{r_i} \tag{39}$$

In general for a set of point charges q_j the potential at a point r_i is given by the algebraic sum of the individual potential, *i.e.*

$$V(r_i) = \frac{1}{4\pi\varepsilon_0}\sum_{j}\frac{q_j}{r_{ij}} \tag{39a}$$

In case of a continuous distribution of charges

$$V(r_i) = \frac{1}{4\pi\varepsilon_0}\int \frac{\rho d\tau_j}{r_{ij}} \tag{39b}$$

Comparing Eq. (39) with Eq. (38), we arrive at the conclusion that the *potential of the field produced by a system of charges equals the algebraic sum of the potentials produced by each of the charges separately*. Whereas the field strengths are added vectorially in the superposition of fields, the potentials are added algebraical. This is why it is usually much simpler to calculate the potentials than the electric field strengths.

Examination of Eq. (37) shows that the charge q at a point of the field with the potential V or ϕ has potential energy

$$V_P = qV = q\phi \tag{40}$$

Hence, the work of the field forces on the charge q can be expressed through the potential difference:

$$W_{12} = V_{P,1} - V_{P,2} = q(V_1 - V_2) = q(\phi_1 - \phi_2) \tag{41}$$

Obviously, the work done on a charge by the forces of a field equals the product of the magnitude of the charge and the difference between the potentials at the initial and final points (*i.e.* the potential decrement).

If the charge q is removed from a point having the potential V or ϕ to infinity (where by convention the potential vanishes), then the work of the field forces will be

$$W_\infty = q\phi = qV \tag{42}$$

Obviously, the *potential numerically equals the work done by the forces of the field on a unit positive charge when the latter is removed from the given point to infinity*. Work of the same magnitude must be done against the electric field forces to move a unit positive charge from infinity to the given point of a field.

2.13. RELATION BETWEEN ELECTRIC FIELD STRENGTH AND POTENTIAL

An electric field can be described either with the aid of the vector quantity $\boldsymbol{E}$, or with the aid of the scalar quantity V or ϕ. There must evidently be a definite relation between these quantities. If we bear in mind that $\boldsymbol{E}$ is proportional to the force acting on a charge and ϕ to the potential energy of the charge, it is easy to see that this relation must be similar to that between the potential energy and the force.

The force $\boldsymbol{F}$ is related to the potential energy by the expression

$$\boldsymbol{F} = -\nabla V_P \tag{43}$$

For a charged particle in an electrostatic field, we have

$$\boldsymbol{F} = q\boldsymbol{E}$$

and

$$V_p = q\phi = qV$$

Introducing these values in Eq. (43), one obtains

$$qE = -\nabla(q\phi) = -q\nabla(\phi)$$

or
$$E = -\nabla\phi = -\text{grad}\phi \tag{44}$$

This is the relation between the field strength and potential. Since E has a finite value at any point of the field, the potential ϕ is a continuous function of the coordinates of points of the field. The rectangular components of the electric field E are given by

$$E_x = -\frac{\partial\phi}{\partial x},\ E_y = -\frac{\partial\phi}{\partial y} \text{ and } E_z = -\frac{\partial\phi}{\partial z} \tag{45}$$

Thus

$$E = -\left[\hat{i}\frac{\partial}{\partial x} + \hat{j}\frac{\partial}{\partial y} + \hat{k}\frac{\partial}{\partial z}\right]\phi \tag{45a}$$

In general, the component along the direction corresponding to a displacement ds is

$$E_s = -\frac{\partial\phi}{\partial s} \tag{46}$$

Eqs. (45) or (46) are used to find the electric potential ϕ or V when the field E is known, and conversely.

Let us consider a simple case of a uniform electric field (Fig. 2.29).

The first of Eqs. (45) gives, for an X-axis which is parallel to the field, $E = -\frac{d\phi}{dx}$. Since E is constant and we assume $\phi = 0$ at $x = 0$, by integration, one obtains

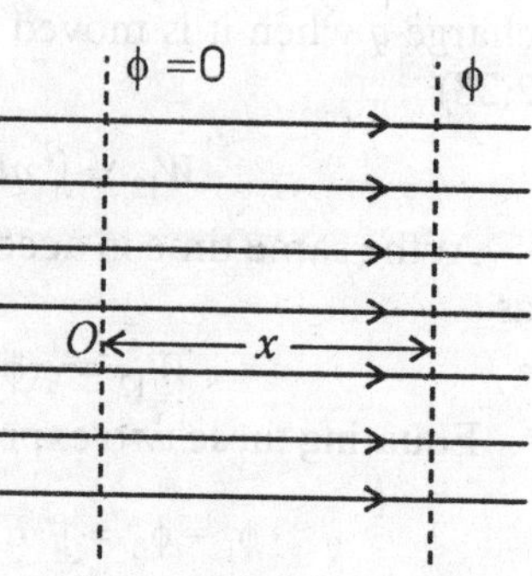

Fig. 2.29: Uniform electric field

$$\int_0^{\phi} d\phi = -\int_0^x E dx = -E\int_0^x dx \text{ or } \phi = -Ex \tag{47}$$

Eq. (47) is very useful relation represented graphically in Fig. 2.30.

We may note that, because of the negative sign in Eq. (46) or Eq. (47), the electric field points in the direction in which the electric potential decreases. When we consider two points x_1 and x_2, Eq. (47) gives $\phi_1 = -Ex_1$ and $\phi_2 = -Ex_2$. Subtracting we have $\phi_2 - \phi_1 = -E(x_2 - x_1)$; or calling $d = x_2 - x_1$, we obtain

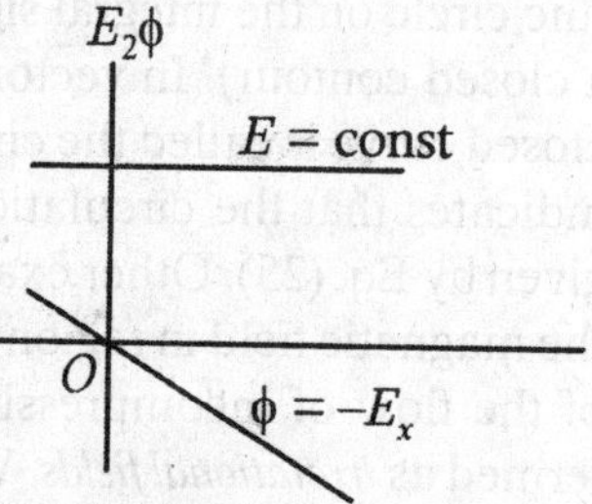

$$E = -\frac{\phi_2 - \phi_1}{d} = \frac{\phi_1 - \phi_2}{d} \tag{48}$$

Fig. 2.30: Variation of E and ϕ for a uniform electric field

Although Eq. (48) is a valid only for uniform electric fields, it can be used to estimate the electric field between two points separated by a distance d, if the potential difference $\phi_1 - \phi_2$ between them is known. If the potential difference $\phi_1 - \phi_2$ is positive, the field points in the direction from x_1 to x_2, and if it is negative it points in the opposite direction. Eq. (48) [or in fact also Eq. (45) or Eq. (46)] indicates that the electric field can also be expressed in volts/metre, a unit which is equivalent to the Newtons/Coulomb given before. One can easily verify it in the following way.

$$\frac{\text{Volt}}{\text{meter}} = \frac{\text{Joule}}{\text{Coulomb - m}} = \frac{\text{Newton - m}}{\text{Coulomb - m}} = \frac{\text{Newton}}{\text{Coulomb}}$$

By common usages, the term volt/m, abbreviated Vm^{-1} is preferred to NC^{-1}.

We have seen that Eq. (44) allows us to find the field strength at every point from the known values of ϕ. One can also solve the reverse problem, *i.e.*, find the potential difference between two arbitrary points of a field according to the given values of E. For this purpose, one can take advantage of the circumstance that the work done by the forces of a field on the charge q when it is moved from point 1 to point 2 be calculated as (Fig. 2.28)

$$W_{12} = \int_1^2 qEdl$$

At the same time in accordance with Eq. (41), this work can be written as

$$W_{12} = q(\phi_1 - \phi_2)$$

Equating these two expressions, one obtains

$$\phi_1 - \phi_2 = \int_1^2 \mathbf{E} \cdot dl \tag{49}$$

The integral can be taken along any lines joining points 1 and 2 because the work of the field forces is independent of the path for circumvention along closed contour $\phi_1 = \phi_2$, and Eq. (49) becomes

$$\oint \mathbf{E} \cdot dl = 0 \tag{50}$$

(the circle on the integral sign indicates that integration is performed over a closed contour). In vector calculus the line integral of a vector round a closed curve is called the circulation or the curl of the vector field. Eq. (50) indicates that the circulation of the electric field is zero. Its sources are given by Eq. (25). Other examples of such fields are the gravitational field, the magnetic field in region free from electric current and the velocity field of the flow of incompressible fluids free from viscosity. Such fields are termed as *irrotational fields*. We must note that relation (50) is not applicable to the field of moving charges (*i.e.*, a field changing with time), *i.e.* this field is not a potential one.

Now, we take curl of vectors in Eq. (44) one obtains

$$\nabla \times \boldsymbol{E} = -\nabla \times \nabla\phi = 0 \tag{50a}$$

Using Stokes' theorem, one can transform line integral in Eq. (50) into a surface integral

$$\oint \boldsymbol{E} \cdot d\boldsymbol{l} = \int_s \nabla \times \boldsymbol{E} \cdot \hat{a}_n dS$$

$$\therefore \quad \int_s \nabla \times \boldsymbol{E} \cdot \hat{a}_n dS = 0$$

and since S is arbitrary, $\nabla \times \boldsymbol{E} = 0$. Thus, one obtains the following two important laws of electrostatics:

$$(i) \qquad \nabla \cdot \boldsymbol{E} = \frac{\rho}{\varepsilon_0} \tag{51}$$

$$(ii) \qquad \nabla \times \boldsymbol{E} = 0 \tag{52}$$

Eq. (51) follows from Coulomb's inverse square law. Eq. (52) does not depend upon this law. All predictions of electrostatics follow from these laws and hence Eq. (51) and Eq. (52) can be considered fundamental laws of electrostatics.

2.14 EQUIPOTENTIAL SURFACES

A locus of the points in an electrostatic field having the same potential is called an equipotential surface. Its equation has the form

$$\phi(r) = c$$

where c is a constant, represents an equipotential surface. Different values of c generate a family of such surfaces.

The potential does not change in movement along an equipotential surface over the distance $dr (d\phi = 0)$. Hence, according to Eq. (46), the tangential component of the E to the surface equals zero. Let us consider a displacement $d\boldsymbol{r}$ on an equipotential surface

$$\phi(\boldsymbol{r}) - \phi(\boldsymbol{r} + d\boldsymbol{r}) = 0$$

$$i.e. \qquad \phi(\boldsymbol{r}) - \left\{\phi(\boldsymbol{r}) + \frac{\partial \phi}{\partial r} \cdot d\boldsymbol{r}\right\} = 0$$

$$\therefore \qquad \frac{\partial \phi}{\partial r} \cdot d\boldsymbol{r} = 0$$

$$i.e. \qquad \nabla\phi \cdot d\boldsymbol{r} = 0$$

$$\text{or} \qquad \boldsymbol{E} \cdot d\boldsymbol{r} = 0 \tag{53}$$

This shows that if $d\boldsymbol{r}$ is on the surface, $\boldsymbol{E}$ must be normal to the surface, *i.e.*, the field lines are normal or orthogonal to the equipotential surface.

An equipotential surface can be drawn through any point of a field, consequently one can construct an infinitely great number of such surfaces.

They are conventionally drawn so that the potential difference for two adjacent surfaces is the same everywhere. Thus the density of the equipotential surfaces allows us to assess the magnitude of the field strength. Indeed the denser are the equipotential surfaces, the more rapidly does the potential change when moving along a normal to the surface. Hence $\nabla\phi$ is at the given place, and therefore $\boldsymbol{E}$ is greater too.

Figure 2.7 shows equipotential surfaces (more exactly, their intersections with the plane of the drawing) for the field of a point charge. Figures 2.8 and 2.9 shows the equipotential surfaces along with the lines of force for two equal but opposite charges and two identical charges respectively. In accordance with the nature of the dependence of E or r equipotential surfaces become the denser, the nearer we approach a charge.

Equipotential surfaces for a homogeneous field are a collection of equispaced planes right angles to the direction of field.

2.15 ELECTROSTATIC ENERGY

The energy of electrostatic systems is solely the potential energy arising from the interaction between the charges.

Let us consider a charge q_1, be situated at a certain point in space and charge q_2 which was initially at infinity, be brought up to a distance r_{12} from q_1. We have the potential of the charge q_1 at r_{12} is $\frac{q_1}{4\pi\varepsilon_0 r_{12}}$. Obviously, this is the work done in bringing a unit charge from infinity at the charge q_2 from infinity to the point specified by r_{12} is $\frac{q_1 q_2}{4\pi\varepsilon_0 r_{12}}$.

Now a third charge q_3 is added to the system. We will have to perform work against the field of q_1 and q_3. If r_{13}, r_{23} are the distances of q_3 from q_1 and q_2 respectively, then the additional contribution to the potential energy is as follows.

$$\frac{q_1 q_3}{4\pi\varepsilon_0 r_{13}} + \frac{q_2 q_3}{4\pi\varepsilon_0 r_{23}}$$

Proceeding to build up the assembly in this way, one finds the total energy of the assembly is

$$W = \frac{q_2}{4\pi\varepsilon_0}\left(\frac{q_1}{r_{12}}\right) + \frac{q_3}{4\pi\varepsilon_0}\left(\frac{q_1}{r_{13}} + \frac{q_2}{r_{23}}\right) + \frac{q_4}{4\pi\varepsilon_0}\left(\frac{q_1}{r_{14}} + \frac{q_2}{r_{24}} + \frac{q_3}{r_{34}}\right) + \ldots$$

$$= \frac{1}{4\pi\varepsilon_0}\sum_i q_i \sum_{j<i} \frac{q_i}{r_{ij}} \qquad (54)$$

The restriction $j < i$ ensures that the interaction between every pair is counted only once. One can write Eq. (54) in a different form, *i.e.*

$$W = \frac{1}{4\pi\varepsilon_0}\frac{1}{2}q_1\left(\frac{q_2}{r_{12}}+\frac{q_3}{r_{13}}+\ldots\right)+\frac{1}{4\pi\varepsilon_0}\frac{1}{2}q_2\left(\frac{q_1}{r_{21}}+\frac{q_3}{r_{23}}+\ldots\right)$$

$$+\frac{1}{4\pi\varepsilon_0}\frac{1}{2}q_3\left(\frac{q_1}{r_{31}}+\frac{q_2}{r_{32}}+\frac{q_4}{r_{34}}+\ldots\right)+\ldots$$

$$= \frac{1}{4\pi\varepsilon_0}\frac{1}{2}\sum_{i=1} q_i \sum_{j=i}\frac{q_j}{r_{ij}};\ i \neq j \tag{55}$$

The factor ½ appears in Eq. (55) because in this expression each pair is counted twice.

Since $= \frac{1}{4\pi\varepsilon_0}\sum_{j\neq i}\frac{q_j}{r_{ij}}$ is the potential ϕ_i, produced by all the charges of the system except ith one at the point at which q_i is situated, we have

$$W = \frac{1}{2}\sum_{i} q_i\phi_i \tag{56}$$

When there is a continuous distribution of charges and charges are not localized, we have

$$W = \frac{1}{2}\int \phi_\rho d\tau \tag{57}$$

ILLUSTRATIVE EXAMPLES

Example 1

A point charge of 2×10^{-6} C is placed at the centre of a hypothetical cube with the side 0.1 m each. Determine the electric flux through each face.

Solution

A cube has six equal faces. The charges is situated at a centre of the cube. Obviously, equal amount of flux will be passing through each face.

Electric flux from charge q is equal to $= q/\varepsilon_0 = \frac{2\times10^{-6}}{\varepsilon_0}$

Therefore flux from each face $= \frac{2\times10^{-6}}{6\times8.85\times10^{-12}}$

$= 37.7 \times 10^3$ N-m²/C.

Example 2

Find the magnitude and direction of electric field to keep an electron in equilibrium in air.

Solution

$$Ee = mg$$

or
$$E = \frac{mg}{e} = \frac{3.1 \times 10^{-31} \times 9.8}{1.6 \times 10^{-19}}$$
$$= 55.74 \times 10^{-12} \text{ N/C}$$

Example 3

A rectangular surface 0.2 m × 0.3 m is placed in uniform electrostatic field 2000 N/C, such that normal to the surface makes an angle of 30° with direction of electrostatic field. Find the flux density through this surface.

Solution

$$d\phi = \mathbf{E} \cdot \mathbf{ds} = Eds\cos\theta$$
$$= 0.06 \times 2000 \times \frac{\sqrt{3}}{2}$$
$$= 60\sqrt{3} = 103.92 \frac{\text{N - m}^2}{\text{C}}$$

Example 4

A large metal plate has a uniform surface charge density $\sigma = -2 \times 10^{-6}$ C/m^2. From what distance should the electron with energy 100 eV be fired so that it just fails to strike the plate?

Solution

Suppose that the electron be thrown towards the plate from a distance *d*. Electric field due to plate
$$E = \sigma / \varepsilon_0$$
Force experienced by electron due to electric field
$$F = Ee = \frac{\sigma e}{\varepsilon_0}$$
Work done on electron in bringing him up to the plate
$$F = \frac{e\sigma d}{\varepsilon_0}$$
If the work done is greater than electron energy, the electron will not able to reach the plate. Thus
$$\frac{e\sigma d}{\varepsilon_0} = 100 \text{ eV} = 100 \times 1.6 \times 10^{19} \text{ J}$$
or
$$d = \frac{100 \times 1.6 \times 10^{-19} \times 2 \times 8.852 \times 10^{-12}}{1.6 \times 10^{-19} \times 2 \times 10^{-6}}$$
$$= 4.42 \times 10^{-4} \text{ m} = 0.442 \text{ mm}$$

Clearly, if the electron is being thrown greater than 4.43×10^{-4} m distance, it will not be able to reach the plate.

Example 5

Two small spheres are positively charged. The amount of charge on each sphere is different. The combined charge on both the spheres is 5×10^{-5} C. When the spheres are 2 m apart, each sphere is repelled from the other by a force of 1 N. Find, how the charges are distributed between the spheres.

Solution

Let the charges on both the spheres be q_1 and q_2 C.

$\therefore$ $$q_1 + q_2 = 5 \times 10^{-5} \text{ C} \qquad (i)$$

$$Fe = 1 = \frac{q_1 q_2}{4\pi\varepsilon_0 (2)^2} = \frac{q_1 q_2 \times 9 \times 10^9}{4}$$

or $$q_1 q_2 = \frac{4}{9} \times 10^{-9} \qquad (ii)$$

From Eqs. (*i*) and (*ii*), we have

$$q_1 - q_2 = \left[(q_1 + q_2)^2 - 4q_1 q_2\right]^{1/2}$$

$$= \left[25 \times 10^{-10} - \frac{16}{9} \times 10^{-9}\right]^{1/2}$$

$$= \left(0.72 \times 10^{-9}\right)^{1/2} = 2.68 \times 10^{-5} \qquad (iii)$$

Solving Eqs. (*i*) and (*iii*), one obtains

$$q_1 = 38.4 \times 10^{-6} \text{ C and } q_2 = 11.6 \times 10^{-6} \text{ C.}$$

Example 6

An infinite line charge has a charge density 2 micro Coulomb/m. Calculate the electric field at a point which is at a normal distance of 10 cm from the line charge.

Solution

We have $$\boldsymbol{E} = \frac{\lambda}{2\pi\varepsilon_0 r}\hat{r}$$

$$= \frac{2 \times 2 \times 10^{-6} \times 9 \times 10^9}{(0.1)^2}\hat{r}$$

$= 36 \times 10^4$ N/C along the perpendicular direction of line charge.

Example 7

Two large metal plates of area 1 m^2 face each other. They are 5 cm apart and carry equal and opposite charges on their inner surfaces. If E between the plates is 55 N/C, what is the charge on the plates?

Solution

We have $$E = \frac{\sigma}{\varepsilon_0} = \frac{q}{A\varepsilon_0}$$

$\therefore$ $$q = EA\varepsilon_0 = 55 \times (1)^2 \times 8.55 \times 10^{-12}$$

$$= 0.49 \times 10^{-9}\text{ C}$$

Example 8

A spherical charge distribution is given by

$$r = r_0(1 - r/a) \text{ when } r \gtrless a$$

$$r = 0 \qquad \text{when } r > a_0$$

where a is the radius of the sphere. Find the electric field intensity at points outside and inside. Find the value of r for which E is maximum.

Solution

Electric field at external point

$$E = \frac{\rho_0}{\varepsilon_0}\frac{a^3}{12r^2} \qquad (i)$$

Electric field at internal point

$$E = \frac{\rho_0}{\varepsilon_0}\left(\frac{r}{3} - \frac{r^2}{4a}\right) \qquad (ii)$$

E is maximum inside the charge distribution at a distance is given by

$$r = 2a/3.$$

Example 9

A thin electric shell of metal has a radius of 0.25 micro Coulomb. Calculate the electric intensity at a point (*i*) inside the shell (*ii*) just outside the shell, and (*iii*) 3 m from the centre of the shell.

Solution

(*i*) The electric intensity at a point inside a charged shell is zero.

(*ii*) For external points the shell behaves as if its entire charge is concentrated at the centre. Thus, if R be the radius of the shell, the intensity at a point just outside the shell is

$$E = \frac{1}{4\pi\varepsilon_0}\frac{q}{R^2} \qquad (i)$$

Here $q = 0.2$ micro Coulomb $= 0.2 \times 10^{-6}$ C, $R = 0.25$

$$\therefore \quad E = 9 \times 10^9 \times \frac{0.2 \times 10^{-6}}{(0.25)^2} = 2.88 \times 10^4 \text{ N/C}$$

(*iii*) The intensity at an external point at a distance *r* from the centre of the shell is given by

$$E = \frac{1}{4\pi\varepsilon_0}\frac{q}{r^2}$$

$$r = 0.3 \text{ m}$$

$$= 9 \times 10^9 \times \frac{0.2 \times 10^{-6}}{(3)^2} = 200 \text{ N/C}$$

Example 10

The electric field in a region of space is given by $\mathbf{E} = 8x\hat{i} - 4y\hat{j} - 4z\hat{k}$, where $\hat{i}, \hat{j}$ and $\hat{k}$ are unit vectors along *x*, *y* and *z* axes respectively. Find the equations of line of force in the plane $z = 0$.

Solution

The *x*, *y* and *z* components of electric field are $E_x = 8x$, $E_y = -4y$ and $E_z = -4z$. In the plane $z = 0$, $E_z = 0$, so that the electric field is confined to the *x*-*y* plane. At any point (x, y) in this plane, the shape of the line of force is

$$\frac{dy}{dx} = \frac{E_y}{E_x} = -\frac{4y}{8x} = -\frac{y}{2x}$$

$$\therefore \quad \frac{2dy}{y} = -\frac{dx}{x}$$

Integrating, we obtain

$$\log y^2 = -\log x + \log c$$

where *c* is constant of integration.

Thus $\quad y^2 = c/x$

or $\quad xy^2 = c$

This is the equation of the line of force. For different values of *c*, we can obtain different lines of force.

REVIEW QUESTIONS

1. State Coulomb's law in electrostatics. What do you mean by the flux of an electric field?
2. State and prove Gauss's law in electrostatics. Apply it to calculate the electric field intensity due to a uniformly charged sphere (non conducting) at points (*i*) outside the sphere (*ii*) at the surface of the sphere and (*iii*) inside the sphere.

3. Using Gauss theorem discuss the electric field created by a line charge distribution.
4. Use Gauss's law to find the field intensity at a point near an infinite sheet of charge. How is the result modified if the charge lies only on one side of the sheet?
5. Deduce the Gauss's law in differential form, *i.e.*, $\nabla \cdot \mathbf{E} = \rho/\varepsilon_0$ [BE]
6. Define electric field and intensity.
7. Derive Coulomb's law from Gauss's law. Would Gauss's law hold if the exponent in Coulomb's law were not exactly two?
8. Establish on the basis of Gauss's law the fact that when an excess charge is given to an insulated conductor, the charge reside entirely on the outer surface of the conductor.
9. Apply Gauss's theorem to show that the electric field near a charged conductor is twice to that due to a none conducting thin sheet of charge, with the same surface charge density as the conductor. [BE]

PROBLEMS

1. Point charges $+4q$ and $-q$ are separated by a distance 'a'. Show that the only position where a third charge $+q$ could be in equilibrium is along the line joining the first two and at a distance 'a' from $-q$.
2. An infinite line charge has a charge density 2 micro Coulomb/m. Calculate the electric field at a point which is at a normal distance of 10 cm from the line charge. [36×10^4 N/C]
3. The flux entering a closed surface is 2×10^3 N-m^2/C and flux leaving the closed surface is 8×10^3 N-m^2/C. Calculate the charge enclosed by the surface. [5.31×10^{-8} C]
4. A thin spherical shell of metal has a radius of 0.25 m and carries a charge of 0.2 μC. Calculate the electric field at a point (*i*) 3 m from the centre of the shell (*ii*) just outside the shell, and (*iii*) inside the shell. Given $\varepsilon_0 = 8.85 \times 10^{-12}$ C^2/N^{-1}m^2. [BE] [(*i*) 200 N/C (*ii*) 2.88×10^4 N/C (*iii*) 0]
5. If the electric field near the earth's surface be 300 V/m directed downward. What is the surface density of charge on earth's surface? [2.65×10^{-9} C/m^2]
6. An insulated soap bubble of radius 10 cm is given a charge of 4.8×10^{-8} C. Find the increase in radius due to this charge, neglecting the surface tension effect. Given atmospheric pressure = 10^5 N/m^2. [B.E] [2.74×10^{-9}m]

SHORT ANSWER QUESTIONS

1. Can an electric field may be represented by lines of force. [Yes]
2. How many electrons are there in one Coulomb charge?

$$\left[\frac{1}{1.602 \times 10^{-19}} = 6.25 \times 10^{18}\right]$$

3. Does Coulomb's law of electric force obey Newton's third law of motion? [Yes]
4. Write the dimensional formula for ε_0. [$M^{-1}L^{-3}T^4A^2$]
5. Can a point in space have a non-zero potential even when the electric field in the space is zero?

[Yes, electric intensity at a given point is given by $E = -\frac{d\phi}{dx}$. If potential ϕ is constant in space, the potential gradient in any direction is zero, hence the intensity is zero. Obviously, potential can exist where there is no electric field]

OBJECTIVE QUESTIONS

1. The work done in moving a positive charge on an equipotential surface is __________. [zero]
2. The number of lines of force that radiate outward from one Coulomb of positive charge is __________. [$1/\varepsilon_0 = 1.13 \times 10^{11}$]
3. Two thin infinite parallel plates have uniform charge densities $+\sigma$ and $-\sigma$. The electric field in the space between them is __________. [σ/ε_0]
4. When placed in a uniform field a dipole experiences only a __________. [torque]
5. The work done in carrying a charge q once sound a circle of radius r with a charge Q at the centre is __________. [zero]

Magnetostatics and Electrodynamics

3.1 INTRODUCTION

An important kind of interaction among the fundamental particles composing matter is the one called electromagnetic interaction. It is associated with a characteristic property of each particle called its *electric charge*. To describe the electromagnetic interaction, one will have to introduce the notion of *electromagnetic field*, characterized by two vectors, the electric field $\boldsymbol{E}$ and the magnetic field $\boldsymbol{B}$, such that the force on an electric charge is given by

$$\boldsymbol{F} = q(\boldsymbol{E} + \boldsymbol{V} \times \boldsymbol{B}) \quad (1)$$

The electric and magnetic fields $\boldsymbol{E}$ and $\boldsymbol{B}$ are in turn determined by the position of charges themselves and by their motions (or currents). The separation of the electromagnetic field into its electric and magnetic components depends on the relative motion of the observer and the charges producing the field. Also the fields $\boldsymbol{E}$ and $\boldsymbol{B}$ are directly correlated with each other by the Ampere-Maxwell and Faraday-Henry laws. All these relations are expressed by four laws viz. Gauss's theorem, Bio-Savart law, Ampere's theorem and Faraday's law. Maxwell put these fundamental laws of electromagnetism into the form of four equations which have now become famous Maxwell's equations. The theory of the electromagnetic field is condensed into these four laws. These laws constitute the basic framework of the theory of electromagnetic interactions. The electric charge q and the current I are called the sources of the electromagnetic field since, given q and I, Maxwell's equations allow us to compute $\boldsymbol{E}$ and $\boldsymbol{B}$.

According to Maxwell's theory, the rate of propagation of electric and magnetic interaction is equal to the velocity of light in the given medium. Maxwell's theory reveals the electromagnetic nature of light. The theoretically predicted velocity of these electromagnetic waves agrees with the experimentally determined velocity of light in vacuum. It is also found that the other types of waves, e.g. *X*-rays, ultraviolet rays and the infrared rays possess the same velocity and are essentially of the same nature as light waves. Maxwell's electromagnetic theory explains fairly satisfactorily most of the properties of these radiations. Apart from the mathematical deduction of equations, Maxwell proposed the idea of *displacement currents.* We know that the empty space is a non-conductor of electricity. Maxwell proposed that electric currents in empty space can exist in the form of "*displacement currents*". However, the term displacement current is misleading and Maxwell's picture unnecessary, since there is no such current between the plates of the capacitor (see sec 3.2).

3.2 DISPLACEMENT CURRENT

Maxwell proposed the idea of displacement current while he was critically examining electromagnetic theory. He put forward this fact that the magnetic fields are generated not only due to circulating current but also due to time dependent charges in the electric field. This type of time varying field is equivalent to a current and Maxwell interpreted this as a *displacement current*.

To understand the concept of displacement current let us consider two plates *A* and *B* of a parallel plate capacitor separated by a dielectric of permittivity ε (Fig. 3.1). The plate *P* of the capacitor is charged positively with a surface charge density σ.

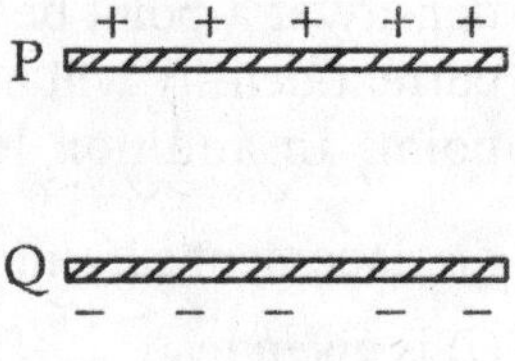

P + + + + +

Q − − − − −

Fig. 3.1

Plate *Q* is charged negatively and its surface density is also σ. If the area of either plate is *A*, then the total charge on either plate will be $A\sigma$. Now join the two plates by a conducting wire. Obviously, electrons will flow from *Q* to *P*. In other words, this is equivalent to flow of a conventional current from *P* to *Q*. The magnitude of the resulting current in the connecting wire $= \frac{d}{dt}(A\sigma)$. Maxwell assumed that currents are always circuital. Obviously, a current equal to the conduction current in the connecting wire flows through the dielectric in the capacitor. Maxwell interpreted this as *displacement current.*

We have the electric intensity in dielectric

$$E = \frac{\sigma}{\varepsilon} \tag{2}$$

or $\sigma = E\varepsilon$

$\therefore$ Displacement current $= \frac{d}{dt}(A\sigma) = A\frac{d}{dt}(\varepsilon E)$

But $\varepsilon E = D$ is electric displacement and hence displacement current $= A\frac{dD}{dt}$

Displacement current flowing per unit area, *i.e.* $\frac{dD}{dt}$ is called the *displacement current density*. Since E is a vector and hence D (displacement) is also a vector. Obviously, one can write

$$\text{Displacement current} = \frac{dD}{dt} \tag{3}$$

From the above discussion it is obvious that, the current in the circuit becomes continuous (even between the plates of the capacitor) if one accepts the idea of displacement current.

We must remember that the term "*displacement current*" is purely conventional. In essence, the displacement current is time varying electric field. The only reason for the quantity given by Eq. (3) a "current" is that the dimensions of this quantity coincides with that of current density. Of all the physical properties belonging to a real current, a displacement current has only one the ability of producing a magnetic field.

3.3 TOTAL CURRENT

Suppose that there is a volume distribution of charge and let the charge density at a point be ρ. Now, if the charges are in motion the resulting current density will be eV, where V is the velocity of the charges at that point. In addition to the conduction current $J(= \sigma V)$, if there is a displacement current $\left(\frac{\partial \boldsymbol{D}}{\partial t}\right)$ at the concerned point, then the total current (I) is given by

$$I = \frac{\partial \boldsymbol{D}}{\partial t} + \boldsymbol{J} \tag{4}$$

The partial differentiation sign indicates that $\boldsymbol{D}$ may be a function of both space and time.

3.4 CONTINUITY EQUATION

Let us consider a closed surface S enclosing a volume V. If J is the current density at some point in the surface then the current passing out of the surface is $\int_s \boldsymbol{J} \cdot \hat{n} ds$, where $\hat{n}$ is the unit outward drawn normal at the point. We know that the charge is conserved and hence the above current must be equal to the rate of decrease of charge within the volume V, *i.e.* $-\frac{\partial}{\partial t}\int_s e dV$ where ρ is the charge density within the volume. For conservation of charge, we have

$$\int_s \boldsymbol{J}\cdot\hat{n}ds = -\frac{\partial}{\partial t}\int \rho dV = -\int_v \frac{\partial}{\partial t}(\rho dV)$$

Making use of Gauss's theorem, one obtains

$$\int_s \boldsymbol{J}\cdot\hat{n}ds = \int_v \nabla\cdot\boldsymbol{J}dV = -\int_v \frac{\partial\rho}{\partial t}dV$$

$$\therefore \qquad \nabla\cdot\boldsymbol{J} = -\frac{\partial\rho}{\partial t}$$

or $$\nabla\cdot\boldsymbol{J} + \frac{\partial\rho}{\partial t} = 0 \qquad (5)$$

Eq. (5) is known as the *equation of continuity* and it emphasises the conservation of charge and energy.

In the steady state $\frac{\partial\rho}{\partial t} = 0$

$$\therefore \qquad \nabla\cdot\boldsymbol{J} = 0 \qquad (6)$$

This is valid in the region which does not contain a source or sink of current.

3.5 GAUSS'S THEOREM APPLIED TO ELECTROSTATICS

Let us consider a closed surface S enclosing a volume V and let the surface enclose an electric charge of volume density ρ. Let dV be an elementary part of the total volume V enclosed by S, then the total charge enclosed by surface $S = \int_v \rho dV$. Let $\boldsymbol{D}$ be the electric induction at some point in S, then the normal magnetic flux across an element dS of the surface S is $\boldsymbol{D}\cdot\hat{n}ds$, where $\hat{n}$ is the unit vector drawn outward normal to the surface at the point considered. Obviously, the total outward normal flux across the surface $S = \int_s \boldsymbol{D}\cdot\hat{n}ds$. According to Gauss's theorem this must be equal to the total charge enclosed by surface S. Thus

$$\int_s \boldsymbol{D}\cdot\hat{n}ds = \int_v \rho dV$$

But $$\int_s \boldsymbol{D}\cdot\hat{n}ds = \int_v \nabla\cdot\boldsymbol{B}dV \qquad \text{(Guss's theorem)}$$

$$\therefore \qquad \int_v \nabla\cdot\boldsymbol{D}dV = \int_v \rho dV$$

Obviously, the integrals on both sides of the above equation are volume integrals taken over the same volume and hence the two integrands must be equal, *i.e.*

$$\nabla\cdot\boldsymbol{D} = \rho \qquad (7)$$

In this form, the Gauss's theorem shows that the sources of electric displacement are free electric charges.

The application of Gauss's theorem in calculating $\boldsymbol{D}$ consists in choosing a closed surface of such shape that the electric flux ϕ can be readily calculated by simple means.

3.6 BIOT-SAVART LAW

This law establishes the magnitude and direction of vector of magnetic induction $d\boldsymbol{B}$ at any arbitrary point P of a magnetic field set up in vacuum by an element of a conductor of length dl carrying a current I (Fig. 3.2).

$$dB = \frac{\mu_0}{4\pi} Idl \frac{\sin\theta}{r^2} \tag{8}$$

Fig. 3.2

Let the position vector of P be $\boldsymbol{r} \cdot \boldsymbol{dl}$ is the vector of the element of conductor numerically equal to dl and of same direction as the current. $dl\sin\theta$ can be written as

$$\boldsymbol{dl} \times \hat{r} = \frac{\boldsymbol{dl} \times \boldsymbol{r}}{r}$$

$$\therefore \qquad d\boldsymbol{B} = \frac{\mu_0}{4\pi} \frac{I\boldsymbol{dl} \times \boldsymbol{r}}{r^3}$$

The total induction at P due to the entire conductor RS is given by

$$\boldsymbol{B} = \frac{\mu_0}{4\pi} \int \frac{I\boldsymbol{dl} \times \boldsymbol{r}}{r^3} \tag{9}$$

We must note that the direction of $\boldsymbol{B}$ is that of $\boldsymbol{dl} \times \boldsymbol{r}$. Taking the divergence on both sides of Eq. (9), one obtains

$$\nabla \cdot \boldsymbol{B} = \frac{\mu_0 I}{4\pi} \int \nabla \cdot \frac{\boldsymbol{dl} \times \boldsymbol{r}}{r^3}$$

But $$\nabla \cdot \left(\frac{\boldsymbol{dl} \times \boldsymbol{r}}{r^3}\right) = \frac{\boldsymbol{r}}{r^3} \cdot \nabla \times \boldsymbol{dl} - \boldsymbol{dl} \cdot \nabla \times \left(\frac{\boldsymbol{r}}{r^3}\right)$$

Since at the field point P, $\boldsymbol{dl}$ is independent of the coordinates of P and hence $\nabla \times \boldsymbol{dl} = 0$.

$$\nabla \cdot \boldsymbol{dl} \times \left(\frac{\boldsymbol{r}}{r^3}\right) = -\boldsymbol{dl} \cdot \nabla \times \left(\frac{\boldsymbol{r}}{r^3}\right)$$

$$= -\boldsymbol{dl} \cdot \nabla \times \nabla\left(\frac{1}{r}\right) = 0$$

Hence $$\nabla \cdot \boldsymbol{B} = 0 \tag{10}$$

This is the first law in magnetostatics corresponding to the relation $\nabla \times \boldsymbol{E} = 0$ in electrostatics. The relation (10) shows that the magnetic field is *solenoidal* in contrast to the electric field which is irrotational. Eq. (10) is one of Maxwell's equations for an electromagnetic field.

3.7 MAGNETIC VECTOR POTENTIAL

We know that $\nabla \cdot \boldsymbol{B} = 0$. As the divergence of the curl of a vector vanishes, we can write the megnetic induction in the form

$$\boldsymbol{B} = \nabla \times \boldsymbol{A} \tag{11}$$

where the vector field $\boldsymbol{A}$ is called the *magnetic vector potential*. As $\nabla \times \boldsymbol{B} = \mu_0 \boldsymbol{J}$, we have

$$\nabla \times (\nabla \times \boldsymbol{A}) = \mu_0 \boldsymbol{J} \tag{12}$$

Using the vector expansion

$$\nabla \times (\nabla \times \boldsymbol{A}) = \nabla(\nabla \cdot \boldsymbol{A}) - \nabla^2 \boldsymbol{A} \tag{13}$$

and choosing

$$\nabla \cdot \boldsymbol{A} = 0 \tag{14}$$

we get

$$\nabla^2 \boldsymbol{A} = -\mu_0 \boldsymbol{J}. \tag{15}$$

The equations for the cartesian components of $\boldsymbol{A}$ are

$$\nabla^2 A_x = -\mu_0 J_x,\ \nabla^2 A_y = -\mu_0 J_y, \text{ and } \nabla^2 A_z = -\mu_0 J_z. \tag{16}$$

Each of these component equations has the form of Poisson's equation $\nabla^2 \phi = -(\dot{\rho}/\varepsilon_0)$ in electrostatics, which has the solution

$$\phi(\boldsymbol{r}) = \frac{1}{4\pi\varepsilon_0} \int \frac{\rho(\boldsymbol{r}_0)\, dV_0}{|\boldsymbol{r} - \boldsymbol{r}_0|} \tag{17}$$

Using this as a guide, the solution for A_i, where i stands for x, y, and z, is written as

$$A_i(\boldsymbol{r}) = \frac{\mu_0}{4\pi} \int \frac{J_i(\boldsymbol{r}_0)\, dV_0}{|\boldsymbol{r} - \boldsymbol{r}_0|}$$

Thus we can write

$$\boldsymbol{A}(\boldsymbol{r}) = \frac{\mu_0}{4\pi} \int \frac{\boldsymbol{J}(\boldsymbol{r}_0)\, dV_0}{l} \tag{18}$$

where $l = |\boldsymbol{r} - \boldsymbol{r}_0|$.

For a line current I flowing in a wire, $\boldsymbol{J} dV_0$ is replaced by $I d\boldsymbol{l}_0$, where $d\boldsymbol{l}_0$ is a differential length vector along the wire in the direction of the current. Thus for a line current, we get

$$\boldsymbol{A}(\boldsymbol{r}) = \frac{\mu_0}{4\pi} \int \frac{I(\boldsymbol{r}_0)\, d\boldsymbol{l}_0}{l}. \tag{19}$$

For expressions for the magnetic vector potential at a point due to a long straight current-carrying wire and at a point outside and inside a uniformly long solenoid, refer examples 1 and 2.

3.8 ELECTROMAGNETIC INDUCTION

Laws of Electromagnetic Induction

Faraday observed experimentally that whenever the magnetic flux linked with an electric circuit changed, a current was induced in the circuit—a phenomenon, known as the *electromagnetic induction*. The results of Faraday's experiments led to the development of two useful laws:

1. **Neumann's Law:** *The induced electromotive force (e.m.f.) in a circuit is equal to the time rate of change of the magnetic flux linked with the circuit.*
2. **Lenz's law:** *The direction of the induced e.m.f. or the current is such that it will oppose the change of flux producing it.*

If Φ is the flux linked with a circuit at time t, then $d\Phi/dt$ gives the time rate of variation of flux. The combination of the two laws of electromagnetic induction yields

$$\varepsilon = -\frac{d\Phi}{dt} \tag{20}$$

where ε is the induced e.m.f. The negative sign signifies that the e.m.f. ε opposes the changes of flux. Hence the induced e.m.f. is often referred to as the back *e.m.f.* If R is the resistance of the circuit, then the induced current is

$$i = \frac{\varepsilon}{R} = -\frac{1}{R}\frac{d\Phi}{dt} \tag{21}$$

In S.I. units, ε is expressed in volt, R in ohm, i in ampere, Φ in Weber, and t in second.

Lenz's law ensures that perpetual-motion machines are not possible to construct. It indicates that if we have a closed curve in a conductor, the induced current will oppose the change of flux through the curve. If the negative sign in Eq. (20) were absent, the induced current would have enhanced the flux change, which in turn, would have increased the induced e.m.f. further. The process would have been regenerative leading to a runaway. Lenz's law does not lead to this runaway; rather it implies a tendency to stability. This law is a special case of a very general principle, called *Le Chatelier's principle*, which states that a physical system always opposes a change impressed upon it externally.

Equation (20) is often referred to as the integral form of *Faraday's law of electromagnetic induction* and is a fundamental equation of electromagnetism.

Let C be a closed circuit binding an open surface S, placed in a magnetic field $\boldsymbol{B}$. The flux Φ through S is given by

$$\Phi = \int_s \boldsymbol{B} \cdot \hat{n} da \tag{22}$$

The induced e.m.f., generated around C, is given by Eq. (20). If the electric field is $\boldsymbol{E}$, the induced e.m.f. around the curve C, is given from definition by

$$\varepsilon = \oint_C \boldsymbol{E} \cdot d\boldsymbol{l} \tag{23}$$

With the aid of Eqs. (22) and (23), Eq. (20) can be written as

$$\oint_C \boldsymbol{E} \cdot d\boldsymbol{l} = -\frac{d}{dt}\left(\int_S \boldsymbol{B} \cdot \hat{\boldsymbol{n}} da\right) \tag{24}$$

If the circuit is fixed, the time derivative can be moved inside the integral, whence it becomes a partial derivative, $\boldsymbol{B}$ being a function of position as well. Furthermore, using Stokes' theorem, the left-hand side of Eq. (24) can be converted into a surface integral. The result is

$$\int_S (\nabla \times \boldsymbol{E}) \cdot \hat{\boldsymbol{n}} da = -\int_S \frac{\partial \boldsymbol{B}}{\partial t} \cdot \hat{\boldsymbol{n}} da \tag{25}$$

Equation (25) is true for any arbitrary fixed surface S

$$\therefore \qquad \nabla \times \boldsymbol{B} = -\frac{\partial B}{\partial t} \tag{26}$$

which is the *differential form* Faraday's law.

As an application of Faraday's law of electromagnetic induction, consider a long straight wire XY (Fig. 3.3) carrying a current $I(t)$. The flux throught a rectangular wire $abcd$ in the plane of the wire XY can be readily calculated. The magnetic induction at a distance r from the wire XY is given by

$$B = \frac{\mu_0 I(t)}{2\pi r}$$

Fig. 3.3: Induced e.m.f. in a rectangular wire in the plane of a current-carrying wire

The flux through an elementary area within $abcd$, shown shaded in Fig. 3.3, is $Bldr$, where $l = ad = bc$, the area of the element being ldr. The total flux Φ through $abcd$ is given by

$$\Phi = \frac{\mu_0 l I(t)}{2\pi} \int_{r=r_1}^{r_2} \frac{dr}{r} \tag{27}$$

where r_1 and r_2 are the distances of the sides ad and bc, respectively, from the wire. Thus

$$\Phi = \frac{\mu_0 l I(t)}{2\pi} \ln\left(\frac{r_2}{r_1}\right) \tag{28}$$

The induced e.m.f. around the curve *abcd* is

$$\varepsilon = -\frac{d\Phi}{dt} = -\frac{\mu_0 l}{2\pi}\left(\ln\frac{r_2}{r_1}\right)\frac{dI}{dt} \tag{29}$$

If dI/dt is positive, the induced current in *abcd* flows in the counter clockwise direction, as shown in Fig. 3.3. If dI/dt is negative, the direction of the induced current in *abcd* will be clockwise.

Eddy Currents

When ***B*** changes with time, currents are induced in the conducting materials placed in the region where ***B*** is nonzero. These currents are termed *Foucault currents* or *eddy currents.* They heat up the conducting materials through the Joule losses and are therefore undesirable in electromagnetic devices. Eddy currents are minimised by using laminated conductors, as explained below.

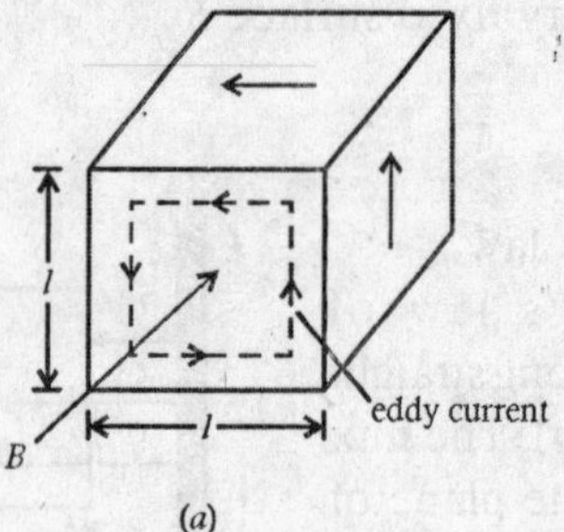

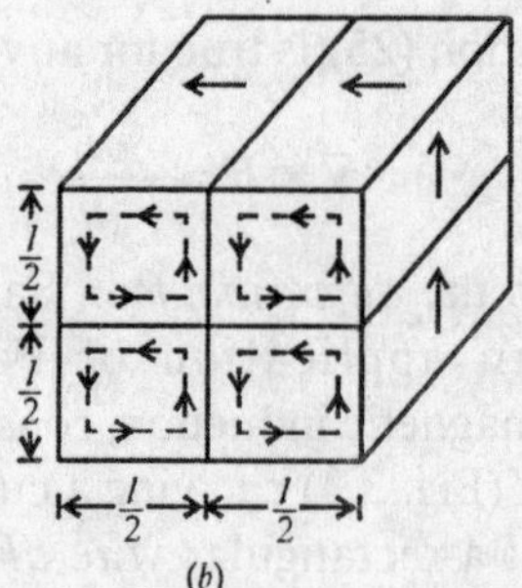

Fig. 3.4

Consider a block of a conducting material placed in a changing magnetic field $B(t)$ [Fig. 3.4]. The Joule loss due to the induced circulating current in the block is proportional to l^4. Let us now divide the block into four equal parts [as shown in Fig. 3.4 (*b*)] insulated from one another. The Joule loss in each part will now be proportional to $(l/2)^4$. The total Joule loss in the four blocks will hence be proportional to $4(l/2)^4 = l^4/4$. The Joule loss is thus reduced by a factor of 4 when the block is divided into four parts. In general, if a cross-section (normal to ***B***) of the conductor is laminated into *n* equal parts, the Joule losses due to eddy currents are reduced by a factor of *n*. Transformer cores are laminated to reduce eddy currents in the core.

The magnetic fields produced by the eddy currents weaken the original magnetic field causing the eddy currents. Some magnetic shielding is therefore obtained for magnetic fields changing with time.

Eddy currents are utilized in divices like *induction heaters.* Here the Joule heat produced by eddy currents is employed to raise the temperature of an object without disturbing it otherwise. Thus *food may be cooked* uniformly by placing it in a region fo strong changing magnetic fields. Eddy currents are also used *in melting metals* by placing them in a verying magnetic field of high frequency, say, a few tens of kilo hertz. Some *alloys* which can not be prepared otherwise, are prepared with the help of the eddy-current effect, the advantage being that the metals are not oxidized since the melting process can be carried out in an oxygen-free chamber. As eddy-currents oppose the relative motion between a conductor and the magnetic field, they have been exploited in *eddy-current brakes.* The working of the *speedometer* of a car is also based on the eddy-current effect. An aluminium drum rests on a pivot around a magnet which rotates at a rate proportional to the speed of the car. So eddy currents are induced in the drum and hence the latter experiences a torque due to the magnetic field. The torque tends to rotate the drum in the direction of the magnet, and is a measure of the speed of the car. A pointer attached to the drum rotates over a speedometer dial to read the speed. A common application of eddy currents is in the *damping of oscillations* of pointers of some electriacal instruments.

Conducting Bar Moving on Tracks through & Constant Magnetic Field

Let us consider a conducting bar *ab* move at a constant velocity v on fixed conducting tracks through a constant magnetic field $\boldsymbol{B}$ (Fig. 3.5). We assume that v is much less than c, the speed of light in free space, so that the relativistic corrections are unimportant.

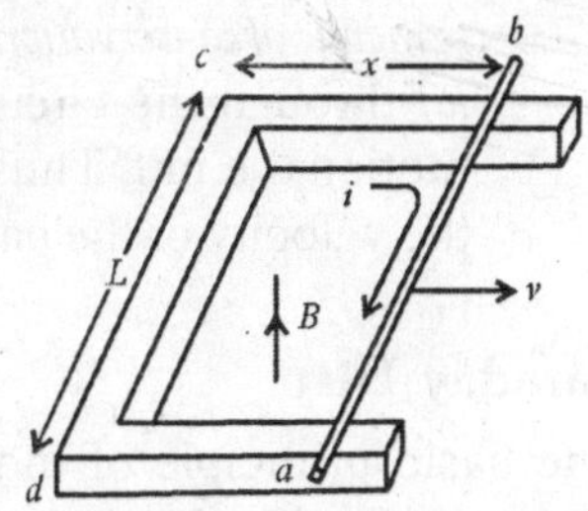

Fig. 3.5: A conducting bar moving on stationary tracks

Let at time t, the distance $cb = da = x$. The area of the loop *abcd* at this instant is Lx. If $\boldsymbol{B}$ is perpendicular to the loop, the instantaneous flux through the loop is $\Phi = BLx$. The induced e.m.f. in the loop is given by

$$\varepsilon = \frac{d\Phi}{dt} = -BL\frac{dx}{dt} = -BLv \tag{30}$$

If R is the resistance of the loop then the induced current is

$$i = \frac{\varepsilon}{R} = \frac{BLv}{R} \tag{31}$$

Let us assume here that the induced current i is so small that the magnetic field produced by it is negligible comparad to B. This implies that the resistance R of the circuit is sufficiently large.

The negative sign in Eq. (31) indicates that the induced current flows in the clockwise direction for the direction of the magnetic field shown in Fig. 3.5. If the direction of B is reversed, the induced current will flow in the counter-clockwise direction.

If the magnetic field B is not perpendicular to the plane of the loop but makes an angle θ with it, B in Eqs. (30) and (31) will be replaced by $B\sin\theta$. The e.m.f. induced in this case depends on the velocity and not on the position of the bar. It is, therefore, called a *motional e.m.f.*

One can also determine the motional e.m.f. from either of the following two considerations:

(1) In Fig. 3.5, the current carriers in the sliding bar ab move with the velocity v. The force experienced by a current carrier of change q is qvB, directed along the length of the sliding bar. The force per unit charge, i.e. vB is the electric field set up in the moving bar of length L. The induced e.m.f. (or the motional e.m.f.) is the work done in taking a unit charge completely round the circuit and is given by the product of the electric field and the length L, i.e., by BvL.

(2) If i is the induced current in Fig. 3.5, the force experienced by the sliding bar of length L is $F = BiL$. If the bar moves through a distance dx in time dt, the work done is $Fdx = BiLdx$. This work, by the *principle of conservation of energy*, is expended in circulating the charge idt through the circuit, and is represented by εidt, where ε is the motional e.m.f. Thus $BiLdx = \varepsilon idt$, or $\varepsilon = BLv$, where $v(= dx/dt)$ is the velocity of the bar.

Faraday Disc

The basic priniciple of induction generators is illustrated with the help of a Faraday disc, shown in Fig. 3.6. The Faraday disc is essentially a conducting disc of radius a rotated about its axis with a constant angular velocity ω. The disc is placed in a uniform magnetic field $\boldsymbol{B}$ perpendicular to its plane. The free electrons in the disc at a distance r from its axis will experience a Lorentz force—$e\boldsymbol{v} \times \boldsymbol{B}$, the charge on an electron being—e. Here v is the linear velocity at a distance r from the axis, i.e., $v = \omega r$. The Lorentz force is directed towards the centre of the disc. Thus, there will be an accumulation of electrons towards the centre, and a deficiency of electrons in the outer edge of the disc. This charge separation produces an electric field towards the centre which will oppose further charge separation. An equilibrium is reached when the electric field $\boldsymbol{E}$ caused by the charge separation will balance the Lorentz force. Thus,

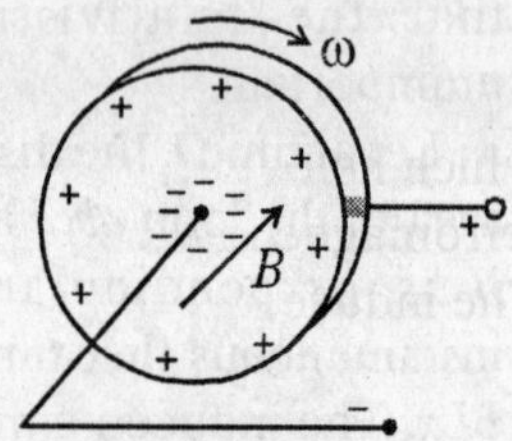

Fig. 3.6: A Faraday disc

$$\boldsymbol{E} + \boldsymbol{v} \times \boldsymbol{B} = \boldsymbol{0} \text{ or, } E = -\omega r B$$

The potential difference between the rim of the disc and its centre is

$$\phi = -\int_0^a E dr = \frac{\omega B a^2}{2} \tag{32}$$

which is the e.m.f. induced in the disc.

If a wire is externally connected between the rim and centre of the disc, a current will flow in the wire due to this induced e.m.f. This is essentially the principle of an induction generator.

If instead of rotating the disc, a current is allowed to flow from the axis to the rim, the disc would experience a torque due to $\boldsymbol{B}$ and would rotate. The generator would then become a motor.

Self-Inductance

When a current flows in a circuit, the magnetic flux produced by the current depends on the geomety of the circuit and, for nonferromagnetic materials, it is proportional to the current. Thus if I is the current flowing in a circuit the attendant magnetic flux can be written as

$$\Phi = LI \tag{33}$$

where L is the *self-inductance*, or simply, the inductance of the circuit. Equation (33) defines the self-inductance as the *total flux linked with the circuit for unit current flowing in the circuit.*

For a rigid stationary circuit, the changes in the flux is caused by changes in the current. Thus

$$\frac{d\Phi}{dt} = \frac{d\Phi}{dI}\frac{dI}{dt} \tag{34}$$

which holds even when Φ is not linearly related to I, as in the the case of ferromagnetic materials where the permeability depends on the current. The induced e.m.f. in the circuit is given by

$$\varepsilon = -\frac{d\Phi}{dt} = -L\frac{dI}{dt} \tag{35}$$

where
$$L = \frac{d\Phi}{dI} \tag{36}$$

Equation (35) defines the self-inductance L as the *e.m.f. induced in the circuit for a unit rate of change of current in it.* If Φ is not linearly related to I, the inductance defined by Eq. (36) is termed the *incremental inductance* to distinguish it from the inductance defined by Eq. (33). However, when Φ is proportional to I, the two definitions lead to the same result, since $\Phi/I = d\Phi/dI$ in that case.

The negative sign in Eq. (35) indicates that the induced e.m.f. opposes the flow of current. Thus *the self-inductance measures the ability of a circuit to oppose the variation of current in it.*

In SI units, Φ, I, t, ε and L are expressed in Weber, ampere, second, volt, and henry, respectively.

$$1 \text{ henry} = \frac{1 \text{ Weber}}{1 \text{ ampere}} = \frac{1 \text{ volt - second}}{1 \text{ ampere}} = 1 \text{ V} \cdot \text{A}^{-1} \cdot \text{s}$$

As Φ has the dimensions $[ML^2T^{-2}I^{-1}]$, the dimensions of L are $[ML^2T^{-2}I^{-2}]$.

Calculation of Self-Inductances

(*a*) A Long Solenoid: The expression for the self-inductance of an air-cored very long solenoid can be found by noting that the magnetic induction at a point inside it is given by

$$B = \frac{\mu_0 NI}{l}$$

where μ_0 is the permeability of free space (the relative permeability of air being unity to a very good approximation), N is the total number of turns, and l is the length of the solenoid. If A is the area of cross-section of the solenoid, the flux linking each turn is

$$\Phi_1 = BA = \frac{\mu_0 NIA}{l} \tag{37}$$

and the total flux linking N turns is

$$\Phi = N\Phi_1 = \frac{\mu_0 N^2 A}{l} I \tag{38}$$

The self-inductance of the solenoid is

$$L = \frac{\Phi}{I} = \frac{d\Phi}{dI} = \frac{\mu_0 N^2 A}{l} \tag{39}$$

Equation (39) also gives the self-inductance of an air-cored toroidal coil of N number of turns of mean length l, and cross-sectional area A.

If the solenoid is not very long, the magnetic induction B at an axial point is given by

$$B = \frac{\mu_0 NI}{2L} \int_{\theta_1}^{\pi-\theta_2} \sin\theta d\theta$$

$$= \frac{\mu_0 NI}{2L} (\cos\theta_1 + \cos\theta_2)$$

Now we have

$$\cos\theta_1 = z_0/\sqrt{a^2 + z_0^2} \text{ and } \cos\theta_2 = (l - z_0)/\sqrt{a^2 + (l - z_0)^2}$$

where l is the length of the solenoid. Thus

$$B = \frac{\mu_0 NI}{2l}\left[\frac{z_0}{\sqrt{a^2 + z_0^2}} + \frac{l - z_0}{\sqrt{a^2 + (l - z_0)^2}}\right]$$

In a small length dz_0, the number of turns is $(N/l)dz_0$. The flux linking these turns is

$$d\Phi = \frac{B(\pi a^2)N}{l} dz_0$$

the variation of the magnetic flux over the cross-sectional area being neglected. The total flux linking N turns is

$$\Phi = \int_{z_0=0}^{l} \frac{B(\pi a^2)N}{l} dz_0 = \frac{\mu_0 N^2 IA}{2l^2}\int_0^l \left[\frac{z_0}{\sqrt{a^2 + z_0^2}} + \frac{l - z_0}{\sqrt{a^2 + (l - z_0)^2}}\right] dz_0$$

$$= \frac{\mu_0 N^2 IA}{2l^2}\left[\sqrt{a^2 + z_0^2} - \sqrt{a^2 + (l - z_0)^2}\right]_0^l$$

$$= \frac{\mu_0 N^2 A}{l^2}\left(\sqrt{a^2 + l^2} - a\right) I$$

where $A = \pi a^2$ = area of cross-section. The self-inductance is therefore given by

$$L = \frac{\mu_0 N^2 A}{l^2}\left(\sqrt{a^2 + l^2} - a\right) \tag{40}$$

Equation (40) goes over to Eq. (39) when $l \gg a$.

(b) Two Long Parallel Wires: Let P and Q be two long parallel wires separated by a distance d in air (Fig. 3.7). The lengths of the wires are much larger than d. The two wires carry the same current I in the opposite direction. The two wires thus constitute an open-air transmission line. The radius a of each wire is assumed to be much less than d, so that the flux inside the wires themselves can be neglected. The flux is concentrated

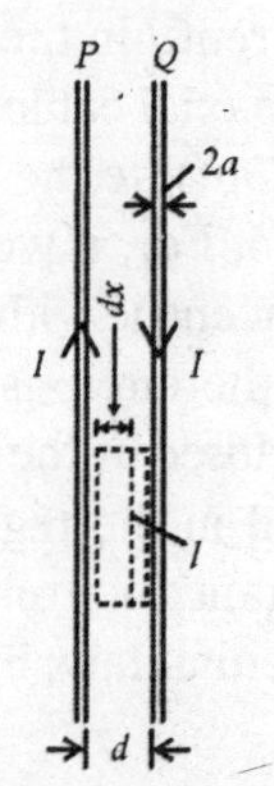

Fig. 3.7: Two long parallel wires

in the region between the two wires. Consider a rectangular area of length l and width dx in the region between the wires. If B is the magnetic induction in this region, the flux through the rectangle is $Bldx$. If the rectangular element is at a distance x from the wire P, then,

$$B = \frac{\mu_0 I}{2\pi}\left(\frac{1}{x} + \frac{1}{d-x}\right)$$

where the contributions due to the wires P and Q are added. The total flux linkage is

$$\Phi = \int_{x=a}^{d-a} Bldx = \frac{\mu_0 Il}{2\pi}\left[\ln\frac{x}{x-d}\right]_a^{d-a} = \frac{\mu_0 Il}{\pi}\ln\left(\frac{d-a}{a}\right)$$

The self-inductance L is given by

$$L = \frac{\Phi}{I} = \frac{d\Phi}{dI} = \frac{\mu_0 l}{\pi}\ln\left(\frac{d-a}{a}\right) \tag{41}$$

Since $d \gg a$, the self-inductance per unit length is

$$\frac{L}{l} = \frac{\mu_0}{\pi}\ln\frac{d}{a} \tag{42}$$

If the wires are very close together, $(d - a) \approx a$. Then Eq. (41) gives $L \approx 0$. This is the principle of *noninductive winding* employed in the construction of resistance coils. An insulated wire is doubled on itself before it is wound on the format.

(c) Coaxial Cylinders: Consider two coaxial cylinders of radii a and b, respectively ($b > a$), as shown in Fig. 3.8. The two cylinders carry the same current I in the opposite directions; they thus form a coaxial cable.

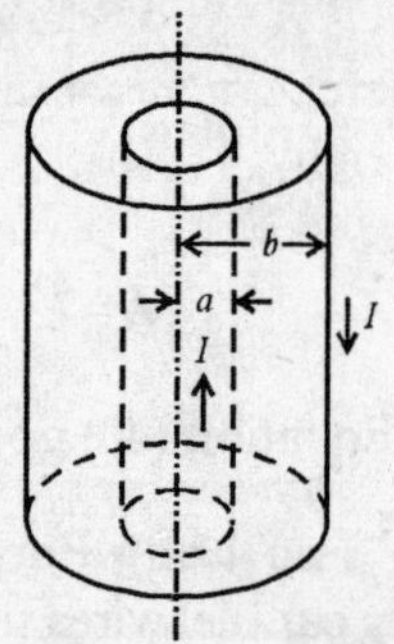

Fig. 3.8: Two coaxial cylinders

Outside the two cylinders, the magnetic field is zero. For, if we consider a circle of radius r ($r > b$) concentric with the cable, the total current enclosed by the circle is $I - I = 0$. If $a < r < b$, the current enclosed by the circular path is I. Hence the magnetic field in the region between the two cylinders at a distance r from the axis is given, from Ampere's circuital law, by

$$B = \frac{\mu_0 I}{2\pi r}$$

If we imagine two coaxial cylinders of radii r and $r + dr$, and of unit length, the flux in the region between the two cylinders is $B \times (1 \times dr) = Bdr$. The total flux is

$$\Phi = \int_{r=a}^{b} Bdr = \frac{\mu_0 I}{2\pi} \int_a^b \frac{dr}{r} = \frac{\mu_0 I}{2\pi} \ln \frac{b}{a}$$

The inductance per unit axial length is

$$L = \frac{\mu_0}{2\pi} \ln \frac{b}{a} \tag{44}$$

In the above derivation we have neglected the flux inside the materials of the two cylinders. This is justified when $b \gg a$ and the thickness of the outer cylinder is very small compared to b.

If, instead of air, the space between the two cylinders is filled with a medium of permeability μ, μ_0 in Eq. (44) has to be replaced by μ.

Mutual Inductance

Suppose that there are N number of circuits designated by 1, 2, ..., N. When the circuits carry currents, the flux produced by the current in one circuit may link with another circuit. The flux Φ_i linked with the current i is therefore determined by adding up the contributions from each circuit. Thus

$$\Phi_i = \Phi_{i1} + \Phi_{i2} + \ldots + \Phi_{ii} + \ldots + \Phi_{iN}$$

$$= \sum_{j=1}^{N} \Phi_{ij} \tag{45}$$

where Φ_{ij} is the flux through the ith circuit due to the jth circuit. For non-ferromagnetic materials, Φ_{ij} is proportional to the current I_j in the jth circuit. Hence one can write

$$\Phi_{ij} = M_{ij} I_j \qquad (i \neq j), \tag{46}$$

where M_{ij} is referred to as the *mutual inductance* between the circuits i and j. *The mutual inductance between two circuits is therefore defined as the flux linked with one circuit due to unit current in the other.* Observe that Φ_{ii} is the self flux and Φ_{ii}/I_i is the self-inductance L_i of the ith circuit.

The induced *e.m.f.* in the ith circuit, ε_i, is given by

$$\varepsilon_i = -\frac{d\Phi_i}{dt} = -\sum_{j=1}^{N} \frac{d\Phi_{ij}}{dt} \tag{47}$$

For rigid fixed circuits, the changes in Φ_{ij}'s are caused by changes in the currents. Thus

$$\frac{d\Phi_{ij}}{dt} = \frac{d\Phi_{ij}}{dI_j} \cdot \frac{dI_j}{dt} = M_{ij} \frac{dI_j}{dt} \tag{48}$$

where $$M_{ij} = \frac{d\Phi}{dI_j} \quad (i \neq j). \tag{49}$$

Therefore the e.m.f. induced in the ith circuit due to the change of current in the jth circuit is

$$\varepsilon_{ij} = -M_{ij} \frac{dI_j}{dt} \tag{50}$$

Thus the *mutual inductance between the two circuits may also be difined as the e.m.f. induced in one circuit due to the unit rate of change of current in the other.* The two definitions lead to the same result when the fluxes are linearly related to the currents.

We shall below that $M_{ij} = M_{ji}$, so that there is no ambiguity in the subscripts. Note that $d\Phi_{ii}/dI$ is the self-inductance L_i of the ith circuit. The unit of mutual inductance is the same as that of self-inductance, i.e. henry.

Neumann's Formula

Consider two loops C_1 and C_2, as shown in Fig. 3.9. The loop C_1 carries a current I_1. The flux Φ_{21} linked with C_2 due to the current in C_1 is written as

$$\Phi_{21} = \int_{s_2} \boldsymbol{B}_{21} \cdot \hat{n} dS_2 \tag{51}$$

Using the vector potential $\boldsymbol{A}_{21}$ corresponding to $\boldsymbol{B}_{21}$, we can write with the help of Stoke's theorem

$$\Phi_{21} = \oint_{C_2} \boldsymbol{A}_{21} \cdot d\boldsymbol{l}_2 \tag{52}$$

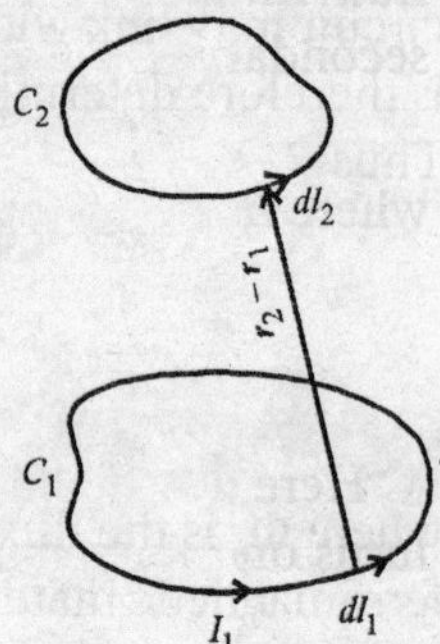

Fig. 3.9: Mutual inductance between loops C_1 and C_2

The vector potential $\boldsymbol{A}_{21}$ can be expressed from Eq. (51) as

$$\boldsymbol{A}_{21} = \frac{\mu_0}{4\pi} \oint_{C_1} \frac{I_1 d\boldsymbol{l}_1}{|\boldsymbol{r}_2 - \boldsymbol{r}_1|} \tag{53}$$

$$\Phi_{21} = \frac{\mu_0 I_1}{4\pi} \oint_{C_2} \oint_{C_1} \frac{d\boldsymbol{l}_1 \cdot d\boldsymbol{l}_2}{|\boldsymbol{r}_2 - \boldsymbol{r}_1|} = M_{21} I_1$$

where $$M_{21} = \frac{\mu_0}{4\pi} \oint_{C_2} \oint_{C_1} \frac{d\boldsymbol{l}_1 \cdot d\boldsymbol{l}_2}{|\boldsymbol{r}_2 - \boldsymbol{r}_1|} \tag{54}$$

Since $|\boldsymbol{r}_1 - \boldsymbol{r}_2| = |\boldsymbol{r}_2 - \boldsymbol{r}_1|$, the order of the integration may be interchanged without affecting the final result. It immediately follows that

$$M_{21} = M_{12} = M = \frac{\mu_0}{4\pi} \oint_{C_1} \oint_{C_2} \frac{d\boldsymbol{l}_1 \cdot d\boldsymbol{l}_2}{|\boldsymbol{r}_1 - \boldsymbol{r}_2|} \tag{55}$$

which is called *Neumann's formula*, and *M*(without the subscripts), is simply referred to as the mutual inductance between the loops.

Calculation of Mutual Inductance

(a) Two Solenoids: Consider a short solenoid *S* wound over a long narrow solenoid *P* at its centre (Fig. 3.10). The two solenoids *P* and *S* are referred to as the primary and the secondary solenoids, respectively. By connecting the terminals *a* and *d* to a source of electrical energy, a current *I* is passed through the primary. If N_1 is the total number of turns on the primary solenoid of cross-sectional area *A* and length *l*, then the magnetic flux linked with each turn of the secondary is

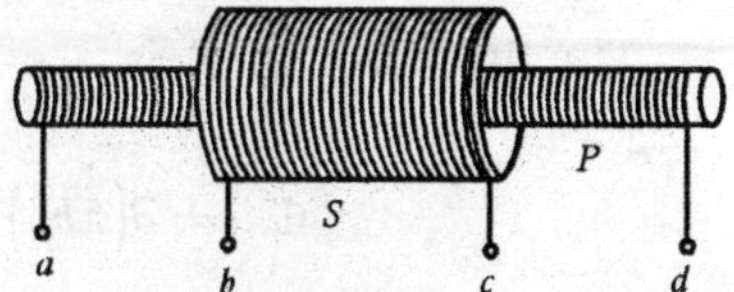

Fig. 3.10: Two solenoids

$$\Phi = BA$$

where *B* is the magnetic induction given by,

$$B = \frac{\mu N_1 I}{l}$$

Here μ is the permeability of the core material. If N_2 is the number of turns on the secondary solenoid, the flux linked with the secondary is

$$N_2\Phi = \frac{\mu N_1 N_2 A I}{l}$$

Therefore, the mutual inductance between the primary and the secondary is

$$M = \frac{\mu N_1 N_2 A I}{l} \tag{56}$$

The same expression will be obtained when a short solenoid is wound around a toroid.

If the current through the primary solenoid increases at the rate of *dI*/*dt*, the induced e.m.f. in the secondary across the terminals *b* and *c* is −*MdI*/*dt*. This induced e.m.f. causes a flow of current in the secondary circuit when the terminals *b* and *c* are connected through an external circuit.

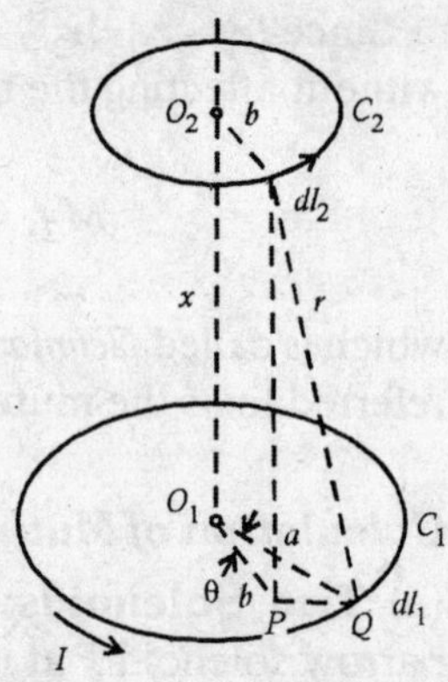

Fig. 3.11: Two parallel circular coaxial loops

(*b*) Two Parallel Coaxial Circular Loops: Suppose C_1 and C_2 are two parallel circular coaxial loops with centres O_1 and O_2, and readii a and b, respectively. Let x be the axial separation O_1O_2 (Fig. 3.11). If the coil C_2 is so small that the flux through it can be assumed to be uniform, the mutual inductance between C_1 and C_2 can be calculated as follows.

If I is the current in C_1, the magnetic induction at O_2 is given by

$$B = \frac{\mu_0 I a^2}{2\left(a^2 + x^2\right)^{3/2}}$$

The flux through C_2 is

$$\Phi_{21} = B\left(\pi b^2\right) = \frac{\pi \mu_0 a^2 b^2}{2\left(a^2 + x^2\right)^{3/2}} I \tag{57}$$

Therefore, the mutual inductance between the loops is

$$M = \frac{\Phi_{21}}{I} = \frac{\pi \mu_0 a^2 b^2}{2\left(a^2 + x^2\right)^{3/2}} \tag{58}$$

If the loop C_2 is not so small, the mutual inductance can be determined by applying Neumann's formula as shown below.

Consider two elementary lengths dl_1 and dl_2 on C_1 and C_2, respectively, as shown in Fig. 3.11. If we project dl_2 on the plane of C_1, then $dl_1 \times dl_2 = dl_1 dl_2 \cos\theta$, where$\theta$ is the angle shown in Fig. 3.11. Also, $dl_1 = a d\theta$. So that $dl_1 \times dl_2 = a\cos\theta d\theta dl_2$. The distance r between dl_1 and dl_2 is given by

$$r^2 = x^2 + (PQ)^2 = x^2 + (a^2 + b^2 - 2ab\cos\theta)$$

$$= x^2 + a^2 + b^2 - 2ab[2\cos^2(\theta/2) - 1]$$

$$= x^2 + (a + b)^2 - 4ab\cos^2(\theta/2)$$

$$= h^2\left[1 - \frac{4ab\cos^2(\theta/2)}{h^2}\right]$$

where $h^2 = x^2 + (a + b)^2$. Hence

$$\frac{1}{r} = \frac{1}{h}\left[1 - \frac{4ab\cos^2(\theta/2)}{h^2}\right]^{-1/2}$$

$$= \frac{1}{h}\left(1 + \frac{2ab}{h^2}\cos^2\frac{\theta}{2} + \frac{6a^2b^2}{h^4}\cos^4\frac{\theta}{2} + \ldots\right)$$

From Neumann's formula [Eq. (55)], the mutual inductance between the loops is writen as

$$M = \frac{\mu_0}{4\pi}\oint_{C_1}\oint_{C_2}\frac{dl_1 \cdot dl_2}{r} = \frac{\mu_0}{4\pi h}a\oint_{C_2} dl_2 \int_0^{2\pi}\cos\theta$$

$$\left(1 + \frac{2ab}{h^2}\cos^2\frac{\theta}{2} + \frac{6a^2b^2}{h^4}\cos^4\frac{\theta}{2} + \ldots\right)d\theta$$

Now,

$\int_0^{2\pi}\cos\theta d\theta = 0$, $\int_0^{2\pi}\cos\theta\cos^2\theta/2d\theta = \pi/2$, $\int_0^{2\pi}\cos\theta\cos^4\theta/2d\theta = \pi/2$, etc.

Also $\oint dl_2 = 2\pi b$. Therefore,

$$M = \frac{\mu_0 ab}{2h}\left(\frac{\pi ab}{h^2} + \frac{3\pi^2b^2}{h^4} + \ldots\right)$$

$$= \frac{\mu_0\pi a^2b^2}{2h^3}\left(1 + \frac{3ab}{h^2} + \ldots\right) \qquad (59)$$

If $a \gg b$, Eq. (59) reduces to Eq. (58).

Inductances in Series and Parallel

Series Connection

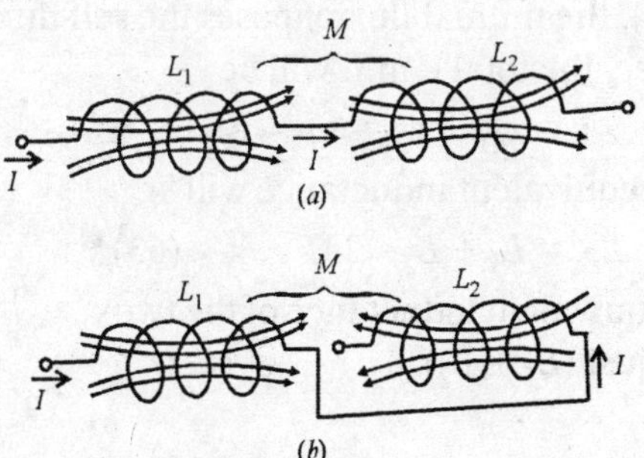

Fig. 3.12: Two coils connected in series (*a*) Mutual flux aiding self flux (*b*) Mutual flux opposing self flux

Consider two coils of self inductances L_1 and L_2 connected in series. Let a current *I* pass through the two coils at time *t*. In Fig. 3.12 (*a*), the mutual flux augments the self flux of the coils. The induced e.m.f. in coil 1 due to its self inductance is

$$\varepsilon_{11} = -L_1 \frac{dI}{dt}$$

while the e.m.f. induced in coil 2 due to the current I in coil 1 is

$$\varepsilon_{21} = -M \frac{dI}{dt}$$

where M is the mutual inductance between the coils. The e.m.f. induced in coil 2 due to its self inductance is

$$\varepsilon_{22} = -L_2 \frac{dI}{dt}$$

and the e.m.f. in coil 1 due to the current I in coil 2 is

$$\varepsilon_{12} = -M \frac{dI}{dt}$$

As the fluxes aid one another, the total e.m.f. is

$$\varepsilon = \varepsilon_{11} + \varepsilon_{22} + \varepsilon_{12} + \varepsilon_{21} = -(L_1 + L_2 + 2M)\frac{dI}{dt} \quad (60)$$

If L_{eq} is the equivalent inductance, then

$$\varepsilon = -L_{eq} \frac{dI}{dt} \quad (61)$$

Comparing Eqs. (60) and (61), we obtain

$$L_{eq} = L_1 + L_2 + 2M \quad (62)$$

In Fig. 3.12 (*b*), the mutual flux opposes the self flux of the two coils in series. In this case, the total e.m.f. will be

$$\varepsilon = \varepsilon_{11} + \varepsilon_{22} - \varepsilon_{12} - \varepsilon_{21}$$

Therefore the equivalent inductance will be

$$L_{eq} = L_1 + L_2 - 2M \quad (63)$$

If $M = 0$, the equivalent inductance of the two coils in series is just $(L_1 + L_2)$.

Parallel Connection

Figure 3.13 shows two coils of self inductances L_1 and L_2 connected in parallel. Let the total current I be divided into the parts I_1 through coil 1 and I_2 through coil 2. Then, $I = I_1 + I_2$.

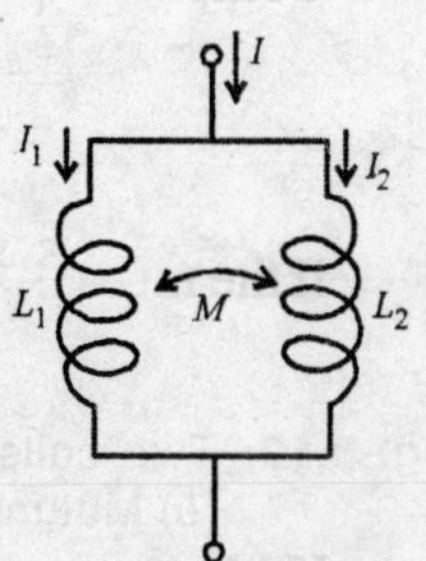

Fig. 3.13: Two inductances connected in parallel

Assuming that the mutual flux aids the self flux, the total e.m.f. induced in coil 1 is

$$\varepsilon_1 = -L_1 \frac{dI_1}{dt} - M \frac{dI_2}{dt}$$

where M is the mutual inductance between the coils. Similarly, the total e.m.f. for coil 2 is

$$\varepsilon_2 = -L_2 \frac{dI_2}{dt} - M \frac{dI_1}{dt}$$

Since the two coils are in parallel, $\varepsilon_1 = \varepsilon_2 = \varepsilon$ (say). Therefore,

$$L_1 \frac{dI_1}{dt} + M \frac{dI_2}{dt} = -\varepsilon \quad \text{and} \quad M \frac{dI_1}{dt} + L_2 \frac{dI_2}{dt} = -\varepsilon$$

By Cramer's rule,

$$\frac{dI_1}{dt} = \frac{-\varepsilon(L_2 - M)}{L_1 L_2 - M^2} \quad \text{and} \quad \frac{dI_2}{dt} = \frac{-\varepsilon(L_1 - M)}{L_1 L_2 - M^2}$$

Therefore,

$$\frac{dI}{dt} = \frac{dI_1}{dt} + \frac{dI_2}{dt} = \frac{-\varepsilon(L_1 + L_2 - 2M)}{L_1 L_2 - M^2}$$

or

$$\varepsilon = -\frac{L_1 L_2 - M^2}{L_1 + L_2 - 2M} \frac{dI}{dt} \tag{64}$$

If L_{eq} is the equivalent inductance, then

$$\varepsilon = -L_{eq} \frac{dI}{dt} \tag{65}$$

Comparing Eqs. (64) and (65), one obtains

$$L_{eq} = \frac{L_1 L_2 - M^2}{L_1 + L_2 - 2M} \tag{66}$$

If the mutual flux opposes the self flux, the equivalent inductance will be

$$L_{eq} = \frac{L_1 L_2 - M^2}{L_1 + L_2 + 2M} \tag{67}$$

If $M = 0$, the equivalent inductance of two inductances in parallel will be $L_1 L_2/(L_1 + L_2)$.

Ceofficient of Coupling

If we consider two loops, the mutual flux between them can be less than or at best equal to the self fluxes of the two loops. That is, $\Phi_{12} \le \Phi_{22}$ and $\Phi_{21} \le \Phi_{11}$. We can write $\Phi_{12} = k_1\Phi_{22}$ and $\Phi_{21} = k_2\Phi_{11}$, where k_1 and k_2 are two numbers less than or equal to unity. Thus we get

$$MI_2 = \Phi_{12} = k_1\Phi_{22} = k_1L_2I_2$$

since $\Phi_{22} = L_2I_2$

Therefore, $$M = k_1L_2 \tag{68}$$

Again, $$MI_1 = \Phi_{21} = k_2\Phi_{11} = k_2L_1I_1 \text{ or } M = k_2L_1 \tag{69}$$

From Eqs. (68) and (69), we have

$$M^2 = k_1k_2L_1L_2 \text{ or, } M = k\sqrt{L_1L_2}, \tag{70}$$

where $0 \le k \le 1$. The geometrical constant k is known as the *coefficient of coupling** of the loops. The coefficient of coupling may be varied by varying the geometry, offering a means of constructing a variable inductance (we may note that L_{eq}'s of Eqs. (62), (63), (66) and (67) depends on M).

3.9 FARADAY'S LAW OF INDUCTION

The phenomenon of electromagnetic induction is the development of an induced electromotive force (ε) in a conducting circuit placed in a varying magnetic field. If the circuit is closed, a current called an *induced current*, is produced in it. Faraday and Henry independently studied the phenomenon of electromagnetic induction and gave a law. According to this law of electromagnetic induction, the e.m.f. induced in a circuit is equal to the negative rate of increase of magnetic flux enclosed by the circuit. Negative sign takes account of Lenz's Law. Let us consider a surface S which encloses a closed circuit C as its boundary (Fig. 3.14). Let $\boldsymbol{E}$ be the electric intensity at point P in the circuit C, then the work done in taking a unit charge thorough a small segment dl of the current at point P is $\boldsymbol{E} \cdot dl$. Obviously, the total workdone in taking a unit positive charge once round the closed circuit

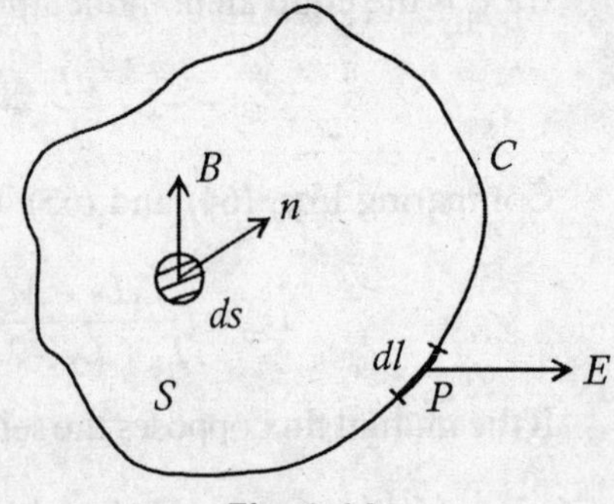

Fig. 3.14

*Sometimes k is allowed values in the range $-1 \le k \le 1$ to accommodate positive as well as negative values of M. The positive values of M imply that the mutual flux is aiding the self flux, while the negative value of M imply that the mutual flux is opposing the self flux.

$\oint \boldsymbol{E} \cdot d\boldsymbol{l}$. This is the induced e.m.f. in the circuit, *i.e.*

$$\varepsilon = \oint_c \boldsymbol{E} \cdot d\boldsymbol{l} \tag{71}$$

Let *ds* be a small element of the surfaces. If the magnetic flux density at this point is $\boldsymbol{B}$, then the normal flux across *ds* will be $\boldsymbol{B} \cdot \hat{n} ds$, where $\hat{n}$ is unit normal outward to *ds*. Obvioulsy, the total magnetic flux enclosed by *C* is given by

$$\phi = \oint_s \boldsymbol{B} \cdot \hat{n} ds \tag{72}$$

But Faraday's law is

$$\varepsilon = -\frac{\partial \phi}{\partial t}$$

$$\therefore \qquad \oint_c \boldsymbol{E} \cdot d\boldsymbol{l} = -\frac{\partial}{\partial t}\int_s \boldsymbol{B} \cdot \hat{n} ds \tag{73}$$

Time *t* and area *S* are independent variable quantities and hence can be interchanged, *i.e.*

$$\oint_c \boldsymbol{E} \cdot d\boldsymbol{l} = -\int_s \frac{\partial \boldsymbol{B}}{\partial t} \cdot \hat{n} ds$$

But from Stoke's curl theorem,

$$\oint_c \boldsymbol{E} \cdot d\boldsymbol{l} = -\oint_s (\nabla \times \boldsymbol{E}) \cdot \hat{n} ds$$

$$\therefore \qquad \oint_s \nabla \times \boldsymbol{E} \cdot \hat{n} ds = -\oint_s \frac{\partial \boldsymbol{B}}{\partial t} \cdot \hat{n} ds$$

Obviously, on either side of the above equation we have a surface integral taken over the same surface. Thus the two integrands must be equal.

$$\nabla \times \boldsymbol{E} = -\frac{\partial \boldsymbol{B}}{\partial t} \tag{74}$$

Eq. (74) is the differential form of Faraday's laws of electromagnetic induction.

The Eq. (74) and Eq. (7) show that the electric field has a non conservative part due to changing magnetic flux density as well as a conservative part due to electric charge density. Eq. (74) or its equivalents express relations that must exist between the time rate of change of the magnetic field at a point and the electric field existing at the same point of space. It illustrates in a very obvious way the close inter-relationship between the electric and magnetic components of an electromagnetic field. Eq. (74) is one of the Maxwell's equations for an electromagnetic field.

3.10 AMPERE'S LAW FOR MAGNETOMOTIVE FORCE

A magnetic field is one of the forms of an electromagnetic field. Its distinguishing feature is that it acts only on moving particles and bodies having an electric charge, as well as on magnetized bodies regardless of their state of motion. A magnetic field is produced by current carrying conductors, by moving electrically charged particles and bodies, by magnetized bodies, or by a variable electric field (by displacement current).

The force exerted by the magnetic field on the current carrying conductor is called *Ampere's force*. The relation between the magnetic field and the electric current is called *Ampere's law*. This law states that the line integral of the magnetic flux density taken round any closed path is equal to μ_0 times the enclosed by the path.

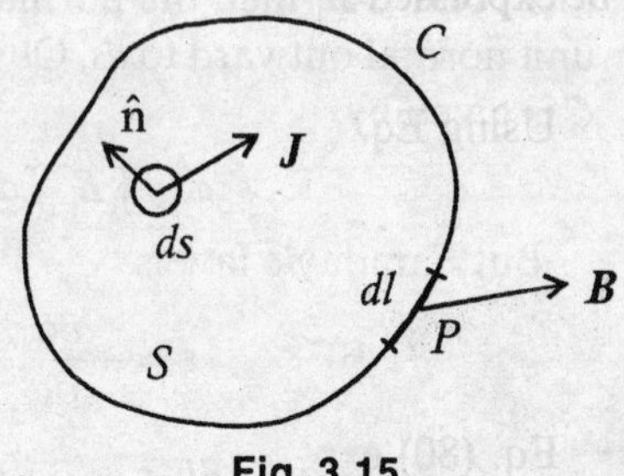

Fig. 3.15

Let us consider a closed circuit C as shown in Fig. 3.15. S is shown a surface which has C as its boundary.

Let $\boldsymbol{B}$ be the magnetic flux density at some point P in the circuit and dl be an elementary segment of the circuit at P. Let ds be an elementary part of the surface S. Let there is a current of density J at ds, then one finds that the normal current passing through ds will be $\boldsymbol{J} \cdot \hat{n} ds$, where $\hat{n}$ is unit normal drawn outward to ds. Obvioulsy the total current enclosed by the circuit C will be

$$I = \oint_s \boldsymbol{J} \cdot \hat{n} ds \tag{75}$$

We have seen that there exist a relationship between the magnetic field and the electric field in the same space (Eq. 73 (*a*)). The close relationship that exists between the electric and magnetic field suggests that an analogous relation should exist between the time rate of charge of an electric field and a magnetic field at the same place. Eq. 73 (*a*) relates the circulation of the electric field to the time rate of change of the flux of the magnetic field. We might expect that a similar expression must relate the circulation of magnetic field to the time rate of change of flux of electric field. Thus

$$I = \oint_c \boldsymbol{B} \cdot d\boldsymbol{l} = \mu_0 \oint_s J \cdot \hat{n} ds$$

We have from Stoke's theorem, the tangential line integral of any vector taken round a closed circuit is equal to the surface integral of the curl of that vector taken over any surface which has that circuit as its boundary. Hence

$$\int_c \boldsymbol{B} \cdot d\boldsymbol{l} = \oint_s \hat{n} \cdot \nabla \times \boldsymbol{B} ds$$

or $$\oint_s \hat{n} \cdot \nabla \times \boldsymbol{B} ds = \mu_0 \oint_s \boldsymbol{J} \cdot \hat{n} ds$$

Since we have taken surface integrals on both the sides over the same surface and hence

$$\nabla \times \boldsymbol{B} = \mu_0 \boldsymbol{I} \tag{76}$$

This is the second law of magnetostatics corresponding to $\nabla \times \boldsymbol{E} = \rho/\varepsilon_0$ of electrostatics. The integral form of this law is obtained as

$$\oint \boldsymbol{B} \cdot dl = \mu_0 \boldsymbol{I} \tag{77}$$

This is known as A*mpere's circuital law*. Since $\boldsymbol{B} = \mu_0 \boldsymbol{H}$, Eq. (76) can also be expressed as

$$\nabla \times \boldsymbol{H} = I \tag{78}$$

Using Eq. (4), one obtains

$$\nabla \times \boldsymbol{H} = \frac{\partial \boldsymbol{D}}{\partial t} + \boldsymbol{J} \tag{79}$$

$$\nabla \times \boldsymbol{B} = \mu_0 \left(\boldsymbol{J} + \varepsilon_0 \frac{\partial \boldsymbol{E}}{\partial t} \right) \tag{80}$$

Eq. (80) expresses a relation between the electric current at a point in space and the electric and magnetic fields at the same point. In empty space, where there are no currents, $\boldsymbol{J} = 0$, and Eq. (80) becomes

$$\nabla \times \boldsymbol{B} = \mu_0 \varepsilon_0 \frac{\partial \boldsymbol{E}}{\partial t} \tag{81}$$

Physical Meaning of $\nabla \cdot \boldsymbol{E} = 0$, $\nabla \times \boldsymbol{E} = -\frac{\partial \boldsymbol{B}}{\partial t}$, and $\nabla \times \boldsymbol{B} = 0$

If the divergence and the curl of a vector field are known everywhere, that field can be calculated at any point. $\nabla \cdot \boldsymbol{E} = 0$ and $\nabla \times \boldsymbol{E} = -\frac{\partial \boldsymbol{B}}{\partial t}$ enable us to do this for the electric field. Obvioulsy, they constitute two of the Maxwell's equations (out of four) describing the electromagnetic field. $\nabla \cdot \boldsymbol{B} = 0$ is valid for both steady and time varying fields and states. This signifies that a conduction current as well as a changing flux produces a magnetic field.

3.11 MAXWELL'S EQUATIONS

The theory of electromagnetic field is condensed into four fundamental laws. *i.e.* (*i*) Gauss's law for electric field (*ii*) Gauss's law for magnetic field (*iii*) Faraday's law of electromagnetic interaction (*iv*) Ampere's law of circuital current, and these are called Maxwell's equations. These equations may be written down, both in their differential and integral forms, as in Table 3.1 and 3.2 respectively

It has been found that above set of Maxwell's laws are in agreement with experimental results in all situations. Maxwell's equations are used in integral or differential form, depending on the problem to be solved. The solution of the equations may not be simple in all cases.

Gauss's Law of electrostatics is a basic law in electrostatics. The differential form of Gauss's law is $\nabla \cdot \boldsymbol{D} = \rho$. In this form, the law shows

TABLE 3.1: MAXWELL'S EQUATIONS IN DIFFERENTIAL FORM

Law	*Valuum*	*Free Space*	*Dielectric medium*	*Constant Field (with time)*
Gauss's law for electric field	$\nabla . \boldsymbol{E} = \frac{\rho}{\varepsilon_0}$	$\nabla . \boldsymbol{E} = 0$	$\nabla . \boldsymbol{E} = \rho/\varepsilon$	$\nabla . \boldsymbol{E} = \rho/\varepsilon$
Gauss's law for magnetic field	$\nabla . \boldsymbol{B} = 0$	$\nabla . \boldsymbol{B} = 0$	$\nabla . \boldsymbol{B} = 0$	$\nabla . \boldsymbol{B} = 0$
Faraday's law	$\nabla \times \boldsymbol{E} = -\frac{\partial \boldsymbol{B}}{\partial t}$	$\nabla \times \boldsymbol{E} = -\frac{\partial \boldsymbol{B}}{\partial t}$	$\nabla \times \boldsymbol{E} = -\frac{\partial \boldsymbol{B}}{\partial t}$	$\nabla \times \boldsymbol{E} = 0$
Ampere's circuital law	$\nabla \times \boldsymbol{B} = \mu_0 \left(\boldsymbol{J} + \varepsilon_0 \frac{\partial \boldsymbol{E}}{\partial t} \right)$	$\nabla \times \boldsymbol{B} = \mu_0 \varepsilon_0 \frac{\partial \boldsymbol{E}}{\partial t}$	$\nabla \times \boldsymbol{B} = \mu \left(\boldsymbol{J} + \varepsilon \frac{\partial \boldsymbol{E}}{\partial t} \right)$	$\nabla \times \boldsymbol{B} = \mu \boldsymbol{J}$

TABLE 3.2: MAXWELL'S EQUATIONS IN INTEGRAL FORM

Law	*Vacuum*	*Free Space*	*Dielectric medium*	*Constant Field (with time)*
Gauss's law for electric field	$\oint_S \mathbf{E}.d\mathbf{s} = \int_V \frac{\rho}{\varepsilon_0} dv$	$\oint_S \mathbf{E}.d\mathbf{s} = 0$	$\oint_S \mathbf{E}.d\mathbf{s} = \oint_v \frac{\rho}{\varepsilon} dv$	$\oint_S \mathbf{E}.d\mathbf{s} = \int_v \frac{\rho}{\varepsilon} dv$
Gauss's law for magnetic field	$\oint_S \mathbf{B}.d\mathbf{s} = 0$	$\oint_S \mathbf{B}.d\mathbf{s} = 0$	$\oint_S \mathbf{B}.d\mathbf{s} = 0$	$\oint_S \mathbf{B}.d\mathbf{s} = 0$
Faraday's law	$\int_C \mathbf{E}.d\mathbf{l} = -\int_S \frac{\partial \mathbf{B}}{\partial t}.d\mathbf{s}$	$\int_C \mathbf{E}.d\mathbf{l} = -\int_S \frac{\partial \mathbf{B}}{\partial t}.d\mathbf{s}$	$\int_C \mathbf{E}.d\mathbf{l} = -\int_S \frac{\partial \mathbf{B}}{\partial t}.d\mathbf{s}$	$\int_C \mathbf{E}.d\mathbf{l} = 0$
Ampere's circuital law	$\int_C \mathbf{B}.d\mathbf{l} = \mu \int_{ds} \left(\mathbf{J} + \varepsilon_0 \frac{\partial \mathbf{E}}{\partial t} \right).d\mathbf{s}$	$\int_C \mathbf{B}.d\mathbf{l} = \mu_0 \varepsilon_0 \int_s \frac{\partial \mathbf{E}}{\partial t}.d\mathbf{s}$	$\int_C \mathbf{B}.d\mathbf{l} = \mu \int_s \left(\mathbf{J} + \varepsilon \frac{\partial \mathbf{E}}{\partial t} \right).d\mathbf{s}$	$\int_C \mathbf{B}.d\mathbf{l} = \mu \int_s \mathbf{J}.d\mathbf{s}$

that the sources of electric displacement are free electric charges (since $\rho \neq 0$), *i.e.* the electric flux lines either leave or enter a closed surface. Obvioulsy, within the closed surface, therefore, there must be either a source or a sink of flux lines. We know from convention or a sink of flux lines start from positive charge as sources and terminate in negative charges as sinks. This shows that positive and negative charges exist independently.

The Gauss law for magnetic field in differential form, $\nabla \cdot \boldsymbol{B} = 0$ means that the magnetic flux lines are continuous, *i.e.* they form loops. Obviously, there are no sources or sinks. This shows that magnetic poles always occur in pairs.

Maxwell's equations are compatiable with the principle of relativity in the sense that they remain invariant under a Lorentz transformation. That is, their form does not change when the coordinates x, y, z and time t are transformed according to the Lorentz transformation.

The synthesis of electromagnetic interactions as expressed by Maxwell's equations is one of the greatest achievements in physics, and that is what places these interactions in a unique position. They are the best understood of all interactions and the only ones, so far, that can be expressed in a closed, consistent, mathematical form.

We must note, however, that Maxwell's equations, as they have been presented, have their limitations. They work very well when dealing with electromagnetic interactions between large aggregates of charges, such as radiating antennas, electric circuits, and even beams of ionized atoms or molecules. But it has been found that the electromagnetic interactions between fundamental particles (especially at high energies) must be treated in a somewhat different way, and according to the laws of quantum mechanics, constituting a technique called *quantum electrodynamics.* This will not be considered in this book.

Maxwell's equations are also consistent with the law of conservation of energy. For example, we consider Maxwell's fourth equation,

$$\nabla \times \boldsymbol{H} = \boldsymbol{J} + \frac{\partial \boldsymbol{D}}{\partial t}$$

Let us take divergence of both sides of this equation, we obtain

$$\nabla \times (\nabla \times \boldsymbol{H}) = \nabla \cdot \left(\boldsymbol{J} + \frac{\partial \boldsymbol{D}}{\partial t} \right) = \nabla \cdot \boldsymbol{J} + \frac{\partial}{\partial t}(\nabla \cdot \boldsymbol{D})$$

but $\qquad \nabla \cdot \boldsymbol{H} = 0 \therefore \nabla \cdot \nabla \times \boldsymbol{H} = \nabla \times \nabla \cdot \boldsymbol{H} = 0$

Thus one obtains

$$\nabla \cdot \boldsymbol{J} + \frac{\partial}{\partial t}(\nabla \cdot \boldsymbol{D}) = \nabla \cdot \boldsymbol{J} + \frac{\partial \rho}{\partial t} = 0$$

Which is the equation of continuity. Since equation of continuity is based on the law of conservation of charge and hence the Maxwell's equation under consideration is consistent with the laws of conservation of energy.

3.12 MAXWELL'S EQUATIONS UNDER STATIC CONDITIONS

One can obtain Maxwell's equations under static conditions by equating time rates to zero. Thus one obtains

$$\left.\begin{aligned} \nabla \cdot \boldsymbol{D} &= \rho \\ \nabla \times \boldsymbol{E} &= 0 \end{aligned}\right\} \text{(Electrostatics)} \tag{82}$$

$$\left.\begin{aligned} \nabla \cdot \boldsymbol{B} &= 0 \\ \nabla \times \boldsymbol{H} &= \boldsymbol{J} \end{aligned}\right\} \text{(Magnetostatics)} \tag{83}$$

Eq. (82) involve only electric fields whereas Eq. (83) involve only magnetic fields.

3.13 MORE ABOUT MAXWELL'S EQUATIONS

One can see the interdependence of Maxwell's equations and can judge that the equations involving the curls are more fundamental than those involving the divergence. Let us verify these statements.

Let us take divergence on both sides of Eq. (80), one obtains

$$\nabla \cdot \nabla \times \boldsymbol{H} = \nabla \cdot \boldsymbol{J} + \nabla \cdot \frac{\partial \boldsymbol{D}}{\partial t} = \nabla \cdot \boldsymbol{J} + \frac{\partial}{\partial t}(\nabla \cdot \boldsymbol{D})$$

But $\nabla \cdot \nabla \times \boldsymbol{H} = 0$ and from equation of continuity, we have $\nabla \cdot \boldsymbol{J} = -\frac{\partial \rho}{\partial t}$

$$\therefore \quad -\frac{\partial \rho}{\partial t} + \frac{\partial}{\partial t}(\nabla \cdot \boldsymbol{D}) = 0$$

or

$$\frac{\partial \rho}{\partial t} = \frac{\partial}{\partial t}(\nabla \cdot \boldsymbol{D})$$

$$\therefore \quad \nabla \cdot \boldsymbol{D} = \rho$$

Now we take the divergence on both side of Eq. (74), one obtains

$$\nabla \cdot \nabla \times \boldsymbol{E} = -\nabla \cdot \frac{\partial \boldsymbol{B}}{\partial t} = -\frac{\partial}{\partial t}(\nabla \cdot \boldsymbol{B})$$

But $\quad \nabla \cdot \nabla \times \boldsymbol{E} = 0$ and hence $\frac{\partial}{\partial t}(\nabla \cdot \boldsymbol{B}) = 0$

or $\quad \nabla \cdot \boldsymbol{B} = \text{constant}.$

The above must hold for every kind of magnetic field and must be true for a region where $\boldsymbol{B}$ = constant or zero. Obvioulsy, constant must be zero. This means

$$\nabla \cdot \boldsymbol{B} = 0$$

This shows that Eq. (7) can be derived from Eq. (80) and Eq. (10) can be derived from Eq. (74).

Physical Meanings of $\nabla \times \boldsymbol{E} = -\frac{\partial \boldsymbol{B}}{\partial t}$, $\nabla \cdot \boldsymbol{E} = 0$, $\nabla \cdot \boldsymbol{B} = 0$ and $\nabla \times \boldsymbol{H} = \partial \boldsymbol{D}/\partial t$

We have seen that Maxwell's electromagnetic equations are a series of classical equations that govern the behaviour of electromagnetic waves in all practical situations. They connect vector quantities applying to any point in a varying electric or magnetic field.

From these equations, Maxwell demonstrated that each field vector obeys a wave equation. Maxwell further showed that where a varying electric field exist, it is accompanied by a varying magnetic field induced at right angles, and *vice versa,* and the two form an electromagnetic field that could propagate as a transverses wave. $\nabla \cdot \boldsymbol{E} = \rho/\varepsilon_0$ and $\nabla \cdot \boldsymbol{B} = 0$ are simply differential form of Gauss's law in electrostatics. $\nabla \cdot \boldsymbol{B} = 0$ also signifies the non existence of magnetic molecule. $\nabla \times \boldsymbol{E} = -\frac{\partial \boldsymbol{B}}{\partial t}$ represents the differential form of Faraday's law of electromagnetic induction. $\nabla \times \boldsymbol{B} = 0$ or $\nabla \times \boldsymbol{H} = \boldsymbol{J} + \frac{\partial \boldsymbol{B}}{\partial t}$ is the modified form of Ampere's circuital law. Obviously, these field equations, represent generalisations of fundamental laws of electricity and magnetism.

$$\nabla \times \boldsymbol{E} = -\frac{\partial \boldsymbol{B}}{\partial t}$$

This describe the coupling between the electric and magnetic field vectors and their interaction with matter in space and time. This relation together with $\nabla \times \boldsymbol{H} = \frac{\partial \boldsymbol{D}}{\partial t}$ (where $\boldsymbol{D} = \varepsilon_0 \boldsymbol{E}$) in free space shows that time-variable $\boldsymbol{E}$ and $\boldsymbol{H}$ field cannot exist independently, e.g. if $\boldsymbol{E}$ is a function of time, then $\boldsymbol{D} = \varepsilon_0 \boldsymbol{E}$ will also be a function of time so that $\partial \boldsymbol{D}/\partial t$ will be non zero. Consequently $\nabla \times \boldsymbol{H}$ is non zero, and so a non zero $\boldsymbol{H}$ must exist. In a similar way one can show that if $\boldsymbol{H}$ is a function of time, then there must be an $\boldsymbol{E}$ field present.

These relations also show that a static $\boldsymbol{E}$ field can exist in the absence of a magnetic field $\boldsymbol{H}$; e.g. a capacitor with a static charge Q furnishes an example. Like wise a conductor with a constant current I has a magnetic field $\boldsymbol{H}$ without an $\boldsymbol{E}$ field.

3.14 ELECTROMAGNETIC WAVES

Let us consider an isotropic and homogenous medium in which there are no charges and currents, *i.e.* $\rho = 0$ and $J = 0$. Maxwell's equations reduce to

$$\nabla \cdot \boldsymbol{D} = 0 \text{ or } \nabla \cdot \boldsymbol{E} = 0 \tag{84}$$

$$\nabla \cdot \boldsymbol{B} = 0 \text{ or } \nabla \cdot \boldsymbol{H} = 0 \tag{85}$$

$$\nabla \times \boldsymbol{E} = -\frac{\partial \boldsymbol{B}}{\partial t} = -\mu \frac{\partial \boldsymbol{H}}{\partial t} \tag{86}$$

$$\nabla \times \boldsymbol{H} = \frac{\partial \boldsymbol{D}}{\partial t} = \varepsilon \frac{\partial \boldsymbol{E}}{\partial t} \tag{87}$$

Take curl on both sides of Eq. (86),

$$\nabla \times \nabla \times \boldsymbol{E} = -\nabla \times \mu \frac{\partial \boldsymbol{H}}{\partial t} = -\mu \frac{\partial}{\partial t}(\nabla \times \boldsymbol{H})$$

But $\nabla \times \nabla \times \boldsymbol{E} = \nabla(\nabla \cdot \boldsymbol{E}) - \nabla^2 \boldsymbol{E}$

We have from Eq. (84), $\nabla \cdot \boldsymbol{E} = 0$

$\therefore$
$$\nabla^2 \boldsymbol{E} = \mu \frac{\partial}{\partial t}(\nabla \times \boldsymbol{H}) = \varepsilon\mu \frac{\partial}{\partial t}\left(\frac{\partial \boldsymbol{E}}{\partial t}\right)$$

or
$$\nabla^2 \boldsymbol{E} = \varepsilon\mu \frac{\partial^2 \boldsymbol{E}}{\partial t^2} \tag{88}$$

Similarly starting with Eq. (87), one can obtain the equation

$$\nabla^2 \boldsymbol{H} = \varepsilon\mu \frac{\partial^2 \boldsymbol{H}}{\partial t^2} \tag{89}$$

By expanding the L.H.S. of Eq. (88) and Eq. (89), one can easily see that these are differential equations of wave motion, obviously, the velocity of these waves is given by

$$c = \frac{1}{\sqrt{\varepsilon\mu}} \tag{90}$$

This reveals that the fields generated by moving charges can leave the source and travel through space in the form of waves. For free space $\varepsilon = \varepsilon_0 = 8.8542 \times 10^{-12}\ \text{C}^2/\text{N-m}^2$ and

$\mu = \mu_0 = 4\pi \times 10^{-7}\ \text{H/m}\ (= \text{N-s}^2/\text{C}^2)$

Thus,
$$c = \frac{1}{\sqrt{8.8542 \times 10^{-12} \times 4\pi \times 10^{-7}}} = 2.99794 \times 10^8\ \text{m/s}$$

represents the speed of propagation of electromagnetic waves in free space. Obviously, the theoretical value of the speed of electromagnetic waves in free space agrees very closely with the experimental value of the velocity of light in free space. This result is of immense importance in physics. This unites the subjects of optics and electromagnetism. It also gives the idea of the electromagnetic spectrum. Electromagnetic waves cover a wide rang of frequencies of wavelengths, and may be classified according to their main source. The classification does not have very sharp boundaries, since different sources may produce waves in overlapping ranges of frequencies. Fig. 3.16 relates the various sections of the electromagnetic spectrum in terms of energy, frequency and wavelength. Light is simply a form of electromagnetic radiation. *X*-rays, ultraviolet rays, infra red rays, radiowaves and microwaves are all electromagnetic radiations. They differ from each other only in the order of magnitude of their wavelength, *i.e.* frequency. Obviously, all these travel with the same speed, *i.e.* speed of light in free space (Fig. 3.16).

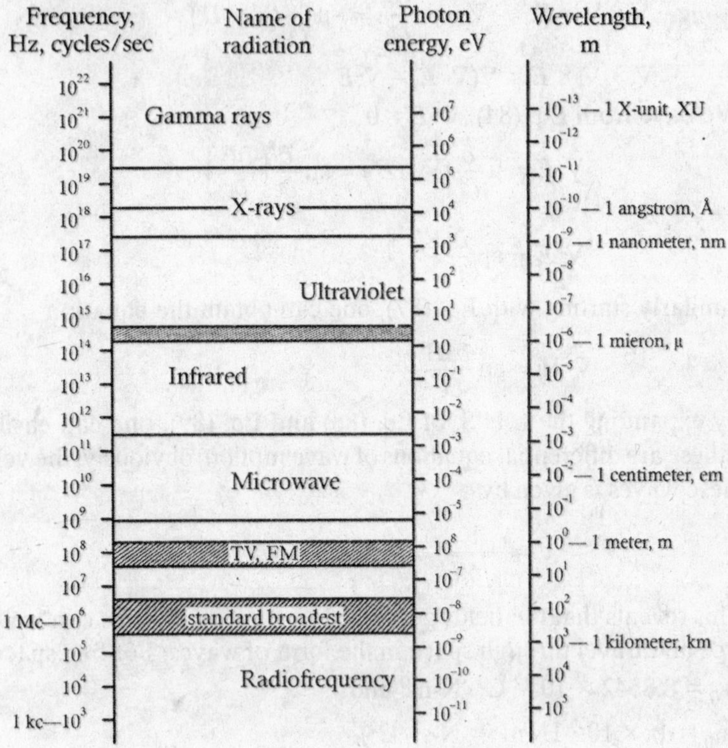

Frequency, Hz, cycles/sec
Name of radiation
Photon energy, eV
Wevelength, m
10^{22}
10^{21}
10^{20}
10^{19}
10^{18}
10^{17}
10^{16}
10^{15}
10^{14}
10^{13}
10^{12}
10^{11}
10^{10}
10^{9}
10^{8}
10^{7}
1 Mc—10^{6}
10^{5}
10^{4}
1 kc—10^{3}
Gamma rays
X-rays
Ultraviolet
Infrared
Microwave
TV, FM
standard broadest
Radiofrequency
10^{7}
10^{6}
10^{5}
10^{4}
10^{3}
10^{2}
10^{1}
10
10^{-1}
10^{-2}
10^{-3}
10^{-4}
10^{-5}
10^{-6}
10^{-7}
10^{-8}
10^{-9}
10^{-10}
10^{-11}
10^{-13} — 1 X-unit, XU
10^{-12}
10^{-11}
10^{-10} — 1 angstrom, Å
10^{-9} — 1 nanometer, nm
10^{-8}
10^{-7}
10^{-6} — 1 mieron, μ
10^{-5}
10^{-4}
10^{-3}
10^{-2} — 1 centimeter, em
10^{-1}
10^{0} — 1 meter, m
10^{1}
10^{2}
10^{3} — 1 kilometer, km
10^{4}
10^{5}

Fig. 3.16

3.15 PLANE WAVES IN A DIELECTRIC

A wave is said to be plane if all the wave fronts in it are plane and normal to the direction of propagation. Let us choose a system of rectangular coordinate axes such that the direction of propagation of the wave is along the *Z*-axis as shown in Fig. 3.17.

X
E
B
O
Y
Z
Direction of
wave propagation

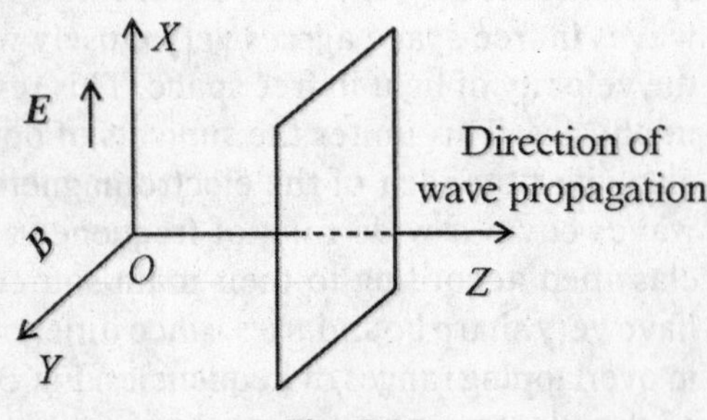

Fig. 3.17

The components of E are E_x, E_y and E_z and similarly the components of H are H_x, H_y and H_z.

Let us consider that the wave is in a plane and direction of propagation is along the Z-axis. Obviously, the value of any component of either H or E must have the same value at all points in a plane normal to the Z-axis. This means all differential coefficients with respect to x and y must vanish for a given value of z.

In the light of this and expressing L.H.S. of $\nabla \times \boldsymbol{H} = \varepsilon \frac{\partial \boldsymbol{E}}{\partial t}$ in the determinant form, one obtains

$$\begin{vmatrix} \hat{i} & \hat{j} & \hat{k} \\ 0 & 0 & \partial/\partial z \\ H_x & H_y & H_z \end{vmatrix} = \varepsilon\left(\frac{\partial E_x}{\partial t}\hat{i} + \frac{\partial E_y}{\partial t}\hat{j} + \frac{\partial E_z}{\partial t}\hat{k}\right) \tag{91}$$

where $\hat{i}$, $\hat{j}$, $\hat{k}$ are the unit vectors along x, y and z axis respectively. Equating the coefficients of these unit vectors on unit side of Eq. (91), one obtains

$$-\frac{\partial H_y}{\partial z} = \varepsilon\frac{\partial E_x}{\partial t} \tag{92}$$

$$\frac{\partial H_x}{\partial z} = \varepsilon\frac{\partial E_y}{\partial t} \tag{93}$$

$$0 = \varepsilon\frac{\partial E_z}{\partial t} \tag{94}$$

Similarly from (86), one obtains

$$\begin{vmatrix} \hat{i} & \hat{j} & \hat{k} \\ 0 & 0 & \partial/\partial z \\ E_x & E_y & E_z \end{vmatrix} = -\mu\left(\hat{i}\frac{\partial H_x}{\partial x} + \hat{j}\frac{\partial H_y}{\partial y} + \hat{k}\frac{\partial H_z}{\partial z}\right) \tag{95}$$

Equating the coefficients of unit vectors on both the sides, one obtains

$$\frac{\partial E_y}{\partial z} = \mu\frac{\partial H_x}{\partial x} \tag{96}$$

$$\frac{\partial E_x}{\partial z} = -\mu\frac{\partial H_y}{\partial t} \tag{97}$$

$$0 = -\mu\frac{\partial H_z}{\partial t} \tag{98}$$

From Eq. (94) we note that E_z is either a constant or zero. Similarly Eq. (98) reveals that H_z is either zero or a constant. However, the constant values of electric and magnetic fields are not relevant in the context of wave motion. Thus $E_z = 0$ and $H_z = 0$, *i.e.*, the electric and magnetic vectors have no components along the direction of propagation. This means that they are at right angles to the direction of propagation. Thus the waves are transverse (Fig. 3.18).

The vectors **E** and **H** are at right angles to each other. Thus if the direction of propagation is along the z-axis and if **E** is assumed to point in the *x*-direction then **H** will point in the *y*-direction (Fig. 3.18).

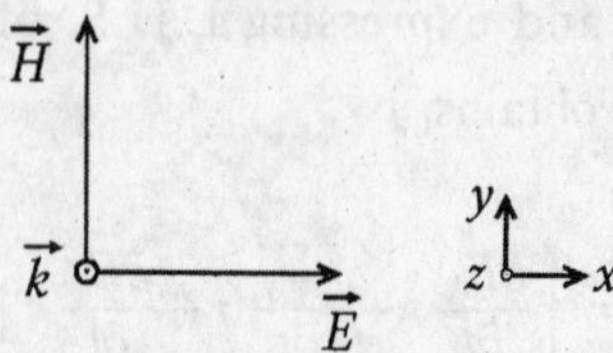

Fig. 3.18: If a plane wave is propagating in the *z*-direction (which is coming out of paper) and if at any instant of time the electric vector is along the *x*-axis then the magnetic vector will be along the y-axis

3.16 RELATION BETWEEN *E* AND *H*

Let us express

$$E_x = f(z - ct) \tag{99}$$

where *c* is the velocity of the wave. Differentiating Eq. (99) with respect to *t*, one obtains

$$\frac{\partial E_x}{\partial t} = -cf'(z - ct) \tag{100}$$

Using Eq. (100) Eq. (92) becomes

$$\frac{-\partial H_y}{\partial z} = -c\varepsilon f'(z - ct) \tag{101}$$

Integrating (101) with respect to *z*, one obtains

$$H_y = c\varepsilon f(z - ct) = c\varepsilon E_x \tag{102}$$

we have $c = \dfrac{1}{\sqrt{\varepsilon\mu}}$

$$\therefore \quad H_y = \frac{1}{\sqrt{\varepsilon\mu}}\varepsilon E_x = \sqrt{\frac{\varepsilon}{\mu}}E_x \tag{103}$$

We can easily see that Eq. (97) also leads to the same result. This result shows that the electric and magnetic vectors are in phase with each other, *i.e.*, **E** and **H** both attain the maximum and minimum values simultaneously. Fig. 3.19 represent a sinusoidal electromagnetic wave.

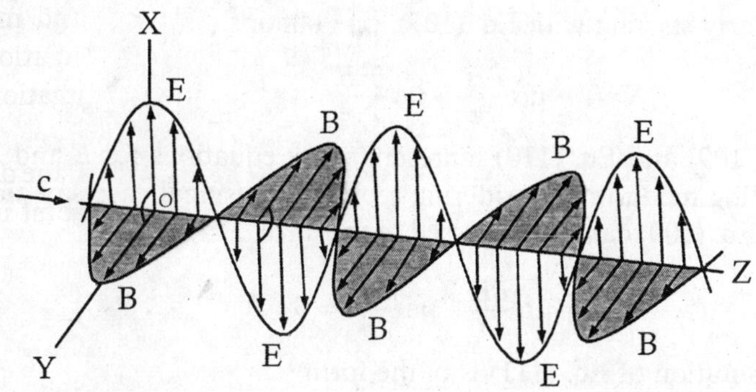

Fig. 3.19: Electric and magnetic fields in a harmonic plane electromagnet wave

This theoretical prediction about the Maxwell's equations has been amply confirmed by experiment, and it results in several phenomena. Besides plane wave solutions to Maxwell's equations there are also cylindrical and spherical electromagnetic waves. At a large distance from the source a limited portion of a cylindrical or a spherical wave can practically be considered as plane, and in this case the electric and magnetic fields are also perpendicular to each other and to the direction of propagation (*i.e.*, radial).

3.17 THE WAVE EQUATION IN A CONDUCTING MEDIUM

If there are no free charges in a medium, than $\rho = 0$. If the conductivity of the medium is σ then for a conducting medium, we have $\boldsymbol{J} = \sigma \boldsymbol{E}$. Thus Maxwell's equations become

$$\nabla \cdot \boldsymbol{E} = 0 \tag{104}$$

$$\nabla \cdot \boldsymbol{H} = 0 \tag{105}$$

$$\nabla \times \boldsymbol{E} = -\mu \frac{\partial \boldsymbol{H}}{\partial t} \tag{106}$$

$$\nabla \times \boldsymbol{H} = \sigma \boldsymbol{E} + \varepsilon \frac{\partial \boldsymbol{E}}{\nabla t} \tag{107}$$

Taking the curl of Eq. (106), on obtains

$$\nabla \times \nabla \times \boldsymbol{E} = -\mu \nabla \times \frac{\partial \boldsymbol{H}}{\partial t}$$

$$\nabla(\nabla \cdot \boldsymbol{E}) - \nabla^2 \boldsymbol{E} = -\mu\sigma \frac{\partial \boldsymbol{E}}{\partial t} - \mu\varepsilon \frac{\partial^2 \boldsymbol{E}}{\partial t^2} \tag{108}$$

Using Eq. (104), we get

$$\nabla^2 \boldsymbol{E} - \mu\sigma \frac{\partial \boldsymbol{E}}{\partial t} - \mu\varepsilon \frac{\partial^2 \boldsymbol{E}}{\partial t^2} = 0 \tag{109}$$

Similarly starting with Eq. (107), one obtains

$$\nabla^2 \boldsymbol{H} - \mu\sigma\frac{\partial \boldsymbol{H}}{\partial t} - \mu\varepsilon\frac{\partial^2 \boldsymbol{H}}{\partial t^2} \tag{110}$$

Eqs. (109) and Eq. (110) represent wave equations for $\boldsymbol{E}$ and $\boldsymbol{H}$ for a conducting medium. Considering a one dimensional case say along the X-axis, Eq. (109) can be expressed in the form

$$\frac{\partial^2 \boldsymbol{E}}{\partial x^2} - \mu\sigma\frac{\partial \boldsymbol{E}}{\partial t} - \mu\varepsilon\frac{\partial^2 \boldsymbol{E}}{\partial t^2} = 0 \tag{111}$$

The solution of Eq. (111) is of the form

$$E = E_0 \exp[i(\omega t - kx)] \tag{112}$$

Substituting for E from Eq. (112) in Eq. (111) and moving the common factor, one obtains

$$-k^2 + \varepsilon\mu\omega^2 - 2\omega\mu\sigma = 0$$

or

$$k^2 = \mu\varepsilon\omega^2\left[1 - \frac{i\sigma}{\varepsilon\omega}\right]$$

$$k = \omega\sqrt{\varepsilon\mu}\left[1 - \frac{i\sigma}{\varepsilon\omega}\right]^{1/2} \tag{113}$$

If $\sigma/\varepsilon\omega <<$ (say $\gtrsim 0.01$), the medium can be classified as a dielectric and if $\frac{\sigma}{\varepsilon\omega} >> 1$ (say $\gtrsim 100$), the medium can be classified as a conductor. For

$$0.01 \lesssim \frac{\sigma}{\varepsilon\omega} \lesssim 100$$

the medium is said to be a quasi-conductor. Thus depending on the frequency, a particular material can behave as a dielectric or as a conductor. For example, for fresh water $\varepsilon/\varepsilon_0 \approx 80$ and $\sigma \approx 10^{-3}$ mhos/m (both ε and σ can be assumed to be constants at low frequencies), one finds

$$\frac{\sigma}{\varepsilon} = \frac{10^{-3}}{80 \times 8.85 \times 10^{-12}} \approx 1.4 \times 10^6 \text{ sec}^{-1}$$

For $\omega = 2\pi \times 10$ sec^{-1}, $\frac{\sigma}{\omega\varepsilon} \approx 2 \times 10^4$ and for $\omega \approx 2\pi \times 10^{10}$ sec^{-1}, $\frac{\sigma}{\omega\varepsilon} \approx 2 \times 10^{-5}$. Thus fresh water behaves as a good conductor for $\nu \lesssim 10^3$ sec^{-1} and as a dielectric for $\nu \gtrsim 10^7$ sec^{-1}. For copper, one may assume $\varepsilon \approx \varepsilon_0$ and $\sigma \approx 5.8 \times 10^7$ mhos/m and for $\omega \approx 2\pi \times 10^{10}$ sec^{-1}, one obtains $\frac{\sigma}{\omega\varepsilon} \approx \frac{5.8 \times 10^7}{10^{10} \times 8.9 \times 10^{-12}} \approx 10^8$. Obviously, for such frequencies copper behaves as an excellent conductor.

For good conductor, neglecting 1 in comparison to $i\sigma/\varepsilon\omega$ Eq. (113) reduces to

$$k = \omega\sqrt{\varepsilon\omega}\left(-\frac{i\sigma}{\varepsilon\omega}\right)^{1/2} = \omega\sqrt{\varepsilon\mu}\left(\frac{\sigma}{2\varepsilon\omega}\right)^{1/2}(-2i)^{1/2}$$

$$\because \quad (1-i)^2 = -2i \qquad \therefore\ (-2i)^{1/2} = 1-i$$

or
$$k = \left(\frac{\mu\omega\sigma}{2}\right)^{1/2}(1-i) = \left(\frac{\mu\omega\sigma}{2}\right)^{1/2} - i\left(\frac{\mu\omega\sigma}{2}\right)^{1/2} \tag{114}$$

Obviously, k is imaginary, writing

$k = \alpha - i\beta$, one obtains

$$\alpha = \beta = \left(\frac{\mu\omega\sigma}{2}\right)^{1/2}$$

Eq. (112) becomes

$$E = E_0\exp[i(\omega t - \alpha x + i\beta x)]$$

$$= E_0\exp(-\beta x)\exp[i(\omega t - \alpha x)] \tag{115}$$

Eq. (115) represents an attenuated wave. The attenuation is due to a Joule-loss. Obviously the amplitude of the wave decreases with distance x exponentially. The constant β is called the *absorption coefficient*. From Eq. (115), it can be easily seen that the field decreases by a factor e in traversing a distance

$$\delta = \frac{1}{\beta} = \left(\frac{2}{\mu\omega\sigma}\right)^{1/2} \tag{116}$$

which is know as the *penetration depth* or *skin depth*.

The relation (116) indicates that the skin depth goes to zero as the conductivity approaches infinity and is small for good conductors at high frequency currents. Thus for copper $\sigma = 58 \times 10^6$ mhos/m, the skin depth at various frequencies is

Frequency	*Skin depth* (δ)
60 Hz	8.5×10^{-3}m
1 MHz	6.6×10^{-5}m
30 GHz	3.8×10^{-7}m

A poor conductor can be made a good conductor with a thin coating of silver or copper. The rapid attenuation of waves indicates that in high frequency circuits currents flows only on the surface of the conductor. The relatively high δ value in the case of sea water (10^{-1}m at 30 kHz) explains why the radio communication with submarines becomes difficult at depth of several metres.

3.18 THE POYNTING VECTOR

Let us consider Eqs. (74) and (79)

$$\nabla \times \boldsymbol{E} = -\frac{\partial \boldsymbol{B}}{\partial t}$$

$$\nabla \times \boldsymbol{H} = J + \frac{\partial \boldsymbol{D}}{\partial t}$$

Now $\nabla \cdot (E \times \boldsymbol{H}) = H \cdot \nabla \times \boldsymbol{E} - \boldsymbol{E} \cdot \nabla \times \boldsymbol{H}$

$$= -\boldsymbol{H} \cdot \frac{\partial \boldsymbol{B}}{\partial t} - \boldsymbol{J} \cdot \boldsymbol{E} - \boldsymbol{E} \cdot \frac{\partial \boldsymbol{D}}{\partial t} \qquad (117)$$

For a linear material, we have

$$\boldsymbol{H} \cdot \frac{\partial \boldsymbol{B}}{\partial t} + \boldsymbol{E} \cdot \frac{\partial \boldsymbol{D}}{\partial t} = \mu \boldsymbol{H} \cdot \frac{\partial \boldsymbol{H}}{\partial t} + \varepsilon \boldsymbol{E} \cdot \frac{\partial \boldsymbol{E}}{\partial t}$$

$$= \frac{1}{2}\mu \frac{\partial}{\partial t}(\boldsymbol{H} \cdot \boldsymbol{H}) + \frac{1}{2}\varepsilon \frac{\partial}{\partial t}(\boldsymbol{E} \cdot \boldsymbol{E})$$

$$= \frac{1}{2}\frac{\partial}{\partial t}(\boldsymbol{B} \cdot \boldsymbol{H} + \boldsymbol{D} \cdot \boldsymbol{E})$$

Thus Eq. (117) can be rewritten in the form

$$\nabla \cdot \boldsymbol{S} + \frac{\partial u}{\partial t} = -\boldsymbol{J} \cdot \boldsymbol{E} \qquad (118)$$

where $\boldsymbol{S} \approx \boldsymbol{E} \times \boldsymbol{H}$ (119)

is known as *Poynting vector*. The vector $\boldsymbol{S} = \boldsymbol{E} \times \boldsymbol{H}$ has the dimensions of $\frac{\text{energy}}{\text{area} \times \text{time}}$ and, hence, it represents the energy that flows out of the boundary per unit time.

Also $u = \frac{1}{2}(\boldsymbol{B} \cdot \boldsymbol{H} + \boldsymbol{D} \cdot \boldsymbol{E})$ (120)

Eq. (118) resembles the equation of continuity and for a physical interpretation one can note that if a charge q (moving with velocity v) is acted on by an electromagnetic field then the work done by the field in moving it through a distance ds would be $\boldsymbol{F} \,.\, ds$. thus the work done per unit time would be

$$\boldsymbol{F} \cdot \frac{ds}{dt} = \boldsymbol{F} \cdot \boldsymbol{V} = [q\boldsymbol{E} + q\boldsymbol{V} \times \boldsymbol{B}] \cdot \boldsymbol{V}$$

$$= q\boldsymbol{E} \cdot \boldsymbol{V} \qquad (121)$$

If there are N charged particles per unit volume, each carrying a charge q, then the work done per unit volume would be

$$Nq\boldsymbol{V} \cdot \boldsymbol{E} = \boldsymbol{J} \cdot \boldsymbol{E} \qquad (122)$$

where $\boldsymbol{J}$ is the current density. The energy appears in the form of kinetic (or heat) energy of the charged particles. Thus the term $\boldsymbol{J} \cdot \boldsymbol{E}$ represents the familiar Joules loss and, therefore, the quantity $\boldsymbol{J} \cdot \boldsymbol{E}$ appearing on R.H.S. of Eq. (118) would represent the rate at which energy is produced per unit volume per unit time. Obviously, one may interpret Eq. (118) as an equation of continuity for energy with u representing energy per unit volume. The quantities ½$(\boldsymbol{D} \cdot \boldsymbol{E})$ and ½$(\boldsymbol{B} \cdot \boldsymbol{H})$ represent the electrical and magnetic energies per unit volume respectively.

The direction of $\boldsymbol{E} \times \boldsymbol{B}$ for a plane electromagnetic wave is perpendicular to the wave-front and is therefore pointing in the direction of propaga-tion of the wave (Fig. 3.20). The magnitude of poynting vector

$$|\boldsymbol{E} \times \boldsymbol{B}| = EB = \frac{1}{2} E^2$$

Thus the vector $c\boldsymbol{E} \times \boldsymbol{B}$ has the magnitude E^2. Then the flux $c^2\varepsilon_0 \boldsymbol{E} \times \boldsymbol{B}$ (poynting vector) across a surface S is given by

$$\int_s c^2 \varepsilon_0 (\boldsymbol{E} \times \boldsymbol{B}) \cdot u_N ds = \frac{d(\text{Energy})}{dt} \tag{123}$$

is the energy crossing the area S per unit time, and for that reason it has been designated by $\frac{d(\text{Energy})}{dt}$.

Since electromagnetic radiation propagates with velocity c, one may use the relation between energy and momentum $p = \frac{v(\text{Energy})}{c^2}$, ($v = c$) to obtain the momentum p per unit volume associated with an electromagnetic wave. Thus

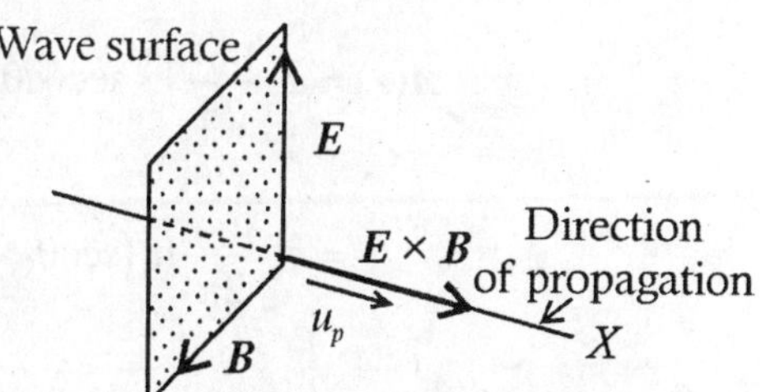

Fig. 3.20: Definition of the direction of energy flow in an electro-magnet wave

$$p = \frac{\text{Energy}}{c} = \frac{\varepsilon_0 E^2}{c} = \varepsilon_0 |\boldsymbol{E} \times \boldsymbol{B}| \tag{124}$$

Since momentum must have the direction of propagation, one may express the above equation in vector form as

$$\boldsymbol{p} = \frac{E}{c} \boldsymbol{u} = \varepsilon_0 \boldsymbol{E} \times \boldsymbol{B} \tag{125}$$

where $\boldsymbol{u}$ is the unit vector in the direction of propagation.

If an electromagnetic wave has momentum, it also has angular momentum. The angular momentum per unit volume is

$$\boldsymbol{L} = \boldsymbol{r} \times \boldsymbol{p} = \varepsilon_0 \boldsymbol{r} \times (\boldsymbol{E} \times \boldsymbol{B}) \tag{126}$$

This is called the "orbital" angular momentum of radiation. In addition, electromagnetic radiation possesses an intrinsic angular momentum, or spin. Concluding, one can say that an electromagnetic wave carries momentum and angular momentum as well as energy.

ILLUSTRATIVE EXAMPLES

Example 1

Determine the magnetic vector potential at a distance r from a very long thin straight wire carrying a current I. Hence find the corresponding magnetic field $\boldsymbol{B}$.

Solution

If $\hat{a}_z$ is a unit vector in the direction of the current I, from Eq. (19) we write for the magnetic vector potential

$$\boldsymbol{A}(\boldsymbol{r}) = \hat{a}_z \frac{\mu_0 I}{4\pi} \int_{-\theta_1}^{\theta_2} \sec\theta d\theta$$

As the length L of the wire is very large compared with r, we have $\theta_1 = \theta_2$. So,

$$\boldsymbol{A}(\boldsymbol{r}) = \hat{a}_z \frac{\mu_0 I}{2\pi} \int_{0}^{\theta_1} \sec\theta d\theta$$

$$= \hat{a}_z \frac{\mu_0 I}{2\pi} \ln(\sec\theta_1 + \tan\theta_1)$$

Since $L \gg r$, we have $\sec\theta_1 \simeq \tan\theta_1 = \frac{L}{2r}$. Hence

$$\boldsymbol{A}(\boldsymbol{r}) = \hat{a}_z \frac{\mu_0 I}{2\pi} \ln\left(\frac{L}{r}\right)$$

Using cylindrical coordinates (r, θ, z), we note that

$$A_z = \frac{\mu_0 I}{2\pi} \ln\left(\frac{L}{r}\right), \; A_r = 0, \text{ and } A_\theta = 0.$$

Hence, $\boldsymbol{B} = \nabla \times \boldsymbol{A} = -\hat{a}_\theta \frac{dA_z}{dr} = \hat{a}_\theta \frac{\mu_0 I}{2\pi r}$, where $\hat{a}_\theta$ is a unit vector in the θ-direction.

Example 2

An infinitely long solenoid of radius 'a' having n_0 number of turns of wire per unit length carries the current I. Find the magnetic vector potential at a distance $r(>a)$ from the axis of the solenoid.

Solution

Equation (19) shows that $\boldsymbol{A}$ is in the direction of the current I. Since I flows along the circumference of the solenoid, $\boldsymbol{A}$ is perpendicular to $\boldsymbol{r}$ (Fig. 3.21). By symmetry, $\boldsymbol{A}$ has the same magnitude at all points at an axial distance r from the solenoid.

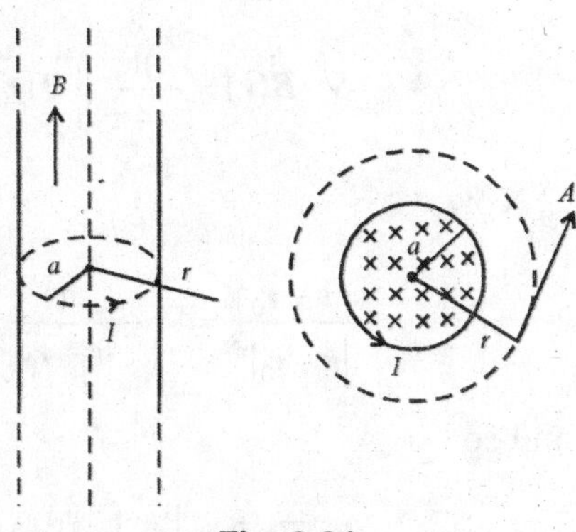

Fig. 3.21

Taking the line integral of $\boldsymbol{A}$ along the circumference of the circle of radius r and relating it to the surface integral of $\nabla \times \boldsymbol{A}$ by Stokes' theorem of vector analysis, we have

$$\oint \boldsymbol{A} \cdot d\boldsymbol{l} = \int_s (\nabla \times \boldsymbol{A}) \cdot \hat{\boldsymbol{n}} ds \quad (i)$$

where S is the surface bounded by the circle. The integral on the right-hand side of (i) is just the flux of $\boldsymbol{B}$ through the cross-section of the solenoid, the magnitude of $\boldsymbol{B}$ outside the solenoid being zero.

Since $\boldsymbol{B} = \mu_0 n_0 I$ inside the solenoid and $\boldsymbol{B}$ is directed along the axis of the solenoid, Eq. (i) gives

$$2\pi r A = (\mu_0 n_0 I)\pi a^2$$

or

$$A = \frac{\mu_0 n_0 I a^2}{2r}.$$

$\boldsymbol{A}$ is always perpendicular to the vector $\boldsymbol{r}$. Note that $A \neq 0$ outside the solenoid although B is zero there.

Note: For a point inside the solenoid ($r < a$), Eq. (i) gives $2\pi r A = (\mu_0 n_0 I)\pi r^2$, or $A = \mu_0 n_0 I r/2$.

Example 3

The magnetic field $\boldsymbol{B}(\boldsymbol{r})$ due to a volume current is, by Biot-Savart law,

$$\boldsymbol{B}(\boldsymbol{r}) = \frac{\mu_0}{4\pi} \int_V \boldsymbol{J}(\boldsymbol{r}_0) \times \frac{\boldsymbol{r} - \boldsymbol{r}_0}{|\boldsymbol{r} - \boldsymbol{r}_0|^3} dV_0$$

show that $\nabla \cdot \boldsymbol{B}(\boldsymbol{r}) = 0$, where ∇ denotes differentiation with respect to r.

Solution

Using the vector identity

$$\nabla \cdot (\boldsymbol{A} \times \boldsymbol{B}) = -\boldsymbol{A} \cdot \nabla \times \mathbf{B} + \boldsymbol{B} \cdot \nabla \times \boldsymbol{A}$$

and identifying $\boldsymbol{A}$ with $\boldsymbol{J}(\boldsymbol{r}_0)$ and $\boldsymbol{B}$ with $\dfrac{\boldsymbol{r} - \boldsymbol{r}_0}{|\boldsymbol{r} - \boldsymbol{r}_0|^3}$, we get

$$\nabla \cdot \boldsymbol{B}(\boldsymbol{r}) = -\frac{\mu_0}{4\pi} \int_V \boldsymbol{J}(\boldsymbol{r}_0) \cdot \nabla \times \frac{\boldsymbol{r} - \boldsymbol{r}_0}{|\boldsymbol{r} - \boldsymbol{r}_0|^3} dV_0$$

Now,

$$\frac{\boldsymbol{r} - \boldsymbol{r}_0}{|\boldsymbol{r} - \boldsymbol{r}_0|^3} = \nabla\left(-\frac{1}{|\boldsymbol{r} - \boldsymbol{r}_0|}\right)$$

and so

$$\nabla \cdot \boldsymbol{B} = \frac{\mu_0}{4\pi} \int_V \boldsymbol{J}(\boldsymbol{r}_0) \cdot \nabla \times \nabla \frac{1}{|\boldsymbol{r} - \boldsymbol{r}_0|} dV_0$$

= 0, since curl of gradient is zero.

Example 4

(*a*) Determine the e.m.f. developed between the terminals of a straight conductor of length l moving with a constant velocity $\boldsymbol{v}$ at right angles to a uniform magnetic field $\boldsymbol{B}$.

(*b*) An aeroplane is moving due north with a velocity of 450 km per hour. Find the potential difference in volt between the ends of its wings, distant 35 meter from each other. Given, the earth's magnetic intensity is 0.35 *Oe*.

Solution

(*a*) Each electron inside the conductor will experience a Lorentz force $-e\boldsymbol{v} \times \boldsymbol{B}$, $-e$ being the charge on an electron. The free electrons will, as a result, be deflected towards the end a of the conductor. There will be a charge separation leading to the development of an electric field $\boldsymbol{E}$ from the end b to the end a. In the steady state,

$$-e\boldsymbol{E} - e\boldsymbol{v} \times \boldsymbol{B} = \boldsymbol{0} \text{ or, } \boldsymbol{E} = -\boldsymbol{v} \times \boldsymbol{B}$$

The total e.m.f. induced in the conductor is

$$\varepsilon = |E| l = vBl.$$

Fig. 3.22

(*b*) The induced e.m.f. is

$$\varepsilon = vBl$$

where v = velocity = 450 km/hour $= \frac{450 \times 10^3}{3600}$ m/s, B = magnetic induction = $\mu_0 H = 4\pi \times 10^{-7} \times 0.35/(4\pi \times 10^{-3}) = 0.35 \times 10^{-4}$ tesla, and l = distance between the wings = 35 m. Therefore,

$$\varepsilon = \frac{450 \times 10^3}{3600} \times 0.35 \times 10^{-4} \times 35 = 0.153 \text{ Volt}.$$

Example 5

Show that for a long thin copper wire carrying a time-varying current $I = t$ ampere, the induced electric field on its surface is independent of the radius of the wire and equal to 10^{-7} V/m. Take the induced field along the axis of the wire to be zero.

Solution

Let the radius of the wire be R. Consider a rectangle $abcd$ in the wire with the side ab along the axis. The magnetic induction at a distance r from the axis ($r < R$) is

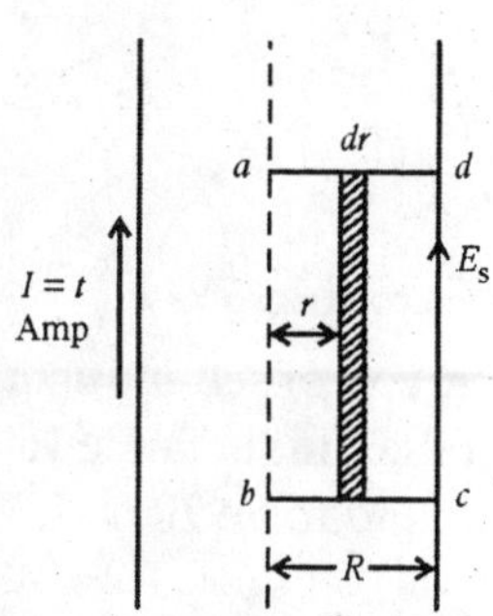

Fig. 3.23

$$B = \frac{\mu_0 I r}{2\pi R^2}$$

The flux through the elementary shaded area within $abcd$ is $Bldr$, where $l = ab = cd$. The flux Φ through $abcd$ is

$$\Phi = \int_0^R \frac{\mu_0 I r}{2\pi R^2} l dr = \frac{\mu_0 I l}{4\pi}$$

The induced e.m.f. along the curve $abcd$ is

$$|\varepsilon| = \frac{d\phi}{dt} = \frac{\mu_0 l}{4\pi}\frac{dI}{dt} = \frac{\mu_0 l}{4\pi} \qquad (i)$$

The wire being thin, $cd \gg da$. If E_a is the induced electric field at the surface and E_0 that along the axis, then $|\varepsilon| = (E_a - E_0)l$. Since $E_0 = 0$, we obtain from (*i*)

$$E_a = \frac{\mu_0}{4\pi} = 10^{-7} \text{ V/m}$$

Clearly, E_a is indepindent of the radius of the wire.

Example 6

An air-cored solenoid has a diameter of 2.5 cm and 500 turns wound over a length of 30 cm. Calculate the self-inductance of the solenoid.

Solution

The self-inductance L of the solenoid is

$$L = \frac{\mu_0 N^2 A}{l}$$

where $\mu_0 = 4\pi \times 10^{-7}$ H/m, N = total number of turns = 500, A = area of cross-section = $\pi(2.5 \times 10^{-2}/2)^2$ m^2 $= \frac{\pi}{4} \times 6.25 \times 10^{-4}$ m^2, and l = length of the solenoid = 30×10^{-2} m. Therefore

$$L = \frac{4\pi \times 10^{-7} \times 25 \times 10^4 \times \pi \times 6.25 \times 10^{-4}}{4 \times 30 \times 10^{-2}} \text{ H}$$

$$= 0.514 \text{ mH.}$$

Example 7

A coil of self-inductance 100 mH is connected in series with another coil of self-inductance 169 mH. The effective inductance of the combination is found to be 70 mH. Determine the coefficient of coupling.

Solution

Since the given effective inductance is less than the sum of the self-inductances of the two coils, the mutual flux must oppose the self-inductance fluxes. Therefore we can write for the effectuve dnductance

$$L_{\text{eff}} = L_1 + L_2 - 2M$$

Given, L_1 = 100 mH, L_2 = 169 mH, and L_{eff} = 70 mH

Therefore $$M = \frac{L_1 + L_2 - L_{\text{eff}}}{2} = 99.5 \text{ mH}$$

The coefficient of coupling is

$$k = \frac{M}{\sqrt{L_1 L_2}} = \frac{99.5}{\sqrt{100 \times 169}} = 0.765$$

Example 8

A conducting disc of conductivity σ has radius a and thickness δ. A uniform magnetic field $\boldsymbol{B}$ is applied in a direction perpendicular to the plane of the disc. If the magnetic field changes with time at the rate of dB/dt, calculate the power dissipated in the disc due to the induced current.

Solution

Consider two circles of radii r and $r + dr$ concentric with the disc ($0 < r < a$) [Fig. 3.24]. The induced e.m.f. in the circular path of radius r is

$$\varepsilon = -\frac{d}{dt}\left(\pi r^2 B\right) = -\pi r^2 \frac{dB}{dt}$$

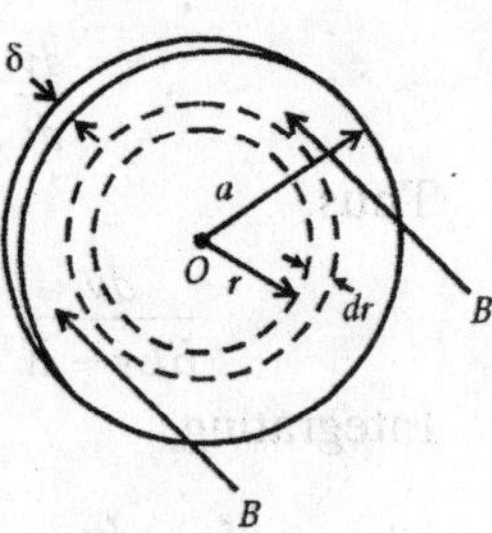

Fig. 3.24

The resistance of the circular path between radii r and $r + dr$ is

$$R = \frac{1}{\sigma}\frac{2\pi r}{\delta dr},$$

the length of the path being $2\pi r$ and the cross-sectional area of current flow being δdr. The power dissipated inside this path is

$$dP = \frac{\varepsilon^2}{R} = \frac{\pi\delta\sigma}{2}\left(\frac{dB}{dt}\right)^2 r^3 dr$$

The total dissipated power P is

$$P = \frac{\pi\delta\sigma}{2}\left(\frac{dB}{dt}\right)^2 \int_0^a r^3 dr = \frac{\pi\delta\sigma a^4}{8}\left(\frac{dB}{dt}\right)^2$$

Note that the disspated power is proportional to the square of the area of the disc. This is a general property of eddy currents.

Example 9

Two long parallel rails of zero resistance separated by a distance L are joined to a cell of e.m.f. E at one end. An external uniform magnetic field B is applied normal to the plane of the rails. A conducting bar of mass m and resistance R is placed across the rails. The bar can slide freely parallel to itself always remaining perpendicular to the rails. Find the velocity of the sliding bar as a function of time. What is the terminal velocity of the bar? What is the current flowing through it at the terminal velocity? Interpret the results and say how the energy supplied by the cell is consumed.

Solution

Let v be the velocity of the conducting bar at any instant of time t. The motional e.m.f. opposing the cell e.m.f. is BLv. The net e.m.f. in the circuit is thius $(E - BLv)$. The current flowing in the circuit is $i = (E - BLv)/R$. The force on the conducting bar is BiL and its accelerations is

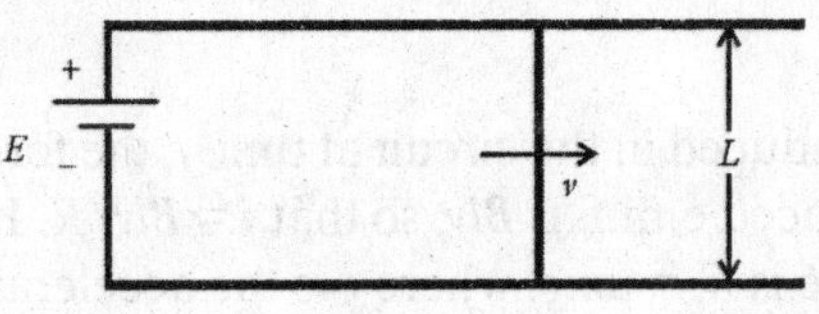

Fig. 25

$$\frac{dv}{dt} = \frac{BiL}{m} = \frac{BL}{mR}(E - BLv)$$

Thus,

$$\frac{dv}{BLv - E} = -\frac{BL}{mR}dt$$

Integrating,

$$\int \frac{dv}{BLv - E} = -\frac{BL}{mR}\int dt, \quad \text{or} \quad \frac{1}{BL}\ln\left(\frac{BLv - E}{K}\right) = -\frac{BL}{mR}t$$

where K is the integration constant.

So
$$BLv - E = Ke^{-B^2L^2t/(mR)}$$

At $t = 0$, $v = 0$. Therefore, $K = -E$, and

$$v = \frac{E}{BL}\left(1 - e^{-B^2L^2t/(mR)}\right)$$

which gives the velocity as a function of time. The terminal velocity v_T of the bar is obtained by letting $t \to \infty$. Thus, $v_T = E/BL$. At the terminal velocity, $E = BLv_T$ and so the current is zero.

At first, the cell e.m.f. E drives a current through the bar which moves outward due to the force exerted on it. The motion of the bar induces a back e.m.f. which reduces the current and hence the acceleration of the bar. The accelertion becomes zero at the terminal velocity where the cell e.m.f. is balanced by the back e.m.f.

The energy supplied by the cell is partly dissipated as Joule heat in the circuit and the rest is used up in moving the bar.

Example 10

A rod PQ of mass m and length l can slide on two conducting parallel rails lying in a horizontal plane and connected to a resistance R, as shown in Fig. 3.26. A uniform magnetic field of induction B is applied normal to the plane of the rails. At time $t = 0$, the rod is momentarily pushed outwards along the rails with a velocity v_0. If μ is the coefficient of friction between the rod and the rails, find an expression for the velocity v of the rod as a function of time t. Also, determine the time that elepses before the rod comes to rest.

Solution

If i is the current induced in the circuit at time t, the force F exerted on the rod is Bil. The induced e.m.f. is Blv, so that $i = Blv/R$. Hence $F = B^2l^2v/R$. The frictional force is $F_r = \mu mg$, where g is the acceleration due to gravity. The equation of motion of the rod PQ is

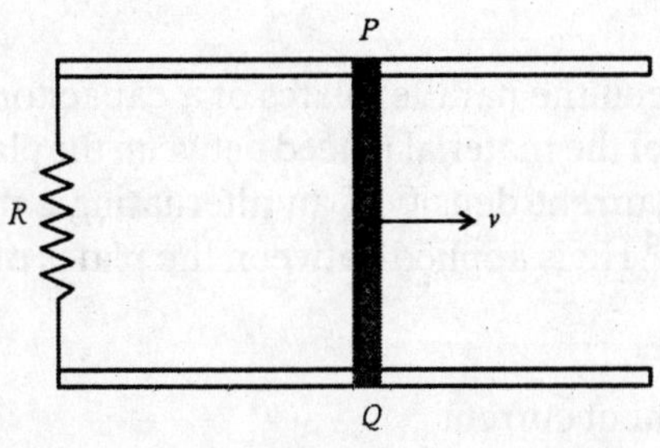

Fig. 26

$$m\frac{dv}{dt} = -F - F_r$$

The negative signs imply that F and F_r are opposite to the direction of v. Thus,

$$\frac{dv}{dt} = -\frac{B^2l^2v}{mR} - \mu g$$

or
$$\int_{v_0}^{v} \frac{dv}{\left(B^2l^2v/mR\right) + \mu g} = -\int_0^t dt$$

Putting $x = (B^2l^2v/mR) + \mu g$, we obtain

$$\frac{mR}{B^2l^2}\int_{\frac{B^2l^2v_0}{mR}+\mu g}^{\frac{B^2l^2v}{mR}+\mu g} \frac{dx}{x} = -t$$

or
$$v = \left(v_0 + \frac{mR\mu g}{B^2l^2}\right)e^{-\frac{t}{r}} - \frac{mR\mu g}{B^2l^2}$$

where
$$r = \frac{mR}{B^2l^2}.$$

If t_0 is the time taken by the rod to come to rest, we have $v = 0$ at $t = t_0$. So

$$0 = \left(v_0 + \frac{mR\mu g}{B^2l^2}\right)e^{-\frac{t_0}{r}} - \frac{mR\mu g}{B^2l^2}$$

or
$$t_0 = r\ln\left[\frac{v_0 + mR\mu g/\left(B^2l^2\right)}{mR\mu g/\left(B^2l^2\right)}\right].$$

Example 11

The distance between the parallel plates of a capacitor is 0.6 mm and the dielectric constant of the material placed between the plates is 15. Calculate the displacement current density if an alternating e.m.f. of 10 volts and frequency 1.5×10^9 Hz is applied between the plates of capacitor.

Solution

We have displacement current

$$J_d = \frac{\partial D}{\partial t}$$

Magnitude of displacement current density for alternating field

$$J_d = \omega D = \omega \varepsilon_0 \varepsilon_r E$$

$\therefore$ Electric field $E = \dfrac{V}{d}$

Here V is potential difference and d is the distance between the plates of the capacitor.

$$\therefore \quad J_d = \omega \varepsilon_0 \varepsilon_r \frac{V}{d} = \frac{2 \times 3.14 \times 1.5 \times 10^9 \times 8.85 \times 10^{-2} \times 15 \times 100}{6 \times 10^{-4}}$$

$$= 2.08 \times 10^5 \text{ amp/m}^2$$

Example 12

Show that the following field vectors in free space ($\mu = \mu_0$, $\varepsilon = \varepsilon_0$, $\sigma = 0$, $\rho = 0$, $J = 0$) satisfy all the Maxwell's equations. $\boldsymbol{E} = E_0 \cos(\omega t - \beta z)\,\hat{i}$ and $\boldsymbol{H} = \dfrac{E_0}{\eta}\cos(\omega t - \beta z)\,\hat{j}$.

Solution

From Faraday's law, we have

$$\nabla \times \boldsymbol{E} = -\mu_0 \frac{\partial \boldsymbol{H}}{\partial t}$$

or

$$\begin{vmatrix} \hat{i} & \hat{j} & \hat{k} \\ \dfrac{\partial}{\partial x} & \dfrac{\partial}{\partial y} & \dfrac{\partial}{\partial z} \\ E_x & E_y & E_z \end{vmatrix} = -\mu_0 \frac{\partial \boldsymbol{H}}{\partial t}$$

Then, with $E_y = E_z = 0$, and $H_x = H_z = 0$, we have

$$\left(\frac{\partial E_{0z}}{\partial y} - \frac{\partial E_{0y}}{\partial z}\right)\hat{i} + \left(\frac{\partial E_{0x}}{\partial z} - \frac{\partial E_{0z}}{\partial z}\right)\hat{j} + \left(\frac{\partial E_{0y}}{\partial x} - \frac{\partial E_{0x}}{\partial y}\right)$$

$$= -\mu_0 \frac{E_0}{\eta} \frac{\partial}{\partial t}\left[\cos(\omega t - \beta z)\right]\hat{k}$$

or $$\frac{\partial E_x}{\partial z} a_y = \mu_0 \frac{E_0}{\eta} \omega \sin(\omega t - \beta z)\hat{j}$$

or $$\beta E_0 \sin(\omega t - \beta z)\hat{j} = \omega \mu_0 \frac{E_0}{\eta} \sin(\omega t - \beta z)\hat{j}$$

This yields $$\beta = \omega\, \mu_0/\eta \qquad (i)$$

From Ampere's law, we have

$$\nabla \times \boldsymbol{H} = J_0 + \varepsilon_0 \frac{\partial \boldsymbol{E}}{\partial t} \qquad (ii)$$

Expanding the curl, we obtain

$$\nabla \times \boldsymbol{H} = \frac{\partial H_z}{\partial z}\hat{i} = -\frac{\beta E_0}{\eta} \sin(\omega t - \beta z)\hat{i} \qquad (iii)$$

Using Eq. (*iii*), Eq. (*ii*) become

$$-\frac{\beta E_0}{\eta} \sin(\omega t - \beta z)\hat{i} = -\varepsilon_0 \omega E_0 \sin(\omega t - \beta z)\hat{i}$$

This gives $$\beta = \omega \varepsilon_0 \eta \qquad (iv)$$

Combining Eq. (*i*) and Eq. (*iv*), one obtains

$$\frac{\omega \mu_0}{\eta} = \omega \varepsilon_0 \eta \quad \text{or} \quad \eta^2 = \frac{\mu_0}{\varepsilon_0}$$

$$\therefore \qquad \eta = \pm\sqrt{\frac{\mu_0}{\varepsilon_0}} \qquad (v)$$

Using (*v*), one obtains from (*i*) or (*iv*)

$$\beta = \pm\, \omega \sqrt{\mu_0 \varepsilon_0} \qquad (vi)$$

Let us now check Gauss's laws. We have

$$\nabla \cdot \boldsymbol{D} = \varepsilon_0 \nabla \cdot \boldsymbol{E} = 0$$

or $$\frac{\partial E_x}{\partial x} + \frac{\partial E_y}{\partial y} + \frac{\partial E_z}{\partial z} = 0$$

which we observe to be true since the only component of E, E_x is independent of x.

Now, $$\nabla \cdot \boldsymbol{B} = \mu_0 \nabla \cdot \boldsymbol{H} = 0$$

or $$\frac{\partial H_x}{\partial x} + \frac{\partial H_y}{\partial y} + \frac{\partial H_z}{\partial z} = 0$$

This is also seen to be true since the only component of H, H_y is independent of y obviously, the given fields are valid ones but only if η and β satisfy constraints in (*v*) and (*vi*)

Example 13

Compare the conduction and displacement current densities in copper ($\varepsilon = \varepsilon_0$, $\mu = \mu_0$) and $\sigma = 5.8 \times 10^7$ S/m at a frequency of 1 MHz.

Solution

Let us assume the sinusoidal variation of the electric field in the material, we have

$$E = E_0 \sin\omega t \ \text{V/m}$$

Here $\omega = 2\pi\nu$ and $\nu = 10^6$ c/s, the conduction current density is

$$J_c = \sigma E = \sigma E_0 \sin\omega t \ \text{A/m}^2$$

The displacement current density is

$$J_d = \frac{\partial D}{\partial t} = \varepsilon \frac{\partial E}{\partial t} = \omega \varepsilon E_0 \cos \omega t \quad \text{A/m}^2$$

The ratio of the magnitudes of these currents is

$$\frac{|J_c|}{|J_d|} = \frac{\sigma}{\omega\varepsilon}$$

For copper at 1 MHz, we find

$$\frac{\sigma}{\omega\varepsilon} = \frac{5.8 \times 10^7}{2\pi \times 10^6 \times \left(\frac{1}{3\pi}\right) \times 10^{-9}} = 1.04 \times 10^{12}$$

This is very large value and, therefore, even if the frequency is raised to an extremely high value (~100 GHz or 10^{11} Hz), the conduction current dominates the displacement current by an enormous amount. Obviously, for copper and other conductors it is reasonable to neglect displacement current.

Example 14

For a medium, conductivity $\sigma = 5$ mho/m and dielectric constant = 1. If an electric field $E = 250 \sin(10^{10} t)$ is applied then find conducting current and displacement current in the medium. At what frequency both the currents will be equal?

Solution

We have the conducting current density

$$J_c = \sigma E$$

$$\sigma = 5 \text{ mho/m}$$

$$E = 250\sin(10^{10}t)$$

$$\therefore \quad J_c = 5 \times 250\sin(10^{10}t)$$

$$= 1250\sin(10^{10}t) \text{ amp/m}^2$$

Displacement current density

$$J_d = \varepsilon \frac{\partial E}{\partial t} = \varepsilon_0 \varepsilon_r \frac{\partial E}{\partial t}$$

$$\therefore \quad J_d = \varepsilon_0 \varepsilon_r \frac{\partial}{\partial t} 250 \sin(10^{10}t)$$

$$= \varepsilon_0 \varepsilon_r 250 \times 10^{10} \cos(10^{10}t)$$

$$= 8.85 \times 10^{-12} \times 1 \times 250 \times 10^{10}\cos(10^{10}t)$$

$$= 22.1 \cos(10^{10}t) \text{ amp/m}^2$$

For $|J_c| = |J_d|$, we have

$$\sigma E = \varepsilon_0 \varepsilon_r \omega E \therefore \omega = \frac{\sigma}{\varepsilon_0 \varepsilon_r} = \frac{5}{8.85 \times 10^{-12} \times 1}$$

$$= 5.65 \times 10^{11} \text{ radian/s}$$

REVIEW QUESTIONS

1. State Faraday's law of electromagnetic induction. Use it to deduce, $\nabla \times E = -\frac{\partial B}{\partial t}$. [BE]

3. (*a*) State the laws of electromagnetic induction. Calculate the self-inductance of a uniformly wound solenoid. [C.U. 1982]

(*b*) State Faraday's law of induction and express it in differential form.

4. (*a*) What do you understand by 'self' and 'mutual' inductances? Find an expression for the mutual inductance between the primary and the secondary of a standard solenoid.

(*b*) What are the dimensions of inductance in terms of (MLTI)?

5. Explain the term 'eddy current'. How would you reduce losses due to eddy currents? Name any practical application of eddy current.

6. Determine the self-inductance per unit length of an open-air transmission line. Explain the principle of non-inductive winding.

7. Calculate the self-inductance per unit length of a coaxial cable, the space between the two cylinders containing a material of permeability μ.

8. (*a*) Obtain Neumann's formula for mutual inductance between two loops.

(*b*) Apply it to calculate the mutual inductance between two parallel coaxial circular loops.

9. There are two parallel circular loops of radii a and b ($a \ll b$); the separation between the centres of the loops is r. Find the flux through the smaller loop due to a current I in the smaller loop? Why?

10. Show that the equivalent inductance of two coils of self-inductances L_1 and L_2, connected in series, is

$$L_{eq} = L_1 + L_2 \pm 2M,$$

where M is the mutual inductance between the coils.

11. Show that the equivalent inductance of two coil of self-inductances L_1 and L_2, connected in parallel, is

$$L_{eq} = \frac{L_1 L_2 - M^2}{L_1 + L_2 \mp 2M}$$

12. A metal disc of radius a rotates with a constant angular velocity ω in a magnetic field of induction B, normal to the plance of the disc. Derive an expression for the potential difference between the centre and the rim of the disc.

13. From the fundamental principle establish the relation

$$\nabla \times \boldsymbol{E} = -\frac{\partial \boldsymbol{B}}{\partial t}$$

14. How was the concepts of displacement current helpful in removing discrepancy in Ampere's law? Prove that $\nabla \times \boldsymbol{H} = \boldsymbol{J} + \boldsymbol{D}$.

15. What is the physical significance of equations (*i*) $\nabla \cdot \boldsymbol{B} = 0$ and (*ii*) $\nabla \cdot \boldsymbol{D} = \rho$?

16. Using Maxwell's equations prove that (*i*) $\nabla \cdot \boldsymbol{B} = 0$ and (*ii*) $\nabla \cdot \boldsymbol{D} = \rho$

17. Deduce the equations: (*i*) $\nabla \times \boldsymbol{E} = -\frac{\partial \boldsymbol{B}}{\partial t}$ and (*ii*) $\nabla \times \boldsymbol{H} = \boldsymbol{J} + \frac{\partial \boldsymbol{D}}{\partial t}$.

18. Prove the continuity equation $\nabla \boldsymbol{J} + \frac{\partial \rho}{\partial t} = 0$.

19. What is displacement current ? Show that the conduction current in a lead wire is identical with the displacement in the gap of the condenser.

20. Show that Maxwell's equation $\nabla \times \boldsymbol{H} = \boldsymbol{J}$ is inconsistent with the law of conservation of energy. How this has been modified in the light of the principle of conservation of energy?

21. Making use of Maxwell's equations, obtain the differential equation for an electromagnetic wave.

22. How electromagnetic waves are propagated? Show that these waves are transverse and travel with the velocity of light in vacuum.

23. Do the magnitudes of electric intensity (*E*) and magnetic intensity (*H*) vectors in the electromagnetic wave become maximum and minimum simultaneously? Explain it clearly.

24. From Maxwell's curl equations derive the wave equation in $\boldsymbol{H}$ for a plane wave travelling in the positive x-direction in a medium with constants $\mu = \mu_0$, $\varepsilon = \varepsilon_0$ and $\sigma = 0$. The electric field is in the y-direction.

25. Show that the energy stored in a magnetic field per unit volume is $\frac{1}{2}B^2/\mu_0$ and the corresponding expression for the energy density in an electric field is $\frac{1}{2}\varepsilon_0 E^2$.

26. Write Maxwell's equations in differential as well as integral form.

27. What is an electromagnetic spectrum? Write the names of its constituent members: Why classification does not have very sharp boundaries.

28. Why do the different parts of the electromagnetic spectrum behave in a different way when propagating through matter?

29. Why are visible light waves are called electromagnetic waves? Justify your answer.

30. Show that when an electromagnetic wave enters a metal gets attenuated. Obtain an expression for the skin depth.

31. What is Poynting vector? Obtain an expression for it.

32. An electromagnetic wave carries momentum and angular momentum as well as energy. Explain and justify your answer.

33. Show that the energy of an electromagnetic wave is equally shared between the electric and magnetic fields.

PROBLEMS

1. A straight piece of wire moves freely on two smooth parallel rails 80 cm apart. A magnetic intensity of 2000 oersted exists perpendicular to the rails and the wire. The wire is always at right angles to the rails. A voltmeter connected between the rails reads 50 volt. Determine the speed of the wire [**Ans.** 312.5 m/s]

2. A metallic disc of radius 12 cm rotates at the rate of 1500 r.p.m. A uniform magnetic field of 30 gauss is app;ied perpendicular to the plane of the disc. Determine the e.m.f. induced between the centre and the edge of the disc.

[**Ans.** 3.39 mV]

3. The electric field of a plane electromagnetic wave in vacuum, using MKSC units, is represented by $E_x = 0$, $E_y = 0.5\cos\left[2\pi\times10^8\left(t-\frac{x}{c}\right)\right]$, $E_z = 0$ (*a*) Determine the wavelength, the state of polarization, and the direction of propagation (*b*) compute the magnetic field of the wave (*c*) calculate the energy flux per unit area.

[(*a*) 3 m, linearly polarized in the *xy*-plane, propagates in the +*x*-direction.

(*b*) $B_x = B_y = 0$, $B_z = \frac{1}{6}\times10^{-8}\cos\left[2\pi\times10^{-8}\left(t-\frac{x}{c}\right)\right]$. (*c*) 3.316×10^{-4} W/m^{-2}.]

4. Electromagnetic radiation from the sun falls on the earth's surface at the rate of 1.4×10^3 W-m^{-2}. Assuming that this radiation can be considered as plane waves, estimate the magnitude of the electric and magnetic field amplitudes in the wave. [1.15×10^3 N/C, 3.84×10^{-6} T]

5. The conductivity of a medium is 2 mho/m and its dielectric constant is 2. An electric field $E = 220\sin 10^6 t$ is applied on it. Calculate the conduction current and displacement current densities.

[$440\sin(10^6 t)$ A/m^2, $5.53\times10^{-2}\cos(10^6 t)$ A/m^2]

6. If distilled water has constant s $\mu_r = 1$, $\varepsilon_r = 81$, and power factor = 0.05 at 1 GHz, calculate (*a*) the $1/e$ depth of penetration and (*b*) the 1 per cent depth of penetration at this frequency. [$\delta(1/e) = 212$ mm, (*b*) $\delta(1\%)977$ mm]

SHORT ANSWER QUESTIONS

1. What is absolute index of refraction?

Ans. $\eta = \frac{c}{v} = \frac{1}{\sqrt{\varepsilon_0\mu_0}}\sqrt{\varepsilon\mu} = \sqrt{\frac{\varepsilon\mu}{\varepsilon_0\mu_0}} = \sqrt{\varepsilon_r\mu_r}$.

2. Is gamma rays are electromagnetic radiations?

Ans. Yes, these waves are of nuclear origin.

3. Are there electromagnetic waves in cosmic radiations?

Ans. Yes, there are electromagnetic waves of even shorter wavelengths or longer frequencies.

4. Why does not one see the portion other than visible one of the electromagnetic spectrum?

Ans. The retina of the eye is sensitive only to colours in the visible region, *i.e.* wavelength lying between 3900 Å and 7800 Å. This region corresponds to visible part of the spectrum.

5. On what factors the velocity of electromagnetic wave depends?

Ans. (*i*) permittivity and (*ii*) permeability, $c = \frac{1}{\sqrt{\varepsilon\mu}}$.

11. Do electromagnetic waves have same velocity in all transparent media?

Ans. No, as the refractive index is different for different media, $\eta = \frac{c}{v}$, v is different for different media.

12. An electromagnetic wave carries momentum. What it signifies?

Ans. An electromagnetic interaction between two electric charges means an exchange of energy and momentum between the charges.

OBJECTIVE QUESTIONS

1. The ratio $\sqrt{\mu/\varepsilon}$ for a medium is called

(a) characteristic impedance (b) reactance
(c) refractive index (d) frequency.

2. Poynting vector signifies

(a) current density vector producing electromagnetic field
(b) power density vector producing electrostatic field
(c) power density vector producing electromagnetic field
(d) current density vector producing electrostatic field.

3. The characteristic equation of plane wave in E independent of two dimensions y and z is

(a) $\frac{\partial^2 E}{\partial t^2} = 0$ (b) $\frac{\partial E}{\partial t} = 0$

(c) $\frac{\partial E}{\partial x} = 0$ (d) $\frac{\partial^2 E}{\partial x^2} = 0$

4. Poynting theorem relating the electric field intensity E, magnetic field intensity H and the rate of energy flow per unit area at a point P is given by

(a) $\boldsymbol{P} = \frac{1}{\boldsymbol{E} \times \boldsymbol{H}}$ (b) $\boldsymbol{P} = \boldsymbol{E} \times \boldsymbol{H}$

(c) $\boldsymbol{P} = \boldsymbol{E}/\boldsymbol{H}$ (d) $\boldsymbol{P} = \boldsymbol{H}/\boldsymbol{E}$.

5. For a electromagnetic wave propagated in a good dialectic having $(\sigma/\omega\varepsilon) >> 1$, the attenuation factor α and phase shift factor β are given by

(a) $\alpha = \frac{\sigma}{2}\sqrt{\frac{\mu}{\varepsilon}}$, $\beta = \sqrt{\omega\mu\varepsilon}$
(b) $\alpha = \sqrt{\frac{\mu}{\varepsilon}}$, $\beta = \frac{1}{2}\sqrt{\omega\mu\varepsilon}$
(c) $\alpha = \sqrt{\frac{\mu}{\varepsilon}}$, $\beta = \sqrt{\frac{\omega\mu}{\varepsilon}}$
(d) $\alpha = \frac{\sigma}{2}\sqrt{\frac{\mu}{\varepsilon}}$, $\beta = \sqrt{\mu\varepsilon}$.

6. The value of $\sqrt{\frac{\mu}{\varepsilon}}$ for a free space is about

(a) 380 Ω
(b) 38 Ω.
(c) 3.8 Ω
(d) 3800 Ω.

7. Poynting vector $\boldsymbol{P} = \boldsymbol{E} \times \boldsymbol{H}$ has the dimensions

(a) Watt-m
(b) Watt–m^2
(c) Watt/m
(d) Watt/m^2.

8. In a perfect dielectric, wave propagation occurs

(a) with small attenuation
(b) with large attenuation
(c) with zero attenuation
(d) none of the above.

9. An electromagnetic wave incident on a perfect conductor is

(a) fully transmitted
(b) entirely reflected
(c) partially transmitted
(d) none of the above.

10. The intrinsic impedance of free space is given by

(a) $z_0 = \mu_0/\varepsilon_0$
(b) $z_0 = \sqrt{\frac{\mu_0}{\varepsilon_0}}$
(c) $z_0 = \mu_0\varepsilon_0$
(d) $z = \sqrt{\mu_0\varepsilon_0}$.

11. In a uniform plane wave, E and H are related by

(a) $\frac{E}{H} = \sqrt{\frac{\mu}{\varepsilon}}$
(b) $\frac{E}{H} = \sqrt{\frac{\varepsilon}{\mu}}$
(c) $\frac{E}{H} = \frac{\mu}{\varepsilon}$
(d) $\frac{E}{H} = \frac{\varepsilon}{\mu}$.

ANSWERS

1. c	2. c	3. a	4. b	5. a	6. a	7. d
8. c	9. b	10. b	11. a			

Wave Optics, Lasers and Fibre Optics

Wave Optics

4.1 NATURE OF LIGHT

From early recorded times, man had wondered about the mysteries of light and imagined its exact nature and relationship to human vision. Today, much is known about the nature of light, and some are still being discovered. Each of these of new discoveries is added to our growing knowledge on light. Based on these observations, laws have been formulated which predict the behaviour of light. Indeed one of the basic questions one needs to answer about light concerns its nature. Just what is light? When we scratch a match or switch on a light bulb in a dark room, we see not only the match or light bulb, but we see other objects too. Moreover, these objects may come in different colours. Colour is a word we use to describe certain physiological and psychological sensation of vision. Just what is it that makes the difference we have between colours? How do these colours combine with one another?

The observed behaviour of light as it travels from one place to another, and its interaction with matter has led to the development of a lot of technological tools. Tools have been developed which allow us to see objects as tiny as single blood cells and bacteria, and also see bodies in galaxies long distances away. We have developed high-speed cameras used for taking photographs of fast-moving objects, and also have tools for taking photos during the night and under the depths of the sea. All these instruments use the intricate behaviour of light as it travels from one place to another, and as it interacts with material objects.

We can understand the nature of light when we relate light to vision. An object becomes visible when light coming form it meets the eye. A body is visible either because it is luminous, or because it is illuminated. Luminous objects produce their own light. Examples include lamps, the stars, and the sun. Most common objects we find around us are non-luminous, and they include such things as wood, concrete, and writing paper. These non-luminous objects are made visible by the reflection of light from their surfaces.

Air, glass, and water readily transmit light. They are said to be **transparent.** Other substances tend to diffuse the light, and scatter it in many directions. The result is that it is barely possible to make out the profiles of objects when viewed through such materials. We are not able to discern details of the object being viewed. These materials are said to be **translucent.** Frosted light bulbs, paper soaked in paraffin or vegetable oil are examples of translucent materials. Substances which do not transmit light at all are said to be **opaque.**

Energy can be transmitted from one point to another by two methods. In one of these, matter may move from one point to the other. This is the **corpuscular theory**. In the other method, a wave disturbance can travel between the two points in the medium. This is the **wave theory**. Two theories have been put forward to help explain the behaviour of light as it travels from one point to another, and as it interacts with matter. These are the **corpuscular theory**, and the **wave theory.** Both theories have been advanced to account for the nature, propagation, and interaction of light with matter.

The Corpuscular Theory

The corpuscular theory assumes light to be made up of particles, or **corpuscles,** originating from the light source, and moving in straight lines through space or through a medium. One of the most famous champions of this theory was Sir Isaac Newton himself. The small corpuscles passing through the eye are thought to stimulate the sensory nerves and produce the sense of vision. The corpuscles were later named **photons** (quantum of energy) the name they have even today.

The greatest success of the quantum theory lies in its account of the **photoelectric effect,** *i.e.* the production of electricity from certain metal surfaces when light of certain colour is incident on the surface. In this, it is thought that the energy of the photon is transferred to an electron in the metal surface, and the electron gets ejected. This leads to the electric current that is measured.

The Wave Theory

Christian Huygens (1629-1695) first conceived light as a wave motion. In his theory, he proposed light as a transverse wave motion. More than one hundred years later, this proposal was backed by the result of electromagnetic wave studies carried out by James Clarke Maxwell. In the wave theory, light is pictured as consisting of transverse waves of progressively changing electric and magnetic fields. The fields act at right angles to each other, and to the direction of propagation of the wave. In the propagation of light in a medium, Huygens proposed that each point on the wavefront acts as a secondary source of new disturbance that spreads out spherically. The new wavefront is the envelope of all the individual wavelets that originate from the old wavefront as shown in Fig. 4.1.

Many studies were carried out on the wave theory of light at the beginning of the nineteenth century as a result of discoveries made by Thomas Young of England, and Augustin Jean Fresnel of France. Young's theoretical work predicted the interference of light waves from two sources, just like surface waves in water. Fresnel carried out both theory and experiment much further, explaining diffraction by straight edges, obstacles and apertures, by a combination of Young's principle of interference, and Huygens's principle of constructing secondary wavefronts.

Wave-Particle Duality

There are two theories needed to explain all the properties and interactions of light. These are the wave theory, and the corpuscular (now, quantum) theory. It is now known that light is a form of transverse waves in which

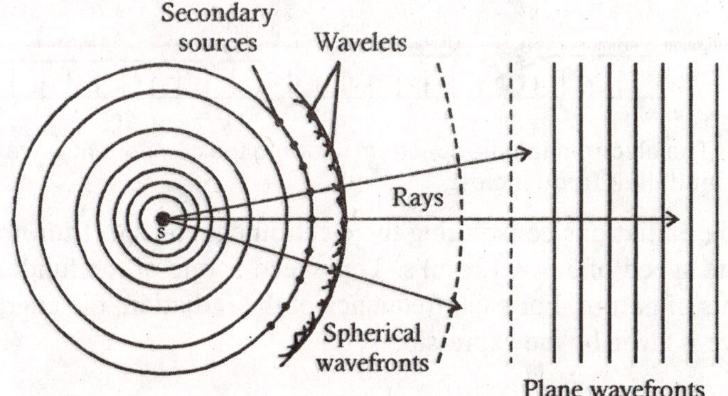

Fig. 4.1: Huygen's construction of secondary wavefronts. Spherical wavefronts from a point source become plane wavefronts at a great distance from the source

the electric and magnetic vibrations are perpendicular to the direction of travel of the wave (see Fig. 4.3). Thus light waves have wavelength and frequency. This wavelength controls the colour of the light. Light also exhibits other phenomena, e.g. interference, diffraction, that are explained only if we assume that light is a wave motion. On the other hand, there are other observed phenomena that can only be satisfactorily explained if light is assumed to be made up of photons. About the most important of these is the photoelectric effect.

From the above, it seems that we thus have two seemingly contradictory ways of accounting for the behaviour of light. The existence of two theories to account for light is referred to as the **wave-particle duality**. In our approach to explaining the nature and behaviour of light, we sometimes assume wave properties, and at times we assume particle properties. In general the wave theory is used to explain phenomena associated with the propagation of light. Phenomena like interference and diffraction are explained using the wave theory. The quantum theory comes into play when it is desired to explain phenomena dealing with the interaction of light with matter.

The Electromagnetic Spectrum

An orderly arrangement of radiation according to wavelength, frequency, or energy is known as a spectrum. The electromagnetic spectrum is such a spectrum. In it, the components of light are arranged in order of frequency and wavelength. The important parts of the electromagnetic spectrum are shown diagrammatically in Fig. 4.2

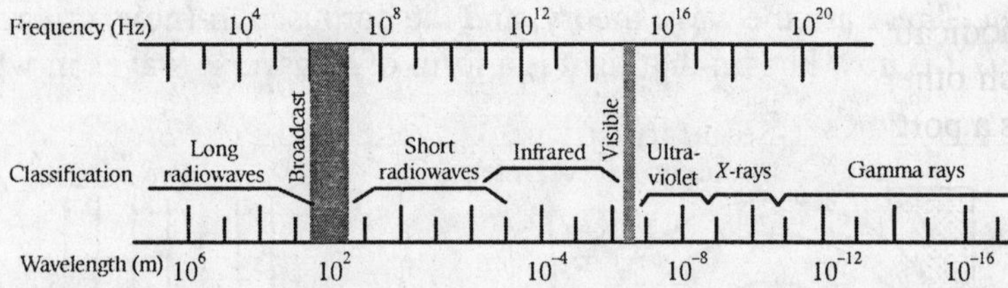

Fig. 4.2: The electromagnetic spectrum showing the component wavelengths and their frequencies.

All the radiations constituting the electromagnetic spectrum travel at the same speed of 3.0×10^8 m/s. This speed is one of the fundamental constants of nature. For each frequency of the radiation, the energy E of the wave is given by the expression

$$E = h\nu$$

where ν is the frequency of the radiation, and h is a constant, known as Planck's constant (h) with numerical value of 6.7×10^{-34} Js.

The most familiar frequency range is *visible light*, the narrow band of electromagnetic radiation of which the human eye is sensitive. This extends from the violet on the high frequency (or short wavelength) side to red on the low frequency (long wavelength) side. *X*-rays are highly energetic type of atomic radiation with high penetrating power. Because of their high penetration, these rays have applications in medical imaging, in non-destructive testing of materials, and in studying the crystal structure of minerals. *X*-rays are produced from specialised tubes and also from radioactive substances. Closely related to *X*-rays are gamma rays. These are highly penetrating radiations produced from certain radioactive sources. Gamma rays in general have more energy than *X*-rays.

To the longer wavelength side of the visible region we start with the infrared radiation, known more commonly as heat radiation. Radio waves used for broadcast and other communication purposes have wavelengths ranging from about 1 mm (for short radio waves) to 10^6 m. The most useful region for normal broadcast has wavelength between 10 m and about 400 m.

Becoming increasingly important in technology are microwaves. These are electromagnetic radiations having frequency in the range 10^9 to 10^{12} Hz, and a wavelength between 1.0 mm to 1.0 m. The popular application of these waves now-a-days is in cooking. A microwave oven is a relatively small, box-like oven that raises the temperature of an electrically non-conducting material (the food and its container) by subjecting the material to a high frequency electromagnetic field. The heating occurs throughout the food, rather than from the outside first—as in most other cookers. This greatly reduces cooking time and energy consumption.

As stated above, the light wave is an electromagnetic wave consisting of periodically varying electric and magnetic fields oscillating at right angles to each other and to the direction of the propagation of waves. Fig. 4.3 shows a portion of electromagnetic wave.

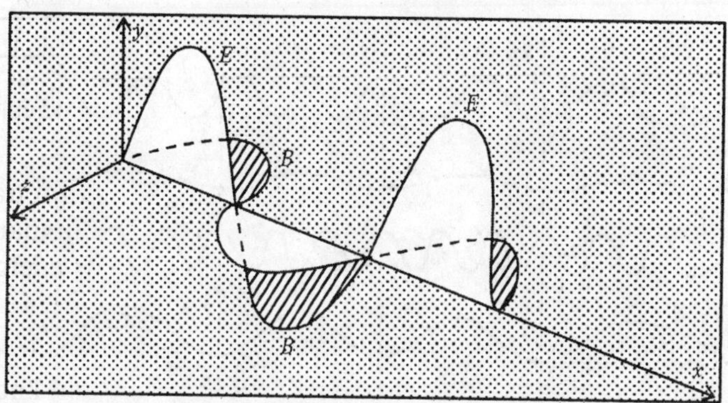

Fig. 4.3: An electromagnetic wave

The electric field in the wave is defined by the electric field strength ***E*** and the magnetic field by the magnetic induction vector ***B***. Vectors ***E*** and ***B*** are mutually perpendicular to each other and also to the direction of propagation. Both vectors ***E*** and ***B*** are of equal importance to the wave. However, a light wave is often represented by ***E*** wave as many of the effects of light are mainly due to the electric field and can be easily explained with the help of vector ***E***. The magnetic field is implied to be oscillating in a plane normal to the plane of the electric field oscillations and we have not shown it specifically in Fig. 4.3.

The light wave depicted as electromagnetic wave in Fig. 4.3 can be represented mathematically by the following expressions

$$E = E_0 \sin(kx - \omega t)$$

$$B = B_0 \sin(kx - \omega t)$$

where $\boldsymbol{k}$ is wave vector, $\omega(= 2\pi\nu)$ is angular frequency and E_0 and B_0 are maximum amplitudes of $\boldsymbol{E}$ and $\boldsymbol{B}$ vectors respectively. We may note that Fig. 5.3 and the above equations represent an ideal plane, monochromatic and harmonic electromagnetic wave which travel with the speed of light $c\ (= 3 \times 10^8\ \text{m/s})$ in vacuum.

4.2 COHERENCE

Conventional light sources produce incoherent light. This means that the light that emerges from a conventional light source is mixture of waves at various frequencies that reinforce or cancel each other in a random fashion. Figure 4.4 depicts this situation. Obviously, the wavefront thus produced varies from point to point and changes from time to time. The wave from a laser is called almost ***coherent*** because it is an orderly wave of one frequency where the whole beam, is ***spatially*** in phase. There are, thus, two independent facets of this coherence: namely, ***temporal coherence*** and ***spatial coherence***.

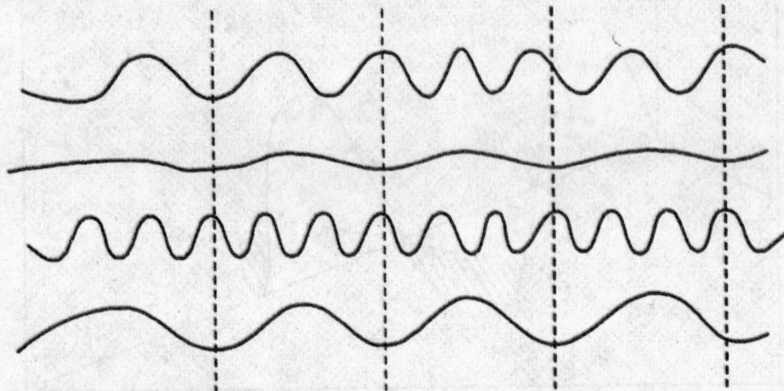

Fig. 4.4: Incoherent waves. There is no relationship between one wave and another

Temporal Coherence

This type of coherence refers to the correlation between the radiation field at a point and the radiation field at the same point at a later time, *i.e.*, the relation between $E(x, y, z, t_1)$ and $E(x, y, z, t_2)$, where these represent the radiation field at the point (x, y, z) at times t_1 and t_2 respectively. If the phase difference between the two radiation fields is constant during the period normally covered by observation (~ few microseconds), the wave is said to have ***temporal coherence***. If the phase difference between two radiation fields changes many times and in an irregular way during the shortest period of observation, the wave is said to be incoherent.

Time coherence comes about when each cycle of the wave takes exactly, the same time to pass a point in space. This really means that the frequency of the wave is not varying and is given the name ***monochromatic***.

Spatial Coherence

Two fields at two different points on a wavefront of a given electromagnetic wave are said to be space coherent if they preserve a constant phase difference over any time t, *i.e.*, space coherency requires that the waves not only are of the same frequency, but they are in phase in space. Figure 4.5 shows these conditions. In Fig. 4.5 (*a*) the waves are monochromatic (time coherent) but are not in sequence in space. This can occur when the source of the wave is physically broad rather than a point source. Fig. 4.5. (*b*) shows a spatially and time-coherent wave. The whole wavefront is in step and each cycle takes the same length of time. The vary nature of laser mechanism produces this coherent signal.

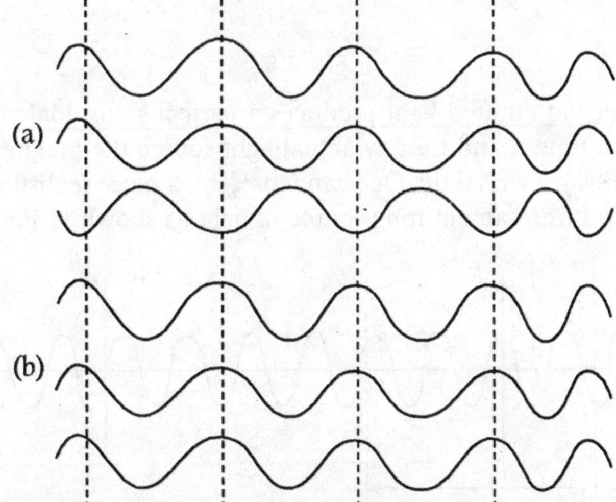

Fig. 4.5: (*a*) Time coherent space incoherent (*b*) Time and space coherent

Spatial coherence is possible even when the two beams are individually time incoherent, as long as any phase change in one of the beams is accompanied by a simultaneous equal phase change in the other beam. With the ordinary light sources, this is only possible if the two beams have been derived from the same part of the source.

Time incoherence is a characteristic of single beam of light, whereas space incoherence concerns the relationship between two separate beams of light. Two beams of light produced in different parts of light source will have been emitted by different groups of atoms. Each beam will be time-incoherent and will suffer random phase changes, as a result of which the phase difference between the two beams will also suffer rapid and random changes. Two such light beams are said to be space incoherent.

Coherence Time and Coherence Length

Let us consider that the radiation field E, from a light source is an ideal sinusoidal function of time t, at any given position can be written as

$$E = a\cos(\omega_0 t + \phi) \tag{1}$$

Here a is the magnitude of the field, ω_0 the angular frequency and ϕ the phase. Figure 4.6 depicts the field represented by Eq. (1)

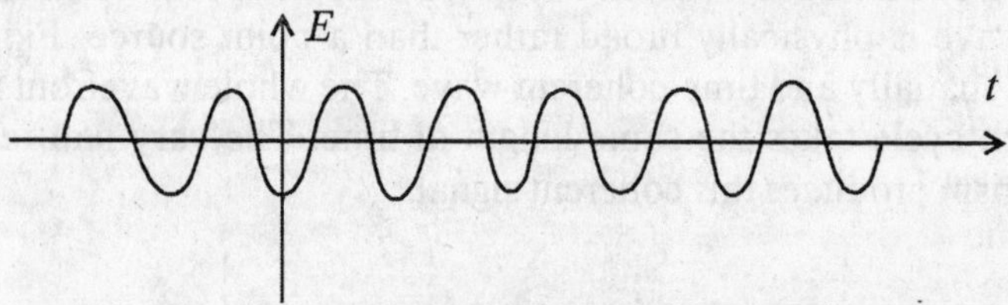

Fig. 4.6

However, no emitted light produces a perfect sinusoidal variation of field E with time t, and for any actual light source the magnitude a and phase ϕ will vary with time. One can represent, a more realistic picture of the radiation from an emitting source of light as shown in Fig. 4.7.

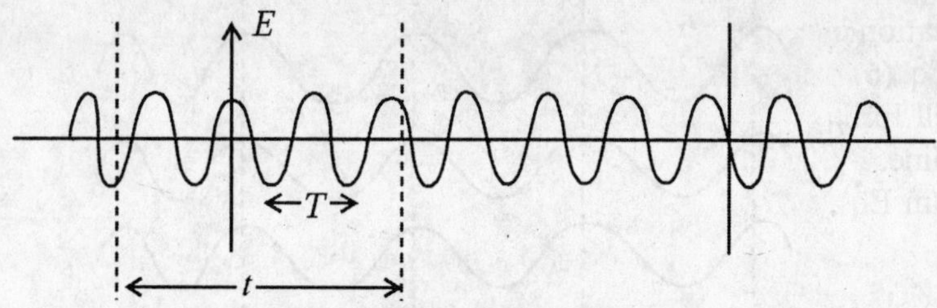

Fig. 4.7

Figure 4.7 represents the average time duration for which an ideal sinusoidal emission occurs. τ is called *coherence time* or *interval of coherence*.

Let T denotes the period of the oscillation, where $T = \frac{\omega_0}{2\pi}$. The spatial dimension, L, for which the light may be considered to be a perfect sinusoidal is given by

$$L \approx cT \approx \frac{\tau}{T}\lambda \tag{2}$$

Here c is the velocity of light and L is called the ***coherence length***. After time τ, there is no correlation between the phases of the waves.

In terms of purity Q of the spectral line ($Q \approx \lambda/\Delta\lambda$, where $\Delta\lambda$ is half width of the spectral line), we have

$$c_\tau \approx L \approx \lambda Q \tag{3}$$

4.3 PRINCIPLE OF SUPERPOSITION

When two or more waves arrive at a point in space simultaneously the disturbance at that point is the *vector sum* of the disturbance caused by those waves taken individually, *i.e.*, the net wave disturbance at any given time and place is the vector sum of all the wave disturbances at that particular place at that particular time. This is called the principle of superposition.

Let Y_1, the displacement at the observation point due to one wave at any instant t is given by

$$Y_1 = a_1\cos\omega t = a_1\cos 2\pi\nu t \tag{4}$$

Let the second wave have the same frequency ν and cause displacement in the same direction, represented by

$$Y_2 = a_2\cos(\omega t + \delta) = a\cos(2\pi\nu t + \delta) \tag{5}$$

where δ represents the *phase difference* between the waves at the point of consideration.

According to *Young's principle of superposition*, when both waves are present simultaneously the resultant displacement Y at that point is given by

$$Y = Y_1 + Y_2 \tag{6}$$

It is worthwhile to mention that principle of superposition is of wider application in which Y_1 and Y_2 could have different directions. Obviously, then Eq. (6) is a vector equation. In discussing interference of light waves, we will limit ourselves to the situations where Y_1 and Y_2 are along the same line.

From Eqs. (4), (5) and (6) one obtains

$$\begin{aligned} Y &= a_1\cos 2\pi\nu t + a_2\cos(2\pi\nu t + \delta) \\ &= a_1\cos 2\pi\nu t + a_2\cos 2\pi\nu t\cos\delta - a_2\sin 2\pi\nu t\sin\delta \\ &= (a_1 + a_2\cos\delta)\cos 2\pi\nu t - (a_2\sin\delta)\sin 2\pi\nu t \\ &= a\cos\phi\cos 2\pi\nu t - a\sin\phi\sin 2\pi\nu t \\ &= a\cos(2\pi\nu t + \phi) \end{aligned} \tag{7}$$

where

$$\left.\begin{aligned} a\cos\phi &= a_1 + a_2\cos\delta \\ a\sin\phi &= a_2\sin\delta \end{aligned}\right\} \tag{8}$$

Squaring and adding the terms of Eq. (8), one obtains

$$a^2 = a_1^2 + a_2^2 + 2a_1a_2\cos\phi \tag{9}$$

and dividing Eq. (8), gives

$$\tan\theta = \frac{a_2 \sin\delta}{a_1 + a_2 \cos\delta} \tag{10}$$

(i) Constructive Superposition

When phase difference $\phi = 0, 2\pi, 4\pi, \ldots = 2n\pi$, where $n = 0, 1, 2, 3, \ldots$, we have from Eq. (9),

$$a^2 = a_1^2 + a_2^2 + 2a_1a_2 = (a_1 + a_2)^2 \tag{11}$$

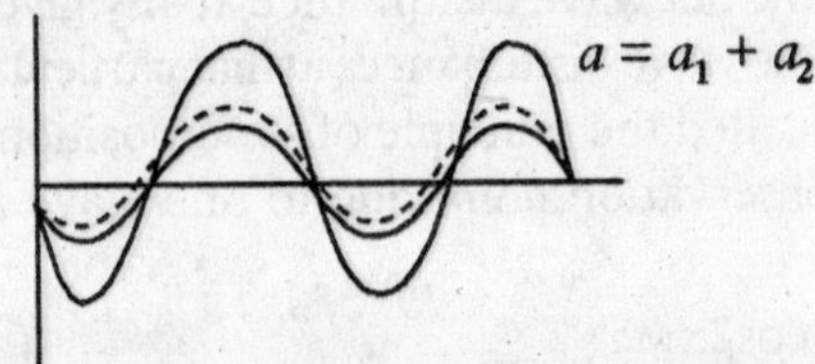

(a) Constructive superposition

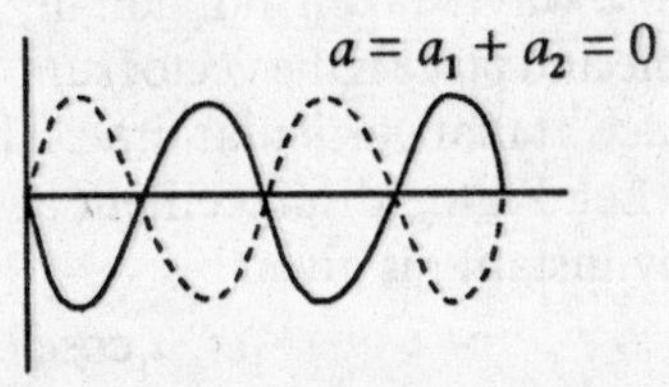

(b) Destructive superposition

Fig. 4.3

This is called the case of ***constructive superposition*** or ***additive interference***. In addition if $a_1 = a_2$ then

$$a^2 = 4a_1^2 + 4a_2^2$$

i.e. intensity become 4 times.

(ii) The Destructive Superposition

If $\phi = \pi, 3\pi, 5\pi, \ldots = (2n + 1)\pi$, where $n = 0, 1, 2, 3, \ldots$ one obtains

$$a^2 = a_1^2 + a_2^2 - 2a_1a_2 = (a_1 - a_2)^2 \tag{12}$$

This shows that the value of a is minimum. If $a_1 = a_2$, then $a = 0$. This is called the ***destructive superposition***, $a = 0$ corresponds to zero intensity. Obviously the change of intensity from maxima to minima is very sharp and can be observed clearly.

We, therefore, conclude that (*a*) the phase difference between two vibrations must be $\delta = 2n\pi$ for constructive superposition (*b*) $\delta = (2n+1)\pi$ for destructive superposition.

We know that path difference of a wavelength distance (λ) corresponding to 2π phase difference. Hence one can express path difference for constructive and destructive superposition as follows:

$$\Delta = \frac{\lambda}{2\pi} \cdot 2n\pi = 2n\lambda / 2 \quad \text{[Constructive superposition]}$$

$$\Delta = \frac{\lambda}{2\pi}(2n + 1) = (2n + 1)\lambda/2 \quad \text{[Destructive superposition]} \tag{13}$$

The above discussion demonstrates that first of all, superposition of waves results in a spatial redistribution of the oscillation energy; in other words the intensity of waves is redistributed in space. Regions with zero intensity and those with intensity above the intensity of overlapping waves are produced. Second, this spatial redistribution of energy is found to be time-independent. This means that a stable pattern of fixed interference fringes is formed.

The phenomenon of wave interference reduces therefore to a spatial redistribution of intensities of waves, which results in formation of a fixed pattern of interference fringes.

It is worthwhile to mention that interference of two plane waves constitutes one of the elementary cases of interference. It is also possible to analyse the cases of interfering plane and spherical waves. Obviously, the more complicated the wave fronts, more complex the interference fringe patterns are.

4.4 INTERFERENCE

A very important characteristic of wave motion is the phenomenon of interference. This occurs when two or more wave motions coincide in *space* and *time*, *i.e.* when two (or more) waves are in the same place at the same time, we say that they *interfere* with each other. The pattern of bright and dark fringes that results is called an *interference pattern*. By "interference", we really mean that the waves are in the same place at the same time and apparently get in each other's way and follow the **principle of superposition**.

We have seen that as result of superposition of two or more waves the redistribution of energy takes place. This modification is called interference.

The interference in which there is an *increase of amplitude* and hence *intensity* is called *constructive interference* and when there is a *decrease* of *amplitude* and hence *intensity* is called the *destructive interference* (Fig. 4.8).

Interference in Light

In the case of light the observation of interference is not so easy as for other types of wave motion. One cannot observe interference pattern on a screen by placing two candles in front of it, *i.e.* by using two conventional sources of light we fail to observe any reinforcement or weakening of the light intensity in the illuminated space. In order to make the interferences in light persistent, *coherent sources* of light must be used. *Two waves are said to be coherent if their phase difference is independent of time.* This condition is met by two monochromatic waves of the same frequency.

The necessary conditions for producing persistent interference in light can be summarized as follows:

(*i*) The two sources should be coherent, *i.e.*, they should emit continuously the waves either in the same phase or with a constant phase difference.

(*ii*) The two sources should emit continuously the waves of the same *wavelength*, *period* or *frequency*. Above two conditions are necessary for sustained interference.

(*iii*) If the interfering waves are polarised than their state of polarisation must be same.

(*iv*) The two coherent sources should lie very close to each other so that the path difference between the waves reaching particular point on the screen is not very large. If this is not so, the dark and bright fringes will not be sharp.

(*v*) The two sources must be very narrow. A broad source is equivalent to a large number of point sources lying side by side. Each pair of such sources will give rise to its own interference pattern resulting in a general illumination due to overlapping of the patterns.

(*vi*) The amplitudes of the waves from the two sources should be equal. If it is not so, the intensity of minima will not be much different from that of maxima. This will result in general illumination.

(*vii*) The background must be dark.

(*viii*) The distance between the two coherent sources and the screen must be suitable.

Young's Double Slit Experiment

The classic experiment that demonstrates interference of light was first performed by Thomas Young in 1802. In the original experiment sunlight was used as the source, but any bright source such as a tungsten filament lamp or an arc would be satisfactory. Light is passed through a pinhole S so as to illuminate an aperture consisting of two pinholes or narrow slits S_1 and S_2 (Fig. 4.9). If a white screen is placed in the region beyond the slits, a pattern of bright and dark interference bands can be seen. The key to the experiment is the use of single pinhole S to illuminate the aperture. This provides necessary mutual coherence between the light from the two slits S_1 and S_2.

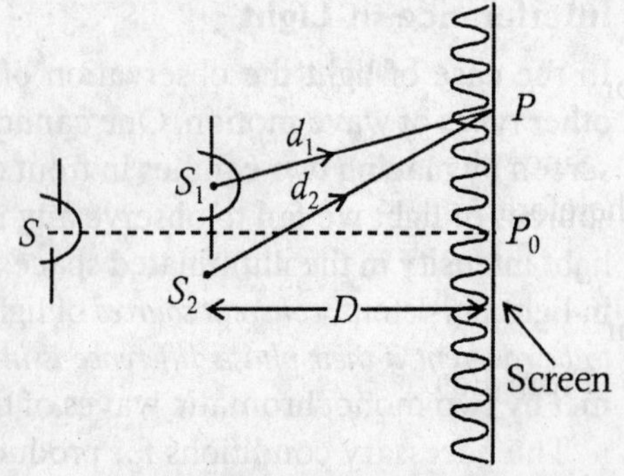

Fig. 4.9: Young's double slit experiment

The usual elementary analysis of the Young's experiment involves finding the difference in phase between the two waves arriving at a given point P over the distances d_1 and d_2 as shown in Fig. 4.9. Now, we shall obtain expressions for path difference and phase difference.

Theory

Consider a source of light S emitting monochromatic waves of wavelength λ. Let S_1 and S_2 be two narrow slits equidistant from the source S separated

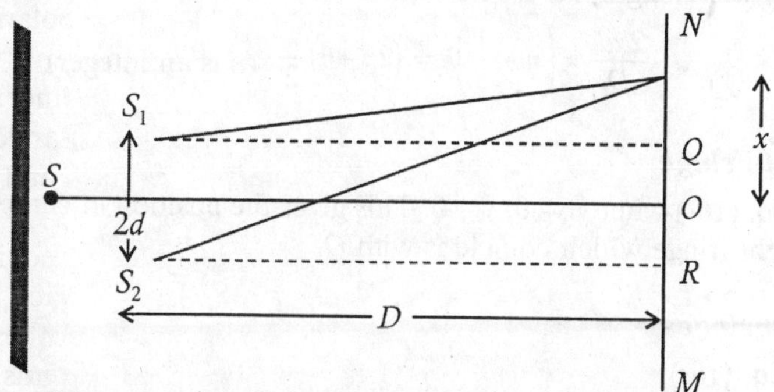

Fig. 4.10: Interference of two waves in the case of the double slit

by a distance $2d$ and acting as coherent sources. Let MN be the screen placed symmetrically to the slit perpendicular to the line SO at a distance D from S_2. Let P be a point distant x from O where the rays are from two slits S_1 and S_2, *i.e.*, S_1P and S_2P. The intensity at P depends upon the path difference $S_2P - S_1P$. For the point O, the path $S_1O - S_2O = 0$, *i.e.* both waves reach at O in the same phase. The point O is, therefore, essentially bright. Whether point P is bright or dark depends upon the path difference $S_2P - S_1P$. From Fig. 4.10, we have

$$PQ = x - d \quad \text{and} \quad PR = x + d$$

$$S_1P^2 = D^2 + (x - d)^2 \tag{14}$$

and
$$S_2P^2 = D^2 + (x + d)^2 \tag{15}$$

Subtracting Eq. (14) from Eq. (15), one obtains

$$S_2P^2 - S_1P^2 = (x + d)^2 - (x - d)^2 = 4xd$$

or
$$S_2P - S_1P = \frac{4xd}{S_2P + S_1P} \tag{16}$$

Since the point P is very nearer to O and hence $D >> x$ and $D >> 2d$, therefore,

$$S_1P \approx S_2P = D$$

or
$$S_1P + S_2P = 2D \tag{16a}$$

From Eqs. (16) and (16a), one obtains

$$S_2P - S_1P = \frac{4xd}{2D} = \frac{2xd}{D} \tag{16b}$$

Now, if S_1 and S_2 be in the same phase, then P will have maximum intensity, *i.e.* $(S_1P - S_2P)$ is a whole number of wavelength, *i.e.* for Maxima

$$\frac{2xd}{D} = n\lambda \qquad (n \text{ is an integer}) \qquad (16c)$$

Also P will have minimum intensity if $(S_2P - S_1P)$ is a whole number plus half wavelength, *i.e.* for minima

$$\frac{2xd}{D} = \left(n + \frac{1}{2}\right)\lambda = (2n+1)\frac{\lambda}{2} \quad (n \text{ is an integer}) \qquad (16d)$$

Central Fringe

From Eq. (16*c*) when $n = 0$, $x = 0$. This gives the position of centre of the first bright fringe which coincides with O.

Bright Fringes

From Eq. (16*c*)

For $n = 1$ $\quad x_1 = \frac{D}{2d}\lambda$

For $n = 2$ $\quad x_2 = \frac{D}{2d}2\lambda$

For $n = 3$ $\quad x_3 = \frac{D}{2d}3\lambda$

For $n = m$ $\quad x_m = \frac{D}{2d}m\lambda$ (17)

If β be the distance between the two nearest bright fringes known as the fringe width, then

$$\beta = X_{n+1} - X_n = X_2 - X_1 = \frac{\lambda D}{2d} \qquad (18)$$

As Eq. (18) is independent of n, therefore, the fringe width (β) for all bright fringes is the same.

For $n = 0$ Let $x = x_0$, then $x_0 = \frac{D}{2d}\frac{\lambda}{2}$

For $n = 1$ $\quad X_1' = \frac{D}{2d}\frac{3\lambda}{2}$

For $n = 2$ $\quad X_2' = \frac{D}{2d}\frac{5\lambda}{2}$

For $n = n$ $\quad X_n' = \frac{D}{2d}(2n+1)\frac{\lambda}{2}$ (19)

$\therefore$ Fringe width $\quad \beta = X'_{n+1} - X'_n = X'_2 - X'_1 = \frac{D}{2d}\lambda$ (20)

This is again the same as that for a bright fringe. Thus one finds that

(*i*) All fringes whether dark or bright are of equal width.

(*ii*) The dark and bright fringes are alternately placed above and below the central bright fringe at *O*.

(*iii*) If D and $2d$ are constants then fringe width $\beta \propto \lambda$. Obviously, fringes produced by light of shorter wavelength will be *narrow* as compared to those produced by longer wavelength.

(*iv*) If λ and D are kept constant, then $\beta \propto 1/2d$. Thus if the distance between the two coherent sources ($2d$) becomes sufficiently large, the fringes will become very narrow and general illumination will result.

(v) If $2d$ and λ are kept constant, then $\beta \propto D$. If D is sufficiently large, the fringes will become very wide and their intensity decreases to such an extent that there is general illumination produced on the screen.

Conditions of Maxima and Minima in Terms of Phase Difference

We know that a path difference of λ corresponds to 2π difference. Hence we write Eqs. (16*c*) and (16*d*) as below:

$$\Delta = \frac{2\pi}{\lambda}\frac{2xd}{D} = n\lambda \qquad \text{[For Maxima]}$$

$$\Delta = \frac{2\pi}{\lambda}\frac{2xd}{D} = \left(n + \frac{1}{2}\right)\lambda \qquad \text{[For Minima]} \tag{21}$$

Shape of the Interference Fringes

We have seen that the fringes formed at any point P on the screen will be bright or dark, if

$$S_2P - S_1P = n\lambda \qquad \text{(bright)}$$

$$S_2P - S_1P = \left(n + \frac{1}{2}\right)\lambda \qquad \text{(dark)}$$

For any value of n, the locus of maximum or minimum intensity point will be $S_2P - S_1P = \Delta$ (Obviously Δ has to be less then $2d$).

The locus of such a point is a hyperbola. Since wavelength of the light is very small and hence the curves of intersection with the plane of the screen will be very nearly straight lines (Fig. 4.11).

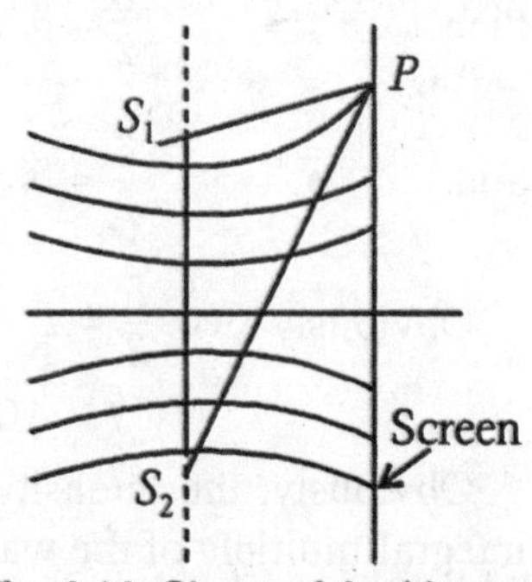

Fig. 4.11: Shape of the fringes

Intensity or Energy Distribution

Consider two waves from coherent sources S_1 and S_2 as

$$Y_1 = a\sin\omega t$$

and $$Y_2 = a\sin(\omega t + \delta)$$

The resultant displacement is given by

$$Y = Y_1 + Y_2 = a\sin\omega t + a\sin(\omega t + \delta)$$

$$= 2a\sin\left(\omega t + \frac{\delta}{2}\right)\cos\frac{\delta}{2}$$

$$\left[\because \sin A + \sin B = 2\sin\frac{A+B}{2}\cos\frac{A-B}{2}\right]$$

or $$Y = 2a\cos\frac{\delta}{2}\sin\left(\omega t + \frac{\delta}{2}\right) \quad (22)$$

Thus the amplitude of the resultant wave is

$$A = 2a\cos\frac{\delta}{2} \quad (23)$$

Now the intensity at a point is proportional to the square of amplitude and if I_0 is the intensity due to the resultant wave and I due to the individual wave, then

$$I \propto 4a^2\cos\frac{\delta}{2} \quad \text{and} \quad I_0 \propto a^2$$

$$\therefore \quad \frac{I}{I_0} = 4\cos^2\frac{\delta}{2}$$

or $$I = 4I_0\cos^2\frac{\delta}{2} \quad (24)$$

For Maxima

When the path difference is 0, λ, 2λ, ..., nλ

then $$\frac{\delta}{2} = 0,\ \pi,\ 2\pi,\ 2(2\pi),\ \ldots,\ n(2\pi)$$

and $$\frac{\delta}{2} = 0,\ \pi,\ 2\pi,\ \ldots,\ n\pi$$

Obviously $\cos\frac{\delta}{2} = I$

$$I = 4I_0, \text{ which is maximum.}$$

Obviously, the intensity is maximum when the path difference is an integral multiple of the wavelength (λ) and the fringes are bright.

For Minima

When the path difference is

$$\frac{\lambda}{2}, \frac{3\lambda}{2}, \ldots, (2n+1)\frac{\lambda}{2}$$

$\therefore$ The phase difference $\delta = \pi, 3\pi, \ldots, (2n+1)\pi$

and $$\frac{\delta}{2} = \frac{\pi}{2}, \frac{3\pi}{2}, \ldots, (2n+1)\frac{\pi}{2}$$

Obviously $\cos\frac{\delta}{2} = 0 \therefore I = 0$

Clearly, the intensity is minimum when the path difference is an odd multiple of $\lambda/2$, *i.e.*, the fringes are dark.

Thus the resultant intensity at a point varies from minimum value 0 to maximum value $4I_0$, *i.e.*, 4 times the intensity due to the individual waves.

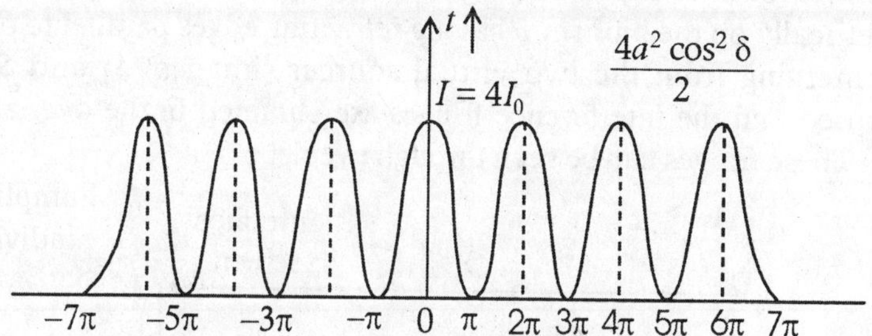

Fig. 4.12: Intensity or energy distribution

The intensity or energy distribution curve is as shown in Fig. 4.12 and the average value of intensity is $\frac{4I_0 + 0}{2} = 2I_0$, which is equal to the sum of intensity of two individual waves. This shows that there is no destruction of energy due to interference and thus the law of conservation of energy is not violated. The energy is merely transferred from places of minimum intensity to those of maximum intensity. As a result of this the brightness at the maxima is greater than the brightness which would have been produced by the simple addition of two light beams without interference. Obviously, ***interference is the phenomenon of redistribution of light energy***.

4.5 TYPES OF INTERFERENCE

The phenomenon of interference is usually divided into two classes viz. Interference due to division of wavefront and interference due to division of amplitude.

(A) Interference Due to Division of Wavefront

The incident wavefront is divided into two parts, either by reflection or refraction. These two parts of the same wavefront travel unequal distances and reunite at small angle on the screen to produce interference fringes or bands. Following devices are based on the principle of division of wavefront to obtain interference.

(*a*) Biprism, (*b*) Fresnel's double mirror, (*c*) Lloyd's single mirror, etc.

Fresnel's Biprism

It is a device for obtaining two coherent sources for producing sustained interference. It may be supposed to be made up of two prisms of very small refracting angles placed base to base. In actual practice it is made from a single glass plate by grinding and polishing so that it becomes a single prism with one of its angles about 179° and other two about 30° each.

A horizontal section of the apparatus for production of interference fringes is shown in Fig. 4.13 in which *S* is a narrow vertical slit illuminated by monochromatic light. The light emerging from *S* is made to fall symmetrically on the biprism *P* having refracting edges parallel to *S*. The light emerging from the two virtual sources (images) S_1 and S_2 are superposed and the interference fringes are obtained in the overlapping region. These fringes can be seen through eyepiece.

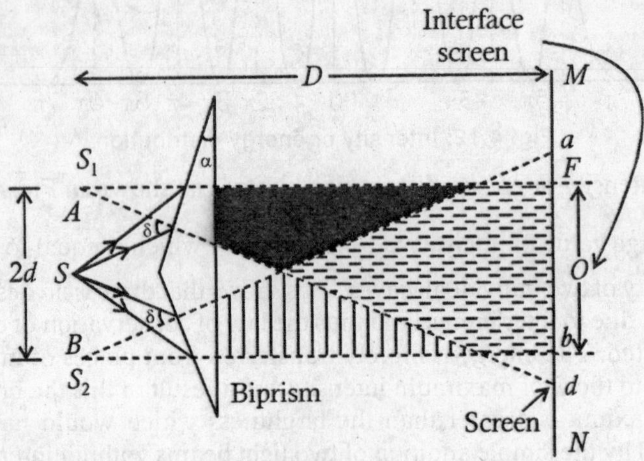

Fig. 4.13

Theory

The complete theory refers to section 4. The point *O* is equidistant from *A* and *B* and hence it has maximum intensity. On both sides of *O*, alternately *bright* and *dark* fringes are produced (Fig. 4.14). The spacing between two adjacent bright (or dark) fringes, *i.e.* fringe width is given by the relation (18) or (20), *i.e.*

$$\beta = \frac{D}{2d}\lambda$$

where β is fringe width, D is distance between source and screen, $2d$ is distance between two coherent sources S_1 and S_2 and λ is wavelength of light used.

The above arrangement provides a very accurate method of finding the wavelength λ of light emitted by a monochromatic source.

Fig. 4.14: Biprism fringes

The relation is

$$\lambda = \frac{d}{2D}\beta \tag{25}$$

From Eq. (25), it is clear that if $2d$, β and D are measured, the wavelength λ can be determined.

Experimental Arrangement for Determination of 2d, β and D

The experimental arrangement consists of an optical bench carrying three stands each for a slit, biprism and a micrometer eyepiece (Fig. 4.15). The biprism and slit can be rotated in their own planes while the biprism and the eyepiece can also be moved at right angles to the length of the optical bench. The eyepiece can be moved with a micrometer screw and the distance moved can be accurately read on the attached circular and linear scales. The slit is illuminated with the monochromatic light whose wavelength is to be determined. The following adjustments are made prior to the observations:

(*i*) The optical bench is levelled.

(*ii*) The eyepiece is focused on the cross wires and one of them is set vertical and other horizontal.

(*iii*) The slit, eyepiece and the biprism are adjusted to the same height.

(*iv*) The slit is made vertical by rotating it in its own plane and then it is made narrow.

(*v*) To make the biprism at right angles to the length of the optical bench, the biprism is moved at right angles to the optical bench until on looking through the biprism along the axis of the beam, the two equally bright images S_1 and S_2 appear in the field of view.

(*vi*) To bring the overlapping region in the field of view the eyepiece is moved at right angles to the length of the optical bench.

(*vii*) Now the lateral shift is removed, so that the line joining slit and the edge of the biprism becomes parallel to the length of the optical bench.

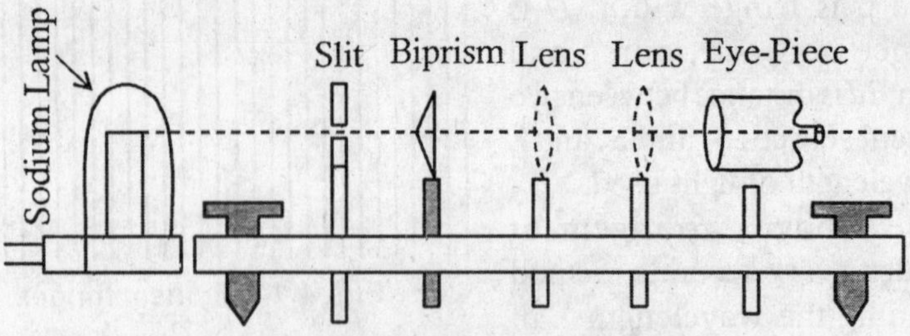

Fig. 4.15: Biprism experiment

Now one can measure β, $2d$, D for determination of λ.

(*i*) **Determination of β:** For determination of fringe width (β), leaving the first few fringes the vertical crosswire of the eyepiece is set on a bright fringe and the position of the eyepiece is noted. Now by giving the eyepiece a lateral motion, the positions of the eyepiece corresponding to successive bright fringes are noted. From these observations fringe width (β) is calculated.

(*ii*) **Determination of *D*:** The distance between S_1, S_2 and the focal plane of the eyepiece, (D) is determined by noting the positions of the slit and eyepiece on the scale of the optical bench. By applying the bench error correction the corrected value of D is obtained.

(*iii*) **Determination of 2*d*:** For determination of $2d$, a convex lens whose focal length is less then ¼th of the distance between the slit and the eyepiece is mounted between the biprism and the eyepiece on the optical bench. Now, by moving the convex lens along the length of the optical bench two such positions as in displacement method are obtained such that the real images S_1 and S_2 are observed in the eyepiece (Fig. 4.16). Let the separation between the two real images in the two cases d_1 and d_2 respectively. In the first position of the lens L_1, by applying magnification formula, we obtain

$$\text{Magnification} = \frac{d_1}{2d} = \frac{V}{U} \tag{26}$$

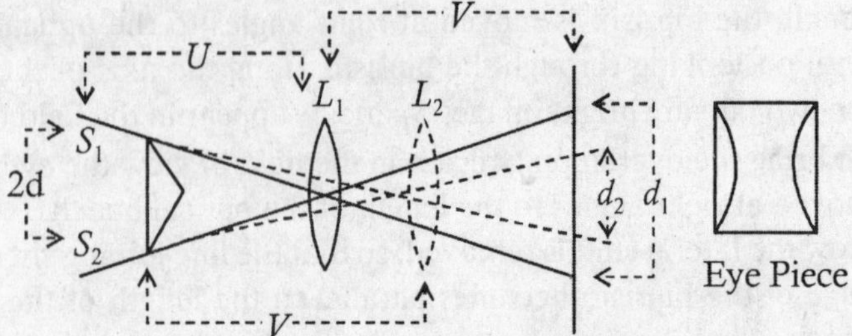

Fig. 4.16: Measurement of distance between virtual coherent sources

For the second position of lens L_2, V and U are interchanged. Again applying the magnification formula, we have

$$\text{Magnification} = \frac{U}{V} = \frac{d_2}{2d} \tag{27}$$

Multiplying Eqs. (26) and (27), we obtain

$$2d = \sqrt{d_1 d_2} \tag{28}$$

For different positions of the eyepiece several pairs of values of d_1 and d_2 are obtained and the mean value of $2d$ is found.

Location of Zero Order Fringes or White Light Fringes in Fresnel's Biprism Experiment

With monochromatic light, all bright fringes are of the same colour. Therefore, it is impossible to locate the zero order fringe. But we know that with white light source zero order fringe is white because the path difference for each wavelength of light reaching from the coherent sources at O (Fig. 4.13) is same. Therefore, by replacing monochromatic light by white light one can easily place the crosswire in the position of zero order fringe. Now replace the white light source by monochromatic light source, the crosswire in the eyepiece will be in the position of zero order band.

When monochromatic source of light is replaced by a white light source, a fringe system appears in the field of view with central white fringe surrounded by a few coloured fringes on both sides. The fringes of mixed colours are obtained due to overlapping of the maxima and minima of different colours at one place. The outer edges of the bands are violet whereas the inner edges are red. This is due to the fact that the fringe width for violet is minimum and maximum for red. Therefore, the first dark band of violet is obtained first and that of the red the last on either side of the central zero order band. The inner edge of the first minimum of violet receives sufficient light intensity from red because maximum of red falls in its vicinity and hence the edge is reddish. Similarly the first maximum of violet falls close to the inner edge of the minimum of red and hence the edge appears violet. For points at large distances from the centre, the maxima and minima due to large number of wavelengths overlap and uniform illumination is observed.

The distance of nth fringe from the centre (with monochromatic light) for blue and red eyes will be

$$X_b^n = \frac{n\lambda_b D}{a(\mu_b - 1)\alpha} \tag{29}$$

$$X_v^n = \frac{n\lambda_v D}{a(\mu_v - 1)\alpha} \tag{30}$$

Determination of the Thickness of a Thin Sheet of Transparent Material

Let a plate of thickness t and refractive index μ be introduced in the path, S_1P as shown in Fig. 4.17. The path S_1P is travelled by light partly in air and partly in plate. The path travelled in air = $S_1P - t$ and path travelled in plate = t. The optical path travelled by a ray is defined as

Optical path = Distance travelled × Refractive index of the medium

$$\therefore \quad \text{Optical path along } S_1P = (S_1P - t) \times 1 + t\mu$$

$$= S_1P + t(\mu - 1)$$

and optical path along $\quad S_2P = S_2P$

$$\therefore \text{ The optical path difference } = S_2P - S_1P$$

$$= S_2P - S_1P + (\mu - 1)t$$

Due to the introduction of a transparent plate of thickness, the optical path difference changes. The change of optical path difference causes fringe system to displace. White light fringe system is used to note this displacement of fringe system. Initially the cross-wires are focused on the central white without the plate G. Next the plate of thickness t is introduced in the path S_1P and the displacement is noted with the help of micrometer. It is worthwhile to mention that the thickness t of the plate should be small enough to displace fringes within the field of view.

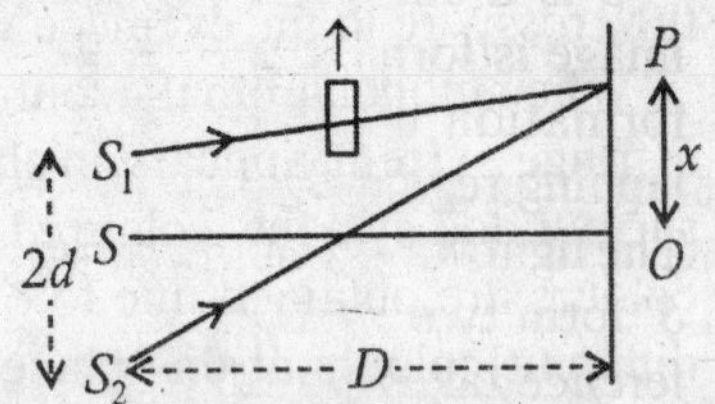

Fig. 4.17: Determination of the thickness of thin transparent film

In Fig. 4.17, O represents the position of zero order, *i.e.* central white fringe. Due to introduction of sheet G the central white fringe shifts from O to P. Therefore, the condition satisfied is

$$S_2P = (S_1P - t) + \mu t = S_1P + (\mu - 1)t \tag{31}$$

Lateral displacement $x = OP$ is measured.

Now we replace white light source by monochromatic light source. If β is the fringe width, then x/β represents the number of fringes of wavelength λ; which would have shifted due to introduction of plate in the path S_1P. The shift of one fringe corresponds to path difference λ and hence the path difference due to sheet, *i.e.*

$$S_2P - S_1P = \frac{x}{\beta}\lambda = (\mu - 1)t$$

But $$\beta = \frac{D}{2d}\lambda$$

$$\therefore \quad (\mu - 1)t = x\frac{2d}{D} \tag{32}$$

Knowing x, $2d$, D and μ thickness t of the sheet each can be calculated using Eq. (32).

From $$x = \frac{D}{2d}(\mu - 1)\,t$$

If displacement $x = n$ fringe width, than $x = n\beta$

$$\therefore \qquad n\beta = \frac{D}{2d}(\mu - 1)\,t$$

or $$t = \frac{n\beta}{\mu - 1}\frac{2d}{D} = \frac{n\lambda}{\mu - 1} \qquad (33)$$

One can also calculate thickness of the plate knowing λ, μ and n.

The method is not suitable for determination of thickness of thick sheets due to (*i*) variation of μ with wavelength, and (*ii*) large lateral displacement.

Interference by Lloyd's Single Mirror

Lloyd's mirror is a plane, blackened glass plate about 30 cm in length and 6 to 8 cm in breadth (Fig. 4.18). Light incident on it is reflected from the front face. The light entering the plate is absorbed and hence there will be no reflection from the rear portion of the plate.

S is a slit source of light whose image is formed at S' by M_1M_2. The formation of image (S) and overlapping region is shown in Fig. 4.18. The light waves diverging from S and S' form two coherent sources. Interference takes place between the two waves diverging from S and S'. It is evident from Fig. 4.18 that the interference fringes are restricted to the region FA of the screen lying between the rays reflected from the two ends of the mirror.

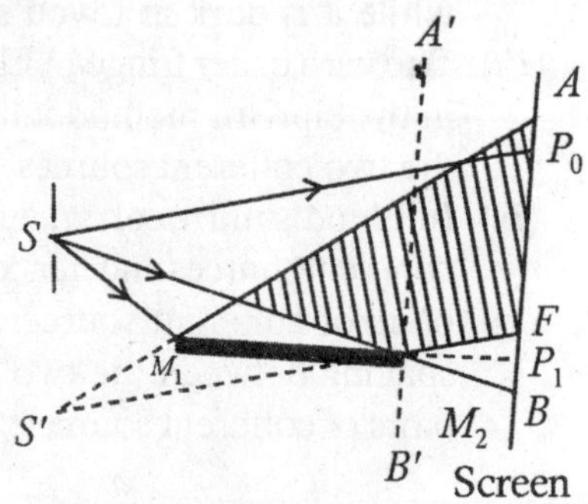

Fig. 4.18: Interference by Lloyd's single mirror

We know that when light waves are reflected at the surface of a denser medium, while travelling from a rarer to a denser medium, they suffer a phase change of π. Obviously, the waves diverging from S and S' will have a permanent phase difference of π, in addition to the phase difference due to path difference. If the path difference between the interfering waves is $n\pi$ ($n = 0, 1, 2, \ldots$), then they meet in **opposite** phases (due to the permanent phase difference π) and cause **destructive** interference. If the path difference is $\left(n + \frac{1}{2}\right)\lambda$, then the phase difference, odd multiple of π due to this path difference, gets added to the permanent phase difference π, so that the resultant phase difference is even multiple of π. Obviously, the waves are then in phase, and they cause **constructive interference**. Clearly, the conditions for darkness and brightness on the screen are just the opposite of those in the case of biprism (if the conditions are expressed in terms of path difference).

Central Fringe

The central fringe which would be obtained at P_0 (Fig. 4.13) is not visible because the fringe system is restricted to the region FA. At P_0 the difference $SP_0 - S'P_0 = 0$ and hence to central fringe would be dark. However, zero order can be obtained by introducing a transparent sheet of suitable thickness in the path of the actual source S.

Comparison of Lloyd's Mirror and Biprism Experiment

The fringes obtained from Lloyd's mirror differ from those obtained with Biprism in the following respects:

(*i*) In Biprism experiment full interference pattern is obtained while in Lloyld's mirror less then half the interference pattern is obtained.

(*ii*) With white light source the zero order fringe is white in Biprism while it is dark in Llyod's mirror.

(*iii*) The zero order fringe in Llyold's mirror method is very sharp while in the biprism method it has a width equal to the width of either of the two coherent sources.

(*iv*) In Llyod's mirror arrangement $2d$ is different from different pairs of coherent sources and hence fringe width β is also different for different pairs of coherent sources. In the case of biprism experiment, the spacing between the two consecutive maxima is the same for all pairs of coherent sources.

Achromatic Fringes

Bright and dark fringes obtained by using white light source are called achromatic fringes. The fringe width is given by

$$\beta = \frac{D}{2d}\lambda \tag{34}$$

This shows that fringe width is different for different wavelengths. However, β remains constant in heterogeneous light provided $\lambda/2d$ is kept constant. This means

$$2d \propto \lambda$$

In the case of biprism, the condition (34) is not fulfilled because the distance of the corresponding point is constant. In the case of Llyods's single mirror, the distance between the corresponding points is different for different pairs and hence it can be employed to obtain achromatic fringes.

Fresnel's Double Mirror

One can obtain two close coherent sources by the method of reflection by using Fresnel's double mirror as shown by M_1NM_2 in Fig. 4.19. The parts M_1N and M_2N of double mirror are inclined with each other at very small angle. In actual practice double mirror is made by grinding a single glass plate whose front surface is polished and silvered and back surface is blackened to avoid multiple-reflection. S is a source, whose images are formed at S_1 and S_2 are the virtual coherent sources. Since the two mirrors M_1N and M_2N are inclined at small angle and hence the two virtual images S_1 and S_2 are quite close to each other. As is evident from Fig. 4.19, S, S_1 and S_2 lie on a circle with the pole of the mirror as centre. From Fig. 4.19, we have

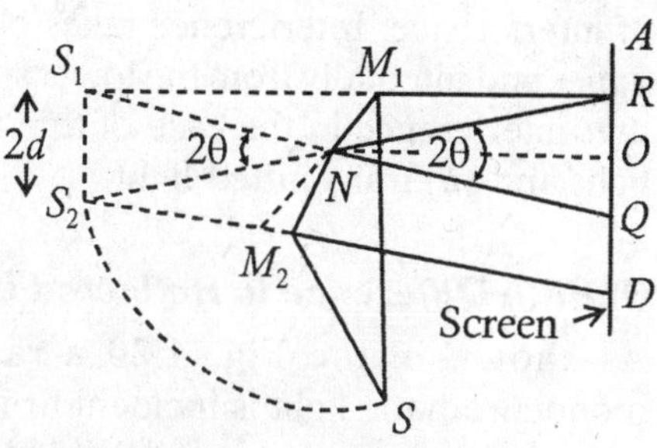

Fig. 4.19: Fresnel's double mirror

$$2d = a\cdot 2\theta \tag{35}$$

Here a is the distance of N from S_1, S_2 or S.

In actual experiment, the distance of S from N is kept quite small, so that $2d$ may be small. If b be the distance of N from screen, then

$$D = a + b \tag{36}$$

where D is called the distance between the source and the focal plane of the eyepiece.

One can obtain the wavelength of the monochromatic light by the following relation

$$\lambda = \frac{2a}{(a+b)}\theta\beta \tag{37}$$

(B) Interference Due to Division of Amplitude (Thin Films)

The amplitude of the incoming beam of light from the monochromatic source is divided into two parts either by partial reflection or refraction. These divided beams reunite after traversing different paths and interfere constructively or destructively. Due to use of broad sources of light, diffraction effects are minimised. The devices based on this principle are: (*a*) Newton's rings, (*b*) Michelson's interferometer, (*c*) Parallel and wedge shaped films, and (*d*) Fabry Perot Interferometer. Truly speaking, (*a*) and (*b*) are examples of (*c*).

Interference in Thin Films

Thin films of transparent materials like soap babbles, or of a drop of oil spread on the surface of water show brilliant colours when exposed to an extended source of light. The phenomenon can be explained on the basis of interference. Interference takes place between rays reflected from the upper and internally from the lower surface of the film. It has been observed that interference in the case of thin films takes place due to (*i*) reflected light and (*ii*) transmitted light.

(*i*) *Path Difference in Reflected Light*

As shown in the Fig. 4.20 a ray of monochromatic light is incident from an extended source on a parallel sided film of thickness t and refractive index μ. At B, the ray divides into a reflected ray BR_1 and a refracted ray BC. At C the ray is again partly reflected and partly refracted. Similar reflections and refractions occur at D, E, etc. Thus one obtains a set of parallel reflected rays and parallel refracted rays. Let us calculate the effective path difference between the reflected rays BR_1 and DR_2. Let DM be $\perp'$ to the reflected ray BR_1. Obviously, the difference between BR_1 and DR_2

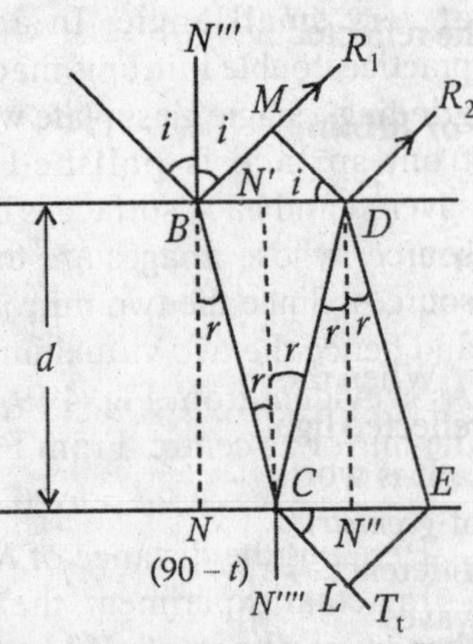

Fig. 4.20: Interference in thin films

$$= \text{Path } BCD \text{ in medium} - \text{Path } BM \text{ in air}$$

$$= \mu(BC + CD) - BM$$

$$= 2\mu BC - BM \quad (\because BC = CD)$$

From Fig. 4.20, we have

$$BC = BN \sec r^t / \cos r$$

and
$$BM = BD \cos(90 - i) = BD \sin i$$

$$= 2BN' \sin i = 2t \tan r \sin i$$

$$= 2t \frac{\sin r}{\cos r} \frac{\sin i}{\sin r} \sin r = 2t\mu \sec r \sin^2 r$$

$\therefore$ The path difference

$$= 2\mu \frac{t}{\cos r} - \frac{2\mu t \sin^2 r}{\cos r}$$

$$= 2\mu \frac{(1 - \sin^2 r)}{\cos r} = 2\mu t \cos r$$

The reflected ray BR_1 suffers a phase change of π on reflection at B and as we know that a phase change of π is equivalent to a path difference of $\lambda/2$, one obtains the effective path difference between BR_1 and DR_2

$$= 2\mu t \cos r + \lambda/2$$

For Maxima

$$2\mu t \cos r + \lambda/2 = n\lambda \tag{38}$$

or $$2\mu t \cos r = (2n - 1)\lambda/2$$

In general $$2\mu t \cos r = (2n \pm 1)\lambda/2 \tag{39}$$

When the condition for maxima is satisfied, the film appears bright in the reflected light.

For Minima

$$= 2\mu t \cos r + \frac{\lambda}{2} = (2n + 1)\frac{\lambda}{2}$$

or $$2\mu t \cos r = n\lambda \tag{40}$$

(where $n = 0, 1, 2, 3, \ldots,$ etc.)

When the condition for minima is satisfied, the film appears dark in the reflected light.

It is worthwhile to mention that conditions (39) and (40) are in terms of *geometrical path difference* and not in terms of the total optical path difference. These conditions hold good for every successive pair of reflected waves.

When the *thickness of the film* (t) *is very small as compared to the wavelength* (λ) *of the light, i.e.*

$$t << \lambda$$

the total optical path difference is $\lambda/2$ only. Obviously, the rays will produce destructive interference and film will appear **dark** by reflected light.

Note: The interference pattern will not be perfect because the intensities of the rays will not be same and moreover their amplitudes are different. The amplitude depends on the amount of light reflected and transmitted through the film. It has been observed that for normal incidence, about 4% of the incident light is reflected and rest *i.e.*, 96% is BR_1 and DR_2. Obviously, the intensity never vanishes completely and perfect dark fringes will not be observed.

Path Difference in Transmitted Light

Path difference in the transmitted light

$$= \text{Path } CDE \text{ in film} - \text{Path } CL \text{ in air}$$

$$= \mu(BD + DE) - CL = 2\mu BC - CL$$

From Fig. 4.20, we have

$$CL = CE\cos(90 - i) = CE\cos i = BD\sin i$$

∴ Path difference between CT_1 and ET_2

$$= 2\mu BC - BD\sin i \qquad BD\sin i = 2t\tan r\sin i$$

$$= 2\mu BC - 2t\tan r\sin i \qquad \sin i = \mu\sin r$$

$$= \frac{2\mu t}{\cos r} - 2\mu t\sin r\tan r$$

$$= \frac{2\mu t}{\cos r} - 2\mu t\frac{\sin^2 r}{\cos r}$$

$$= 2\mu t\left(\frac{1-\sin^2 r}{\cos r}\right) = 2\mu t\cos r$$

In the case of transmitted light there will be no phase change due to reflection at C or D because in either case light is travelling from denser to rarer medium. Obviously, the conditions for maxima and minima in transmitted light are

$$2\mu t\cos r = n\lambda \qquad \text{(Maxima)} \qquad (41)$$

$$2\mu t\cos r = (2n+1)\lambda/2 \qquad \text{(Minima)} \qquad (42)$$

A comparison of the conditions for maxima and minima in the reflected and transmitted light reveals that they are complementary.

Interference fringes in the case of transmitted light are less distinct because the difference in amplitude between CT_1 and ET_2 is very large. However, when the angle of incidence is nearly 45°, the fringes are more distinct.

Colours of Thin Films

When white light is incident on a thin film, the light which comes from any point from it will not include the colour whose wavelength satisfies the condition

$$2\mu t\cos r = n\lambda$$

in the reflected light. Obviously, the film will appear coloured and the colour will depend upon the thickness (t) and the angle of inclination. If t and r are constant then the colour will be uniform. In the case of water or oil different colours are seen because t and r vary.

Classification of Fringes Exhibited by Thin Films

The optical phase difference (δ) for a parallel thin film is given by

$$\delta = \frac{2\pi}{\lambda} = 2\mu t\cos r \qquad (43)$$

This shows that δ depends upon (*i*) λ, (*ii*) μt (*iii*) r. On this basis the fringe systems exhibited by thin films are classified into three classes, e.g.

(*i*) Haidenger fringes or fringes of equal inclination, (*ii*) Fringes of equal chromatic order and (*iii*) Fizeu fringes or fringes of equal thickness.

(*i*) Haidenger Fringes or Fringes of Equal Inclination: In this class, fringe of whatever shape it might be, corresponds to a particular value of r, while μt is constant. Constancy of μt automatically demands that light used must be almost monochromatic, *i.e.* λ be constant. Every fringe of the system corresponds to one value of r, *i.e.* characteristic of definite inclination. Familiar example of Haidenger fringes is circular fringes in Michelson interferometer with monochromatic light. These fringes are also called as **isocline fringes.**

(*ii*) Fringes of Equal Chromatic Order: When μt and r are kept practically constant, then each fringe corresponds to a particular wavelength λ. Obviously a coloured pattern of fringes is observed. These fringes are known as **fringes of equal chromatic order** (*FECO*).

(*iii*) Fizeau Fringes or Fringes of Equal Thickness: If a parallel or almost parallel pencil of rays of monochromatic light falls on a film whose thickness t varies at different places, then dark and bright interference fringes are seen in the reflected light at the top surface of the film. These are called **fringes of equal thickness**, or **isopachic fringes**, because each one passes through the points with the same values of t. Equal thickness fringes, localized on the surface of the film, can be observed on a screen as of a converging lens. Circular fringes as Newton's rings are the familiar example of this class. In white light, a system of differently coloured equal-thickness fringes is observed.

Wedge-shaped Film

Let us suppose that the thin film is not parallel-sided but wedge-shaped, *i.e.* its surfaces make some angle with each other. One can obtain such film by placing two accurately plane pieces of glass in contact at one end, the far end being kept apart by a very thin object like a human hair or a cigarette paper. Such a film, when illuminated with a parallel beam of monochromatic light will produce evenly-spaced linear dark and bright fringes running parallel to the apex of the wedge. When such a film is illuminated with white light, coloured fringes are observed. If the light comes from an extended source, the interference pattern can be seen only at the portion of the wedge near the edge.

Consider a wedge-shaped film *ABC* of refractive index μ and whose surfaces are inclined at an angle θ (Fig. 4.21). The light from a sodium lamp is made to fall normally on the system by using a plane glass plate G and a travelling microscope is focused on the fringes formed. Straight fringes parallel to the line of contact of the two glass plates will be observed through the microscope.

Rays of light reflected from the upper and lower faces of the air film between the plates suffer interference. The path deference between the interfering rays is given by

$$\Delta = 2\mu t \cos r + \lambda/2 \tag{44}$$

Fig. 4.21: Interference from thin wedge-shaped film

here t is the thickness of the air film at the point of the incidence of the incident ray. Since light is incident normally, we have $r = 0$ and $r = 1$. For air, $\mu = 1$. Using $r = 0$ and $r = 1$, Eq. (44) reduces to

$$\Delta = 2t + \lambda/2 \tag{45}$$

The minimum thickness of the film is $t = 0$. The thickness varies as we move from A to B. At those points where t is such that

$$\Delta = 2n\lambda \qquad (n = 1, 2, 3, \ldots)$$

there will be brightness, and at points where t is such that

$$\Delta = \left(n + \frac{1}{2}\right)\lambda \qquad (n = 0, 1, 2, 3, \ldots)$$

there will be darkness. Since the thickness t remains constant at all points on a straight line parallel to the wire, one obtains alternately dark and bright straight fringes. These are called '**fringes of equal thickness**'. These fringes are formed at the upper surface of the air film and hence these are **localised**.

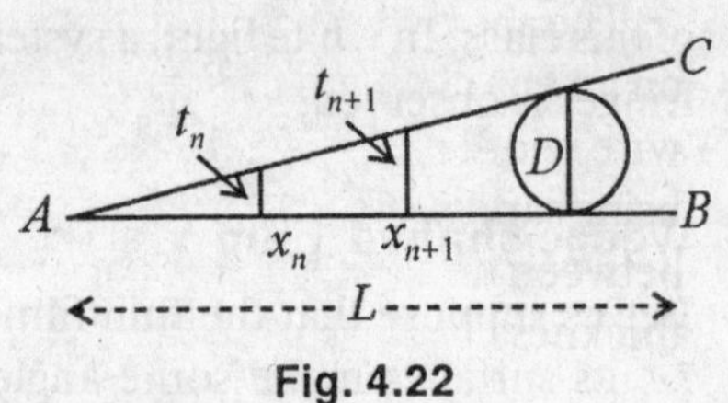

Fig. 4.22

The nth dark fringe is formed where the thickness t_n of the air film satisfies the following condition

$$\Delta = 2t_n + \frac{\lambda}{2} = \left(n + \frac{1}{2}\right)\lambda \tag{46}$$

$(n + 1)$th dark fringe is formed when the thickness of the air film t_{n+1} satisfies the following condition

$$\Delta = 2t_{n+1} + \frac{\lambda}{2} = \left((n+1) + \frac{1}{2}\right)\lambda \tag{47}$$

Subtracting Eq. (46) from Eq. (47), one obtains

$$2(t_{n+1} - t_n) = \lambda \tag{48}$$

If the distance between A and the space r is L and the thickness of the space r is D then from similar triangles (Fig. 4.22), one obtains

$$\frac{t_{n+1}}{X_{n+1}} = \frac{t_n}{X_n} = \frac{D}{L} \tag{49}$$

where X_{n+1} and X_n are the distances of the $(n + 1)$th and nth dark fringes from A. We have

$$t_{n+1} = \frac{D}{L} X_{n+2} \text{ and } t_n = \frac{D}{L} X_n \tag{50}$$

From Eqs. (49) and (50), one obtains

$$2\frac{D}{L}(X_{n+1} - X_n) = \lambda$$

or $$2\frac{D}{L}\beta = \lambda \qquad [\beta = (X_{n+1} - X_n) \text{ is fringe width}]$$

or $$\beta = \frac{L\lambda}{2D} \tag{51}$$

The total width of a known large number of fringes is measured with the travelling microscope and the fringe width (β) is calculated. The distance L between the end A and the space r is measured. Knowing the wavelength of light used the thickness D of the space r is calculated.

The above method can be used to determine the thickness of the insulation on an enamelled copper wire. First, the experiment is performed using the wire width enamel as the space r. Then the enamel insulation is removed using sand paper and the experiment is repeated. Obviously, the difference between the diameters of the wire obtained in the two cases gives twice the thickness of the insulation.

Testing the Optical Flatness of Planes

If the two surfaces AB and AC (Fig. 4.22) are perfectly plane, the air film gradually varies in thickness from A to B. Equal thickness fringes will be observed if a beam of parallel monochromatic light is made to fall on the film because each fringe is the locus of the points at which the thickness of the film has a constant value. If the fringes are not of equal thickness it means the surfaces are not optically flat. The standard method is to take an optically plane surface AB and the surface to be tested AC. The fringes are observed through microscope and if they are of equal thickness, the surface AC is plane. If not, the plane AC is not optically flat. The surface

AC is polished and the process is repeated. When the fringes observed are of equal thickness, it means that the plate *AC* is optically flat.

If a beam of parallel monochromatic light is made to fall on a film of absolutely uniform thickness then t being the same at all points the film will appear to be uniformly tinted at all points and no interference fringes will be observed. This result provides the manufacturer of optical flats with a method of testing minute deviations from true optical flatness of the surface.

Necessity of an Extended Source

Let us consider that a thin film is viewed in reflected light from a point source S (Fig. 4.23). From Fig. 4.23 (*a*) it is evident that for each one angle of incidence, the interference rays are parallel to each other but for different angles of incidence the different pairs of interfering rays are obtained along widely different angles in accordance with the laws of reflection. Since the size of the pupil of the eye is limited, the interfering rays only forming a small portion of the film can enter the eye. Obviously, it is clear that with a point source the entire film cannot be viewed by placing the eye in one position.

However, when the film is viewed with a broad source [Fig. 4.23 (*b*)], the rays reflected from different parts of the film converge and hence by placing the eye in a suitable position, one can see the entire film simultaneously. It is worthwhile to mention that a narrow source is used to observe interference fringes in the case of a Fresnel's biprism, double mirrors and Llyod's single mirror.

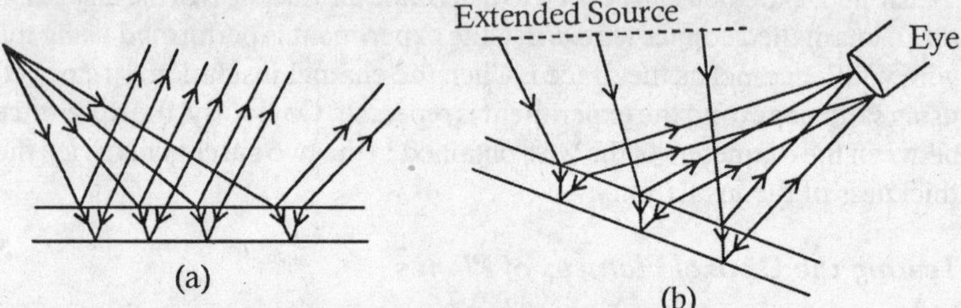

Fig. 4.23: A thin film illuminated by extended source of light

Newton's Rings

When a convex lens of large radius of curvature is placed on a plane glass plate, an air film is formed between the upper surface of the plate and the lower surface of the lens. If monochromatic light be allowed to fall normally on this film, system of alternate bright and dark concentric rings with their centre dark is formed in the air film. As we move away from the centre the spacing between the rings goes on decreasing. These are named as Newton's rings. These rings are formed as a result of the interference between the lens and the plate.

The experimental arrangement for observing Newton's rings is shown in Fig. 4.24. L is the lens kept in contact with the plane glass plate E. To get good fringes the radius of curvature of the face of the lens in contact with the glass plate must be fairly large, say about 1 m. GG is a glass plate fixed above the lens at 45° to the vertical. The light from a monochromatic source S (usually sodium lamp) is rendered parallel by a condensing lens L_1 and made to fall on G in a horizontal direction. The light is practically reflected downwards towards the lens plate combination. After passing through the lens (L), part of it is reflected at the lower lens surface and part at the top surface of the glass plate (E). Then the light travels upwards through the glass plate G and is observed through a travelling microscope (M). There will be a path difference between the light reflected at the glass surface and the lens surface. As a result a maximum or minimum will be observed according as the path difference is an even multiple or an odd multiple of half of wavelength ($\lambda/2$). The locus of all points which have the same thickness is a circle. Therefore, circular fringes alternately bright and dark will be observed. The dark fringes are found to be more sharp than the bright fringes and therefore easier to measure.

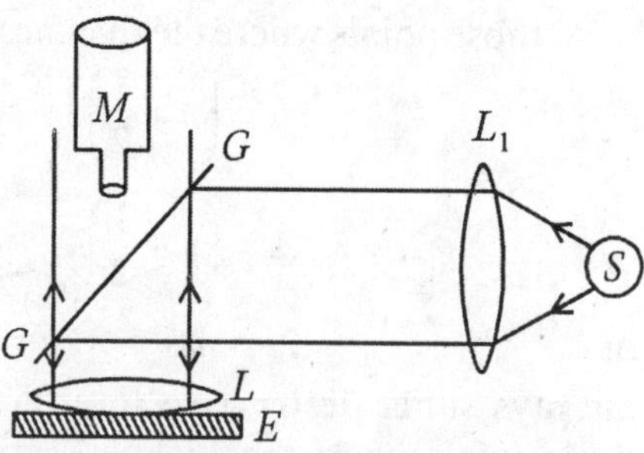

Fig. 4.24: Experimental arrangement for Newton's rings experiment

Mathematical Analysis

In Fig. 4.25 AB is an incident ray and corresponding to it C and D are interfering reflected rays. The path difference between the reflected interfering rays in given by

$$\Delta = 2\mu t \cos(r + \theta) + \lambda/2 \quad (52)$$

where μ is refractive index of the film. For air $\mu = 1$, t is the thickness of the air film at the point in question. r is the angle of refraction in the air film and θ is the angle of wedge. The experimental setup is adjusted so that $r = 0$ and θ is also very small.

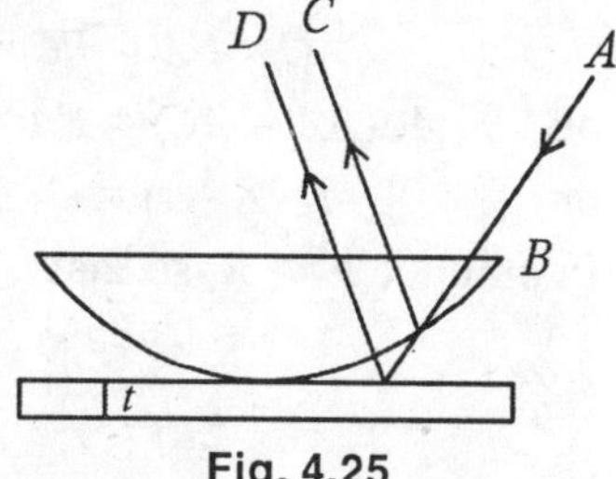

Fig. 4.25

$$\Delta = 2t + \lambda/2 \quad (53)$$

At the point of contact of the lens and the plate, $t = 0$, hence $\Delta = \lambda/2$. This is the condition for minimum intensity. Hence the central spot is dark.

At those points where the thickness of the air film satisfies the condition

$$\Delta = \left(n + \frac{1}{2}\right)\lambda \tag{54}$$

or $$2t + \lambda / 2 = \left(n + \frac{1}{2}\right)\lambda$$

or $$2t = n\lambda \quad (n = 1, 2, \ldots) \text{ for dark rings} \tag{54a}$$

the rays suffer **destructive interference** and there will be **darkness**. At those points where the thickness satisfies the condition

$$\Delta = n\lambda \tag{55}$$

or $$2t + \frac{\lambda}{2} = n\lambda \quad (n = 1, 2, \ldots) \text{ for bright rings}$$

or $$2t = \left(n - \frac{1}{2}\right)\lambda \tag{56}$$

the rays suffer **constructive interference** and there will be **brightness**.

If we imagine concentric circle in the plane of the lower glass plate, with their centre at the point of contact between the plate and the lens, then the thickness of the air film remains constant for every circle. Obviously, the points of brightness and points of darkness lie along concentric circle. The interference fringes, therefore, consist of alternately bright and dark circle. These are called **Newton's rings**. One can easily see from Fig. 4.25 that the interfering rays intersect somewhere near the top of the air film. Therefore, Newton's rings are **localised** and can be seen through a microscope.

Diameter of Dark and Bright Rings

Let R be the radius of curvature of the surface PAQ of the lens and r be the radius of Newton's ring PCQ with centre C (Fig. 4.26). From the geometry of the figure, we have

$$AC \times CA' = PC \times CQ$$

or $$AC(AA' - AC) = PA \times CQ$$

or $$t\,(2R - t) = r \times r = r^2$$

In practice $t >> R$, so that

$$2R - t \approx R$$

$\therefore$ $$2Rt = r^2$$

or $$2t = \frac{r^2}{R} \tag{57}$$

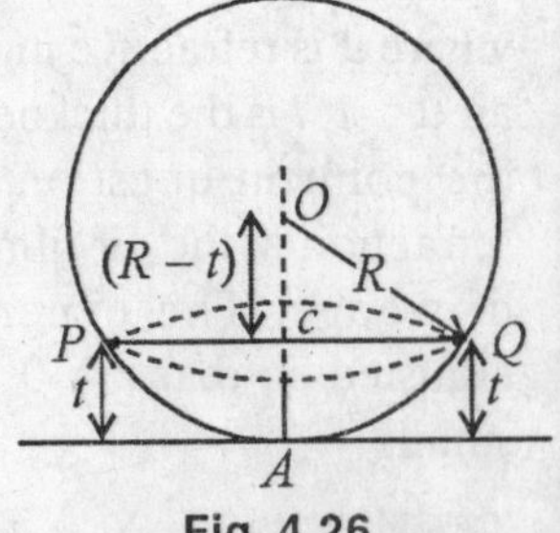

Fig. 4.26

∴ Path difference

$$\Delta = 2t + \frac{\lambda}{2} = \frac{r^2}{R} + \frac{\lambda}{2}$$

Condition for minima, i.e. darkness

$$\frac{r^2}{R} + \frac{\lambda}{2} = (2n+1)\frac{\lambda}{2}$$

or $$\frac{r^2}{R} = n\lambda \qquad (n = 0, 1, 2, \ldots) \tag{58}$$

Eq. (58) gives the radii r of the dark rings. The first dark ring is obtained with $n = 0$. Obviously, the radius of first dark ring is zero. Therefore, the central spot at the point of contact of the lens with the glass plate is dark. The first, second, ..., nth dark rings are obtained with $n = 1, 2, \ldots, n$. Therefore, the radius of nth dark ring is given by

$$\frac{r_n^2}{R} = n\lambda \tag{59}$$

or $$r_n = \sqrt{Rn\lambda} \tag{60}$$

If D_n the diameter of nth dark ring ($D_n = 2r_n$), then

$$\frac{D_n^2}{4R} = n\lambda$$

or $$D_n = \sqrt{4R\lambda}\sqrt{n}$$

or $$D_n \propto \sqrt{n} \tag{61}$$

Therefore **diameters of dark rings are proportional to the square roots of the natural numbers.**

Condition for maxima (bright rings)

$$\Delta = \frac{r^2}{R} + \frac{\lambda}{2} = n\lambda$$

or $$\frac{r^2}{R} = \left(n - \frac{1}{2}\right)\lambda = (2n-1)\frac{\lambda}{2} \qquad (n = 1, 2, 3, \ldots)$$

∴ Radii of nth bright ring

$$r_n^2 = (2n-1)R\frac{\lambda}{2}$$

∴ $$r_n = \sqrt{\frac{R\lambda}{2}}\sqrt{(2n-1)}$$

or diameter of nth bright ring ($D_n = 2r_n$)

$$\therefore \qquad D_n = \sqrt{2R\lambda}\sqrt{2n-1} \tag{62}$$

or

$$D_n = \sqrt{2n-1} \tag{63}$$

Therefore, **the diameters of the bright rings are proportional to the square roots of odd numbers.**

Determination of Wavelength

Eq. (59) suggests a method of finding the wavelength of a given radiation if R is known or of finding R if the wavelength λ is known. If D_n be the diameter of nth dark ring then from Eq. (59)

$$D_n^2 = 4Rn\lambda \tag{64}$$

Similarly the diameter of the $(n + m)$th ring is given by

$$D_{n+m}^2 = 4R(n+m)\lambda \tag{65}$$

Subtracting Eq. (64) form Eq. (65), one obtains

$$\lambda = \frac{D_{n+m}^2 - D_n^2}{4Rm} \tag{66}$$

After setting up the rings the central spot must be observed to be dark. If however it is bright the lens is tapped gently with a pencil till it becomes dark. The microscope is focused on the rings. The diameters of the successive rings are measured. From Eq. (64) it is evident that the **fringe width decreases with the order of the fringe and the fringes (rings) get closer with increase in their order**. The wavelength λ is then calculated using Eq. (66). The radius of curvature R of the lower surface of the lens is found by Boy's method, or, if R is too large, by using a telescope.

If medium other then air forms the film, then Eq. (66) takes the form

$$\lambda = \frac{D_{n+m}'^2 - D_n'^2}{4Rm}\mu \tag{67}$$

where μ is the refractive index of the medium and D'_{n+m} and D'_n are the diameters of $(n + m)$th and nth rings respectively.

Determination of Refractive Index of a Liquid using Newton's Rings

Newton's rings method can be used for the determination of refractive index (μ) of a liquid available in small quantity. The transparent liquid whose refractive index is to be determined is placed between the lens L and the glass plate E (Fig. 4.24). If liquid is rarer than glass, a phase change of π will occur at the time of reflection from the lower face of the liquid film. If the liquid is denser than glass, then a phase change of π will occur due to reflection at the upper face of the film. Hence in either case, a path difference of $\lambda/2$ is introduced due to phase change. Therefore path difference between the interfering waves in reflected light will be

$$\Delta = 2\mu t \cos r + \frac{\lambda}{2}$$

where μ is the refractive index of the liquid used. Since $r = 0$ and $\cos r = 1$ for normal incidence, we have

$$\Delta = 2\mu t + \frac{\lambda}{2}$$

But $$2t = \frac{r_2}{R} \qquad \text{(see Eq. (57))}$$

$\therefore$ Path difference $$\Delta = \frac{\mu r^2}{R} + \frac{\lambda}{2}$$

For minima, i.e. dark rings

$$\frac{\mu r^2}{R} + \frac{\lambda}{2} = (2n+1)\frac{\lambda}{2} \qquad (n = 0, 1, 2, \ldots)$$

or $$\frac{\mu D^2}{4R} = n\lambda \qquad [\text{where } D = 2r]$$

or $$D^2 = \frac{4Rn\lambda}{\mu}$$

$\therefore$ For nth dark ring

$$D_n^2 = \frac{4Rn\lambda}{\mu} \tag{68}$$

For $(n + p)$th dark ring

$$D_{n+p}^2 = \frac{4R(n+p)\lambda}{\mu} \tag{69}$$

Subtracting Eq. (68) from Eq. (69), one obtains

$$D_{n+p}^2 - D_n^2 = \frac{4Rp\lambda}{4} \tag{70}$$

or $$\mu = \frac{4Rp\lambda}{D_{n+p}^2 - D_n^2} \tag{71}$$

Keeping the same value of p for air, one obtains

$$1 = \frac{4Rp\lambda}{D'_{n+p} - D'^2_n} \qquad (\mu = 1 \text{ for air}) \tag{72}$$

From Eq. (71) and Eq. (72), one obtains

$$\mu = \frac{D'^2_{n+p} - D'^2_n}{D_{n+p}^2 - D_n^2} \tag{73}$$

The expression shows that $D'_{n+p} > D_{n+p}$ and $D'_n > D_n$. Using Eq. (73) refractive index of the liquid can be determined.

Newton's Rings Due to Transmitted Light

Arrangement for observing Newton's rings due to transmitted light is shown in Fig. 4.27. The condition for **maxima** and **bright** rings is

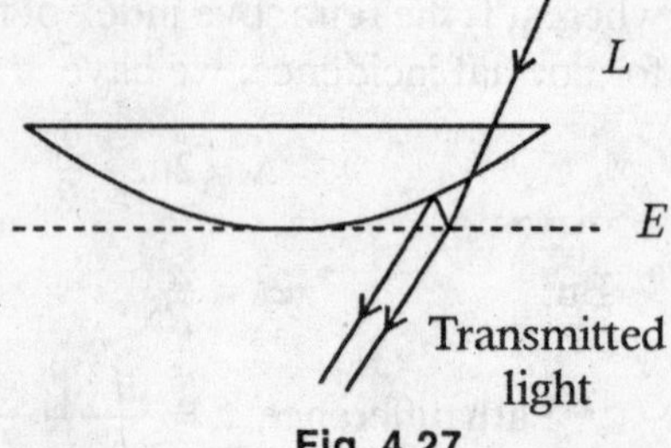

Fig. 4.27

$$2\mu t \cos r = n\lambda$$

$\therefore \quad r = 0, \cos r = 1$ and $\mu = 1$ for air

$\therefore \quad 2t = n\lambda (n = 0, 1, 2, 3, \ldots)$

When $n = 0$, for bright rings, $r = 0$.

Obviously, in the case of Newton's rings due to transmitted light, the central ring is bright, *i.e.* just opposite to the rings due to reflected light.

For dark rings or **minima**, we have

$$2\mu t \cos r = (2n+1)\frac{\lambda}{2}$$

$\therefore \quad r = 0, \cos r = 1$ and $\mu = 1$ for air

$$\therefore \quad 2t = (2n+1)\frac{\lambda}{2},$$

Radius of **bright rings** is given by

$$r_n^2 = n\lambda R$$

For **dark rings**

$$r_n^2 = \frac{(2n+1)\lambda R}{2}$$

Newton's Rings with White Light

We have seen that with monochromatic light, Newton's rings are alternately dark and bright. The diameter of the ring depends upon the wavelength (λ) of the light used. When in place of monochromatic source of light, white light source is used, the diameter of the rings of the different colours will be different and coloured rings are observed. Only the first few rings are clear and after that due to overlapping of the rings of different colours of white light, one cannot see them.

Newton's Rings with both Curved Surfaces

In place of plane glass plate below the plane convex lens in Newton's rings experiment, one can employ curved spherical surfaces as shown in Fig. 4.28. In Fig. 4.28 (*a*) a concave surface is in contact with the convex surface and in Fig. 4.28 (*b*) two convex surfaces are shown in contact with each other. In the former case, it is essential that the radius of curvature of the concave surface (say R_2), is greater than the radius of curvature of the convex surface (say R_1). In the second case, radii of curvatures of the two lenses in contact should be large to enclose a thin film in between.

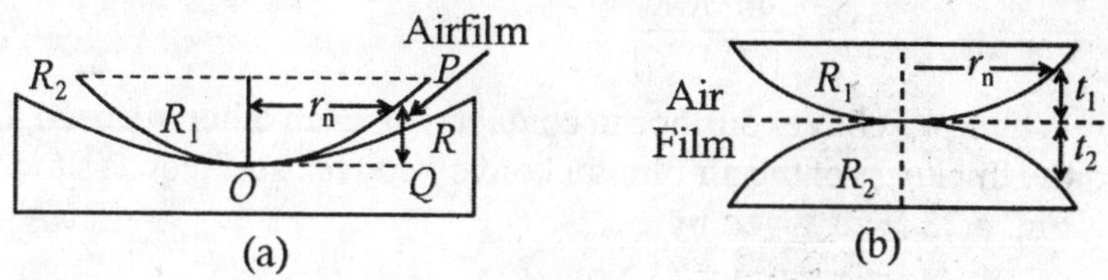

Fig. 4.28

(*i*) In Fig. 4.28 (*a*), a convex surface of radius of curvature R_1 is in contact with a concave surface R_2 at point O ($R_2 > R_1$). Let us consider a Newton's ring of radius r_n where the thickness of the air film is t. From the figure, it is evident that

$$t = PQ - QR = \left[\frac{r_n^2}{2R_1} - \frac{r_n^2}{2R_2}\right] \tag{74}$$

For thin film and normal incidence, the path difference is given by

$$\Delta = 2\mu t = 2\mu\left[\frac{r_n^2}{2R_1} - \frac{r_n^2}{2R_2}\right] \tag{75}$$

For **maxima** or **bright rings**, the path difference given by Eq. (75) must satisfy the condition

$$\mu r_n^2\left[\frac{1}{R_1} - \frac{1}{R_2}\right] = (2n+1)\frac{\lambda}{2}$$

or $$\mu D_n^2\left[\frac{R_2 - R_1}{2R_1R_2}\right] = (2n+1)\frac{\lambda}{2}$$

When $D_n = 2r_n$ is the diameter of the nth ring

or $$D_n^2 = \frac{2(2n+1)\lambda R_1R_2}{\mu(R_2 - R_1)} \tag{76}$$

For air, $\mu = 1$ and hence Eq. (76) becomes

$$D_n^2 = \frac{2(2n+1)\lambda R_1R_2}{(R_2 - R_1)} \tag{77}$$

For **minima** or **dark rings**, one obtains the diameter of *n*th ring as

$$D_n^2 = \frac{4n\lambda R_1 R_2}{\mu(R_2 - R_1)} \tag{78}$$

For air $$D_n^2 = \frac{4n\lambda R_1 R_2}{(R_2 - R_1)} \tag{79}$$

(*ii*) When two convex surface in contact with each other are used, then the thickness of the air film for convex-convex surfaces as shown in Fig. 4.28 (*b*) is given by

$$t = t_1 + t_2 = \frac{r_n^2}{2R_1} + \frac{r_n^2}{2R_2} \tag{80}$$

Proceeding, exactly for the previous case, one obtains the expression for the diameter of *n*th **bright** ring as

$$D_n^2 = \frac{2(2n+1)\lambda R_1 R_2}{\mu(R_1 + R_2)} \tag{81}$$

For **air put** $\mu = 1$ in Eq. (81).
The diameter of *n*th dark ring is given by

$$D_n^2 = \frac{4n\lambda R_1 R_2}{\mu(R_1 + R_2)} \quad \text{(For air } \mu = 1) \tag{82}$$

One can find the diameter of any order using Eqs. (78), (79), (81) and (82). Wavelength relation can be expressed as

$$D_{n+p}^2 - D_n^2 = \frac{4pR_1 R_2}{\mu(R_2 \pm R_1)} \tag{83}$$

Here $(R_2 - R_1)$ is for concave-convex and $(R_2 + R_1)$ for convex-convex surface. Similarly, one can obtain expression for transmitted patterns.

Michelson Interferometer

Michelson interferometer devised by Prof. A. Michelson, is based on the phenomenon of interference. The apparatus is shown in Fig. 4.29. It consists of an optically flat glass plate G_1 semi-silvered at the back, two front silvered mirrors M_1 and M_2. Glass plate G_1 is set at 45° to the directions AM_1 and AM_2 which are perpendicular to each other. The mirror M_1 is mounted on a fine pitch screw (capable of reading upto 10^{-5} cm accurately) with whose help M_1 can be moved exactly parallel to itself.

The mirror M_2 is provided with three levelling screws at its back. The plate G_2 is called the **compensating plate**. It is exactly similar to plate G_1 and kept parallel to G_1. The plane of G_2 can be slightly altered by means of a fine screw. The plate G_2 is essentially required to work with white light source.

The two rays AM_1 and AM_2 after reflection from M_1 and M_2 respectively produce interference when collected by a telescope T.

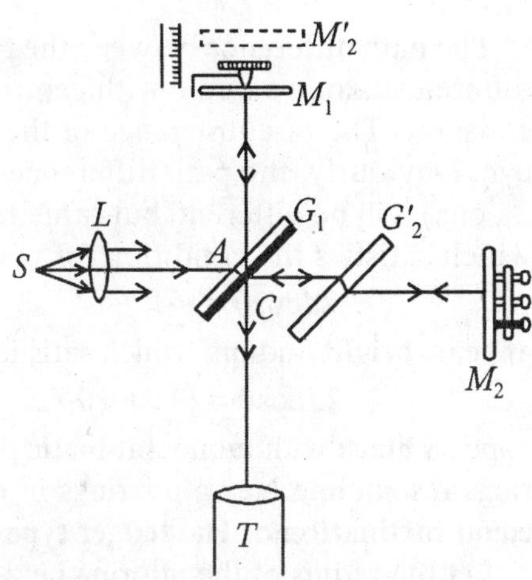

Fig. 4.29: Michelson Interferometer

Working

Light from monochromatic source S after being rendered parallel by convex lens L, falls on G_1. The beam is divided into two parts, one being reflected along AM_1 and the other being transmitted along AM_2. The reflected part going along AM_1 falls on mirror M_1 normally, hence it is reflected back along its original path and then after passing through plate G_1 once again, proceeds along the path AT. The transmitted beam is also incident normally on mirror M_2, hence it is also reflected back along its original path. This suffers reflection at the lower surface of the plate G_1 and then proceeds along AT. Obviously, the telescope T receives two beams which are originally derived from the same single beam SA. Clearly, these two beams are in a position to produce interference fringes in the field of view of telescope T. The ray AM_1 passes through the plate G_1 thrice whereas the ray AM_2 passes through it only once. That is why the second plate G_2 of the same thickness and inclination as G_1 is introduced to compensate the two paths. The whole arrangement equivalent to interference observed from two reflecting faces M_1 and M_2' enclosing an air film, where M_2' is the virtual position between the faces can be varied with the help of micrometer screws with M_1 or M_2.

The path difference between the two interfering rays arises due to the difference in distances of M_1 and M_2 from G_1 and also due to the fact that the rays travelling from M_2 to G_1 suffer a phase change of π which is reflected at G_1. Two beams incident on the telescope interfere with each other and fringes are observed in the focal plane of the telescope.

Types of Fringes

(*i*) **Circular Fringes**: Circular fringes are produced when the mirror M_1 and virtual mirror M'_2 which is the image of M_2 are parallel. If the distances of the mirrors M_1 and M_2 from G_1 differ by distance d, then the separation between M_1 and M'_2 will be $2d$ [Fig. 4.30 (*a*)].

The path difference between the two beams will be $2d\cos\theta$. The path difference along a circle with centre O as the foot of the $\perp^r$ from eye is constant. The circumference of the circle subtends an angle θ on the eye. Obviously, the path differences for different values of θ (say θ, θ' ..., etc.) will be different, but same for one value of θ. Therefore, a ring which satisfies the condition for maxima, *i.e.*

$$2d\cos\theta = n\lambda$$

appears **bright** and one which satisfies the condition for **minima**, *i.e.*

$$2d\cos\theta = (2n + 1)\lambda/2$$

appears **black** with monochromatic light and a pattern of bright and dark rings resembling Newton's rings in origin and are known as fringes of **equal inclination** or **Haidenger type of fringes.**

Let the radius of the nth ring be X'_n and distance of the foot of the $\perp^r$ from eye be L, then for small values of θ, we have

$$\theta = \frac{X_n}{L}$$

One can write the optical path difference for bright rings as

$$2d\left(1 - \frac{\theta'}{2}\right) = 2d\left(1 - \frac{X_n^2}{2L^2}\right) = n\lambda$$

Thus for successive rings, one can write

$$X_{n-1}^2 = X_n^2 = \frac{\lambda L^2}{d}$$

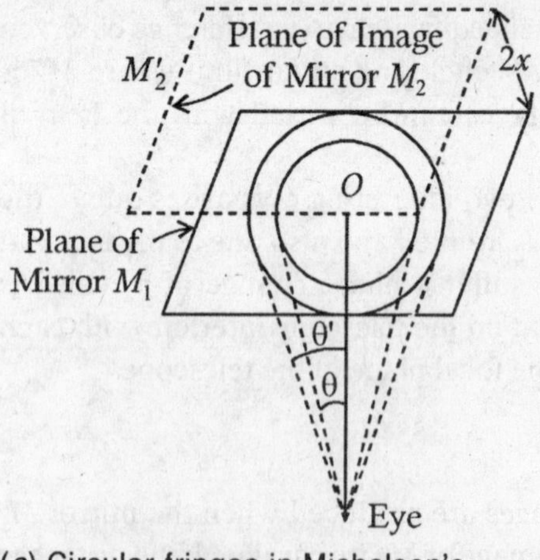

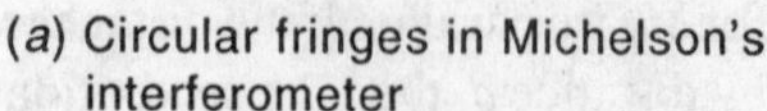
(*a*) Circular fringes in Michelson's interferometer

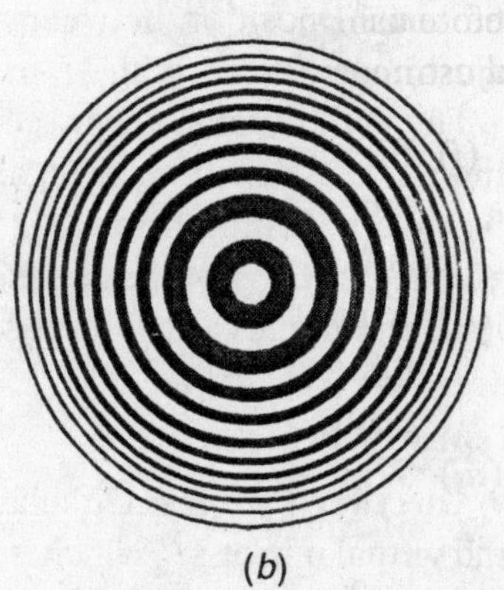
(*b*)

Fig. 4.30

***(ii)* Localized Fringes:** When the mirror M_1 and the virtual image M_2' of mirror M_2 are inclined, the air film enclosed is wedge shaped and localized fringes (straight bands and curved fringes) as shown in Fig. 4.31 are observed. When M_1 actually intersects M_2' in the middle, the fringes are perfectly straight [Fig. 4.31 (*b*)]. In the other positions, the shape of the fringes is curved and always convex towards the thin edge of the wedge [Fig. 4.31 (*a*) and (*c*)]. When the path difference are large, this type of fringes are not observed.

***(iii)* White Light Straight Fringes:** When white light source is used, the fringes are observed only when the path difference is small. Since the different colours overlap on each other hence only the first few coloured fringes are visible. The central fringe is dark or white depends upon whether the plate G_1 is unpolished or polished at the back. White light fringes are useful for the determination of zero path difference, especially in the case of measurement of the standard meter.

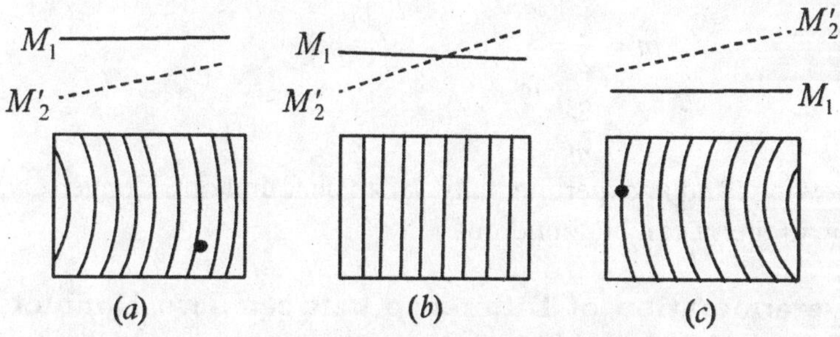

Fig. 4.31: Localized fringes by Michelson interferometer

Setting Michelson Interferometer for Experiment

Before using Michelson interferometer, for any experiment, the following adjustments are made:

(*i*) First the interferometer is illuminated with a broad source of monochromatic light. Two mirrors are kept at different distances. A condensing lens is used.

(*ii*) A pin-hole is placed between the source S and plate G_1. Its four images are observed due to reflections from the silvered and unsilvered portions of the plate G_1 and mirrors M_1 and M_2.

(*iii*) With the help of screws at the back of the one of the mirrors M_1 and M_2, the weaker images are made to coincide with the stronger ones, *i.e.* only two images of the pin-hole remain in the field of view. As soon as this adjustment is complete, the circular rings appear automatically.

If the centre of the fringe system is not in the field of view, then one can bring it in the centre of the field of view with the help of the special screws provided for the purpose at the top of the mirror.

Uses of Michelson's Interferometer

Michelson's interferometer has been used for a variety of purposes, some of the which are described below.

(*i*) Determination of Wavelength of Monochromatic Light: First the fringes are obtained in the field of view with the help of a given monochromatic source of light and Michelson's interferometer. Telescope is focused on one of the fringes. Now, the mirror M_1 is moved parallel to itself, then fringes sweep across the field of view of the telescope. For every displacement of $\lambda/2$ by mirror M_1, one bright fringe goes past the cross-wire of the telescope. Number of fringes is counted. If m fringes move across, when the mirror traverses a known distance l, then

$$m = \frac{\lambda}{2} = l \quad \text{or} \quad 2l = m\lambda$$

$$\therefore \qquad \lambda = \frac{2l}{m} \tag{84}$$

Knowing l and m experimentally, λ can be found out. Conversely, if λ is known, then l can be found out.

(*ii*) Determination of Difference between Two Doublets or Resolution of Spectral Lines: Certain composite spectral lines, which ordinarily could not be easily separated, have been resolved by Michelson's interferometer by studying the change of visibility of fringes with increasing path difference. Suppose the source S of light emits composite light of wavelength λ_1 and λ_2. Both these wavelengths will give their own fringe systems and due to their superposition the fringes are not clearly visible. When mirror M_1 is moved, then these two sets of fringes get in step or out of step alternately. The two sets are said to be in step when bright fringe of one set falls on the bright fringe of the other. In that case the fringes become very distinct and well defined, *i.e.* maximum brightness is produced. The position of mirror M_1 is further moved and a stage is reached when the visibility decreases and becomes least when the two sets are out of step, *i.e.* the dark fringes of one system fall on the bright fringes of the other. On further moving the mirror we again obtain a position of maximum brightness and so on.

If d be the displacement of the mirror M_1 for two consecutive positions of maximum distinctness or indistinctness, the path difference is $2d$. If the number of fringes of wavelength λ_1 lying in this path be n_1 then the number of fringes corresponding to fringes of wavelength λ_2 will be $(n + 1)$.

$$\therefore \quad 2d = n\lambda_1 = (n+1)\lambda_2 = n\lambda_2 + \lambda_2$$

or
$$n = \frac{2d}{\lambda_1}$$

Also
$$n = \frac{2d - \lambda_2}{\lambda_2} = \frac{2d}{\lambda_2} - 1$$

$$\therefore \quad \frac{2d}{\lambda_1} = \frac{2d}{\lambda_2} - 1$$

or
$$2d\left(\frac{1}{\lambda_2} - \frac{1}{\lambda_1}\right) = 1$$

or
$$\frac{\lambda_1 - \lambda_2}{\lambda_1\lambda_2} = \frac{1}{2d}$$

or
$$\lambda_1 - \lambda_2 = \frac{\lambda_1\lambda_2}{2d} \approx \frac{\lambda^2}{2d} \qquad (\because \lambda_1\lambda_2 \approx \lambda^2)$$

or
$$\Delta\lambda \approx \frac{\lambda^2}{2d} \tag{85}$$

where λ is the geometric mean of two wavelengths. This method can also be used for testing whether a given source of light is monochromatic or not.

***(iii)* Determination of Refractive Index or Thickness of thin Films:** For the determination of refractive index of thin transparent plate or determination of thickness of thin transparent film or plate, white light fringes are obtained and the telescope is focused on the central white fringe. The transparent plate or film is introduced in the path of one of the interfering beams, say AM_2. On doing so the position of the central fringe shifts and mirror M_2 is moved through a distance x with the help of the micrometer screw attached to it till the central fringe is again brought back to the initial position. If t is the thickness of the film, then the path of beam is increased by $(\mu - 1)t$. Therefore, the path difference between the two beams becomes $2(\mu - 1)t$. Suppose due to this path difference, m fringes move across the field (in distance x), then

$$2(\mu - 1)t = m\lambda \tag{86}$$

If t, m and λ are known, then μ can be calculated. If μ, m_2 and λ are known, then t can be calculated.

***(iv)* Calibration of Standard Metre:** Michelson used a modified form of the interferometer of determining the number of wavelengths of a monochromatic light contained in the standard meter kept at the International Bureau of Weights and Measures at Sevres near Paris. For this experiment he selected the red cadmium line of wavelength 6438 Å, because it is the simplest and the most homogeneous spectral line.

Fabry-Perot Interferometer

Principle

Sharp Haidinger fringes are produced here by means of multiple reflection from two partially silvered plates placed parallel to each other.

Construction

It essentially consists of two parallel glass plates *A* and *B* separated from each other (Fig. 4.32). Each plate is slightly wedge-shaped (Fig. 4.32(*a*)) to avoid inner reflecting rays to interfere.

The inner surfaces P_1, P_2 of the plates facing each other are optically plane and coated with a fine film of aluminium which reflects about 75% of the incident light. In the experimental arrangement both the plates are fitted on a straight carriage marked linearly and fitted with a micrometer scale. One of the plates *B* facing the observer is fixed and provided with three screws (Fig. 4.32(*b*)) with which the inner reflecting surface can be made exactly parallel to the facing (inner) surface of plate *A*. The plate *A* can be moved along the carriage perpendicularly in its plane so that plate *A* is always parallel to path *B* and the thickness of air film between the two plates can be varied at will.

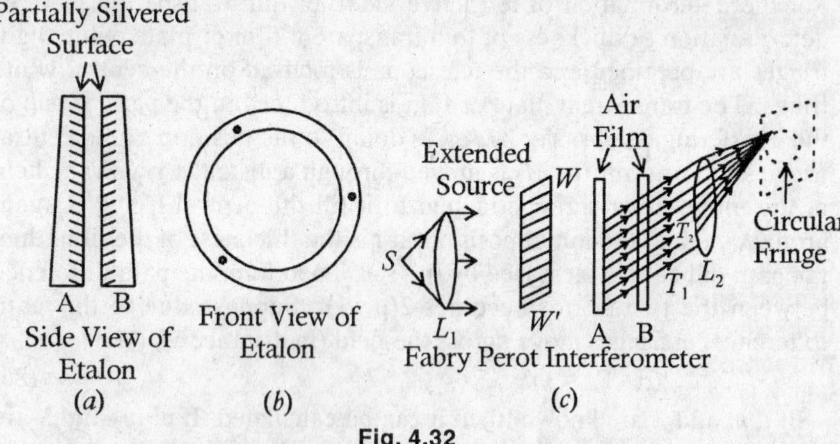

Fig. 4.32

Light from a monochromatic extended source *S* (Sodium Lamp) is rendered parallel by collimating lens L_1 (Fig. 4.32*c*). The parallel light beam suffers multiple reflection in the film enclosed between the two plates

A and B. The transmitted light T_1 T_2 interfere and results into circular fringes (Fig. 4.32(*c*)) of equal thickness called Haidinger fringes, which are formed in the focal plane of the telescope L and can be viewed directly.

Theory of Intensity of Fringes

P_1, P_2 (Fig. 4.33) represent the plane of the two inner reflecting surfaces of the plates A and B. Light from the extended monochromatic source is incident over the plate P_1 at all angles. One such incident ray is shown in Fig. 4.33. Let the angle of incident be θ. Here we have shown ray diagram for convenience. But actually it is the plane wavefront which is incident and undergoes both reflection and (transmission) thereby its amplitude goes on decreasing successively, which is not the case in Michelson interferometer.

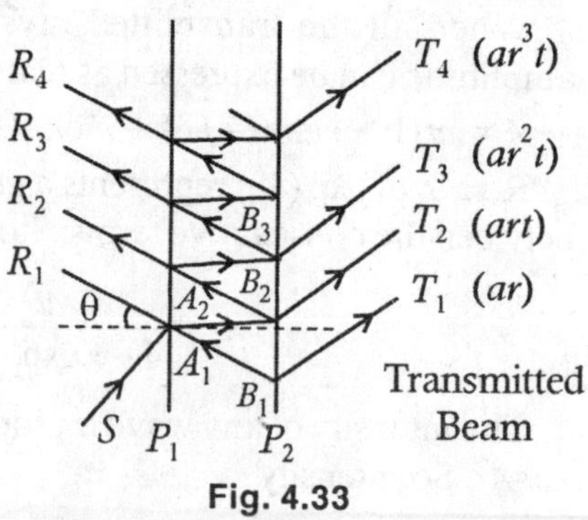

Fig. 4.33

Let SA_1 be the path of a plane wavefront of amplitude '*a*' incident at angle θ. As a result of the multiple reflections, consequent transmissions occur at the two boundary surfaces P_1 and P_2, and we get a set of parallel reflected waves (beam) $A_1R_1, A_2R_2, A_3R_3, \ldots$ and a set of parallel transmitted waves (beam) B_1T_1, B_2T_2, B_3T_3, ..., etc.

Suppose the reflection coefficient of each surface be '*r*' and similarly the transmission coefficient of each surface be '*t*'. Obviously the fractions of amplitudes reflected and transmitted are $\sqrt{r}$ and $\sqrt{t}$ respectively. The amplitude of wave along A_1B_1 is $(a\sqrt{t_1})$ and similarly amplitude of wave moving along B_1T_1 is $a\sqrt{t}\sqrt{t} = at$. The amplitude of wave B_1A_2 is $a\,t\sqrt{r}$ and amplitude of wave A_2B_2 is 'atr'. In a similar way we can calculate the amplitudes of corresponding reflected and transmitted waves as has been shown in Fig. 4.33.

We shall consider the interference among the transmitted beam B_1T_1, B_2T_2, B_3T_3, ..., etc. They are capable of interfering since they have been obtained from same incident beam SA_1, but their respective amplitudes are decreasing successively. Neglecting any phase change due to reflection from silvered surfaces, there is a constant phase difference between any two consecutive transmitted rays, due to path difference '*d*' between them, which is $2\mu d \cos\theta$

Phase difference due to this path difference is

$$S = \frac{2\pi}{\lambda}[2\mu d \cos\theta] = \frac{4\pi}{\lambda}[d \cos\theta] \qquad (87)$$

(since for air $\mu = 1$)

If the phase of first transmitted ray (wave) T_1 is assumed to be zero then it represents the real part of $at \exp(j\omega t)$ and the successive transmitted rays (waves) are represented by the real part of $art\exp(j\omega t - \delta)$, $ar^2t\exp(j\omega t - 2\delta)$ respectively.

When all the transmitted rays (waves) are focused their resultant amplitude can be expressed as real part of

$$A = at[1 + r\exp(-j\delta) + r^2\exp(-2j\delta) + r^3\exp(-3j\delta) + \ldots] \tag{88}$$

R.H.S. of Eq. (88) represents a geometric series having ratio $r[\exp(j\delta)]$ between the consecutive terms. Sum of the terms is real part of amplitude

$$A = \left[\frac{at}{1 - r\exp(-j\delta)}\right] \tag{89}$$

The intensity of any wave is proportional to the square of its amplitude, *i.e.* A^2. So intensity

$$I \propto \frac{a^2t^2}{[1 - r\exp(-j\delta)][1 - r\exp(j\delta)]}$$

or $$I = \frac{ka^2t^2}{1 - r[\exp(-j\delta) + \exp(-j\delta)] + r^2} \quad \text{where } k \text{ is constant}$$

or $$I = \frac{k^2a^2t^2}{1 - 2r\cos\delta + r^2} \tag{90}$$

The above expression can be further expressed as

$$I = \frac{ka^2r^2}{(1-r)^2}\left\{\frac{1}{1 + \frac{4r}{(1-r)^2}\sin^2\left(\frac{\delta}{2}\right)}\right\} \tag{91}$$

If we assume that there is no absorption of energy in the plate then $1 - r = t$ and Eq. (91) can be expressed as

$$I = \frac{ka^2(1-r)^2}{(1-r)^2}\left\{\frac{1}{1 + \frac{4r}{(1-r)^2}\sin^2\left(\frac{\delta}{2}\right)}\right\} \tag{92}$$

Equation (92) represents the expression for intensity in general. From this expression we can get expression for I_{max} and I_{min}

Maximum intensity I_{max} is obtained when

$$\cos\delta = 1$$

or $$\delta = 2n\pi \quad \text{when } n = 0, 1, 2, \ldots$$

So $$I_{max} = I_0 = \frac{ka^2}{(1-r)^2}(1-r)^2 = ka^2 \tag{93}$$

and minimum intensity is obtained when

$$\cos\delta = -1$$

$$\delta = (2n+1) \qquad \text{when } n = 0, 1, 2, \ldots$$

So $$I_{min} = \frac{ka^2(1-r)^2}{(1+r)^2} \tag{94}$$

Using Eq. (92), one obtains the general expression of intensity

$$I = I_{max}\left[\frac{1}{1 + F\sin^2\frac{\delta}{2}}\right] \quad \text{where } F = \frac{4r}{(1-r)^2} \tag{95}$$

and called *coefficient of fineness* or *coefficient of sharpness*

or $$\frac{I}{I_{max}} = \frac{1}{1 + F\sin^2\left(\frac{\delta}{2}\right)} \tag{96}$$

The conditions of maxima and minima in terms of path difference due to air film of thickness '*d*' are

(*i*) $2d\cos\theta = n\lambda$ (Maxima)

(*ii*) $2d\cos\theta = (2n-1)\frac{\lambda}{2}$ (Minima)

As '*d*' the separation between the two plates is constant for a given experimental set-up, so for particular value of *n* and λ, the angle 'θ' must be constant. The locus of all points having same value of θ is circle [Fig. 4.32(*c*)]

Half Fringe Width and Sharpness of Fringes

Half fringe width is the width of fringe in terms of phase difference between points on either side of maxima where intensity falls to half its maximum value. So from Eq. (96), we have

$$\frac{I}{I_{max}} = \frac{1}{1 + F\sin^2\frac{\delta}{2}} = \frac{1}{2}$$

or $$1 + F\sin^2\left(\frac{\delta}{2}\right) = 2$$

or $$F\sin^2\left(\frac{\delta}{2}\right) = 1$$

or $$\sin\left(\frac{\delta}{2}\right) = \sqrt{\frac{1}{F}} = \frac{1}{\sqrt{\{4r/(1-r)^2\}}} = \frac{1-r}{2\sqrt{r}}$$

$\therefore$ $$\delta = 2\sin^{-1}\left[\frac{1-r}{2\sqrt{r}}\right] \qquad (97)$$

If δ is small we have $\sin\left(\frac{\delta}{2}\right) = \frac{\delta}{2}$ (Radians)

Equation (97) can be written as

$$\delta = \frac{1-r}{\sqrt{r}} \qquad (98)$$

Sharpness of maxima depends upon half fringe width. Smaller is half fringe width sharper is fringe.

From Eq. (97 or 98) it is clear that half fringe width decreases as the value of reflection coefficient 'r' increases, and consequently maximas are sharper.

In general here the maxima are sharper (Fig. 4.34) than those obtained in case of Michelson interferometer. Fig. 4.34 (*a*) represents the half width and Fig. 4.34 (*b*) represents the variation of I/I_{max} with δ for different values of r.

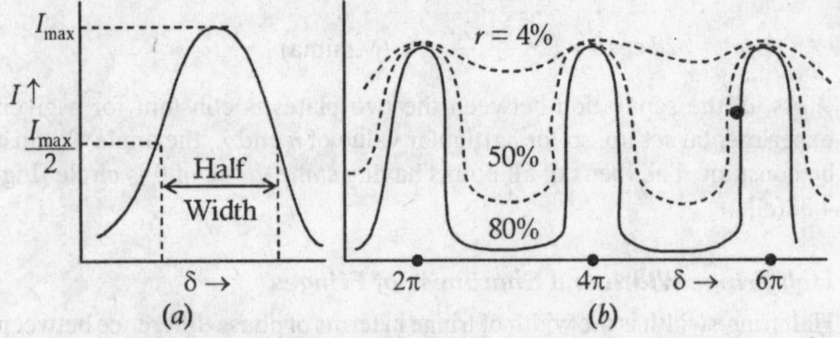

Fig. 4.34: Intensity curves for different values of reflection coefficient

Measurement of Wavelength of a Monochromatic Source of Light

The Fabry Perot etalons are adjusted so that circular fringes are formal in the focal plane or telescope or eye (Fig. 4.32*c*). Let 'n' the order of bright fringe at the centre of fringe system. At the centre the angle $\theta = 0°$. So path difference between two consecutive transmitted beams is

$$2t = n\lambda$$

When the movable plate A is moved parallel to B through a distance $\frac{\lambda}{2}$

on the carriage on which it is mounted, the path difference '2*t*' changes by λ and hence we observe next order bright fringe at the centre of the system.

If the movable plate moves from position x_1 to x_2 on the Carriage Scale and *N* be the number of fringes appearing at the centre during this movement of plate, then

$$N\frac{\lambda}{2} = x_2 - x_1$$

$$\therefore \qquad \lambda = \frac{2(x_2 - x_1)}{N} \qquad (99)$$

Hence we can determine the unknown wavelength of light from Eq. (99).

Measurement of Difference in Wavelength (of two close Spectral Lines)

The difference in wavelength of two close spectral lines (sources) can be determined by the method of coincidence.

Let λ_1 and λ_2 be the wavelengths of two close spectral lines (D_1, D_2 in case of Sodium light). At first the movable plate *A* is brought in contact with plate *B*. In this situation the fringe system due to two sources (wavelength) λ_1 and λ_2 will partially coincide. By moving the plate *A* and parallel to *B*, the path difference '*d*' increases, so that fringe system due to λ_1 and λ_2 will separate out and maximum discordance occurs when the rings system due to λ_2 are half way between those due to λ_1. Let for maximum discordance the separation between the plates be d_1. At the centre of system $\theta = 0°$ and $\cos\theta = 1$ so

$$2d_1 = n_1\lambda_1 = \left(n_1 + \frac{1}{2}\right)\lambda_2 \text{ (here assume } \lambda_1 > \lambda_2\text{)} \qquad (100)$$

or $$n_1\lambda_1 = n_1\lambda_2 + \frac{\lambda_2}{2}$$

or $$n_1(\lambda_1 - \lambda_2) = \frac{\lambda_2}{2}$$

or $$n_1 = \frac{\lambda_2}{2(\lambda_1 - \lambda_2)}$$

Substituting for n_1 in Eq. (100), we get

$$2d_1 = \frac{\lambda_2(\lambda_1)}{2(\lambda_1 - \lambda_2)}$$

$$\therefore \qquad \lambda_1 - \lambda_2 = \frac{\lambda_1 \lambda_2}{4d_1} \qquad (101)$$

Since the two spectral lines are closely situated, so we have

$$\lambda_1 \lambda_2 = \lambda^2$$

$$\therefore \qquad \lambda_1 - \lambda_2 = \frac{\lambda^2}{4d_1} \qquad (101a)$$

If we further increase the separation between the two plates, then the rings system again coincides and then separate out and second maximum discordance appears.

Let the separation between plates be d_2.

$$2d_2 = n_2\lambda_1 = \left(n + \frac{3}{2}\right)\lambda_2 \qquad (102)$$

From Eqs. (101) and (102), one obtains

$$2(d_2 - d_1) = (n_2 - n_1)\lambda_1 = (n_2 - n_1)\lambda_2 + \lambda_1$$

or $(n_2 - n_1)(\lambda_1 - \lambda_2) = \lambda^2$

$$\text{or} \qquad (n_2 - n_1) = \frac{\lambda_2}{\lambda_1 - \lambda_2} \qquad (103)$$

Substituting the value of $(n_2 - n_1)$ from Eq. 103 in Eq. (102), we get

$$2(d_2 - d_1) = \left[\frac{\lambda_2}{\lambda_1 - \lambda_2}\right]\lambda_1$$

$$\text{or} \qquad \lambda_1 - \lambda_2 = \frac{\lambda_1\lambda_2}{2(d_2 - d_1)}$$

$$\therefore \qquad \Delta\lambda = \lambda_1 - \lambda_2 = \frac{\lambda^2}{2(d_2 - d_1)} \qquad (104)$$

Here $\lambda^2 = \lambda_1\lambda_2$ and λ is the geometric mean of two wavelengths.

Hence by noting the separation d_1 and d_2 on the Carriage Scale the difference in wavelength $\Delta\lambda$ can be calculated from Eq. (104).

Chromatic Resolving Power

As explained above the fringes obtained in case of Fabry-Perot interferometer are sharper than those obtained in case of Michelson interferometer on account of this characteristic of fine structure of a spectral line and its half width can be determined more accurately.

The chromatic resolving power $(\lambda/\Delta\lambda)$ of an instrument depends on the fact, that how much the separation $(\Delta\lambda)$ between two consecutive spectral lines, could be determined by knowing the mean wavelength 'λ' of the two lines ($\lambda_1\lambda_2 = \lambda^2$).

According to Fabry-Perot Interferometer

$$\frac{\lambda}{\Delta\lambda} = (-)\frac{m}{\Delta m_0}$$

and this change depends on the sharpness (intensity) of the fringe pattern which is given by

$$I = \frac{I_{max}}{1 + \left[\frac{4r}{(1-r)^2}\right]\sin^2\left(\frac{\delta}{2}\right)} \quad (105)$$

Here I_{max} represents the maximum intensity of the central maxima and I represents the intensity of any other maxima separated from the central maxima by δ.

Rayleigh's criterion for separation of two spectral lines is that the maxima of one should coincide with the minima of the other line as shown in Fig. 4.35.

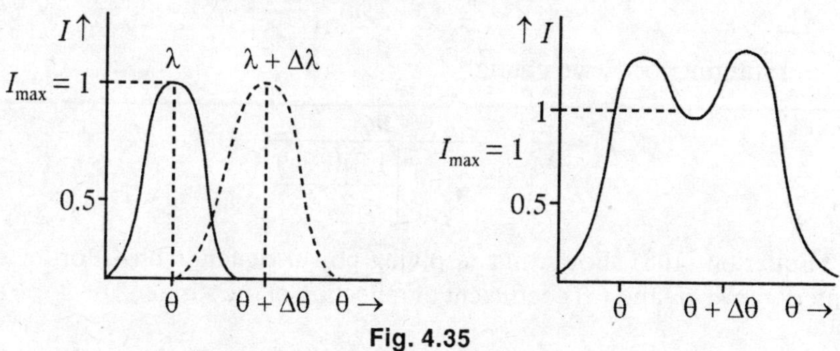

Fig. 4.35

In the above situation the maximum intensity of the envelope in between the two maximas (Fig. 4.35) should be $\frac{8}{\pi^2} I_{max}(\approx 0.8I_{max})$. It means contribution to the intensity of this part of the envelope due to separate spectral line be $\frac{4}{\pi^2} I_{max}(\approx 0.4I_{max})$.

According to the above definition, Eq. (105) can be expressed as

$$\frac{I}{I_{max}} = \frac{4}{\pi^2} = 0.405 = \frac{1}{1 + \frac{4r}{(1-r)^2}\sin^2\left(\frac{\delta}{2}\right)}$$

or $$0.405 + 0.405\left[\frac{4r}{(1-r)^2}\sin^2\left(\frac{\delta}{2}\right)\right] = 1$$

$$\sin^2\frac{\delta}{2} = \frac{0.595}{0.405\left[\frac{4r}{(1-r)^2}\right]} = \frac{1.469}{F} \qquad \text{where } F = \frac{4r}{(1-r)^2}$$

or

or

$$\delta = 2\sin^{-1}\left[\sqrt{\frac{1.469}{F}}\right] \tag{106}$$

or

$$2\delta = 4\sin^{-1}\left[\sqrt{\frac{1.469}{F}}\right] \tag{106a}$$

Hence the separation between two consecutive maxima which can be detected should be

$$\Delta m_0 = \frac{4\sin^{-1}\left[\sqrt{\frac{1.469}{F}}\right]}{2\pi}$$

Now by definition

$$\text{Resolving power } \frac{-\lambda}{\Delta\lambda} = (-)\frac{m_0}{\Delta m_0} = (-)\frac{m_0\pi}{2\sin^{-1}\left[\frac{1.21}{\sqrt{F}}\right]} \tag{107}$$

Substituting for F, we obtain

$$\text{R. P.} = \frac{\lambda}{\Delta\lambda} = -\frac{m_0\pi}{2\sin^{-1}\left[\frac{1.21(1-r)}{2\sqrt{r}}\right]} \tag{108}$$

Equation (108) shows that resolving power depends on (*i*) order of interference m_0 and (*ii*) coefficient of reflection of the surface.

ILLUSTRATIVE EXAMPLES

Example 1

The coherence length for sodium D_1 line is 2.5 cm. Determine (*a*) the spectral width of the line, $\Delta\lambda$, (*b*) the purity factor, Q, (*c*) the coherence time τ. Take wavelength of light 6000 Å.

Solution

(*a*) $$\Delta\lambda \approx \frac{\lambda}{2L} \approx \frac{36\times10^{-10}}{5}\text{ cm} \approx 7\text{ Å}$$

(*b*) $$Q \approx \frac{\lambda}{\Delta\lambda} \approx \frac{6\times10^{-5}}{7\times10^{-8}} \approx 10^3$$

(*c*) $$\tau \approx \frac{L}{C} = \frac{2.5}{3\times10^{10}}\text{ sec} \approx 0.8\times10^{-10}\text{ sec.}$$

Example 2

In a Young's double slit experiment on interference, the distance between the slits is 1 mm; the distance between the slit and the screen is 1 m. The wavelength used is 5893 Å. Compare the intensity at a point distant 1 mm from the centre to that at the centre. Also find the minimum distance from the centre of a point where the intensity is half of that at the centre.

Solution

(*i*) Path difference $\Delta = \dfrac{x \cdot 2d}{D} = \dfrac{0.1 \times 0.1}{100} = 10^{-4}\ \text{cm} = 10^{-5}\ \text{m}$

(*ii*) Path difference $\delta = \Delta \dfrac{2x}{\lambda} = \dfrac{10^{-4} \times 2\pi}{5893 \times 10^{-8}} = 3.39\pi$ radians

(*iii*) $\dfrac{I}{I_0} = \cos^2 \dfrac{\delta}{2} = \cos^2(1.697\pi) = \cos^2(125.5°) = 0.3372$

(*iv*) When the intensity is half of the maximum, if δ is the phase difference, then we have

$$\cos^2 \frac{\delta}{2} = 0.5$$

or
$$\cos \frac{\delta}{2} = \sqrt{0.5} = 0.707$$

∴
$$\frac{\delta}{2} = \frac{\pi^o}{4} \quad \text{or} \quad \delta = \frac{\pi^o}{2}$$

Obviously path difference $\Delta = \delta \dfrac{\lambda}{2\pi} = \dfrac{\lambda}{4}$

(*v*) Distance of the point on the screen from the centre

$$x = \Delta \frac{D}{2d} = \frac{\lambda}{4} \frac{D}{2d} = \frac{5893 \times 10^{-8} \times 100}{4 \times 0.1}$$

$$= 1.47 \times 10\ \text{cm} = 1.47\ \text{m}.$$

Example 3

Two coherent sources whose intensity ratio is 100 : 1 produce interference fringes. Show that the ratio of maximum to minimum intensity in the fringe system is 121 : 81.

Solution

$$\frac{I_{max}}{I_{min}} = \frac{(a_1 + a_2)^2}{(a_1 - a_2)^2}$$

where a_1 and a_2 are amplitudes of two waves.

Given $\frac{I_1}{I_2} = \frac{100}{1}$, where I_1 and I_2 are intensities of two sources

$$\because \quad I_1 \propto a_1^2 \text{ and } I_2 \propto a_2^2$$

$$\therefore \quad \frac{a_1}{a_2} = \frac{10}{1} \text{ or } a_1 = 10a_2$$

$$\therefore \quad \frac{I_{max}}{I_{min}} = \frac{(10a_2 + a_2)^2}{(10a_2 - a_2)^2} = \frac{(11)^2}{(9)^2} = 121 : 81$$

Example 4

The incident faces of a glass biprism of μ = 1.5 makes angles of 2° with the flat face of the prism. The slit is at a distance of 10 cm from the biprism and is illuminated by light of wavelength 5890 Å. Calculate the band width of the fringes formed at a distance of 1 m from the biprism.

Solution

The distance $2d$ between the two virtual sources

$$2d = 2a(\mu - 1)\alpha$$

$$a = 10 \text{ cm}$$

$$\mu = 1.5$$

$$\alpha = 2° = \frac{\pi}{180} \times 2 \text{ radians}$$

$$= \pi/90 \text{ radians}$$

$$D = 100 + 10 = 110 \text{ cm}$$

$$\therefore \quad 2d = 2a \times 10(1.5 - 1) \times \frac{\pi}{90} \text{ cm}$$

$$= 0.349 \text{ cm}$$

Now, Fringe width $\beta = \frac{D\lambda}{2d}$

$$\therefore \quad \beta = \frac{110 \times 5890 \times 10^{-8}}{0.349} \text{ cm} = 0.019 \text{ cm}.$$

Example 5

In Fresnel's biprism experiment, the fringe width of 0.185 cm is observed at a distance of 1 m from the slit. The image of the coherent sources is then produced at the same distance from the slit by placing a convex lens at 30 cm from the slit. Two images are found to be separated by 0.7 cm. Calculate the wavelength of the light used.

Solution

Let U and V be the distances of the virtual sources and their images from the convex lens. Let $2d$ and d_1 be the separation of sources and their images respectively, then we have

$$\frac{U}{V} = \frac{2d}{d_1} \text{ or } 2d = \frac{U}{V} d_1 = \frac{30}{70} \times 0.7 = 0.3 \text{ cm}$$

$$\text{Fringe width } \beta = \frac{D\lambda}{2d}$$

$$\therefore \qquad \lambda = \beta \frac{2d}{D} = \frac{0.0185 \times 0.3}{100}$$

$$= 5550 \times 10^{-8} \text{ cm} = 5550 \text{ Å}.$$

Example 6

A thin slit illuminated by a monochromatic source of light is placed at a distance of 0.5 m from a biprism of $\mu = 1.5$. The distance between two consecutive bands formed on a screen placed at a distance of 1 m from the biprism is found to be 0.12 mm. If the wavelength of the light used be 5890 Å, find the magnitude of obtuse angle of biprism.

Solution

We have $$\beta = \frac{D\lambda}{2a(\mu - 1)x}$$

$$D = 50 + 100 = 150 \text{ cm} = 1.5 \text{ m}$$

$$\lambda = 5890 \times 10^{-8} \text{ cm}$$

$$a = 50 \text{ cm}$$

$$\mu = 1.5$$

$$b = 0.012 \text{ cm}$$

$$\therefore \qquad \alpha = \frac{D\lambda}{2a(\mu - 1)\beta}$$

$$\therefore \qquad \alpha = \frac{150 \times 5890 \times 10^{-8}}{2 \times 50 \times (1.5 - 1) \times 0.012} \text{ radians}$$

$$= \frac{150 \times 5890 \times 10^{-8}}{2 \times 50 \times 0.5 \times 0.012} \times \frac{180}{\pi} \text{ degree}$$

$$= 0.84^{\circ}.$$

Example 7

A Llyold's single mirror has a length 5 cm. It is kept with plane at a distance of 1 mm from a slit illuminated by a monochromatic source of wavelength 5890 Å. The distance between the slit and the nearest edge of mirror is 15 cm. A screen is kept at a distance of 1.2 m from the slit. Find the bandwidth and width of the interference pattern.

Solution

(*i*) $$\beta = \frac{D}{2d}\lambda$$

$$2d = 2 \text{ mm} = 0.2 \text{ cm}$$

$$\lambda = 5890 \times 10^{-8} \text{ cm}$$

$$D = 120 \text{ cm}$$

$$\therefore \quad \beta = \frac{120}{0.2} \times 5890 \times 10^{-8} \text{ cm}$$

$$= 3.5 \times 10^{-2} \text{ cm}$$

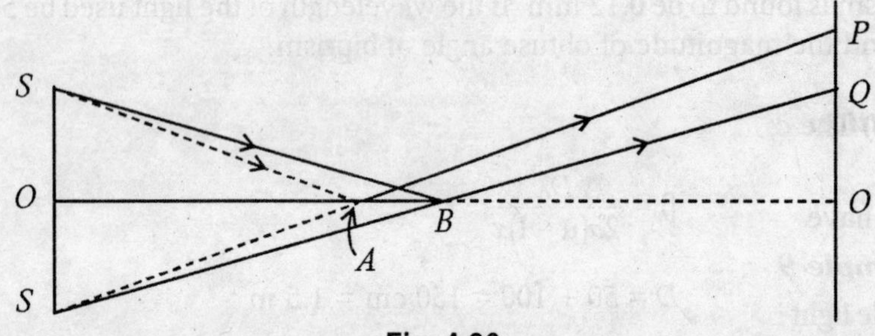

Fig. 4.36

(*ii*) In the Fig. 4.36, we have

$$\Delta OSA \;|||\; \Delta CPA \qquad \therefore \frac{PC}{SO} = \frac{AC}{OA}$$

i.e. $$PC = SO \times \frac{AC}{OA} = 0.1 \times \frac{110}{10} = 1.1 \text{ cm}$$

Similarly from triangle $OSB \;|||\; \Delta CQB$

$$\therefore \quad \frac{OC}{SO} = \frac{BC}{BO} \qquad i.e.\; QC = SO \times \frac{BC}{BO}$$

$$= 0.1 \times \frac{105}{15} = 0.7 \text{ cm}$$

Obviously, the width of the interference pattern

$$QP = PC - QC = 1.1 - 0.7 = 0.4 \text{ cm.}$$

Example 8

The angle between the two planes of Fresnel's double mirror is 20′. The point of intersection of the mirrors is at a distance of 0.3. from the slit illuminated by a monochromatic source of light of wavelength 5460 Å. The eyepiece is mounted at a distance of 120 cm from this point of intersection. Calculate the bandwidth and position of the 10th dark band from the centre.

Solution

(*i*) $$\beta = \frac{D}{2d}\lambda$$

$$\theta = 20' = \frac{\pi}{540} \text{ radians}$$

$$a = 30 \text{ cm}$$

$$\lambda = 5460 \times 10^{-8} \text{ cm}$$

$$D = 30 + 120 = 150 \text{ cm}$$

$$\therefore \quad \beta = \frac{D\lambda}{2a\theta} = \frac{150 \times 5460 \times 10^{-8}}{2 \times 30 \times \pi/540}$$

$$= 0.0235 \text{ cm.}$$

(*ii*) The distance of the 10th dark band from the centre

$$= 9.5\beta = 9.5 \times 0.0235 \text{ cm} = 0.223 \text{ cm.}$$

Example 9

While light is incident on a soap film at an angle $\sin^{-1}(4/5)$ and the reflected light on examination by a spectroscope shows dark bands. Two consecutive dark bands correspond to wavelengths 6.1×10^{-5} cm and 6×10^{-5} cm. If μ of the film be 4/3, calculate its thickness.

Solution

The path difference for dark bands in reflected light is given by

$$2\mu t \cos r = \mu\lambda \tag{1}$$

Let the order of the band width wavelength $\lambda_1 = 6.1 \times 10^{-7}$ m be n, then the order with wavelength $\lambda_2 = 6 \times 10^{-7}$ m will be $(n+1)$.

Thus one can write (*i*) for the dark bands of two wavelengths as below

$$2\mu t \cos r = n\lambda_1 \tag{2}$$

and $$2\mu t \cos r = (n+1)\lambda_2 \tag{3}$$

From (2) and (3), one obtains

$$n\lambda_1 = (n+1)\lambda_2$$

or
$$n = \frac{\lambda_2}{\lambda_2 - \lambda_1} \tag{4}$$

From (2) and (4), we obtain

$$2\mu t \cos r = \frac{\lambda_1 \lambda_2}{\lambda_2 - \lambda_1}$$

or
$$t = \frac{\lambda_1 \lambda_2}{\lambda_2 - \lambda_1} \frac{1}{2\mu t \cos r} \tag{5}$$

From Snell's law, we have

$$\mu = \frac{\sin i}{\sin r} \quad \text{or} \quad \sin r = \frac{\sin i}{\mu}$$

$$\therefore \qquad \cos^2 r = 1 - \sin^2 r = 1 - \frac{\sin^2 i}{\mu^2} = \frac{\mu^2 \sin^2 i}{\mu^2}$$

or
$$\mu \cos r = \sqrt{\mu^2 - \sin^2 i} \tag{6}$$

Substituting the value of $\mu \cos r$ in (5), one obtains

$$t = \frac{\lambda_1 \lambda_2}{\lambda_1 - \lambda_2} \frac{1}{2\sqrt{\mu^2 - \sin^2 i}} \tag{7}$$

$$\lambda_1 = 6.1 \times 10^{-7} \text{ m}, \lambda_2 = 6 \times 10^{-7} \text{ m}, \mu = 4/3 \text{ and } \sin i = 4/5$$

$$\therefore \qquad t = \frac{6.1 \times 10^{-7} \times 6 \times 10^{-7}}{(6.1 \times 10^{-7} - 6 \times 10^{-7})2\sqrt{\left(\frac{4}{3}\right)^2 - \left(\frac{4}{5}\right)^2}} \text{ m}$$

$$= \frac{6.1 \times 6 \times 10^{-7} \times 3 \times 5}{0.1 \times 2 \times 4 \times 4} \text{ m} = 171.5 \times 10^{-7} \text{ m}$$

Example 10

Intereference fringes are produced with monochromatic light falling on a wedge-shaped film of cellophane whose index of refraction is 1.4. The angle of the wedge is 40 seconds and the distance between the successive fringes is 1.25 mm. Calculate the wavelength of the light used.

Solution

The spacing between consecutive bands in a wedge viewed normally is given by

$$\Delta = \frac{\lambda}{2\mu\theta}$$

$$\Delta = 0.125 \text{ cm}$$

$$\mu = 1.4$$

$$\theta = 40'' = \left(\frac{10}{60 \times 60}\right)^0$$

$$= \frac{40}{3600} \times \frac{\pi}{180} \text{ radian}$$

or $$\lambda = \Delta 2\mu\theta$$

$\therefore$ $$\lambda = 0.125 \times 2 \times 1.4 \times \frac{40}{3600} \times \frac{\pi}{180}$$

$$= 0.125 \times 2 \times 1.4 \times \frac{1}{90} \times \frac{22}{7 \times 180} = 7 \times 10^{-5} \text{ cm}$$

Example 11

Two pieces of plane glass are placed together with a piece of paper between the two at one edge. Find the angle, in seconds, of the wedge-shaped air film between the plates, if on viewing the film normally with monochromatic (blue) light of wavelength 4800 Å there are 18 band per cm.

Solution

Let space occupied by *n*th band be $(x_2 = x_1)$, then the condition can be expressed as

$$2\mu\theta(x_2 = x_1) = n\lambda$$

$$\lambda = 4800 \text{ Å}$$

$$n = 18$$

$$\mu = 1 \text{ (air)}$$

$$x_2 - x_1 = 1 \text{ cm}$$

$$\theta = ?$$

$\therefore$ $$2 \times 1 \times \theta \times 1 = 18 \times 4800 \times 10^{-8}$$

or $$\theta = 9 \times 4800 \times 10^{-8} \text{ radian}$$

$$= 9 \times 4800 \times 10^{-8} \times \frac{180}{\pi} \text{ degree}$$

$$= 9 \times 4.8 \times 10^{-5} \times \frac{180}{22} \times 7 \times 3600 \text{ second}$$

$$= 89.1 \text{ second}$$

Example 12

A vertical soap film of length 10 cm is viewed by reflected monochromatic sodium light of wavelength 5893 Å. Just before the film breaks there are (*i*) 12 black and 11 bright fringes, (*ii*) 12 black and 12 bright fringes. If the refractive index of a soap solution is 1.33, find the angle of wedge formed and thickness of the film at the base just before it breaks.

Solution

(*i*) The bandwidth = 10/11, as the length of 10 cm is divided into 11 spacings. The first and the last fringe being black and in between fringes bright.

$$\therefore \quad \beta = \frac{\lambda}{2\mu\theta}$$

$$\lambda = 5893 \times 10^{-8} \text{ cm}, \ \mu = 1.33$$

$$\beta = \frac{10}{11}$$

$$\therefore \quad \theta = \frac{\lambda}{2\mu\beta}$$

$$\therefore \quad \theta = \frac{5893 \times 10^{-8} \times 11}{2 \times 10 \times 1.33} = 2.44 \times 10^{-5} \text{ radian}$$

when the film breaks, obviously, the thickness of the top is zero.

$\therefore$ Thickness at the bottom $= x_0\theta = 10 \times 2.44 \times 10^{-5}$ cm

$$= 2.44 \times 10^{-4} \text{ cm}$$

(*ii*) Band width $\beta = \frac{10}{11.5}$ cm (∵ spacing = 11.5)

$$\therefore \text{ Angle of wedge } = \frac{\lambda}{2\beta\mu} = \frac{5893 \times 10^{-8} \times 11.5}{2 \times 10 \times 1.33}$$

$$= 2.55 \times 10^{-5} \text{ radian}$$

$\therefore$ Thickness of the film at the base

$$x_0\theta = 10 \times 2.55 \times 10^{-5} \text{ cm}$$

$$= 2.55 \times 10^{-4} \text{ cm.}$$

Example 13

Newton's rings are obtained by a source emitting λ_1 = 6000 Å and λ_2 = 4500 Å. If *n*th ring of λ_1 coincides with (*n* + 1)th ring of λ_2, then find the diameter of *n*th ring of λ_1 wavelength. The radius of curvature of the curved surface is 90 cm.

Solution

Let *D* be the diameter of *n*th ring for wavelengths λ_1 and λ_2, then we have

$$D^2 = 4n_1R = 4(n+1)\lambda_2R \quad (1)$$

(Here we have assumed that rings are formed in air for which μ = 1)

$$\lambda_1 = 6000 \times 10^{-8} \text{ cm}, \lambda_2 = 4500 \times 10^{-8} \text{ cm}$$

$\therefore$ $$4n \times 6 \times 10^{-5} R = 4(n+1) \times 4.5 \times 10^{-5} R$$

or $$6n = 4.5(n+1) \therefore n = 3$$

Now $$D = \sqrt{4n\lambda_1 R} = \sqrt{4 \times 3 \times 6 \times 10^{-5} \times 90} = 0.2545 \text{ cm}.$$

Example 14

A thin plano-convex lens of focal length 1.8 m and of refractive index 1.6 is used to obtain Newton's rings. The wavelength of the light used is 5890 Å. Calculate the radius of the 10th dark ring by (*i*) reflection and (*ii*) transmission.

Solution

We have for the lens $\frac{1}{f} = (\mu - 1)\left(\frac{1}{R_1} - \frac{1}{R_2}\right)$

$$f = 180 \text{ cm}, \quad \mu = 1.6$$

$$R = ?$$

$\therefore$ $$\frac{1}{180} = (1.6 - 1)\frac{1}{R}$$

or $$R = 0.6 \times 180 = 108 \text{ cm} = 1.08 \text{ m}$$

$\therefore$ Radius of 10th dark ring by **reflection**

$$D_n = \sqrt{4nR\lambda} = \sqrt{4 \times 10 \times 108 \times 5890 \times 10^{-8}} = 0.504 \text{ cm}$$

Radius of 10th dark ring by **transmission**

$$D'_n = \sqrt{\left(n - \frac{1}{2}\right)4R\lambda} = (9.5 \times 4 \times 108 \times 5890 \times 10^{-5})^{1/2} = 0.49 \text{ cm}$$

Example 15

A thin mica sheet of $\mu = 1.4$ is placed in one arm of Michelson's interferometer. If shift of 7 fringes occur for light of wavelength 5890 Å, calculate the thickness of mica sheet.

Solution

Let t be the thickness of the film. When it is put in the path of one of the interfering beams of Michelson's interferometer an additional path difference of $2(\mu - 1)t$ is introduced. If N be the number of fringes shifted, we have

$$2(\mu - 1)t = N\lambda$$

$$N = 7$$

$$\lambda = 5890 \times 10^{-8} \text{ cm}$$

$$\mu = 1.4$$

$$\therefore \quad t = \frac{N\lambda}{2(\mu - 1)}$$

$$\therefore \quad t = \frac{7 \times 5890 \times 10^{-8}}{2(1.4 - 1)} = 7.74 \times 10^{-4} \text{ cm}$$

Example 16

Michelson's interferometer is used with sodium discharge tube which gives light of wavelength 5890 Å. It is observed that interference pattern disappears and reappears periodically as one moves the movable mirror. Calculate the change in path difference between the two successive reappearances of interference pattern.

Solution

If d be the distance moved by movable mirror between two successive positions of maximum distances or indistinctness, we have

$$\lambda_1 - \lambda_2 = \frac{\lambda_1 \lambda_2}{2d}$$

$$\lambda_1 = 5890 \times 10^{-8} \text{ cm}$$

$$\lambda_2 = 5890 \times 10^{-8} \text{ cm}$$

or

$$d = \frac{\lambda_1 \lambda_2}{2(\lambda_1 - \lambda_2)}$$

or

$$d = \frac{5896 \times 5890 \times 10^{-16}}{2(5890 - 5890) \times 10^{-8}} = 0.028 \text{ cm}$$

Example 17

The path difference between the mirrors of a Michelson's interferometer is 0.6 cm. When the light of wavelength 6000 Å is used, calculate the angular radius of the 10th bright fringe in the circular pattern of fringes obtained.

Solution

The condition to obtain bright fringe at the centre is

$$2d = n_1 \lambda \tag{1}$$

where d is the distance between the two mirrors and n_1 is an integer. Along a direction inclined at an angle θ, a bright fringe is obtained if

$$2d\cos\theta = n_2 \lambda \tag{2}$$

where θ is the angular radius of the circular bright fringe obtained and n_2 is another integer. Subtracting (2) from (1), one obtains

$$2d(1-\cos\theta) = (n_1 - n_2)\lambda = n\lambda$$

$$n = 0.6 \text{ cm}$$

$$n = 10$$

$$\lambda = 6000 \times 10^{-8} \text{ cm}$$

$\therefore$ $$1-\cos\theta = \frac{n\lambda}{2d}$$

or $$1-\cos\theta = \frac{10\times 6\times 10^{-5}}{1.2} = 5\times 10^{-4}$$

$\therefore$ $$\cos\theta = 1 - 0.0005 = 0.9995$$

$\therefore$ $$\theta = 1.812$$

Example 18

A Fabry-Perot interferometer is used to determine the difference between the wavelengths of two closely separated spectral lines, whose average wavelength is 6400 Å. It is found that in a given point near the centre of fringe pattern of rings coincides and it again coincides when the separation is changed by 1.024 mm. Show that the difference between the wavelengths is 2 Å.

Solution

We have $$\Delta\lambda = \lambda_2 - \lambda_1 = \frac{\lambda^2}{2(t_2 - t_1)}$$

$$\lambda = 6400 \text{ Å}$$

$$= 6400 \times 10^{-10} \text{ m}$$

$$t_2 - t_1 = 1.024 \text{ mm}$$

$$= 1.024 \times 10^{-3} \text{ m}$$

$$= \frac{6400\times 6400\times 10^{-20}}{2\times 1.024\times 10^{-3}}$$

$$= 2 \text{ Å}$$

REVIEW QUESTIONS

1. Give the theory of interference fringes. What are the conditions to get sustained interferences pattern?
2. Illustrate the formation of interference fringes in Fresnel's biprism. Show, how would you see it to determine the wavelength of monochromatic light?

3. Calculate the displacement if fringes when a thin transparent plate is introduced in the path of one of the interfering beams of a biprism. Show how this method is used to find the thickness of mica sheet.
4. Describe Fresnel's biprism method of producing interference fringes and determining the wavelength of light.
5. Explain the formation of Newton's rings in reflected light. Prove that the diameters of the dark rings are proportional to the square root of the natural numbers.
6. How do the fringes obtained with Lloyd's mirror differ from those obtained with Fresnel's biprism ? Explain the formation of fringes with white light in case of Lloyd's mirror.
7. (*a*) Explain the colours of thin films. Why should the film be thin in order to see the colours?

   (*b*) Explain what you mean by

   (*i*) Fringes of equal thickness

   (*ii*) Fringes of equal inclination (Haidenger Fringes).
8. (*a*) Describe the formation of colours in thin films and show that with monochromatic light, the interference patterns of reflected and transmitted light are complementary.

   (*b*) With a suitable diagram explain why a broad source of white light is essential to observe colours of thin films. Prove that an excessively thin film appears black in reflected light.
9. Explain the formation of Newton's rings. What will happen if

   (*a*) White light us used?

   (*b*) The lens is lifted slowly from the flat plate?
10. Explain the colours when a thin film of transparent material is observed in reflected white light using Fresnel's biprism. Why are colours not observed in case of thick films?
11. (*a*) Describe with necessary theory, the experiment to determine the wavelength of sodium light using Fresnel's biprism. Why is the angle kept so small?

    (*b*) How can you use the biprism to find the thickness of a thin sheet of mica?
12. Discuss the formation of Newton's rings by (*i*) transmitted light (*ii*) reflected light. Derive the expression for the radius of *n*th dark ring in the case of reflected light.
13. Briefly explain how the Michelson's interferometer is used to determine the difference in wavelength of the two spectral lines (D_1, D_2) of sodium.
14. How do the fringes obtained with Lloyd's mirror differ from those obtained with Fresnel's biprism ? Explain the formation of fringes with white light in case of Lloyd's mirror.
15. Describe the construction with diagram and outline the theory of Michelson's interferometer. Discuss the nature of interference pattern produced.
16. Explain the phenomena of interference for transmitted beam from an air film placed between two plain plates. What are the factors which determine the sharpness of these rings?
17. A plain parallel thin film is illuminated by a parallel beam of white light. Explain with necessary theory, the colour exhibited by the film in reflected light.

18. Describe, with a near sketch, Michelson's interferometer and explain how it may be used to measure the wavelength of monochromatic light.

19. (*a*) Give the theory and the experimental arrangement of Lloyd's mirrors to measure the wavelength of monochromatic light.

(*b*) Distinguish between the fringes produced by Lloyd's mirror and Fresnel's biprism.

20. (*a*) Describe coherent source. Discuss why two independent sources of light of same wavelength cannot show interference?

(*b*) What is the role of compensating glass plate in Michelson's interferometer?

21. Describe an interference method for the measurement of radius of curvature of low power convex lens.

22. Describe and explain the formation of Newton's rings in reflected monochromatic light. How can these be used to determine the wavelength of light. Derive the formula used.

23. Discuss the formation and location of interference fringes in a thin wedged shaped film illumination by monochromatic light.

24. Describe the construction with diagram and outline the theory of Michelson interferometer. How may it be used to measure the wavelength of monochromatic light?

25. Distinguish between 'Temporal coherence' and 'spatial coherence'. Discuss why two independent sources of light of the same wavelength cannot produce interference fringes.

26. Explain the method of finding the wavelength of a monochromatic light with the help of a biprism. Give the theory of the method and adjustments.

27. Describe the construction and working of Michelson's interferometer. How would you use it to measure the wavelength of monochromatic light?

28. Give theory of the measurement of thickness of a thin film transparent sheet by Fresnel's biprism.

29. Newton's rings are formed by reflected light of wavelength λ using a plano-convex lens of radius of curvature R. Derive expression for radius of nth bright ring. If Newton's ring lens and plate do not touch at the centre (due to, say, dust particles) such that the lens is 'Δ' height above the plate at the centre, what is the radius of the ring in this case?

30. (*a*) How shall you measure thickness of thin silicon chip using Biprism experiment? Derive the formula used.

(*b*) White light is incident on a soap film at an angle $\sin^{-1}(4/5)$. In the reflected light two consecutive dark bands corresponding to wavelengths 6.1×10^{-5} cm and 6×10^{-5} cm are observed. If refractive index for the film be $\mu = 4/3$, what is thickness of the film?

(*c*) The coherent time for Na-light of wavelength 5890 Å is $\tau_c = 10^{10}$ s. What is the maximum thickness of film that could be measured using interference of Na-light? [Raj. BE, 2003]

31. (*a*) Draw labelled diagram of Michelson interferometer, how is it used to find the wavelength of light? What is the order of the central fringe?

(*b*) In a biprism experiment zero order and tenth order maxima are formed at 12.34 mm and 14.73 mm respectively on the screen, when its slit is illuminated with light of wavelength 6000 Å. Calculate the position of zero order and twentieth order fringes when illuminated with light of wavelength 5000 Å. Other arrangements remaining the same. [Raj. BE, 2003]

32. (*a*) Prove that the for thin films, the interference patterns in reflected and transmitted light are complementary.

(*b*) Explain why the excessively thin films seen by reflected light appear dark.

(*c*) Light containing two wavelengths λ_1 and λ_2 falls normally on a plano-convex lens of radius of curvature R resting on a glass plate. If the nth dark ring due to λ_1, coincides with the $(n + 1)$th dark ring due to λ_2, prove that the radius of the nth dark ring of λ_1, is $= \sqrt{\left(\frac{\lambda_1 \lambda_2 R}{\lambda_1 - \lambda_2}\right)}$. [Raj. BE, 2002]

33. (*a*) A double slit of separation 1.5 mm is illuminated by while light (4000 Å- 8000 Å). On a screen 120 cm away coloured interference pattern is formed. If a pinhole is made on this screen at a distance of 3.00 mm from the central white fringe, what wavelength will be absent in the transmitted light?

(*b*) Explain with diagram the formation of Newton rings. What will happen if white light is used?

(*c*) In what respects Haidinger's fringes in Michelson interferometer distinct in character from fringes in thin films (Newtons Rings). [Raj. BE, 2001]

34. (*a*) With schematic diagram explain the working of Michelson's interferometer. How shall you use it to measure wavelength separation between two closely spaced lines say D_1 and D_2 lines of sodium lamp?

(*b*) Evacuated tubes of 5 cm length are in the arms of Michelson's interferometer set to observe fringes of equal inclination using sodium light of wavelength 589.3 mm. If it is observed that ten finger pass a given point in the field of view as air is introduced into one of the tubes. Determine pressure inside the tube, give that refreactive index of air increases as pressure increases as per relation $\mu = 1 + 3 \times 10^{-4} P$ where P is pressure in atoms. [Raj. BE, 2001]

35. (*a*) Define special and temporal coherence.

(*b*) Explain the formation of the Newton's rings in reflected light. Prove that the diameter of dark rings are proportional to the square root of the natural numbers.

(*c*) In a biprism experiment the micrometer readings of zero order fringe and tenth order fringe are 2.37 mm and 3.55 mm respectively. Wavelength of light used is 5890 Å. Determine.

(*i*) The position of zeroth and 10th order fringes when λ changes to 7500 Å.

(*ii*) The position of zeroth order fringe when mica sheet of thickness 0.02 mm and $f = 1.5$ is introduced to cover half of the biprism. [Raj. BE, 2000]

36. How will you use Fabry-Perot interferometer to determine the wavelength difference between the two spectral lines of sodium. Why the fringes formed in the case of Fabry-Perot interferometer are sharper than those obtained in case of Michelson interferometer.

PROBLEMS

1. A Fresnel biprism has an angle of 1° and index of refraction 1.5. Interference fringes are formed on a screen 8 m away from the biprism by the slit situated at the distance of 0.3 m away from the biprism. Find the fringe width of the fringes produced for the light of wavelength 6000 Å.

2. A soap film ($\mu = 4/3$ and $t = 2.26 \times 10^{-5}$ cm) is illuminated by white light incident at an angle of 45°. The light reflected by it is examined by spectroscope in which in found a dark band corresponding to wavelength of 6×10^{-5} cm. Calculate the order of interference of dark band.

3. Calculate the thickness of the thinnest film ($\mu = 1.4$) in which interference of violet component ($\lambda = 4000$ Å) of incident light can take place by reflection.

4. A biprism has the refracting angle for each part of 1.2° and refractive index 1.54. The distance between the slit and the biprism is 10 cm and that between the biprism and the eyepiece is 90 cm. If the wavelength of light is 5890 Å, calculate (*i*) the bandwidth (*ii*) the distance of the 4th dark band from centre and (*iii*) the distance between 10th bright band on one side and 10th dark band on the other side of the centre. [**Ans.** (*i*) 0.026 cm (*ii*) 0.09 cm (*iii*) 0.5 cm]

5. A Lloyd's mirror has length of 5 cm. It is kept with its plane at a distance of 1 mm from a slit illuminated by a monochromatic source of wavelength 5890 Å. The distance between the slit and the nearest edge of the mirror is 10 cm. A screen is kept at a distance 1.2 m from the slit. Find the bandwidth obtained. Also calculate the width of the interference pattern obtained. [**Ans.** 0.4 cm]

6. A glass plate of thickness 1.2 micron is placed in the path of interfering beams in a biprism experiment. The wavelength of the source is 6000 Å. The central band is found to shift by one band width. Calculate the refreactive index of the glass.

7. A man, whose eyes are 15 mm above the oil film on water surface, observes greenish colour at a distance of 1 m from his feet. Calculate the probable thickness of the film. $\lambda_{\text{green}} = 5000$ Å, $\mu_{\text{oil}} = 1.4, \mu_w = 1.5$.

   [**Hint:** $2\mu t = (2n+1)\lambda/2$]

8. Two plane surfaces of glass in contact along one edge are separated at the opposite edge by a thin wire. 30 interference fringes are observed between the two edges when normally reflected sodium light is used. Find the diameter of the wire used. Wavelength 5893 Å. [**Ans.** 8×10^{-6} m]

9. In a Newton's ring experiment a parallel beam of light of wavelength 6560 Å and 5248 Å is incident normally on the arrangement. (*i*) It is found that the *n*th dark ring of the first coincides with $(n + 1)$th dark ring of the second, calculate *n*. (*ii*) If the radius of curvature of the convex surface is 0.8 m, calculate the diameter of this ring. [**Ans.** (*i*) $n = 4$ (*ii*) 0.2892 cm]

10. In Newton's ring experiment the diameter of the 5th ring is reduced to half of its value after introducing a liquid below the convex surface. Calculate the refreactive index of the liquid.

11. A Newton's ring arrangement is used with a source emitting $\lambda_1 = 6 \times 10^{-7}$ m and $\lambda_2 = 4.5 \times 10^{-7}$ m and it is found that the *n*th dark ring of λ_1 coincides with $(n + 1)$th dark ring of λ_2. If radius of curvature of curved surface be 0.9 m, find the diameter of the *n*th dark ring for λ_1.

12. The radii of two consecutive bright rings obtained in Newton's ring experiment by reflection are 4.8 mm and 5.2 mm. If the wavelength used is 5460 Å, calculate the radius of curvature of the convex surface used. [**Ans.** 183 m]

13. Newton's rings are obtained by reflection in the usual way. When a liquid is introduced in between the lens surface and plane glass plate the radius of the 10th dark ring reduces by 0.3 mm. If the radius of curvature of the surface is 1 m and the wavelength used is 5600 Å, calculate the refractive index of the liquid.

   [**Ans.** 1.3]

14. A plane beam of monochromatic light falls normally on a thin film of oil which covers a glass plate. The λ of source can be varied continuously. Complete destructive interference of reflected light observed for $\lambda = 5 \times 10^{-7}$ m and 7×10^{-7} m and for no wavelength in between. If μ of oil and glass are 1.3 and 1.5 respectively, find the thickness of the film.

15. In a Michelson's interferometer the path difference between the two mirrors is 4 mm and the wavelength of the light used is 6400 Å. Calculate the angular radius of he 5th bright fringe in the circular pattern of fringes obtained.

    [**Ans.** 1.69°]

16. In an experiment with Michelson's interferometer the positions of the mirror for maximum distinctness give the reading on the scale 0.5828 mm and 0.8773 mm. If the mean wavelength is 5893 Å for sodium (light), calculate the difference between the wavelengths of the two components. [**Ans.** 5.9 Å]

17. In the experiment to measure the refractive index of air using a Michelsons's interferometer a wavelength of 6000 Å is used. It is observed that the mirror has to be shifted by 100 fringes when the air inside the tube is completely removed. If the length of the tube is 20 cm, calculate the refractive index of air.

    [**Ans.** 1.00015]

18. A thin film of glass of refractive index 1.5 is inserted in one arm of Michelson's interferometer. Find the thickness of the film if a shift of 20 fringes is obtained. The wavelength is 5460 Å. [**Ans.** 1.09×10^{-5} m]

19. White light is incident at an angle of 30° to the normal on two parallel glass plates separated by an air film of 0.001 cm thickness and reflected light is examined by the spectroscope. Find the number of dark bands seen in the spectrum between the wavelength of 4×10^{-5} cm and 7×10^{-5} cm.

20. In moving the mirror of the Michelson's interferometer through a distance of 0.141 mm, 500 fringes are counted. Calculate the wavelength of light

21. Newton's rings are formed with reflected light of wavelength 5890 Å with a liquid between the plane and the curved surfaces. The diameter of 5th dark ring is 0.32 cm and radius of curvature of curved surface is 120 cm. Calculate the refractive index of the liquid. [Jodhpur BE, 97]

22. White light is incident on a soap film at an angle of $\sin^{-1}(4/5)$. In the reflected light two consecutive dark bands corresponding to wavelengths 6.1×10^{-5} cm and 6×10^{-5} cm are observed. If the μ of the film is 4/3, calculate its thickness.

23. In a biprism experiment the 5th order and 10th order maximum fall at micrometer readings 13.535 mm and 14.730 mm respectively when wavelength of 6000 Å is used for producing interference pattern. If wavelength is changed to 5000 Å, calculate the positions of zero order and 20th order fringes, other arrangements remaining same.

24. Interference fringes are produced by monochromatic light falling normally on a wedge shaped film of colophon whose refractive index is 1.4. If the angle of the wedge is 20 seconds of arc and the distance between successive fringes is 0.25 cms, calculate the wavelength of light.

25. The convex surface of a plano-convex glass lens with curvature radius $R = 40$ cm comes into contract with a glass plate. A certain ring observed in reflected light has radius r =2.5 mm. Watching the given ring, the lens was gradually

raised from the plate by a distance $\Delta h = 5.0\ \mu m$. What has the radius of that ring become equal to? [Raj. BE, 97]

26. In biprism experiment fringes are first observed with sodium light ($\lambda = 5893$ Å) and fringe width was measured as 347 mm. Then two thin transparent sheets A and B of thickness of 0.016 mm and 0.02 mm and refractive indices 1.65 and 1.45 respectively were introduced in two beams. Calculate the shift of fringe system. Is the shift toward A or B? [Raj. BE, 98]

27. Newton's rings are observed between a spherical surface of radius of curvature 120 cm and a plane plate. The diameters of 5th and 16th bright rings are 0.314 cm and 584 cm. Calculate the diameter of 25th and 37th bright rings and also the wavelength of light used. [BE, 98]

28. Biprism fringes are produced using a source of wavelength 5893 Å. The biprism is of refractive index 1.50 and refracting angles 1.04° and 1.23°. The distance of focal plane of eyepiece from the biprism is 56.1 cm and the distance form slit to biprism is 12.4 cm. Calculate (*i*) separation of the coherent slit images and (*ii*) the fringe width. [BE, 99]

29. Michelson interferometer is adjusted to form circular fringes using light of wavelength 5000 Å. When the difference of path lengths between the mirrors of interferometer is 2.5 mm the fringe pattern having bright fringe at the centre, and we say it the first bright fringe. (*i*) What is the angular radius of the 10th bright mirrors slowly, 60 fringes cross the centre. How much path length is changed? [BE, 2000]

SHORT ANSWER QUESTIONS

1. Can two electric bulbs with point like filament of the same material each 15 W and lying close to each other produce interference?

Ans. As the two electric bulbs are two independent sources, *i.e.* they are not coherent sources. Obviously, they cannot produce intereference.

2. What becomes of the energy of light waves in destructive interference?

Ans. The energy in the destructive interference is transferred to the region where intensity is further increased.

3. The phenomenon of interference can be observed only when the two sources have common characteristics? What are these characteristics?

Ans. The two sources be monochromatic, vibrate with same amplitude and must bear constant phase difference.

4. The central part in Newton's rings seen in reflected light appears dark why?

Ans. In Newton's rings experimental arrangement at the point of the contact between the lens and the glass plate the thickness of the air film is zero. Therefore, there is no path difference between the interfering rays due to difference in path lengths. But one of the rays suffers a phase change of π on reflection at the surface of the glass plate, *i.e.* denser medium. This is why the rays suffer destructive interference and the centre appears dark.

5. How is the interference pattern in Young's double slit experiment affected if one of the slits, say S, is covered with black opaque paper?

Ans. There is a bright central slit but no interference pattern is obtained.

OBJECTIVE QUESTIONS

1. The two sources of light are said to be -----------, if they emit continuously the waves either in the same phase or with a constant phase difference.
[coherent]
2. Colour of thin film is due to ---------- of light. [interference]
3. In Young's double slit experiment, the fringes produced by light of ---------- wavelength will be narrow as compared to those by ---------- wavelength.
[shorter, longer]
4. In the transmitted system the film will appear bright if $2\mu t \cos r$ is an ---------- multiple of $\lambda/2$ and it will appear ----------- if it is and add multiple of λ.
[even, dark]
5. In the case of a transverse wave an additional phase change of ---------- or a path difference of ---------- is introduced when reflection takes place at the surface of denser medium. [π, $\lambda/2$]
6. In Young's double slit experiment, the fringes become ---------- if one of the slits is covered with cellophane paper. [indistinct]

Polarisation

5.1 INTRODUCTION

The phenomenon of interference and diffraction prove that light is a form of wave motion but they do not reveal the character of this wave motion, *i.e.* whether it is (*i*) longitudinal or (*ii*) transverse. The new phenomenon called polarisation proved conclusively that light is a transverse wave motion, because ***polarisability*** is characteristic of transverse waves.

According to electromagnetic theory, light is an electromagnetic disturbance in which the electric and magnetic fields are in the transverse plane and the two fields vary continuously and rapidly with time. The propagation of an alternating electromagnetic field constitutes a phenomenon called electromagnetic waves. Electromagnetic waves are transverse since the electric and magnetic intensity vectors ***E*** and ***B*** of the wave fields are mutually perpendicular and lie in a plane perpendicular to the velocity vector ***v*** of wave propagation. Vectors ***v***, ***E*** and ***B*** form a right handed system (Fig. 5.1) where from the head of vector ***v*** the smallest angle of rotation from vector ***E*** to ***B*** is counter clockwise.

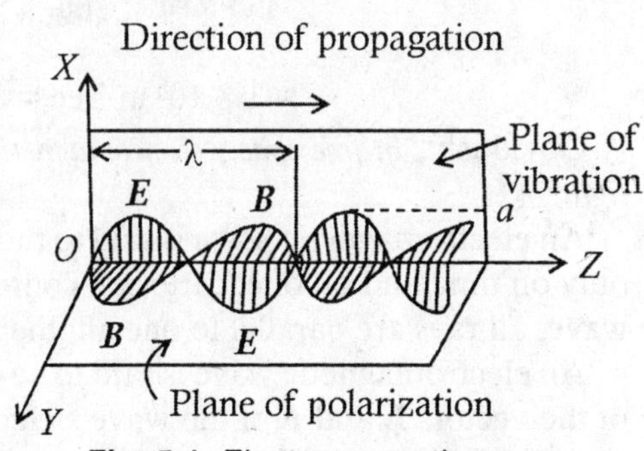

Fig. 5.1: Electromagnetic wave

A ray is a line whose tangent at each point coincides with the direction of wave propagation at that point, *i.e.* with the direction of energy transfer.

The relation between ***E*** and ***B*** in an electromagnetic wave propagated in nonconducting medium are determined by Maxwell's equations.

$$\nabla \cdot \boldsymbol{D} = \rho$$

$$\nabla \cdot \boldsymbol{B} = 0$$

$$\nabla \times \boldsymbol{E} = -\frac{\partial \boldsymbol{B}}{\partial t}$$

$$\nabla \times \boldsymbol{B} = \boldsymbol{J} + \frac{\partial \boldsymbol{D}}{\partial t} \tag{1}$$

Here ρ and ***j*** refer respectively to the charge density and current density respectively. ***D*** and ***E***, ***B*** and ***H*** and ***J*** and ***E*** are related with each other for an isotropic and homogeneous medium as

$$\boldsymbol{D} = \in \boldsymbol{E},\ \boldsymbol{B} = \mu \boldsymbol{H} \text{ and } \boldsymbol{J} = \sigma \boldsymbol{E}$$

Solving Maxwell's equations for a homogeneous isotropic non-conducting medium, one obtains

$$\nabla^2 \boldsymbol{B} = \mu \in \frac{\partial^2 \boldsymbol{E}}{\partial t^2}$$

and
$$\nabla^2 \boldsymbol{B} = \mu \in \frac{\partial^2 \boldsymbol{B}}{\partial t^2} \tag{2}$$

Eq. (2) represents a 3-dimensional vector wave equation where velocity is given by

$$v = \frac{1}{\sqrt{\mu \in}} \tag{3}$$

If the medium is supposed to be vacuum, then $\in = \in_0$ and $\mu = \mu_0$ and we have

$$v_0 = \frac{1}{\sqrt{\mu_0 \in_0}} = \frac{1}{\sqrt{4\pi \times 10^{-7} \times \left(4\pi \times 9 \times 10^9\right)^{-1}}}$$

$$= 3 \times 10^8 \text{ m/sec} = c \text{ (velocity of light)}$$

Obviously, *in free space, electromagnetic waves travel with the velocity of light.*

An electromagnetic wave is said to the plane if vectors ***E*** and ***B*** depend only on time and on one Cartesian coordinate, for instance *x*. In a plane wave, all rays are parallel to one another.

An electromagnetic wave is said to be *monochromatic* if the components of the vectors ***E*** and ***B*** of the wave's electromagnetic field oscillate with sinusoidal motion of the same frequency, which is called the *frequency of the wave.* A monochromatic wave is unlimited in space and time.

Experimentally, it has been found that it is the electric vector ***E*** (or electric field *E*), that produces the optical polarization. In a beam of ordinary light the vibrations are equally probable in all directions at right angles to the direction of propagation. There is thus perfect symmetry around the direction of propagation (Fig. 5.2). Such light waves are said to be ***unpolarised*** waves. If such perfect symmetry is lacking, ***i.e.*** asymmetry exists in any beam of light, the beam is said to be polarised. When we speak of 'vibrations' in the light we refer, then, to the variations of the electric vector ***E*** confined in planes transverse to the direction of propagation (such waves are said to be linearly polarised).

The plane in which the vibrations take place is called the ***vibration plane*** and the plane at right angles to the ***vibration plane*** is called the ***plane of polarisation*** (Fig. 5.3).

Figure 5.3 shows that light after passing through tourmaline crystal *T* has vibrations only in one plane. Such a light is called ***plane polarised light***. We shall discuss about elliptically and circularly polarised light later.

ABCD is called the *plane of vibration*. This contains the crystallographic axis and the vibrations of transmitted light lie in this plane.

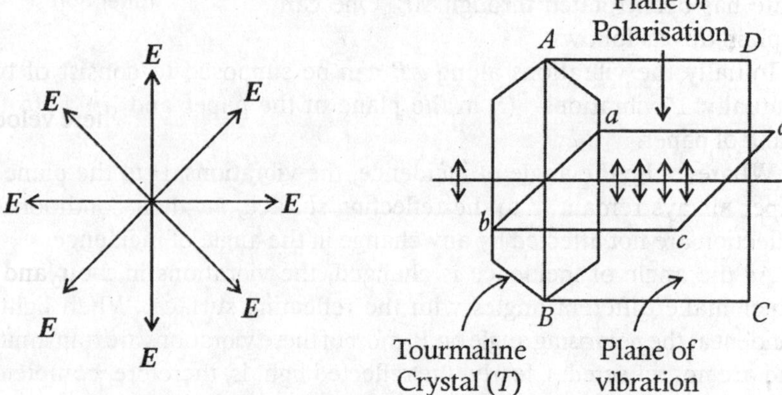

Fig. 5.2: Unpolarised light

Fig. 5.3: Plane of polarisation and plane of vibration

The plane *abcd* perpendicular to the plane of vibration in which no vibrations lie is called the *plane of polarisation.*

5.2 METHODS OF PRODUCING PLANE POLARISED LIGHT

There are following three important methods of producing plane polarised light:

(*i*) Polarisation by Reflection

Malus in 1808 discovered that when natural light is incident in air on the glass surface, the reflected beam is partially polarised. The transmitted beams are also partially polarised. The degree of polarisation of the reflected beam depends upon the angle of incidence and is maximum for a certain value called the *polarising angle* for the medium. The angle of maximum polarisation varies with wavelength. The angle of polarisation for glass of refractive index 1.51 is 57.5°. Complete polarisation is possible with monochromatic light and not with white light.

Let us consider two polished glass plates inclined equally to the horizon. A monochromatic beam of light *AB* incident at *B* is reflected along *BC* and after striking the second plate at *C* is reflected along *CD* at an angle of 57.5° (Fig. 5.4). If we now rotate the upper plate *C* about *BC* as an axis then we observe that the intensity of the reflected beam along *CD* gradually decreases and becomes zero when the plate has been rotated through 90°. One can explain this as follows:

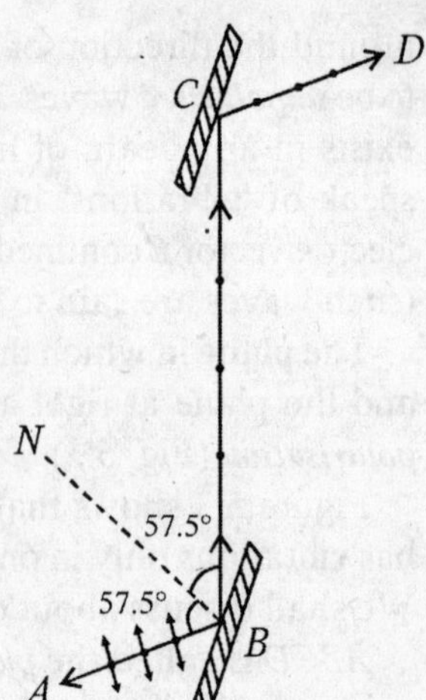

Fig. 5.4: Polarisation by reflection

Initially the vibrations along *AB* can be supposed to consist of two mutually $\perp^r$ vibrations—(*i*) in the plane of the paper and (*ii*) $\perp^r$ to the plane of paper.

Whatever be the angle of incidence, the vibrations $\perp^r$ to the plane of paper always remain $||^l$ to the reflecting surface, *i.e.* their conditions of reflection are not affected by any change in the angle of incidence.

As the angle of incidence is changed, the vibrations in the plane of paper make different angles with the reflecting surface. When light is incident at the polarising angle on *B*, most of these vibrations are transmitted and are not reflected. Clearly, the reflected light is, therefore, completely polarised in the plane of incidence. It is worthwhile to note that the transmitted light is never completely plane polarised as some of the vibrations $\perp^r$ to the plane of incidence are also transmitted.

Brewster's Law

Sir David Brewster, a Scottish physicist, has shown that ***when light is incident at the polarising angle, the reflected ray and the refracted ray are*** $\perp^r$ to each other. This result is known as ***Brewster's law***. If in this position the angle of incidence is *i* and the angle of refraction is *r* then by ***Snell's law***.

$$\mu = \frac{\sin i}{\sin r} \quad (1)$$

But according to Brewster's law

$$i + r = 90° \quad (2)$$

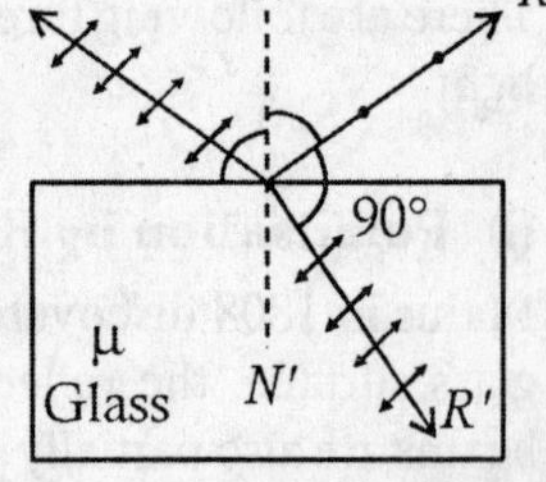

Fig. 5.5

or $r = (90 - i)$

$$\therefore \quad \sin r = \sin(90 - i) - \cos i \tag{3}$$

From Eqs. (1) and (3), we have

$$\mu = \frac{\sin i}{\cos i} = \tan i \tag{4}$$

(*ii*) Polarisation by Refraction—Piles of Plates

We have seen that with natural light incident at polarising angle on a glass plate, the reflected light is completely polarised, in the plane of incidence, *i.e.*, the reflected light consists of only $\perp^r$ vibrations. However, the intensity of the reflected beam is low because only 15% of $\perp^r$ vibrations are reflected and the remaining 85% are refracted. The reflected beam is not completely plane polarised but consists of 85% of the $\perp^r$ vibrations and all parallel vibrations.

However, by using a pile of glass plates, instead of a single glass plate, refracted beam by repeated reflections takes place.

The pile of plates consists of a number of thin glass plates placed one over the top of the other inside a wooden tube (Fig. 5.6). The plates are in contact with each other and are inclined at an angle of 32.5° (90 – 57.5°) to the axis of the tube. In this position the light would be incident at an angle of 57.5° which is the polarising angle for glass. Obviously, this arrangement will give two beams of plane polarised light—one by reflection consisting of vibrations $\perp^r$ to the plane of incidence and the other by refraction consisting of vibrations $||^l$ to the plane of incidence. Clearly, both the beams are plane-polarised but with their planes of polarisation at right angles to each other.

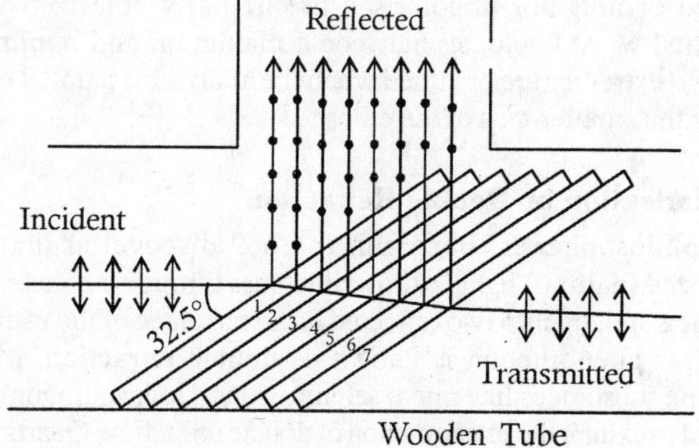

Fig. 5.6: Pile of plates

Degree of Polarisation

To obtain the degree of polarization or proportion of polarisation (p), let I_v and I_d denote the intensities of components with vibrations $||^l$ and $\perp^r$ to the plane of incidence respectively in the transmitted light, then

$$p = \frac{I_p - I_d}{I_p + I_d} \tag{5}$$

Note that $p = 1$ for linearly polarised waves and that $p = 0$ for unpolarised waves. Provostaye and Desains have shown that Eq. (5) finally reduces to

$$p = \frac{n}{n + \left(2\mu/1 - \mu^2\right)^2} \tag{6}$$

where μ is the refractive index of the material of plates and n is the number of plates.

5.3 LAWS OF MALUS

Malus found that the intensity of the transmitted beam varies with the angle between the plane of transmission and the plane of the polarizer. The ***Cosine Law of Malus*** can be stated as follows:

The intensity of light transmitted by the analyser is proportional to the square of the cosine of the angle between the planes of transmission of analyser and polarizer. We have the relation

$$I = a^2\cos\theta = I_0\cos^2\theta \tag{7}$$

where I_0 is the intensity to the incident wave and I that of the transmitted wave. When $\theta = 0$, or π, the intensity of transmitted light is maximum; when $\theta = \pi/2$ or $3\pi/2$, it is zero. Therefore, when the analyser is rotated, the intensity of the transmitted wave fluctuates between 0 and I_0. This, for example, affords a means of determining whether a wave, such as light, is polarised or not. For unpolarised or circularly polarized waves, the transmitted wave fluctuates between a maximum and minimum wave. These two extremes are obtained when the analyser is parallel either to the larger or the smaller axis of the ellipse.

(*ii*) Polarisation by Double Refraction

Dutch philosopher E. Bartholin in 1669 discovered that when an unpolarized (natural) light is allowed to pass through a calcite (or quartz) crystal, it is split up into two reflected beams in place of the usual one as in glass. This phenomenon is known as double refraction. Many other crystalline substances like mica, selenite, sugar, topaz, aragonite, ice, etc. are found to exhibit the phenomenon of double refraction. Quartz and calcite are particularly important as they are used extensively in the manufacture of special instruments.

To demonstrate the phenomenon of double refraction make an ink dot on a sheet of paper and view through a calcite crystal. We observe two images of the dot. Now rotate the crystal slowly as shown in Fig. 5.7(*a*), we notice that one of the two images remains stationary while the second image rotates around the first. The image which remains stationary is called as the ***ordinary image*** and the ***refracted ray*** which produces this image is called as *ordinary ray* denoted by *O*-ray. *O*-ray obeys the ordinary laws of refraction. The second image which rotates round the first is called as the ***extraordinary image*** and the ray which produces this image is called ***extraordinary ray*** and denoted by *E*-ray. *E*-ray does not obey the ordinary laws of refraction.

Let us consider a beam of light *AB* from a point source is incident at an angle *i* on a calcite crystal [Fig, 5.7 (*b*)]. It is split up into *O* and *E* rays inside the crystal. The vibrations in the ordinary ray are at right angles to the principal plane. Those in the extraordinary ray are at right angles to the principal plane. Obviously the *O*-ray is plane polarized in the principal plane and the extraordinary ray is plane polarised at right angles to the principal plane.

The refractive indices for *O*-ray and *E*-ray are respectively

$$\mu_O = \frac{\sin i}{\sin r_1} \quad \text{and} \quad \mu_E = \frac{\sin i}{\sin r_2} \tag{8}$$

where r_1 and r_2 are angles of refraction which *O*-ray and *E*-ray made inside the crystal respectively. For calcite crystal $r_1 < r_2$ and hence $\mu_O > \mu_E$, inside the calcite crystal the velocity of light for the *O*-ray is less then for the *E*-ray.

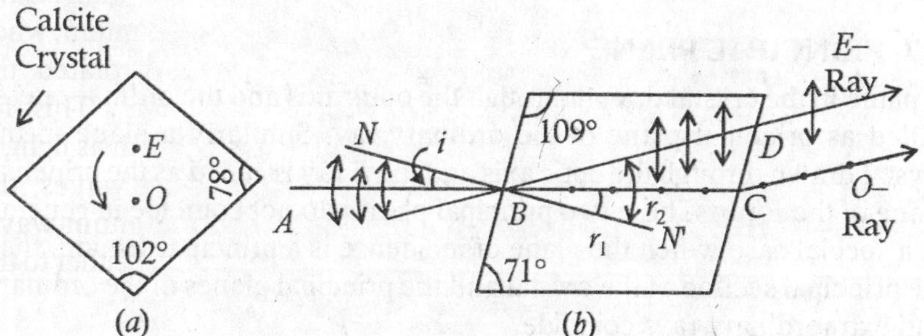

Fig. 5.7: Double refraction through calcite crystal

5.4 CALCITE CRYSTAL

Calcite (Iceland spar) is crystalised calcium carbonate ($CaCO_3$). It exists in nature in several forms but it cleaves very perfectly along three directions forming rhombohedra or rhombus. The six faces of the rhombohedra are parallelograms each having angles of 102° and 78° or more

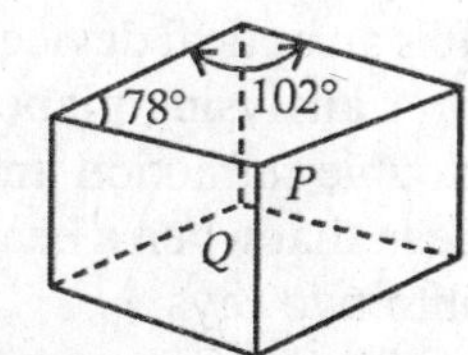

Fig. 5.8: Calcite crystal

accurately 101°53′ and 78° 3′ (Fig. 5.8). At the two opposite corners *P* and *Q* of the rhombus, three angles of 102° meet, whereas at the other angles of 72° meet. The corners where three obtuse angles meet are called as blunt corners.

5.5 OPTIC AXIS

A line passing through any one of the blunt corners and equally inclined with the three faces, which meet at this corner, represents the direction of the crystallographic axis of the crystal (Fig. 5.9) and any direction parallel to it is called ***optic axis***. In Fig. 5.9, *xy* represents the optic axis. Any line parallel to *xy* is also an optic axis. It may be emphasized here that (*i*) *optic axis is a direction and not a particular ray and* (*ii*) *if a ray of light falls along the optic axis or in a direction parallel to it, then it does not break up into O and E rays, i.e., there is no double refraction along optic axis.*

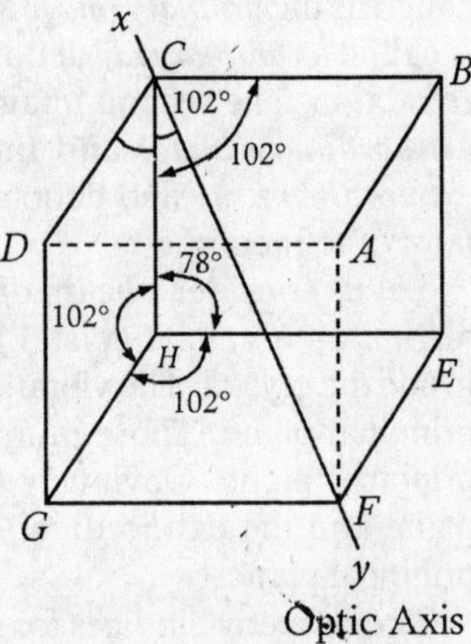

Fig. 5.9: Optic axis

5.6 PRINCIPAL SECTION

A plane which contains the optic axis and ⊥′ to the opposite faces of the crystal is called the ***principal section*** of the crystal. Since a crystal has six faces, therefore, for every point there are ***three principal sections***. A principal section of calcite crystal is a parallelogram having angles of 71′ and 109°.

5.7 PRINCIPAL PLANE

A plane in the crystal drawn through the optic axis and the ordinary ray is called as principal plane of the ordinary ray. Similarly a plane in the crystal drawn through the optic axis and the *E*-ray is called as the principal plane of the *E*-ray. These two principal planes do not coincide in general. In a special case, when the plane of incidence is a principal section, then the principal section of the crystal and the principal planes of the ordinary and extraordinary rays coincide.

5.8 NICOL PRISM

It is an optical device made from calcite and frequently used for producing and analysing plane-polarised light. It is based on the phenomenon of double refraction and was invented by William Nicol in 1828. We have seen that when a beam of light is transmitted through Calcite, it breaks up into two rays:

(*i*) the *O*-ray which has its vibrations ⊥′ to the principal section of the crystal and (*ii*) the *E*-ray which has its vibrations ||' to the principal section. The *O*-ray is cut off by total internal reflection leaving the *E*-ray alone.

Construction of Nicol

Nicol prisms are constructed from a calcite rhombohedra, the edges of end faces of which are equal but one-third the length of the other longer edges [Fig. 5.10 (*a*)]. Let *A'BCDEFG'H* represent such a crystal having *A'* and *G'* as its blunt corners where three obtuse angles of the three faces meet. *A'CG'E* is one of the principal sections of the crystal with *A'CG'* = 71°.

First the two short end faces *A'BCD* and *EFG'H* are grounded in such a way that the angle *ACG* becomes 68° instead of 71°. The rhomb is then cut along the plane *AKGL* [Fig. 5.10 (*a*)]. The two surfaces are grounded and polished optically flat and then cemented together in their original position with Canada balsam, which is a transparent cement with a refractive index about midway between the refractive index of *O*-ray and that of *E*-ray (Fig. 5.10 (*b*)). For sodium light of wavelength λ = 5893 Å, refractive index for Canada balsam is 1.55, μ_O = 1.658, μ_E =1.486.

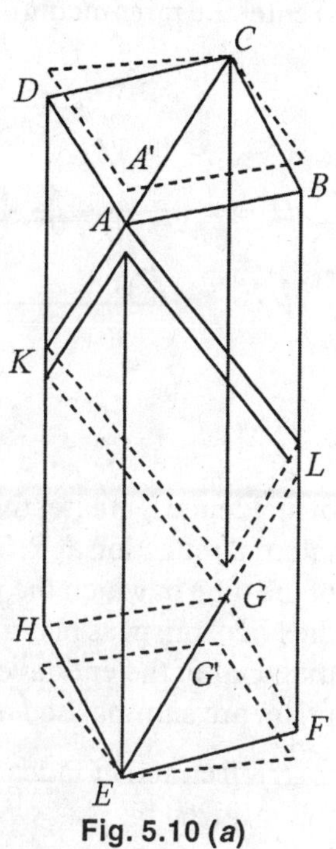

Fig. 5.10 (*a*)

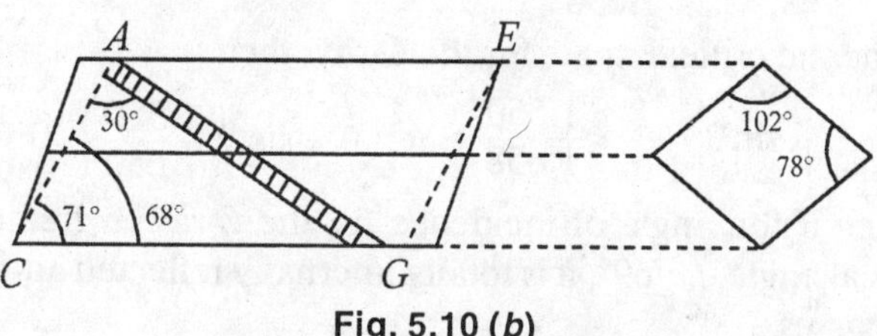

Fig. 5.10 (*b*)

The sides of Nicol prism are blackened to absorb any light incident on the sides.

Action of Nicol Prism

Figure 5.11 represents the section *ACGE* of the Nicol prism. The diagonal *AG* represents the Canada balsam layer. Canada balsam layer acts as a rarer medium for an *O*-ray and as a denser medium for the *E*-ray. When a beam of natural light enters one of the end faces in a direction parallel to the long side of the Nicol, it is doubly refracted and is decomposed into *O*-plane polarised beam and *E*-ordinary plane polarised beam. The *E*-ray, in passing from the calcite into Canada balsam, passes into a denser medium and emerges from the Nicol prism along $S'T'$. Its vibration will be ||^l to the shorter diagonal of the end face EG. The O-ray on the other hand tries to enter the rarer medium, *viz.* Canada balsam from calcite. Since the angle

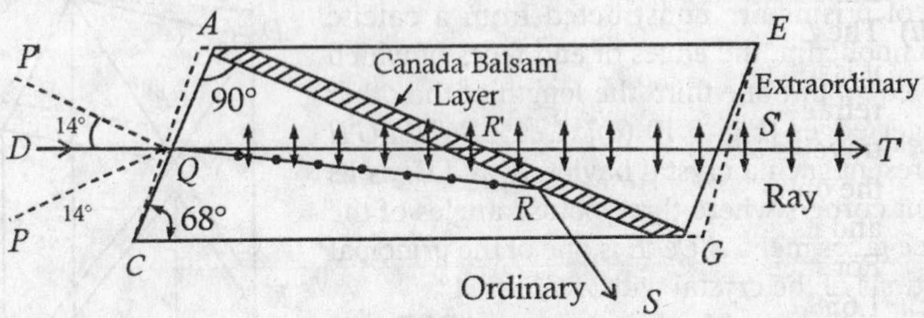

Fig. 5.11: Nicol prism

of incidence is greater than the critical angle and hence it experiences total internal reflection at R. The O-ray is absorbed by the black internal surface of the tube in which the prism is mounted. Obviously, the light that leaves the Nicol prism as plane polarised with the vibration plane ||^l to the shorter diagonal of the end face. The salient features of the working of the Nicol prism are summarised as follows:

(*i*) Refractive index (μ) for O-ray with respect to Canada balsam is given by

$$\mu = \frac{1.658}{1.550}$$

If θ be the critical angle for the O-ray, then

$$\sin\theta = \frac{1}{\mu} = \frac{1.550}{1.658} \quad \text{or} \quad \theta = \sin^{-1}\left(\frac{1.550}{1.650}\right) = 69^{\circ}$$

Hence if the angle of incidence for the O-ray is greater than the critical angle, *i.e.* 69°, it is totally internally reflected and only the E-ray passes.

(*ii*) If the angle of incidence is less then the critical angle for the O-ray then it will also be transmitted through the Nicol prism.

Limitations of Nicol Prism as a Polarising Device

(*i*) As a polarising device Nicol prism is only effective with a slightly convergent or slightly divergent beam of light. Nicol fails to be effective polarizer with too convergent or too divergent incident beam.

(*ii*) If the incident ray makes an angle much smaller than $\angle CQP$ with the face CA, the O-ray will strike the Canada balsam layer at an angle less than the critical angle (69°) and hence *O*-ray will also be transmitted and light emerging from Nicol will not be plane polarised. It is observed that the angle of incidence on the face *CA* is limited to ∠*PQP*" which is nearly 14° beyond which the *O*-ray is transmitted through the Nicol.

(*iii*) The *E*-ray also has a limit to the angle of incidence beyond which it is totally internally reflected by the Canada balsam surface. The refractive index for the *E*-ray, *i.e.* μ_E = 1.486 when the *E*-ray is travelling at right angles to the direction of the optic axis. Along the optic axis, the *E*-ray travels with the same velocity as the *O*-ray and its refractive index is same as that of *O*-ray, *i.e.* $\mu_E = \mu_o = 1.658$. For intermediate angles μ_E lies between the two limits 1.486 and 1.658. Obviously, for a particular angle μ_E may be greater then 1.55 and angle of incidence greater than the critical angle, then the *E*-ray will also be totally internally reflected at the Canada balsam surface. Clearly, no light will then emerge out of the Nicol.

(*iv*) Nicol prism cannot be used for highly convergent or divergent beams.

Uses of Nicols

(i) Two Parallel Nicols

When two Nicol prisms are lined up coaxially one behind the other, as shown in Fig. 5.12, one of them acts as a polarizer and the other acts as an analyser. The first Nicol P_1 which produces the plane polarised light is called the ***polarizer*** and the second Nicol P_2 which is used for examining the state of polarisation of transmitted light is called the ***analyser***. The combination of two such nicols is called a ***polariscope***. When the two nicols are placed with their principal sections ||^l to each other as shown in Fig. 5.12 (*a*), then the *E*-ray transmitted by P_1 is freely transmitted by P_2.

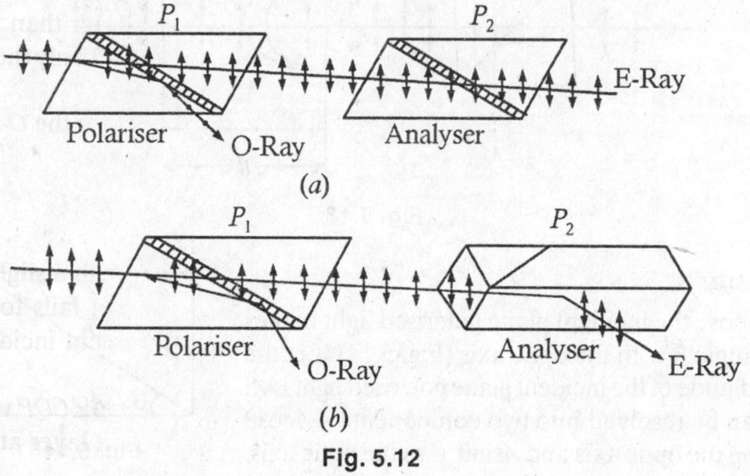

Fig. 5.12

If we rotate one of the two nicols, the intensity of the transmitted beam decreases till it is reduced to zero.

However, if one of the two nicols is further rotated so that it has turned through an angle of 180°, *E*-ray would be transmitted.

If I_0 be the intensity of transmitted beam when the principal sections of P_1 and P_2 are parallel and I be the intensity when the principal sections of P_1 and P_2 are inclined at an angle θ, then according to cosine law of Malus we have

$$I = I_0 \cos^2\theta \tag{9}$$

One can utilise the above facts for detecting the plane polarised light. If the given light on examination through a rotating Nicol (say P_2) exhibits a variation in intensity with minimum intensity zero, the given light is clearly plane polarised.

5.9 ELLIPTICALLY AND CIRCULARLY POLARISED LIGHT

Let monochromatic light be passed through Nicol prism N_1. The outcoming beam is plane polarised (Fig. 5.13). Now, this plane polarised light is allowed to fall normally on a uni-axial doubly refracting calcite or quartz plate *P*, whose faces have been cut ||^l to the optic axis. On entering the doubly refracting calcite crystal, the plane polarised light is split up into two components *O* and *E*. Both these rays travel inside the plate along the same path but with different velocities. For calcite, $V_E > V_O$ and hence the *O*-ray falls behind the *E*-ray and a phase difference δ is introduced between them after transversing a thickness *d* of the calcite plate.

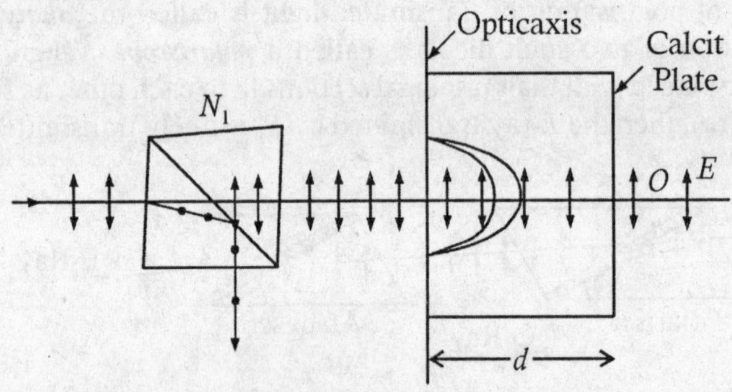

Fig. 5.13

Theory

Suppose the incident plane polarised light makes an angle θ with the optic axis (Fig. 5.14). Let the amplitude of the incident plane polarised light is *A*. It can be resolved into two components—$A\cos\theta$ along the optic axis and $A\sin\theta$ $\perp^r$ to the optic axis.

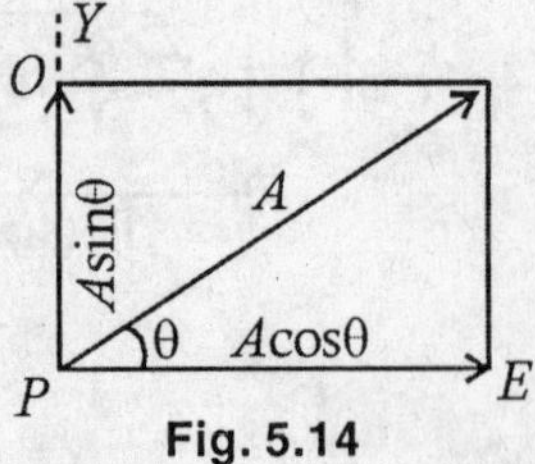

Fig. 5.14

We can represent the two vibrations by two simple harmonic motions at right angles to each other as

$$x = A\cos\theta \sin(\omega t + \delta) \qquad (E\text{-wave vibrations}) \quad (10)$$

$$y = A\cos\theta \sin\omega t \qquad (O\text{-wave vibrations}) \quad (11)$$

Put $A\cos\theta = a$ and $A\sin\theta = b$ (12)

Using Eqs. (12) and (10), Eq. (11) becomes

$$x = a\sin\omega t + \delta \tag{13}$$

$$y = b\sin\omega t \tag{14}$$

One can obtain the nature of resultant vibration by eliminating t from Eqs. (13) and (14), we have from Eq. (14)

$$\frac{y}{b} = \sin\omega t \quad \text{and} \quad \cos\omega t = \sqrt{1 - y^2/b^2} \tag{15}$$

Writing Eq. (23) as

$$\frac{x}{a} = \sin\omega t\cos\delta + \cos\omega t\sin\delta$$

or

$$\frac{x}{a} = \frac{y}{b}\cos\delta + \sqrt{1 - y^2/b^2}\,\sin\delta$$

or

$$\frac{x}{a} - \frac{y}{b}\cos\delta = \sqrt{1 - y^2/b^2}\,\sin\delta$$

Squaring and rearranging, we obtain

$$\frac{x^2}{a^2} + \frac{y^2}{b^2} - \frac{2xy}{ab}\cos\delta = \sin^2\delta \tag{16}$$

This equation represents an ellipse described in a rectangle of sides $2a$ and $2b$. Obviously, the emergent light from the crystal is generally elliptically polarised. We now consider some particular cases of interest.

Case I. When $\delta = 0$, sins$\delta = 0$, $\cos\delta = 1$. Eq. (16) gives

$$\frac{x^2}{a^2} + \frac{y^2}{b^2} - \frac{2xy}{ab} = 0$$

or

$$\left(\frac{x}{a} - \frac{y}{b}\right)^2 = 0$$

or

$$y = +\frac{b}{a}x \tag{17}$$

Eq. (17) is the equation of a straight line passing through the origin and having slope $+b/2$. Obviously, the emergent light is ***plane polarised*** with vibrations in the same plane as in the incident light [Fig. 5.15(a)].

Case II. When $\delta = \pi/4$, Eq. (26) reduces to

$$\frac{x^2}{a^2} + \frac{y^2}{b^2} - \frac{\sqrt{2}xy}{ab} = \frac{1}{2}$$

This is an equations of oblique ellipse [Fig. 5.15(b)].

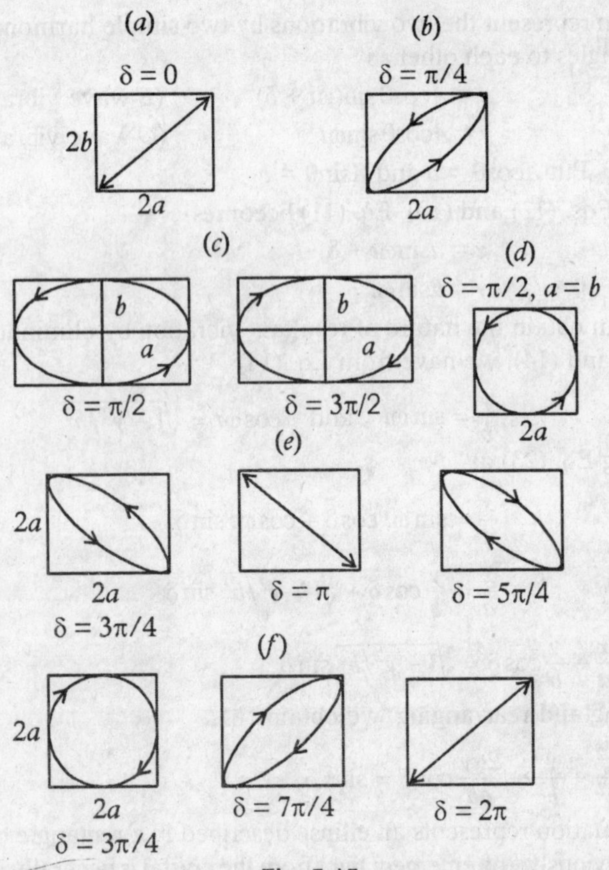

Fig. 5.15

Case III. (*i*) When the thickness (d) of the crystal plate is such that $\delta = \pi/2$, $3\pi/2$, then $\cos\delta = 0$, $\sin\delta = 1$ and Eq. (16) reduces to

$$\frac{x^2}{a^2} + \frac{y^2}{b^2} = 1$$

which represents an ellipse with the major and minor axes parallel to the two privileged directions of the crystal. Obviously the emergent light is elliptically polarised ($a \neq b$) and the plane of the ellipse being normal to the direction of propagation [Fig. 5.15 (*c*)].

(*ii*) If an addition to δ being an odd multiple of $\pi/2$ suppose $a = b$, *i.e.* let the incident plane vibration be equally inclined to the two privileged directions of the crystal ($\theta = 45°$), then the emergent light can be expressed by the equation

$$x^2 + y^2 = a^2 \tag{19}$$

which obviously represents a circle of radius a [Fig. 5.15 (*d*)]. Thus the emergent light is circularly polarised. Such a crystal plate is called a ***quarter wave plate***, which satisfies the general condition $\delta = (2K + 1)\pi/2$, where K is an integer.

Case IV. If $\delta = K\pi$ (K can have odd values), then $\cos\delta = -1$ and $\sin\delta = 0$. Eq. (16) reduces to

$$\frac{x^2}{a^2} + \frac{y^2}{b^2} + \frac{2xy}{ab} = 0$$

or
$$\left(\frac{x}{a} + \frac{y}{b}\right)^2 = 0$$

or
$$y = -\left(\frac{b}{a}\right)x \qquad (20)$$

Eq. (20) is again the equation of straight line passing through origin but with a slope $(-b/a)$ in the negative direction of X-axis [Fig. 5.15(*e*)]. Obviously, the emergent light is plane polarised with the vibration direction making an angle $2\theta = 2\tan^{-1}(b/a)$ with the incident light. The crystal plate behaves as a half wave plate.

Case V. For $\delta = \pi/4$ or $7\pi/4$, Eq. (16) represents the general equation of ellipse [Fig. 5.15 (*f*)] with the major axis inclined in the positive and minor axis in the negative direction of the X-axis respectively.

Case VI. For all other values of δ, the nature of emergent vibrations will be as shown in Fig. 5.15.

From the above analysis we conclude as follows:

(*i*) Fig. 5.15 represents the shape of the resultant transmitted vibrations for different values of δ.

(*ii*) Elliptic vibrations represent the general case when the two components of vibrations have any value of δ other than a multiple of π except for the special case of circular vibrations. Obviously, the plane polarised and circularly polarised light are the special cases of elliptically polarised light.

(*iii*) $0 < \delta < 2\pi$, the ellipse is described anticlockwise and light is said to be ***left handed elliptically polarised light***.

(*iv*) $\pi < \delta < 2\pi$, the ellipse is described in the clockwise direction and the light is said to be ***right handed elliptically polarised***.

(*v*) $\delta = 0$ and $\delta = 2\pi$, the plane of vibrations are parallel.

5.10 PHASE RETARDATION PLATES

Electromagnetic theory of light reveals that plane waves of light can pass through a crystal only if its vibration plane is $||^l$ to one of the two privileged directions which are mutually $\perp^r$ to each other. These directions are also known as ***azimuthal directions***. The theory further shows that the velocities of propagation for these two possible vibrations are different. The direction along which the velocity is smaller is called the ***slow axis*** and the other is called the ***fast axis***. Let the velocities along these axes be c' and c'' and velocity in vacuum is c. Obviously, the two refractive indices corresponding to the vibrations are $\mu' = c/c'$ and $\mu'' = c/c''$. If we cut the crystal with its faces $||^l$ to the optic axis then the two privileged directions will be those of the ***ordinary*** **(*O*)** and ***extraordinary*** **(*E*)** vibrations.

Now consider a thin crystal plate cut with optic axis is $||^l$ to the faces of the plate (Fig. 5.16). Let a plane polarized wave is incident normally on such a plate, the vibrations in the wave break at first face into one vibration along the optic axis and the other vibration $\perp^r$ to it. These components travel thought the crystal plate in the same direction along the normal but with different velocities v_O and v_E. If the thickness of the crystal plate is d, the difference of time taken in traversing the plate by the two beams is

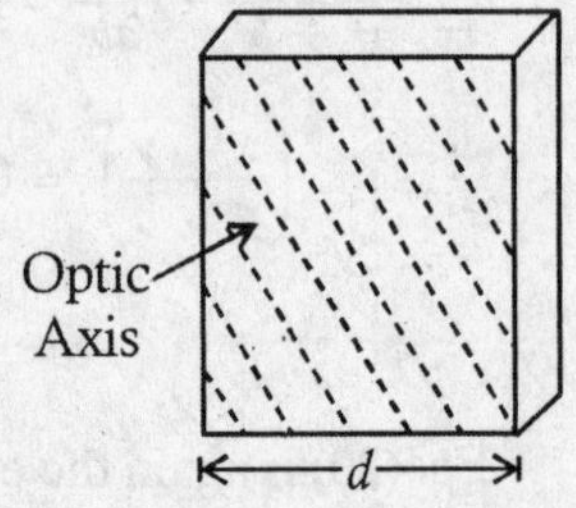

Fig. 5.16: A phase retardation plate

$$|\Delta t| = d\left(\frac{1}{v_O} - \frac{1}{v_E}\right) \tag{21}$$

in terms of equivalent length in vacuum, the path difference or the relative retardation is

$$\delta \text{ or } \Delta x = c|\Delta t| = d|\mu_O - \mu_E| \tag{22}$$

Obviously, a relative phase retardation $\frac{2\pi}{\lambda}\Delta x = \frac{2\pi}{\lambda}(\mu_O - \mu_E)d$ is created between two mutually $\perp^r$ vibrations. Due to this reason, such plates are called ***phase retardation plates***. If we choose the thickness of the plate to make the retardation in path equal to $\lambda/4$ or $\lambda/2$, the plates are then called ***quarter wave plate*** or ***half-wave plate*** respectively.

(*i*) Quarter Wave Plate

If the path difference in Eq. (22) satisfies the condition

$$d|\mu_O - \mu_E| = \left(n \pm \frac{1}{4}\right)\lambda \tag{23}$$

where n is an integer, the crystal plate is called quarter wave plate. Its thickness can be determined from

$$d = \frac{\left(n \pm \frac{1}{4}\right)\lambda}{|\mu_O - \mu_E|} \tag{23(a)}$$

If the plane polarised light whose plane of vibration is inclined at an angle of 45° to the optic axis is incident on a quarter wave plate, the emergent light is circularly polarised (Fig. 5.17).

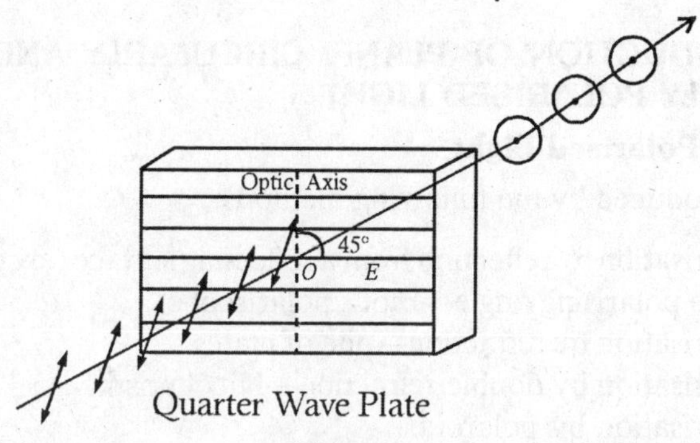

Quarter Wave Plate

Fig. 5.17

(ii) Half Wave Plate

If the path difference (Eq. 22) satisfies the condition

$$d|\mu_O - \mu_E| = \left(n + \frac{1}{2}\right)\lambda \tag{24}$$

the crystal plate is called half wave plate. Its thickness is given by

$$d = \left(n + \frac{1}{2}\right)\lambda \Big/ |\mu_O - \mu_E| \tag{24a}$$

In practice the quarter wave plate and half wave plate have to be supported between glass plates. When plane-polarised light falls on a half wave plate with its plane of vibration making an angle θ than the emergent light is also plane-polarised with the only difference that its plane of vibration from that of the incident light (Fig. 5.18).

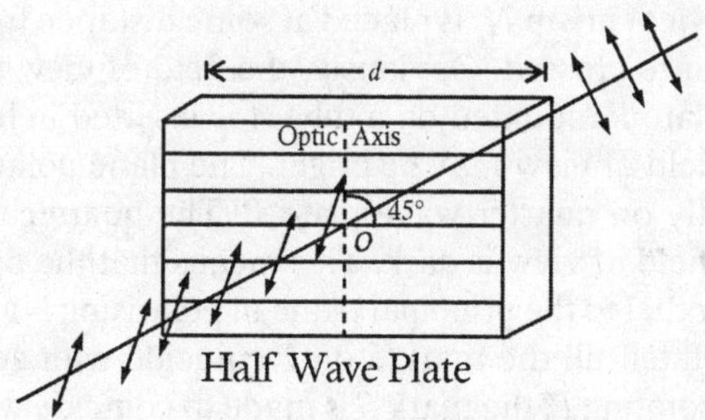

Half Wave Plate

Fig. 5.18

The *QW* and *HW* are usually made of thin sheets of split mica or of quartz cut ||′ to optic axis. A major disadvantage of these retardation plates is that they are true for one particular wavelength because their thickness (d) depends upon wavelength (λ). Obviously, the same plate will not be equally effective for all the colours.

5.16 PRODUCTION OF PLANE, CIRCULARLY AND ELLIPTICALLY POLARISED LIGHT

(a) Plane Polarised Light

It can be produced by the following methods:

(*i*) Polarisation by reflection from a reflecting surface (except metallic) at the polarising angle—Biot's polariscope.
(*ii*) Polarisation by refraction—pile of plates
(*iii*) Polarisation by double refraction—Nicol prism.
(*iv*) Polarisation by polaroid.

Usually, for producing plane polarised light, a Nicol prism is used. A beam of monochromatic light is passed through a Nicol prism. The beam is split up into *E*-ray and *O*-ray is totally internally reflected back at the Canada balsam layer while the *E*-ray passes through the prism. The emergent light beam is plane polarised.

(b) Circularly Polarised Light

One can produce circularly polarised light by the two waves of equal amplitudes vibrating $\perp^r$ to each other and having a path difference of $\lambda/4$ or phase difference of $\pi/2$. For this purpose plane polarised light is allowed to fall normally on a quarter wave plate such that vibrations in the incident light make in angle of 45° with the optic axis of the plate. The emergent light will be circularly polarised.

Circularly polarised light can also be produced by the arrangement shown in Fig. 5.19. A parallel beam of light is allowed to fall on Nicol prism N_1. The Nicol prism N_2 is placed at some distance from prism N_1 so that N_1 and N_2 are crossed. Obviously, the field of view will be dark. A quarter wave plate P mounted on a tube A is inserted in between N_1 and N_2 so that the field of view may be bright. The plane polarised light from N_1 falls normally on quarter wave plate P. The quarter wave plate P is rotated till the field of view is dark. This means that the optic axis of $\lambda/4$ plate is now parallel to the principal plane of polarising Nicol N_1. Keeping P fixed, A is rotated till the mark S on P coincide with zero mark on A. Afterwards by rotating P, the mark S is made to coincide with 45° on A. In this position the vibrations of plane polarised light incident on the $\lambda/4$ plate make an angle 45° with the direction of the optic axis of the $\lambda/4$ plate. The plane polarised light entering the $\lambda/4$ plate is split up into O and E components of equals amplitudes and time period and as a result the emergent light from P is circularly polarised. If N_2 is rotated at this stage, the field of view shows no change in intensity. This is just smaller to the case when ordinary light passes through the Nicol prism.

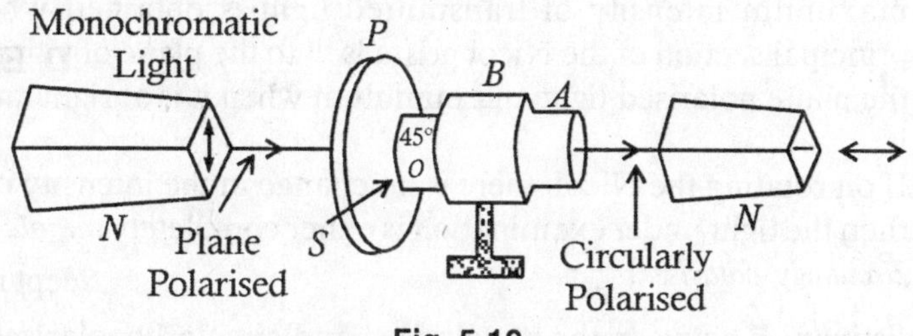

Fig. 5.19

(c) Elliptically Polarised Light

To produce elliptically polarised light, the two waves of unequal amplitudes vibrating at right angles to each other should have a path difference of $\lambda/4$ or a phase difference of $\pi/2$. One can use the arrangement of Fig. 5.19. A parallel beam of monochromatic light is allowed to fall on N_1 and N_1 and N_2 are crossed, *i.e.* field of view is dark. The plane polarised light from N_1 falls normally on $\lambda/4$ plate P. The field of view is illuminated and light emerging out of $\lambda/4$ plate P is elliptically polarised. In this case, one will have to be careful that the vibrations of the plane-polarised light emerging out of $\lambda/4$ plate P should not make an angle of 45° with the optic axis of P. On rotating N_2, we will observe that the intensity of illumination of the field of view varies between a maximum and minimum. This just resembles the case when a mixture of plane polarised and unpolarised light is circularly polarised by rotating Nicol N_2.

5.12 ANALYSIS OR DETECTION OF PLANE, CIRCULARLY AND ELLIPTICALLY POLARISED LIGHT

Suppose we are asked to find the state of polarisation of the supplied light, *i.e.* whether the light is linearly or elliptically or circularly polarised. We can adopt the following procedure for this purpose. Hold a Nicol prism in front of the light and rotate it about the incident light as the axis. The following three situations may arise:

(*i*) On rotating the Nicol prism, the light intensity may vary between maximum value and zero. It is so, then can say that the given light is ***plane polarised***.

(*ii*) If light intensity varies between maximum and minimum values but never becomes zero, the light under analysis is either ***elliptically polarised*** or ***partially polarised***, *i.e.* it is mixture of plane polarised and unpolarised light. In the case of elliptically polarised light, maximum light would be transmitted when the principal section of the Nicol prism is parallel to the major axis and minimum, when it is ||^l to the minor axis. In the case of partially polarised light, maximum intensity of transmitted light is obtained when the principal section of the Nicol prism is ||^l to the plane of vibration of the plane polarised light and minimum when it is at right angles to it.

(*iii*) If on rotating the Nicol, there is no change in the intensity of light, then the light under examination is either completely ***unpolarised*** or *circularly polarised.*

To distinguish between the unpolarised and circularly polarised light, we introduce a $\lambda/4$ plate between the light and analysing Nicol. If the light under examination is circularly-polarised, then it would emerge from the $\lambda/4$ plate as a plane polarised light which would be extinguished completely twice during one complete rotation of the Nicol prism.

If, on the other hand, the light under examination is ***unpolarised***, then it would not become plane polarised on passing through $\lambda/4$ plate and would, therefore, not be extinguished for any position of the Nicol prism.

To distinguish between ***elliptically-polarised*** and ***partially polarised light*** we again introduce a $\lambda/4$ plate between the light and the Nicol prism. If the original light is ***elliptically-polarised*** than after passing it through $\lambda/4$ plate, it will become plane polarised and would be completely extinguished twice during each complete rotation of the Nicol prism. If the light under examination is ***partially polarised***, then on passing it through the $\lambda/4$ plate, the plane polarised in the mixture would become elliptically polarised. Obviously, on rotating the Nicol prism, intensity of the transmitted light would again vary between maximum and minimum values and will never become zero in any orientation of the Nicol prism. The above results have been summarised in Chart 5.1.

5.13 OPTICAL ACTIVITY

Arago in 1811 discovered that when a quartz plate is introduced between two crossed Nicol prism, light is partially restored even if the crystal is cut with its faces normal to the optic axis. Arago further observed that by simply rotating the analyser Nicol through a suitable angle, the field of view could again made dark. Obviously, the quartz is found to rotate the plane of polarisation. This property is called ***optical activity*** and it is exhibited by many other crystals, e.g. sugar crystal, potassium phosphate crystal, etc.

The substances like quartz, sugar in solution, cinnabar, sulphate of strychnine, etc. which rotate the plane of vibration (and also plane of polarisation) are known as *optically active substances.* All the ***optically active substances*** can be classified into two classes as mentioned below.

Chart 5.1
Incident Light

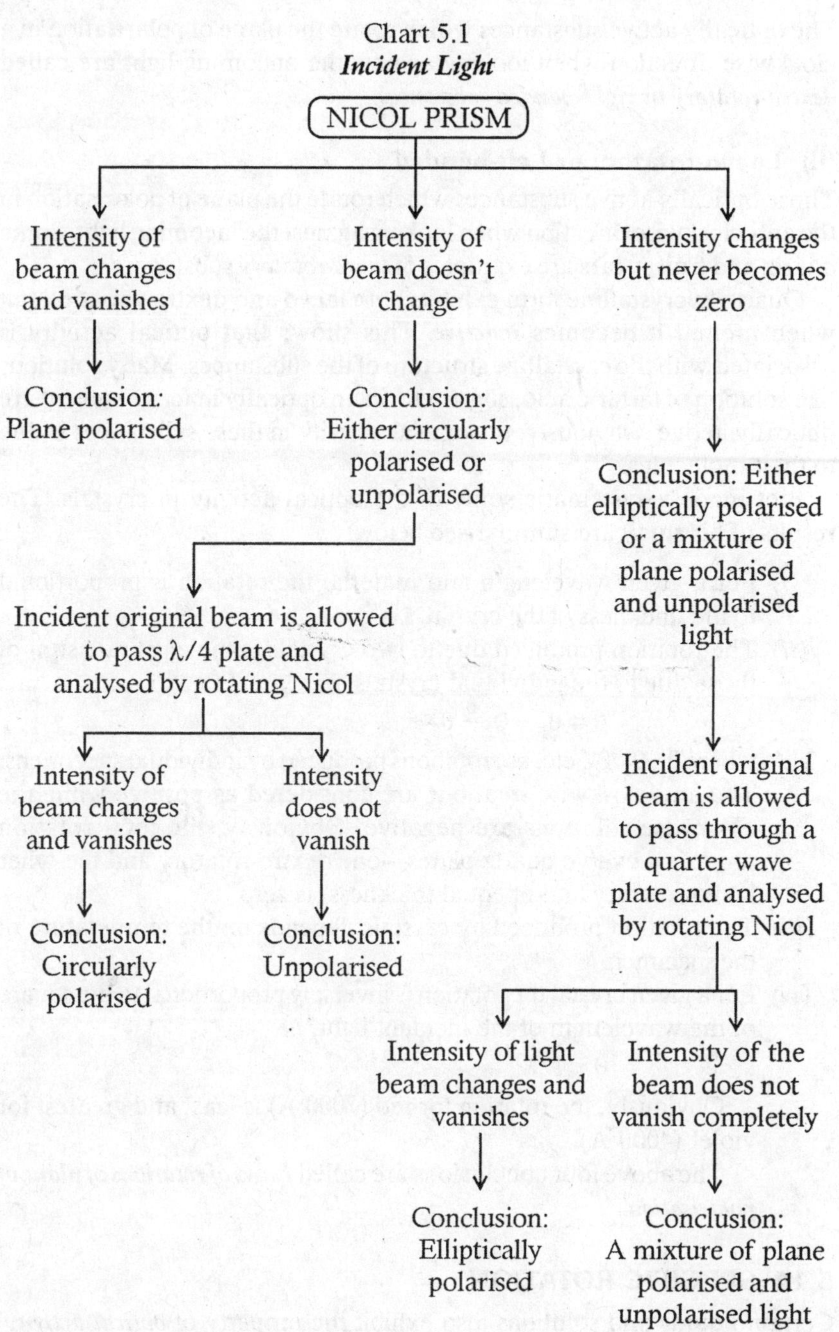

(*i*) Dextro-rotatory or Right-handed

The optically active substances which rotate the plane of polarisation in a clockwise direction when looking against the oncoming light are called ***dextro-rotatory*** or ***right-handed substances.***

(*ii*) Laevo-rotatory or Left-handed

Those optically active substances which rotate the plane of polarisation in the anti-closkwise direction when looking against the incoming light. Some quartz and fruit sugars are examples of ***laevo-rotatory*** substances.

Quartz in crystalline form exhibits both laevo and dextro varieties, but when melted it becomes ***inactive***. This shows that optical activity is associated with the crystalline structure of the substances. Many solutions like solution of tartaric acid, sugar, etc. in an optically inactive solvent are optically active. Obviously, the optical activity in these substances is due to their molecules.

Biot made a systematic study of the optical activity in crystals. The results of his study are summarised below:

(*i*) For a given wavelength and material the rotation is proportional to the thickness of the crystal, *i.e.* $\theta \propto l$.

(*ii*) The rotation produced due to two crystals is the algebraic sum of those due to the individual crystals, *i.e.*

$$\theta = \theta_1 + \theta_2 + \theta_3 + \ldots$$

where θ_1, θ_2, θ_3, etc. are rotations produced by individual specimens. The anti-closkwise rotations are considered as positive while the closkwise rotations are negative. Obviously, the total rotation produced by two quartz plates—one dextro-rotatory and the other laevo-rotatory and of equal thickness, is zero.

(*iii*) The rotation produced by crystals depends on the temperature of the specimen.

(*iv*) For a given crystal the rotation is inversely proportional to the square of the wavelength of the incident light, *i.e.*

$$\theta \propto 1/\lambda^2$$

Obviously, the rotation for red (7000 Å) is least and greatest for violet (4000 Å).

The above four conclusions are called ***Laws of rotations of plane of polarisation.***

5.14 SPECIFIC ROTATION

Certain liquids and solutions also exhibit the property of ***optical activity***. For a given wavelength the rotation produced by a solution at a given temperature is proportional to the length of the solution, *i.e.*

$$\theta \propto l \text{ and } \theta \propto C$$

$$\therefore \qquad \theta \propto lC \text{ or } \theta \propto SlC \tag{45}$$

Here l is the length of the solution column in decimetres, C the concentration in gm/cc of the solution and S is a constant which depends upon the nature of the active substance and is known as the ***specific rotation*** of the solution. Thus

$$S = \frac{\theta}{lC} = \frac{\text{Rotation in degrees}}{\text{Length in decimeters} \times \text{Concentration in gm/cc}} \tag{46}$$

$$= \frac{\theta}{l(m/V)} = \frac{\theta V}{lm} \tag{47}$$

Here m (in gms) is the mass of the active substance present in V *cc* of solution. Thus *specific rotation of a solution for a given wavelength of light at a temperature t is defined as the rotation produced by solution of unit length (diameter) containing a unit mass* (1 gm) *of the optically active substance per unit volume* (1 cc) *of the solution.*

From the above definition of specific rotation it is clear that it is not a constant but it varies with the wavelength of light, the nature of the inactive solvent, the concentration of the solution and its temperature. Clearly, the specific rotation is greater for light of shorter wavelength and *vice versa.*

For quartz for which the rotation is very large, the specific rotation is defined as the rotation produced by 1 mm thick plate.

Molecular Rotation

The product of specific rotation and the molecular weight of the substance is called the ***molecular rotation.***

5.15 POLARIMETERS

These are the instruments used for finding the optical rotation of different solutions. When these are used for finding the optical rotation of sugar, they are called ***saccharimeters***. Polarimeter can be used to find the specific rotation of sugar solution or if specific rotation is known, they can be used to find its concentration. The following polarimeters are widely used.

(*i*) Laurent's Half Shade Polarimeter

Figure 5.21 is a schematic representation of this polarimeter. N_1 and N_2 are two Nicol prisms. N_1 is a polarizer and N_2 is an analyser. Behind Nicol prism N_1 there is a half wave plate of quartz Q which covers one half of the field of view while the other half G is a glass plate. G absorbs the same amount of light as the quartz plate Q. T is a glass tube, containing the optically active substance and closed at the ends by cover slips and metal covers. This tube is mounted between the two Nicols N_1 and N_2. It is worthwhile to mention than there will be no air bubbles in the path of the light. The air bubbles (if any) will appear at the upper portion of the wide bore (T_1) of the tube T. S is a circular slit illuminated with monochromatic light.

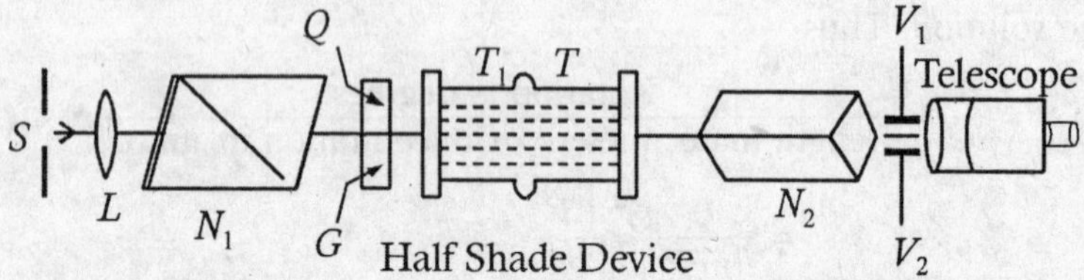

Fig. 5.21: Laurent's half shade polarimeter

The lens L renders the light from slit S parallel and directs it towards the polarising Nicol N_1. The light emerging from N_1 plane polarised with its vibrations in the principal plane of the Nicol. This plane polarised light passes through Laurent's half shade device and then through the tube T containing the sugar solution (active substance). The emergent light, on passing through the analysing Nicol prism N_2, is viewed through a Galilean telescope to which is attached a scale V_1 V_2.

Half Shade Device

Laurent's half shade device consists of a semicircular half wave plate of quartz Q cut with optic axiz $||^l$ to its faces and at right angles to the vertical diameter. This is cemented together with another semicircular plate of glass G which is of such a thickness that its absorption of light is same as that of the quartz. Thus a composite circular plate is formed (Fig. 5.22). Suppose a beam of plane polarised light with its vibration plane parallel to CD is incident on the half shade. The light passing through the glass portion is unaffected. The light passing through the quartz half plate Q is split into two vibrations, one vertical and the other horizontal. Vertical component called E-component is parallel to the optic axis YY' and the other O-component $\perp^r$ to it, *i.e.* along XX'.

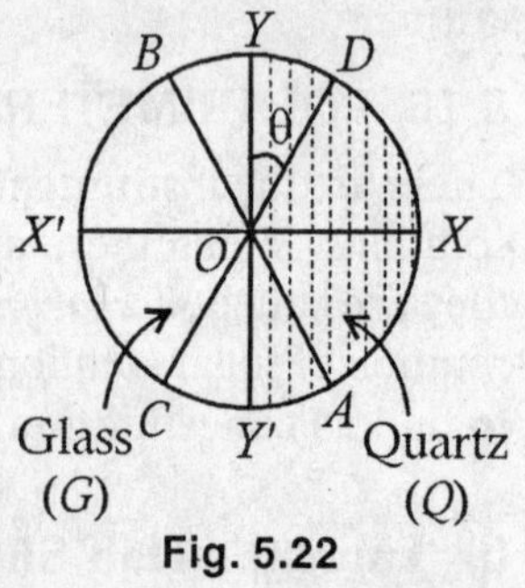

Fig. 5.22

Inside quartz $V_O > V_E$ and hence on emergence the O-component will gain a phase of π over the E-component. Due to this phase difference of π, the direction of O-component is reversed, *i.e.* if OM is the initial direction then its final direction will be ON (Fig. 5.23). Thus if the component of initial incident vibrations along OD in quartz plate are along OL and OM, the emergent wave will be the resultant of the vibrations along OL and ON. Obviously, the resultant of OL and ON is OB which makes the angle θ with Y-axis as incident vibrations along AB *do* but on the other side, *i.e.*

$$\angle DOL = \angle BOL = \theta$$

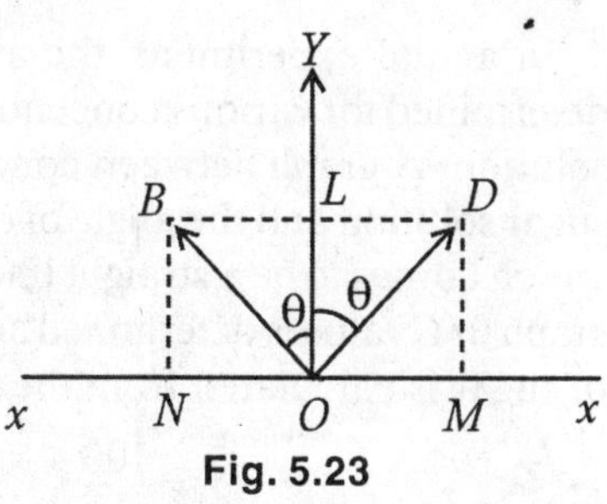

Fig. 5.23

This shows that the plate *rotates the plane of polarisation by an angle* 2θ.

Thus we notice that there are two plane polarised beams (*i*) one beam emerging from glass half plate with vibrations in the plane OD and (*ii*) other beam from quarter half with vibrations in the plane OB. If the analyser N_2 has its principal plane or section along YY', *i.e.* along the direction which bisects $\angle AOC$. The amplitudes of light incident on the Nicol N_2 from both the halves will be equal. Obviously, the field of view will be equally bright [Fig. 5.24 (*a*)].

If N_2 is rotated to the right of YY', then the right half will be brighter as compared to the left half portion as shown in Fig. 5.24 (*b*). If N_2 is rotated to the left YY' then the left half is brighter as compared to the right half [Fig. 5.24 (*c*)]. Obviously, the *half shade serves the purpose of dividing the field of view into two halves.* The half shade device is so accurate that if the analysing Nicol prism is rotated through even small angle from the position of equal brightness, a marked change in the intensity of the two halves of the field very accurately if they are simultaneously presented to the eye.

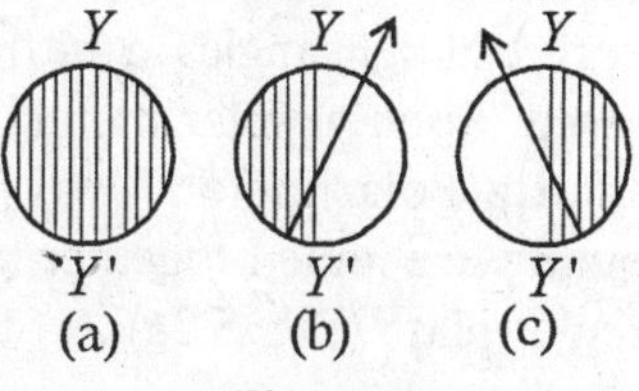

Fig. 5.24

The disadvantage with half shade device is this that it can be used only for special wavelength for which it has been designed.

Determination of Specific Rotation of Sugar

To determine the specific rotation of an optically active substance (say, sugar solution), N_2 (analyser) is set in the position for equal brightness of the fields of view, first with water in the tube T. The readings of the verniers V_1 and V_2 are recorded. Now, the water in the tube T is replaced by sugar solution. Due to this the vibrations from quartz half and glass half are rotated. Obviously, the field of view is not equally bright. The analyser is rotated in the clockwise direction and is brought to a position so that the entire field of view is equally bright. The new positions of the verniers V_1 and V_2 on the scale are again recorded. The angle (θ) through which the analyser has been rotated gives the angle through which the plane of vibration has been rotated by the sugar solution.

In actual experiment, the angle of rotation is determined for various concentrations of the sugar solution. A graph between concentration C of the sugar solution and the angle of rotation θ is plotted which comes to be a straight line (Fig. 5.25). From graph θ/C value is determined and specific rotation of sugar is calculated from the relation.

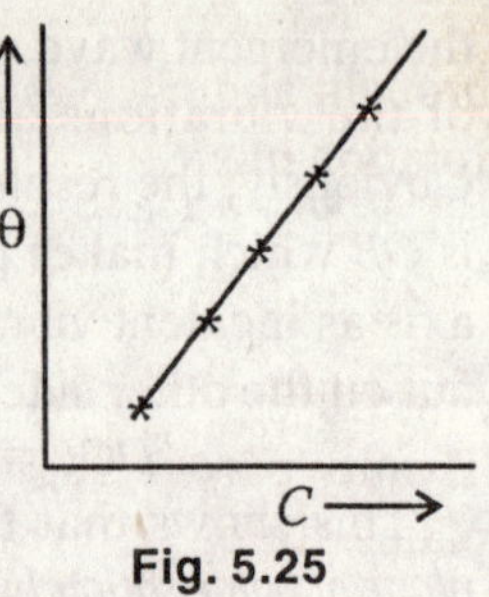

Fig. 5.25

$$S = \frac{100}{lc}$$

5.16 BIQUARTZ POLARIMETER

Instead of a half shade plate, a biquartz plate consisting of two semicircular pieces of quartz (each 3.75 mm in thickness, one left handed and other right handed, which are cut so that their faces are at right angles to the optic axis) is used. Such a polarimeter is called Biquartz Polarimeter. Two plates of biquartz are joined together to form a circular plate (Fig. 5.26).

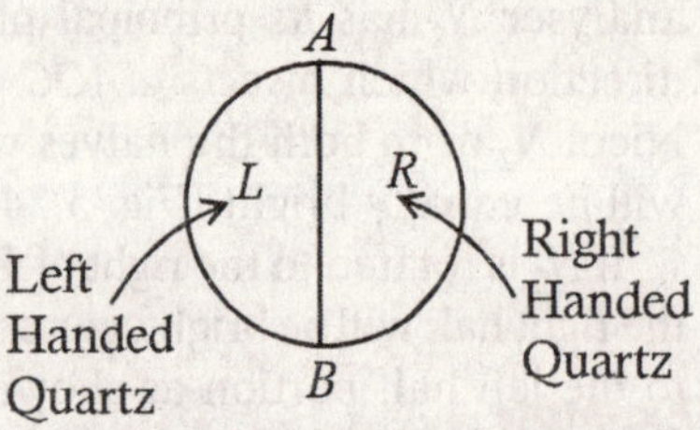

Fig. 5.26

Action of the Biquartz Plate

Each half portion of the biquartz circular plate produces equal rotation of the light incident normally upon them but in opposite directions. This device is placed just behind the polarising Nicol. If the different angles and the emergent light passes through the analysing Nicol. Waves of different refractibilities will be obstructed in the two halves, and the two semicircles will appear of different colours. If, however, the principal planes of polarising and analysing Nicols are parallel, the light of the same colour will be extinguished in both the halves and other colours will be present in same proportion; the two parts will thus appear grey-violet coloured. This colour is named as the ***tint of a passage*** or ***sensitive tint***. If the analyser is now rotated through a very small angle, one of the quartz piece becomes pink and other blue. This shows that a marked change is brought about in the appearance of the biquartz plate on slight rotation of the analyser.

Comparison with Half Shade Device

In comparison to half shade device, the biquartz device is much more sensitive because on slight rotation of the analyser a marked change is brought about in the appearance. For accurate determination of specific rotation biquartz device is used.

ILLUSTRATIVE EXAMPLES

Example 1

Twenty parallel glass plates of refractive index 1.5 are used in a pile of plates. Calculate the degree of polarisation of the transmitted beam.

Solution: Given that $n = 20, \quad \mu = 1.5$

$$P = \frac{n}{n-(2\mu/\mu^2-1)}$$

$$= \frac{20}{\left[20+\dfrac{2\times 1.5}{(1.5)^2-1}\right]} = \frac{20}{20+\left(\dfrac{3}{1.25}\right)^2}$$

$$= 0.776.$$

Example 2

The principal planes of two Nicol prisms form an angle of 30°. Prove that the intensity of the transmitted light diminished in the ratio of 3/2, if the angle between principal planes becomes 45°.

Solution

When the angle between the principal planes is 30°, intensity of the transmitted light is

$$I_1 = I_0 \cos^2 30° = \frac{3}{4} I_0 \qquad \left(\because \cos 30° = \sqrt{3}/2\right) \qquad (1)$$

When the angle between the principal planes is 45°, the intensity of the transmitted light is

$$I_2 = I_0 \cos^2 45° = \frac{I_0}{2} \qquad (2)$$

From Eqs. (1) and (2), we have $\dfrac{I_1}{I_2} = \dfrac{3}{2}$.

Example 3

Two Nicols are first crossed and then one of them is rotated through 60°. Calculate the percentage of incident light transmitted.

Solution

Let us suppose that the intensity of incident unpolarized light beam is $2a^2$. When this beam enters the first Nicol, then it is broken up into O and E

rays, each of amplitude a, *i.e.* intensity a^2. From the first Nicol, O-ray is lost by total internal reflection and only E-ray of amplitude a and intensity a^2 is transmitted.

When one of the two Nicols is rotated through 60° from the crossed positions, the angles between the principal planes of two Nicols will be

$$90^\circ \pm 60^\circ = 150^\circ \text{ or } 30^\circ$$

The intensity of finally emergent light from the second Nicol can be calculated from Malus law as

$$I = I_0 \cos^2 \theta = a^2 \cos^2 30^\circ = \frac{3}{4} a^2$$

$\therefore$ Percentage of light transmitted from the system is

$$= \frac{a^2 \cos^2 30}{2a^2} \times 100 = \frac{1}{2}\left(\frac{\sqrt{3}}{2}\right)^2 \times 100 = \frac{3}{8} \times 100 = 37.5\% .$$

Example 4

The refractive index of a crystal is 1.5533 for the extraordinary ray and 1.5442 for the ordinary ray. Calculate the minimum thickness of the crystal plate which will behave like a quarter wave plate for radiation of wavelength 5000 Å.

Solution

We know that for a crystal plate to behave like a quarter wave plate the difference in the optical paths for the ordinary and extraordinary rays through the crystal must differ by an odd multiple of a quarter wave plate.

Therefore, if the minimum thickness of this to happen is t, then we have

$$(1.5533 - 1.5442)t = \frac{\lambda}{\Delta} = \frac{5 \times 10^{-5}}{4}$$

or $$t = \frac{5 \times 10^{-5}}{4 \times 0.0089} \text{ cm} = 0.0011 \text{ cm}$$

The next larger thickness for which the crystal plate behaves like a quarter wave plate, will be

$$= 3t = 3 \times 0.0011 = 0.0033 \text{ cm.}$$

Example 5

Plane polarised light falls normally on a quarter wave plate. Explain what will be the nature of the emergent light if the plane of polarisation of the incident light makes the following angles with the principal plane of the quarter wave plate 0°, 30°, 45°, 90°.

Solution

When a plane polarised light falls normally on a quarter wave plate and the optic axis is parallel to the crystal face, then the plane of polarisation of the incident polarised light is assumed to make 45° with the principal plane and the emergent light is used to be circularly polarised. The incident beam can be represented by

$$E_x = E_s = E_0 \sin 45^\circ \sin \omega t = \frac{E_0}{2} \sin \omega t$$

and
$$E_y = E_f = E_0 \sin 45^\circ \sin \omega t = \frac{E_0}{2} \sin \omega t$$

Here *s* and *f* represent slow and fast components. After the wave emerges out of the quarter wave plate, the slow component will be ahead of phase by $\pi/2$. Obviously, the emergent beam is given by

$$E_x = \frac{E_o}{\sqrt{2}} \sin(\omega t + \pi/2) = \frac{E_o}{\sqrt{2}} \cos \omega t$$

and
$$E_y = \frac{E_o}{\sqrt{2}} \sin \omega t$$

represents a circularly polarised wave. Similarly, one can analyse for $\delta = 0$, 30°, $\pi/2$.

Example 6

For calcite $\mu_O = 1.658$, $\mu_E = 1.486$. For sodium light (λ = 5893 Å), calculate thickness of a quarter wave plate of calcite for this light.

Solution

$$t = \frac{\lambda}{4(\mu_o \sim \mu_e)} = \frac{5893 \times 10^{-8}}{4(1.658 \sim 1.4860)}$$
$$= 8.58 \times 10^{-3} \text{ cm}$$

Example 7

A beam of linearly polarised light changes to a circularly polarised beam by passing it through a crystalline plate 0.03 mm thick. Assuming this thickness to be minimum, calculate the difference in refractive indices of two rays in the crystal for light of wavelength 6000 Å.

Solution

A linearly polarized light can become circularly polarised when the slice is quarter wave plate, *i.e.*, we have

$$t = \frac{\lambda}{4(\mu_o \sim \mu_e)}$$

or $$\mu_o - \mu_e = \frac{\lambda}{4t}$$

$$= \frac{6000 \times 10^{-8}}{4 \times 0.003} = 0.005.$$

Example 8

The faces of quartz plate are parallel to the optic axis of the crystal. (*a*) What is the thinnest possible plate that would serve to put the ordinary and extraordinary rays of λ = 5890 Å a half wave apart on their exit? (b) What amplitudes of this thickness would give the same result? The indices of refraction of quartz are μ_E = 1.553, μ_O = 1.544.

Solution

(*a*) According to the question the quartz plate behaves like a half wave plate, and hence its thickness is given by

$$t = \frac{\lambda}{2(\mu_E \sim \mu_O)} = \frac{5890 \times 10^{-8}}{2(1.553 - 1.544)} = 3.27 \times 10^{-3} \text{ cm}$$

(*b*) The plates which would introduce a path difference of $\lambda/2$ or its odd multiples between the ordinary (*O*) and extraordinary (*E*) rays will produce the same effect. Obviously, the thickness of such plates is given by

$$= (2n + 1)t \qquad \text{where } n = 0, 1, 2, \ldots$$

$$= t, 3t, 5t, \ldots$$

$$= 3.27 \times 10^{-3} \text{ cm } 9.81 \times 10^{-3} \text{ cm}, \ldots$$

Example 9

The refractive index of calcite for ordinary ray is 1.6584, and the refractive index of Canada balsam is 1.550. A ray of light is incident parallel to the length side. Calculate the angle of incidence of the ordinary ray on the Canada balsam layer and show that it will suffer total internal reflection at that layer.

Solution

When the incident ray is parallel to the length side as shown in Fig. 5.27, then the angle of incidence is 22°.

Now, $$\mu_O = \frac{\sin i}{\sin r}$$

$\therefore$ $$\sin r = \frac{\sin i}{\mu_O} = \frac{\sin 22}{1.6584} = 0.2259$$

$\therefore$ $$r = 13.05°.$$

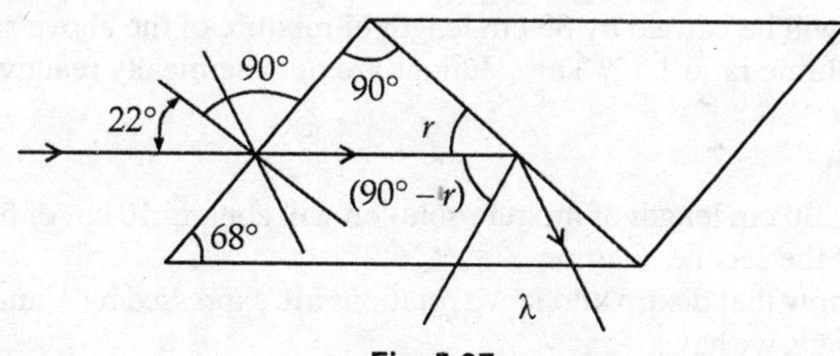

Fig. 5.27

Obviously, the angle of incidence on the Canada balsam layer = 90 – *r* = 90 – 13.05 = 76.95°.

Now, the critical angle of calcite with Canada balsam for ordinary ray,

$$C = \sin^{-1}\frac{1.550}{1.6584} = 69.17°$$

This shows that ordinary ray will suffer internal reflection at the Canada balsam layer.

Example 10

Find the specific rotation of a given sample of sugar solution if the plane of polarisation is turned through 26.4°. The length of the tube containing 20% solution is 20 cm.

Solution: Given that θ = 26.4°, *l* = 2 decimeter *C* = 20% = 0.2 gm/cc

$$S = \frac{\theta}{lC}$$

$$\therefore \quad S = \frac{26.4}{2 \times 0.2} = 66°$$

Example 11

A solution of camphor and alcohol in a tube 0.2 m long is found to rotate the plane of polarisation of light by 27°. What is the mass of the camphor in unit volume of the solution. The specific rotation of camphor is 54° decimeter/gm/cc.

Solution: Given that θ = 27°, *S* = 54°, *l* = 0.2 m = 2 decimeter

$$\frac{\theta}{lC}$$

$$\therefore \quad C = \frac{\theta}{lC} = \frac{27}{54 \times 2} = \frac{1}{4} = 0.25 \text{ gm / cc} = 25\%$$

Example 12

If 20 cm length of a certain solution causes right handed rotation 38°, and 30 cm of another solution causes left handed rotation 24°, what optical rotation will be caused by 30 cm length of mixture of the above solution in the volume ratio 1 : 2? The solutions are not chemically reactive.

Solution

As given, 30 cm length of mixture solution will contain 10 cm of first and 20 cm of the second solution.

We know that dextro and laevo rotations are expressed by + and – sign respectively, we have

Optical rotation produced by first solution

$$\theta_1 = +\frac{42^\circ}{20} \times 10 = +21^\circ \text{ (\textit{i.e.} right handed rotation)}$$

Optical rotation produced by second solution

$$\theta_2 = +\frac{27^\circ}{30} \times 20 = -18^\circ \text{ (\textit{i.e.} left handed rotation)}$$

∴ Total rotation produced is given by

$$\theta = \theta_1 + \theta_2 = +21^\circ - 18^\circ = +3^\circ$$

Clearly, the mixture will cause right handed rotation of 3°.

Example 13

The optical rotation produced by a particular material is found to be 30° per mm at $\lambda = 5500$ Å and 50° per mm at $\lambda = 4000$ Å. In the visible region of the spectrum, the rotation of the plane of polarisation is given by $\theta = a + b\lambda^{-2}$. Find the values of the constants a and b for the given material. Calculate the possible thickness of the material which when interposed between two crossed Nicols will produce maximum transmission for $\lambda = 5500$ Å.

Solution

As per problem, the rotation of plane of polarisation is given by

(*i*)
$$\theta = a + \frac{b}{\lambda^2} \qquad (1)$$

$$\lambda = 5000 \times 10^{-8} \text{ cm} = 5 \times 10^{-4} \text{ mm}$$

$$\theta = 30^\circ/\text{mm}$$

∴
$$30^\circ = a + \frac{b}{(5 \times 10^{-4})^2} \qquad (2)$$

(*ii*) When $\lambda = 4000 \times 10^{-8}$ cm $= 4 \times 10^{-4}$ mm, $\theta = 50^\circ$/mm, then

$$50^\circ = a + \frac{b}{(4 \times 10^{-4})^2} \qquad (3)$$

From Eqs. (2) and (3), one obtains

$$a = \frac{50}{9}, \quad \text{and} \quad b = \frac{8}{9} \times 10^{-5} \qquad (4)$$

Substituting the values of a and b in Eq. (1), one obtains the value of optical rotation for $\lambda = 5500$ Å $= 5.5 \times 10^{-4}$ mm as

$$\theta = -\frac{50}{9} + \frac{8000 \times 10^{-8}}{9 \times \left(5.5 \times 10^{-4}\right)^2} = 24°/\text{mm}$$

Clearly, 1 mm thick plate produces a rotation of 24°. To obtain maximum transmission between two crossed Nicols, the plane of polarisation should be rotated through 90°. The desired thickness of the plate for the said purpose will be

$$t = \frac{90}{24} \text{ mm} = 3.75 \text{ mm}.$$

REVIEW QUESTIONS

1. What is double refraction? Describe how Huyghen explained it. Describe the construction and action of a Nicol prism.
2. Describe the construction and working of a Nicol prism. Discuss how do you obtain a plane polarised beam with it?
3. Describe a Nicol prism. Show clearly how it is constructed. Explain its action. How can you use it as a polarizer and analyser?
4. Give Huyghen's construction of the wave fronts for the ordinary and extraordinary waves when a beam of light is reflected through a doubly refracted crystal. Optic axis as in the plane of incidence and is $\perp^r$ to the crystal surface.
5. Thin plates are cur from a negative doubly refracting crystal with various positions of the optic axis as described below. Explain your observation giving diagrams.
   (*i*) Optic axis lying in plane of incidence but parallel to upper face.
   (*ii*) Optic axis lying in plane of incidence but $\perp^r$ to upper face.
   (*iii*) Optic axis lying in plane of incidence but inclined to upper face.
   (*iv*) Optic axis $\perp^r$ to plane of incidence but parallel to upper face.
6. Explain Huyghen's theory of double refraction in uni-axial crystals. Distinguish between negative and positive crystals?
7. Give the wave theory of the origin of the double refraction of light at a calcite crystal for normal and oblique incidence.
8. Define polarisation. Are sound waves polarised ? What are elliptically and circularly polarised light? Plane polarised light is normally incident on a $\lambda/4$ plate. Sate conditions under which circularly or elliptically polarised light can be obtained.
9. Describe how (*i*) plane polarised (*ii*) elliptically polarised and (*iii*) circularly polarised light can be produced and detected.
10. How can you produce elliptically and circularly polarised light? Explain fully how will you distinguish between (*a*) unpolarised light, and circularly polarised

light, (*b*) partially polarised light and unpolarized light. State the polarisation in the emergent beam. What changes would occur if plate is $\lambda/8$ plate?

11. Light passes through a Nicol polarizer with its shorter diagonal vertical. The analyser is set for minimum intensity. How will you set a half wave plate between the two Nicols so that maximum intensity of light is seen through the analyser?
12. Unpolarised light falls on two polarising sheets so oriented that no light is transmitted. If third polarising sheet is placed between them, can light be transmitted? Explain (*i*) how do you find the polarising direction? (*ii*) How do you find optic axis and polarising direction of a $\lambda/4$ plate?
13. What is quarter wave plate? Deduce its thickness for a given wavelength in terms of its refractive index.
14. How would you produce the following with the help of a Nicol prism and quarter wave plate: (*i*) plane polarised light (*ii*) circularly polarised light and (*iii*) elliptically polarised light?
15. Give two methods of producing elliptically polarised light. How would you distinguish it from partially polarised light.
16. Elliptically polarised light falls normally on a quarter wave plate. Explain the nature of emergent light if the major axis of the ellipse makes the following angles angles with the principal plane of the quarter wave plate (0°, 30°, 90°).
17. (*a*) Describe the construction and working of a phase retardation plate. What are quarter and half wave plates?

    (*b*) Explain how you will produce and detect plane, circularly, and elliptically polarised light.
18. Define specific rotation. Describe the construction and working of Laurent half-shade polarimeters?
19. Define specific rotation. How will you determine it with the half of a half-shade polarimeter?
20. Give with full details the construction and working of Laurent's half shade polarimeter.
21. What is meant by rotatory polarisation? How can a right circularly polarized light be transformed to left circularly polarised light?
22. Discuss the phenomenon of rotation of the plane of polarisation of light by optically active materials. Give the necessary theory. Show that the rotation of plane of vibration

$$= \frac{\pi d}{\lambda}(n_A - n_C)$$

    where n_A and n_C are the refractive indices of the crystal in the direction of the optic axis for anti-clockwise circular polarised light and d is the thickness of the crystal lattice.
23. (*a*) What do you understand by a quarter-wave plate and half-wave plate?

    (*b*) Explain the functions of various parts of Laurent's half shade polarimeter.
24. Distinguish between circularly and elliptically polarized light. How would you produce and detect circularly polarized light?
25. Define specific rotation. Describe the construction and principle of a half shade polarimeter.

26. (*a*) Describe the construction and working of Nicol prism. How can you use it as a polarizer and analyser?

(*b*) What is quarter-wave plate?

27. Describe how you will produce elliptically polarized light and distinguish it from partially polarised light.

28. A transparent plate is given. Using two nicol prism (or polaroides) how would you find whether the plate is a quarter wave plate, a half wave plate or simple glass plate? [Raj. BE, 2000]

29. Write short notes on:

(*a*) Specific rotation and its measurement. [Raj. BE, 2000]

(*b*) Laurent's half shade polarimeter. [Raj. BE, 2001]

30. (*a*) What do you mean by optical rotation? Discuss, how shall you measure optical rotation of sugar solution using a bi-quartz device.

(*b*) Discuss the state of polarizations of the following light waves:

(*i*) $\mathbf{E} = \hat{j}A\cos(kx - \omega t) + \hat{k}2A\cos(kx - \omega t + \pi/4)$

(*ii*) $\mathbf{E} = \hat{j}A\cos(kx - \omega t) + \hat{k}A\sin(kx - \omega t)$ [Raj. BE, 2003]

31. (*a*) How will you distinguish plane polarized, circularly polarised and elliptically polarised light?

(*b*) A phase retardation plate of quartz has thickness 0.1436 mm. For what wavelengths in the visible region will it act as quarter wave plate?

(*c*) What is bi-quartz device. [Raj. BE, 2002]

32. (*a*) Describe the construction of a 'Nicol prism' and explain its action in converting unpolarised into plane polarised light.

(*b*) Plane polarised light passes through a quartz plate with its optic axis parallel to the faces. Calculate the least thickness of the plate for which the emergent beam will be.

(*i*) Plane polarised

(*ii*) Circularly polarised

Given: $\mu_E = 1.5533$, $\mu_O = 1.5442$ and $\lambda = 5000$ Å [Raj. BE, 2002]

33. (*a*) Explain the working & construction of Nicol prism. Mention its limitation.

(*b*) Show that linearly polarised light can be represented as a superposition of two circularly polarised lights of suitable amplitude and phases.

(c) What is optical activity? Mention the laws of optical rotation.

[Raj. BE, 2001]

34. (*a*) How shall you differentiate the following?

(*i*) Three sources of light having the same physical appearance: partially polarized, elliptically polarized and mixture of unpolarized and circularly polarized light?

(*ii*) Half wave plate and quarter wave light?

(*b*) How shall you measure specific rotation of glucose solution using a biquartz device? [Raj. BE, 2001]

35. Define specific rotation? Describe the construction and working of Laurent's half shade polarimeter.

36. A transparent plate is given. Using two Nicol prisms (or) how would you find whether the plate is a quarter wave plate, a half-wave plate or a simple glass plate? [Raj. BE, 2000]

PROBLEMS

1. The refractive index of quartz is 1.5443 for the ordinary ray and 1.5334 for the extraordinary ray. Calculate the thickness of a quartz plate which will act as a quarter wave plate for light of wavelength 6000 Å. [**Ans.** 0.0014 Å]

2. The refractive index of calcite is 1.648 for the ordinary ray and 1.486 for the extraordinary ray. A plane parallel plate of calcite is cut of thickness 0.01 mm. For what wavelengths in the visible range will this plate behave as (*i*) a quarter wave plate (*ii*) a half-wave plate?
[**Ans.** (*i*) 7200, 5899, 4985 and 4320 Å, (*ii*) 6480 and 4629 Å]

3. How much a polarizer and an analyser be oriented so that a beam of natural light is reduced to (*i*) 0.5 (*ii*) 0.25 (*iii*) 0.75 (*iv*) 0.125 of its initial value?
[**Ans.** 45°, 60°, 30°, 69°]

4. The critical angle of glass is 40° and that of a liquid is 50°. Find the angle of polarisation when light is incident on glass from the liquid. [**Ans.** 50°]

5. Calculate the thickness of quarter wave plate from the following data $\mu_O =$ 1.6247, $\mu_E = 1.5672$ and $\lambda = 5893$ Å [**Ans.** 2.56 micron]

6. What is the maximum thickness that a crystal plate should have if it is to act as a quarter wave plate for radiation of wavelength 6000 Å. The refractive index of crystal is 1.553 for the ordinary ray and 1.554 for the extraordinary ray.
[**Ans.** 0.0014 cm]

7. Two nicols are oriented with their principal planes making on angle of 30°. What percentage of incident unpolarized light will pass through the system?
[**Ans.** 37.5%]

8. Calculate the thickness of a quarter wave plate for wavelength 5890 Å. The refractive indices of quartz (for λ = 5890 Å) for *E*-ray and *O*-ray are equal to 1.5533 and 1.5443 respectively. [**Ans.** 1.636×10^{-3} cm]

9. Find the thickness of a quarter wave plate when the wavelength of light is equal to 5880 Å and $\mu_O = 1.5818$, $\mu_E = 1.5508$. [**Ans.** 1.474×10^{-3} cm]

10. The thickness of a quarter wave plate for a certain wavelength of light is 2 micron. What will be the nature of light obtained when a plane polarised beam of light passes through it? What would happen if the thickness is 3 micron?
[**Ans.** (*i*) $x^2 + y^2 = a^2$, circularly polarised (*ii*) If $x^2 + y^2 + \sqrt{2}xy = a^2/2$ elliptically polarised]

11. A cellophane sheet behaves as a half wave plate for light of wavelength 4000 Å. How would this sheet behave for light of wavelength of 8000 Å? It may be assumed that refractive indices do not change with wavelength.
[**Ans.** $\lambda/4$ plate]

12. A plane polarised beam of light passes through a polarizer with its allowed direction inclined at 45° with the plane of vibration of the incident beam. A second polarizer is kept after this with its allowed direction at 45° with that of the first. Compare the intensity emergent from the second with that of incident on first. What will happen if the first polarizer is removed? [**Ans.** 0.25, zero]

13. The specific rotation of sucrose is 66.4. Calculate the rotation of a column of 20 cm of solution of concentration 12.5 gm of sucrose in 100 cc of water.

[**Ans.** 16°36′]

14. On inserting a polarimeter tube 25 cm long containing a sugar solution of unknown strength, it is found that the plate of polarization its rotated through 10°. Find the strength of sugar rotation in per cc, given specific rotation of sugar rotation to be 60° per decimeter for unit concentration. [**Ans.** 0.066 gm/cc]

15. A solution of 0.3 gm of camphor in 10 cc of alcohol in a 20 cm long tube produces a rotation of 30°. Find the specific rotation of camphor. [**Ans.** 55]

16. A solution of an optically active solute produces a rotation of 20° of the plane of polarization in a path length of 10 cm when the concentration is 20 gm/litre. What is the concentration in a solution which produces a rotation of 30° in a path length of 5 cm? [**Ans.** 60 gm/litre]

17. A tube of sugar solution 20 cm long is placed between crossed nicols and illuminated with light of wavelength 6000 Å. If the optical rotation produced is 13° and the specific rotation is 65°, determine the strength of the solution.

[**Ans.** 0.1 gm/cc]

SHORT ANSWER QUESTIONS

1. Why light waves can be polarised while sound waves cannot be polarised?

Ans. Sound waves are longitudinal in nature whereas light waves are transverse in nature. Only transverse waves can be polarised.

2. How polarisation of light establishes that light waves are transverse in nature?

Ans. We know that longitudinal waves can have vibrations only along the direction of propagation of wave, *i.e.* one state of polarisation. Experiments reveal that light can have several different states of polarisation, which means that the vibrations are in plane perpendicular to the directions of the wave.

3. How does the polarisation of light afford a convincing evidence of transverse nature of light?

Ans. If light waves were not transverse in nature then the vibrations would have passed through two tourmaline crystals with their optic axes perpendicular to each other, *i.e.* in crossed position.

4. What is the angle between the direction of propagation and plane of polarisation in the propagation of electromagnetic waves?

Ans. 0°

5. A calcite crystal is placed over a dot on a piece of paper and rotated. What will you observe through the calcite crystal?

Ans. One dot rotating about the other.

OBJECTIVE QUESTIONS

1. Nicol prism is an optical device made from the bifringent calcite crystal and is used both as a __________ and as an __________. (polariser, analyser)

2. Along the optic axis, the velocities of the ordinary and extraordinary ray are the __________. (same)

3. Polarising angle is the angle of incidence for which the reflected light is completely ________ with vibrations perpendicular to the plane of incidence. (plane polarised)
4. If the natural light falls on a uniaxial crystal, the intensities of the ordinary and extraordinary wave as they enter the crystal are ________. (same)
5. According to cosine low of Malus, the intensity of polarised light emerging from an analyser is proportional to the square of the ________ of the angle between the optic axes of the polariser and the ________. (cosine, analyser)

6

Lasers

6.1 INTRODUCTION

We live in an age where laser light shows are common and lasers light up the sky. Within the past few years, lasers have come into wide spread use in science, medicine, manufacturing, and in many other fields. The term laser stands for '*Light Amplification by Stimulated Emission of Radiation*'. Laser produces a highly intense, concentrated and parallel beam of monochromatic and coherent light. In the words of C. H. Townes, inventor of Maser and Laser,' *what the laser does essentially is to give us electronic type control over light. It is a marriage of optics and electronics.*

Historically the laser is the outgrowth of 'MASER' a device that uses microwaves instead of light waves. The word 'maser' is an acronym for Microwave (or Molecular) Amplification by Stimulated Emission of Radiation and is also known as *optical maser*. The principles on which lasers operate undoubtedly place them in the family of masers, but in structure properties and applications, lasers are quite different from masers. The major difference between the two devices is that of working frequency; the typical operating frequencies of masers are nearly 10^{10} Hz, while the frequencies in which laser operate are of the order of 10^{15} Hz.

The basic idea behind the laser and maser is the process of stimulated *emission* that was first postulated by Einstein in his derivation of Planck's radiation law in 1917. If his considerations has been applied to non-equilibrium situations, the operating principle of the laser and maser could have been proposed at that time. In fact, it took about 35 years for others to come up with the basic idea and make this idea work.

Before 1960 the laser was just a twinkle in the eye of scientists. Today it is a brighter than the sun reality. The laser (it rhymes with razor) promises to be in measurement techniques, communications, military weaponry, and the conquest of space. It also figures in industrial processes, computers, medicine, chemistry and other fascinating possibilities. In brief the applications of laser light range from straight forward (such as surveying and welding) to the sophisticated (such as optical communications and holography) and it may be that more interesting applications are still to come.

6.2 IMPORTANT PROPERTIES OF LASER LIGHT

The salient features of a laser light are (*i*) *directionality* (*ii*) *intensity* (*iii*) *monochromaticity* (*iv*) *coherence.* These features are discussed below.

(*i*) Laser Light is Highly Directional

A laser beam is highly directional, *i.e.* it spreads very little. A laser beam departs from strict parallelism only because of diffraction at the exit aperature of the laser. For example, a laser pulse used to measure the distance to the Moon generates a spot on the Moon's surface with a diameter of only a few metres. Light from an ordinary bulb can be made into an approximately parallel beam by using a lens, but the beam divergence is much greater than for laser light and a parallel beam of light from a torch on the earth shall spread in an area of few square kilometres. Each point on a light bulb's filament forms its own separate beam, and the angular divergence of the overall composite beam is set by the size of the filament. The directionality of a beam can be measured by the *full angle beam divergence* which is two times the angle subtended by the outer edge of the beam with the beam axis. The outer edge of the beam is considered to be the line where the beam intensity decreases to 1/exp (*I*) times its value at the axis. Let us consider that an aperature of diameter *d* radiates a beam having a plane wavefront. Let the beam travels as a parallel beam through the distance of Rayleigh range, which is of the order of d^2/λ, where λ is the radiated wavelength. Obviously, the beam begins to spread with distance owing to diffraction as shown in Fig. 6.1. In general, the aperature diameter (*d*) control the angular spread $\Delta\theta$, radians of the far-field beam. The relationship between $\Delta\theta$ and *d* is given by

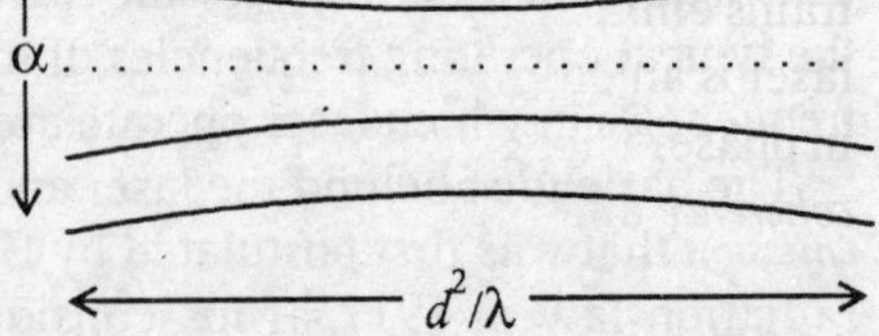

Fig. 6.1: Spreading of beam beyond Rayleigh range

$$\Delta\theta = \frac{\lambda}{d} \tag{1}$$

Hence, the shorter the wavelength and larger the aperature diameter, the smaller the angular divergence of the beam. For a typical laser beam, the beam divergence is less then 10^{-5} radian, exhibiting that the beam spreading is less than 10^{-5} m for every meter.

(*ii*) Laser Light is Highly Intense, *i.e.* it can be Sharply Focused

If two light beams transport the same amount of energy, the beam that can be focused to the smaller spot will have the greater intensity at that spot. Lasers emit narrow beam of light, so that the energy is concentrated in a small region. The spatial and spectral confinement of energy in laser beam accounts for its high intensity. For laser light, the focused spot can be so small that an intensity of 10^{17} W/cm^2 is readily obtained. An oxyacetylene flame, by contrast, has an intensity of only about 10^3 W/cm^2. Laser beams can generate power densities millions of times greater than those on the surface of the sun. It is found that when a laser beam is focused by a lens, the tremendous intensity at the focal point can produce a radiation pressure of the order of 10^6 kg per sq. cm.

(*iii*) Laser Light is Highly Monochromatic

Light from an ordinary incandescent light bulb is spread over a continuous range of wavelength and is certainly not monochromatic. The radiation from a fluorescent neon sign is monochromatic, to about 1 part in about 10^6. However, the sharpness of definition of laser light can be many times greater, as much as 1 part in 10^{15}.

(*iv*) Laser Light is Highly Coherent

Individuals long wave (wave trains) for laser light can be reversed hundred kilometres long. When two separated beams that have travelled such distances over separate paths their common origin and are able to form a pattern of interference fringes. The corresponding *coherence length* for wave trains emitted by light bulb is typically less than a metre. The wave from a laser is an orderly wave of one frequency when the whole beam is *spatially* in phase. There are two independent facts of coherence namely (*a*) *temporal coherence* and (*b*) *spatial coherence.*

(*a*) *Temporal Coherence*

This refers to the correlation between the radiation field at a point and the radiation field at the same point at a later time, *i.e.* the relation between $E(x, y, z, t_1)$ and $E(x, y, z, t_2)$, where these represent the radiation field at the point (x, y, z) at times t_1 and t_2 respectively. If the phase difference between the two radiation fields is constant during the period normally covered by observations (~few microseconds), the wave is said to have *temporal coherence.* If the phase difference between two radiation fields changes many times and in an irregular way during the shortest period of observation (μs), the wave is said to be incoherent.

Time coherence comes about when each cycle of the wave takes exactly the same time to pass a point in space. This really means that the frequency of the wave is not varying and is given the name monochromatic.

(b) Spatial Coherence

Two fields at different points on a wavefront of given electromagnetic wave are said to be space coherent if they preserve a constant phase difference over any time *t*, *i.e.* space coherence requires that the waves not only are of same frequency, but that they are in phase in space.

Spatial coherence is possible even when the two beams are individually time incoherent, as long as any phase change in one of the beam is accompanied by a simultaneous equal phase change in the other beam. With the ordinary light sources, this is only possible if the two beams have been derived from the same part of the source.

Coherence Time and Coherence Length

Let us consider that the radiation field E, from a light source is an ideal sinusoidal function of time t, then at any given position one can write

$$E = a\cos(\omega_0 t + \phi) \tag{2}$$

here a is the magnitude of the field, ω_0 the angular frequency and ϕ the phase. Figure 6.2 depicts the field.

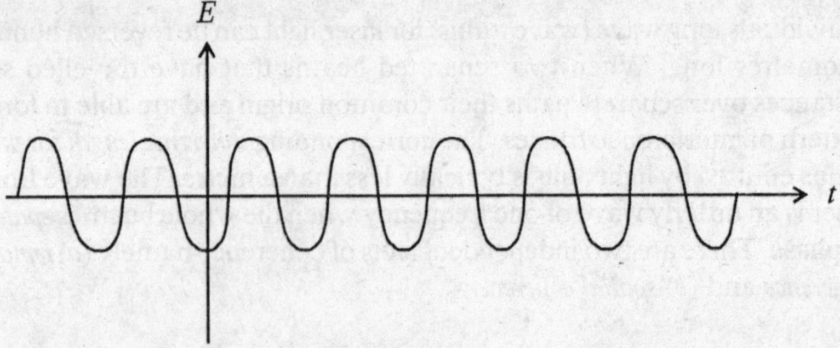

Fig. 6.2

However, no emitted light produces a perfect sinusoidal variation of field E with time t, and for any actual light source the magnitude a and phase ϕ will very with time. One can represents a more realistic picture of the radiation from an emitting source of light as shown in Fig. 6.3. In Fig. 6.3, τ represents the average time vibration for which an ideal sinusoidal

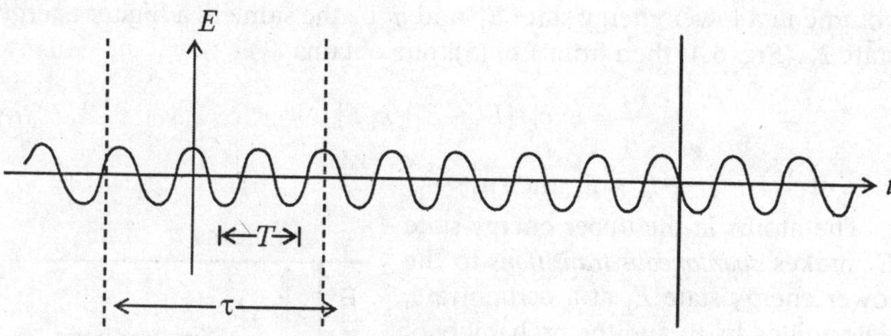

Fig. 6.3

emission occurs. τ is called *coherence time* or *interval of coherence.* Let T denotes the period of the oscillation, where $T = \frac{\omega_0}{2\pi}$. The spatial dimension L, for which the light may be considered to be a perfect sinusoidal is given by

$$L \approx cT \approx \frac{\tau}{T}\lambda \tag{3}$$

Here c is the velocity of light and L is called the *Coherence length.* After time τ, there is no correlation between the phases of the waves.

It terms of purity Q of the spectral line ($Q \approx \frac{\lambda}{\Delta\lambda}$, where $\Delta\lambda$ is half width of the spectral line), we have

$$c\tau \approx L \approx \lambda Q \tag{4}$$

6.3 EINSTEIN'S THEORY OF ATOMIC TRANSITIONS AND EINSTEIN'S *A* AND *B* COEFFICIENTS

Transition between the atomic energy states is a statistical process and as such it is not possible to predict which particular atom will make a transition from one state to another at a particular instant. However, in an assembly of a very large number of atoms, it is possible to calculate the irradiative transition between two states, based on the laws of probability. Einstein in 1917, first calculated the probability of such transitions assuming the atomic system to be in equilibrium with electromagnetic radiation.

Let us consider an assembly of atoms at an absolute temperature T in which the atoms may be in different energy states. Let n_0 be number of atoms per unit volume in the ground state ($E = 0$) then the number of atoms n per unit volume in an excited state of energy E is given by the Boltzmann distribution law

$$n = n_0 \exp[-E/k_B T] \tag{5}$$

where k_B (= 1.38×10^{-23} J/K) is the Boltzmann constant. The maximum number of atoms in an assembly of large number of atoms at an absolute temperature T is in ground state. Let n_1 be the number of atoms per unit volume in a lower energy state E_1, and n_2 be the same in a higher energy state E_2 (Fig. 6.4) then from Eq. (5), one obtains

$$\frac{n_2}{n_1} = \exp[-(E_2 - E_1)/k_B T] \tag{6}$$

If $E_2 < E_1$, then $n_2 < n_1$.

The atoms in the upper energy state E_2 makes *spontaneous transitions* to the lower energy state E_1 at a certain rate, determined by n_2 and the probability of spontaneous transition per unit time A_{21} from E_2 to E_1. Hence the number of spontaneous transition per unit volume per unit time is $A_{21}n_2$. These transitions lead to the emission of electromagnetic radiation of energy given by

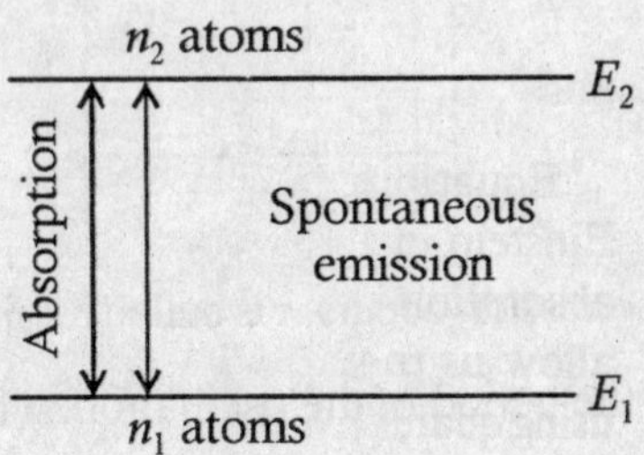

Fig. 6.4: Atomic transition

$$h\nu = E_2 - E_1 \tag{7}$$

where h is Planck's constant and ν is frequency of the electromagnetic radiation. We must not that $1/A_{21}$ is a measure of the lifetime of the upper state against spontaneous decay to the lower state.

Now assume that the atoms are in equilibrium with lectromagnetic radiation of frequency $\nu = (E_2 - E_1)/h$ having energy density u_ν. Due to absorption of energy from this radiation, some of the atoms in the lower energy state E_1 make upward transitions to the higher energy state E_2. The rate of such transitions is determined by the number n_1 and the energy density u_ν of the radiation and is given by $B_{12}n_1u_\nu$ per unit volume per second. Obviously B_{12} is the probability of the *radiation induced transitions* per unit from the lower state E_1 to the upper state E_2.

In addition to the above two types of transitions, there may also be induced transitions of atoms from E_2 to E_1 caused by interactions with the electromagnetic radiation. We call it as a *stimulated* or *induced emission* of radiation of frequency $\nu = (E_2 - E_1)/h$. The number of such induced or stimulated transitions per unit volume per unit time is $B_{21}n_2u_\nu$, where B_{21} is the probability from E_2 to E_1 per unit time.

The coefficients A_{12}, B_{12} and B_{21} are known as Einstein's A and B coefficients. Under equilibrium, the time rate of transitions from E_2 to E_1 must be equal to that from E_1 to E_2. So one can write

$$A_{21}n_2 + B_{21}n_2u_\nu = B_{12}n_1u_\nu$$

or $$u_\nu = \frac{A_{21}n_2}{B_{12}n_1 - B_{21}n_2} = \frac{A_{21}}{B_{21}} \frac{1}{\left(\frac{B_{12}}{B_{21}}\right)\frac{n_1}{n_2} - 1} = \frac{A_{21}}{B_{21}} \frac{1}{\left(\frac{B_{12}}{B_{21}}\right)\exp\left(\frac{h\nu}{k_BT}\right) - 1} \tag{8}$$

Eq. (8) must agree with Planck's radiation formula

$$u_\nu = \frac{8\pi h\nu^3}{c^3}\frac{1}{\left[\exp(h\nu/k_BT)-1\right]} \tag{9}$$

Comparison of Eqs. (8) and (9) yields

$$B_{12} = B_{21} \tag{10}$$

and

$$\frac{A_{21}}{B_{21}} = \frac{8\pi h\nu^3}{c^3} \tag{11}$$

Equations (10) and (11) are called Einstein's relations obtained by Einstein in 1917. The relation (10) shows that the induced emission and absorption probabilities per unit time are equal. The calculation does not allow us to obtain the values of A_{21}, B_{21} and B_{12}. They must be derived using quantum mechanical considerations. Using these relations, one obtains

$$\frac{A_{21}}{B_{21}u_\nu} = \exp\left(\frac{h\nu}{k_BT}\right) - 1 \tag{12}$$

which is the ratio of the number of spontaneous to stimulated transitions. Therefore if $h\nu >> k_BT$, spontaneous emission is much more probable than induced emission, which can be completely neglected. This holds true in the case of electronic transitions in atoms and molecules and in the case of radiative transitions in nuclei. But if $h\nu << k_BT$, as it is in the microwave region of the spectrum, induced or stimulated emission may become important.

Induced emission is the result of the action of the incoming radiation on the atoms (or molecules) of the substance. Hence the forced atomic oscillations bear a constant difference relative to the incoming radiation. This means that all atoms radiate in phase and therefore *induced emission is coherent*. On the other hand, spontaneous transitions occur at random, with no correlation between the times at which atoms undergo transitions. Therefore the phases of atomic radiations are distributed randomly. Obviously, spontaneous emission is incoherent.

6.4 PRINCIPLE OF LASER

The principle of laser exploits the phenomenon of stimulated transitions discussed in section 6.3. We know that light is emitted from a source when the atoms in the source make transitions from an excited (E_2) state to a lower energy state (E_1) spontaneously. Normally, the mean life of excited atoms before spontaneous emission occurs is about 10^{-8} s. However, for some excited states, this mean life is perhaps as much as 10^5 times longer. We call such long-lived states *metastable*, they play an important role in laser operation. The energy of the emitted photon in the *spontaneous transitions* is

$$h\nu_{12} = E_2 - E_1 \tag{13}$$

on other hand, if the atom is placed in an electromagnetic field that is alternating at frequency ν, the atom can absorb an amount of energy $h\nu_{12}$ (photon) from that field and move to the higher energy state E_2. This process is called *absorption*.

Besides absorption and spontaneous emission, a third type of transition from an upper energy state E_1 may be induced by an incident photon of energy $h\nu_{12} = E_2 - E_1$ giving rise to the *stimulated* emission of radiation. This process is called stimulated because the event is triggered by the external photon. The emitting photon is in every way identical to the stimulating photon. Thus the waves associated with photons have the same energy, polarization, and direction of travel. All the above three types of transition are shown in Fig. 6.5.

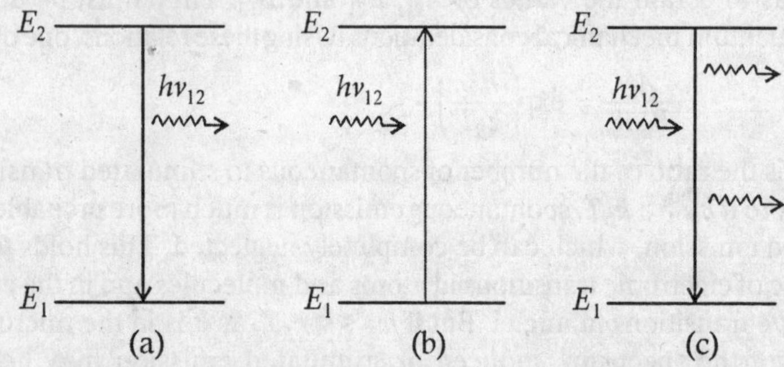

Fig. 6.5: (*a*) Spontaneous transition from an excited state E_2 to a lower state E_1, with an emission of photon $h\nu_{12}$ (*b*) absorption (*c*) Induced or stimulated transition from an excited level E_2 to a lower state E_1. A light wave with photon energy $h\nu_{12}$ causes the atom, to emit a photon of the same energy, increasing the energy of light wave

From Fig. 6.5 (*c*) it is obvious that we get two photons in this case. If these two photons are now incident on two other atoms in the excited state E_2, then it will result in the induced emission of two more photons so that there will be four coherent photons of the same energy result. The process goes on in a chain, yielding a large number of coherent photons of energy $h\nu_{12}$. This will increase the intensity of the coherent radiation enormously and amplification due to induced emission is achieved (Fig. 6.6).

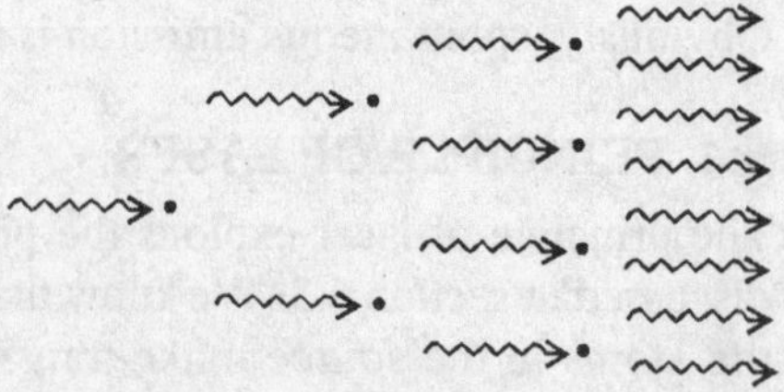

Fig. 6.6: Amplification due to induced emission

The necessary condition for the said amplification of the light intensity by the induced emission of radiation is that we must have more photons emitted than absorbed, *i.e.* we must have a situations in which stimulated emission dominates. The direct way to bring this about is to start with more atoms in the excited state than in the ground state. Normally the number of atoms in the excited upper energy state (E_2) is much lower than in the lower energy state (E_1), *i.e.*, $n_2 << n_1$ due to thermal equilibrium. As a result the rate of downward induced transition $B_{12}n_2u_\nu$ is much less than that of the upward transitions $B_{12}n_2u_\nu$ between the same two energy levels due to absorption of photons. To increase the rate of induced emission from E_2 to E_1, it is necessary to achieve $n_2 >> n_1$. This is known as *population inversion*. The states with an inverse population are sometime referred to as states with a negative temperature. However, since such a population inversion is not consistent with thermal equilibrium, we must think up clever ways to set up and maintain one.

6.5 POPULATION INVERSION BY PUMPING

The physical mechanisms of populating and depopulating energy levels are many and diverse. Accordingly there exist various means of rendering the medium active, collectively known as the processes of pumping. Few of the population inversion processes are as follows:

(*i*) Optical Pumping

In this process, the predominant population of the upper level is achieved by means of light energy delivered from approximately selected sources such as gaseous discharged flash tubes, or continuously burning tubes.

(*ii*) Electrical Pumping

This is accomplished by means of a sufficiently intense electrical discharge in the medium and is particularly suited to gas media. The discharge converts the gas into a plasma where active centres collide inelastically with free electrons and cause the predominant populations of the upper pumping level. Inelastic collisions of active centres with other atoms and molecules purposefully introduced into the gas are also of importance for pumping as they provide resonance energy exchange.

(*iii*) Chemical Pumping

This raises active centres into the higher level by means of suitable exothermal chemical reactions in the active material.

(*iv*) Heat Pumping

In this scheme of pumping the active material at first brought to a high temperature and then rapidly cooled down.

6.6 OPTICAL RESONATORS

The selections of some photon states and the suppression of other states can be realized by means of an *optical resonator*, a principal component of each laser. In its simplest version the optical resonant cavity is a pair of mirrors set on an optic axis which defines the direction of the laser beam. The active material is placed in between these mirrors. Solid active materials are often in the shape of a cylinder whose axis is aligned with the axis of the optic resonator, the length of the cylinder being about ten times its diameters. At least one of the mirrors of the resonator cavity is made semitransparent to serve as an output element passing the light out of the resonator.

Figure 6.7 shows the schematic arrangement of a simple resonator. The photons produced spontaneously in the OO' direction sufficiently close to it will travel within the active material a relatively long way, which is elongated by multiple reflections from the resonator's mirrors. These photons interact with excited active centres to eventually initiate a powerful avalanche of stimulated photons constituting the laser beam. The photons emitted in other directions (and their stimulated avalanches) will traverse a relatively short path length in the materials and die out soon.

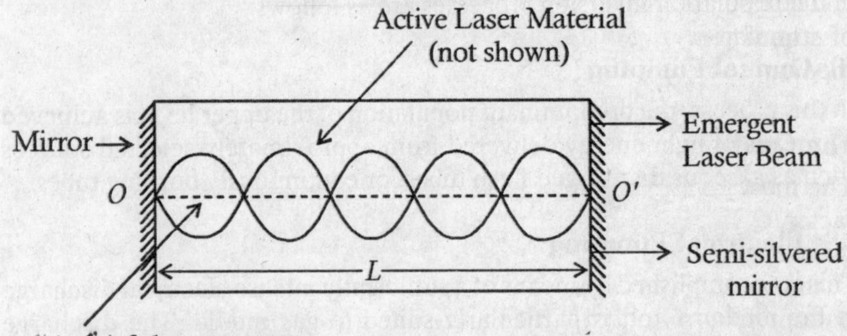

Fig. 6.7: The optical resonator determines the direction of laser action (along the resonator axis) and favours the process of stimulated emission in this direction

Thus the optical resonant cavity provides the desired selectivity of photon states by primarily confining the possible direction of photon propagation. As a result, lasing action occurs in this direction.

Let L be the length of the resonator cavity. The change in one round trip between the mirrors is given by

$$\delta = \frac{2\pi}{\lambda} \cdot 2L + 2\gamma$$

where γ is the phase change at reflection of each mirror and λ is the wavelength of light. For resonance, *i.e.* reinforcement, δ must be a integral multiple of 2π. So, if m is an integer, one obtains

$$\delta = \frac{4\pi L}{\lambda} + 2\gamma + 2\pi m$$

or $$\nu = \frac{c}{\lambda} = \frac{mc}{2L} - \frac{\gamma c}{2\pi L} \tag{14}$$

where c is the speed of light. Obviously, the light having frequencies that satisfy the condition (14) will resonate and set up stationary waves in the cavity as shown in Fig. 6.7. Each mode corresponds to one integral value of m. Since one of the mirrors of the resonator is partially transmitting, *i.e.* half silvered, some of the radiations leaks out, thus providing the collimated laser *output*.

The optical resonator can provide selectivity in other properties of radiation. Of course, the selectivity in photon energy is secured by the choice of active centres with a system of levels is much more complicated than the laser scheme shown in Fig. 6.7. Actual active centres can have a few lasing transitions rather than one. To exclude undesired transitions, the resonant cavity may be provided with mirrors whose reflecting power is a function of frequency, so that the undesired transitions will be damped out, and the necessary selectivity will be ensured for photon states.

The resonant cavity plays, therefore, a key role as if guides the processes of stimulated emission induced by spontaneous photons in the active medium so that a laser radiation of high coherence properties results.

The Quality Factor (Q)

The most important characteristics of the optical resonator is its quality factor, Q. In the standing wave pattern, the reflectors give two nodes. The wavelength of light λ being much smaller the optical resonator cavity length L, the number of half waves between the two mirrors is very large. One obtains from Eq. (14) for the difference in frequencies between two consecutive modes as

$$\Delta\nu = \nu_{m+1} - \nu_m = \frac{c}{2L}$$

Obviously, there is a spread in the frequency of the laser beam, which is controlled by the spectral character of the emission line contributing to the gain. The sharpness of the cavity resonance is determined by Q, a dimensionless quantity. Q is used to show the ability of the cavity to store energy. The slower the energy of a radiation field falls off, *i.e.* the smaller the losses in the resonator, consequently, the higher the resonator's Q, Q of a cavity is defined as

$$Q = 2\pi \frac{\text{Maximum energy stored per cycle in the mode}}{\text{Energy dissipated per cycle in the mode}}$$

In a passive resonant cavity, the light beam irradiance gradually declines with travel distance owing to various losses in the medium, namely, absorption, scattering, radiation through side surfaces, etc. This decreases

Q and reduce the sharpness of resonance. If $d\nu$ is the line width of the cavity mode, then one can easily show that

$$Q = \nu / d\nu$$

The smaller the line width, the higher the Q of the cavity, and the sharper the resonance. In the case of the He-Ne laser with a cavity of length 1 m, we have $\Delta\nu = c/2L = 1.5 \times 10^8$ Hz. Obviously, this is a small fraction of the mean frequency of the output radiation at 4.75×10^{14} Hz. This reveals the high degree of monochromaticity of the laser radiation.

6.7 AMPLIFICATION OF LIGHT AND THRESHOLD CONDITION

Let us consider a system of atoms through which a quasi-monochromatic light beam of frequency ν propagates in the Z-direction. Let S_1 and S_2 represents the two planes of area A perpendicular to the Z-direction and located at z and $z + dz$, respectively, as shown in Fig. 6.8. The volume of the active laser material between the planes S_1 and $S_2 = Adz$. If n_1 is the concentration of the atoms in the lower state, u_ν is the energy density of the radiation and B_{12} is the Einstein's B coefficient, then the time rate of stimulated absorption $= B_{12}n_1u_\nu Adz$. Obviously, the energy absorbed per unit time in the volume Adz is therefore equal to $B_{12}n_1u_\nu h\nu Adz$. Similarly, one finds the energy gained per unit time in the volume Adz due to stimulated emission is $B_{12}n_2u_\nu h\nu Adz$, where n_2 is the concentration of atoms in the upper state. Here we have ignored the radiation due to spontaneous emission because such radiations are directed at random and are lost. Now, the net energy gain per unit time in volume Adz in the frequency range ν and $\nu + d\nu$ is

$$(B_{21}n_2 - B_{12}n_1)u_\nu h\nu Adz = B_{12}(n_2 - n_1)u_\nu h\nu Adz \qquad (\because B_{12} = B_{21})$$

Let us consider that I_ν be the intensity of the beam at S_1. Obviously, the energy entering the volume Adz per unit time through S_1 is $I_\nu(z)A$, Similarly, if $I_\nu(z + dz)$ be the intensity at S_2 then the energy moving out of the volume Adz per unit time through S_2 is

$$I_\nu(z + dz)A = \left(I_\nu(z) + \frac{\delta I_\nu}{\delta z} dz \right) A$$

obviously, net energy leaving the volume per unit time is $\frac{\delta I_\nu}{dz} Adz$. This must be equal to the energy gained by volume Adz due to stimulated transitions. Thus, one obtains

$$\frac{\delta I_\nu}{dz} Adz = B_{12}u_\nu h\nu (n_2 - n_1)Adz$$

$$\frac{\delta I_\nu}{dz} = B_{12}h\nu (n_2 - n_1)u_\nu \qquad (15)$$

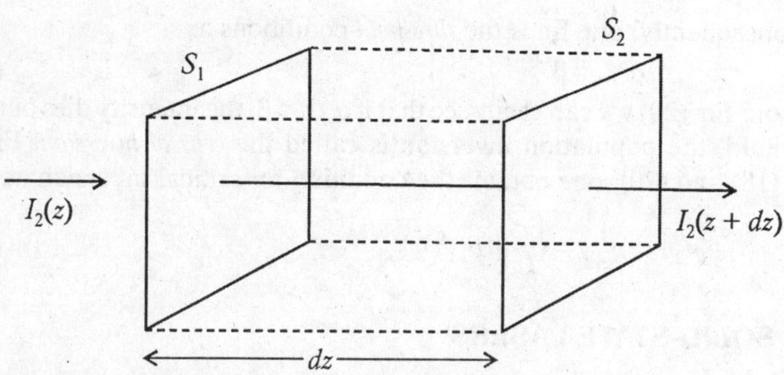

Fig. 6.8: Amplification of a light beam passing the plane S_1 and S_2 located at z and $z + dz$ respectively

Also, we have

$$I_\nu = \nu u_\nu = \frac{c}{n} u_2 \tag{16}$$

where $\nu = c/n$ is the velocity of light in the medium having refractive index n. Using Eq. (16), Eq. (15) reduced to

$$\frac{I}{I_\nu}\frac{dI_\nu}{dz} = \alpha_\nu \tag{17}$$

where
$$\alpha_\nu = \frac{B_{12} h \nu n(n_2 - n_1)}{c} \tag{18}$$

On integration, Eq. (17) gives

$$I_\nu(z) = I_\nu(0)\exp(\alpha_\nu z) \tag{19}$$

Eq. (19) represents the growth of the intensity of the light it traverses a distance z through the active medium of laser. Obviously, the growth results from the population inversion created by pumping and stimulated transitions. The quantity α_ν is called the gain constant having the dimension of reverse length. We can see that the gain constant α_ν increases with the degree of population inversion $n_2 > n_1$ and hence α_ν given by Eq. 18 is positive. This means that the intensity increases exponentially with distance z.

Let us consider that r_1 and r_2 be the reflection coefficients of two mirrors of the optical cavity. Obviously, in one complete round trip, the intensity of light decreases to $I_\nu(0)r_1r_2$ where $I_\nu(0)$ is the initial intensity. Let us write $r_1r_2 = \exp(-2\beta)$ where β is positive and can be considered to be a measure of the loss of intensity of light as it traverses from one mirror to the other. This means that on traversing a length L, the intensity drops to $I_\nu \exp(-\beta)$, We may assume that the reflection losses are the only losses present in the cavity. Thus the intensity in a mirror to mirror passage becomes $I_\nu(0)\exp(\alpha_\nu L - \beta)$. Thus for a substained laser action the amplification must compensate for energy losses. Obviously, for the lasing action to build up, we must have

$$\exp(\alpha_\nu L - \beta) > 1 \quad \text{or} \quad \alpha_\nu L > \beta$$

Consequently, one finds the *threshold* conditions as

$$\alpha_\nu L = \beta \tag{20}$$

From Eq. (20) we can easily see that if $\alpha_\nu L < \beta$, the intensity dies out. At threshold, the population inversion is called the *critical inversion*. From Eqs. (18) and (20), one obtains the condition for critical inversion as

$$(n_2 - n_1)_{\text{crit}} = \frac{\beta c}{L}(\mu B_{12} h\nu)^{-1} \tag{21}$$

6.8 SOLID-STATE LASERS

(*i*) Ruby Laser

The first successful operation of the laser was achieved by T. H. Maiman of the Mughes Aircraft Company in July, 1960, using ruby crystals. Ruby is aluminium oxide in which a few of the aluminium atoms have been replaced by chromium atoms ($Al_2O_3 : Cr^{3+}$), the chromium atoms being the ones that take part in a laser action. Maiman used 0.05% Cr^{3+} to get a pink colour, which is due to the fact that chromium absorbs, ultraviolet green and yellow, and transmits only red and blue. This absorption raises chromium atoms to excited states from which two steps are required to carry them back to the ground energy level. In the first step chromium atoms give up some of their energy to lattice and fall in *metastable* state, characterized by a comparatively long life time. This transition is a non-radiative transition. In the second step, chromium atoms emit radiation of wavelength 6943 Å in the red region of the spectrum and fall to ground level. The metastable has a life time of about 10^{-3} second compared to the extremely small life time of 10^{-7} second of the higher energy level. Thus the necessary time delay between the useless relaxation transition and the useful stimulated transition for the laser action is readily obtained. The energy levels in the ruby laser are shown in Fig. 6.9.

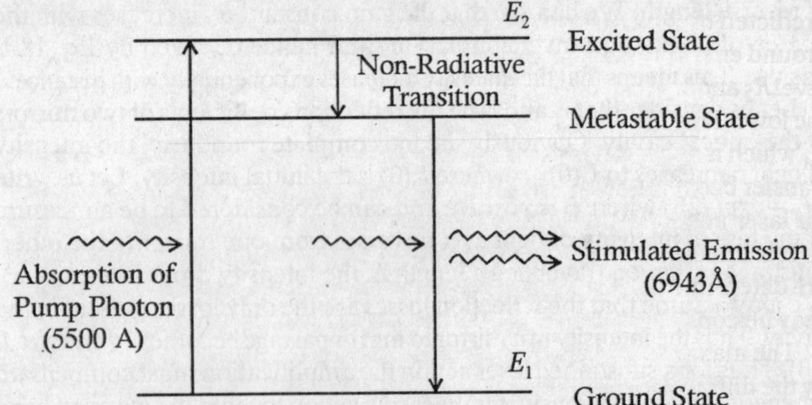

Fig. 6.9: Energy-band scheme for laser action in ruby [$Al_2O_3 : Cr^{3+}$]

Ruby optical laser apparatus is shown in Fig. 6.10. A ruby is machined into a rod typically 5 cm long and 0.5 cm in diameter with polished end which are parallel and optically flat. This is placed near an electronic flash tube which provides the pumping energy. The ruby is cooled using liquid nitrogen. Upto a certain flash intensity, ordinary luminescent emission dominates, and a powerful beam of red light flashes out from the ends of ruby rod. This red light has the enormous power output of more than 10^4 Watts over a beam of cross-section 10^{-4} square metre. The emitted band is within a wavelength interval of about 0.02 Å. The beam from the end face of the ruby rod has an angular spread is less then one degree. Typical output of a ruby laser is 10^{-20} Kilowatts in milli second pulses.

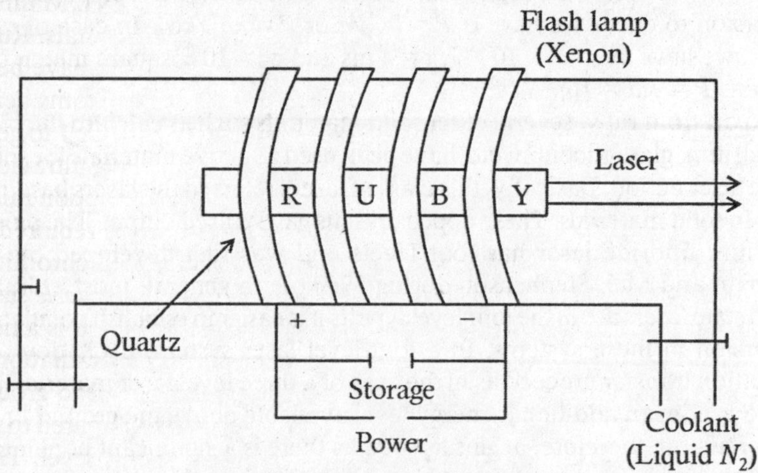

Fig. 6.10: Pulse ruby optical laser

The stimulated emission predominates over spontaneous emission when the population of metastable exceeds considerably over the population predicted by Boltzmann distribution relation and in comparison with the ground energy state. The metastable state consists of closely spaced double level R_2 and R_1. The larger contribution to laser action, however, comes from the lower level, *i.e.* R_1. This is due to the fact that the spontaneous life time for R_1 which is about 4×10^{-2} sec exceeds the thermal relaxation time for energy transfer between R_2 and R_1. The minimum energy required to be pumped for laser action to start, is of the order of 50 to 100 J of absorbed energy.

It is important to note that this type of oscillator action has losses which are determined by the flatness and parallelism of the end mirror. The losses may be considerable, and may reduce the efficiency of the laser.

The maximum space coherence (directionality) obtainable is determined by the diffraction of the mirrors. The light-beam spread angle $\delta\theta$, is given by

$$\delta\theta \cong \frac{\lambda}{D} \qquad (22)$$

where λ is the wavelength of the emitted radiation and D is diameter of the source, which is the mirror in this case.

Pumping Power

One can roughly estimate the pumping power necessary for the population inversion as follows:

Let us consider that P be the power of the pumping light of frequency ν_p incident on a unit area of the active material. Let σ_p be the absorption cross-section of the pump light. Thus the pump power absorbed by the material = $\sigma_P P$. Obviously, this power is used in raising the atoms to the excited state. Let τ be the life time of the upper (excited state), the power consumed in the transition to the lower state is $h\nu_P/\tau$. For population inversion to occur we have $\sigma_P P = h\nu_P/\tau$ or $P = h\nu_P/\tau\sigma_P$. In case of ruby laser, we have $h\nu_P = 3 \times 10^{-19}$ J, $\tau = 4$ ms and $\sigma_P = 10^{-23}$ square meter, one obtains $P = 7.5 \times 10^6$ W/s^2.

Apart from ruby, several other solid materials such as calcium fluoride-samarium, glass-niobium, etc. have been used as active materials for solid-state laser action. Basically, this class of lasers refers to the lasers based on ions in solid materials. These doped crystals gave pulsed output. The pulsed calcium fluoride laser has four levels and was first developed by P.P Sorokin and M.J. Stephenson during 1960-61. In general, most ordinary solid state lasers are of the four level type as it is easier to establish population inversion in these systems. In a four-level laser system, excitation and radiation transfer proceed as in the case of a three level laser material, but there is now an additional energy level available above the ground level. Laser action, therefore, begins as soon as there is a significant population of the level above this additional level, which in turn increase efficiency.

Next to ruby, Nd^{3+} is the most commonly used solid material for lasers. It is capable of producing infrared radiations over a large range, the most useful wavelength being 1.064 μm. Rare earth materials used as solid laser give coherent radiation in the range 0.6 to 2.6 μm. Another common solid laser is the YAG laser which has four levels instead of the three levels in ruby. The materials consists of Ytterium Aluminium Garnet ($Y_3Al_5O_{12}$) doped with ND^{3+} ions in place of Y. On pumping optically by flash lamps, the ND^{3+} ions are excited to their characteristic bands. The excited ions decay non-radiatively into the metastable state $4F_{3/2}$ which serves as the upper laser level and has a lifetime of about 0.25 milli second. The radiative (laser) transistions occur from this state predominantly to the $4I_{11/2}$ state below giving the 1.064 μm beam.

(*ii*) Nd: YAG Laser

This is a four-level solid state laser system in which Yttrium Aluminium Garnet ($Y_3Al_5O_{12}$) commonly known as YAG doped with neodymium ions Nd^{3+} acts as *active medium*. The active medium, Yttrium Aluminium

Garnet is taken in the form of a crystal and is drawn into a rod. The neodymium ions (Nd^{3+}) acts as the *active centers.*

The end faces of the active medium, *i.e.* Nd : YAG rod are ground polished and silvered to act as the *optical resonator mirrors* or one can form the optical cavity by using tow external reflecting mirrors M_1 and M_2 (Fig. 6.11).

A xenon flash lamp or a krypton flash lamp is used as a *optical pumping source.*

Figure 6.11 shown the schematic diagram of a Nd : YAG four level solid state laser. The active medium Nd : YAG rod and a krypton flash lamp are enclosed inside an elliptical reflector (Fig. 6.12). To achieve that the entire flash radiations get focused on the laser rod, the Nd : YAG active material rod is placed at one focal axis and the flash lamp at the other focal axis of the ellipsoidal reflector.

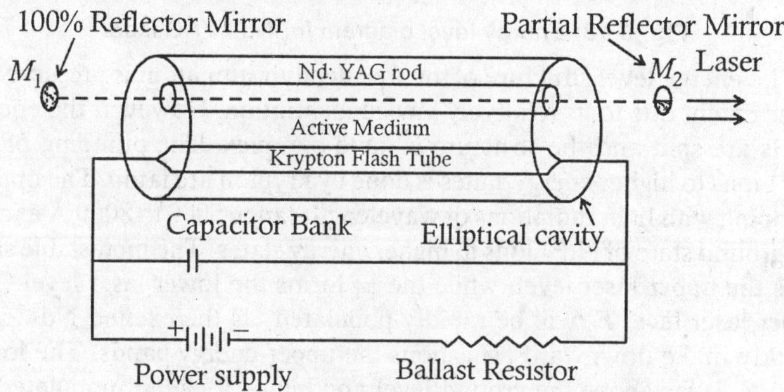

Fig. 6.11: Nd : YAG four level solid state laser

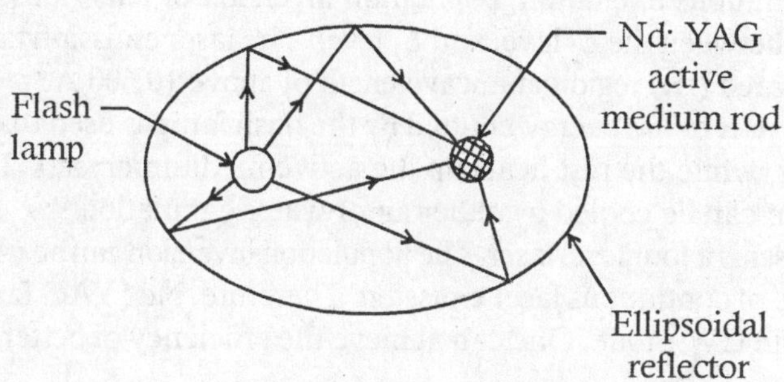

Fig. 6.12: Schematic of a Nd : YAG laser housed in an elliptical reflector

Working

When the flashed lamp is switched on, the optical pumping excites the Nd^{3+} ions from the ground energy state E_0 to the higher energy level E_3

and E_4 by absorbing radiations of wavelengths 0.80 μm and 0.73 μm respectively. Figure 6.13 shows the energy level diagram.

Nd^{3+} ions from the excited state then make a transition from these energy levels. Energy level E_2 is metastable state and the transition from the energy level E_4 to E_2 is a *non-radiative* transition.

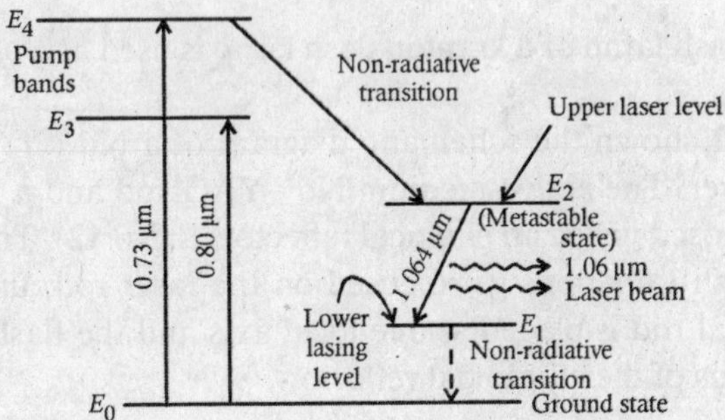

Fig. 6.13: Energy level diagram for Nd : YAG laser

The energy level structure of the free neodymium atom is preserved to some extent due to its relatively low concentration. However, the energy levels are split and the structure is quite complex. The pumping of the Nd^{3+} ions to higher energy states is done by krypton arc lamp. The optical pumping with light radiations of wavelength range 5000 to 8000 Å excites the ground state of Nd^{3+} ions to higher energy states. The metastable state E_2 is the upper laser level, while the E_1 forms the lower laser level. The upper laser level E_2 will be rapidly populated, as the excited Nd^{3+} ions quickly make downward transitions the upper energy bands. The lower level E_1 is far above the ground level and hence it cannot populated by Nd^{3+} ions through thermal transitions from the ground level E_0. Therefore, upon continuous excitation, population inversion of Nd^{3+} ions is readily achieved between the E_2 level and E_1 level. The laser emission takes place in the infrared (IR) region at a wavelength of above 10,600 Å (= 1.06 μm).

Only a part of the energy emitted by the flash lamp is used to excite the Nd^{3+} ions, while the rest heats up the active medium crystal. Therefore, the system can be cooled by either air or water circulation.

This laser is a four level laser. The population inversion can be maintained in the face of continuous laser emission. Therefore, Nd : YAG laser can be operated in CW mode. One can achieve the efficiency of better than 1%.

Applications

Nd : YAG lasers find many applications. The major applications are:

(*i*) in the study of inertial confinement of fusion.

(*ii*) used in military as range finder and as target designations.

(*ii*) used for cutting, drilling, welding and surface hardening of the industrial products.
(*iv*) used in medical field for cataract surgery, to treat gastrointestinal bleeding and gall bladder surgery.
(*v*) used in long haul communication.

6.9 SEMICONDUCTOR (PN JUNCTION) LASERS

Coherent laser radiation ranging from the infrared, through the ultraviolet regions of the spectrum has been obtained from semiconductors and semiconductor junction diodes. Laser radiation can be obtained by pumping a solid intrinsic semiconductor or a PN junction diode. The method of pumping varies with the type of semiconductor used. The most common type of semiconductor used us a III–V compound. Semiconductor lasers are different from other lasers in the following respects:

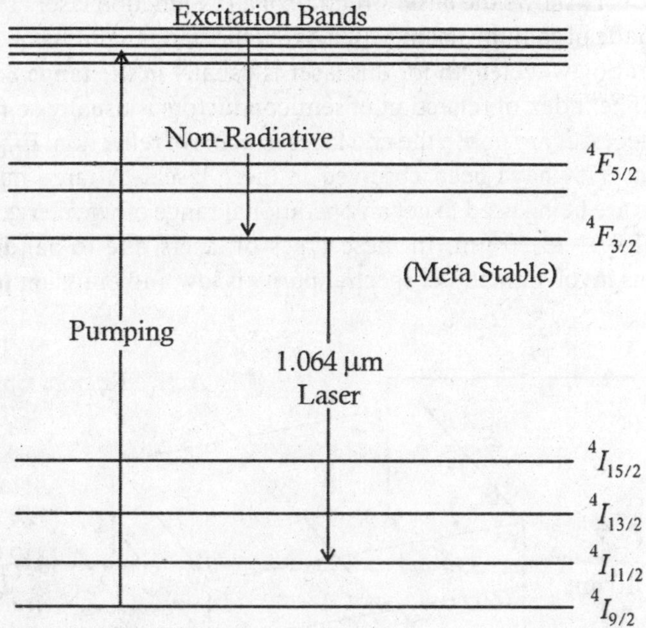

Fig. 6.14: Energy level diagram of Nd in YAG laser

(*i*) Transition occurs between energy bands rather than between discrete energy levels, the emission being a result of electron transitions form the conduction to the valence bands.
(*ii*) The laser is very small in size, typically 50 μm × 250 μm × 50 μm.
(*iii*) The characteristics of the laser beam are strongly influenced by the properties of the junction material.
(*iv*) In PN junction lasers, population inversion occurs in the very narrow region about the junction. The pumping of the PN junction

laser is accomplished by the application of a forward bias to the diode.

It is a property of semiconductors that the downward transitions of electrons recombining with holes may or may not result in the emission of optical energy. Based on this property, semiconductors can be classified as *indirect and direct*. This classification is related to the energy band structure: silicon and germanium are indirect types, whereas most of the III–V compounds are direct. In direct types, radiative transitions take place faster than non-radiative and impurity transitions. In indirect semiconductors, radiative recombinations occur faster than radiative recombinations, so that in essence photon emission is suppressed. In fact, some radiation is emitted by almost all semiconductor PN junctions, but in junctions using the indirect type of semiconductors, this radiation is very inefficient.

Figure 6.15 shows the basic structure of a PN junction laser. The diode laser is made of gallium arsenide (GaAs) by the standard diffusion method. The operating wavelength for this laser is usually in the range 8400 Å to 8500 Å. The index of reflection of semiconductors is usually so high that it is not necessary to polish the end faces to achieve reflection. Efficiencies as high as 70% have been observed in these lasers. A large number of materials are being used to get an operational range of wavelengths in the region of 0.33 to 30 μm. In these types of lasers due to band-to-band transitions involvement, the spectral purity is low and emission pattern is broad.

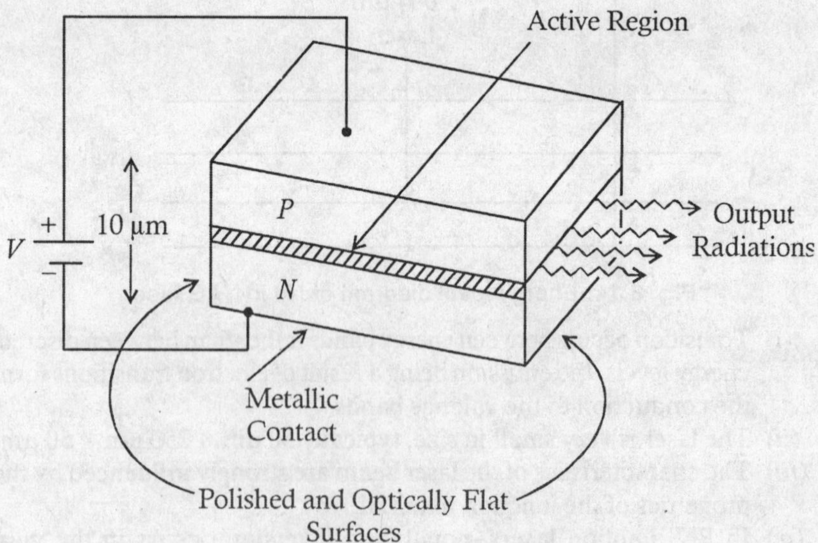

Fig. 6.15: Semiconductor laser

An important difference between a solid state laser and a semiconductor laser is that while in the former the bulk of the material is the host and the active material is as low as nearly one per cent, whereas in the latter the entire material is participating in the laser action. It is important to mention that the semiconductor diode laser is essentially an electroluminescent diode made from a degenerated direct band-gap semiconductor like GaAs, operating under a heavy forward bias.

Population Inversion in a Semiconductor Laser

One can obtain the population inversion in a PN junction by heavily doping both the P and N regions and applying a forward bias. The heavy doping pushes the Fermi level E_F in the conduction band in the N-side and in the valence band in the P-side [Fig. 6.16 (*a*)]. The system is in thermodynamic equilibrium when there is no applied voltage and obviously there is no population inversion. To create population inversion we will have to disturb the thermodynamic equilibrium, *i.e.* a forward bias has to be applied. For an applied forward voltage V, the band diagram of the on junction is shown in Fig. 6.16 (*b*).

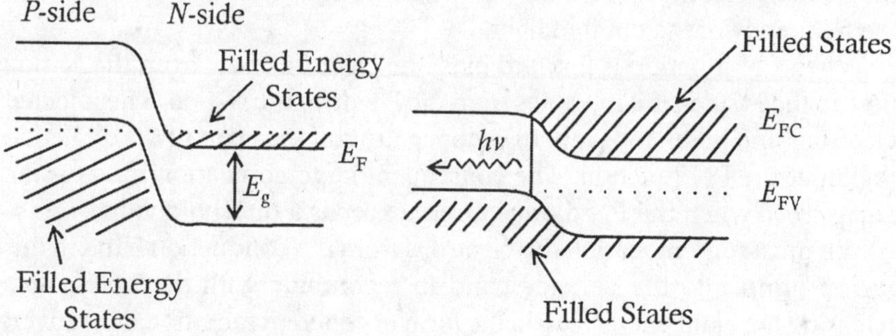

Fig. 6.16: (*a*) Energy band diagram of a PN junction diode (unbiased heavily doped) used as a laser

Fig. 6.16: (*b*) Energy band diagram of an heavily doped PN junction diode (for an applied voltage *V*)

Let us assume that the charge carriers in the two bands are in thermal equilibrium with each other, This means that the population distributions in the conduction and in the valence bands are described by the quasi Fermi levels E_{FC} and E_{FV} respectively one can express the occupation probability in the conduction band by

$$f_c(E_c) = \frac{1}{1 + \exp(E_c - E_{Fc})/k_B T} \tag{23}$$

and that in the valence band is expressed as

$$f_v(E_v) = \frac{1}{1 + \exp(E_v - E_{Fv})/k_B T} \tag{24}$$

For a light beam to be incident, one finds that the time rate of absorption of quanta R_a is proportional to (*i*) the probability per unit time of the transition from the valence band to the conduction band B_{vc} (*ii*) the density of incident radiation $u(\omega)$, (*iii*) the probability of occupancy of the valence bands $f_v(E_v)$ and (*iv*) the probability of vacancy of the conduction $[1 - f_c(E_c)]$. Thus, one can write

$$R_a = CB_{vc}u(\omega)f_c(E_v)[1 - f_c(E_c)] \tag{25}$$

Here C (constant of proportionality) includes the density of states in the two bands. Similarly, one can write the expression for the time rate of emission of the quanta by stimulated emission R_e as

$$R_e = CB_{cv}u(\omega)f_c(E_c)[1 - f_v(E_v)] \tag{26}$$

For laser action, the emission must exceed absorption, we have $R_e > R_a$, *i.e.*

$$f_c = (E_c)[1 - f_v(E_v)] > f_v(E_v)[1 - f_c(E_c)] \tag{27}$$

We can easily see that Einstein's B coefficients B_{cv} and B_{vc} are equal. Using Eqs. (23) and (24), Eq. (27) reduces to

$$E_{FC} - E_{FV} > E_C - E_V = h\nu \tag{28}$$

Since $E_{FC} - E_{FV} = eV$, where V is the applied forward bias, the inequality (28) reveals that the applied forward voltage must exceed $h\nu / e$, where ν is the frequency of the emitted light.

When PN junction is forward bias, the electrons flow from the N-side flow to the P-side, and the holes from the P-side to the N-side. The injected electrons and holes appear in high concentrations in a part of the depletion region at the PN junction. The consequent large population initiates the *lasing action* when the PN diode current exceeds a threshold value.

We must remember that the electrons from the conduction band in this region jump into the valence band to recombine with the holes, thus releasing the band energy gap in the form of coherent radiation. Moreover, the P and N regions of the diode absorbs the laser radiation which interacts with the lattice photons. Obviously, the laser action is restricted to the very thin junction region of the diode.

Semiconductor lasers are an important light source for fibre-optic communications and are key components of such common appliances as compact disc players, supermarket scanners, laser printers, fax machines, and laser pointers. The lasers in these applications are so called double heterostructure lasers, essentially diodes consisting of an active semiconductor region sandwitched between doped semiconductor cladding layers, one N-type, the other P-type. The cladding regions supply electrons and holes to the active region when an appropriate bias voltage is applied. They also have a higher band gap and lower refractive index than the active layer, so that the injected electrons and holes as well as the photons generated by their annihilation are confined to the active region.

Double heterostructure lasers have demonstrated high performance and have been successfully commercialized for wavelengths ranging from blue to the near infrared (IR), upto about 1.6 μm. Unfortunately, few semiconductor materials emitting in the mid-IR (2-20 μm) are reliable, easily processed, and insensitive to temperature cycling—the repeated heating and cooling associated with laser operation.

6.10 GAS LASERS

Laser action can also be made to occur in some gases, such as CO_2 or a mixture of He and Ne. The credit for the discovery of first gas laser goes to A. Javan, W. Bennet and D. R. Herriot of Bell Labs. They used the electric discharge through a gas for ionisation and pumping action, *i.e.* resonant transmission of excitation energy in a gas discharge. Sometimes excitation or pumping may be provided externally by a separate electrode. The cavity is about 30 to 50 cm long having a tube diameter of about 5 mm and two concave reflecting mirrors at the ends. A steady potential of a few thousand volts is used for getting a discharge. The main advantage of such laser resides in that they operate continuously, through some gas lasers are also capable of pulsed working. These lasers display exceptionally high monochromaticity, most pure spectrum and high stability of frequency. All these features make these lasers extremely useful in various branches of science and engineering. The tube used in these lasers is made up of glass or quartz. The ends of the tube terminate in flat glass windows inclined at Brewster' angle to eliminate reflections.

He-Ne gas lasers were the first to be developed and used. They operate in three distinct ranges, in the red at 6328 Å, in the near infrared around 1.15 μm and in the infrared at 3.39 μm. The partial energy levels diagram of He-Ne is shown in Fig. 6.17 (*b*), which explains the origin of these lines. With the help of a dispersive prism, these lines can be separated.

A gaseous mixture of He and Ne (about 10 : 1) is used as active medium. When the helium atoms in the mixture are excited to the metastable state by means of a *dc* or radio-frequency source, they collide with the ground state neon atoms and resonant energy transfer takes place. One of the excited states of the neon has nearly the same excitation energy as He metastable state. Hence the Ne atoms are excited in the collision to a specific energy level as the helium atom returns to the ground-level. This causes a population inversion in the energy states of neon and laser beam is obtained. He-Ne gas laser is shown in Fig. 6.17 (*a*).

Carbon Dioxide (CO_2) Laser

This is a very useful and efficient laser. CO_2 laser is four level molecular laser and operates in *far infrared* region

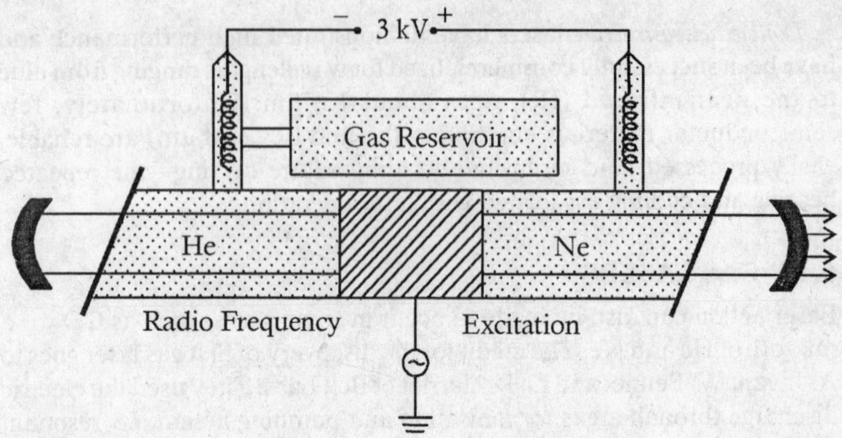

Fig. 6.17 (*a*): He-Ne gas laser

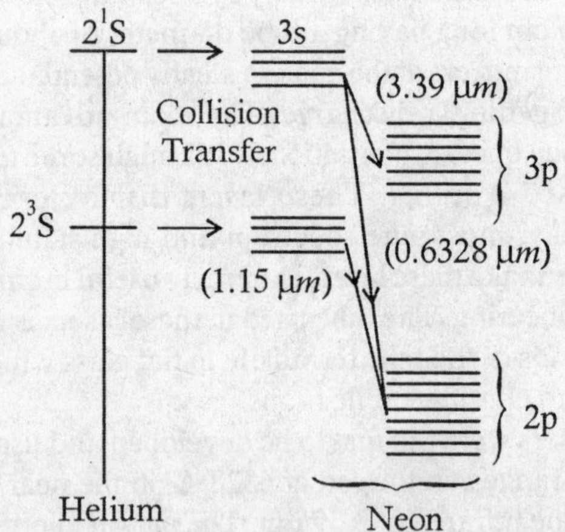

Fig. 6.17 (*b*): Partial energy levels of He-Ne gas laser

***Active Medium*:** The active medium for CO_2 laser is a mixture of CO_2, N_2 and He gases. In this laser, the transitions between various vibrational levels of same electronic states of CO_2 molecule leads to the production of laser beams.

Energy Levels of CO_2 Molecule

The energy levels of an isolated atom are discrete and narrow, but in the case of molecules the energy spectrum is complicated due to several additional features. Each electronic level is associated with nearly equally spaced vibrational levels and each vibrational level in turn has a number of rotational levels.

The Vibrational Models of CO_2 Molecule

CO_2 molecule is a linear molecule consisting of central carbon (C–) atom with two oxygen (O–) atoms attached on either side. A molecule of this type undergoes three independent vibrational oscillations known as the *vibrational modes* as shown in Fig. 6.18. These vibrational degrees of freedom are quantized. At any one time, a CO_2 molecule can vibrate in a linear combination of three fundamental modes. The three states of vibration of the molecule is represented by three quantum numbers ($m\ n\ q$). These quantum numbers represent the amount of energy associated with each mode, e.g. (0 2 0) indicates that the molecule in this energy state is in the pure bending modes with two units of energy. The three different vibrational modes are:

(*i*) **Symmetric Stretching:** The oxygen atoms oscillate along the axis of the molecule, by simultaneous moving away (departing) and approaching the carbon atom in between, which is stationary [Fig. 6.18 (*a*)]. The corresponding frequency is termed as symmetric *stretching frequency* and is denoted by m.

(*ii*) **Bending Mode:** All the three atoms of CO_2 molecule undergo vibrational oscillations perpendicular to the axis of the CO_2 molecule. However, the movement of oxygen atoms and the carbon atom will be in the opposite directions [Fig. 6.18 (*b*)]. The corresponding frequency is termed as the bending frequency and is denoted by n.

(*iii*) **Asymmetric Stretching Mode:** All the three atoms of CO_2 molecule oscillate along the axis of the molecule with both the oxygen atoms moving in the same direction while the carbon atom moving in the oppositge direction [Fig. 6.18 (*c*)]. The corresponding frequency is termed as the stretching frequency and denoted by quantum number q.

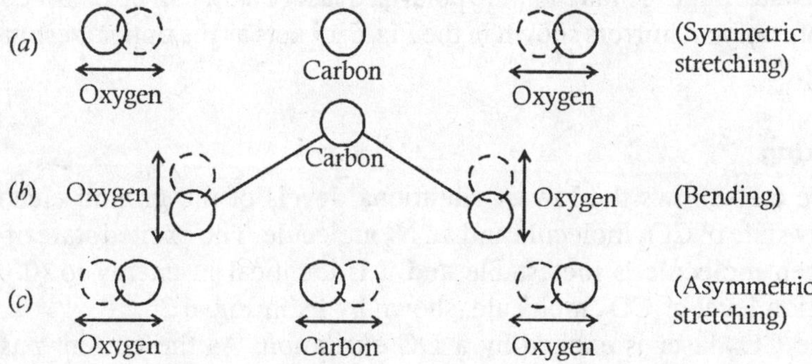

Fig. 6.18: The vibration modes of CO_2 molecule

We may note that each vibrational state is associated with rotational states corresponding to the rotation of CO_2 molecule about its centre of mass. The seperations between energy vibrational-rotational states are much smaller on the energy scale compared to the separation between electron energy levels. At any instant, the CO_2 molecule can be vibrating in a linear combination of above three fundamental modes. The N_2 molecule is also characterised by similar vibrational levels.

Construction

Figure 6.19 shows the schematic of typical CO_2 laser. It is basically a discharge tube of 2.5 cm diameter and 5 m length. The discharge tube is filled with a mixture of CO_2, nitrogen and helium gases in 1 : 2 : 3 proportions respectively. Other additives, e.g. water vapour is also added. CO_2 moles acts as active centres. Discharge is produced by DC excitation (Fig. 6.19).

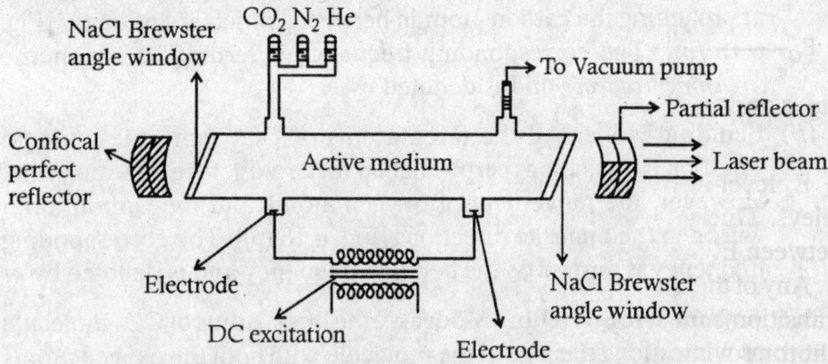

Fig. 6.19: Schematic of carbondioxide laser device

At the end of discharge tube, sodium chloride windows are placed inclined at Brewster angle to the resonator axis. The windows are inclined at Brewster angle so that a plane polarized laser beam can be obtained. A pair of confocal mirrors shown in the Fig. 6.19 acts as the optical resonator cavity.

Working

Figure 6.20 shows the lowest vibrational levels of the ground electron energy state of CO_2 molecule and an N_2 molecule. The excited state of the nitrogen molecule is metastable and it is identical in energy to (0 0 1) vibration level of CO_2 molecule, shown by E_4 in Fig. 6.20.

The CO_2 laser is excited by a DC excitation. As the current passes through the mixture of gases, the N_2 molecules get excited to the metastable state, i.e.

$$N_2 + e^- \rightarrow N_2^*$$

The excited N_2 molecules cannot spontaneously lose their energy and consequently, the number of N_2 molecules at that level keeps on increasing.

These N_2 molecules return to ground state through inelastic collisions with ground state CO_2 molecules. Thereby the CO_2 melecules get excited to the level E_4 (Fig. 6.20).

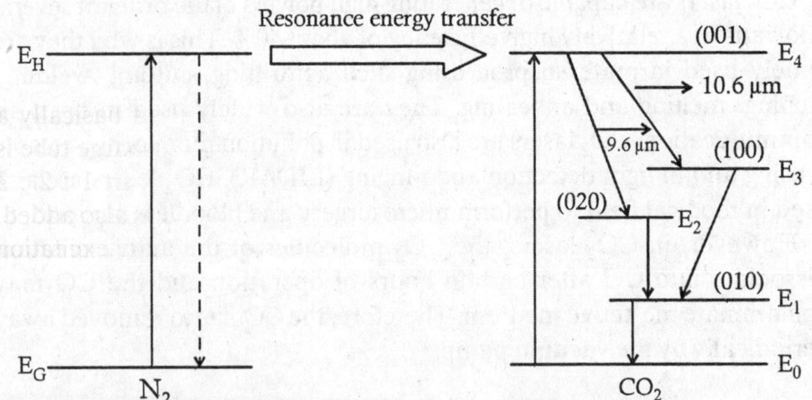

Fig. 6.20: Energy levels of nitrogen and CO_2 molecules and transition between the energy levels

E_4 level is the upper laser level and E_3 and E_2 levels act as the lower lelevs. Due to continuous excitation, population inversion is achieved between E_4 level and levels E_3 and E_2.

Any of the spontaneously emitted photon make the molecules to undergo transition between $E_4 \rightarrow E_3$ thereby emitting a stimulated photon. These photons while travelling along the resonator axis undergo multiple reflections at the optical resonator mirrors. As a result, the transition $E_4 \rightarrow E_3$ yields an intense and coherent laser beam output of wavelength 10 µm (1,06,000 Å). The lasing transition between $E_4 \rightarrow E_2$ levels produces far IR radiation at 9.6 µm (96,000 Å) wavelength. Both the wavelengths clearly belong to far IR region.

The transition between $E_3 \rightarrow E_1$ and $E_2 \rightarrow E_1$ takes place due to elastic collisions between the excited CO_2 molecules with the unexcited CO_2 molecules. Finally, the CO_2 molecule by elastic collision with helium atoms make the transition $E_1 \rightarrow E_0$, i.e. helium helps in depopulating the lower levels. Moreover, helium helps to conduct heat away to the walls of the discharge tube keeping CO_2 cold. This is achieved due to high thermal conductivity of helium.

N_2 gas helps to increase the population of the upper level of CO_2.

CO_2 laser operates in CW mode and the power output of this laser is 50-60 W/m, when there is longitudinal flow of gases.

(The energy states of CO_2 molecule is represented by three quantum numbers (*m n q*). Quantum number (0 1 0) means that, the CO_2 molecule in this energy state is in a pure bending mode of vibrations. One can provide similar explanations for other quantum numbers, i.e. (1 0 0) and (0 0 1)).

When the flow of gases is perpendicular to the discharge tube, the power output can be raised to 10 kW/m. Such type of CO_2 lasers are called as Transversely Excited Atmospheric Pressure Laser or TEA LASER.

CO_2 lasers are capable of generating high powers of the order of several kilowatts at a relatively high efficiency of about 40%. This is why they are widely used in material processing such as drilling, cutting, welding, etching, melting and annealing. They are also widely used in open air communication. CO_2 lasers are also used in pollution monitoring, remote sensing and in light detection and ranging (LIDAR). CO_2 lasers are also used in medical field to perform microsurgery and bloodless surgery.

However, in CO_2 lasers, the CO_2 molecules of the active medium dissociate into CO after certain hours of operation and the CO may contaminate the active medium. Therefore, the CO has to removed away periodically by the vacuum pump.

6.11 CHEMICAL LASERS

In chemical lasers, population inversion results from a chemical reaction. The high energy density that can be released in exothermic chemical reactions has led to the interest in converting this energy to coherent optical energy. If one could use the energy of such chemical reactions to obtain a population inversion directly, one could build compact and powerful lasers. Such lasers may also become a powerful tool in the study of kinetics of chemical reactions.

The first chemical laser was developed by J. V. Kasper and G. C. Pimentel in 1965. The active molecule was vibrationally excited hydrogen chloride (HCl^*) molecule, formed by the following chemical reaction

$$\left.\begin{array}{l} Cl + H_2 \rightarrow HCl^* + H \\ H + Cl_2 \rightarrow HCl^* + Cl \end{array}\right\} \qquad (29)$$

The initial chlorine molecule atoms were produced by flash photolysis according to the following reaction:

$$Cl_2 + h\nu \rightarrow 2Cl \qquad (30)$$

corresponding to the several rotation lines of the HCl molecule the output is 3.7 to 3.84 μm.

Since the discovery of H_2—Cl_2 flash photolysis laser many pulsed and continuous lasers based on the hydrogen isotopes and halogens have been developed. The powers of some of these chemical lasers go as high as several hundred watts. In all these chemical lasers some external means such as electric discharge, flash photolysis or shock heating starts the chemical reaction. However in 1969 T. Cool and R. Stephens of Cornell developed a purely chemical laser based on the following chemical reactions:

$$\left.\begin{aligned} &F_2 + NO \rightarrow NOF + F \\ &F + D_2 \rightarrow DF + D \\ &D + F_2 \rightarrow DF + F \\ &(DF)^* + CO_2 \rightarrow DF + CO_2 \end{aligned}\right\} \quad (31)$$

In the first chemical reaction F_2 is dissociated to give free fluorine atoms which on mixing with D_2 yield vibrationally excited DF. The excited DF transfers its energy to CO_2 and laser action results at the wavelength 10.6 microns. The power output from this laser is about 500–600 watts.

6.12 DYE LASERS

P. P. Sorokin and J.R. Lankard of I.B.M. in 1966 obtained laser action in organic molecules by their efficient fluorescence. They also produced laser action in many dye solutions which were pumped with microsecond long flashes from high intensity lamps. As shown in the energy level diagram 6.21, the optical excitation and emission in organic dyes take place between the rational vibrational levels of the ground singlet state S_0 and the excited singlet state S_1.

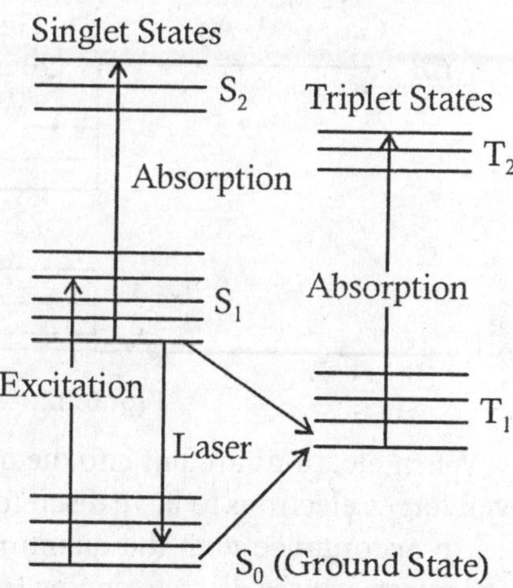

Fig. 6.21: Optical excitation and level emission between vibrational S_0 and S_1 in a dye laser

The active medium of an organic dye laser is an organic fluorescent material dissolved in a common solvent. These substances derive their colour from strong absorption in the visible region of spectrum. Sorokin obtained the laser action by using a solution containing a dye-chloro-aluminium pthalocyanine at 7650 Å. Dye laser can operate at room temperature and moreover they are tunable.

6.13 X-RAY AND γ-RAY LASERS—GRASER

Lasers generating wavelengths shorter than 1 Å would have many important applications. In principle, laser action could be achieved at very short wavelengths using radiation arising from nearly forbidden transitions. At these short wavelengths there is no competition from non-radiative transitions, hence proposals have been made for achieving X-ray or γ-ray lasers in solids than in gases. Soft X-ray lasers, which a decade ago were generally conceded to be impossible, are now a reality. The possibility of a nuclear laser, often termed a 'graser' (for "gamma ray laser") was recognized early in 1960's. A graser may be possible if ways can be found to achieve Mossbauer transitions and population inversion simultaneously in nuclear isomers. Looking to its military importance efforts are going on in various laboratories of the world to achieve graser.

6.14 QUANTUM WELL LASER (*QWL*)

Quantum wells in which a smaller band gap semiconductor (such as GaAs) some tens of *A* thick, sandwiched between the layers of a larger band-gap difference between the two materials have been produced due to recent development in crystal growth. There is an offset of the conduction and the valence band edges across the quantum well interfaces due to the band gap difference between the two materials as shown in Fig. 6.22.

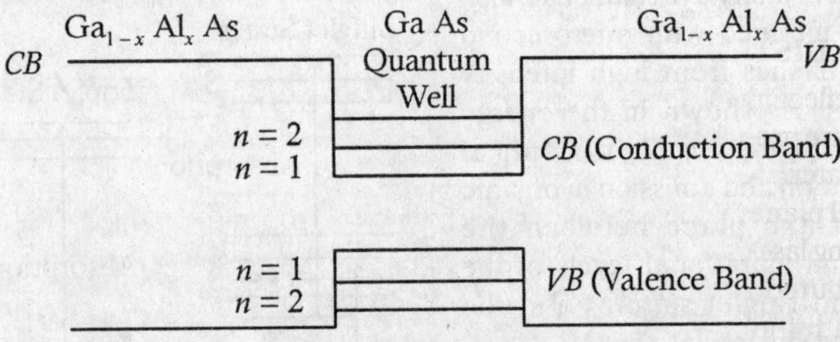

Fig. 6.22: Quantum well

When electrons are put into the quantum well, the confinement in the well forces electrons to have discrete energy levels designated by $n = 1, 2, \ldots$. In accordance with the quantum principle, obviously, the quantum size effects causes discrete energy levels in the direction perpendicular to the interfacial planes. A semiconductor laser based on the principle of quantum well is called as a *quantum well laser*.

The gain becomes larger and sharper for a quantum well laser than that for an ordinary semiconductor laser due to the formation of the discrete levels which reduces the spread of the electron. The threshold current of the quantum well laser is low and moreover, one can control the output wavelength by the well thickness. One finds that quantum well laser has an excellent temperature characteristic and a narrow spectral gain width.

QWL is an improved version of the ordinary *double-hetro structure* laser shown in Fig. 6.23. One can see that the thickness *d* of the active layer of GaAs contained between the outer layers of N-type (Ga, Al) As and P-type (Ga, Al) As and is about 200 nm. Electrons from the N-type (Ga, Al) As and holes from P-type (Ga, Al) As move into the conduction band and valence band of the active layer respectively. This causes a population inversion for lasing action. The light is guided between the outer layers due to the larger refractive index of the active layer. This offers a narrow beam. We can easily see that this is an additional feature over the confinement of the carriers. We can also see that *d* is greater in the ordinary double hetro structure laser than in *QWL*. This means that the energy level quantization in the valence and conduction bands in the double hetro structure laser is of no significance.

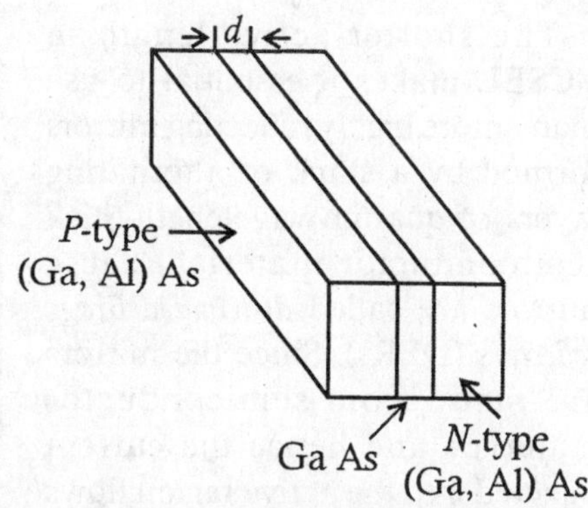

Fig. 6.23: Double hetro structure laser

6.15 DISTRIBUTED FEEDBACK (DBF) EMITTING LASER

The reflecting mirror facets in the conventional edge-emitting semiconductor laser are produced at either end of the optical cavity by cleaving or etching the material along two parallel crystal planes. In the *DBF* edge emitting laser, the reflections are distributed along the cavity length by producing periodic index varaitions which individually reflects a small fraction of the photons as shown in Fig. 6.24.

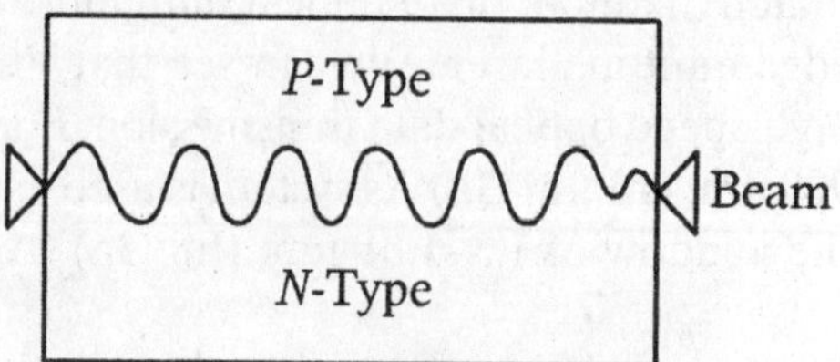

Fig. 6.24: A DFB edge-emitting laser

We must note that collective reflection in *DBF* edge emitting laser is more or less the same as the reflection from the facets of a conventional laser. However, the feedback mechanism of the *DBF* laser is much more sensitive to the wavelength in comparison to the conventional laser. Obviously, the emitted light of the *DBF* laser has a much narrower spectral line width. One can efficiently couple the coherent output beam to an optical fibre.

A thin wide and relatively long cavity is required in either type of edge-emitting laser. This means that the device size is relatively large and it required some tens of *mA* to operate.

6.16 VERTICAL-CAVITY SURFACE-EMITTING LASER (*VCSEL*)

VCSEL developed in 1990s, requires an active light emitting region contained between two mirrors. In this device, light is recirculated vertically in an optical cavity so that emission occurs normally from the chip surface. In VCSEL, the length of the active region is considerably shorter than in other laser devices.

The shorter active length in VCSEL makes it essential to use many more highly reflecting mirrors formed by a stack of alternating layers of quarter-wavelength-thick semiconductor material. These mirrors are called *distributed Bragg reflectors* (DBRs). Since the mirrors are made from semiconductor materials and hence the current injected into the active region flows directly through these mirrors. One can form a PN junction with current flowing vertically through the device by doping one DBR-P-type and other DBR-N-type as shown in Fig. 6.25.

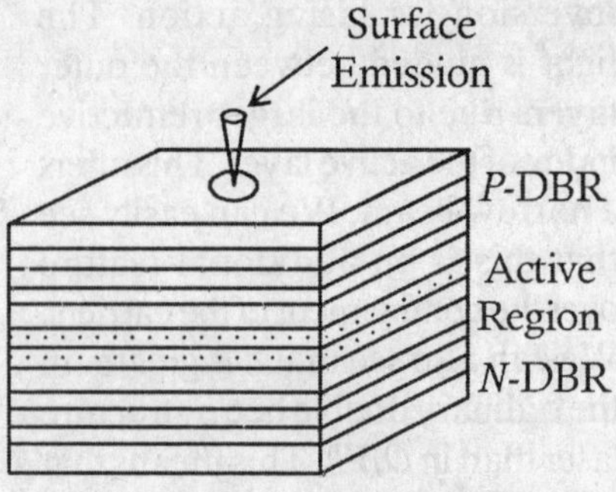

Fig. 6.25: A VCSEL Laser

A VCSEL has compact size and the design flexibility of the output aperature. The small size of VCSEL ensures that the current required for lasing action is quite small. The low capacitance of the device makes the operating speed inherently high. One can tailored the emission pattern to match the input of an optical fibre more effectively than possible with an edge emitting laser. We can see that VCSEL's form an ideal choice for high-speed optical data transmission. For the wavelength window of 780–980 nm, (Al, In, Ga) As system is used as material for VCSEL whereas for the window of 1.3–1.55 μm, (In, Ga) (As, P) system is used.

6.17 QUANTUM DOT LASERS

Quantum dot lasers offer the potential to yield better properties as compared to quantum lasers. These emit light at wavelengths that are determined by the energy levels of dots. The research in nano-crystal quantum dots has opened the door for developing novel optical and opto-electronic devices, such as tunable lasers, optical amplifiers and light emitting diodes from assemblies of these invisibly small particles. Quantum dot lasers work like any other semiconductor lasers such as those found in home audio compact disk players. The goal of a quantum dot laser is to manipulate the material into a high energy state. The result is the net release of energy which emerges as a photon. European group of scientists have launched a project called 'BIG BAND' to develop ultra-wide band InP based quantum dot lasers amplifiers for telecom applications.

The primary goal of 'BIG BAND' is to produce QD-based semiconductor optical amplifiers (SAOs) operating over the entire telecommunication window. SOAs are seen as potential replacement for erbium-doped fibre amplifiers in optical networks. Another goal is to push PnP device emission to longer wavelengths for gas sensing applications.

6.18 APPLICATIONS OF LASERS

With the discovery of laser, man's control of light has been and will continue to extend to an unpredictable age and diverse number of applications. The applications discussed follow from one or more of the special properties of the laser beam, namely: coherency, power, or high intensity directionality, and narrow frequency age. It would be practically impossible to list all the potential uses of the laser, but some of the important uses and applications of lasers are mentioned below. The technological uses of lasers are enormous, and their development is but example of many of the interplay between basic science and technology.

(*i*) Surgery

The powerful, well collimated beam emitted by the laser shows promise in many applications. Exploration is being conducted in such fields as tooth drilling, eye surgery, cancer treatment, and cardiovascular surgery. In most of these cases the laser is like an *optical knife* which would be more accurate, less painful, and faster than the scalpel.

(*ii*) Welding and Metal Cutting

Experiments have shown that the laser can be used effectively for welding. By passing the laser beam through a lens system, an extremely high energy density can be focused in a very small area. The coherency of the light beam permits this high degree of focusing. It appears that the laser will be most useful for spot welding, with the electronic beam welder or resistance welder continuing to perform beam welds. Some of the advantages in using the laser for spot welding are: (*a*) no physical contact with the material being welded, (*b*) welding can be accompanied in air, (*c*) very little rise in temperature in the area surrounding the weld, and (*d*) a variety of metals have been successfully welded. Welding transistor junctions, thermocouples, electronic micro-modules are examples of the precise work which the laser can perform. Another interesting aspect of welding with the laser is that it is possible to weld a joint even after the joint has seen scaled inside a glass envelope.

With some minor changes in control, the same types of laser that perform welding can be used for cutting or burning metals. Experiments have succeeded in punching holes of the order of 0.005 inch diameter in sheets of metal. The laser has also succeeded in cutting holes in diamond, one of the hardest materials.

(*iii*) Communication

The application of the laser to the field of communication is receiving a large amount of technological attention. Since the light from the laser is coherent, it can theoretically carry messages in the same manner as low-frequency carriers.

The frequency of the laser is so high that each message or band of frequencies is a very small percentage of one carrier frequency (in this case it is the frequency of the laser). For instance, if a carrier is 1,10,000 Hz and we were allowed to modulate this with messages up to 10 per cent of the this frequency, then 1,00,000 Hz would be out from total allowable message band width. However, at 1,00,000,000 Hz carrier frequency, the allowable 10 per cent would give 10,000,000 Hz total message band width. Obviously, many more messages can be sent at higher carrier frequencies.

Powerful though laser light may be, it is light and is impeded to some extent by out atmosphere even under good conditions.

(*iv*) Lasers in Industry

The precision property of laser light has been of immense help in industry, particularly in testing the quality of optical components, *e.g.* prisms, lenses, etc. The use of lasers in industry has considerably increased the accuracy in the measurement of the size of the physical quantities.

(*v*) Lasers in Biology

The ability of laser beams to concentrate high power density of light at a focal point has opened a new area of micro Raman spectroscopic analysis. This enables one to extend such studies to biological and biomedical samples available only in very small quantities. The radiation pressure of a laser beam has been made to act on viruses or bacteria as well as on atoms, but in a different way. In this case pair of laser beams are used, which act as *optical tweezers.* Laser beams can also be used to study the forces that curl up long molecules such as DNA.

(*vi*) Laser in Astronomy

Radio astronomers have found lasers extremely valuable for amplifying very faint radio signals from space. Radio telescopes have now an additional accessory—a laser (ruby), with the help of which the astronomers have been able to extend their range of observation.

(*vii*) Military Applications of Lasers

Prediction have been made that the laser beam could be used to disintegrate a striking missile in space. A tremendously large amount of energy would have to be produced to create a heat beam that would actually disintegrate its target. Along these lines, it might be possible to create a truth death-ray gun to be used against humans.

A completely different use of the laser in a military application involves a communication in secrecy. Since the laser beam is well collimated and because of this, a communication link could be established that would be hard to detect unless one were in direct link between the points of communication.

(*viii*) Laser-Fusion

Fusion reactions occur when two light nuclei such as Deuterium (D), Tritium (T), or Helium (^3He) collide and rearrange themselves so as to form two other nuclei of smaller mass with a consequent release of energy. These reactions take place only at very high energy, *i.e.* high temperature (~10^8 K). At this extreme temperature the gas atoms are fully ionized and fully ionized gas is called *plasma.* For the release of sufficient amount of thermonuclear energy and to sustain fusion reaction, it is necessary to confine the hot plasma, so that the fusion reaction can sustain for a long time. The laser has the potential to generate extreme high temperatures and pressure required to initiate a fusion reaction and concentrate large amount of energy in a very small volume. Obviously, lasers can be an extremely useful tool in controlled fusion research.

In the laser-fusion process a solid, spherical pellet of D-T is envisioned to be irradiated by a short pulse, high energy laser beam, which accomplishes a number of tasks: (*a*) the leading portion of the laser beam first creates a plasma surrounding the pellet such that the remainder of the beam is nearly fully absorbed; (*b*) the main portion of the beam then delivers energy to the pellet which not only vaporizes, ionizes, and heats the D-T mixture, but it also induces an implosion, which theoretically results in an ultimate compression of 1000–10,000 times the solid density; (*c*) the resultant highly compressed hot D–T plasma then supports fusion reactions at an extremely high rate in the short time (~10^{-12} secs) before the pellet flies apart due to its high internal pressure.

(*ix*) Controlling and Trapping of Atoms

At room temperatures, a gas of atoms moves with speeds that one can estimate from equipartition to be given by $mv_{rms}^2 / 2 = \frac{3}{2}kT$. For sodium atoms, for e.g., $m = 22 \times 1.7 \times 10^{-27}$ kg, and then one find from this formula that $v_{rms} \approx 0.6 \times 10^3$ m/s. With the help of laser, one can stop atoms and study their properties at leisure.

Atoms that are trapped at low temperatures may be used for a variety of purposes. Trapped atoms can be allowed to fall freely. Their excitations can be measured as they are subsequently hit with a series of short laser pulses of different frequencies. These excitations change with the motion of the atoms, because there is a Doppler shift that is determined by the speed of the atoms. In this way the weaking of gravity due to a rise of as little as 3 cm has been measured! Such an improvement in the measurement of local variations in *g* promises to be of great use in oil and mineral exploration.

(*x*) *Lasers have important applications in non-linear optics and in holography.*

HOLOGRAPHY

8.19 INTRODUCTION

Holography is a way of recording and then reconstructing waves invented by Dennis Gabor in 1948. The waves may be of any kind—light, sound, *X*-ray, corpuscular waves, etc.

The word 'holography' originates from the Greek 'holos' meaning 'the whole'. By using this word, the inventor of holography wanted to stress that it records complete information about a wave—both about its amplitude and its phase.

In conventional photography, only the distribution of the amplitude (more exactly, of its square) is recorded in two-dimensional projection of an object onto the plane of the photograph. For this reason, when examining a photograph from various directions, we do not obtain new angles of approach, and we cannot see, for instance, what is happening behind objects in the foreground.

A hologram, on the contrary, regenerates not a two-dimensional image of an object, but the field of the wave which it scatters. By changing our point of observation within the confines of this wave field, we see the object from different angles, sensing its three dimensional and realistic nature.

It is worthwhile to mention that Gabor did not have a laser when he formulated the idea of holography. He performed his first experiments with mercury arc lamp as source of light. The technique of preparing holography became a practical proposition only after the advent of lasers. In 1963, E. Leith and J. Upatneiks prepared laser holograms for the first time. This opened the way to major advances in holography.

The principle of holography can best be explained in two steps: (*i*) recording of the hologram and (*ii*) reconstructing the image.

A schematic diagram showing how holograms are recorded is depicted in Fig. 6.26.

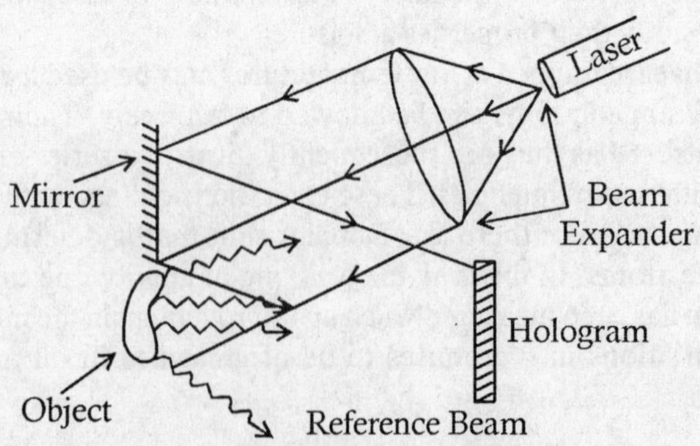

Fig. 6.26: Recording a hologram

(*i*) **The Object Whose Hologram is to be Illuminated with a Laser Beam:** The scattered light wave impinges on a photographic plate. The reference beam-part of the light from the same laser reflected from a mirror—impinges on the same plate. The superposition of these two beams produces an interference pattern and this is recorded on that plate. The developed plate is called as the *hologram*. Since the interference pattern of a hologram is generally very fine, a developed hologram does not differ in appearance from a uniformly exposed plate. A hologram often contains rings and strips. They are due to the diffraction of light on particles of dust getting onto the mirrors and lenses, and have nothing in common with the interference microstructure containing the record of the light wave scattered by the object. The hologram contains, however, enough information to provide a complete reconstruction of the object.

(*ii*) **Reconstruction of the image from a Hologram:** Remove the object and put the hologram is the place where it was when formed then switch on the laser (Fig. 6.27). Now look through the hologram like through a window, you will see the object at its previous place, as if it were not removed, *i.e.* when light from a laser is directed at the hologram from the same direction as the references beam, part of light is 'bent' so that it appears to come from the place once occupied

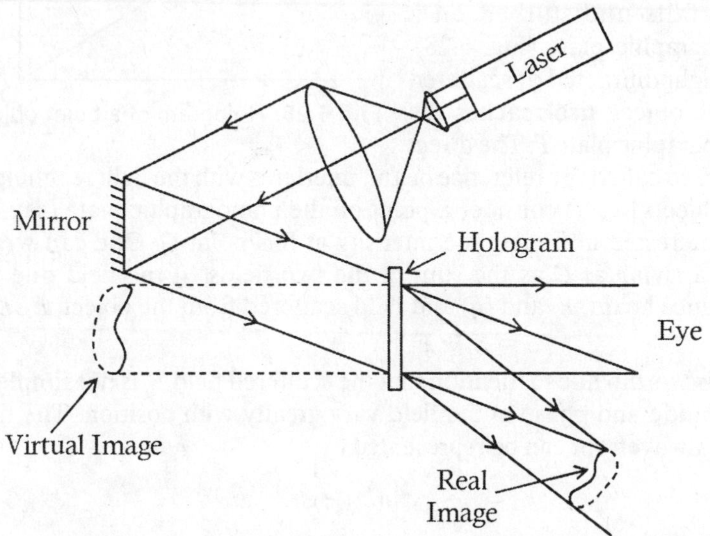

Fig. 6.27: Reconstruction of the image from a hologram

by the object and the result is remarkably realistic 3-dimensional image. The visible object seems to real that we can detect parallax by changing the position of our head. When looking at the near and far parts of the object, we have to accommodate our eyes differently, and if we want to photograph it, we will have to choose such an aperture, as in ordinary photography, which will ensure an adequate field depth. If this is not done, some parts of the object will be sharp on the photograph, and other ones blurred.

From Fig. 6.27 it is clear that two images are produced when the laser beam, which is now broken as the *readout wave*, interacts with the interference pattern on the holographic plate and two images are produced by the diffracted waves. One of them appears at the original position occupied by the object called the *virtual* image and the other is called the *real image* which can be photographed directly without using a lens. Since, the real image reverses foreground and background and hence the interest of the observer lies in the virtual image.

Theory

To understand the various characteristics of holograms it is essential no understand the theory on which recording of a hologram and reconstruction of the image is based.

Let us consider a small object O. Let it be illuminated by a source of light, but most of the light falls undisturbed on a photographic plate (Fig. 6.28). The light diffracted or scattered by the object also reaches the photographic plate P. The direct beam, so called the reference beam, interferes with the light reaching from the object. Let us consider a point on the photographic plate (say Q) we are interested in finding the intensity at this point Q. One can write the field arriving at Q as the sum of the two fields (*i*) the field due to the reference beam, E, and (*ii*) and field scattered from the object, E_s, *i.e.*

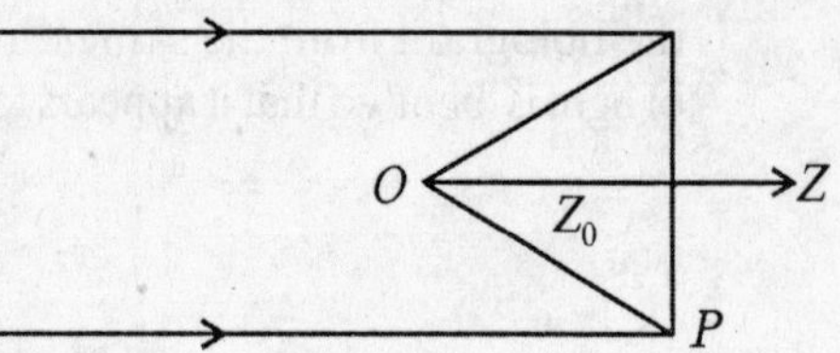

Fig. 6.28: Hologram of a point object O

$$E = E_r + E_s \tag{32}$$

It is worthwhile to mention that the scattered field E_s is not simple, both amplitude and phase of the field vary greatly with position. The field of such a wavefront can be represented by

$$E_s = \frac{A_0}{r_0} \exp\left[i(kr_0 - \omega t)\right] \tag{33}$$

Here $r_0 = PQ$ (Q is not shown in Fig. 6.28).

The reflected wavefronts are spherical and concentric around the point of origin. One can represent the field E_r by plane wave as follows:

$$E_r = A_r \exp[i(kZ_0 - \omega t)] \tag{34}$$

where Z_0 is the distance of the photographic plate from O.

The intensity at the point Q is given by

$$I = |E_r + E_s|^2$$

Substituting the values of E_s and E_r from Eq. (33) and (34) respectively, one obtains

$$I = |A_r|^2 + \frac{|A_0|^2}{r_0^2} + \frac{A_0 A_r^*}{r_0} \exp[ik(r_0 - Z_0)] + \frac{A_0 * A_r}{r_0} \exp[ik(Z_0 - r_0)] \quad (35)$$

Choosing the constants k and ϕ suitably, one can combine last two terms in Eq. (35) and obtain

$$I = |A_r|^2 + \frac{|A_0|^2}{r_0^2} + k\frac{\cos(k(r_0 - Z_0) + \phi)}{r_0} \quad (36)$$

Eq. (36) shows that the blackening of the photographic plate depends on the three terms, the last term being the most important as it gives the effect of the functions A_0 and ϕ. Obviously, the cosine term shows that the total intensity I as function of r_0 shows a series of maxima and minima. The interference of the plane wave E_0 with the spherical wave E_s thus produces a set of circular interference fringes on the photographic plate, which when developed forms the hologram.

If one assumes that the response is proportional to the intensity I, the power transmission of the plate represented by T^2 is given by

$$T^2 = 1 - I\alpha \quad (37)$$

where α is a constant. Eq. (37) can be approximated as

$$T \cong 1 - \frac{1}{2} I\alpha \quad (38)$$

Reconstruction

Let us consider that this hologram is illuminated by a beam which is identical to the reference beam, the field of the transmitted wave will be given the expression.

$$E_1 = TE_r\left(1 - \frac{\alpha}{2} I\right) A_r \exp[i(kZ_0 - \omega t)]$$

$$= \left[1 - \frac{\alpha}{2}|A_r|^2 - \frac{\alpha}{2}\frac{|A_0|^2}{r_0^2}\right] A_r \exp[i(kZ_0 - \omega t)]$$

$$- \frac{\alpha}{2}\frac{A_0 A_r^*}{r_0} \exp[ik(r_0 - Z_0)] A_r \exp[i(kZ_0 - \omega t)]$$

$$- \frac{\alpha}{2}\frac{A_0 * A_r}{r_0} \exp[ik(Z_0 - r_0)] A_r \exp[i(kZ_0 - \omega t)] \quad (39)$$

$$= \left[1 - \frac{\alpha}{2}|A_r|^2 - \frac{\alpha}{2}\frac{|A_0|^2}{r_0^2}\right] A_r \exp[i(kZ_0 - \omega t)] - \frac{\alpha A_0 * |A_r|^2}{2r_0} \exp[i(kr_0 - \omega t)]$$

$$- \frac{\alpha}{2}\frac{A_0 * |r|^2}{r_0} \exp[i(ikZ_0 - kr_0 - \omega t)]$$

One can distinguish the following three terms on the R.H.S. of Eq. (39):

(*i*) The first term represents the attenuated incident wave.

(*ii*) The second term represents the wave surface identical to the wave surface emitted by the object except for a constant factor. Thus, this wave surface when projected back toward the illuminating source, seems to emanate from an apparent object located from a place where the original object was located. We say that these waves produce a virtual image, and an observer who looks through the plate would see the object in three dimensions (Fig. 6.29). As stated earlier, one can take a photograph of the image by a camera.

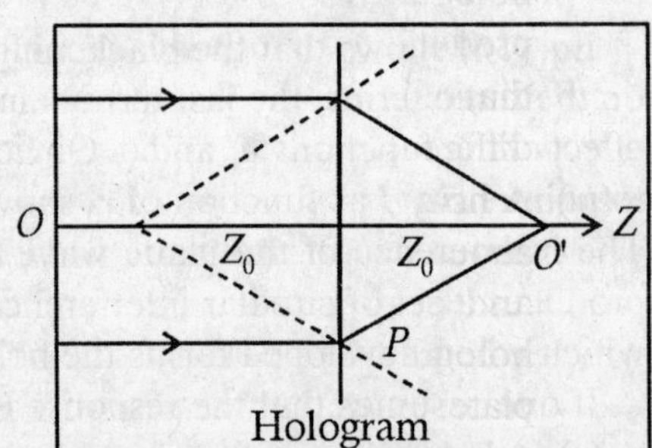

Fig. 6.29: Reconstruction of the image from a hologram

(*iii*) The third term represents a wave surface which is also a replica of the regional wave, except that they have conjugate, or reversed curvature; originally diverging spherical waves from an object are converted into converging spherical waves. It converges at a point *O'* and produces a real image at this point. One can photograph this real image without a lens. Thus a hologram produces a real image *O'* and a virtual image *O*.

Gas lasers have high coherence and continuous wave mode and hence they are often used for holography. Moreover, the emitting power of gas lasers is low and hence the time of exposure is large. Therefore, using gas lasers moving objects cannot be holographed. To prepare the holograph of moving objects and to record the development of a process in time, pulsed solid state lasers are often used. With the use of pulsed solid state lasers one can cut down the exposure time to about 10^{-3} sec.

Some Important Properties of a Hologram

(*i*) The hologram acts somewhat like a simple periodic diffraction grating. It produces a beam direct and two first-order diffracted beams on either side of the direct beam. The zero order of the diffracted light is the direct beam, while the two first orders on either side comprise the virtual and the real images.

(*ii*) The virtual image produced by a hologram appears in complete three-dimensional form. Each section of a hologram is capable o reconstructing an image of the entire object. As the pieces become smaller, the quality of the reconstructed image will become poorer, small details will vanish, and a characteristic speckled structure will appear.

(*iii*) In holography the viewer always sees a positive image whether a positive or a negative photographic transparency is used for the hologram. The reason for this that a negative hologram merely produces a wave field that is shifted 180° in phase with respect to that of a positive hologram. Since the eye is insensitive to this phase difference, the view seen by the observer is identical in the two cases.

(*iv*) A hologram is a reliable method of data storage. The information about a point object is recorded over the whole area of the hologram and each such point of a real object is recorded on the whole hologram and not in one of its points as is the case in photographic plate.

(*v*) A hologram of a complex object can be considered as an interference (coherent) superposition of holograms from individual points or more complex portions of the object. The amplitudes of light waves are summated with account of the phase relations between them in such superposition.

We can also imagine a hologram that is an incoherent superposition of holograms of different objects or parts of the same object. In this case, the photolayer summates the illumination they create. If the number of such consecutive superpositions is not too great, the hologram will simultaneously reconstruct several consecutively recorded light waves without any appreciable distortions. This property of hologram is used for consecutive recording of the waves from several objects or several states of the same object on a single hologram.

(*vi*) The curious property of the wavefront reconstruction process is that it does produce negative. The hologram itself would normally be regarded as a negative but the image it produces is a positive. A hologram copied by contact printing would be reversed in the sense that opaque areas would now become transparent and *vice versa*. The image reconstructed from the copy, however, would remain a positive and would be indistinguishable from the image produced by the original except for the small degradation in quality.

Applications of Holography

Holography has wide range of applications in science and technology. The main applications of holography are listed below:

(*i*) Three-dimensional images: Holographic, Cinematography and Television.
(*ii*) Holographic interferometry
(*iii*) Spatial filtration and character recognition
(*iv*) Spectroscopy
(*v*) In production processes and optical engineering.

The conventional holography needs coherent light. This is a drawback. Recent advances in techniques, e.g. *sandwich holography* offer a new method of storing confidential data. The *rainbow* (white light transmission) holography is a significant achievement because it uses while light.

ILLUSTRATIVE EXAMPLES

Example 1

A laser bean has a power of 59 mW. It has an aperature of 5×10^{-3} m and it emits light of wavelength 7200 Å. The beam is focused with lens of focal length 0.1 m. Show that the area spread and the intensity of the image are 2.074×10^{-10} m² and 2.4×10^{8} W/m² respectively.

Solution

$$d\theta = \frac{\lambda}{d}$$

$$\lambda = 7200 \times 10^{-10} \text{ m}$$

$$d = 5 \times 10^{-3} \text{ m}$$

$$d\theta = \frac{7200 \times 10^{-10}}{5 \times 10^{-3}} = 1.44 \times 10^{-4} \text{ radian}$$

(*i*) Area spread = $(d\theta f)^2$

Here $f = 0.1$ m

$\therefore$ Areal spread = $(1.44 \times 10^{-4} \times 0.1)^2 = 2.074 \times 10^{-4}$ m²

(*ii*)
$$\text{Intensity} = \frac{\text{Power}}{\text{Area}} = \frac{\text{Power}}{\text{Area spread}}$$

$$= \frac{50 \times 10^{-3}\text{W}}{2.074 \times 10^{-10}\text{ m}^2} = 2.411 \times 10^{8} \text{ W/m}^2$$

Example 2

A laser beam has a wavelength of 8×10^{-3} m. The laser beam is sent to moon. The distance of the moon is 4×10^{5} km from the earth's surface.

Determine (*i*) the angular spread of the beam and (*ii*) the axial spread when it reaches the moon's surface.

Solution

We have $d\theta = \lambda / d$

$$\lambda = 8 \times 10^{-7} \text{ m}$$

$$d = 5 \times 10^{-3} \text{ m}$$

$$\therefore \quad \text{Angular spread } d\theta = \frac{8 \times 10^{-7}}{5 \times 10^{-3}} = 1.6 \times 10^{-4} \text{ radian}$$

$$\text{Areal spread} = (\Delta d\theta)^2 \,|\, \Delta = 4 \times 10^5 \text{ km} = 4 \times 10^8 \text{ m}$$

$$= (4 \times 10^8 \times 1.6 \times 10^{-4})^2 = 4.096 \times 10^9 \text{ m}^2$$

Example 3

A laser beam of wavelength 6000 Å on earth is focused by a lens (or mirror) of diameter 2 m on to a crater on the moon's surface. The distance of the moon from earth is 4×10^8 m. Neglecting the éffect of earth's atmosphere show that the angular and areal spreads of the spot on the moon will be 3×10^{-7} radian and 1.44×10^4 m^2 respectively.

Solution

(*i*) Given that

$$\lambda = 6000 \times 10^{-10} \text{ m}$$

$$d = 2 \text{ m}$$

$$\Delta = 4 \times 10^8 \text{ m}$$

$$\text{Angular spread } d\theta = \frac{\lambda}{d} = \frac{6 \times 10^{-7}}{2} = 3 \times 10^{-7} \text{ radian}$$

Example 4

A laser beam has aperature $d = 1.8 \times 10^{-2}$ m and emits radiation of wavelength $\lambda = 5 \times 10^{-7}$ m. Calculate (*i*) the semi angle of the cone of its beam (*ii*) the solid angle of this cone.

Solution

The semi-angle of the cone of the laser beam

$$\theta = \frac{\lambda}{\alpha} = \frac{5 \times 10^{-7} \text{ m}}{1.8 \times 10^{-2} \text{ m}} = 2.72 \times 10^{-5} \text{ radian}$$

(*ii*) Solid angle of this cone

$$\phi = \pi \frac{\Delta^2 \theta^2}{\Delta^2} = \pi\theta^2 = \frac{22}{7} \times \left(2.72 \times 10^{-5}\right)^2 \approx 2.5 \times 10^{-9} \text{ starad}$$

Example 5

In example 4, calculate the areal spread of the beam as it reaches the moon's surface. The distance of the moon from the earth is 4×10^8 m.

Solution

$$\text{Areal spread} = \Delta^2\phi = 4 \times 10^8 \text{ m}$$

Example 6

The coherence length for sodium light is 2.945×10^{-2} m. The wavelength of sodium light is 5890 Å. Determine (*i*) the number of oscillations corresponding to coherence length and (*ii*) the coherence time.

Solution

(*i*) Given that $\lambda = 5890 \times 10^{-10}$ m

$c = 3 \times 10^{10}$ m/s

$L = 2.945 \times 10^{-2}$ m

Number of oscillations in coherence length L,

$$n = \frac{L}{\lambda} = \frac{2.945 \times 10^{-2}}{5890 \times 10^{10}} = 5 \times 10^4$$

(*ii*) Coherence time $= L/c = \dfrac{2.945 \times 10^{-2}}{3 \times 10^8} = 9.8 \times 10^{-11}$ s

Example 7

The coherence length for sodium D_1 line is 2.5 cm. Determine (*i*) spectral width of the line $\Delta\lambda$, (*ii*) the purity factor, Q and (*iii*) the coherence time τ. The wavelength of light is 6000 Å.

Solution

(*i*) $$\Delta\lambda = \frac{\lambda}{2L} = \frac{36 \times 10^{-10}}{5} \text{ cm} \approx 7A$$

(*ii*) $$Q \approx \frac{\lambda}{\Delta\lambda} \approx \frac{6 \times 10^{-5}}{7 \times 10^{-8}} \approx 10^3$$

(*iii*) $$\tau = \frac{L}{C} = \frac{2.5}{3 \times 10^{10}} \text{ sec} \cong 0.8 \times 10^{-10}\text{s}$$

REVIEW QUESTIONS

1. Explain the terms absorption, spontaneous emission and stimulated emission of radiation. Obtain a relation between transition probabilities of spontaneous and stimulated emissions.
2. What is a laser ? Explain the salient features of a laser radiation. Mention some of its principal uses.

3. What are the important features of stimulated emission? Discuss the essential requirements for producing laser action?
4. What is difference between temporal and spatial coherence? Explain coherence time and coherence length.
5. Give an elementary theory of the laser.
6. Describe the construction and working of a ruby laser with necessary diagrams.
7. Describe a He-Ne gas laser. How is population inversion achieved in this type of laser?
8. What are Einstein's A and B coefficients. Obtain a relationship between them.
9. Explain the role of an optical resonator in a laser. How population inversion is achieved in a laser?
10. Explain what do you understand by the modes of a cavity resonator. Explain quality factor.
11. Explain the principle and working of a semiconductor laser. Derive the condition for the predominance of the stimulated emission over absorption for this laser device.
12. Explain the principle of holography and discuss its salient features. Mention its uses.
13. Explain the basic principle of holography. What are salient features of holography. Mention some important uses of holography.
14. (*a*) Distinguish between spontaneous and induced emissions. How does induced emission dominates in He-Ne laser?

    (*b*) How does an optical fibre function in transporting electromagnetic energy? Show that numerical aperture of a step index fibre is given by.

$$NA = \frac{\sqrt{2\Delta}}{n_0}$$

    where symbols have their usual meanings.

    (*b*) Holography [Raj. BE, 2003]
15. (*a*) Describe the construction and working of He-Ne laser. How is population inversion achieved in such a laser?

    (*b*) A laser beam has a wavelength of 700 mm and aperature of 5 mm. The laser beam is sent to moon. The distance of moon is 4×10^5 km from Earth. Calculate (*i*) the angular spread of the beam and (*ii*) and areal spread when it reaches the moon. [Raj. BE, 2002]
16. (*a*) In He-Ne laser, what is the function of the He atoms? Explain the answer with the help of energy level diagram for He-Ne. Describe with a neat sketch the working of He-Ne laser.

    (*b*) Derive the relation between Einstein's coefficients and discuss the results. [Raj. BE, 2002]
17. (*a*) Describe a ruby laser & explain its working, showing the effective energy levels.

    (*b*) Described the method of recording the hologram & reconstruction of image from it. [Raj. BE, 2001]
18. (*a*) What are essential requirements to produce a laser? How are these fulfilled in a He-Ne laser?

    (*b*) What is an optical fiber? What do you mean by numerical aperture of an optical fiber? Find an expression for the numerical aperture of a stop index optical fiber.

19. Explain the essential requirements for producing laser action. How are these requirements obtained in case of Ruby laser? Draw a diagram to represent the component of Ruby laser. [Raj. BE, 2000]

20. (*a*) Explain how population inversion is obtained in ruby laser.
(*b*) Explain the basic principles of holography. [Raj. BE, 2000]

21. Describe the laser action in Helium Neon laser with energy diagram. Describe various applications of laser. [Raj. BE, 1999]

22. Write short notes on the following:
(*a*) Ruby laser (*b*) Gas laser
(*c*) Semiconductor laser (*d*) Einstein's *A* and *B* coefficient
(*e*) Optical resonator (*f*) Optical fiber
(*g*) Holography [Raj. BE, 2002] (*h*) Quantum-well laser.

23. Explain the working of Nd : YAG laser. Draw its energy level diagram. Mention its applications.

24. How CO_2 laser works. Describe its working and draw energy level diagram.

PROBLEMS

1. Determine the minimum difference between the two arms of a Michelson interferometer that seeks to measure two closely spaced wavelengths using He-Ne laser for which the coherence length is 600 km. (**Ans.** 300 km)

[Hint: $2d = n_1\lambda_1 = n_2\lambda_2$ or $2d\left(\frac{1}{\lambda_2} - \frac{1}{\lambda_2}\right) = n_2 - n_1 = m$, where n_1, n_2 and m are integers. Putting $\Delta\lambda = \lambda_1 - \lambda_2$, $m = \frac{2d\Delta\lambda}{\lambda^2} = \frac{2d\Delta\nu}{\nu\lambda} = \frac{2d}{L_c}$ where L_c is coherence length. Obviously, the minimum difference between the two atoms of the interferometer is $d = \frac{L_c}{2} = \frac{600}{2} = 300$ km]

2. The output power of a gas laser is 1 mW and the emitted wavelength is 630 nm. Show that the number of photons emitted per second is about 3×10^{15} per second.

3. The coherence length of Sodium D_2 line is 2.5 cm. If the velocity of light is 3×10^8 m/s, then show that the coherence time is 0.8×10^{-18} sec.

4. Calculate the spot size of a He-Ne laser beam at the centre of a confocal cavity if the length of the cavity is 1.24 m. The wavelength is 633 nm. (**Ans.** 0.5 mm)

5. Calculate the inversion density for He-Ne laser aperature at 633 nm if the temperature of the discharge is 150° C and the gain constant is 3 per cent per metre. The life time of the upper state for spontaneous transition is 10^{-7} s. (**Ans.** $3.1 \times 10^{14}/\text{m}^2$)

6. The bandwidth of a He-Ne laser is 500 Hz. Obtain the coherence time and the longitudinal coherence length. (**Ans.** (*i*) 2 ms (*ii*) 600 km)

7. Calculate the gain constant β of a laser having the following parameters : wavelength = 650 nm, inversion density $(n_2 - n_1) = 5 \times 10^{22}\ /\text{m}^3$, life time for spontaneous emission 2×10^{-4} and line width $\Delta\lambda = 15$ Å. (**Ans.** 3.95)

8. Show that for a normal optical source with temperature about 10^3 K and wavelength 6000 Å, the emission is predominantly due to the spontaneous transitions.

9. A laser beam of wavelength 6000 Å, power 10 mW, and angular spread 5×10^{-5} rad is focused by a lens of focal length 10 cm. Calculate the radius and power density of the image. What is lateral coherence width?
(**Ans.** (*i*) radius = 2.5×10^{-4} cm (*ii*) Power density = 5.1×10^4 W/cm^2 (*iii*) lateral coherence width = 1.2 cm).

10. For sodium line, the wavelength is 5890 Å and coherence time is 10^{-10} s. Show that the monochromaticity of the source is (5890 ± 0.0578) Å.

SHORT ANSWER QUESTIONS

1. Write salient features of laser radiation.
2. What is the relation between the degree of nonmonochromaticity (ζ) and longitudinal coherence length (L_c) or simply coherence length?

Ans. $\zeta = c/L_c \nu_0$

3. What is the relation between the coherence time (τ_c) and frequency spread of the wave ($\Delta\nu$)?

Ans. $\tau_c = 1/\Delta\nu$

4. In terms of quantum picture, when a light beam is said to be a perfectly coherent?

Ans. When all the constituent photons of a light beam have the same energy, the same direction of momentum and identical polarization, the light beam is said to be perfectly coherent.

5. What is the relation between Einstein's A_{21} and B_{21} coefficients?

Ans. $A_{21}/B_{21} = 8\pi h\nu^3/c^3$

6. What do you understand by a metastable state?

Ans. Normally, the mean life of excited atoms before the spontaneous emission occurs is about 10^{-8} s. However, for some excited states, this mean life is perhaps as much as 10^5 times longer. We call such long-lived states metastable; they occupy an important role in laser action.

7. Mention any three methods of pumping for creating population inversion.

Ans. (*i*) Photon excitation (*ii*) Electron excitation and (*iii*) Inelastic collision between atoms.

8. What is a hologram?

Ans. A hologram contains the encoded image of the object in the form of the interference pattern.

9. What is fundamental basis of laser operation?

Ans. Photons are bosons. Atomic decays accompanied by the emission of radiation are enhanced if there are already photons presents that are identical to the emitted photon. This is the basis of laser operation.

10. What is the active medium in CO_2 laser?

Ans. The active medium in CO_2 laser is mixture of CO_2, N_2 and He gases.

11. What is the active medium in Nd : YAG laser?

Ans. An Nd : YAG laser Yttrium Aluminium Garnet doped with Nd^{3+} ions is the active medium.

OBJECTIVE QUESTIONS

1. A laser light is highly monochromatic and __________. [directional]
2. The two *B* coefficients of Einstein are __________. [equal]
3. The order of pumping power necessary to achieve the population inversion in a ruby laser is __________. [10^7 W/m^2]
4. Laser light is considered to be coherent because it consists of coordinated waves of exactly the __________ wavelength. [same]
5. An hologram is an encoded __________ of the object recorded on a photographic plate. [image]
6. A gyro laser detects slow rotational motions, down to __________ and less. [μ rad/s]

Fibre Optics

7.1 INTRODUCTION

Optical fibre is a wave guide through which light can be transmitted with very little leakage through the side walls. Optical fibres transport light signals from place to place as metallic conductors transport electrical signals. Optical fibres can guide light around bends and can carry light for long distances with very little attenuation (loss of light power). Telecommunications by modulating a beam of light (LED or Laser as a light source) and using optical waveguides has the inherent advantages of very large bandwidth, small dimensions, low losses bandwidth, and insensitivity to electrical interference.

An optical fibre is a glass or plastic conduits as thin as a human hair, designed to guide light waves along their length. Usually an optical fibre is a cylindrical wave guide system through which the optical wave can propagate. The principle by which light wave travel through the fibre is '**total internal reflection**' without loss of incident intensity.

An important structure used in optical system is the layered structure or the **wave guide structure**. As the name implies, these structures are used to confine the optical waves in a well defined region and guide their propagation. One can make the layered structures from non-crystalline materials or from crystalline materials. For example glass is used to produce optical fibres used in optical communication, whereas semiconductor wave guides are used in semiconductor laser.

An optical fibre is a very thin and flexible medium of cylindrical shape. It has three principal sections: (*i*) **the core** (*ii*) **the cladding** (*iii*) **the jacket**. The inner most section of the fibre is referred as the cladding. We must note that the optical properties of the cladding are different from those of the core. The outermost section of the fibre is known as the jacket. Jacket is made of plastic or polymer and other materials and this protects the fibre structure from moisture, abrasion, mechanical shocks, and other environmental hazards.

The core forms the actual working structure of the optical fibre. The radiation (say light) entering the core, suffering a number of familiar phenomenon of total internal reflection at the core-cladding interface. Obviously, the interface between core and cladding acts as a mirror at which total internal reflection of the transmitted light takes place.

With the advent of optical fibres, an era of *photonics* has been ushered in. Apart from its use as communication channel, optical fibres find wide uses in other areas. Sensors for detecting electrical, mechanical, thermal energies are made using optical fibres. Copying machines, inexpensive and simple display systems, etc. also utilizes fibre optics. In medical diagnostics, fibroscopes are used in a variety of forms.

7.2 TOTAL INTERNAL REFLECTION

When a ray of light is incident at an angle i in denser medium (refractive index n_1), the refracted rays will bend away from the normal in rarer medium (refractive index n_2) and we have

$$\text{Snell's law} \quad \frac{\sin i}{\sin r} = \frac{n_2}{n_1}$$

If the angle of incidence say i_c when the angle of refraction $r = 90°$, then

$$\sin i_c = \frac{n_2}{n_1} \qquad (\sin r = \sin 90° = 1) \tag{1}$$

The angle i_c is known as **critical angle.** The value of the critical angle (i_c) depends upon the refractive index of the denser medium (n_2). When the light in denser medium is incident at an angle i_c ($i > i_c$), then light will be reflected back into the denser medium. This is known as **total internal reflection** and this phenomenon is used in optical fibre communication. When light enters one end of the fibre, it undergoes successive total internal reflections from side walls and travels down the length of the fibre along the zigzag paths. A small fraction of light may escape through side walls but a major fraction emerges out from other end of the fibre.

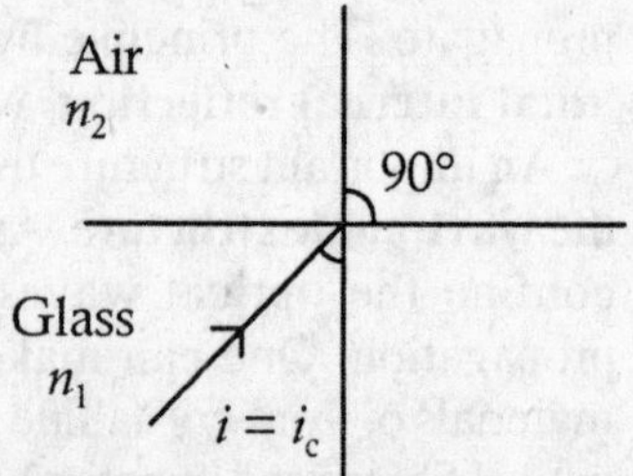

Fig. 7.1: Total internal reflection

7.3 THE OPTICAL FIBRE

Figure 7.2 (*a*) shows a glass fibre consisting of a (cylindrical) central core cladded by a material of slightly lower refractive index. One can represent the corresponding refractive index distribution (in the transverse direction) as given by

$$\begin{aligned} n(r) &= n_1 \qquad 0 < r < a \qquad \text{Core} \\ &= n_2 \qquad r < a \qquad \text{Cladding} \end{aligned}$$

and has been shown in Fig. 7.2 (*b*). Obviously n_1 is the refractive index of the core of a fibre, and n_2 that of the cladding ($n_1 > n_2$). A ray of light *OA*, incident on one end—face of the core, is refracted into the core along *AB*. If n_0 is the refractive index of air, we have from Snell's law

$$\frac{\sin i}{\sin \theta} = \frac{n_1}{n_0} \tag{2}$$

The ray *AB* is incident on the core-cladding interface at angle $\phi = (90 - \theta)$. We can see that as *i* decreases θ also decreases so that ϕ increases. When this ray has to suffer total internal reflection at the core-cladding interface.

$$\sin \phi \, (= \cos \theta) > n_2/n_1 \tag{3}$$

Thus

$$\sin \theta < \left[1 - \left(\frac{n_2}{n_1}\right)^2\right]^{\frac{1}{2}} \tag{4}$$

and one obtains

$$\sin i_m < \frac{n_1}{n_0}\left[1 - \left(\frac{n_2}{n_1}\right)^2\right]^{\frac{1}{2}} = \left[\left(n_1^2 - n_2^2\right)/n_0^2\right]^{1/2}$$

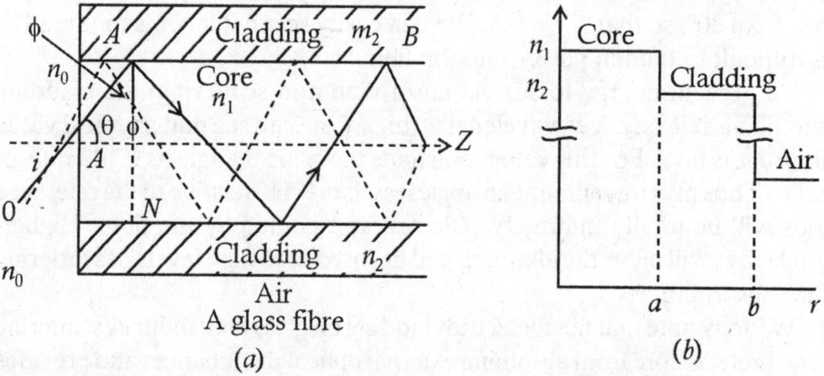

A glass fibre

(*a*)

(*b*)

Fig. 7.2: (*a*) Refraction of a ray of light into core (*b*) Distribution of refractive index of a cladded optical fibre which consists of a cylindrical glass structure surrounded by a material of slightly lower refractive index

If $\left(n_1^2 - n_2^2\right) \geq n_0^2$, then for all values of i, total internal reflection will occur. Assuming $n_0 = 1$, one obtains the maximum value of sin$i$ for a ray to be guided is

$$\sin i_m = \left(n_1^2 - n_2^2\right)^{1/2} \quad \text{when} \quad n_1^2 < n_2^2 + 1$$
$$= 1 \quad \text{when} \quad n_1^2 > n_2^2 + 1 \tag{5}$$

Obviously, if a core of light is incident on one end of the fibre, it will be guided through it provided the semi angle of the core is less then i_m. This angle is a measure of the light gathering power of the fibre. The core of the semi vertical angle θ_m is termed as the acceptance core and the angle θ_m is called the acceptance angle of the fibre ($\theta_m = \sin^{-1}[n_1^2 - n_2^2]$). Acceptance angle may be defined as the maximum angle that a light ray can have relative to the axis of the fibre and propagation down the fibre.

The *numerical aperature* (*NA*), sometimes referred as the *figure of merit* of an optical fibre, and is defined as

$$NA = \sin\theta_m = \frac{1}{n_0}\sqrt{n_1^2 - n_2^2} = \sqrt{n_1^2 - n_2^2} \qquad (\because n_0 = 1) \tag{6}$$

Usually $(n_1 - n_2)$ is not large and therefore, one can

have (writing $n_1 - n_2 = n_1\left(1 - \dfrac{n_2}{n_1}\right) = n_1\sqrt{2\Delta}$)

$$NA = n_1\sqrt{2\Delta} \tag{7}$$

where $$\Delta = \frac{\delta_n}{n_1} = \frac{n_1^2 - n_2^2}{2n_1^2} = \frac{(n_1 - n_2)(n_1 + n_2)}{2n_1^2} = \frac{n_1 - n_2}{n_1} \tag{8}$$

The quantity Δ is usually referred as the fractional difference of the refractive indices of the core and the cladding. Δ is also called *relative refractive index difference when* $n_1 = n_2$. For $n_1 = 1.48$ and $n_2 = 1.46$, one obtains $NA \approx 0.24$. This give $i_m \approx 14°$. Obviously, the acceptance angle i_m is less than 30°, so that $NA < 0.5$. We can easily see that if NA is too small, it is difficult to launch power into the fibre.

To prevent energy losses via absorption and scattering, the cladding should be at least a few wavelengths thick. One can see that a typical value for n_2/n_1 is 0.99. For this value, one finds the critical angle as 0.142 rads or 8.11°. Thus rays travelling at an angle less than 8.11° relative to the reference axis will be totally internally reflected and guided by the fibre. Higher-angle rays will enter the cladding and be lost due to high levels of scattering and absorption.

We may note that the dielectric cladding on glass core reduces scattering loss, protects core from absorbing external optical disturbances and provides mechanical strength to main core glass fibre. Sometimes there is buffer coating over cladding which adds further strength to main fibre and protects fibre from mechanical vibrations and impact.

7.4 CLASSIFICATION OF FIBRES

Usually the optical fibres are grouped under three categories, viz:

(*i*) Stepped index multimode fibre
(*ii*) Stepped index monomode fibre
(*iii*) Graded index multimode fibre.

(*i*) Stepped Index Multimode Fibre

Stepped index is the simplest type of an optical fibre consisting of a thin cylindrical structure of transparent glassy material of uniform refractive index n_1 surrounded by a cladding of another material of uniform but slightly lower refractive index n_2 (Fig. 7.2). These fibres are referred to as stepped index fibres due to the step discontinuity of the index profile at the core-cladding interface. The light rays incident on the end—face of the core at angles of incidence within the acceptance angle are guided along the core in different paths. It is observed that light rays of different colours of frequencies are also separated as they travel inside the core. This means multimode propagation is possible in the fibre as shown in Fig. 7.3.

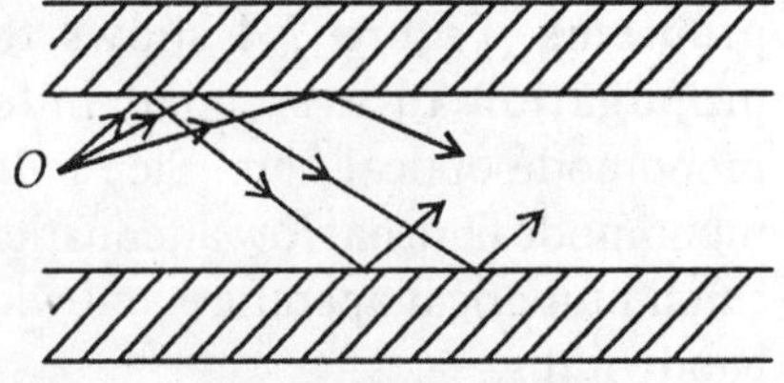

Fig. 7.3: Multimode propagation in a fibre

From Fig. 7.3 we can see that the time taken by the different light rays propagating along paths of different lengths upto the destination are different. This means the variations of the successive pulses of light may overlap, thus distorting the information being carried. Such distortions are usually called as the transit-time dispersion.

One can show that the time taken by a ray is function of the angle θ made by the ray with the *Z*-axis which leads to pulse dispersion and given by (Fig. 7.2)

$$\tau = \frac{n_1 L}{c \cos\theta} \tag{9}$$

where $L = AB$. If we assume that all rays lying between 0 and θ_c (critical angle) are present than the time taken by the rays corresponding to $\theta = \theta_c \approx \cos^{-1}(n_2/n_1)$ would be given by

$$\tau_{min} = \frac{n_1 L}{c} \quad \text{and} \quad \tau_{max} = \frac{n_1^2 L}{n_2 c} \tag{10}$$

Thus, if all the input rays were excited simultaneously, the rays would occupy a time interval at the output end of duration

$$\Delta\tau = \tau_{max} - \tau_{min} = \frac{n_1 L}{c}\left(\frac{n_1}{n_2} - 1\right) \tag{11}$$

For a typical fibre, one can assume $n_2 = 1.46$ (pure silica),

$$\Delta = \frac{n_1 - n_2}{n_1} \Rightarrow n_1 \approx 1.4746$$

one obtains $\Delta\tau \cong 50 \times 10^{-9}$ s/km

$$\cong 50 \text{ ns/km}.$$

This shows that an impulse after traversing through the fibre of length 1 km broadens to a pulse dispersion. For this purpose two alternative solutions exist one involving the use of graded index fibres and the other involving single mode fibres.

(*ii*) Stepped Index Monomode Fibre

One can make core vary thin, *i.e.* the core diameter is almost of the same order as the wavelength for the light to be propagated. This helps in minimizing the transit-time dispersion problems. Figure 7.4 shows the propagation in a stepped index monomode optical fibre. Step index monomode fibre has low attenuation, small numerical aperature and wide bandwidths.

Fig. 7.4: Propagation in a stepped index monomode optical fibre

One can show that when $V < 2.405$, only the LP_{01} mode can propagate, *i.e.* fibre becomes single mode fibre. Theoretically, LP_{01} mode will propagate no matter how small the V value, *i.e.* no matter how small the core radius a.

$$\because \quad V = \frac{2\pi}{\lambda} a(NA)$$

$$\therefore \quad a = \frac{V\lambda}{2\pi(NA)}$$

When $V < 2.405$ then

$$a < \frac{2.405\lambda}{2\pi(NA)}$$

or

$$a < \frac{2.405\lambda}{2\pi\sqrt{n_1^2 - n_2^2}}$$

for single mode fibre.

We may note that in general in single mode fibres cores, tend to be only a few microns in diameter, but it has an advantage from several points of view if core can be made as large as possible. One can easily note that this may be achieved by reducing the NA value, *i.e.* by making core and cladding R.I. vary close to each other. One of the major advantage of using single mode fibre is the inter model dispersion no longer exists, and so they are capable of carrying very large bandwidth signal for long distance communication.

(iii) Graded Index Multimode Optical Fibre

In comparison to stepped index monomode fibres, graded index multimode fibre is quite cost effective and offer a less expensive method of overcoming the problem of transit-time dispersion. It is observed that in such fibres, the refractive index of the core falls of gradually from the axis towards the core-cladding interface. This means that the individual light rays are gradually refracted in the core itself. Figure 7.5 shows the wave propagation at different angles of incidence, travelling different distances from the central axis before suffering reflections to recross the central axis.

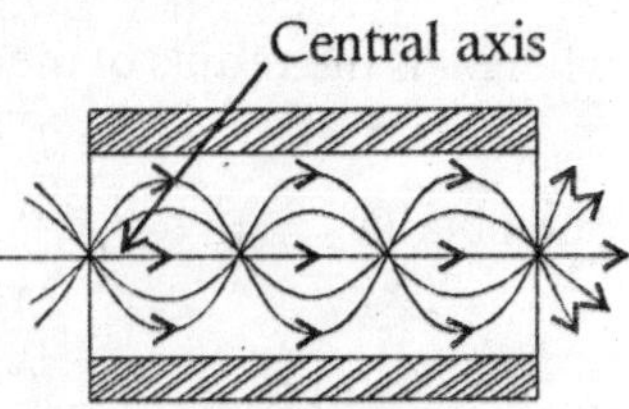

Fig. 7.5: Propagation in a graded index multimode optical fibre

We may note that the number of guided modes is dependent upon the physical parameters, *i.e.* the refractive index difference between core and cladding (*i.e.* $N.A. = \sqrt{n_1^2 - n_2^2}$), core radius of the fibre a and the wavelength (λ) of the transmitted light, are all included in the normalized frequently V' for the fibre, where $V = 2\pi/\lambda a (N.A.)$ = Normalized frequency. One finds that the total number of guided modes for step index fibre is related to V by the following approximate expression.

$$M_s = \frac{V^2}{2}$$

From figure it is obvious that the light rays with large angles of incidence transverse greater paths than those with smaller angles of incidence. However, the velocity of propagation increases with a drop of the refractive index and thus all the rays recross the central axis at nearly the same instant of time. This helps in reducing considerably the transit-time dispersion.

Fibre optic systems use optical fibres to transmit information, in the form of coded pulses or fragmented images (using bundles of fibres). Over moderate distances they are used in telecommunications for which purpose they are becoming competitive with electric cables. In fibre communication, two bands are extensively used: 800 nm to 900 nm and 1200 nm to 1400 nm. The losses in the fibres for the said two bands are very low. Lasers and *LED*'s (light emitting diodes) serve as light sources for fibres. Photodetectors are employed for processing the signal at the receiver and of the optical fibre link. Photodetectors convert the light signals into electrical waveforms. Optical fibres are also used in medical instruments (fiberscopes) to examine internal body cavities, such as stomach and bladder.

V-Number: An optical fibre is characterized by one more important parameter, called as V-number. This is more generally called as *normalized frequency* of the optical fibre and given by the relation

$$V = \frac{2\pi a}{\lambda}\sqrt{n_1^2 - n_2^2} \tag{12}$$

where a is the radius of the core and λ is the free space wavelength. Using

$$NA = \sqrt{n_1^2 - n_2^2} = n_1\sqrt{2\Delta}$$

one obtains

$$V = \frac{2\pi a}{\lambda}(NA) \tag{13}$$

$$V = \frac{2\pi a}{\lambda} n_1\sqrt{2\Delta} \tag{14}$$

The maximum bar of modes N_m supported by a Stepped index (SI) fibre is determined by

$$N_m \cong \frac{1}{2}V^2 \tag{15}$$

Thus for $V = 10$, $N_m = 50$

For $V < 2.405$, the optical fibre can support only one mode and is classified as SMF. MMFs can have values $V > 2.405$ and can support many modes simultaneously.

The wavelength corresponding the value of $V = 2.405$ is known as the *cut off wavelength* (λ_c) of the optical fibre. We have

$$\lambda_c = \frac{\lambda V}{2.405} \tag{16}$$

From Eq. (14), one can deduce that single mode properties can be realized by decreasing the core diameter and/or decreasing Δ such that $V > 2.405$. We may note that either a larger core and/or a larger Δ will result in multimode properties.

In case of GRIN fibres, for large values of V, we have

$$N_m \cong \frac{V^2}{4} \tag{17}$$

7.5 ADVANTAGES OF OPTICAL FIBRES

Fibres optic systems offer many advantages. Few of them are listed below:

(*i*) Optical fibres are light in weight and moreover occupy less space than coaxial (bundles twisted pair) cables.

(*ii*) Optical fibres are composed of dielectric materials and therefore they are totally immune to extraneous interfering electromegnetic signals.

(*iii*) There is virtually no signal leakage from optical fibres and hence cross-talks between neighbouring is not possible, whereas it is quite frequent in conventional metallic systems.

(*iv*) The attenuation of signals in optical fibres is much smaller than in coaxial cables or twisted pair.

(*v*) Optical fibres are immune to electromagnetic interference and do not pick up line currents. Obviously, these can be safely used in high voltage environments, e.g. in a power station, and can be laid alongside metallic power cables.

(*v*) The basic raw materials used in the fabrication of low loss fibres is silica, which is abundantly available in nature (although in an impure form), whereas copper constitutes the basic raw material in coaxial cables. Obviously, fibres are cost effective, *i.e.* more economical.

(*vii*) Optical fibres can withstand environmental hazards better and also have longer life compared to copper cables.

(*viii*) Optical fibres are nonconductive and noninductive. This means there is no radiation and interference with other communication systems.

(*ix*) The speed of transmission through optical fibres is very fast because the signals are carried by light.

(*x*) The channel capacity of optical fibres is very large, compared to the audioband (5 to 10 Hz), the frequency of the carrier light wave is very high ($\sim 10^{14}$ Hz). Obviously, one can accommodate a large number of speech signal channels in the fibres whose bandwidth is larger than that of a wire transmission line.

7.6 DISADVANTAGES OF OPTICAL FIBRE SYSTEMS

(*i*) Optical fibre systems are *virtually useless themselves.* To be practical, they must be connected to standard electronic facilities which often requires expensive interfaces.

(*ii*) Optical fibres by themselves have a significantly lower tensile strength than coaxial cable. This can be improved by coating the fibre with standard Kevlar and a protective jacket of *PVC.*

(*iii*) Occasionally it is necessary to provide electrical power to remote interface or regenerating equipment. This can not be accomplished with the optical cable so additional metallic cables must be included in the cable assembly.

(*iv*) Optical fibre cable systems are relatively new and have not had sufficient time to prove their reliability.

(*v*) Optical fibres require special tools to splice and repair cables and special test equipment to make routine measurements. Repairing fibre cables is also difficult and expensive and techniques working on optical fibre cables also require special skills and training.

In spite of all the attractive features that fibres optical communication systems possess, a fully developed fibre optical communication system started becoming available in the commercial market only by 1975. Ever since the technology in this area is progressing well. The attainment of very low pass fibre has also stimulated forcefully efforts in the development of supporting technologies line long life semiconductor light sources, low loss optical fibre cable, fibre optic connectors, low pass splices, and highly sensitive fast response solid state photodetectors.

7.7 TYPES OF RAYS

The plane containing the central axis meridional rays remain the meridian plane which can be divided in to two categories: (*a*) bound rays that propagate along axis (Fig. 7.6 *a*, *b*) and (*b*) unbound rays that are not bound along axis and are refracted out of the core and absorbed in cladding or buffer.

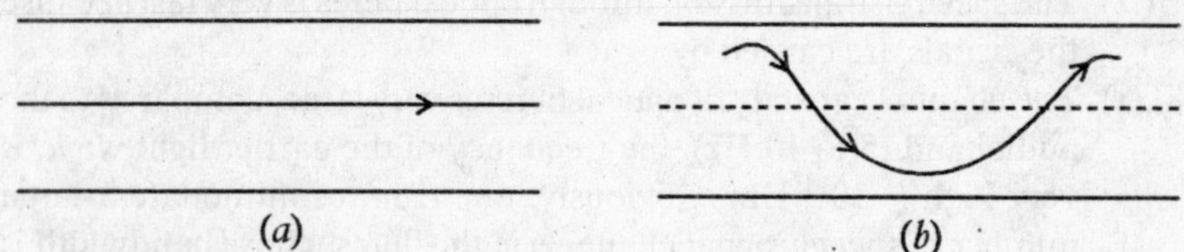

Fig. 7.6: Meridional rays propagation through optical fibre cable light

Skew Rays

These are the rays which enter the optical fibre obliquely and not confined to a plane and follow a helical path around the axis. These rays are soon scattered out of the core at bends and irregularities and thus don't contribute significantly in optical communication.

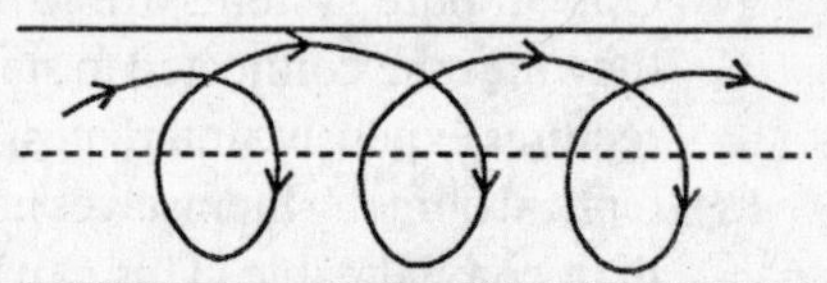

Fig. 7.7: Skewed ray propagation through optical fibre cable light

7.8 MODES OF PROPAGATION IN OPTICAL FIBRE

Light propagation through straight and curve step index optical fibre with cladding is shown in Fig. 7.8.

Here we have shown only one ray and therefore this is called as ray diagram. However, in reality an infinite number of such rays (Fig. 7.9 *a*, *b*)

all slightly displaced from each other can be propagated at a time which are called different modes of propagation. At time of propagation they will follow different paths in the optical fibre (Fig. 7.10).

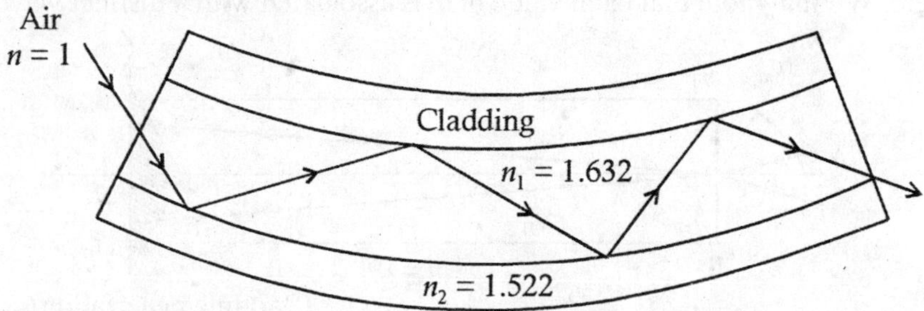

Fig. 7.8: Propagation of light through straight and curve. Step index optical fibre with cladding

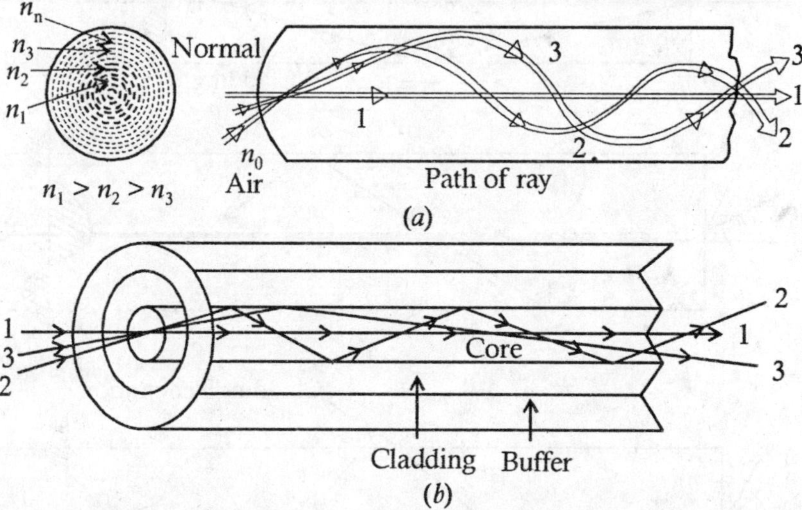

Fig. 7.9: Propagation of light in multimode graded index and step index fibre

We have read that light can be described as an electromagnetic wave. It can be seen that the frequency spectrum of electromagnetic waves extends from the subsonic (a few Hz) to cosmic rays (10^{22} Hz). The light frequency spectrum can be divided into three general bands: infrared, visible and ultraviolet. Electromagnetic wave consists of a periodically varying electric field $\boldsymbol{E}$ and magnetic field $\boldsymbol{H}$ which are perpendicular to each other. The transverse modes are shown in Fig. 7.10. These modes illustrates the case when the electric field $\boldsymbol{E}$ is perpendicular to the direction of propagation (z) and hence $\boldsymbol{E}_z = 0$, but the corresponding component of magnetic field $\boldsymbol{H}$ is in the direction of propagation. These modes are therefore called transverse electric (TE_m) modes. Similarly, when a component of electric field $\boldsymbol{E}$ is in the direction of propagation but $\boldsymbol{H}_z = 0$, the modes so formed are called transverse magnetic (TM_m) modes. We have incorporated the mode number m into this nomenclature by referring to TE_m or TM_m modes etc. We may note that each value of m is associated with a distinct wave

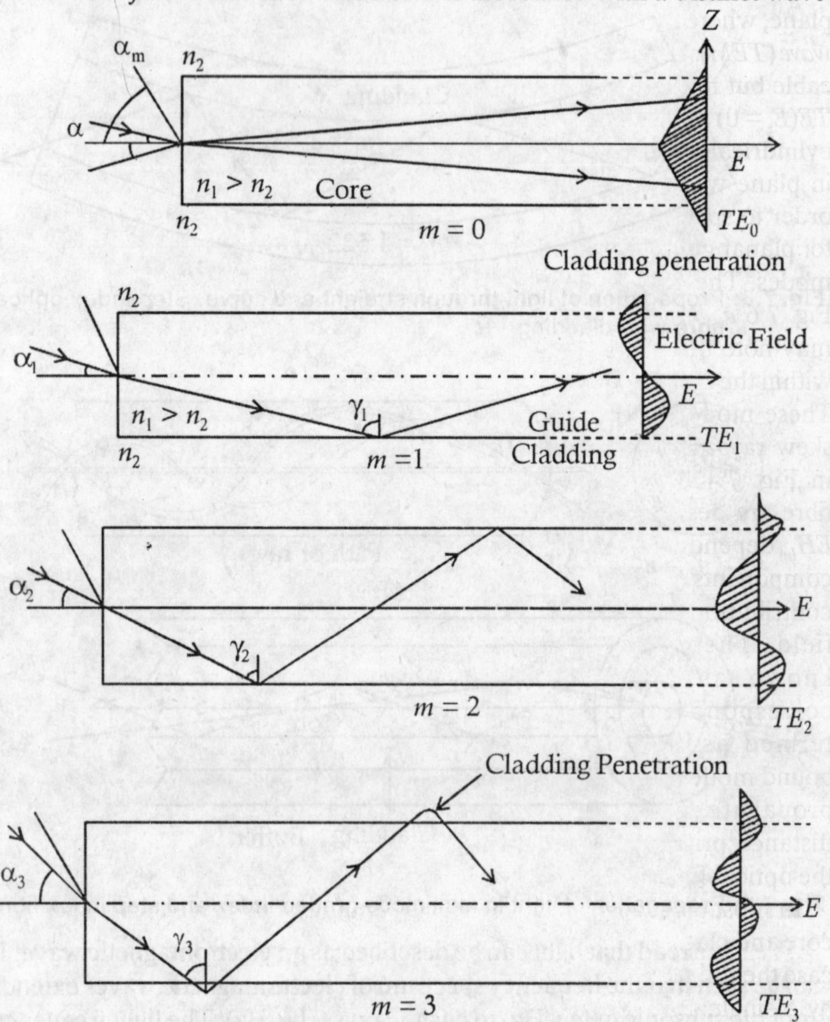

Fig. 7.10: Propagation of light in different modes through step index optical fibre, where α_m = acceptance angle, r_c = critical angle, and $\pi/2 > r_1 > r_2 > r_3 > \ldots r_c$, whereas $\alpha_0 < \alpha_1 < \alpha_2 < \alpha_3 \ldots < \alpha_m$

pattern or mode as shown in Fig. 7.10, within the wave guide they are following different paths. We also find that for each value of m there is a value of γ (γ_m), γ_m can take values in the range γ_c to $\pi/2$, which implies that the smaller the value of γ_m, *i.e.* γ_m closer to γ_c, higher the mode number and *vice versa.* We can see from Fig. 7.10 that the higher order modes, carry along with them a higher portion of mode energy within the cladding. This part of the wave which is carried in the cladding is generally referred as the "evanescent wave". We may note that (*i*) For free space communication, the total electromagnetic wave field lies in the transverse plane, where both $\boldsymbol{E}_z = 0$, and $\boldsymbol{H}_z = 0$. This is called *transverse electromagnetic wave* (*TEM*). However, *TEM* wave occur in metallic conductor like coaxial cable but not in optical wave guide (*ii*) for planar wave guide, generally $TE(E_z = 0)$ and $TM(H_z = 0)$ modes are obtained. (*iii*) As in optical fibre the cylindrical wave guide is bounded in two dimensions, rather than one as in plane wave guide. Obviously two integers l and m are necessary in order to specify the modes, in contrast to the single integer (m) required for planar guide. Thus, for cylindrical wave guide we refer TE_{lm} and TM_{lm} modes. The TE_{lm} and TM_{lm}, modes corresponds to meridional rays (see Fig. 7.6 *a*, *b*) which are travelling within the optical fibre. However, we may note that hybrid modes where $\boldsymbol{E}_z$, and $\boldsymbol{H}_z$ are non-zero also occur within the cylindrical wave guide. These modes which results from skew rays propagation as shown in Fig. 7.11, within the optical fibre are designated as HE_{lm} and EH_{lm} depending upon whether the components of $\boldsymbol{H}$ or $\boldsymbol{E}$ make larger contribution to the transverse field. These skew rays are known as *leaky rays* and corresponding modes are termed as *leaky modes*, not bound modes because they can propagate only appreciable distance prior being lost from the optical fibre.

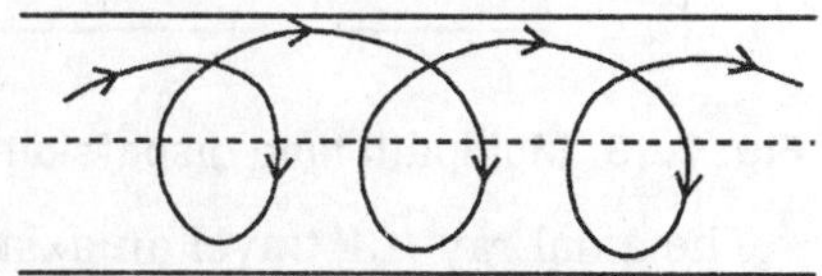

Fig. 7.11: Propagation of skewed rays through optical fibre cable light

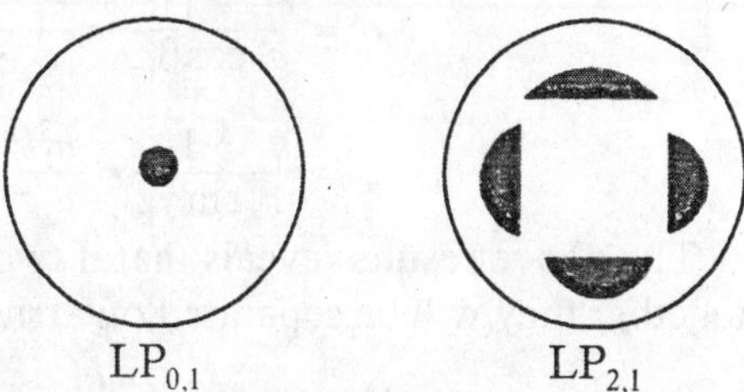

Fig. 7.12: Linearly polarised (LP_{lm}) modes

In most practical wave guides, it is found that the refractive indices of core and cladding differ from each other by only a few per cent and in that case the full set of modes (*i.e. EH*, *HE* and *TE*) can be easily approximated by a single set called linearly polarized (LP_{lm}) modes. An LP_{lm} mode in general has m field maxima along a radius vector and $2l$ field maxima round a circumference (Fig. 7.12).

7.9 DISPERSION—INTER MODEL DISPERSION

(*a*) For Multimode Stepped Index Fibre

When the ray of light of different modes follow different paths, like some rays (shown by AA', Fig. 7.13) which propagate along the optical fibre axis and some of which propagate along the trajectories of the oblique rays, for example CC' or BB' (Fig. 7.13). Obviously, there will be difference in time taken for different modes to travel a distance l. We know that refractive index (R.I.) of a medium is simply a measure of velocity of propagation v of the light in medium. Thus we can write

$$\text{R.I.}(n) = \frac{v_a}{v_m} \quad \text{or} \quad v_m = \frac{v_a}{n} = \frac{c}{n}$$

where $c\,(=v_a)$ is the velocity of light in air and v_m is the velocity of light in medium.

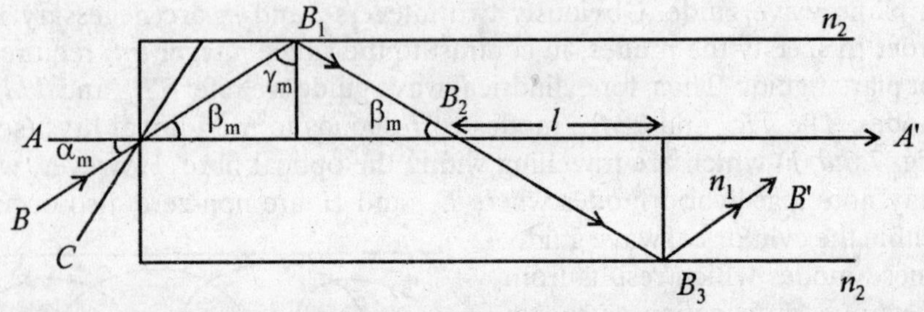

Fig. 7.13: Multipath time dispersion for multimode step index optical fibre

The axial ray AA' travel an axial distance l along the optical fibre in time whereas the oblique ray BB' travel the same axial distance l along the path B_2B_3 in the time

$$t_2 = \frac{n_1}{c}\frac{1}{\cos\beta_m} = \frac{n_1}{c}\frac{1}{\cos(90-\gamma_c)} \qquad \left(\because \sin\gamma_c = \frac{n_2}{n_1}\right)$$

$$= \frac{n_1}{c}\frac{1}{\sin\gamma_c} = \frac{n_1^2 l}{n_2 c}$$

The above results reveals that if two rays (AA' and BB') are launched together they will be separated on arrival of the other end by ΔT, where

$$\Delta T = t_2 - t_1 = \frac{n_1 l}{c}\left(\frac{n_1}{n_2} - 1\right)$$

$$= \frac{n_1 l}{n_2}\cdot\frac{l}{c}\Delta n \qquad (\text{where } \Delta n = n_1 - n_2)$$

Obviously, a pulse containing rays at all possible angles spreads out during propagation by an amount given by

$$\frac{\Delta T}{l} = \frac{n_1}{n_2}\cdot\frac{\Delta n}{c}$$

This is called as **multipath time dispersion** of the step index fibre and this causes pulse broadening.

(*b*) For Graded Index Multimode Fibre

Different modes will follow different paths, sinusoidally. We may note the rays making larger angles (Fig. 7.14) traverse longer path but they do so in lower refractive index area and hence they travel with higher speed of propagation. However refractive index changes gradually in graded index optical fibre, so there is a self focusing mechanisms that leads to a smaller value of pulse dispersion than that of step index optical fibre. This shows that for long distance communication graded index optical fibre will be more useful than step index optical fibre which causes less time dispersion.

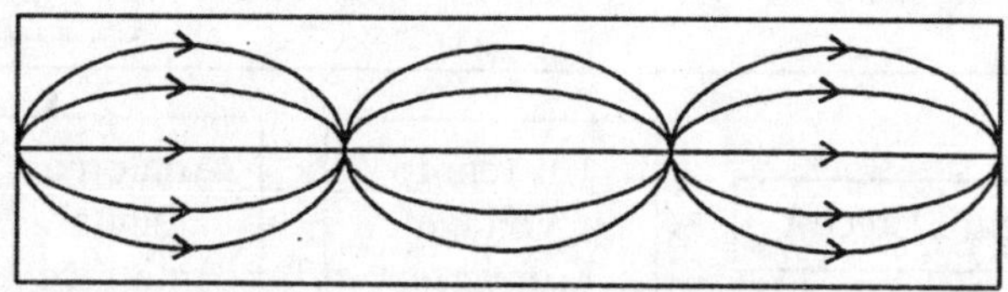

Fig. 7.14: Less multiple time dispersion for graded index optical fibre

7.10 LOSSES IN OPTICAL FIBRE CABLE

The losses in optical fibre, *i.e.* attenuation are mainly caused by absorption and scattering of rays. We may note that

(*i*) Absorption of rays is basically a material property and it occurs at all wavelengths when electronic transitions takes place within material and are followed by non-radiative relaxation processes. This causes an increase of thermal energy in the material.

(*ii*) **Scattering** is also partly a material property but in optical fibres it is caused by the imperfection in their geometries. Scattering occurs when the mode of propagation of light is changed, such that some of the optical energy leaves the optical fibre.

(*iii*) **Losses Due to Tight Bending of Fibres:** Tight bends of the optical fibre cause some of the light not to be internally reflected but to propagate into the cladding and be lost.

(*iv*) **Loss Due to Imperfection in Fibre:** The presence of imperfection in optical fibre material also cause power losses from the optical fibre cable.

(*v*) **Losses Due to Connectors and Splices:** Power losses are also caused for long distance communication cables due to connectors and splices which required to connect two optical fibres temporarily.

7.11 OPTICAL FIBRE COMMUNICATION SYSTEM

Figure 7.15 shows a simplified block diagram of an optical fibre communications link. The three primary building blocks of the link are the transmitter, the receiver, and the fibre guide.

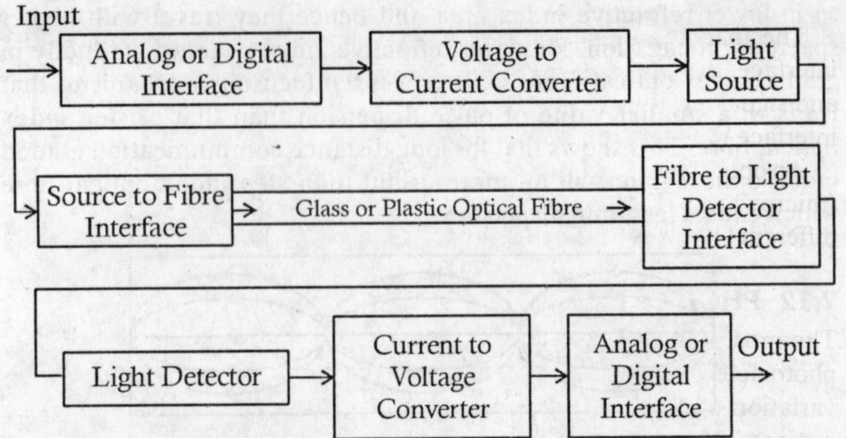

Fig. 7.15: Simplified fibre optic communication link

The transmitter consists of an analog or digital interface, a voltage to current converter, a light source, and a source to fibre light coupler. The fibre guide is either an ultra pure glass or plastic cable. The receiver includes a fibre to light detector coupling device, a photo detector, a current-to-voltage converter, an amplifier, and an analog or digital interface.

In an optical fibre transmitter, the light source can be modulated by a digital or an analog signal. For analog modulation, the input interface matches impedances and limits the input signal amplitude. For digital modulation, the original source may already be in digital form or, if in analog form, it must be converted to a digital pulse stream. For the latter case, an analog-to-digital converter must be included in the interface.

The voltage-to-current converter serves as an electrical interface between the input circuitry and the light source. The light source is either a light-emitting diode (LED) or an injection laser diode (ILD). The amount of light emitted by either an LED or an ILD is proportional to the amount of drive current. Thus the voltage to current converter converts an input signal voltage to a current that is used to drive the light source.

The source-to-optical fibre coupler (such as a lens) is a mechanical interface. Its function is to couple the light emitted by the source into the optical fibre cable. The optical fibre consists of a glass or plastic fibre core, a cladding and a protective jacket. The fibre-to-light detector coupling device is also a mechanical coupler. Its function is to couple as much light as possible from the fibre cable into the light detector.

The light detector is very often either a PIN (*p*-type-intrinsic-*n*-type) diode or an APD (avalanche photodiode). Both the APD and the PIN diodes convert light energy to current. Consequently a current to voltage converter is required. The current to voltage converter transforms changes in detector current to changes in output signals voltage.

The analog to digital interface at the receiver output is also an electrical interface. If analog modulation is used, the interface matches impedances and signal levels to the output circuitry. If digital modulation is used, the interface must include a digital-to-analog converter.

An optical communication system is quite similar in concept to electronic (microwave) communication system (10^9 Hz) only the frequency range is different (10^{15} Hz) and optical fibre instead of metallic fibre.

7.12 PHOTO DETECTOR

The most common semiconductor photo detector is the photo diode. A photo detector detects the optical power incident upon it and converts the variation of this optical power into a correspondingly varying electric current. The most commonly used system in optical fibre communication is *IM/SS*, where intensity of the optical carrier, modulated by biasing current in *LED/* diode laser (*IM*) and that variation of optical power directly detected in photo-detector (*DD*).

(*i*) Photo-Diode Mechanism

The basic structure and extrinsic P-N junction silicon photo diode is shown in Fig. 7.16 (*a*). In normal operation a sufficiently large reverse bias voltage is applied across the diode, so that depletion region as well as barrier potential will be large. Now the light incident on the reverse biased P-N junction diode, cause a photon ($h\nu$) to be absorbed. If the energy of the photon is more than the band gap energy (E_g), *i.e.* $h\nu > E_g$, then the absorbed photon excite an electron from valence band (*VB*) to conduction band (*CB*) and leaving a hole in *VB*. The electron-hole pair created due to incident light that is absorbed in the depletion region of the photo-diode is called photo-carriers.

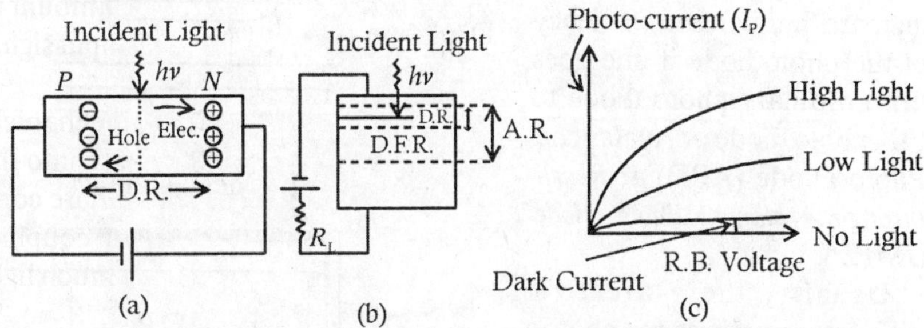

Fig. 7.16: Mechanism of P-N junction silicon photo-diode

Under reverse bias voltage, electron and hole flow in the depletion region (DR) in opposite direction, some of the electron-hole pairs will recombine and hence lost. The distance they will traverse prior to their recombination is called diffusion length (L_n and L_p for electrons and holes respectively) and the time it takes for an electron and hole to recombine is known as carrier life time (T_n and T_p). The relation between the carrier life time and diffusion length can be expressed as

$$L_n = (D_n T_n)^{1/2},\ L_p = (D_p T_p)^{1/2}$$

where D_n and D_p represents the electron and hole diffusion coefficients respectively, which are in cm^2/sec.

We may note that photons can be absorbed in depletion region (DR) as well as in diffusion region (DFR) too, which is usually indicated as absorption region (AR) (see Fig. 7.16 *b*). Obviously, this region becomes a void of free photo carriers. Now, within the deflection region those pairs of electron-hole remains which cannot recombine because of the existence of main electrostatic field due to the depletion region voltage. Any photon induced electron-hole pair there in will be separated and drawn in opposite directions by the combined effect of equivalent depletion region voltage as well as the reverse bias voltage. We may note that the polarities of these two voltages are in the same direction. This effect leading to flow of photo-current in the external circuit (Fig. 7.16 *c*), as photo carriers drift away across the depletion region. Figure 7.16 *c* shows typical output characteristics for reverse bias (RB) P-N photo diode. Different operating conditions are indicated from no light to high level light. When there is no incident light, then also a small reverse current will be there. This is called as dark current.

If the electron-hole pairs are produced outside the absorption region (AR) then there will be higher probability available and in that case no photo current will be available. Obviously this means that we have to make the top layer as thin as possible and effective depletion region as wide as possible in order to maximize the efficiency of photodiode as phote detector, which is termed as *quantum efficiency*. One can increase the quantum efficiency of the photodiode if one goes from ordinary photo diode to PIN photo diode or Avalanche Photo Diode (APD) ar *Reach through Avalanche Photo Diode (RAPD)*.

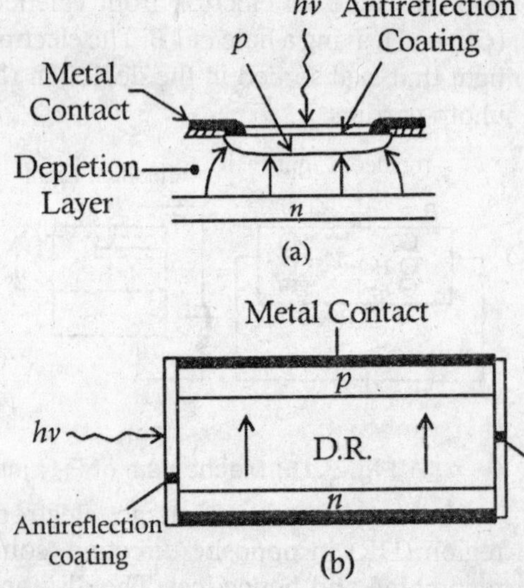

Fig. 7.17: Two different possible structures of photo diode

Usually, there are two different structures for photo-diodes possible as shown in Fig. 7.17. Figure 7.17 (*a*) shows the photodiode with front illuminated operation at 0.8 to 0.9 μm range, whereas

Fig. 7.17 (*b*) shows that side illuminated photodiode structure operating at higher wavelength 1.09 μm range.

(ii) Quantum Efficiency (η) and Responsivity (R)

Quantum efficiency (η) and responsivity or response speed (*R*) for a photo detector are two important characteristics. These two parameters, *i.e.* η and *R* depend on the material band gap, operating wavelength, percentage of doping and thickness of P and N regions. In order to achieve high quantum efficiency, the depletion layer must be thick enough to permit a large fraction of the incident light to be absorbed. This is why PIN or APD photo-diode is better than ordinary photo-diode. One can define quantum efficiency and responsivity as follows:

$$\text{Quantum efficiency, } \eta = \frac{\begin{array}{c}\text{Rate of generation of electron}\\ \text{-hole pair (electron/sec.)}\end{array}}{\begin{array}{c}\text{Rate of photons incident}\\ \text{(Photons/sec.)}\end{array}}$$

or
$$\eta = \frac{I_p/e}{P_0/h\nu}$$

$$\text{Responsivity,} \quad R = I_p/P_0 = \eta e/h\nu = \frac{ne\lambda}{hc} \text{ A/watt}$$

where $I_p \rightarrow$ photo current, $e \rightarrow$ electron charge, $P_0 \rightarrow$ incident optical power, $h\nu \rightarrow$ photon energy and $c \rightarrow$ velocity of light.

(*iii*) Long Wave Length Cut-off

It is essential when considering intrinsic absorption process, that the energy of the incident photon ($h\nu$) should be greater than the band gap energy (E_g) of the material used as photo diode, otherwise electron-hole pair will not generate, *i.e.*

$$h\nu \geq E_g \text{ or } hc/\lambda \geq \lambda \leq \lambda_c \leq hc/E_g$$

Obviously, there will be threshold wavelength for detection, which is commonly known as long wavelength cut-off (λ_c). If $\lambda > \lambda_c$, *i.e.* the incident wavelength of photon (λ) is more than λ_c, then photo detector will not work.

7.13 OPTICAL FIBRE CABLE CONSTRUCTION

A single optical fibre can transmit light energy very effectively (Fig. 7.8 and 7.9). However, in order to transmit an image of an object one requires assembly or bundles of fine glass fibres or plastic fibres because each fibre transmits rays from a very small region of the object (Fig. 7.18). The quality of the image depends upon the diameter of the optical fibre. It can be as small as 10^{-6} m. A bundle of the optical fibre may consist of thousands of individual optical fibres.

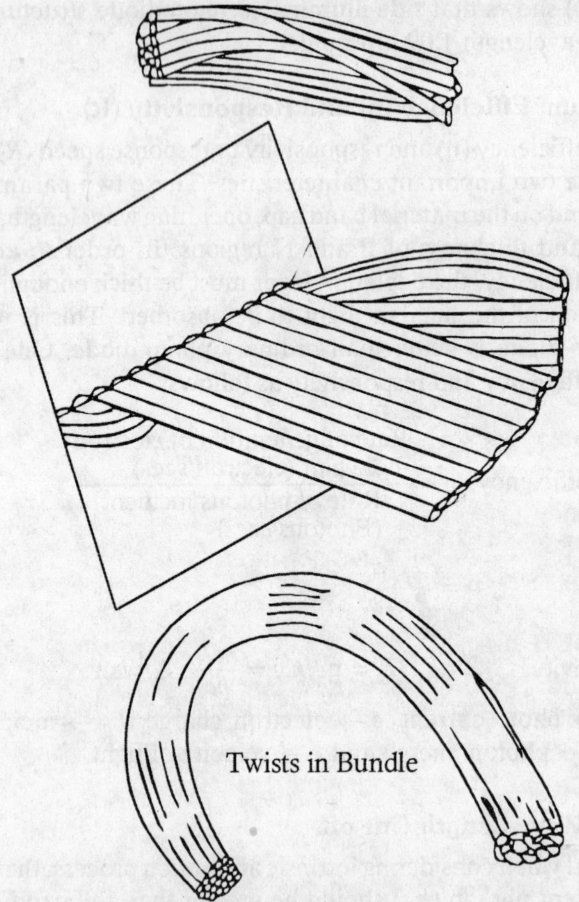

Fig. 7.18: Bundle of optical fibre

Essentially, three varieties of optical fibres are available. All these three varieties are constructed of either glass, plastic or a combination of glass and plastic. The three varieties are:

(*i*) Plastic core and cladding

(*ii*) Glass core with plastic cladding (often called PCS fibre, plastic-clad silica)

(*iii*) Glass core and glass cladding (often called SCS, silica-clad silica)

There is also a possibility of using fourth variety that uses a nonsilicate substance, zinc chloride. Investigations have shown that fibres made of this substance will be as much as 1000 times as effective as glass—their silica based counterpart.

Plastic fibres have several advantages over glass fibres. First, plastic fibres are more flexible and consequently, more rugged than glass. They are easy to install, can better withstand stress, are less expensive, and weigh approximately 60% less than glass. The disadvantage of plastic fibres is their high attenuation characteristic. They do not propagate light as efficiently as glass. Consequently plastic fibres are limited to relatively short runs, such as within a single building or a building complex.

Fibres with glass cores exhibit low attenuation characteristics, however, PCS fibres are slightly better than SCS fibres. PCS fibres also less affected by radiation and, therefore, are more attractive to military applications. SCS fibres have the best propagation characteristics and they are easier to terminate than PCS fibres. Unfortunately, SCS cables are the least rugged, and they are more susceptible to increase in attenuation when exposed to radiation.

The selection of a fibre for a given application is a function of specific system requirements.

Cable Construction

There are many different cable designs available today. Fig. 7.19 shows examples of several fibre-optic cable configurations. Depending upon the configuration, the cable may include a **core**, a **cladding**, a **protective tube**, **buffers**, **strength numbers** and one or more **protective jackets**.

7.14 FIBER OPTIC SENSORS

This is a transducer which can canvert various input variables (physical quantity) into an electrical signal in a measurable form.

In comparison to traditional sensors, the optical fiber sensors are less costly and improved quality and these find wider applications. The optical fiber sensors are used for measuring and sensing various parameters, e.g., pressure, temperature, magnetic field, electric field, humidity, acoustic vibrations, current, etc.

The optical fibers have following major advantages in sensing applications:

(*i*) Optical fiber sensors are light in weight and small in size

(*ii*) Optical fiber sensors have good geometrical flexibility

(*iii*) Optical fiber sensors are made of silica and therefore they are free from the risk of fire and sparks

(*iv*) Optical fiber sensors are immune to electromagnetic interference, *i.e.* they are electrically passive and moreover they do not distort the surrounding electrical and magnetic fields

(*v*) Optical fiber sensors have large bandwidth and are high sensitive

(*vi*) The chemical and enviornmental ruggedness of optical fiber sensors is more.

The optical sensors are non cantact and generally high accuracy devices and systems. In optical fiber sensors, the optical wave acts as information carrier and sensor.

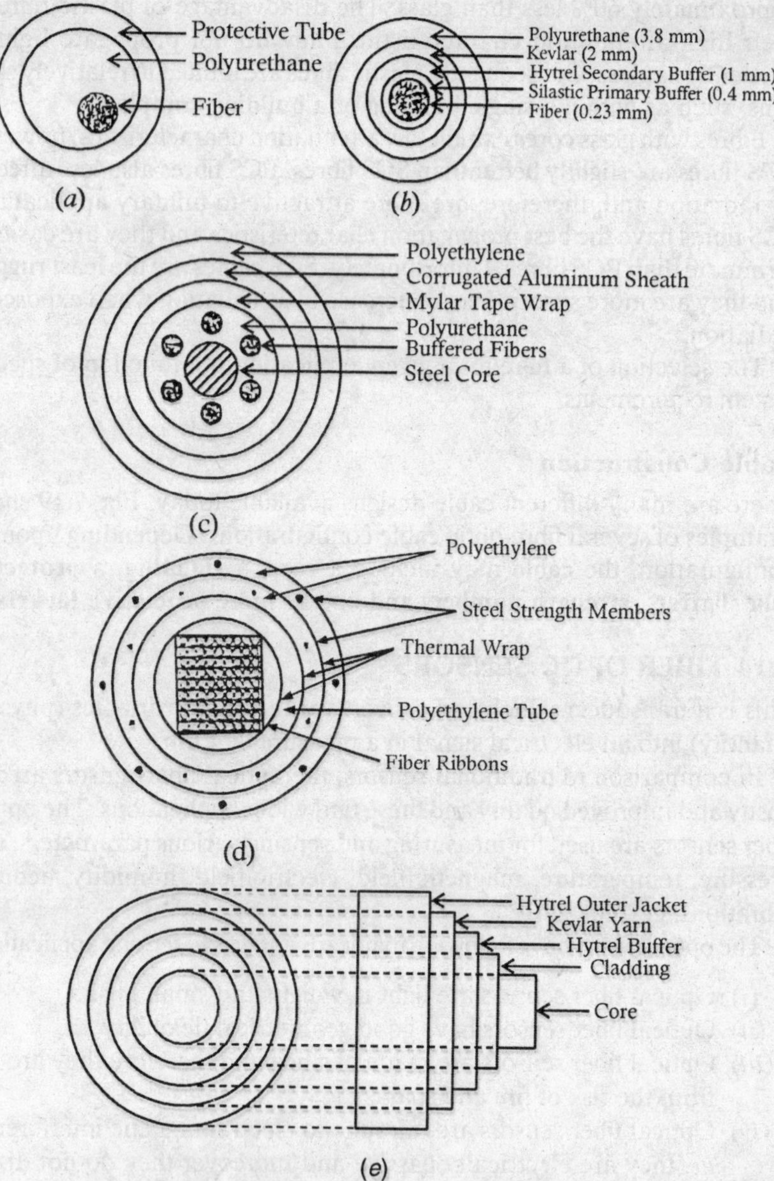

Fig. 7.19: Fibre optic cable configuration: (*a*) loose tube construction (*b*) Constrained fibre; (*c*) multiple strands; (*d*) telephone cable; (*e*) plastic-clad silica cable

Whenever optical sensors are used for measuring a physical parameter, any one of the characteristic like phase, amplitude, intensity, frequency, polarisation and direction of propagation of the wave gets modulated by the measured quantity.

However, the detected quantity is intensity and thus on demodulation it results in change of intensity and thereby relating to the measured quantity.

Types of Optical Fiber Sensors

Optical fiber sensors are of following two types:

(*a*) **Extrinsic or Hybrid Fiber Optic Sensors (Passive Sensors):**

(*i*) In these sensors, the interaction between the light and the measurand (the quantity under measurement) takes place outside the optical fiber.

In extrinsic sensors, the fiber acts merely as a wave guide. This type of sensors have a sensor head, and the sensed optical signal will be transferred to the measurement point with low attenuation and increased mechanical stability for signal processing.

Extrinsic sensors can be used for the measurement of voltages, current, pressure, temperature, force, displacement, etc.

(*i*) **Intrinsic or all Fiber Sensors (Active Sensors):** In these sensors, the measurand acts directly on the fiber itself and produces and change in the transmission characteristics. Examples of this type of sensors are liquid level sensor, Faraday sensor or gyroscope.

Following table gives a few of the physical measurand with their modulation effects in optical fibers.

TABLE 7.1: PHYSICAL MEASURAND WITH THEIR CORRESPONDING MODULATION EFFECTS IN OPTICAL FIBERS

Physical Measurand	Modulation Effects in Optical Fibers
1. Temperature	Thermoluminescence
2. Nuclear radiations	Radiation induced luminescence
3. Pressure	Piezoelectric effect
4. Mechanical force	Stress birefringence
5. Electric field	Electro optic effect
6. Electric current	Electroluminescence

Now, we briefly explain the *displacement sensors* and *pressure sensor.*

(*iii*) **Interferometric Sensors:** These sensors are widely used to measure temperature, pressure, displacement, magnetic field and electric field.

Interferometric sensors are mainly based on the principle of interference pattern produced due to a phase shift on being superimposing and reference beam and a sensing beam.

(b) Displacement Sensor

(i) Based on Intensity Measurement: To make a measurement of the displacement, the intensity modulation of the transmitted light is used. This is the principle of this sensor.

The arrangement of displacement sensor is shown in Fig. 7.20. One end of the transmitting fibre is coupled to a laser source. The other free end is made to face a target located at a distance *x* metre. Similarly, one end of the receiveing fibre is connected to a detector and its free end is made to face the same target.

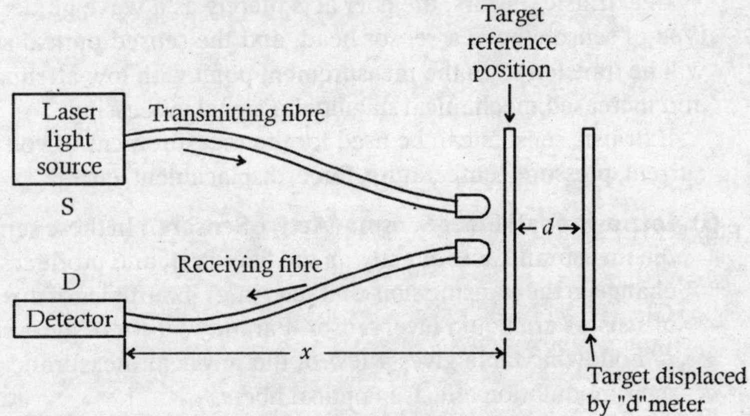

Fig. 7.20: Displacement sensor

A light beam is allowed to incident over the target at a distance *x* meter from the transmitting end of the fibre. After getting reflected from the target, the light is collected by the detecting fiber and detected by the detector *D*. Obviously, the displacement of the object is related to the intensity of light reaching at the detector. Clearly, there will be either an increase in intensity or decrease in intensity with respect to the decrease or increase in distance of the target. Obviously, we can measure the position of the target using the optical fiber detector.

In order to enhance the sensity of the sensor, the axes of the transmitting fiber and receiving fiber should be placed at a specific angle with respect to each other and also to the target.

(ii) Displacement Measurement with the Help of Diffraction Pattern: The arrangement of an optical fiber displacement sensor based on phase modulation is shown in Fig. 7.21.Fig.

A collimated laser light beam is allowed to fall over butted gratting arrangement (Fig. 7.21). The light beam after getting transmitted through the gratings are received by a detector. Clearly, when any one of the grating is transversely displaced, it causes the diffraction angle of the light beam to vary.

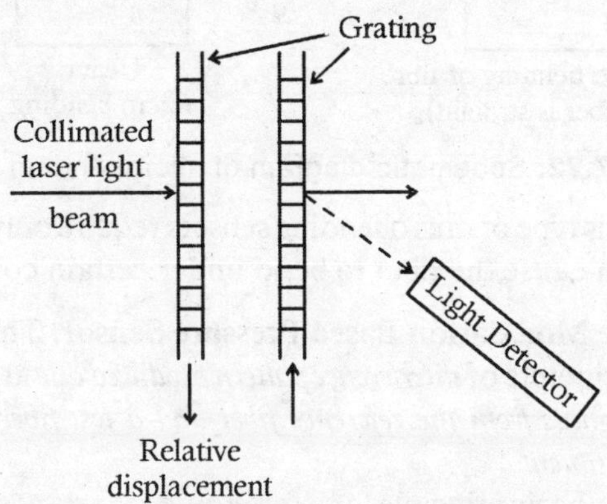

Fig. 7.21: Displacement measurement based on phase modulation

Obviously, the detector senses this as an intensity variation. Clearly, the displacement of the grating results in a high and low intensity beam on the digital detector, thereby providing a direct measurement of the displacement.

(*c*) **Pressure Sensor:** For the downhole measurement in oil wells, optical fiber sensors for the measurement of temperature and pressure have been developed.

These sensors are used in different fields like process control, oceanography, meteorology, hydrology, aerospace, laboratory instrumentation and energy pressure exploration. This sensor is also used to measure blood gas parameters and dosage levels.

There are different types of pressure sensors. A brief description of these are presented below:

(*i*) **Microbending Sensor:** These sensors are based on measurement of the light losses caused by controlled microbending.

The basic principle of a microbending sensor which can be used for measuring pressure, temperature and displacement is shown in Fig. 7.22.

A fibre is mounted in between a pair of plates containing parallel grooves as shown in Fig. 7.22. When we apply pressure over the plates, it causes microbends in the fiber and therefore results in leakage of light thereby reducing the power of the transmitted light. Thus, the measurement of the output power provides the pressure acted over the fiber.

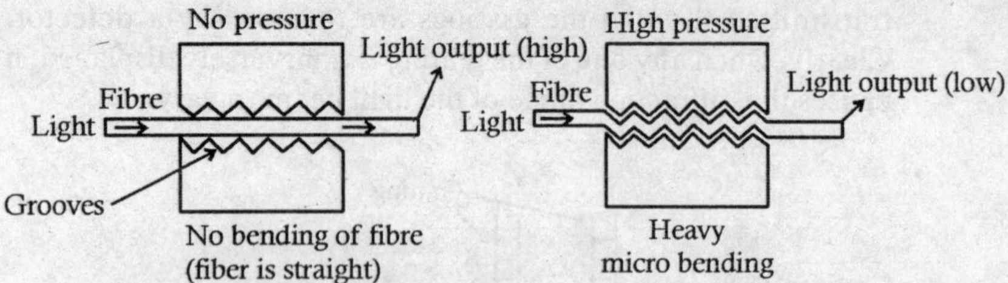

Fig. 7.22: Schematic diagram of microbending sensor

This type of microbending sensors require only an enviornment which cause the fiber to bend under certain conditions.

(ii) Phase Modulation Based Pressure Sensor: These are based on the principle of *interference pattern produced due to phase shift between the signals from the reference fiber and a test fiber in the measuring enviornment.*

The basic principle for interferometric arrangement to measure pressure is schematically illustrated in Fig. 7.23.

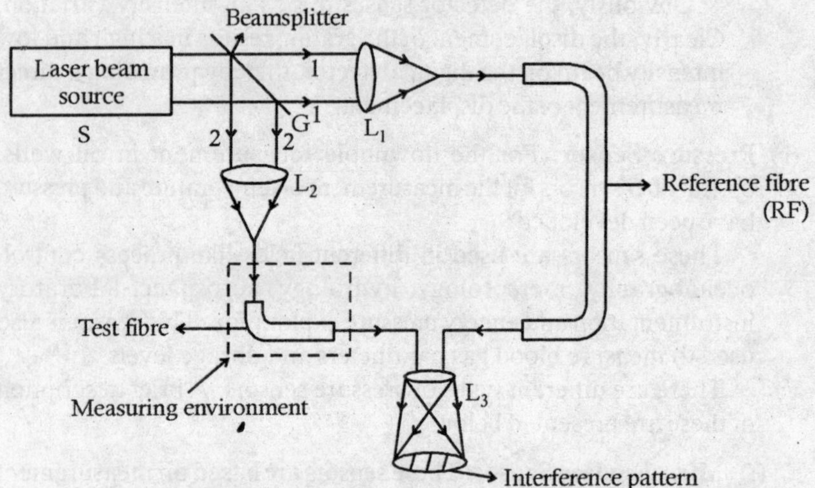

Fig. 7.23: Schematic illustration of pressure sensor based on phase modulation

The light from a source of laser beam (S) is allowed to pass through a beam splitter. This beam splitter is a glass plate G, which is inclined at an angle of 45° to the direction of the laser beam. Obviously, the beam splitter G divides the single beam incident on it into two parts travelling at right angles to each other, marked by 1, 1 and 2, 2.

The main or direct beam 1, 1 passes through lens L_1 and then through a reference fiber which is isolated from any influence of external enviornment. Finally, the beam from the other end of this reference fibre is allowed to pass through lens L_3.

The splitted beam 2, 2 at right angles to the main beam is allowed to pass through a lens L_2 and then through a test fibre which is kept in an varying environment to be measured. Clearly, an enviornmental effect like pressure or temperature when acts over this test fiber, then a path difference is produced due to the change in fiber length, core diameter, and refractive index with respect to the beam from the reference fibre. Finally, this beam after getting transmitted through this test fiber is allowed to pass through the lens L_3.

Due to the phase difference between the two beams from the reference fiber and test fiber produce interference pattern. With the help of this sensor, a very accurate measurement of pressure or temperature may be obtained from this interference pattern.

(*iii*) Photoelastic Pressure Sensor Based on Polarisation: In this type of pressure sensor, the phenomenon of *photoelastic* effect is used to measure the pressure.

Figure 7.24 shows the schematic diagram of a multimode fibre optic sensor, which is used for measuring pressure.

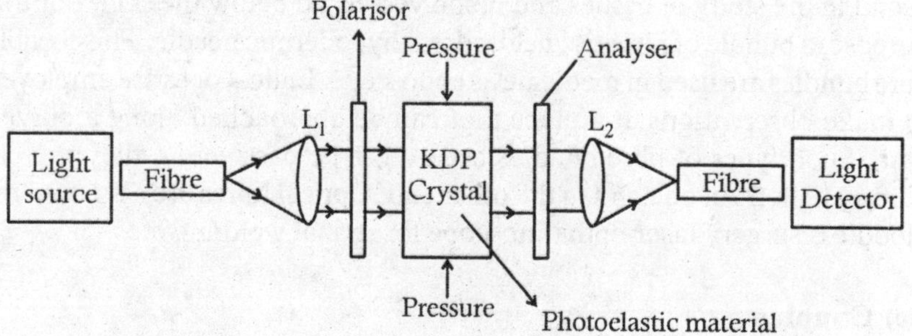

Fig. 7.24: Photoelastic pressure sensor based on polarisation

The photoelastic effect in the device is induced by a mechanical stress on photoelastic material like KDP crystal which is placed between the two cross polarisers. Due to the applied stress over the crystal upon passing a laser beam through it simultaneously the crystal material gets transformed into an optically anisotropic material. This causes a variation in intensity of the transmitted light thrugh the optical fiber sensor. Obviously, by measuring the intensity of the transmitted light beam, the pressure acted over the photoelastic material can be known.

7.15 APPLICATIONS OF OPTICAL FIBRES

Fibre optics technology has matured as a powerful alternative for information transport and sensing. Few important applications of optical fibres are:

(*i*) Communication

Optical fibres are used to carry signals in optical communications due to their unique features, viz. large channel capacity, wide band width, good electrical isolation, no cross talks, etc. Now a days telephone, television, satellite and computer links of advanced communication systems rely heavily on this technology.

(*ii*) Scanning

Optical fibres can be used to measure the distribution of light intensity over an illuminated area. For this purpose, the input end of the optical fibre scans the area, and the light output from the optical fibres is fed to a photodetector. The variation of the photodetector output reveals the light intensity distribution.

(*iii*) Use in Medical Instruments

Optical fibres are used in medical instruments (fibres copes) to examine internal body cavities such as stomach and bladder. Optical fibres are also useful in the study of tissues and blood vessels far below the skin. For this purpose, a bundle of fibres is enclosed in a hypodermic needle. This flexible fibre bundles are used in medicine as endoscope. Endoscopes are employed to make observations at a place that can be approached along a curved path. Such types of fibre bundles are said to be passive, *i.e.* they merely convey light from one end to the other end. Optical fibres are also used for bloodless surgery laser opthalmoscope for retinal welding, etc.

(*iv*) Coupler

One can use the optical fibre for coupling two circuits without introducing a direct electrical link. An LED forms a load in the primary circuit, so that a light output is obtained. Light guide collects the light and transmit to a photodetector in the secondary circuit. The secondary circuit gives an electrical signal in this circuit corresponding to that in the first.

(*v*) Display and Illumination

Optical fibres are employed to carry light to the display units. They are also used to illuminate dials when measuring instruments are used in dim light.

(*vi*) Transfer of Energy

Optical fibres are employed in the ultraviolet region for spectroscopic work and also used to transfer infrared energy from the source to the point of application of heat.

ILLUSTRATIVE EXAMPLES

Example 1

For a 3 μm diameter optical fibre with core and cladding indexes of réfraction of 1.545 and 1.510 respectively; determine the cut-off wavelength.

Solution

$$\lambda_c = \frac{2\pi a n_1 \sqrt{2\Delta}}{2.405} \ \mu\text{m}$$

where

λ_c - Cut-off wavelength (μm)

n_1 - Core index of refraction (unitless)

n_2 - cladding index of refraction (unitless)

a - core radius (μm)

$$\Delta = \frac{n_1 - n_1}{n_1}$$

$$\Delta = \frac{1.545 - 1.510}{1.545} = 0.023$$

$$a = \frac{3\mu m}{2} = 1.5\ \mu\text{m}$$

$$\lambda_c = \frac{(2\pi)(1.5\ \mu\text{m})(1.545)\sqrt{2(0.023)}}{2.405}$$

$$= 1.29\ \mu\text{m}.$$

Example 2

For a multimode index fibre $n_1 = 1.53$, $n_2 = 1.50$ and $\lambda = 1$ μm. If the core radius is 50 μm then calculate the normalized frequency of the fibre (ν) and the number of guided mode.

Solution

Normalized frequency, $\nu = \dfrac{2\pi a(NA)}{\lambda}$

$$= \frac{2 \times 3.14 \times 50 \times 10^{-6} \times \sqrt{(1.53)^2 - (1.50)^2}}{1 \times 60^{-6}}$$

$$= 94.72$$

Total number of guided mode $= M_s = \frac{\nu^2}{2} = \frac{(94.72)^2}{2}$

$= 4486.$

Example 3

For a multimode stepped index fibre with a glass core ($n_1 = 1.5$) and a fused quartz cladding ($n_2 = 1.46$), determine the critical angle (θ_c), acceptance angle (θ_m) and numerical apperature (*NA*). The source to fibre medium is air.

Solution

Critical angle $\theta_c = \sin^{-1}\frac{n_2}{n_1} = \sin^{-1}\frac{1.46}{1.5}$

$= 76.7°$

$$\theta_m = \sin^{-1}\frac{\sqrt{n_1^2 - n_2^2}}{n_0}$$

and $\theta_{m(\max)} = \sin^{-1}\sqrt{n_1^2 - n_2^2}$

$$= \sin^{-1}\sqrt{(1.5)^2 - (1.46)^2}$$

$$= 20.2°$$

$$NA = \sin\theta_m = \sin 20.2 = 0.344.$$

The maximum diameter a single-mode optical fibre can have is proportional to the wavelength of the light ray entering the cable and the numerical aperature of the fibre. Mathematically, the maximum radius of the core of a single mode fibre is

$$r_{\max} = \frac{0.383\,\text{Å}}{NA}$$

where $r_{\max}$ is maximum core radius in meters.

$NA \rightarrow$ numerical aperature (unitless) and

$\lambda \rightarrow$ light ray wavelength (meters)

Example 4

The numerical aperature (*NA*) of an optical fibre is 0.5 and core refractive index 1.54. Determine (*i*) refractive index (RI) of cladding (*ii*) change in core cladding refractive index per unit RI of the core.

Solution

(*i*) $NA = \sin\theta_m = \sqrt{n_1^2 - n_2^2}$

$$\therefore (0.5)^2 = (1.54)^2 - n_2^2 \quad \text{or } n_2 = 1.456$$

(*ii*) RI of the core $= \dfrac{n_1 - n_2}{n_1} = \dfrac{1.54 - 1.456}{1.54} = 0.0542$.

REVIEW QUESTIONS

1. Define a optical fibre system. What is numerical aperature of an optical fibre. What is its significance?
2. Obtain the expression for *NA* is terms of *RI* of the core and cladding of an optical fibre.
3. Outline the primary building blocks of a fibre optic system.
4. Draw block diagram for optical communication through fibre optical cable. Mention the advantages of optical communication?
5. Explain the working of monomode and multimode optical fibres. For long distance communication which optical fibre is preferred and why?
6. Contrast glass and plastic fibre cables.
7. Explain the principle and working of fiber optic sensors.
8. Explain the working of displacement sansor based on: (*i*) intensity measurement and (*ii*) diffraction pattern.
9. Explain the working of a pressure sensor based on: (*i*) Phase modulation and (*ii*) Photoelastic effect.
10. Briefly explain the principle and working of different pressure sensors.
11. Explain the principle and working of microbending sensor.
12. Explain the principle and working of photoelastic pressure sensor.
13. Write short notes on
    (*i*) Fiber optic cable
    (*ii*) Step index fiber
    (*iii*) Graded index fiber
    (*iv*) Fiber optic communication system
    (*v*) Fiber optic sensors

PROBLEMS

1. A multimode step index fibre has $n_1 = 1.53$, $n_2 = 1.50$ and $\lambda = 1\ \mu m$. In order to make this fibre single mode, what will be core radius?

   [Hint : $a \leq \dfrac{2.405\lambda}{2\pi(NA)}$, $NA = \sqrt{(1.53)^2 - (1.50)^2}$

   $$a \leq \frac{2.405 \times 1 \times 10^{-6}}{2 \times 3.14 \times \sqrt{(1.53)^2 - (1.50)^2}}, \therefore a < 1.27\ \mu m]$$

2. Numerical aperature of a optical fibre is 0.5 and core refractive index is 1.48. Show that cladding refractive index is 1.393 and acceptance angle is 30°.

   [Hint : $NA = \sqrt{n_1^2 - n_2^2} = 0.5$, R.I. of core $= n_1 = 1.48$ that is $n_2 = 1.393$ that is R.I of cladding $= n_2 = 1.393$. Acceptance angle $\theta = \sin^{-1}(NA) - 30°$]

3. A step index fibre in air has a NA of 0.16, a core refractive index of 1.45 and core diameter of 60 cm. Show that the normalized frequency for the fibre when light at a wavelength of 0.9 μm transmitted is 3.35×10^5. [BE]

[Hint : $$\nu = \frac{\pi d}{\lambda_0}\sqrt{n_1^2 - n_2^2} = \frac{3.143 \times 0.6\,\text{m}}{9 \times 10^{-7}\,\text{m}} \times 0.16 = 3.35 \times 10^5]$$

4. An optical fibre has a NA of 0.20 and a cladding refractive index of 1.59. Show that the acceptance angle for the fibre in water which has a refractive index of 1.33 is 8.6°.

[Hint : $$NA = \frac{\sqrt{n_1^2 - n_2^2}}{n_0},$$

when the fibre is in air, $n_0 = 1$, and $NA = \sqrt{n_1^2 - n_2^2} = 0.20$

$$\therefore \quad n_1 = \sqrt{(NA)^2 + n_2^2} = \sqrt{(0.20)^2 + (1.59)^2} = 1.6025$$

when the fibre is in water, $n_0 = 1.33$

$$\therefore \quad NA = \frac{\sqrt{n_1^2 - n_2^2}}{n_0} = \frac{\sqrt{(1.6025)^2 - (1.59)^2}}{1.33} = 0.15$$

$$\therefore \quad \theta_0(\text{max}) = \sin^{-1}(NA) = \sin^{-1}(0.15) = 8.6°]$$

5. Show that the core radius neccessary for single mode operation at 850 nm S.I. fibre with $n_1 = 1.480$ and $n_2 = 1.47$ is 1.89 μm. Also show that the NA and maximum acceptance angle of this fibre are 0.1717 and 9° 53 ' 12" respectively.

[Hint : (*i*) $$\nu = \frac{\pi d}{\lambda_0}\sqrt{n_1^2 - n_2^2}$$

$$2.405 = \frac{\pi d}{450 \times 10^{-9}\,\text{m}} \times 0.1717$$

or $$d = 3.79\,\text{mm} \qquad \therefore\ r = 1.89\,\text{mm}$$

(*ii*) $$NA = \sqrt{n_1^2 - n_2^2} = \sqrt{(1.48)^2 - (1.47)^2} = 0.1717$$

(*iii*) $\sin\theta_0$ (max) NA

$$\therefore \quad \theta_0\,(\text{max}) = \sin^{-1}(NA) = \left(n_1^2 - n_2^2\right)^{1/2}$$

$$= \sin^{-1}(0.1717) = 9°\ 53'\ 12"$$

6. A fibre cable has index of refraction core 1.52 and cladding 1.31. Show that the numerical aperature is 0.77.

SHORT ANSWER QUESTIONS

1. To prevent energy losses via absorption and scattering, what should be the order of the cladding in an optical fibre?

Ans. At least a few wavelengths thick.

2. How numerical aperature (NA) of an optical fibre is defined?

Ans. $NA = \sqrt{n_1^2 - n_2^2}$ where n_1 is the refractive index of the core of a fibre and n_2 that of cladding ($n_1 > n_2$) : One can define the numerical aperature of a fibre as the physical area that light must enter in order to propagate through the fibre. Similarly, sources have a numerical aperature that describes the output pattern of the light. The numerical aperatures for light sources can be calculated in terms of angle involved with the output.

3. How will you define acceptance angle (θ_A) for an optical fibre?

Ans. $\theta_A = \sin^{-1}\sqrt{n_1^2 - n_2^2}$

4. What is major economic benefit offered by fibre optics in information technology?

Ans. Very high information transmission rate at a low cost per circuit-km.

5. What type of reflection is expected from an optical fibre?

Ans. Total internal reflection.

6. How is an optical fibre fabricated?

Ans. To fabricate fibre, one can use either different glasses with different refractive indices for core and cladding or use fused silica (quartz glass) whose refractive index is modified by doping. The type of glass used will also vary with application, but the purity (lack of any foreign materials as well as lack of imperfections such as cracks or bubbles) must also be kept very high.

7. What is a kevlar?

Ans. This is a yarn type material having high tensile strength. This is used in fiber construction to provide additional strength to the cable.

8. Explain the relationship between information capacity and bandwidth.

Ans. The information carrying capacity of a communication system is directly proportional to its bandwidth, *i.e.* under the bandwidth, the greater is its information carrying capacity.

9. Explain the purpose of cladding in optical fiber.

Ans. Due to its lower refractivge index than core, the cladding helps in retaining the light with the core. Moreover, it provides mechanical strength, reduces scattering loss and also protects the core from absorbing surface contaminants with which it could come in contact.

10. Mention the advantages of optical fiber communication over metallic cable communication.

Ans. (*i*) Extremely wide bandwidth

(*ii*) Light in weight and small in size

(*iii*) Immune to electrostatic interference

(*iv*) More secured, greater safety and there is no cross talk interference

11. Can a light propagate through a optical fiber cable with the angle of incidence at the entrance end greater than acceptance angle?

Ans. No. Only when the light enters at an angle less than the acceptance angle, light can propagate through the cable by undergoing total internal reflection.

12. What is an optical fiber sensor?

Ans. This is a transducer which can convert various input variables (physical quantity) into an electrical signal in a measurable form.

13. What are the advantages of optical sensors in sensing applications?

Ans. (*i*) They are electrically passive

(*ii*) They have good geometrical flexibility

(*iii*) They are light in weight and are small in size

(*iv*) The chemical and enviornmental ruggedness is more.

14. What are the characteristics of the wave that may get modulated in fiber optic sensor?

Ans. Amplitude or intensity, phase, polarisation, frequency and direction of propagation may get modulated.

OBJECTIVE QUESTIONS

1. Optical fibre is __________ carrier and immune to __________ interference. [non-electrical, electromagnetic]

2. Optical fibre does not exit any __________ that might interfere with signals. [noise]

3. Light has much __________ frequency than radio waves or modulated signals in copper wire. [higher]

4. Optical fibre does not rust or corrode and is much __________ and __________ than copper cable. [smaller, larger]

5. Scattering mechanism cause the transfer of __________ from one propagating mode into a different mode. [optical power]

6. Major implementation of optical fibre transmission involves some form of digital __________. [modulation]

7. Optical fibres are __________ pipes that guide light through them on the principle of total __________ . [light, internal reflection]

8. The refractive index of __________ region is always less than that of the core region. [cladding]

9. Optical fibres are divided into different groups based on their __________ profile. [modulation]

Unit III

Quantum Physics and Applications

Quantum Physics

8.1 INTRODUCTION

The dynamical properties of the observable objects have been successfully explained by the classical mechanics as there was no restriction on the values of the dynamical property. But the concepts of classical mechanics failed to describe the actual behaviour of atomic and molecular dimensions as experimental observations for such systems show that only a well defined set of values are observable, e.g. an electron moving around the nucleus of an atom has only a discrete set of values for energy. This discreteness has been termed as quantization. The explanation of this quantization required the development of a new system of mechanics called quantum mechanics.

8.2 BLACK BODY RADIATIONS

Radiations is emitted by every substance as a result of the vibration of its particles and the character of radiation depends on the nature and temperature of the substance. The ability of a body to radiate is closely related to its ability to absorb radiation as at any constant temperature, the body will be in thermal equilibrium with its surroundings and must be absorbing radiation at the same rate as it limits.

Different solids emit radiations at different rates at the same temperature; the rate is maximum when the solid is perfectly black, *i.e.* behaves as a black body. A black body can absorb all the radiations that fall on it. For practical purposes an isothermal cavity with a small aperture through which radiations from out side may be admitted and absorbed completely due to repeated reflections inside the enclosure is considered a black body. Radiations emerging from the small hole of such a hollow enclosure is, therefore, called 'black body radiations'. The intensity of radiations, *i.e.*

total radiations emitted per unit surface area, from a black body depends only on temperature (T) and is independent of the nature of the solid.

8.3 SPECTRAL DISTRIBUTION OF ENERGY IN THE BLACK BODY RADIATION

The radiation emitted by a black body is not confined to a single wavelength but spreads over a wide spectrum of wavelength. The experimentally observed dependence of the energy density on wavelength (λ) at different temperatures is shown in Fig. 8.1. A close investigation reveals the following important facts:

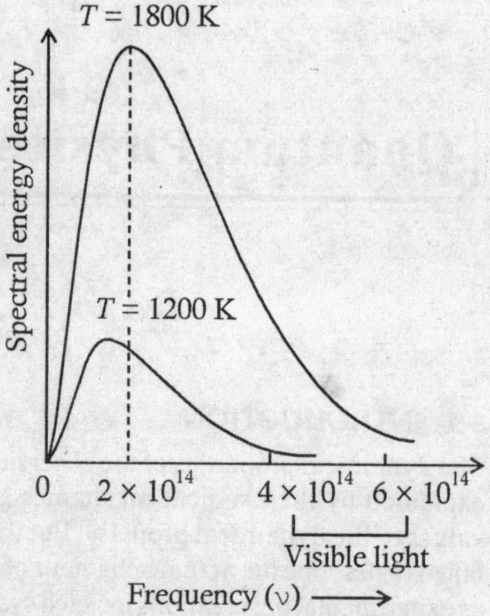

Fig. 8.1: Experimental dependence of the energy density on frequency and temperature

(*i*) At a given temperature, the energy is not uniformly distributed in the radiation spectrum.

(*ii*) At a given temperature the density of radiant energy increases with increase in wavelength and becomes maximum at a particular wavelength (λ_m). With further increase in wavelength, the density of radiant energy decreases.

(*iii*) An increase in temperature increases the energy of all spectral components. The short wavelengths cutoff advance towards the origin as the temperature increases and the peak of the curve shifts to shorter wavelength, *i.e.* (λ_m), (the wavelength for which energy emitted is maximum) decreases.

The shift of the peak of the curve was found to obey the following empirical relationship, commonly called *Wien displacement law*

$$\lambda_m T = \text{Constant} \tag{1}$$

With the help of classical thermodynamics, Wien has further shown that the amount of energy contained in the spectral region included within the wavelength λ and $\lambda + d\lambda$ emitted by a black body is given by

$$E_\lambda d\lambda = \frac{A}{\lambda^5} e^{-B/\lambda T} d\lambda \tag{2}$$

where A and B are constants.

Wien law works well only for short wavelengths. There are considerable deviations from Wien's law at long wavelength and high temperatures. From Eq. (2), one obtains finite energy even for $T = \infty$ (infinite). Rayleigh argued that it is unlikely that energy (E) should be finite for $T = \infty$.

Later, Rayleigh and Jeans adopted a more rigorous method based on classical physics and came up with the following relation

$$E_\lambda d\lambda = C\lambda^{-4}\, Te^{-A/\lambda T} \tag{2a}$$

where C is another constant.

This *Rayleigh-Jeans law* describes the experimental curve quite well in the long wavelength region but the energy density E_λ increases rapidly as λ decreases and approaches infinity for shorter wavelengths. This failure, for explaining black body radiation, was such a crushing blow to classical physics that it is historically referred as the *ultraviolet catastrophe.* It led Max Planck to discover that only if light emissions is a quantum phenomenon then only the correct formula for the energy density of black body radiation be obtained.

The failure of Rayleigh-Jeans law to explain the spectral energy curves of black body radiation indicated that the fundamental assumptions of classical physics required suitable modifications.

8.4 PLANCK'S LAW OF RADIATION

Being well aware of the shortcomings of both Rayleigh-Jeans and Wien Radiation Laws, Planck in October 1900, proposed remarkably well relation for the observed energy curves of the black body radiation. The formula could be written as

$$\rho(\nu)d\nu = \frac{a\,\nu^3}{\exp(b\nu) - 1} \tag{3}$$

where a and b are numerical constants. This formula had no theoretical background and was obtained expirically by the trial and error method to explain the observed results. However, a few weeks later, *i.e.*, in December 1900 Planck announced the theoretical derivation of his radiation law (Eq. 3). For this, he proposed that a black body chamber was filled not only with radiation but also with molecules of a perfect gas, and dipole oscillators of molecular dimensions constituting the black body absorbed energy from the radiation and transferred it partly or wholly to the gas molecules when the latter collided with them. This was a round about way of explaining the emission and absorption of radiation by black body but there was no other way as the mechanism of emission and absorption of radiation by atoms and molecules by direct exchange was not known at that time. Planck calculated that the number of oscilIators per unit volume was $8\pi\nu^2/c^3$, so that the radiation density was given by

$$\rho(\nu) = \frac{8\pi\nu^2}{c^3} E(\nu) \tag{4}$$

where $E(\nu)$ is the mean energy of the oscillator emitting radiation of frequency ν. Now, if $E(\nu)$ were to be determined by equipartition law, *i.e.* $E(\nu) = k_B T$, then equation (4) would lead to the Rayleigh-Jeans law. Planck gave up the hypothesis of continuous emission of radiation by oscillators and assumed that they emitted energy only when they acquired certain minimum energy E or an integral multiple of it, nE. That is to say an oscillator having an amount of energy nE could emit only after it had absorbed the amount $(n + 1)E$, and after emitting the amount of energy E it reverted back to the previous state of energy nE. Thus the radiation of energy E could be emitted by only those oscillators which has energy E, $2E, 3E\ldots\ nE$. Further E was directly proportional to ν, *i.e.*

$$E = h\nu \tag{5}$$

where the constant of proportionality, h, was called the Planck's constant. The above idea can be expressed in the form of following revolutionary postulates which have become the foundation of quantum theory of radiation.

(*i*) The amount of energy emitted or absorbed by an oscillator is proportional to its frequency. Calling the constant of proportionality h, one can write for the change in oscillator energy

$$\Delta\varepsilon = h\nu$$

(*ii*) An oscillator cannot have an arbitrary energy but must occupy one of a discrete set of energy states given by

$$\varepsilon_n = nh\nu$$

where n is an integer or zero. It was assumed that the ground state corresponds to the zero energy state. The value of $h = 6.62619 \times 10^{-34}$ Joule-sec. Planck's constant (h) is a universal constant which plays important role in all quantum phenomena.

The previous picture of a continuum of oscillator states is now replaced by a discrete set of "quantized" states. Furthermore, the amount of energy emitted or absorbed is also quantised, since each quantum must correspond to the energy difference between two states of a given oscillator. Each quantum of electromagnetic energy is called a photon. Using Plank's quantum hypothesis, we now derive the famous Plank's radiation formula.

Let N be the total number of Planck's oscillator and E_T is their total energy, then the average energy per oscillator $\bar{\varepsilon}$, given by

$$\bar{\varepsilon} = \frac{E_T}{N} \tag{6}$$

Let $N_0, N_1, N_2, \ldots N_n, \ldots$, etc. be the number of oscillators having energies $0, \varepsilon, 2\varepsilon, 3\varepsilon, \ldots n\varepsilon$, etc. respectively, then

$$N = N_0 + N_1 + N_2 + , \ldots + N_n + \ldots = \sum_{n=0}^{\infty} N_0 \tag{7}$$

and

$$E_T = 0 + \varepsilon N_1 + 2\varepsilon N_2 + \ldots + n\varepsilon N_n + \ldots = \sum_{n=0}^{\infty} n\varepsilon N_0 \tag{8}$$

According to Maxwell-Boltzmann distribution function, the number of resonators or oscillators having energy nE will be

$$N_n = N_0 \exp\left(-\frac{n\varepsilon}{k_B T}\right) \tag{9}$$

Obviously, the higher energy states are thus less likely to be populated, and putting

$$y = \exp\left(\frac{-\varepsilon}{k_B T}\right)$$

$$N = \sum_{n=0}^{\infty} N_0 y^n$$

$$= N_0(1 + y + y^2 + \ldots) = N_0\left(\frac{1}{1-y}\right) \tag{10}$$

Similarly $\quad E_T = EN_0 \sum_{n=0}^{\infty} n_0 y^n$

$$= EN_0(y + 2y^2 + 3y^3 + \ldots)$$

$$= EN_0 y(1 + 2y + 3y^2 + \ldots)$$

$$= \frac{EN_0 y}{(1-y)^2} \tag{11}$$

Hence $\quad \bar{\varepsilon} = \frac{yE}{1-y} = \frac{E\exp(-\varepsilon/k_B T)}{\exp(\varepsilon/k_B T) - 1} \tag{12}$

Using Eq. (12) for E, the energy density relation (5) becomes

$$\rho(\nu)d\nu = \frac{8\pi\nu^2}{c^3 \exp(\varepsilon/k_B T) - 1} d\nu$$

Since $\quad E = h\nu$

$$\rho(\nu)d\nu = \frac{8\pi h\nu^3}{c^3} \frac{1}{\exp(c\nu/k_B T)} d\nu \tag{13}$$

This is famous Planck's radiation law. This has the same from as Eq. (3) if a is put equal to $8\pi h/c^3$ and b equal to h/k_B. One can express Eq. (13) in terms of wavelengths by using the relation $\nu = c/\lambda$ and hence

$$d\nu = -\frac{c}{\lambda^2}d\lambda, \text{ as}$$

$$\rho(\lambda)d\lambda = \frac{8\pi hc}{\lambda^5}\frac{1}{\exp(ch/\lambda k_B T)-1}d\lambda \tag{14}$$

For small temperatures and short wavelengths, λT is small and hence $\exp(ch/\lambda k_B T) >> 1$. Obviously, 1 in the denominator of Eq. (14) can be neglected and hence Eq. (14) reduces to

$$\rho(\lambda)d\lambda = \frac{8\pi hc}{\lambda^5}\exp(-ch/\lambda k_B T)d\lambda \tag{15}$$

This is Wein's law with $a = 8\pi ch$ and $b = ch/\lambda k_B$. For higher temperatures and larger wavelengths.

$$\exp\left(\frac{ch}{\lambda k_B T}\right) = 1 + \frac{hc}{\lambda k_B T} \tag{16}$$

Using Eq. (16), Eq. (14) becomes ($\therefore$ T is large)

$$\rho(\lambda)d\lambda = \frac{8\pi k_B T}{\lambda^4}d\lambda \tag{17}$$

This is Rayleigh Jean's Law with $a = 8\pi k_B$.

Obviously, the Planck's theory of radiation incorporates all that is valid from the older theories as special cases. It thus serves an excellent example of a conceptual advancement which opened exciting new frontiers while still preserving much of the older physics.

Planck's derivation introduced a new constant h, the value of which was obtained by fitting the formula (17) to the observed curves for black body radiation. The value of $h = 6.626 \times 10^{-34}$ Js.

The most important aspect of Planck's hypothesis was the new idea of discontinuous emission of radiation. According to classical physics, a system can absorb or emit any amount of energy whereas in Planck's hypothesis energy emitted by a black body is restricted by the relation $E = h\nu$. To this definite amount of energy, $h\nu$, Planck gave the name "quantum". Thus, while in classical physics energy is continuous variable, in quantum physics it is "quantized".

Let us now reflect for a moment on the significance of Planck's postulate. It states that the energy of a harmonic oscillator expressed in terms of its frequency of vibration, may on the average occur in units no smaller then

($E = nh\nu$ where $n = 0, 1, 2, \ldots$). It does not say what is the energy of the oscillator at any instant, nor does it say how the value of h is to be calculated.

From this we may conclude that:

(*i*) Planck's postulate is empirical, *i.e.* it contains a parameter h whose value cannot be calculated except through the comparison of some expression in which it appears with the appropriate experimental results. There is no theory of Planck's constant. The only physical reason we can give for its existence is that it is invaluable for the elucidation of microscopic phenomena.

(*ii*) The postulate is statistical in that it refers only to the time-averaged behaviour of the microscopic oscillator. Nothing is said about energy except that it varies directly with the frequency if evaluated over period of time long compared with ν^{-1}.

(*iii*) The postulate does not in itself suggest that a flaw exists in the Maxwell's theory of electromagnetism. It states only what energy an oscillator must have on the average if it is to remain in thermodynamic equilibrium with a radiation field. We should not conclude from this that the latter must necessarily exhibit discrete energies when interacting with matter. We have verified the quantum hypothesis only for the emission of radiation by a heated solid!

8.5 PHOTOELECTRIC EFFECT

The importance of the Planck's hypothesis of energy quantization was quickly realised by Albert Einstein. Einstein extended this idea to all kinds of electromagnetic radiations and explained the important phenomenon of photoelectric effect.

When ultraviolet light of *X*-rays strike a piece of metal it frees electrons from its surface, and it is experimentally possible to count those electrons and to measure their energy. Quantitative investigation of the effect shows that the light intensity determines the number of electrons thus freed but has no influence on their energy. This is contrary to what we would except from Maxwell's electromagnetic theory. Since the intensity of a light wave is proportional to the square of the amplitude of its electric wave is proportional to the square of the amplitude of its electric field vector (E). The energy of the electrons is, surprisingly, determined only by the colour of the light, *i.e.* its frequency.

The following facts must be explained by any satisfactory theory of the photoelectric effect:

(*i*) If the incident light has no frequency component above a certain threshold frequency, no photoelectrons will be emitted.

(*ii*) The threshold frequency appears to depend upon the properties of the metal.

(*iii*) Provided that the threshold frequency is exceeded, photoelectrons will be emitted instantly, regardless of how low the intensity of light source is.

(*iv*) The photoelectric current (number of electrons emitted per second) increases with the intensity of light but is independent of the frequency.

(*v*) The maximum kinetic energy of the photoelectrons is independent of the intensity of light and depends only upon its frequency.

Attempts to explain the above phenomena by means of classical radiation theory met with defeat. Classically, an electron would be ejected as soon as it could accumulate enough energy to overcome the force which binds it to the metal. This energy is called the *work function* of the metal. Since an electromagnetic wave has an energy density associated with it, one would except that the metal should accumulate enough energy and ultimately, to eject an electron, regardless of the frequency or the intensity of the source. However, the classical picture was completely at a loss to explain all of the above experimental facts except number four. From the classical point of view it was reasonable to expect the photoemission to increase when the energy density of the incident radiation is increased.

Einstein used Planck's idea of energy quantization to explain the above mentioned observations. He proposed that electromagnetic radiation (light) could be considered to have properties of both waves and particles. He called the particles of light as 'photons'. Each photon of light possesses an energy equal to $h\nu$ (quantum of light). Before discussing the Einstein's photoelectric effect, we would like to study details of the experimental study of the photoelectric effect.

Photoelectric Effect

When a radiation of high frequency, e.g. ultraviolet light or *X*-rays is incident on a clean metal surface, electrons are emitted from the surface of the metal. The emitted electrons are called photoelectrons and the phenomenon is called the photoelectric effect. The entire range of electromagnetic radiation from gamma rays and *X*-rays to the ultraviolet, the visible and the infrared produce this effect. The photoelectric effect is also observed in solids, liquids and gases. This effect was experimentally observed by Hertz in 1887.

Experimental Arrangement to Study the Photoelectric Effect

A simple experimental arrangement for the study of photoelectric effect is shown in Fig. 8.2.

The emitting surface, *i.e.* photosensitive plate *A* is mounted opposite a metal plate *B* in a highly evacuated glass tube *C*. The two plates *A* and *B* form two electrodes, to which a variable potential difference can be applied. The evacuated glass tube is fitted with quartz window D, through which ultraviolet light from the source S is transmitted and allowed to fall in the plate A.

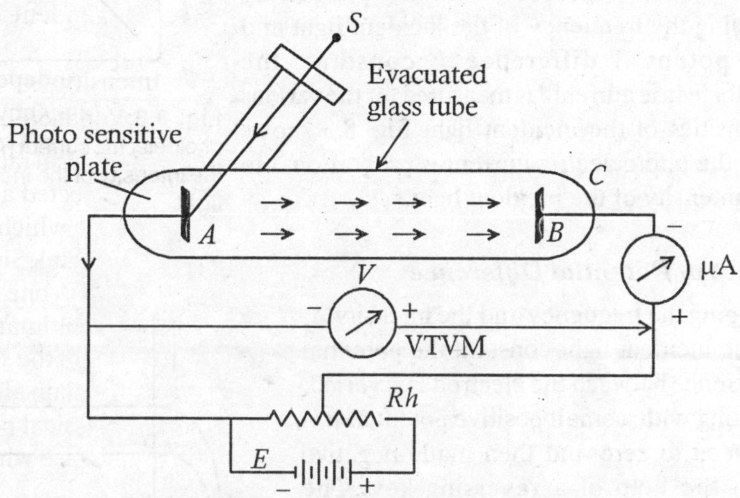

Fig. 8.2: Experimental arrangement to study the photoelectric effect

When the applied potential difference between the plates A and B is such that A is at negative potential with respect to B, the photoelectrons emitted from A are accelerated towards plate B.

The resulting photoelectric current I flowing in the circuit is measured by the μA (micro ammeter) and the accelerating potential difference V is measured by vacuum tube voltmeter. If the potential difference is reversed so that plate B is now at negative potential with respect to A, the photoelectrons are repelled towards A by the retarding potential. Consequently, the photoelectric current is reduced. It is found that the photoelectric current depend on the following factors: (*i*) the frequency of the incident radiation (*ii*) the intensity of incident radiation (*iii*) the potential difference between the electrodes and (*iv*) the nature of the emitting surface.

The important experimental observations are as follows:

Effect of Frequency on the Photoelectric Current

The collector plate B is made sufficiently positive with respect to the emitter photosensitive plate A. The surface of plate A is illuminated with monochromatic light of different frequencies. It is observed that the photoelectric current is produced only when the frequency ν of the incident light is greater than a certain minimum value ν_0. This minimum value of the frequency is called the *threshold frequency* for the given surface.

Effect of the Intensity of the Incident Light

To study the effect of the intensity of the incident light, the collector plate B is made sufficiently positive with respect to the emitter plate A. By keeping the frequency of the incident light and the potential difference V constant, the photoelectric current I is measured for the various intensities of the incident light Fig. 8.3 shows that the photoelectric current is proportional to the intensity of the incident light.

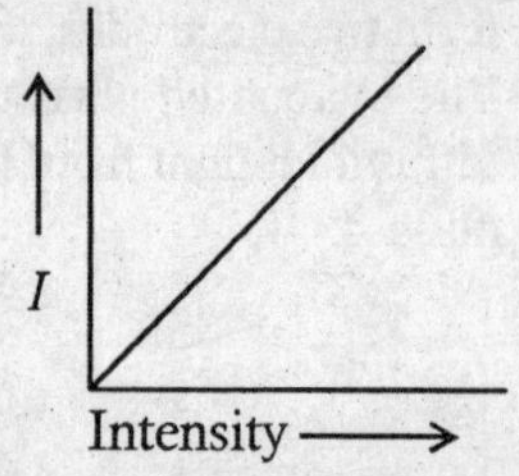

Fig. 8.3: Variation of the photoelectric current (I) with the intensity of the incident light

Effect of Potential Difference

Keeping the frequency and the intensity X_1 of the incident light constant the potential difference between the electrodes is varied. Starting with a small positive potential it is reduced to zero and then made negative with the help of a reversing key. The variation of photoelectric current with the potential difference is as shown in Fig. 8.4. When the collector plate B is positive, the photoelectric current remains constant. When the collector plate is made more and more negative, the photoelectric current goes on decreasing until it is stopped entirely. The retarding potential that stops the photo current is called the stopping potential. If the intensity of the incident light is increased to X_2, then initial photo current for potential V_0 is found to be the same for the light of the same frequency.

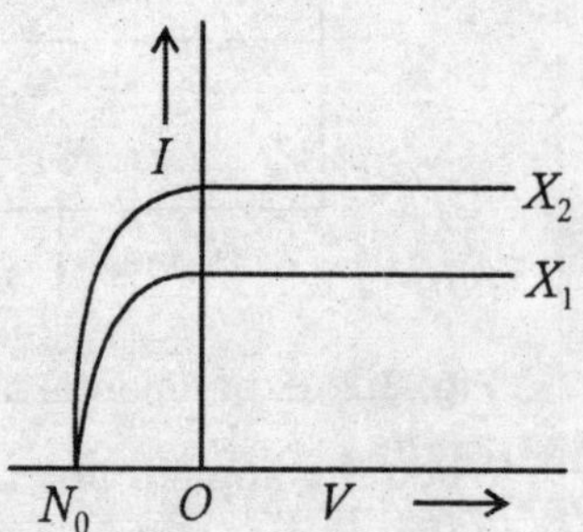

Fig. 8.4: Variation of the photoelectric current (I) with the potential difference (P.D.)

The above experimental observation is explained by assuming that when light on a certain frequency $\nu(\nu > \nu_0)$ is incident on a emitting surface, photoelectrons with kinetic energies ranging from zero to a certain maximum value are emitted from the surface. The photoelectrons which are emitted with the maximum kinetic energy are stopped by the stopping potential V_0 volts. Obviously, the work done by the retarding potential on the faster electrons must be equal to their kinetic energy, *i.e.*

$$eV_0 = \frac{1}{2}mv_{\max}^2 \qquad (18)$$

where m is the mass of the electron, e is the electronic charge and $v_{\max}$ is the maximum velocity of emission of the electrons.

Effect of Frequency on the Stopping Potential

To study the effect of the frequency of the incident light on the stopping potential, the intensities of light of different frequencies are adjusted to produce the same maximum (I_m) value of the photoelectric current when collector B is positive. The potential of collector B is then reduced in steps and made zero. It is then made more and more negative with the help of the reversing key. For the given photocathode, the graphs obtained are as shown in Fig. 8.5. From Fig. 8.5, it is evident that the stopping potential increases as the frequency of the incident light is increased. Thus if the frequencies are in the increasing order, *i.e.* $\nu_1 > \nu_2 > \nu_3$ then the corresponding stopping potentials are also in the increasing order, *i.e.* $V_{01} > V_{02} > V_{03}$.

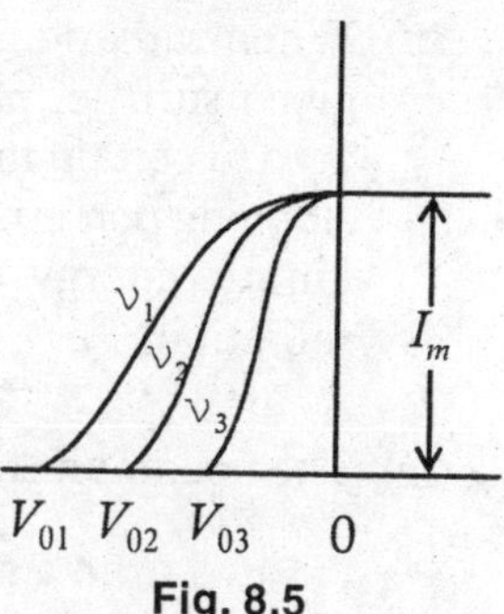

Fig. 8.5

Effect of Frequency on the Maximum Kinetic Energy

Magnitudes of the stopping potential at different frequencies are converted into the maximum kinetic energies. A plot of maximum kinetic energy against the frequency is as shown in Fig. 8.6. A straight line is obtained. The straight line (*a*) meets the frequency axis at a certain frequency ν_0.

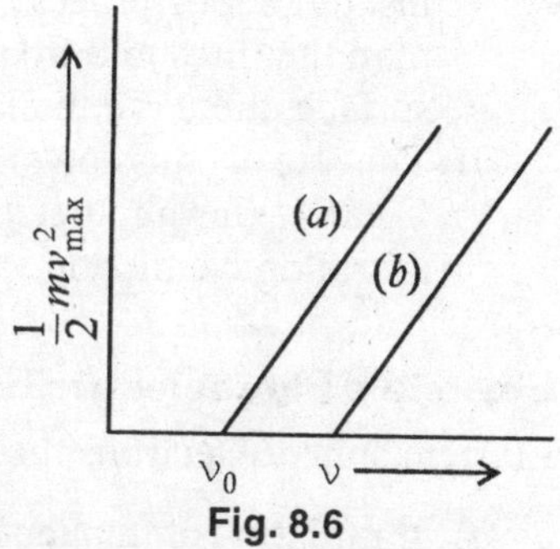

Fig. 8.6

This reveals that ν_0 is the minimum frequency for which the maximum kinetic energy is zero. This minimum value of frequency, is called the threshold frequency for the emitting surface.

Thus the threshold frequency for a given emitting surface defined as minimum frequency of incident light, below which no photoelectrons are emitted, however intense the incident light may be.

From Fig. 8.6 it is evident that for another emitting surface, a straight line (*b*) parallel to (*a*) is obtained.

On the basis of above experimental observations, we can summarize the following important characteristics of the photoelectric effect:

(*i*) For a given photosensitive surface, there is a minimum frequency ν_0 of incident radiation, below which there is no emission of photoelectrons. This frequency (ν_0) is called the threshold frequency for the surface and its value depends on the materials and the nature of the emitting surface.

(*ii*) For a given frequency $\nu(\nu > \nu_0)$, the number of photoelectrons emitted per second from a given surface, under a constant accelerating potential difference is directly proportional to the intensity of incident radiation.

(*iii*) When radiation of a given frequency $\nu(\nu > \nu_0)$ is incident on a given surface, photoelectrons with kinetic energies ranging form zero to certain maximum value, are emitted from the surface. The stopping potential for the photoelectrons and hence the maximum kinetic energy is independent of the intensity of the incident radiation.

(*iv*) The maximum kinetic energy $\left(\frac{1}{2}mv_{\max}^2 = eV_0\right)$ of the photoelectrons emitted from a given surface increases linearly with the frequency of the incident radiation. The straight line so obtained for the given surface meets the frequency axis at the threshold frequency ν_0 for the surface. The slope of the straight line is the same for all emitting surfaces and is equal to Planck's constant h.

(*v*) The emission of photoelectrons from photosensitive surface is an instantaneous process. As soon as radiation of frequency greater than the threshold frequency is incident on a given photosensitive surface, there is emission of photoelectrons.

(*vi*) The rate of emission of the electrons is independent of temperature. Clearly, the photoelectrons phenomenon is entirely different to thermionic emission.

Einstein's Photoelectric Equation

Einstein's photoelectrons theory is based on the following:

(*i*) A radiation of frequency ν consists of a stream of discrete particles, called photons. Energy of each photon or quanta is $h\nu$, where h is Planck's constant. The photons move through space with the speed of light.

(*ii*) When a quanta of energy $h\nu$, *i.e.* a photon is incident on a metal surface, the entire energy of the photon is absorbed by a single electron without any time lag. The probability of absorbing two or more photons at the same time by a metal surface is negligible.

Suppose that radiation of frequency ν is incident on the surface of a metal in vacuum. The entire energy of a photon, *i.e.* $h\nu$ is absorbed by an electron of the metal. We must note that the incident radiation penetrates several atomic diameters into the metal, and electrons at different depths absorb this amount of energy from the incident photons. The energy $h\nu$ absorbed by the electrons in the metal is used in the following two ways:

(*i*) A part of the photon energy $h\nu$ is used up by the electron to do a certain minimum amount of work to overcome attractive forces of positive ions of the metal. The minimum amount of energy is called the *photoelectric work function* and usually denoted by W of the metal. The work function (W) depends on nature of the emitting surface.

(*ii*) The remaining energy ($h\nu - W$) appears as the kinetic energy comes from the surface, it does not suffer any collision with another electron. In this situation, it will be emitted from the surface into vacuum with this energy as maximum kinetic energy $\frac{1}{2}mv_{max}^2$, where m is the mass of the electron and v_{max} its maximum speed for the given frequency. Thus for such an electron, we have

$$\frac{1}{2}mv^2 = h\nu - W \tag{19}$$

The above equation is called Einstein's photoelectric equation. Suppose the frequency of the incident radiation is reduced to a certain value ν_0 such that an electron is just emitted from the surface with zero kinetic energy.

Then we have when $\nu = \nu_0$, $\frac{1}{2}mv_{max}^2 = 0$.

Then from Eq. (19), we have

$$O = h\nu_0 - W$$

or

$$W = h\nu_0 \tag{20}$$

Substituting for $W = h\nu_0$ from Eq. (20) in Eq. (19), we have

$$\frac{1}{2}mv_{max}^2 = h\nu - h\nu_0 = h(\nu - \nu_0) \tag{21}$$

Eq. (21) is another form of Einstein's photoelectric equation. The frequency ν_0 is called the *threshold frequency*. On the basis of Einstein's theory, one can define the threshold frequency as follows:

It is the frequency of incident photon which has just sufficient energy to liberate an electron with zero kinetic energy from the emitting metal surface.

It is worthwhile to mention that the energy of light photon incident on the surface of the metal is not divided, but the whole energy of a photon is used up by the same electron in ejecting from the parent atom and imparting kinetic energy.

Explanation of the characteristics of the photoelectric effect from Einstein's equation

(*i*) Eq. (21) shows that the graph of $\frac{1}{2}mv_{max}^2$ against ν is a straight line for an emitting surface. For all emitting surfaces the straight lines will have the same slope equal to Planck's constant h.

(*ii*) When the intensity of incident radiation is increased, then the increased number of photons in the incident radiation will cause the emission of more photoelectrons per second. Obviously, the photoelectrons current is proportional to the intensity of the incident radiation. The increase of the intensity does not affect the maximum kinetic energy.

(*iii*) From Eq. (20) and Eq. (21), we have

(*a*) For $\nu > \nu_0$ not photoelectrons can be emitted.

(*b*) For $\nu = \nu_0$ photoelectrons are just emitted with zero kinetic energy.

(*c*) For $\nu > \nu_0$ photoelectrons are emitted with kinetic energies ranging from zero to a maximum value.

(*d*) For $\nu > \nu_0$ and the intensity of incident light is increased, the number of photoelectrons emitted per second will increase.

Some recent important observations about the photoelectric effect are:

(*i*) Only a small fraction (~5%) of the incident photon succeeds in ejecting photoelectrons from photo-metal while rest of the photons are absorbed by the system as a whole and generate thermal energy.

(*ii*) Photoelectric effect is also observed for isolated atoms in the form of gas, e.g. Na, K vapour and this process is called photo-ionization.

(*iii*) The energy required for ejecting the electrons may also be provided by heating the metal, which results in the thermionic emission of the electrons.

(*iv*) The advent of lasers capable of emitting coherent radiation at high power levels, two or multi photon photo-emission has now been observed. Theory predicts that the double quantum photo current should be proportional to the square of the power of the incident radiation as opposed to the nearly linear relationship that holds for the single quantum photo-effect. In a multi photon photoelectric effect, the kinetic energy of the photoelectron is given by

$$\frac{1}{2}mv^2 = Nh\nu - eV_s \qquad (22)$$

and the critical frequency is eV_S/Nh which is smaller than the corresponding frequency for single photon process by a factor of $1/N$.

8.6 COMPTON EFFECT

A. H. Compton in 1923, while studying the scattering of *X*-rays by a block of paraffin, observed that the wavelength of the scattered radiation was greater than the wavelength of incident radiation and the shift in wavelength was dependent on the angle of the scattering. The wavelength of the scattered radiation is found maximum at right angles to the incident beam. This effect is called Compton effect. Essentially, it is a demonstration of the scattering of a photon by an electron as a particle-particle scattering.

Theory

The theory of Compton effect was given by Compton and also by Debye at about the same time. Compton considered the radiation to be quantised and assumed the electron of the atom of scattering material to be loosely bound (almost free) and at rest before collision with the photon. Let λ and ν be the wavelength and frequency of the incident radiation. According to the quantum theory of radiation, the energy of each photon associated with this radiation is $h\nu$. When this photon of energy $h\nu$ strikes a loosely bound (almost free) electron at rest, then as a result of collision the electron gains momentum and recoils in a direction making an angle θ, with the direction of incident radiation (Fig. 8.7). At the same time, the photon is scattered at angle ϕ with the direction of the incident radiation with lesser energy $h\nu'(\nu' > \nu)$. It is assumed that law of conservation of energy and law of conservation of momentum holds good.

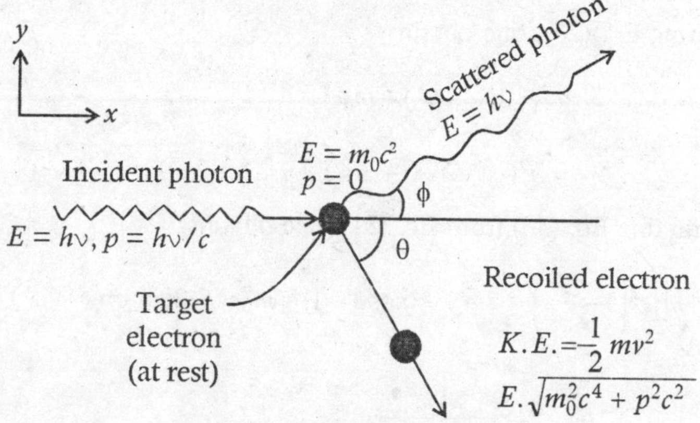

Fig. 8.7: The Compton effect

Let m_0 be the rest mass of the electron and m be the mass of the electron, when it is recoiled with a velocity v, then from the special theory of relativity, we have

$$m = \frac{m_0}{\sqrt{1-(v^2/c^2)}} \tag{23}$$

The momentum of the photon of energy $h\nu$ is given by

$$p = \frac{E}{c} = \frac{h\nu}{c} \tag{24}$$

The energy of the electron (at rest) before collision of will be m_0c^2 and after collision it will be mc^2. From the law of conservation of energy, we have

$$h\nu + m_0c^2 = h\nu' + mc^2 \tag{25}$$

From the law of conservation of momentum, in the *X*-direction

$$\frac{h\nu}{c} = \frac{h\nu'}{c}\cos\phi + mv\cos\theta \tag{26}$$

and in *Y*-direction

$$0 = \frac{h\nu'}{c}\sin\phi + mv\sin\theta \tag{27}$$

From Eqs. (26) and (27), one obtains

$$mv\cos\theta = h(\nu - \nu'\cos\phi) \tag{28}$$

$$mv, c\sin\theta = h\nu'\sin\phi \tag{29}$$

Squaring and adding Eqs. (28) and (29), one obtains

$$m^2v^2c^2 = h^2\left[(\nu - \nu'\cos\phi)^2 + (\nu'\sin\phi)^2\right] \tag{30}$$

Squaring Eq. (25), one obtains

$$m^2c^4 = \left[(\nu - \nu') + m_0c^2\right]^2$$

$$= h^2(\nu^2 - 2\nu\nu' + \nu'^2) + m_0^2c^4 + 2hm_0c^2(\nu - \nu') \tag{31}$$

Subtracting Eq. (30) from Eq. (31), one obtains

$$m^2c^4\left(1 - \frac{v^2}{c^2}\right) = 2h^2\nu\nu'(\cos\phi - 1) + m_0^2c^4 + 2hm_0c^2(\nu - \nu') \tag{32}$$

But $$mc^2 = \frac{m_0c^2}{\sqrt{1 - v^2/c^2}}$$

or $$m^2c^2\left(1 - \frac{v^2}{c^2}\right) = m_0^2c^4 \tag{33}$$

Using Eq. (33), Eq. (32) becomes

$$2h^2\nu\nu'(1 - \cos\phi) = 2hm_0c^2(\nu - \nu') \tag{34}$$

$$h(1 - \cos\phi) = m_0c^2\left(\frac{1}{\nu'} - \frac{1}{\nu}\right)$$

$$= m_0c(\lambda' - \lambda) \tag{34a}$$

Obviously, the shift in wavelength of the scattered photon as compared to that of the incident photon, is

$$\lambda' - \lambda = \Delta\lambda = \frac{h}{m_0 c}(1 - \cos\phi) \tag{35}$$

Also $\lambda' - \lambda = \Delta\lambda = \frac{h}{m_0 c}\sin^2\frac{\phi}{2}$ (35a)

This is Compton expression for the shift in the wavelength of the scattered *X*-rays. h/m_0c is denoted by λ_c and is called *Compton wavelength*. λ_c has the dimensions of length and it has value

$$\frac{h}{m_0 c} = \frac{6.6 \times 10^{-34}}{9.1 \times 10^{-31} \times 3 \times 10^8} = 0.0242 \times 10^{-10} \text{ m}$$

$$= 0.242 \text{ Å}$$

Thus $\Delta\lambda = \lambda_c(1 - \cos\phi)$ (36)

From Eq. (35) it is evident that λ' is always greater than λ as $\cos\phi$ is always less than 1. Furthermore $\Delta\lambda$ is also dependent on $\cos\phi$. Thus maximum value of $(1 - \cos\phi) = [1 - (-1)] = 2$, at $\phi = 180°$ hence the maximum value of the shift in wavelength can be

$$(\Delta\lambda)_{max} = 0.0484 \text{ Å}$$

Thus to have an appreciable relative change in wavelength, *i.e.* $(\Delta\lambda/\lambda)$, the wavelength (λ) of the incident *X*-ray must be quite small. Obviously, the Compton effect is easily observable for smaller wavelengths.

Direction of Recoil Electron

Dividing Eq. (29) by Eq. (28), one obtains

$$\tan\theta = \frac{h\nu'\sin\phi}{h(\nu - \nu'\cos\phi)} = \frac{\nu'\sin\phi}{(\nu - \nu'\cos\phi)} \tag{37}$$

Using Eq. (34), one obtains

$$\frac{1}{\nu'} = \frac{1}{\nu} + \frac{h}{m_0 c^2}(1 - \cos\phi)$$

or

$$= \frac{1}{\nu} + \frac{h}{(m_0 c^2)} 2\sin^2\frac{\phi}{2}$$

or

$$\frac{1}{\nu'} = \frac{1 + \left(\frac{h\nu}{m_0 c^2}\right) 2\sin^2\frac{\phi}{2}}{\nu}$$

or
$$\nu' = \frac{\nu}{1+\left(\frac{h\nu}{m_0c^2}\right)2\sin^2\frac{\phi}{2}} \tag{38}$$

Writing $\beta = h\nu/m_0c^2$, Eq. (38) becomes

$$\nu' = \frac{\nu}{1+2\beta\sin^2\frac{\phi}{2}} \tag{39}$$

Substituting the value of ν' in Eq. (37), one obtains

$$\tan\theta = \frac{\frac{\nu\sin\phi}{[1+2\beta\sin^2\phi/2]}}{\frac{\nu-(\nu\cos\phi)}{(1+2\beta\sin^2\phi/2)}}$$

or
$$\tan\theta = \frac{2\sin\frac{\phi}{2}\cos\frac{\phi}{2}}{2\sin^2\frac{\phi}{2}+2\beta\sin^2\frac{\phi}{2}}$$

or
$$\tan\theta = \frac{2\sin\frac{\phi}{2}\cos\frac{\phi}{2}}{2(1+\beta)\sin^2\frac{\phi}{2}} = \frac{\cot\frac{\phi}{2}}{1+\frac{h\nu}{m_0c^2}} \tag{40}$$

Obviously, the recoil angle θ depends on the scattering angle if $\phi = 0° = 90°$ and if $\phi = 180°$; $\theta = 0°$. Obviously, as ϕ varies from 0 to 180 degree, θ varies from 90° to 0°. This shows that the electron can get recoiled only in the onward direction at angles less then 90°, whereas a photon gets scattered in all directions. Since in most cases $h\nu = 10$ keV, and therefore $h\nu << m_0c^2$, one obtains a simple relation $\theta \approx \frac{1}{2}(\pi - \phi)$, *i.e.* the direction in which the recoiled electron bisects the angle which is supplementary to the angle ϕ made by the scattered photon momentum with the initial photon momentum.

Kinetic Energy of Recoil Electron

The kinetic energy of the recoiled electron also depends on the scattering angle ϕ. One can easily show it as follows:

The kinetic energy of the recoil electron is the difference between the energies of the incident and scattered photons, *i.e.* K. E. of recoiled electron $= h\nu - h\nu'$.

Substituting the value of ν' from Eq. (39), one obtains

$$\text{K.E.} = h\nu - h\left[\nu\Big/1 + 2\beta\sin^2\frac{\phi}{2}\right]$$

$$= h\nu\left[\frac{2\beta\sin^2\dfrac{\phi}{2}}{1+2\beta\sin^2\dfrac{\phi}{2}}\right] \tag{41}$$

Experimental Verification of Compton Effect

The experimental arrangement to demonstrate the Compton effect is shown in Fig. 8.8. A beam of monochromatic *X*-rays of kinetic energy wavelength (λ) is made incident on a target. The distribution of intensity with wavelength of the monochromatic *X*-rays scattered at various angles is measured by *X*-ray spectrometer. The intensity distribution versus wavelengths curves for various scattering angles are shown in Fig. 8.9. Obviously, the curves have two peaks, one corresponding to modified *X*-rays and the other corresponding to unmodified radiations. The difference between two peaks on wavelength axis gives the Compton effect or shift.

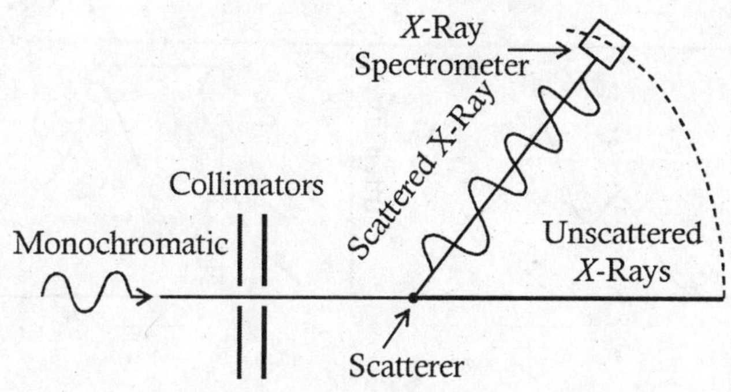

Fig. 8.8

From the various curves it is obvious that greater is the scattering angle, greater is the Compton shift in accordance with Eq. (35). At $\phi = 90°$, $\Delta\lambda = 0.024$ Å is in good agreement with the value obtained from Compton formula given by Eq. (35).

Comparison of Compton Effect and Photoelectric Effect

(*i*) In Compton effect a photon cannot transfer all of its energy to a free electron, whereas in photoelectric effect, an electron absorbs the incident photon completely before it leaves the photometal surface. To understand it clearly, let us consider a Compton collision between a photon and free electron (initially at rest). The incident photon has an energy $h\nu$ and a momentum $h\nu/c$. Let us suppose that this photon transfers its total energy (and momentum) to the electron who acquires a velocity v. From the law of conservation of momentum and energy, we have

$$h\nu = \frac{1}{2}mv^2 \text{ and } \frac{h\nu}{c} = mv$$

Eliminating $h\nu$ between the above two equations, one obtains

$$v = 2c$$

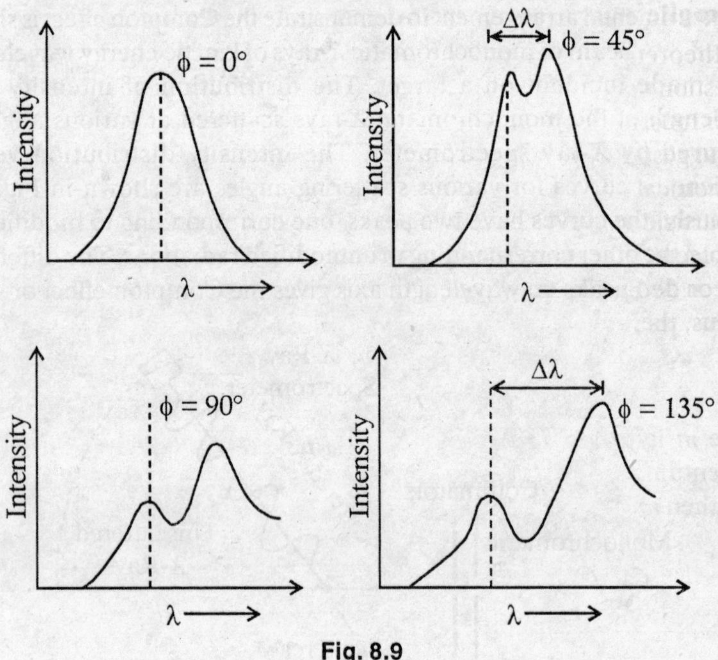

Fig. 8.9

But v cannot exceed c (velocity of light) and hence a free electron cannot absorb all the energy of a photon. In photoelectric effect, an electron absorbs the incident photon completely before it leaves the photometal surface as photo electron. Thus the electron in the photometal surface cannot be free, *i.e.*, it must be bound to an atom otherwise it could not absorb the incident photon completely.

(*ii*) Photoelectric effect can be explained by mere quantum considerations, whereas one has to use relativistic considerations in applying laws of conservation of momentum and energy in case of explaining Compton effect with quantum theory. Obviously, we can say that Compton effect is *relativistic photoelectric effect*.

8.7 MATTER WAVES

The photoelectric effect and Compton effect provided conclusive evidence that electromagnetic radiation possesses corpuscular nature. At the same, the propagation experiments based on interference and diffraction also proved the existence of wave nature for these radiations, Obviously, radiation presents itself in two forms, particle or corpuscle and wave. In view of this, it is rather difficult to accept the conflicting ideas that radiation is a wave which is spread out cover space and also a particle which is localized at a point in space.

de-Broglie Matter Waves

The theoretical physicist, de Broglie, in 1924 put fourth a very bold suggestion that *wave particle duality* is a general property and like radiation, matter must also exhibit dual (*i.e.* particle like and wave like) characteristic. His suggestion was partly based on the intuitive feeling that nature is symmetrical and if radiation can have dual characteristic of being wave and particle then matter particles like electrons, photons, neutrons, atoms or molecules have an associated wave with them. This called as matter wave or de Broglie wave

Thus, the wavelength of the matter wave is

$$\lambda = \frac{h}{p} = \frac{h}{mv} \tag{42}$$

where m is the mass of the inaterial particle, v its velocity and p is momentum. In case, particle velocity is comparable with the velocity of light, then m is replaced by relativistic mass, *i.e.*

$$m = \frac{m_0}{\sqrt{1 - \frac{v^2}{c^2}}}$$

The de Broglie wavelength can be expressed in terms of energy

$$K = \frac{1}{2} mv^2 \text{ or } v = \sqrt{mK}$$

Thus
$$\lambda = \frac{h}{\sqrt{2mK}}$$

If a charge particle carrying a charge q is accelerated through a potential difference of V volts then

$$K = qV$$

or
$$\lambda = \frac{h}{\sqrt{2mqV}}$$

From special theory of relativity, the energy E and momentum p of a particle of rest mass m_0 are related as:

$$E^2 = p^2c^2 + m_0^2c^4 \tag{43}$$

For a photon, $E = h\nu$ and rest mass $m_0 = 0$, hence

$$E^2 = p^2c^2 \text{ or } E = pc \therefore p = \frac{h\nu}{c} = \frac{h}{\lambda} \tag{44}$$

Obviously, Eq. (44) connects the corpuscular mode of description to the wave nature.

de-Broglie Wavelength of Matter Waves

From Planck's theory of radiation, the energy of a photon is given by

$$E = h\nu = hc/\lambda \tag{45}$$

$$\therefore \quad \lambda = hc/E \tag{46}$$

From special theory of relativity

$$E = mc^2 \tag{47}$$

Mass of the photon $m = h\nu/c^2$

Momentum of the photon

$$mc = p\frac{h\nu}{c} = \frac{h}{\lambda}, \quad \therefore \quad p = \frac{h}{\lambda}$$

or

$$\lambda = h/p \tag{48}$$

de Broglie postulated that what is true for an energy packet which we call as quantum or photon is also true for the material particle. Thus for a particle of mass m moving with velocity v, we have $p = mv$ and the de Broglie wave associated with this particle is

$$\lambda = \frac{h}{mv} \tag{49}$$

Here m stands for the relativistic mass. From Eq. (49), it is clear that

(*i*) If m or v is large, the de Broglie wave associated with the particle is small.

(*ii*) The de broglie wave associated with a material particle or photon is independent of any charge associated with it.

The expression for the de Broglie wavelength associated with an accelerated electron in the non-relativistic case, will be

$$\lambda = \frac{h}{\sqrt{2m_oeV}}$$

$$m_0 = 9.1 \times 10^{-31} \text{ kg}, h = 6.6 \times 10^{-34} \text{ Js}, e = 1.6 \times 10^{-19} \text{ C}$$

$$\therefore \qquad \lambda = \frac{6.6 \times 10^{-34}}{\sqrt{2 \times 9.1 \times 10^{-31} \times V}}$$

$$= \frac{12.27 \times 10^{-10}}{\sqrt{V}} = \frac{12.27}{\sqrt{V}} \text{ Å} = \sqrt{\frac{150}{V}} \text{ Å} \qquad (50)$$

If electrons are accelerated by 100 volts, then the wavelengths of de Broglie waves associated with electrons will be 1.227 Å. The wavelength is of the same order of magnitude as the distance between the atomic planes in crystals. Crystals have been widely used for studying the diffraction of *X*-rays whose wavelengths are of the order of 1 Å. This immediately suggests that crystals can be used as diffraction grating to detect the existence of matter waves associated with electrons. Davisson and Germer in 1927 carried out a series of experiments and were first to demonstrate the existence of matter waves.

8.8 THE DAVISSON AND GERMER EXPERIMENT

The experimental arrangement used by Davisson and Germer is shown in Fig. 8.10. It consists of an electron gun which comprises a tungsten filament *F* heated by a low tension battery. The electron emitted by the filament are accelerated in an electric field of known potential difference from a high tension battery. The electrons are collimated to a fine beam by allowing them to pass through a system of suitable slits. This entire arrangement to produce a fine beam of electrons accelerated to a desired velocity is called as electron gun.

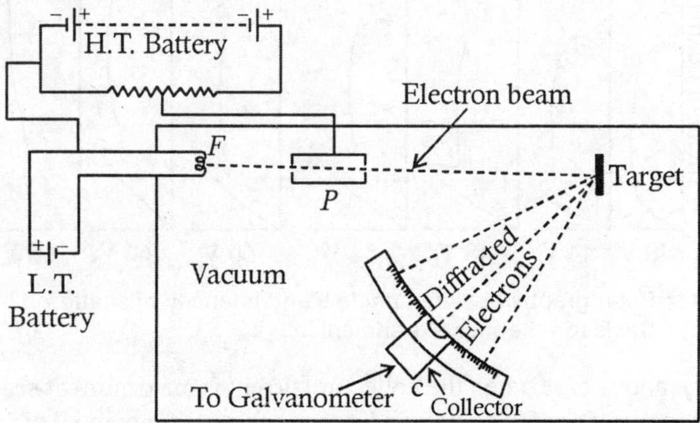

Fig. 8.10: Davisson and Germer experiment

The collimated beam of electrons is made to strike a nickel target capable of rotating about an axis perpendicular to the plane of the paper. The electrons are scattered in all directions by the atoms of nickel crystal. The intensity of the electron beam scattered in a given direction is measured by allowing it to enter in a Faraday cylinder called the collector, which can be rotated about the same axis as the target. The collector is connected to a galvanometer whose deflection is proportional to the intensity of the beam entering the collector. The whole apparatus is enclosed in an evacuated chamber

In one set of observations, the (111) face of the target nickel crystal was presented to the incident beam of electrons. The potential difference between the filament F and the electron gun was adjusted at 40 volts and then gradually increased in small steps. For each value of potential difference, polar graphs were drawn between angle θ (the angle between the incident beam and the beam entering the collector, known as *colatitude*) and the collector current. In these graphs, the length of the radius vector corresponding to a certain angle was drawn directly proportional to the collector current. For the potential difference of 40 volts, a smooth curve was obtained. As the potential difference was gradually increased, a little bump began to appear on the curve which increased to a maximum for $V = 54$ volts and $\theta = 50°$ (Fig. 8.11) Beyond 54 volts, the bump again decreased to become insignificant at about 68 volts. The pronounced maxima observed corresponding to accelerating potential of 54 volts at an angle of 50° with the electron beam provided a convincing proof that electron were associated with waves which after scattering from the regularly spaced atoms of the nickel crystal, gave rise to a constructive interference.

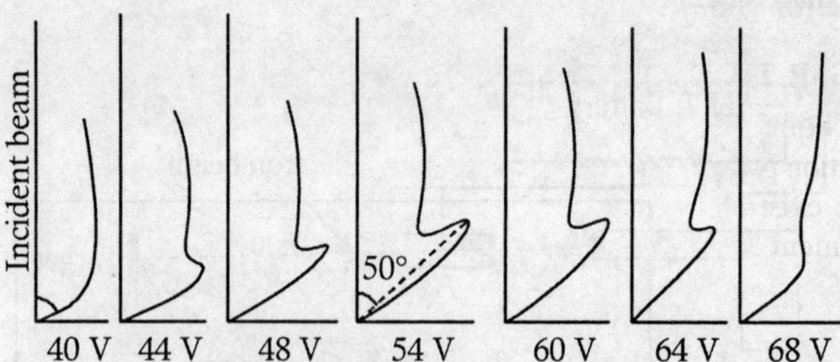

Fig. 8.11: Polar graphs between angle θ and intensity of scattered beam in Davisson-Germer experiment

In the above case when the collector current is maximum at scattering (colatitude) angle of 50° corresponding to accelerating potential of 54 volts, the incident and the scattered beam evidently makes an angle of 65° with the family of Bragg's planes as shown in Fig. 8.12. *X*-ray diffraction studies reveal that the spacing of planes in this family is 0.91 Å. Taking

$n = 1$, $\theta = 65°$, one can calculate the wavelength of the incident wave by applying Bragg's law, $2d\sin\theta\ n\lambda$, we have

$$2 \times 0.91 \times \sin 65° = 1\ \lambda$$

or, $\lambda = 2 \times 0.91 \times 0.9063 = 1.662$ Å

Now applying de Broglie's relation for matter waves, we have

$$\lambda = \frac{h}{mv} = \frac{12.27}{\sqrt{V}}$$

Since $V = 54$ volts, $\lambda = \dfrac{12.27}{\sqrt{54}} = 1.673$ Å

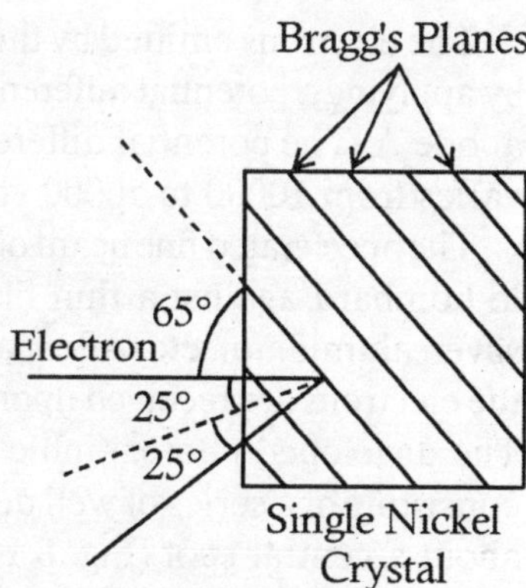

Fig. 8.12

Obviously, there is an excellent agreement between the two result. Thus Davisson and Germer's experiment may be considered to be a verification of the de Broglie hypothesis of matter waves.

The analysis of the Davisson-Germer experiment is actually less straight forward, since the energy of an electron increases when it enters a crystal by an amount equal to the work function of the surface. Hence the electron's speed in the experiment were greater inside the crystal and the corresponding de Broglie wavelengths shorter than the corresponding values outside. An additional complication arises from interference between waves diffracted by different families of Bragg planes, which restricts the occurrence of maxima to certain combinations of electron energy and angle of incidence rather than merely to any combination they obey the Bragg equation.

8.9 G. P. THOMSON'S EXPERIMENT

G. P. Thomson, using *X*-ray powder diffraction method, obtained electron diffraction patterns. These results provided another experimental evidence for the existence of matter waves. A schematic diagram of Thomson's experiment is shown in Fig. 8.13.

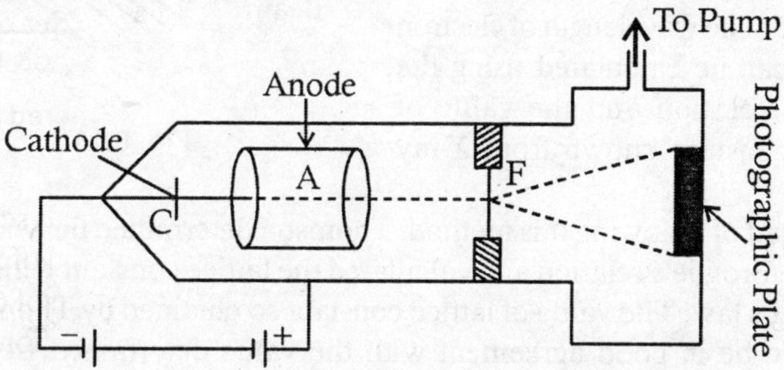

Fig. 8.13: Thomson's experiment for electron diffraction

The electrons emitted by the cathode C are accelerated and channelised by applying a potential difference between the cathode C and the cylindrical anode A. The potential difference is applied by a high tension battery and varies from 10000 to 50000 volts.

The accelerated fine beam of electrons is then made to bombard against a thin film of metal like gold, silver, aluminium, etc. After passing through the film, the electrons are received upon a photographic plate. The developed photographic plate shows a pattern consisting of a series of well defined concentric rings about a central spot (Fig. 8.14). It is worthwhile to mention that the diffraction pattern can be disturbed by a magnetic field, showing that the pattern was produced by electrons and not by any radiation.

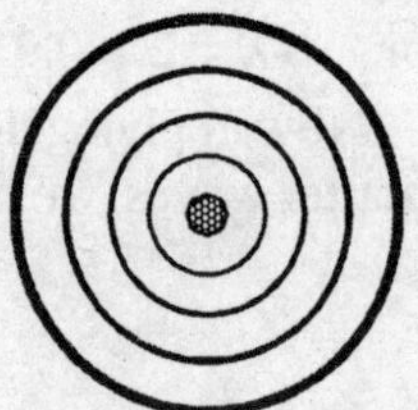

Fig. 8.14: Electron diffraction pattern

In Thomson's experiment, one can imagine the metallic film to be composed of a very large number of tiny crystals oriented at random. Due to the random orientation of these millions of micro-crystals, a given electron beam is intercepted by a particular micro crystal, this set of crystal planes is having suitable angle of orientation so as to satisfy Bragg's law ($2d\sin\theta = n\lambda$) and this gives rise to constructive interference between the reflected rays. But since the crystals are oriented at random, this particular set of crystal planes shall have all possible orientations about the incident beam. Due to this, crystal planes will lie on the surface of a cone of semi-vertical angle 2θ, where θ is the angle which the beam makes with this set of crystal planes as shown in Fig. 8.15. Other sets of crystal planes suitably oriented so as to satisfy Bragg's condition will reflect the beam along the surfaces of cones of different semi-vertical angles. The intercepts of these semi-vertical cones with the photographic film or plate give rise to diffraction pattern as shown in Fig. 8.15. The wavelength of electron waves can be calculated using the Bragg's relation and the value of lattice constant known from X-ray data.

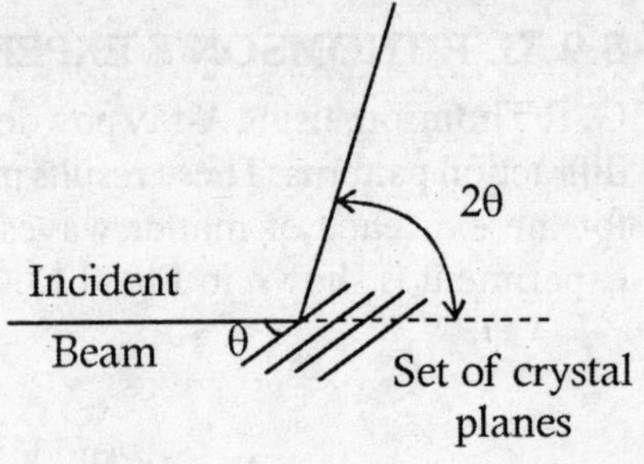

Fig. 8.15

Instead of following this method. Thomson determined the wavelength from de Broglie's relation and calculated the lattice constant with the help of Bragg's law. The value of lattice constant so obtained by Thomson was found to be in good agreement with the value determined from X-ray measurements, confirming the existence of matter waves.

8.10 PROPERTIES OF MATTER WAVES

Following are the properties of matter waves:

(*i*) Wavelength is universally proportional to the mass of velocity of the particle.

(*ii*) When $v = 0$ then $\lambda = \infty$ wave becomes indeterminate and $v = \infty$ then $\lambda = 0$ thus matter waves are generated by the motion of particles.

(*iii*) $\lambda = h/mv$ is independent of charge, thus these waves are produced irrespective of particle being charged or not. This reveals that matter waves are not electromagnetic waves as electromagnetic waves are produced only through motion of charged particles.

(*iv*) The velocity of matter waves is dependent of the velocity of the matter particle and contrary to electromagnetic waves these don't have constant velocity.

(*v*) The velocity of matter wave is greater than the velocity of light.

A particle in motion associated with matter wave has two different velocities—one going mechanical motion v and another group or wave velocity.

Schrodinger postulated that a moving material particle is equivalent to wave packet rather than a single wave. A wave packet comprises a group of waves, each with slightly different velocity and wavelength, with phases and amplitudes so chosen that they interfere constructively over only a small region of space where the particle can be located, outside of which they produce destructive interference so that the amplitude reduces to zero rapidly. Obviously a wave packet is formed as a result of the superposition of many plane waves of different frequencies grouped around some central frequency. Since the envelope of the packet must have a finite spatial extent, the sum of the amplitudes of all of its plane wave components must be zero everywhere except where the packet is localized. Form of a typical wave packet in one dimension is shown in Fig. 8.16.

The velocity with which the wave packet moves is called the group velocity. The energy carried by the packet as its group velocity is thus analogous to the kinetic energy of a free particle of the same velocity. The individual waves

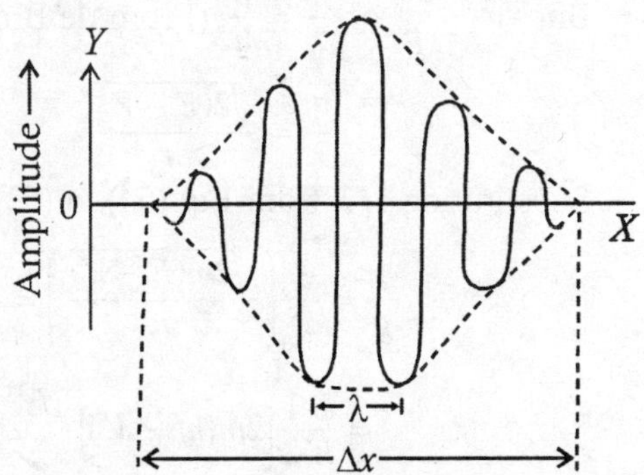

Fig. 8.16: A wave packet in one dimension

forming the wave packet possess an average velocity, u called the *phase velocity*. One can easily show that the velocity of the material particle v is same as *group velocity* of wave packet.

Let us suppose that v is the velocity of the moving particle, g is the group velocity, u is the velocity of an individual wave train comprising the group or wave packet and λ is the wavelength of the individual wave. The group velocity g of a wave packet is found to be given by

$$g = u - \lambda \frac{du}{d\lambda} \tag{51}$$

Writting Eq. (51) as

$$g = \lambda^2\left(\frac{u}{\lambda^2} - \frac{1}{\lambda}\frac{du}{d\lambda}\right)$$

$$= -\lambda^2\left(\frac{1}{\lambda}\frac{du}{d\lambda} - \frac{u}{\lambda^2}\right) = -\lambda^2 \frac{d}{d\lambda}\left(\frac{u}{\lambda}\right)$$

$$= -\lambda^2\left(\frac{df}{d\lambda}\right) \qquad \left[\because f = \frac{u}{\lambda} \text{ (here } f \text{ is frequency)}\right]$$

$$\therefore \quad \frac{1}{g} = -\frac{1}{\lambda^2}\left(\frac{d\lambda}{df}\right) = \frac{d}{df}\left(\frac{1}{\lambda}\right) \tag{52}$$

Let E denote the total energy and V the potential energy of the particle. It m is the mass of the particle, its kinetic energy is given by

$$\frac{1}{2}mv^2 = E - V$$

$$\text{or} \qquad v = \sqrt{\frac{2(E-V)}{m}} \tag{53}$$

$$mv = p = \sqrt{2m(E-V)}$$

But $\lambda = \frac{h}{p}$ or $\frac{1}{\lambda} = \frac{p}{h}$ (From de Broglie's relation)

$$\therefore \quad \frac{1}{\lambda} = \frac{\sqrt{2(E-V)}}{h} \tag{54}$$

Substituting $1/\lambda$ from Eq. (54) in (52), one obtains

$$\frac{1}{g} = \frac{d}{df}\left(\frac{\sqrt{2m(E-V)}}{h}\right) = \frac{1}{h}\frac{d}{df}\sqrt{2m(hf-V)} \qquad [\therefore E = hf]$$

$$= \frac{1}{h}\frac{1}{2}[2m(hf-V)]^{-1/2} 2mh = \sqrt{\frac{m}{2(hf-V)}} = \frac{1}{v} \text{ (from 53)}$$

$$\therefore \quad g = v \tag{55}$$

Obviously, the group velocity g of de Broglie wave group or wave packet associated with a moving particle is the same as the classical velocity v of the particle. However, it is important to note that the motion of particle and motion of a wave packet are equivalent with only one difference, *i.e.* the wave packet has spatial dimension whereas a particle (a mass-point) has not. If the wave packet can be made small enough, the distinction between the two becomes insignificant.

8.11 THE UNCERTAINTY PRINCIPLE

The description of material particles in term of waves and the setting up of a connection between energy and momentum, on one side, and frequency and wavelength, on the other, lead to an indefiniteness that had not been previous suspected in mechanics of particles. This indefiniteness has become famous as the 'uncertainty' principle, or '*indeterminacy principle*' put forward by Heisenberg in 1927. This principle expresses a fundamental limit to the simultaneous determination of certain pairs of variables, such as position and momentum, energy and time. The principle can be stated for different pairs of variables as follows:

The product of uncertainly Δx (or possible error) in the x-coordinate of a particle, in motion, at some instant, and the uncertainly Δp_x, in the x-component of the momentum, at the same instant, is of the order of or greater than $\hbar$, *i.e.*

$$\Delta p_x \Delta x \geq \hbar \left(\hbar = \frac{h}{2\pi} = 1.054 \times 10^{-34} \text{ Js} \right) \qquad (56)$$

An expression of this type also holds good for other components of linear momentum and for angular momentum and angular position, *i.e.*

$$\Delta p_y \Delta y \geq h$$

$$\Delta p_z \Delta z \geq h \qquad (56a)$$

It is important to mention here that there is no restriction on Δx or on Δp_x, but only on their product. Also, there is no restriction on product like $\Delta p_z \Delta y$.

Significance: One can draw following conclusions from (56):

(*a*) If the position coordinate x of a particle in motion is accurately determined at some instant, so that $\Delta x = 0$, then at the same instant the uncertainty Δp_x in the determination of the momentum of the particle becomes infinite.

(*b*) If the momentum p_x of a particle is accurately determined at some instant, so that $\Delta p_x = 0$, then at the same instant the uncertainty Δx in the determination of the position coordinate becomes infinite.

Obviously, if an experiment is designed to measure x or p_x accurately, the other quantity will become completely uncertain.

One can measure both the quantities by means of an experiment, but only within certain limits of accuracy specified by the uncertainty relation.

(*c*) For a particle of mass m moving with velocity v, the product of uncertainty Δx and Δv is given by

$$\Delta x \, \Delta v \geq \frac{\hbar}{m}$$

For a heavy particle $\hbar/m \approx 0$, is very small, and therefore, the product Δx and Δv of the two uncertainties becomes very small. For such particles both x and v can be determined accurately.

For very heavy particles, if m is such that $\hbar/m \approx 0$, the uncertainties vanish and one can measure x and v with perfect accuracy. Obviously, this is the limiting case of classical mechanics. This means, classical mechanics is true for heavy bodies and uncertainties are a characteristic of atomic and subnuclei particles.

The uncertainty relation for the simultaneous measurement of energy E and time t can be expressed as

$$\Delta E \Delta t \geq \hbar \tag{57}$$

Here again there are no restrictions on the accuracy with which E or t can be measured, but only on the product $\Delta E \Delta t$

Significance: The physical significance of the energy-time uncertainty relation is quite different from that of the position momentum uncertainty relation. If ΔE is the maximum uncertainty in the determination of the energy of a system in a particular state, then the minimum time interval for which the system remains in the state is given by

$$\Delta t = \frac{\hbar}{\Delta E}$$

And, if a system remains in a particular state for a maximum time instant Δt, then the minimum uncertainty in the energy of the system, in the state is given by

$$\Delta E = \frac{\hbar}{\Delta t}$$

The uncertainty relation for angular momentum and angular displacement is often expressed as

$$\Delta J \Delta \theta \geq \hbar \tag{58}$$

8.12 EXPERIMENTAL ILLUSTRATIONS

The problem of uncertainty principle is a logical consequence of the dual behaviour of matter. A deeper insight into the nature of the uncertainty principle can be accomplished by the following experiments:

(i) **Localization Experiment** (γ-ray microscope)**:** Let us consider an experiment whereby a particle (say electron) might be observed in a microscope. Light of frequency ν falls upon the particle and is scattered. If it is to be observed, the scattered photon must enter the objective of the microscope as shown in Fig. 8.17. There is, therefore, an uncertainty in its momentum given by $\Delta p = p\sin\alpha$ (From Fig. 8.17), ($\tan\alpha = \Delta p/p$) when α is small, we can write $\tan\alpha = \sin\alpha = \Delta p/p$, since it can go anywhere within a cone of semi-vertical angle α. The particle p recoils with an equal and opposite amount of momentum p in accordance with the Compton effect. The position of the particle is defined by the resolving power of the microscope such that

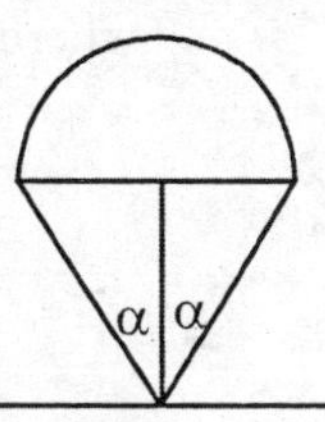

Fig. 8.17

$$\Delta x = \frac{\lambda}{\sin\alpha} \tag{59}$$

Again we have $\sin\alpha = \dfrac{\lambda}{\Delta x} = \dfrac{\Delta p}{p}$ and therefore $\Delta p \Delta x = \hbar$.

To improve the resolving power we can increase the aperture α of the microscope, but in doing so Δp is also increased. Decreasing the wavelength λ would also increase the resolving power, but a corresponding larger amount of momentum would be transferred to the particle from the photon due to a large Compton effect. Similar relationships exist for the momentum along each of the cartesian axes.

(ii) **Determination of position of a particle using a single slit:** Let us suppose that we want to determine the cartesian coordinates of a particle. For simplicity let us determine the x-coordinate only. Let us place a slit in the path of the particle with the width of the slit in the x-direction. If now the particle is observed on the other side of the slit then the particle while crossing the slit must have x-coordinates lying between x_1 and x_2 positions (Fig. 8.18). As a particular example let us suppose that we have a beam of electrons moving along the Z-direction and the width of the slit be d. As regards the passage of an electron through the slit, the x-component of the position of the electron has uncertainty given by

$$\Delta x = x_2 - x_1 = d \tag{60}$$

The wavelength associated with an electron is given by de Broglie relation $\lambda = h/p$ where p is the momentum of the electron.

As the electron emerges from the single slit it is diffracted in an unpredictable direction and acquires certain velocity and momentum in the x-direction. Now, according to the phenomenon of diffraction from a single slit, most of the electrons in the beam will fall inside the central maximum, *i.e.* the diffracted electron has a large probability of hitting the screen in the region BC. From the theory of diffraction, we have

$$AB = \lambda L/d \tag{61}$$

where L is the distance of the screen from the slit.

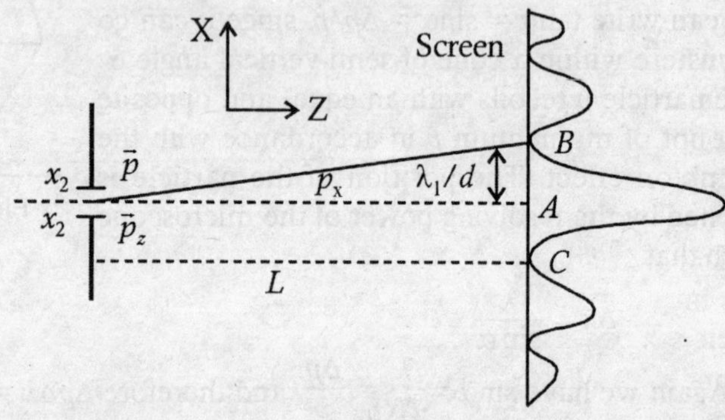

Fig. 8.18

For an electron falling somewhere in the central maximum, the momentum in the x-direction can have any value between p_x and $-p_x$. Obviously, the uncertainty in the x-component of the momentum can be written as

$$\Delta p_x \approx p_z \tag{62}$$

It is worthwhile to note that the uncertainty Δp_x actually be larger than that given by Eq. (62) as electron can fall into other maxima also. From the geometry of the figure, we have

$$\frac{p_x}{p_z} = \frac{\lambda L/d}{L} = \frac{\lambda}{d} \tag{63}$$

Obviously,

$$\Delta p_x \approx p_x \frac{\lambda}{d} p_z \tag{64}$$

Here p_z represents the initial momentum of the electron and it is related to λ as

$$p_z = \frac{h}{\lambda} \tag{65}$$

$$\therefore \quad \Delta p_x \approx \frac{\lambda}{d}\frac{h}{\lambda} \approx \frac{h}{d} \tag{66}$$

$$\therefore \quad \Delta x \Delta p_x \approx d\frac{h}{d} \approx h$$

or

$$\Delta x \Delta p_x > \frac{h}{2x} \text{ (or } \hbar) \tag{67}$$

Obviously, the uncertainty in x can be made as small as desired by reducing the width of the slit but decrease in the width of the slit increases the width of the central maximum, *i.e.* it increases the uncertainty in p_x, so that the product $\Delta x\,\Delta p_x$ is always greater than $\hbar$.

8.13 APPLICATION OF UNCERTAINTY PRINCIPLE

According to uncertainty principle, it is not possible to measure simultaneously the position and momentum of the particle with the desired accuracy. On the other hand, according to classical mechanics, it is possible to know the position and momentum of any particle with desired accuracy provided its initial position and momentum are known. Further, quantum mechanics introduced the laws of probability and therefore probability takes the place of certainty in classical mechanics. Moreover, the quantum mechanics introduces a new concept of particle according to which the term particle is something composed of a very deep and essential unison avoidable uncertainty. The principle of uncertainty explains a large number of facts which could not be explained by classical mechanics. A few of them are discussed below:

(*i*) *Non existence of free electrons in the nucleus*: We know that (*i*) the maximum possible kinetic energy of an electron emitted by a radioactive nuclei is about 4 MeV.

(*ii*) the rest mass of an electron is $m_0 = 9.11 \times 10^{-31}$ kg, and

(*iii*) the diameter of the nucleus is 2×10^{-14} m.

If an electron exists inside the nucleus, it can be anywhere within the diameter of the nucleus. Obviously, the maximum uncertainty Δx in the position of the electron is the same as the diameter of the nucleus, *i.e.*

$$\Delta x = 2 \times 10^{-14} \tag{68}$$

According to Heisenberg uncertainty relation, the product of the uncertainty Δx in the position of the electron and the uncertainty Δp_x in the x component of its momentum is given by

$$\Delta x\,\Delta p_x \geq \hbar$$

The uncertainty in the momentum p_x is

$$\Delta p_x \geq \frac{\hbar}{\Delta x}$$

Minimum uncertainty in the momentum is given by

$$\Delta p_x \geq \frac{\hbar}{\Delta x} = \frac{6.63 \times 10^{-34}}{2\pi \times 2 \times 10^{-14}} = \frac{6.63 \times 10^{-20}}{4\pi}$$

$$= 5.28 \times 10^{-21} \text{ kg-m/sec.} \tag{69}$$

This means that if the electron exists inside the nucleus, then its minimum momentum should be

$$E_{min} = 5.28 \times 10^{-21} \text{ kg-m/sec}$$

According to theory of relativity, the total relativistic energy E of the particle is given by

$$E^2 = p^2c^2 + m_0^2c^4 \quad (70)$$

For the electron having minimum momentum, the minimum energy is given by

$$E_{min}^2 = p_{min}^2 + m_0^2c^4 \quad (71)$$

$$= (5.28 \times 10^{-21} \times 3 \times 10^8)^2 + (9.11 \times 10^{-31})^2 \times (3 \times 10^8)^4$$

$$= 2.5 \times 10^{-24} \times 6.27 \times 10^{-27} \text{ J}$$

Obviously, the second term is much smaller than the first and hence it can be neglected, then

$$E_{min} = \sqrt{2.5 \times 10^{-24}} = 1.58 \times 10^{-12} \text{ J}$$

$$= \frac{1.58 \times 10^{-12}}{1.6 \times 10^{-19}} \text{ eV} = 0.9875 \times 10^7 \text{ eV}$$

$$= 9.875 \times 10^6 \text{ eV} = 9.875 \text{ MeV}$$

Obviously, if a free electron exists in the nucleus, it must have a minimum energy of about 9 MeV.

As stated in the beginning that the maximum kinetic energy which a β-particle (an electron), emitted from a radioactive nuclei can have is 4 MeV. This clearly indicates that free electrons cannot exist inside nuclei.

Ground State Energy of H Atom

Suppose the electron in the hydrogen atom moves around the nucleus, *i.e.* proton in a circular orbit of radius r. The maximum uncertainty in the determination of its position with respect to the proton can be taken equal to r, *i.e.* $\Delta r = r$.

From the uncertainty relation, the minimum uncertainty Δp in the simultaneous determination of its momentum p in the plane of the orbit is given by

$$p = \frac{\hbar}{\Delta r} = \frac{\hbar}{r}$$

Since the momentum p cannot be less than the uncertainty Δp, and hence the minimum possible momentum is

$$p = \frac{\hbar}{r}$$

The kinetic energy K of the electron is given by

$$K = \frac{p^2}{2m} = \frac{\hbar^2}{2mr^2} \tag{72}$$

And the electrostatic potential energy is given by

$$U = \frac{e^2}{4\pi\varepsilon_0 r} \tag{73}$$

Therefore, the total energy of the electron in the hydrogen atom is

$$E = \frac{\hbar^2}{2mr^2} - \frac{e^2}{4\pi\varepsilon_0 r} \tag{74}$$

The ground state energy must be a minimum value of E. Suppose r_1 is the distance of the electron from the proton in the ground state.

For $$E = E_{\min}\left(\frac{dE}{dr}\right)_{r=r_1} = 0$$

$\therefore$ $$-\frac{\hbar^2}{mr_1^3} + \frac{e^2}{4\pi\varepsilon_0 r_1^2} = 0$$

or $$\frac{e^2}{4\pi\varepsilon_0} = \frac{\hbar^2}{mr_1}$$

$\therefore$ $$r_1 = \frac{4\pi\varepsilon_0\hbar^2}{me^2} = \frac{4\pi\varepsilon_0 h^2}{me^2 \times 4\pi^2} = \frac{\varepsilon_0 h^2}{\pi me^2} \tag{75}$$

Eq. (75) is the well known expression for the radii of the Bohr orbit. Using the values:

$$\hbar = 6.63 \times 10^{-31} \text{ Js}, \ m = 9.11 \times 10^{-31} \text{ kg}$$

$$e = 1.6 \times 10^{-19} \text{ C}, \ \varepsilon_0 = 8.85 \times 10^{-12} \text{ C}^2/\text{N-m}^2$$

one obtains $$r_1 = 5.31 \times 10^{-11} \text{ m}$$

This is the radius of the hydrogen atom in the ground state.

The expression for the energy E_1 in the ground state is obtained by substituting

$$r = r_1 = 4\pi\varepsilon_0\hbar^2/me^2$$

From Eq. (74), we have

$$E_1 = \frac{\hbar}{2mr_1^2} - \frac{e^2}{4\pi\varepsilon_0 t_1}$$

$$= \frac{\hbar^2}{2m}\left(-\frac{me^2}{4\pi\varepsilon_0\hbar^2}\right)^2 - \frac{e^2}{4\pi\varepsilon_0}\left(-\frac{me^2}{16\pi\varepsilon_0\hbar^2}\right)$$

$$= \frac{me^4}{32\pi^2\varepsilon_0^2\hbar^2} - \frac{me^4}{16\pi^2\varepsilon_0^2\hbar^2} = -\frac{me^4}{32\pi^2\varepsilon_0^2\hbar^2}$$

$$= \frac{me^4 \times 4\pi^2}{32\pi^2\varepsilon_0^2 h^2} = -\frac{me^4}{8\varepsilon_0^2 h^2} \tag{76}$$

Substituting the numerical values of various constants appearing in Eq. (76) and using 1 eV = 1.6×10^{-19} J, one obtains

$$E_1 = -13.6 \text{ eV}$$

This is the ground state energy of the hydrogen atom.

Ground State Energy of Harmonic Oscillator

The Hamiltonian of a linear harmonic oscillator is given by

$$\hat{H} = \hat{H}(\hat{x}, \hat{p}) = \frac{\hat{p}^2}{2m} + \frac{1}{2}k\hat{x}^2 \tag{77}$$

where $\hat{x}$ and $\hat{p}$ are Hermitian operators corresponding to x and p respectively.

Let ψ represent eigen vector of $\hat{H}$ belonging to eigen value E. Also, let $\hat{A}$ and $\hat{B}$ be, respectively, the operators corresponding to position and momentum, one can write

$$\Delta x = \left\{\left\langle \bar{x}^2 \right\rangle\right\}^{1/2} \text{ and } \Delta p = \left\{\left\langle p^2 \right\rangle\right\}^{1/2} \tag{78}$$

where angular brackets $\langle\ \rangle$ represents the average. Since both x^2 and p^2 are positive for the oscillator, we have

$$\Delta x = 0;\ \Delta p > 0$$

so that $\Delta x\, \Delta p > 0$

According to Heisenberg uncertainty principle

$$\Delta x\, \Delta p \geq \hbar/2, \tag{78a}$$

Now, $E = \left\langle \psi \middle| \hat{H} \middle| \psi \right\rangle = \left\langle \hat{H} \right\rangle$

$$= \frac{1}{2m}\left\langle \hat{p}^2 \right\rangle + \frac{1}{2}k\left\langle \hat{x}^2 \right\rangle$$

$$= \frac{1}{2m}(\Delta p)^2 + \frac{k}{2}(\Delta x)^2$$

$$\geq \frac{1}{2m}(\Delta p)^2 + \frac{\hbar^2 k}{8}\frac{1}{(\Delta p)^2} \tag{78b}$$

The minimum value of E is given by the minimum of the expression,

$$= \frac{1}{2m}(\Delta p)^2 + \frac{\hbar^2 k}{2}\frac{1}{(\Delta p)^2}$$

i.e. $$E_{\min} = E_0 = \left[\frac{(\Delta p)^2}{2m} + \frac{\hbar^2 k}{8(\Delta p)^2}\right]_{\min}$$

$$= \frac{(\Delta p_0)^2}{2m} + \frac{\hbar^2 k}{8(\Delta p_0)^2} \tag{79}$$

where $$(\Delta p_0)^2 = \frac{1}{2}\sqrt{\hbar^2 km} = \frac{m}{2}\hbar\omega_c \tag{80}$$

ω_c being the classical frequency of the oscillator, given by

$$\omega_c = \sqrt{\frac{k}{m}} \tag{81}$$

Thus $$E_0 = \frac{1}{2}\hbar\omega_c \tag{82}$$

The above energy *i.e.* $E_0 = \frac{1}{2}\hbar\omega_c$ is called the '*lowest energy*' or '*Zero point energy*'.

Natural Width of Spectral Line

According to Heisenberg's uncertainty principle of energy and time

$$\Delta E \Delta t \geq \hbar$$

Since the life time of electron in an excited orbit is finite ($\approx 10^{-8}$ sec), therefore the energy levels of the atom given by ($\Delta E \approx h/\Delta t$) must have a finite width. This means that the excited energy levels have a finite energy spread thereby indicating that the radiation emitted when an electron jumps truly monochromatic. In other words the spectral lines can never be infinitely sharp but must have a natural finite width.

8.14 SCHRODINGER'S WAVE MECHANICS

The experiments described in preceding sections tell us that classical physics is inadequate for atomic scale phenomena and for interaction between electron and electromagnetic waves. The idea expressed in Planck's equation $E = h\nu$ and de Broglie's equation $\lambda = h/mv$ are basic to all these problems, but it is not possible to build a complete understanding of atomic phenomena directly upon these ideas.

We require a theory that enables us to predict and explain more complicated atomic phenomena. The theory that accomplishes this is the wave mechanics or quantum mechanics and the heart of the theory is Schrodinger's wave equation. Schrodinger in 1925, using de Broglie's ideas of matter waves set up a wave equation to describe the new mechanics of the particles and successfully applied it to various atomic problems. An equally successful matrix algebra formulation, different in form but basically equivalent to the Schrodinger treatment was developed by Heisenberg. A relativistic treatment of the quantum mechanics was pioneered by P. A. M. Dirac and led to a new and broader picture of nature, including the concept of antimatter. In the present chapter we shall restrict ourselves to the study of the Schrodinger's wave mechanics.

Wave Equation

In classical mechanics, wave equation is a second order differential equation in space and time. Solution of this wave equation represent wave disturbances in a medium. Obviously, a wave equation is the usual basis of mathematical theory of wave motion. As an example of wave equation we can consider the equation of an electromagnetic wave, travelling in the X-direction, *i.e.*

$$\frac{\partial^2 E_y}{\partial x^2} = \frac{1}{c^2} \frac{\partial^2 E_y}{\partial t^2} \tag{83}$$

where E_y is the Y-component of the electric intensity.

We must note that a differential equation for the wave associated with a particle in motion can not be derived from first principles.

One can develop the equation by any one of the following procedures:

(*i*) The classical equations of motion are transformed into a wave equation in accordance with wave properties of matter based on de Broglie's hypothesis.

(*ii*) A complex variable quantity, called the wave function, is assumed to represent a plane simple harmonic wave associated with a free particle, and the classical expression for the total energy is used.

(*iii*) A particle at a given position and at a given time represented by a wave packet which is obtained by superposition of a group of plane waves of nearly the same wavelength, which interfere destructively everywhere except at the wave packet, and the classical expression for the total energy is used.

(*iv*) In the classical expression for the total energy of a particle, the dynamic quantities are replaced by their corresponding operators. These operators allowed to operate upon the wave function.

Any one of the above procedures can be followed to develop the Schrodinger equation.

Wave Function

The variable quantity characterizing de Broglie waves associated with a material particle, is called the *wave function*, denoted by the symbol ψ. The value of the wave function associated with a moving body at the particular point x, y, z in space at time t is related to the likelihood of finding the body there at that time. In quantum mechanics we are concerned with this wave function ψ of the body.

We assume that a wave function is a complex quantity and therefore it cannot have a direct physical meaning. It may be expressed in the form

$$\psi(x, y, z, t) = a + ib \tag{84}$$

where a and b are real functions of the variables (x, y, z, t).

One can obtain the complex conjugate of ψ which is denoted by ψ^* by changing i to $-i$, *i.e.*

$$\psi^*(x, y, z, t) = a - ib \tag{85}$$

Multiplying Eq. (84) by Eq. (85), one obtains

$$\psi(x, y, z, t)\, \psi^*(x, y, z) = a^2 + b^2 \tag{86}$$

On the LHS of Eq. (86) the product of $\psi\psi^*$ is denoted by P, *i.e.*

$$P = \psi(x, y, z, t)\, \psi^*(x, y, z, t) = |\psi(x, y, z, t)|^2 \tag{87}$$

$$\therefore \qquad |\psi(x, y, z, t)|^2 = a^2 + b^2 \tag{88}$$

Obviously, the product of ψ and ψ^* is real and positive if $\psi \neq 0$. Its positive square root is denoted by $|\psi(x, y, z, t)|^2$ and is called the **modulus** of ψ. The quantity $|\psi(x, y, z, t)|^2$ is called the **probability density.** Maxborn has used the result (88) and interpreted $\psi(x, y, z, t)$ as follows:

For the motion of a particle, the quantity

$$PdV = \psi(x, y, z, t)\, \psi^*(x, y, z, t) dV$$

$$= |\psi(x, y, z, t)|^2 dV \tag{89}$$

is the *probability* that the particle will be found in a volume element dV surrounding the point at position (x, y, z) at time t. **Thus the total probability** for finding the particle anywhere in space in question is proportional to the integral of $|\psi(x, y, z, t)|^2$ over all the space. For the motion of a particle in one dimension the quantity

$$Pdx = \psi(x, t) = |\psi(x, t)|^2 dx \tag{90}$$

In this case $|\psi(x, t)|^2$ is called the probability per unit distance.

Thus the total probability is

$$P_x = \int_{-\infty}^{\infty} |\psi(x, t)|^2\, dx \tag{91}$$

For the one dimensional space along the X-axis. For three dimensional space, we have

$$P = \int_{-\infty}^{\infty} |\psi(x, y, z, t)|^2\, dV \tag{92}$$

The above integrals must be finite in order to represent a real particle.

The above discussion clearly reveals that the quantum laws and results of their measurements have probabilitistic interpretation.

The wave function $\psi(r, t)$ is sometimes called a *probability density* for the particle as

$$P(x,t) = \frac{|\psi(x,t)|^2}{\int_{-\infty}^{\infty} |\psi(x,t)|^2 \, dx} \tag{93}$$

where

$$\int_{-\infty}^{\infty} P(x,t)dt = 1 \tag{94}$$

For physical acceptable wave functions it is always possible to introduce an appropriate factor in the wave function such that

$$\int_{-\infty}^{\infty} |\psi(x,t)|^2 \, dx = \int_{\infty}^{\infty} \psi^*(x,t)\psi(x,t)dx = 1 \tag{95}$$

When Eq. (95) is true, wave function ψ is said to be normalised. A wave function which satisfies the condition (95) is said to be normalised (to unity).

The normalising condition for the wave function for the motion of a particle in three dimensions is

$$\int_{-\infty}^{\infty} |\psi(x,y,z)|^2 \, dV = 1 \tag{96}$$

where ψ is a function of the space coordinates (x, y, z).

If a wave function does not satisfy the normalization condition (96), then it must be multiplied by a constant factor, called the normalisation factor. For example,

$$\int \psi^2 dx = N (N \neq 1)$$

The normalisation factor will be $1/\sqrt{N}$ and the normalised wave function will be $\frac{1}{\sqrt{N}}\Psi$. One can easily see that $\frac{1}{\sqrt{N}}\Psi$ will be a good wave function as ψ itself.

From Eq. (93), it is obvious that when a wave function ψ is normalised, the probability density is simply the square of its amplitude. To arrive at a result consistent with physical observation, several additional requirements are imposed on the wave function ψ:

(*i*) ψ must be finite for all values of x, y, z.

(*ii*) ψ must be well behaved, *i.e.*, single valued and continuous every where, obviously, for each set of values of (x, y, z), ψ have one value only.

(*iii*) ψ must be continuous in all regions except in those regions where the potential energy $V(x, y, z) = \infty$.

(*iv*) If a particle can be present in the states described by the wave functions $\psi_1, \psi_2, \psi_3, \ldots, \psi_n$, it can also be in all states described by the wave function constructed from $\psi_1, \psi_2, \psi_3, \ldots, \psi_n$, by the linear transformation

$$\lambda = a_1\psi_1 + a_2\psi + \ldots + a_n\psi_n \tag{97}$$

where $a_1, a_2, a_3, \ldots, a_n$ are arbitrary complex numbers. This is known as *Principle of Superposition.*

(*v*) The wave function ψ must approach zero as x tends to $\pm\infty$.

(*vi*) A wave function satisfying the following condition is said to be an *orthogonal wave function*

$$\int_{-\infty}^{\infty} \psi_i^*(x)\psi_j(x)dx = 0 \quad (i \neq j) \tag{98}$$

The *orthogonality condition* guarantee the non-interference of the wave function representing different states.

(*vii*) Degeneracy: These may, however, be situations when two wave functions ψ_j and ψ_k correspond to the same energy E. This is a case of 'degeneracy.' The system in such an energy state is said to be in a degenerate state. In such a situation, the degenerate wave functions ψ_j and ψ_k will not necessarily be '*orthogonal*'.

(*viii*) Orthonormal set: The normalisation and orthogonality conditions may be combined as follows:

$$f\psi_i\psi_j dv = 1 \text{ if } i = j$$

$$f\psi_i\,\psi_j dv = 0 \text{ if } i \neq j \tag{99}$$

Wave functions satisfying Eq. (99) are said to form "Orthogonal" set of wave functions.

Schrodinger Time Dependent Wave Equation

In quantum mechanics the wave function ψ corresponds to the displacement y of wave motion in a string but ψ is not a measurable quantity and may therefore be complex. We shall therefore assume ψ in the x-direction to be specified by

$$\psi = Ae^{i\omega(t - x/v)} \tag{100}$$

If we replace ω by $2\pi\nu$ and $v = \lambda\nu$ in the above formula, we get

$$\psi = Ae^{-2\pi i\,(\nu t - x/\lambda)} \tag{101}$$

We know that ν and λ can be written in terms of total energy E and momentum p of the particle. Since

$$E = h\nu \text{ and } \lambda = h/p \tag{102}$$

We have

$$\psi = Ae^{-\left(\frac{2\pi i}{h}\right)(Et - px)} \tag{103}$$

This expression for wave function is correct only for freely moving particles while we are interested in situations where the motion of particle is subjected to various restrictions. An important example is an electron bound to an atom by electric field of its nucleus. Let us now obtain a fundamental differential equation for ψ which can be solved under various specific restrictions. Differentiating the above equation twice with respect to x, we get

$$\frac{\partial^2 \psi}{\partial x^2} = -\frac{4\pi^2 p^2}{h^2}\psi \tag{104}$$

and once with respect to t, we get

$$\frac{\partial \psi}{\partial t} = -\frac{2\pi i E}{h}\psi \tag{105}$$

At speeds small compared with that of light, the total energy E of particle is the sum of its kinetic energy $p^2/2m$ and its potential energy V, where V is in general a function of position and time

$$E = p^2/2m + V \tag{106}$$

Multiplying both sides of the equation by ψ, we obtain

$$E\psi = \frac{p^2\psi}{2m} + V\psi \tag{107}$$

But from Eqs. (105) and (104), we get

$$E\psi = -\frac{h}{2\pi i}\frac{\partial \psi}{\partial t} \quad \text{and} \quad p^2\psi = -\frac{h^2}{4\pi^2}\frac{\partial^2 \psi}{\partial x^2} \tag{108}$$

Substituting the value of these in Eq. (107), one obtains

$$\frac{h}{2\pi i}\frac{\partial \psi}{\partial t} = \frac{h^2}{8\pi^2 m}\frac{\partial^4 \psi}{\partial x^2} - V\psi \tag{109}$$

This is the time dependent form of the Schrodinger's equation. In three dimensions the equation is written as

$$\frac{h}{2\pi i}\frac{\partial \psi}{\partial t} = \frac{h^2}{8\pi^2 m}\left(\frac{\partial^2 \psi}{\partial x^2} + \frac{\partial^2 \psi}{\partial y^2} + \frac{\partial^2 \psi}{\partial z^2}\right) - V\psi$$

or

$$-i\hbar\frac{\partial \psi}{\partial t} = \frac{\hbar^2}{2m}\nabla^2\psi - V\psi \tag{110}$$

where $\hbar = \dfrac{h}{2\pi}$ and Laplace operator $\nabla^2 = \left(\dfrac{\partial^2}{\partial x^2} + \dfrac{\partial^2}{\partial y^2} + \dfrac{\partial^2}{\partial z^2}\right)^{2\pi}$

Schrodinger Time Independent Wave Equation

In many problems, the force acting upon the particle and hence the potential energy does not depend explicitly upon time and vary with the position of the particle only. In that case, the Schrodinger equation may be further simplified by eliminating all dependence on time. We know that one dimensional wave function ψ of an unrestricted particle may be written as

$$\psi = Ae^{-\left(\frac{2\pi i}{h}\right)(Et - px)}$$

or
$$= Ae^{-\left(\frac{2\pi iE}{h}\right)t} e^{\left(\frac{2\pi ip}{h}\right)x}$$
$$= \Psi e^{-\left(\frac{2\pi iE}{h}\right)t} \tag{111}$$

This shows that ψ is the product of a time dependent function $e^{-(2piE/h)t}$ and a position dependent function ϕ. Putting the value of ψ in time dependent Schrodinger equation, we get

$$-E\phi e^{-\left(\frac{2\pi iE}{h}\right)t} = \frac{h^2}{8\pi^2 m} e^{-\left(\frac{2\pi iE}{h}\right)t} \frac{\partial^2 \phi}{\partial x^2} - V\phi e^{-\left(\frac{2\pi iE}{h}\right)t} \tag{112}$$

Dividing by the common factor $e^{-\left(\frac{2\pi iE}{h}\right)t}$, we get

$$\frac{\partial^2 \phi}{\partial x^2} + \frac{8\pi^2 m}{h^2}(E - V)\phi = 0 \tag{113}$$

This is the steady-state form of the Schrodinger equation. It can also be written in three dimensions as

$$\frac{\hbar^2}{2m} \nabla^2 \phi + (E - V)\phi = 0 \tag{114}$$

Justification of the Wave Equation

Let us consider a broad wave packet representing a rather sharply defined momentum and moving from a region of space in which the particle is free toward a region in which a force is present. For simplicity we shall restrict to a one-dimensional description. We also suppose that the forces are conservative, so that classically the particle has, and maintains throughout its motion, a constant, energy

$$E = \frac{p^2}{2m} + V \tag{115}$$

Accordingly, the momentum is a function of position. Fourier decomposition of incident wave packet is

$$\psi(x, 0) = \frac{1}{\sqrt{2x}} \int_0^\infty a(k) e^{ikx} dk \tag{116}$$

By omitting the range of integration from $-\infty$ to 0, we have included the assumption that the wave packet moves toward positive x, and contains no negative k components. Using the substitution

$$k = \sqrt{\frac{2mE}{h^2}} \tag{117}$$

one can transform Eq. (117) into an integral over the energy of the free particle, *i.e.*

$$\psi(x, 0) = \int_0^\infty f(E) e^{ik(Ex)} dx \tag{118}$$

where $f(E)$ is assumed to be a smoothly varying positive function of E with a single peak. Obviously, (118) is a rather specialized wave packet. When this wave packet travels in force free space, it spreads in time as follows

$$\psi(x,t) = \int f(E) e^{\frac{i}{\hbar}(px-Et)} dE \tag{119}$$

Now, question arises: how this wave packet will move when it enters the region where a force is present?

Let us consider that the frequency remains unaltered when the wave packet approaches the region of changing potential energy. This is basic assumption and recommends itself by analogy with other types of wave motion. This is further supported by principle of conservation of energy, *i.e.* constant frequency according to the equation $E = h\nu = \hbar\omega$.

In the spirit of above remarks we may compare the motion of a particle through a force field with the propagation of a wave through a strange medium whose detailed microscopic properties we do not propose to analyse here, but which can somehow be set into "sympathetic" vibrations of the same frequency and which thus interacts with the incident wave to modify its properties of propagation, *i.e.*, its velocity of propagation and its wavelength. Thus the form of wave packet moving through a region of force may be represented by

$$\psi(x,t) = \int f(E) e^{\frac{i}{\hbar}[a(x,E)-Et]} dt \tag{120}$$

This is a generalization of Eq. (119) in which $a(x, E)$ is an as yet undetermined function. This function must be chosen so that Eq. (120) represents a wave packet classically. Since $f(E)$ is assumed to be a smooth positive function of E, the peak of the amplitude of Eq. (120) occurs if x and t are chosen so as to make the phase $f(E) - Et$ stationary when $f(E)$ is near its peak. This ensured that the Fourier components near the peak of $f(E)$ are as much in phase as possible and are therefore added constructively. Mathematically, the phase is stationary, if

$$\frac{\partial}{\partial E}\left[a(x,E) - Et\right] = 0 \tag{121}$$

After differentiation, E in this equation is to be evaluated at the energy which corresponds to the peak of $f(E)$. Eq. (121) leads to the condition

$$t = \frac{\partial a}{\partial E} \tag{122}$$

On the other hand, classically for a conservative motion

$$t = \int^{x} \frac{dx'}{V(x')} = \int^{x} \frac{\partial p(x', E)}{\partial E} dx' \tag{123}$$

where $p(x, E)$ is the momentum as defined by Eq. (115). Comparison of Eq. (122) and Eq. (123) gives the possible choice

$$a(x, E) = \int^{x} p(x', E) dx' \tag{124}$$

Thus, we obtain

$$\psi(x, t) = \int^x f(E) e^{\frac{i}{\hbar}\left[\int^x p(x', E)dx' - Et\right]} dx \tag{125}$$

If the particle is free and moves with constant momentum p corresponding to a wavelength $\lambda = h/p$, Eq. (125) is an exact equation instead of an approximation. It is reasonable to assume that Eq. (124) is better approximation, the more nearly $\lambda(x) = h|p(x)|$ is independent of x.

$$\left|\frac{d\lambda}{dx}\right| << 1 \tag{126}$$

If this is true, the wave is quasi-harmonic and $\lambda(x)$ can be considered as its slowly varying wavelengths. Eq. (126) can also be written as

$$\lambda\left|\frac{dp}{dx}\right| << p^2 \tag{127}$$

or

$$\lambda\left|\frac{dp}{dx}\right| << \frac{\left[p(x)\right]^2}{m} \tag{128}$$

This condition implied that the potential energy changes very little over the distance of a wavelength. The particle can then be considered as nearly free over a distance of many wavelengths. For this region $\hbar\frac{dp}{dx}$ can be neglected compared to p^2 in expressions derived from Eq. (125).

One obstacle in the way of applying Eq. (125) extensively is that it may be quite inaccurate in regions where the kinetic energy varies rapidly with x, particularly near the classically turning points of the motion, where $p(x, E) = 0$ and where Eq. (127) can never be satisfied. Hence Eq. (125) does not provide us with an adequate formula for the description of the particle motion, where it is least classical and most interesting.

It is at this point that we ask ourselves whether a more general satisfactory wave function could not be found if we abandoned the explicit construction of wave packets, which represents quite special semi-classical wave forms. A wave packet must always be given in terms of a particular initial wave $\psi(x, 0)$. Let us ask instead of some general relation which $\psi(x, t)$ must satisfy, no matter what the particular initial condition be. Our experience with classical physics suggests that such a region is most likely going to be differential equation for $\psi(x, t)$. In general, we expect that ψ waves are complex quantities and that $\psi(x, 0)$ alone, without knowledge of time derivative, determines the future of the wave. Hence, we shall attempt to obtain a first order differential equation in t. Following this idea, let us differentiate Eq. (125) with respect to time, we have

$$\frac{\partial\psi}{\partial t} \approx -\int f(E)\frac{i}{h}Ee^{\left[\frac{i}{\hbar}(\int pdx' - Et)\right]} dE \tag{129}$$

We would like to eliminate in the initial condition $f(E)$ from RHS of Eq. (129). This can almost be accomplished, if we note

$$\frac{\partial \psi}{\partial t} \approx \frac{i}{\hbar} \int f(E)\, p(x,E) e^{\left[\frac{i}{\hbar}\left(\int^x p dx' - Et\right)\right]} dE$$

and
$$\frac{\partial^2 \psi}{\partial x^2} \approx -\frac{i}{\hbar^2} \int f(E) p^2(x,E)\, e^{\left[\frac{i}{\hbar}\left(\int^x p dx' - Et\right)\right]} dE + \frac{i}{\hbar} \int f(E) \frac{\partial p}{\partial x} e^{\left[\frac{i}{\hbar}\left(\int^x p dx' - Et\right)\right]} dE \tag{130}$$

But according to Eq. (126) the last term is negligible in comparison to first. Using Eqs. (128), (129) and (130), one obtains at least as an approximation

$$i\hbar \frac{\partial \psi}{dt} = \frac{\hbar}{2m} \frac{\partial^2 \psi}{\partial x^2} + V(x)\psi \tag{131}$$

Eq. (131) is valid even when (125) is not. In fact (131) is the quantum mechanical wave equation and thus justifies the Schrodinger's wave equation.

Probability Current Density

Immediately doubts arise over the consistency of the probabilistic interpretation of the wave function as discussed above. If probability of finding the particle in some bounded region of space decreases as time goes on, then the probability of finding it outside this region must increase by the same amount. The probability interpretation of the ψ waves can be made consistently only if this conservation of probability is guaranteed. This requirement is fulfilled owing to Gauss's integral theorem, if it is possible to define a probability current density $\boldsymbol{J}$ which together with the probability density $p = \psi^*\psi$ satisfies a continuity equation

$$\partial \rho / \partial t + \nabla \cdot \boldsymbol{J} = 0 \tag{132}$$

Exactly as in the case of the conservation of matter in hydrodynamics, or conservation of charge in electrodynamics, a relation of the form (132) can easily be deduced from the wave equation. The probability equation

$$p(r)dr = \int_V \psi^* dV \tag{133}$$

gives the probability of finding the particle in a region of space of volume V bounded by the surface A. In order to determine the probability flow of the particle, we must know the changes in the probability with time within the region. Differentiating Eq. (133) with time, one obtains

$$\frac{dp}{dt} = \frac{d}{dt} \int_V \psi^* \psi dV = \int_V \left(\frac{d\psi^*}{dt} \psi + \psi^* \frac{d\psi}{dt} \right) dV \tag{134}$$

Time independent Schrodinger's equation and its complex conjugates are

$$-\frac{\hbar^2}{2m}\nabla^2\psi + V\psi = i\hbar\frac{\partial\psi}{\partial t} \tag{134a}$$

$$-\frac{\hbar^2}{2m}\nabla^2\psi^* + V\psi^* = i\hbar\frac{\partial\psi^*}{\partial t} \tag{134b}$$

Multiplying Eq. (134*a*) by ψ^* and Eq. (134*b*) by ψ and then subtracting Eq. (134*a*) from Eq. (134*b*), one obtains

$$-\frac{\hbar^2}{2m}\psi\nabla^2\psi^* + \frac{\hbar^2}{2m}\psi^*\nabla^2\psi = i\hbar\psi\frac{\partial\psi^*}{dt} - i\hbar\psi^*\frac{\partial\psi}{\partial t}$$

or $$\psi\frac{\partial\psi^*}{dt} + \psi^*\frac{\partial\psi}{\partial t} = -\frac{\hbar^2}{2im}\left(\psi^*\nabla^2\psi - \psi\nabla^2\psi^*\right) \tag{135}$$

From Eqs. (134) and (135), one obtains

$$\frac{dp}{dt} = -\frac{\hbar}{2im}\int_V\left(\psi^*\nabla^2\psi - \psi\nabla^2\psi^*\right)\cdot dV \tag{136}$$

The integral appearing in Eq. (136) can be transformed into integral over the surface *V* by Green's theorem, we have

$$\frac{dp}{dt} = \frac{\hbar}{2im}\int_A(\psi^*\nabla\psi - \psi\nabla\psi^*)\cdot dt \tag{137}$$

or $$\frac{dp}{dt} = \nabla\cdot \boldsymbol{J} = 0 \tag{138}$$

where $\boldsymbol{J} = \hbar/2im\left(\psi^*\nabla\psi - \psi\nabla\psi^*\right)$ is called the probability current density.

Clearly Eq. (138) is analogous to Eq. (132). Therefore it is quite reasonable to interpret *j* as a probability current density. If Eq. (138) is multiplied by charge of the particle then *ep* give the *charge density* and *ej* give the *current density* of charge in quantum mechanical description. One can express Eq. (138) as

$$\frac{dp}{dt} - \nabla\cdot\boldsymbol{J} = -\int_A \boldsymbol{J}\cdot dA \tag{139}$$

This shows that the rate of change of probability of the particle inside a surface is equal to the negative of probability current density through the surface *A*. Eq. (139) is generally useful for the determination of flux of particle in the scattering problems. The idea of probability current density is generally useful for the description of unbounded systems.

We must note that the operator $-i\hbar\nabla$ represents the momentum, even when a force is present. Then the velocity will be represented by the operator

$$-\frac{i\hbar}{m}\nabla \text{ or } \frac{\hbar}{im}\nabla$$

and it is apparent that

$$J = \text{real part of} \left(\psi^* \frac{\hbar}{im} \nabla\psi \right)$$

or real part of $\psi \frac{\hbar}{im} \nabla\psi^*$, while this interpretation of j is suggestive, it must be realised that it is not susceptible to direct measurement in the sense in which P is.

8.15 SIMPLE SOLUTIONS OF SCHRODINGER'S EQUATION

We have developed Schrodinger's wave equation for systems of fundamental particles or atoms in non-relativistic limit: when the total energy of such a system is independent of the time, the solution of Schrodinger equation reduces to the solution of the energy eigen equation ($H\psi = E\psi$). The total solution is then simply the energy eigen function ψE_n multiplied by the time factor, $\exp\left[-(i/\hbar)E_n t\right]$. The energy eigen states are also referred to as stationary state, since the probability density for such a state is constant in time. Further more, the expectation value of any operator that is itself not a function of time is constant for a stationary state. Now we will discuss it in detail.

Schrodinger Equation and Stationary State Solution

Schrodinger's time dependent one dimensional wave equation is

$$-\frac{\hbar^2}{2m}\frac{\partial^2\psi(x,t)}{\partial x^2} + V(x)\,\psi(x,t) = i\hbar\frac{\partial\psi(x,t)}{\partial t} \tag{140}$$

Wave function $\psi(x, t)$ can be expressed as the product of two functions, *i.e.* one involving the time alone while the other coordinate alone, then one can write

$$\psi(x, t) = u(x) f(t) \tag{141}$$

Using Eq. (141), Eq. (140) becomes

$$-\frac{\hbar^2}{2m}\frac{d^2u(x)}{dx^2} f(t) + V(x)\,u(x)\,f(t) = i\hbar\frac{df(t)}{dt} u(x)$$

Dividing by $u(x) f(t)$, one obtains

$$\frac{1}{u(x)}\left[-\frac{\hbar^2}{2m}\frac{d^2u(x)}{dx^2} + V(x)\,u(x)\right] = i\hbar\frac{1}{f(t)}\frac{df(t)}{dt} \tag{142}$$

In the above equation, the left hand side is independent of time but a function of x whereas right hand side is independent of x but a function of time t only. This will be possible only when each side of Eq. (142) will be equal to a same constant, say E. This constant has the dimensions of energy. Thus we have

$$\frac{1}{u(x)}\left[-\frac{\hbar^2}{2m}\frac{d^2u(x)}{dx^2}+V(x)u(x)\right]=E \tag{143}$$

and $$\frac{i\hbar}{f(t)}\frac{df(t)}{dt}=E \tag{144}$$

Equation (144) can be rewritten as

$$\frac{df(t)}{f(t)}=-\frac{i}{\hbar}Edt$$

which on integration yields

$$\ln f(t)=\frac{iE}{\hbar}t+\ln C$$

or $$f(t)=Ce^{-iEt/\hbar} \tag{145}$$

Eq. (143) can be put as follows

$$\frac{d^2u(x)}{dx^2}+\frac{2m}{\hbar^2}\left[E-V(x)\right]u(x)=0 \tag{146}$$

Eq. (146) represents Schrodinger's time independent or steady state equation, whose solution is given by

$$u(x)=\sum_n a_n u_n(x) \tag{147}$$

On substituting Eqs. (145) and (147) in Eq. (141), one obtains

$$\psi(x,t)=\Sigma_n a_n\psi_n(x,t)$$

$$=\Sigma_n a_n\psi_n(x)\exp\left[-\frac{i}{\hbar}E_n t\right] \tag{148}$$

Eq. (148) represents the general solution of the one dimensional time-dependent Schrodinger equation corresponding to various values of the constant energy E of a particle.

Stationary State Solutions

If in a particular state, the probability distribution function $\psi\psi^*$ does not depend upon time, then the state of the system is said to be stationary state.

Suppose we now consider the probability distribution function $\psi\psi^*$ for a system in the state represented by the following wave function

$$\psi(x,y,z,t)=\sum_{n=1}^{\infty}a_n\psi_n(x,y,z)\exp\left(-\frac{i}{\hbar}E_n t\right) \tag{149}$$

The complex conjugate of Eq. (149) is

$$\psi^*(x,y,z,t) = \sum_{n=1}^{\infty} a_n \psi_n^*(x,y,z)\exp\left(-\frac{i}{\hbar}E_n t\right) \tag{150}$$

on multiplying Eqs. (149) and (150), one obtains the probability distribution function $\psi\psi^*$, *i.e.*

$$\psi\psi^* = \left\{\sum_{n=1}^{\infty} a_n \psi_n(x,y,z)\exp\left[\frac{-i}{\hbar}E_n t\right]\right\} \left\{\sum_{m=1}^{\infty} a^*{}_m \psi_m^*(x,y,z)\exp\left[\frac{-i}{\hbar}E_m t\right]\right\}$$

or

$$\psi\psi^* = \sum a_n a_n^* \psi_n(x,y,z)\,\psi_n^*(x,y,z) + \sum_m \sum_n' a_n a_m^* \psi_n(x,y,z)\,\psi_m^*(x,y,z)\exp\left\{\frac{i}{\hbar}(E_m - E_n)t\right\} \tag{151}$$

In the first term in Eq. (151) the terms with $m = n$ have been included while in the second term of the same equation the term with $m = n$ have been excluded, *i.e.*, the prime on the double summation indicates that the terms with $m = n$ are excluded. As time t enters the probability function $\psi\psi^*$ in Eq. (151), it means that ψ expressed by Eq. (149) is not a stationary state solution. If a_n's are zero for all values except for one value of E_n then $\psi\psi^*$ will be independent of time t. In such a case the state represented by wave function ψ would be stationary state. For this reason the wave function will contain only a single term and will be represented by

$$\psi_n(x,y,z,t) = \psi_n(x,y,z)\exp\left(-\frac{i}{\hbar}E_n t\right) \tag{152}$$

The solution represented by (152) is stationary states solution, *i.e.* when the particle is in a particular state of energy, the probability density is independent of time. The probability density for an stationary state or an eigen state is constant in time, and the expectation value for the energy of an eigen state is precisely the energy of the state for all time. One can easily prove that the expectation value of any operator which does not explicitly depend upon time is also independent of time for a stationary state.

Boundary and Continuity Conditions on the Wave Functions

Schrodinger steady state equation in three dimensions is

$$\frac{\partial^2\psi}{\partial x^2} + \frac{\partial^2\psi}{\partial y^2} + \frac{\partial^2\psi}{\partial z^2} + \frac{2m}{\hbar^2}(E - V)\psi = 0 \tag{153}$$

Schrodinger equation (steady state) in one dimension is

$$\frac{\partial^2\psi(x)}{\partial x^2} + \frac{2m}{\hbar^2}(E - V)\psi(x) = 0 \tag{154}$$

The time independent or steady state Schrodinger wave equation is a second order linear differential equation in x [in general in $r(x, y, z)$]. In general, Schrodinger's steady state equation can be solved only for certain values of the energy E. What is meant by this statement has nothing to do with any mathematical difficulties that may be present, but is something much more fundamental. To "solve" Schrodinger's equation for a given system means to obtain a wave function ψ that not only obeys the equation and whatever boundary conditions there are but also fulfils the requirements for an acceptable wave function—namely that it and its derivatives, *i.e.* ψ and $\partial\psi/\partial x$ be continuous, finite and single valued. If there is no such wave function, the system cannot exist in a steady state. Hence, while obtaining the solutions of steady state Schrodinger equation it will be required that the wave function and its gradient are continuous, finite and single valued at every point in space.

Thus energy quantization appears in wave mechanics as a natural element of the theory and energy quantization in the physical world is revealed as universal phenomenon characteristic of all stable systems.

Free Particle System

A free particle is one which is not subjected to any force or potential barrier and is free to move in limitless space. For a linear motion of the particle, the Hamiltonian H is given as

$$H = -\frac{\hbar^2}{2m}\frac{d^2}{dx^2} + V \tag{155}$$

and the wave function ψ (which is also eigen function of H here) is $\psi(x)$.

Since the particle is not subjected to any external force, its potential energy V is constant, which for convenience, may be assumed to be zero. Thus, the Schrodinger steady state equation for the free particle may be written explicitly as

$$\frac{d^2\psi}{dx^2} + \frac{2m}{\hbar^2}E\psi = 0 \tag{156}$$

The general solution of Eq. (156) is

$$\psi = Ae^{ikx} + Be^{-ikx} \tag{157}$$

where $$K = \sqrt{\frac{2m}{\hbar^2}E} \tag{158}$$

and A and B are constants.

Eq. (157) gives the steady state or time independent part of the wave function. The complete wave function for particle is given by

$$\psi(x, t) = \psi(x)e^{-iEt/\hbar}$$

$$= \left(Ae^{ikx} + Be^{-ikx}\right)e^{-iEt/\hbar} \tag{159}$$

$$= Ae^{ikx}\, e^{-i\omega t} + Be^{ikt}e^{-i\omega t} \qquad (\because \omega = E/\hbar)$$

$$= Ae^{-i(\omega t - kx)} + Be^{-i(\omega t + kx)} \tag{160}$$

Eq. (160) represents the sum of two stationary solutions in principle of superposition, *i.e.*, a continuous plane simple harmonic wave.

The first term on the *RHS* of Eq. (160) represents the wave travelling in the positive x-direction. The wavelength of each travelling plane wave is $\lambda = 2\pi/k$. The phase velocity of these waves is

$$v = \nu\lambda = \frac{E}{h}\frac{2\pi}{k} = \frac{E}{\hbar k}$$

According to be-Broglie's hypothesis these waves correspond to a particle of momentum

$$p = \frac{h}{\lambda} = \frac{hk}{2\pi} = \hbar k$$

and energy

$$E\left(= \frac{p^2}{2m} = \frac{\hbar^2 k^2}{2m} = \frac{\hbar^2 2mE}{2m\hbar^2} = E\right)$$

Therefore, for the motion of the particle in the positive x-direction, we may have

$$\psi(x) = Ae^{ikx} \tag{161}$$

or

$$\psi(x, t) = Ae^{-i(\omega t - kx)} \tag{162}$$

The momentum operator, $\frac{\hbar}{i}\frac{\partial}{\partial x}$ operating on the wave function $\psi(x, t)$ gives

$$\frac{\hbar}{i}\frac{\partial\psi}{\partial x} = \frac{\hbar}{i}\left[A(ik)e^{-i(\omega t - kx)}\right]$$

$$= \hbar k A e^{-i(\omega t - kx)}$$

$$= \hbar k\psi = p_x\psi \tag{163}$$

This equation shows that the wave function $\psi(x, t)$ for the particle is an eigen function of the linear momentum operator, and the momentum p_x is the eigen value of the operator. Hence the momentum remains sharp with the value p_x.

Now, as the constant k^2 and hence energy E are completely arbitrary, *i.e.* no constraint has been applied on these, the particle can have any energy between 0 and $+\infty$, *i.e.*, the particle has a continuous energy spectrum. The two sets of waves represented by Eq. (160) are travelling in opposite directions in $+x$ and $-x$ directions. Thus corresponding to the same eigen value E there are two eigen functions as with the energy operator $i\hbar\partial / \partial t$, one obtains

$$i\hbar \frac{\partial}{\partial t}\left(Ae^{ikx}e^{-iEt/\hbar}\right) = E\left(Ae^{-ikx}e^{-iEt/\hbar}\right) \tag{164}$$

and $$i\hbar \frac{\partial}{\partial t}\left(Ae^{-ikx}e^{-iEt/\hbar}\right) = E\left(Ae^{-ikx}e^{-iEt/\hbar}\right) \tag{165}$$

i.e., in the two cases, we have

$$i\hbar \frac{\partial \psi}{\partial t} = E\psi \tag{166}$$

Obviously, ψ has two different values but the eigen value E remains the same. Such eigen functions are termed as "degenerate functions" and in the present case of a free particle the degeneracy is two fold.

The probability p of finding the particle between x and $x + dx$ is given by

$$pdx = \psi(x, t)\psi^*(x, t)dx = A^2dx \tag{167}$$

Therefore, the probability density P for the position of the particle with the definite value of momentum is constant over the X-axis, *i.e.*, all positions of the particle are equally probable. This conclusion is also obtained from the principle of uncertainty.

According to the interpretation of the wave function the probability of finding the particle somewhere in space must be equal to 1, *i.e.*

$$\int_{-\infty}^{\infty} \psi(x,t)\,\psi^*(x,t)\,dx = 1 \tag{168}$$

In this case $\psi(x, t)\ \psi^*(x, t)dx$ = constant = A^2. Hence the wave function for the particle cannot be normalized and A must remain arbitrary. The difficulty arises because we are dealing with an arbitrary case. In practice, we can not have an absolutely free particle. This particle will always be confined within an enclosure in the laboratory, and hence its position can be determined with absolute accuracy.

Particle in Bound States

When the motion of a particle is confined to a limited region such that the particle moves back and forth in the region, the particle is said to be in a bound state. If one dimensional motion of a particle is assumed to take place with zero potential energy over a fixed distance and if the potential energy is assumed to become infinite is the simplest example of all motions in bound states. But in practice such a motion is not possible.

Particle in One Dimensional Box

(Particle in an infinitely deep one-dimensional potential well):

Let us consider the one dimensional motion along X-axis of a particle of mass m in a hollow rectangular box having perfectly rigid wall. Let the origin be at one corner of the box and the X-axis be perpendicular to the parallel opposite walls as shown in Fig. 8.19. Let a be the distance between the walls so that the motion along the X-axis is confined between $x = 0$ and $x = a$. Suppose inside the box, *i.e.* in the region $0 < x < a$, there is no force acting on the particle so that in this region, *i.e.* inside the box, the potential energy $V(x)$ is zero. When the particle collides with the perfectly rigid walls there is no loss of energy so that the total energy E of the particle remains constant. In order to leave the region, the particle will have to do an infinite amount of work. Since this is not possible, it cannot exist outside the box. Hence for $x \geq 0$ *and* $x \geq a$, *i.e.* outside the box, the wave function $\psi(x) = 0$.

The potential energy curve for such a particle is shown in Fig. 8.20 because of its appearance it is called a *square well potential of infinite depth.*

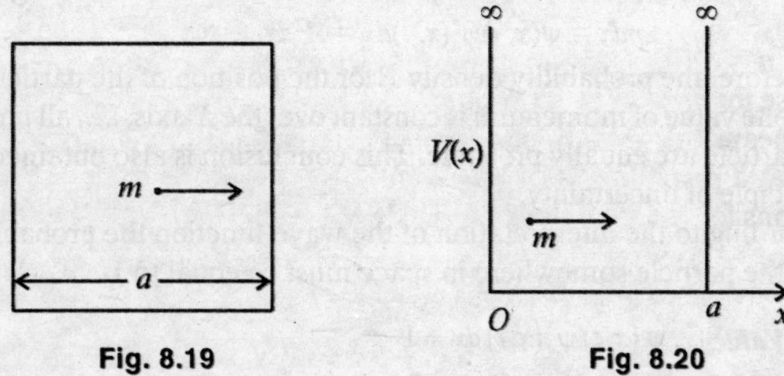

Fig. 8.19 Fig. 8.20

Wave Equation for a Particle

For the motion of the particle along the X-axis inside the box, where $V(x) = 0$, the time independent Schrodinger's wave equation is

$$\frac{d^2\psi}{dx^2} + \frac{2m}{\hbar^2} E\psi = 0 \qquad [\because V(x) = 0] \tag{169}$$

where E is the total energy of the particle. Eq. (169) can be expressed as

$$\frac{d^2\psi}{dx^2} + K^2\psi = 0 \tag{170}$$

where $$K^2 = \frac{2mE}{\hbar^2} \tag{171}$$

For a particular value of the energy, E, K is a constant.

Solution of the Equation

The general solution of Eq. (170) is

$$\psi(x) = A\sin Kx + B\cos Kx \tag{172}$$

where A and B are constants of integration. We have the boundary conditions.

(*i*) At $x = 0$, $\psi(x) = 0$

(*ii*) At $x = a$, $\psi(x) = 0$

From the first condition, one obtains $B = 0$.

Therefore from Eq. (172), one obtains

$$\psi(x) = A \sin Kx = 0 \tag{173}$$

Using second boundary condition, one obtains

$$\psi(a) = A \sin Ka = 0 \tag{174}$$

We cannot take $A = 0$ because there will then be no solution. Hence Eq. (174) is satisfied only when, $Ka = n\,\pi$.

$$K = \frac{n\pi}{a} \tag{175}$$

where n = 1, 2, 3, ... *i.e.*, n is an even integer. We cannot take $n = 0$, because for $n = 0$, $K = 0$, $E = 0$ and hence $\psi(x) = 0$, every where in the box. This means that a particle with zero energy cannot be present inside the box, *i.e.*, a particle in the box, cannot have zero energy. Hence the wave functions for the motion of particle in the region $0 < x < a$ are given by

$$\psi_n(x) = A \sin \frac{n\pi x}{a} \qquad \text{[where } n \text{ is even]} \tag{176}$$

Eigen Values of Energy: Substituting the value of K from Eq. (171) in Eq. (175), one obtains

$$\sqrt{\frac{2mE}{\hbar^2}} = \frac{n\pi}{a}$$

or

$$E = \frac{n^2\pi^2\hbar^2}{2ma^2}$$

$$E_n = \frac{n^2\pi^2\hbar^2}{2ma^2} \tag{177}$$

Eq. (177) represents the eigen values of the energy. These values are called the energy levels of the particle. Using $\hbar = h/2\pi$, one can rewrite Eq. (177) as

$$E_n = \frac{n^2h^2}{8ma^2}$$

From Eq. (177), one can draw the following conclusions:

(*i*) The lowest energy of the particle is obtained by putting $n = 1$ in Eq. (177). One obtains for $n = 1$ as

$$E_1 = \frac{\pi^2\hbar^2}{2ma^2} \tag{178}$$

It is called the ground state energy level of the particle. The value of E_n in terms of E_1 are given by

$$E_n = n^2 E_1 \tag{179}$$

In the above discussion, the n numbers are called quantum numbers while the E values are called energy levels. The quantum states with higher values of n (= 2, 3, 4, ..., etc.) are termed as excited states.

According to quantum mechanics (but not in accordance to classical mechanics) the kinetic energy is not equal to zero in ground state.

(*ii*) The possible values of energy of the particle in the potential box are discrete corresponding to n = 1, 2, 3, ... The classical mechanic does not need any such requirement of discrete energy levels for bound systems. The quantization of energy conclusion is universal, *i.e.*, whenever a particle is confined to move in a limited space, its energy is quantized. For example, the energies of electrons bound to atoms and molecules are quantised.

(*iii*) The Energy Levels: The quantization of energy is shown graphically in Fig. 8.21. The allowed energy levels are shown as horizontal lines. That these lines are straight emphasises the fact that energy (E) is the same irrespective of the position (x) of the particle. Since for a particular value of n, the energy (E) does not change with x, the latter is often termed as "constant of motion", further, since H or E is independent of time (t) the system under consideration is a good and simple example of "conservative system".

The lowest value of E is not zero but $\pi^2 h^2/2ma^2$ (for n = 1). Evidently, the particle in a box will continue to move to and fro in the region 0 to L even at the temperature of absolute zero; for this region it is called the "zero point energy". This conclusion is universal and valid for any system undergoing to and fro motion.

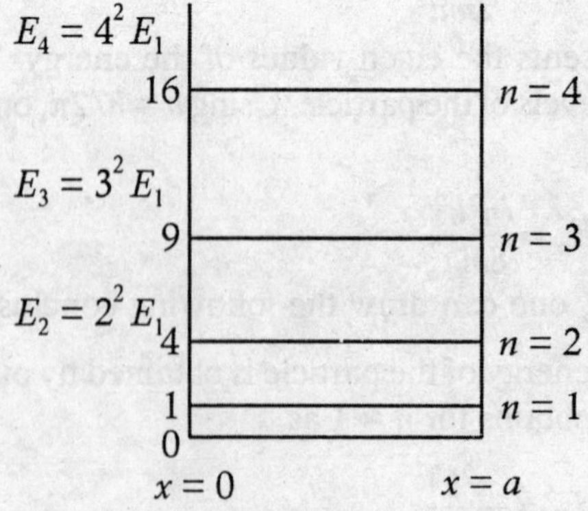

Fig. 8.21

The dependence of E on n^2 results in the increased spacing between energy levels as n increases.

$$\Delta E = E_{n+1} - E_n = \left[(n+1)^2 - n^2\right] E_1 = (2n+1)\, E_1 \qquad (180)$$

Since the energy E_1 is proportional to $1/a^2$, the smaller the box the greater are the energy spacing (ΔE), *i.e.*, the quantization is more pronounced. Alternatively, as the box is made wider ΔE decreases and in the limited case of $L \to \infty$ (free particle) the quantization vanishes.

The wave functions ψ_n corresponding to E_n are called eigen functions.

Normalization of the Wave Functions

The wave functions for the motion of the particle are

$$\psi_n(x) = A \sin \frac{n\pi x}{a} \qquad (181)$$

in the region $0 < x < a$ and $\psi_n = 0$, in the region $x < 0$, $x > a$. This wave function is still incomplete in as much as the constant A is not determined. A can be determined by making the use of the fact that the total probability of the particle is somewhere in the box must be unity, *i.e.*

$$\int_0^a \left|\psi_n(x)^2\right| dx = 1 \qquad (182)$$

or $$\int_0^{a^2} A^2 \sin^2 \frac{n\pi x}{a}\, dx = 1$$

or $$A^2 \int_0^{a^2} \frac{1}{2}\left(1 - \cos \frac{2n\pi x}{a}\right) dx = 1$$

or $$\frac{A^2}{2}\left[x - \frac{a}{2\pi n} \sin \frac{2n\pi x}{a}\right]_0^a$$

The second term of the integrated expression becomes zero at both limits. Therefore,

$$\frac{A^2 a}{2} = 1 \text{ or } A = \sqrt{\frac{2}{a}} \qquad (183)$$

The normalized wave function is thus

$$\psi_n = \sqrt{\frac{2}{a}} \sin \frac{n\pi x}{a} \qquad (184)$$

Note: Actually $\psi = \pm\sqrt{2/a}\, \sin n\, x/a$. However, since probability is given by $|\psi|^2$ and not by ψ, it is immaterial whether positive or negative A is chosen. Customarily, we select the positive series of normalized ψ functions.

Forms of the Wave Functions

The expression for the real form of the wave function can be used conveniently to draw its graphical representations. The wave functions for the first three values of n are shown in Fig. 8.22. It is evident that the wave function ψ_1 has two nodes at $x = 0$, $x = a$. The wave function ψ_2 has three nodes at $x = 0$, $x = a/2$ and at $x = a$. The wave function ψ_3 has four nodes $x = 0$, $x = a/3$, $x = 2a/3$ and at $x = a$. Thus the wave function ψ_n will have $(n + 1)$ nodes. Obviously the increasing number of nodes is consistent with increasing energy of the system.

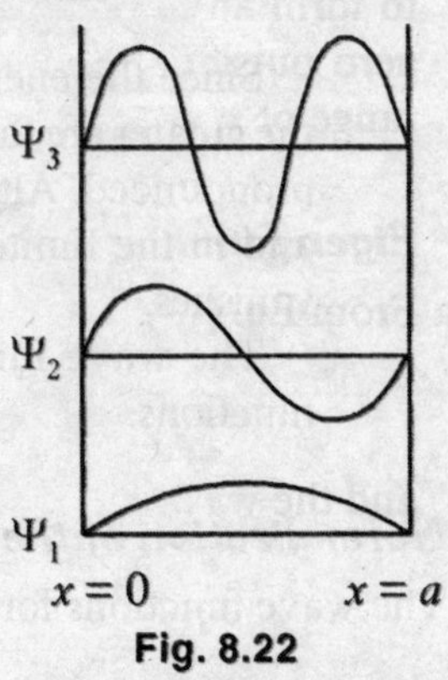

Fig. 8.22

Orthogonality of the Wave Function

Let $\psi_n(x)$ and $\psi_m(x)$ be the normalized wave functions of the particle in the interval $(0, a)$ corresponding to the different energy levels E_n and E_m respectively.

These wave functions are

$$\psi_n(x) = \sqrt{\frac{2}{a}} \sin \frac{n\pi x}{a}$$

$$\psi_m(x) = \sqrt{\frac{2}{a}} \sin \frac{m\pi x}{a}$$

where m and n are integers. In the present case the wave functions are real, therefore,

$$\psi_n^*(x) = \psi_n(x) \text{ and } \psi_m^*(x) = \psi_m(x)$$

Now we have, when $m \neq n$

$$\int_0^a \psi_m^* \psi_n dx = \frac{2}{a} \int_0^a \sin \frac{n\pi x}{a} dx$$

$$= \frac{1}{a} \int_0^a \left[\cos \frac{(m-n)\pi x}{a} - \cos \frac{(m+n)\pi x}{a} \right] dx$$

$$= \frac{1}{a} \left[\frac{a}{\pi(m-n)} \sin \frac{(m-n)\pi x}{a} - \frac{a}{\pi(m+n)} \sin \frac{(m+n)\pi x}{a} \right]_0^a = 0$$

Obviously, the functions are mutually orthogonal in the interval (0, a). These functions are also normalized in this interval. The wave functions which, are normalized and mutually orthogonal in an interval, are said to form an orthogonal set in this interval. Since the wave functions are zero outside the interval (0, a), they are also orthogonal in the whole range of x, *i.e.*, in the interval ($-\infty$, ∞).

Eigen Functions of Momentum

From Eq. (177), we have the possible values of energy as

$$E_n = \frac{n^2\pi^2\hbar^2}{2ma^2} \tag{185}$$

and the wave function of a particle in a box whose energy is E is

$$\psi = A\sin\frac{\pi x}{a} \tag{186}$$

But

$$\sqrt{\frac{2mE}{\hbar^2}} = \frac{x}{a} \tag{187}$$

Substituting the value of a from Eq. (187) in Eq. (186) we have

$$\therefore \quad \psi = A\sin\sqrt{\frac{2mE}{\hbar^2}}\;x \tag{188}$$

Substituting E_n for E yields

$$\psi_n = A\sin\frac{n\pi x}{a} \tag{189}$$

for the eigen function corresponding to the energy eigen values E_n. Substituting $A = \sqrt{2/a}$, we have

$$\psi_n = \sqrt{\frac{2}{a}}\sin\frac{n\pi x}{a} \tag{190}$$

We can easily see that these eigen functions meet all the requirements for each quantum number n, ψ_n is a single valued function of x, and ψ_n and $\partial\psi_n/\partial x$ are continuous. Furthermore, the integral of $|\psi_n|^2$ over all space is finite, as one can see by integrating $|\psi_n|^2 dx$ from $x = 0$ to $x = a$ (since the particle, by hypothesis is confined within these limits):

$$\int_{-\infty}^{\infty}|\psi_n|^2 dx = \int_0^L|\psi_n|^2 dx$$

$$= A^2\int_0^L \sin^2\left(\frac{n\pi x}{a}\right)dx$$

or

$$1 = A^2\frac{a}{2}$$

$$\therefore \quad A = \sqrt{\frac{2}{a}}$$

Eigen Values of the Momentum

The eigen values of the momentum P_n along the X-axis are given by

$$P_n^2 = 2mE_n = 2m\frac{n^2\pi^2\hbar^2}{2ma^2} = \frac{n^2\pi^2\hbar^2}{a^2}$$

or
$$P_n = \frac{n\pi\hbar}{a} = \frac{nh}{2a} \tag{191}$$

Eq. (191) shows that the eigen values of the momentum of the particle are discrete, and the difference between the momenta corresponding to two consecutive energy levels is constant equal to $h/2a$.

Probability of Location of the Particle over a Small Range between x and x+dx

According to quantum mechanics, the probability $P(x)\,dx$ that the particle be found over a small distance dx at position x is given by

$$P(x)\,dx = \int |\psi_n|^2 dx = \frac{2}{a}\sin^2\left(\frac{n\pi x}{a}\right)dx \tag{192}$$

therefore, the probability density for the one dimensional motion is

$$P(x) = \frac{2}{a}\sin^2\left(\frac{n\pi x}{a}\right) \tag{193}$$

The probability density is maximum when

$$\frac{n\pi x}{a} = \frac{\pi}{2}, \frac{3\pi}{2}, \frac{5\pi}{2}$$

or
$$x = \frac{a}{2n}, \frac{3a}{2n}, \frac{5a}{2n} \tag{194}$$

Obviously, for the state defined by $n = 1$, the most probable position of the particle is at $x = a/2$. For $n = 2$, the most probable positions are at $x = a/4$ and $x = 3a/4$. For $n = 3$, the most probable positions are at $x = a/6$, $x = 3a/6$ and $x = 5a/6$. The variation of the probability densities with x for first three values of n are shown in Fig. 8.23.

Thus the quantum mechanical result is quite different from the classical one. According to classical mechanics a particle in such a potential box would travel with a uniform velocity from one wall to the other, and at the walls it would be perfectly reflected. Therefore, the probability of finding the particle within a small distance dx anywhere in the one-dimensional box is the same and is equal to dx/a. The probability $1/a$ is represented by a straight line at height $1/a$ above the X-axis.

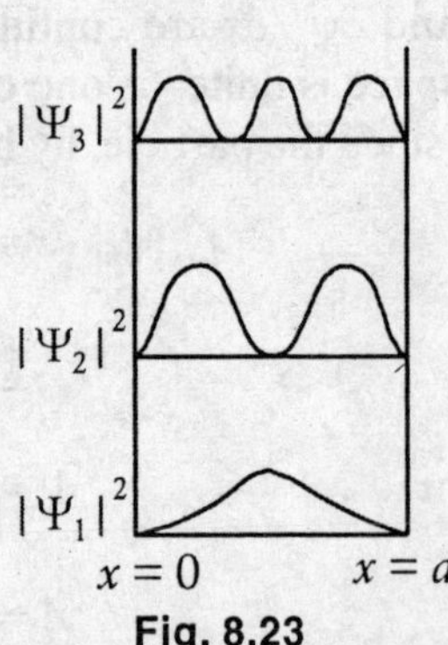

Fig. 8.23

The problem of particle in a one dimensional box is not merely an exercise in mathematics, it can serve as a good model for the calculation of approximate energies in atomic and molecular systems, besides accounting for the effect of quantization.

Particle in a Rectangular Three Dimensional Box

We have seen that a particle moving freely in a one dimensional box (*P.E.* : $V = 0$) serves as a very convenient model for several types of atomic and molecular systems. Calculations, though approximate, agree fairly well with observed results. Electronic motions in atoms and molecules are, however, three dimensional and a three dimensional box model should be more appropriate. Though electronic motions in atoms and molecules are complicated due to some other factors, let us see how far the results of quantum mechanical treatment of a single particle moving in a three dimensional rectangular box are of interest.

Let a particle of mass m be in motion in a rectangular potential box as shown in Fig. 8.24. With sides of lengths a, b, c parallel to the x, y and z axes respectively, suppose there is no force acting on the particle inside the box, so that in the region $0 < x < a$, $0 < y < b$ and $0 < z < c$, the potential energy $V(x, y, z) = 0$ and outside the box $V(x, y, z) = \infty$.

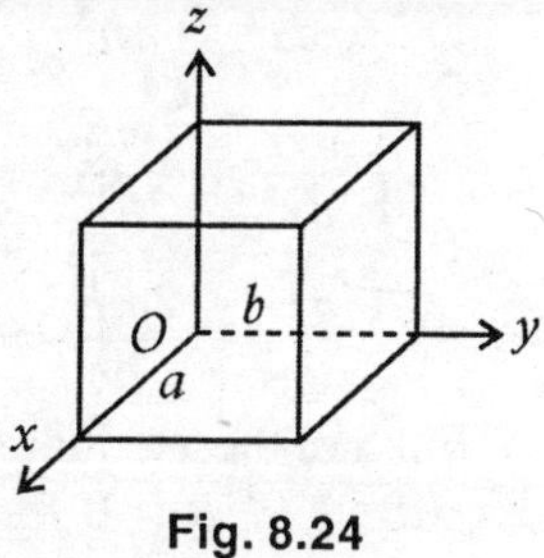

Fig. 8.24

Wave Equation for the Particle

For the motion of the particle inside the box, the Schrodinger time-independent equation reduces to

$$\nabla^2\psi + \frac{2m}{\hbar^2}E\psi = 0 \tag{195}$$

$$\frac{\partial^2\psi}{\partial x^2} + \frac{\partial^2\psi}{\partial y^2} + \frac{\partial^2\psi}{\partial z^2} + \frac{2m}{\hbar^2}E\psi = 0 \tag{196}$$

This is a partial differential equation, where the wave function ψ is a function of x, y and z. The three dimensional equation (196) can be reduced to three ordinary one dimensional equations by the method of separation of variables. We assume that the wave function $\psi(x, y, z)$ is equal to the product of three functions x, y and z each of which is a function of one variable only. Thus we have

$$\psi(x, y, z) = X(x)\,Y(y)\,Z(z) \tag{197}$$

Substituting Eq. (197) in Eq. (196), one obtains

$$YZ\frac{d^2X}{dx^2} + ZX\frac{d^2Y}{dy^2} + XY\frac{d^2Z}{dz^2} + \frac{2mE}{\hbar^2}XYZ = 0 \tag{198}$$

We have used ordinary derivatives instead of partial derivatives because each of the functions x, y and z is a function of one variable only. Dividing Eq. (198) by *XYZ*, one obtains

$$\frac{1}{X}\frac{d^2X}{dx^2}+\frac{1}{Y}\frac{dY^2}{dy^2}+\frac{1}{Z}\frac{d^2Z}{dz^2}+\frac{2m}{\hbar^2}E=0 \tag{199}$$

In Eq. (199) $2\,mE/\,\hbar^2$ is constant for a particular value of kinetic energy. Since the velocity of the particle being a vector quantity, can be resolved into three components along the coordinate axes, the kinetic energy *E* can expressed as the sum of the corresponding terms E_x, E_y and E_z. Hence

$$E=E_x+E_y+E_x \tag{200}$$

From Eqs. (199) and (200), we have following three ordinary differential equations, each comprising a single variable.

$$\frac{1}{X}\frac{d^2X}{dx^2}+\frac{2mE_x}{\hbar^2}=0 \tag{201}$$

$$\frac{1}{Y}\frac{d^2Y}{dx^2}+\frac{2mE_y}{\hbar^2}=0 \tag{202}$$

$$\frac{1}{Z}\frac{d^2Z}{dx^2}+\frac{2mE_z}{\hbar^2}=0 \tag{203}$$

Each of these equations has a form similar to the Schrodinger equation $d^2\psi/dx^2+K^2\psi=0$ for a particle in one dimensional box with similar boundary conditions. In other words, the problem of a particle in three dimensional box is reduced to one of three particles in three separate one dimensional boxes of lengths *a*, *b* and *c*.

The boundary condition applicable to the solution is $X(0) = X(a) = 0$ So the eigen values of E_x are given by

$$E_x=\frac{\pi^2\hbar^2}{2ma^2}n_x^2 \tag{204}$$

where $n = 1, 2, 3, \ldots$, and the corresponding normalized eigen functions are given by

$$X(x)=\sqrt{\frac{2}{a}}\sin\frac{n_x\pi x}{a} \tag{205}$$

The solutions for *y* and *z* are of the same form. One obtains

$$Y(y)=\sqrt{\frac{2}{b}}\sin\frac{n_x\pi y}{b} \tag{206}$$

$$E_y=\frac{\pi^2\hbar^2}{2mb^2}n_y^2 \tag{207}$$

and

$$Z(z)=\sqrt{\frac{2}{c}}\sin\frac{n_z\pi z}{c} \tag{208}$$

$$E_z = \frac{\pi^2\hbar^2}{2mc^2} n_z^2 \tag{209}$$

where n_x, n_x and n_x are now the quantum numbers for the three separate boxes.

Eigen Values of Energy

Substituting the expressions for E_x, E_y and E_z in Eq. (200) one obtains

$$E_{n_x,n_y,n_z} = \frac{\pi^2\hbar^2}{2m}\left[\frac{n_x^2}{a^2} + \frac{n_y^2}{b^2} + \frac{n_z^2}{c^2}\right] \tag{210}$$

where $n_x = 1, 2, 3, \ldots, n_y = 1, 2, 3, \ldots$ and $n_z = 1, 2, 3, \ldots$

Eq. (210) gives the eigen values of the energy of the particle. These values are called the energy-levels of the particle. The lowest (ground state) energy E_1 corresponds to $n_x = n_y = n_z = 1$.

Wave Functions

The total normalized wave functions inside the box for the stationary states are given by

$$\psi_{nx,\,ny,\,nz}(x, y, z) = X(x)\, Y(y)\, Z(z)$$

$$= \sqrt{\frac{8}{abc}} \sin\frac{n_x \pi x}{a} \sin\frac{n_y \pi y}{b} \sin\frac{n_z \pi z}{c} \tag{211}$$

where n_x, n_y and n_z are integers. The wave functions are zero outside the box. It may easily be proved that the wave functions is normalized, because

$$= \frac{8}{abc}\int_0^a \left(\sin\frac{n_x\pi x}{a}\right)^2 dx \int_0^b \left(\sin\frac{n_y\pi y}{b}\right) dy \int_0^c \left(\sin\frac{n_z\pi z}{c}\right)^2 dz = 1$$

From Eqs. (210) and (211), one can draw following conclusions:

(*i*) Three integers n_x, n_y and n_x which are called *quantum numbers* are required to describe each stationary state. If we change the sign of the quantum numbers, there is no change in the energy and in the wave function except that the minus sign appears on RHS of Eq. (211). Obviously, all the stationary states are given by the positive integral values of n_x, n_y and n_z. No quantum number can be zero, because if any one of them is taken zero, then $\psi\,(x, y, z) = 0$ which would mean that the particle does not exist in the box.

(*ii*) The lowest possible energy, *i.e.*, the ground state occurs when $n_x = n_y = y_z = 1$ and it depends on the values of a, b and c.

(*iii*) If the particle is confined in a cubical box in which $a = b = c = 1$. (say), the eigen values of energy are given by

$$E_{n_x,n_y,n_z} = \frac{\pi^2\hbar^2}{2mL^2}\left(n_x^2 + n_y^2 + n_z^2\right) \tag{212}$$

In this case the energy of the particle in the ground state is given by

$$E_{111} = \frac{3\pi^2\hbar^2}{2mL^2} \tag{213}$$

No other state will have this energy, and this state has only one wave function. Therefore, the ground state and the energy-level is said to be non-degenerate. We note that E_{111} is three times that for a particle in one dimensional box of length L.

The corresponding wave function is

$$\psi_{111} = \sqrt{\frac{8}{L^2}} \sin\frac{\pi x}{L} \sin\frac{\pi y}{L} \sin\frac{\pi z}{L} \tag{214}$$

(The subscripts in E and ψ denote the quantum numbers)

(*iv*) Degeneracy: An interesting situation arises in a cubical box. In cubical box, the energy depends on the sum of the squares of the quantum numbers. Consequently the particle having the same energy in an excited state will have several different stationary states, or different wave functions. Such states and energy levels are said to be degenerate.

For example, we consider the first excited state when only one of the quantum numbers is 2, while the others are 1 each, *i.e.*

n_x	n_y	n_z
2	1	1
1	2	1
1	1	2

Obviously there are three independent states corresponding to the following three combinations of n_x, n_y and n_z: (2, 1, 1), (1, 2, 1), and (1, 1, 2). Each state has the same energy

$$E_{211} = E_{121} = E_{112} = \frac{6\pi^2\hbar^2}{2mL^2} \tag{215}$$

but a different wave function

$$\psi_{211} = \sqrt{\frac{8}{L^3}} \sin\frac{2\pi x}{L} \sin\frac{\pi y}{L} \sin\frac{\pi z}{L}$$

$$\psi_{121} = \sqrt{\frac{8}{L^3}} \sin\frac{\pi x}{L} \sin\frac{2\pi y}{L} \sin\frac{\pi z}{L} \tag{216}$$

$$\psi_{112} = \sqrt{\frac{8}{L^3}} \sin\frac{\pi x}{L} \sin\frac{\pi y}{L} \sin\frac{2\pi z}{L}$$

Obviously, the first excited state is triply degenerate or threefold degenerate. The first three energy levels of a particle in a cubical box are shown in Fig. 8.25. The degeneracy arises whenever there is an integral relation among *a, b* and *c*. But if $n_x = n_y = n_z$, the state will be non degenerate. A slight distortion of the system may lead to the breakdown of degeneracy.

E		
$E_{222} = \dfrac{12\pi\hbar^2}{2mL^2}$	222	Non-degenerate
$E_{221} = E_{212} = E_{122} = \dfrac{9\pi^2\hbar^2}{2mL^2}$	221 212 122	Triply-degenerate
$E_{211} = E_{121} = E_{112} = \dfrac{6\pi^2\hbar^2}{2mL^2}$	211 121 112	Triply-degenerate
$E_{111} = \dfrac{3\pi^2\hbar^2}{2mL^2}$	111	Non-degenerate

Fig. 8.25

Degree of Degeneracy: The number of independent wave functions for the stationary states of an energy-level is called the degeneracy of the level.

For a particle in a cubical box a number of energy levels with the corresponding quantum numbers and the degree of degeneracy are given in Table 8.1.

Physical Representation of the Wave Function

The function ψ involves three variables x, y and z and therefore, it cannot be represented graphically. A qualitative understanding of the wave function can be achieved, however, from the nature of the expression (211), the value of the function in the ground state ($n_x = n_y = n_z = 1$) is maximum when, $x = a/2$, $y = b/2$, and $z = c/2$ (*i.e.*, at the box centre) and is positive everywhere. In the excited states, the function changes sign and exhibits modes in some particular regions. For example, for the state ψ_{211} the function vanishes at $x = a/2$ irrespective of the values of y and z, *i.e.* along the yz plane which is, therefore, the "nodal plane". Similarly, the xy and xz planes passing through $b/2$ and $c/2$ are nodal planes for the states ψ_{121} and ψ_{112} respectively.

One can arrive at a better picture by considering $|\psi|^2$ function, *i.e.* probability. For the state ψ_{111} in a cubical box, the probability of finding the particle is maximum at the centre of the box and decreases uniformly towards each side. The probability distribution may be shown pictorially as in Fig 8.26. In the next higher state, the probability distribution exhibits two equal halves and the probability is zero along the vertical planes passing through the box centre (nodal planes) as described above.

TABLE 8.1

Energy Levels	*Quantum Numbers*	*Degree of Degeneracy*
$\frac{3\pi^2\hbar^2}{2mL^2}$	(111)	Non-degenerate
$\frac{6\pi^2\hbar^2}{2mL^2}$	(211), (121), (112)	Threefold degenerate
$\frac{9\pi^2\hbar^2}{2mL^2}$	(221), (212), (122)	Threefold degenerate
$\frac{11\pi^2\hbar^2}{2mL^2}$	(311), (131), (113)	Threefold degenerate
$\frac{12\pi^2\hbar^2}{2mL^2}$	(222)	Non-degenerate
$\frac{14\pi^2\hbar^2}{2mL^2}$	(123), (132), (213) (231), (312), (321)	Sixfold degenerate
$\frac{17\pi^2\hbar^2}{2mL^2}$	(322), (232), (223)	Threefold degenerate
$\frac{18\pi^2\hbar^2}{2mL^2}$	(411), (141), (114)	Threefold degenerate
$\frac{19\pi^2\hbar^2}{2mL^2}$	(331), (313), (133)	Threefold degenerate
$\frac{21\pi^2\hbar^2}{2mL^2}$	(421), (412), (241) (214), (124), (142)	Sixfold degenerate

The state (2, 2, 2) has three vertical nodal planes along each side, *i.e.*, *yz* = *xz* and *xy*. It is apparent that the energy level increases the probability distribution and spread out more and more inside the box and in the limiting case tends to corresponds to the classical prediction.

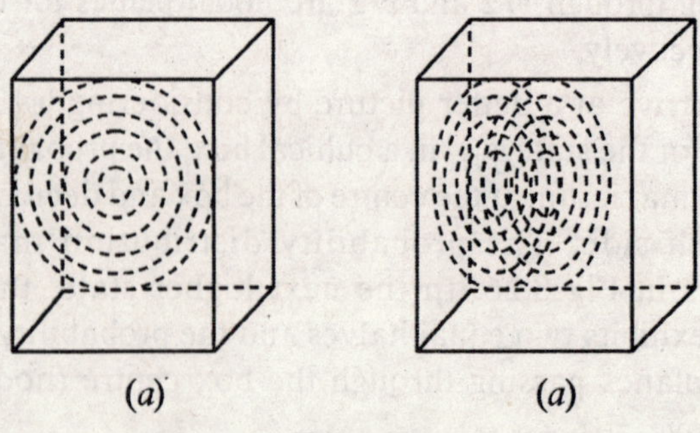

Fig. 8.26

Potential Step (Single Step) Barrier

Let us consider potentials that are constant in time and that are also constant throughout prescribed regions of space. The simplest of these is the step potential shown in Fig. 8.27 in which the one dimensional space is divided into two regions such that $V = 0$ for $x < 0$ and for $V = V_0$ for $x > 0$. When a force field acting on a particle is zero or near zero every where except in a limited region, it is known as step potential. The potential function of a potential step may be represented as

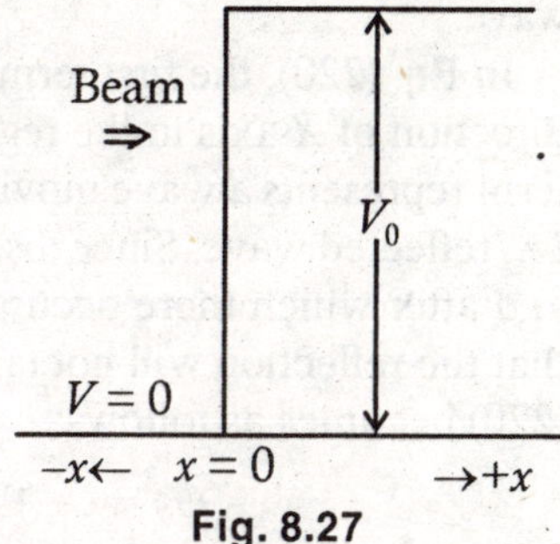

Fig. 8.27

$$V(x) = \begin{cases} 0 \text{ for } x < 0 \\ V_0 \text{ for } x > 0 \end{cases}$$

Suppose the beam of electrons of energy E move from left to right, *i.e.*, along the positive direction of X-axis. Two situations arises, namely $E < V_0$ and $E > V_0$, separately as these lead to different physical situations. We will restrict to one dimension only. We will apply quantum mechanics to this problem according to which electrons behave like a wave when moving from left to right and face a potential hill at $V(x) = V_0$ at $x = 0$.

This problem is analogous to light striking a sheet of glass where the wave is partly reflected and partly transmitted. Here also the electrons at the discontinuity will be partly reflected and partly transmitted. Let us associate a wave function $\psi(x, t)$ with the particle. The wave function has to be obtained by solving the Schrodinger's wave equation.

The Schrodinger wave equation for regions I, where $V(x) = 0$, is

$$\frac{\partial^2\psi}{\partial x^2} - \frac{2m}{\hbar^2}E\psi = 0 \tag{217}$$

The Schrodinger wave equation for the region II is

$$\frac{\partial^2\psi}{\partial x^2} - \frac{2m}{\hbar^2}(E - V_0)\psi = 0 \tag{218}$$

The general solution of Eqs. (217) and (218) will be

$$\psi_1 = Ae^{ip_1x/\hbar} + Be^{-ip_1x/\hbar} \tag{219}$$

$$\psi_2 = Ce^{ip_2x/\hbar} + De^{-ip_2x/\hbar} \tag{220}$$

where ψ_1 and ψ_2 are the wave functions for the regions *I* and *II* respectively, A, B, C and D are constants of integration to be determined by the boundary conditions; p_1 and p_2 are the momenta in the region *I* and *II* respectively and are given by

$$p_1 = \sqrt{2mE}$$

$$p_2 = \sqrt{2m(E - V_0)} \tag{221}$$

In Eq. (219) the first term represents a wave advancing in the positive direction of X-axis in the first region, *i.e.*, incident wave and the second term represents a wave moving in the negative direction of x, *i.e.*, reflected wave.

In Eq. (220), the first term represents a wave advancing in the positive direction of X-axis in the region II, *i.e.*, transmitted wave and the second term represents a wave moving in the negative x-direction in the region II, *i.e.*, reflected wave. Since discontinuity occurs only at $x = 0$ in the region II and after which there occurs no discontinuity in this region. This means that the reflection will not take place in this region, *i.e.*, $D = 0$. Hence Eq. (220) becomes as follows:

$$\psi_2 = Ce^{ip_2x/\hbar} \tag{222}$$

Boundary Conditions

In order for ψ to be finite, E and V must be finite because infinite energies do not occur in nature. Then from Schrodinger's equation, it follows that $\partial^2\psi/\partial x^2$ is not necessarily continuous but is everywhere finite. In order for $\partial^2\psi/dx^2$ to be finite, $\partial\psi/dx$ should be continuous everywhere. By the continuity of $\partial\psi/dx$, we mean

$$\frac{\partial\psi_1}{\partial x} = \frac{\partial\psi_2}{\partial x} \text{ at } x = 0, \textit{ i.e.}$$

$$\left(\frac{\partial\psi_1}{\partial x}\right)_{x=0} = \left(\frac{\partial\psi_2}{\partial x}\right)_{x=0} \tag{222a}$$

This is the first boundary condition. In order for $\partial\psi/dx$ to be continuous, ψ must be continuous. The continuity of ψ implies that $\psi_1 = \psi_2 = 0$ at $x = 0$, *i.e.*

$$(\psi_1)_{x=0} = (\psi_2)_{x=0} \tag{222b}$$

This is the second boundary condition. On applying the boundary condition 222(b) to (219) and (220), one obtains

$$A + B = C \tag{223}$$

On differentiating Eqs. (219) and (222), one obtains

$$\frac{\partial\psi_1}{\partial x} = \frac{ip_1}{\hbar}\left[Ae^{ip_2x/\hbar} - Be^{-ip_2x/\hbar}\right] \tag{224}$$

$$\frac{\partial\psi_1}{\partial x} = \frac{ip_2}{\hbar}Ce^{ip_2x/\hbar} \tag{225}$$

On applying the boundary condition (222a)

$$p_2C = p_1[A + B] \tag{226}$$

Solving Eqs. (223) and (226) for B and C, one obtains

$$C = \frac{2p_1 A}{p_1 + p_2} \tag{227}$$

and $$B = \frac{p_1 - p_2}{p_1 + p_2} = A \tag{228}$$

Obviously, in Eqs. (227) and (228), B anc C denotes the amplitude of reflected and transmitted beams respectively in terms of the amplitude of the incident wave, *i.e.*, A.

At this state, we define *reflectance* (R) and *transmittance* (T) as follows:

Reflectance, Reflectivity or Reflection Coefficient (*R*)

It is defined as the ratio of the magnitude of reflected current to magnitude of indent current, *i.e.*

$$R = \frac{\text{Magnitude of reflected current}}{\text{Magnitude of incident current}} \tag{229}$$

Transmittance (*T*)

At the potential discontinuity the transmittance is defined as the ratio of the magnitude of transmitted current to be magnitude of incident current, *i.e.*

$$T = \frac{\text{Magnitude of reflected current}}{\text{Magnitude of incident current}} \tag{230}$$

Now, we will discuss the problem under following two situations: (*i*) $E > V_0$ and (*ii*) $E < V_0$.

Case I. $E > V_0$

When $E > V_0$, p_2 in Eq. (222) is real. In this case we will now derive the current density in the regions I and II.

In region I, we have

$$\psi_1 = Ae^{ip_1x/\hbar} + Be^{-ip_1x/\hbar} \tag{231}$$

complex conjugate of Eq. (231) will be

$$\psi_1^* = A^* e^{ip_1x/\hbar} + B^* e^{-ip_1x/\hbar} \tag{232}$$

On differentiating Eqs. (231) and (232), with respect to x, one obtains

$$\frac{\partial \psi_1}{\partial x} = \frac{ip_1}{\hbar}\left[Ae^{ip_1x/\hbar} - Be^{-ip_1x/\hbar}\right] \tag{233}$$

$$\frac{\partial \psi_1^*}{\partial x} = -\frac{ip_1}{\hbar}\left[A^* e^{ip_1x/\hbar} - B^* e^{-ip_1x/\hbar}\right] \tag{234}$$

The probability current density is defined as

$$J = \frac{\hbar}{2im}\left[\psi^*\nabla\psi - \psi\nabla\psi^*\right]$$

Applying the above relation to region I, one obtains

$$(J_x)_1 = \frac{\hbar}{2im}\left[\psi_1^*\frac{\partial\psi_1}{\partial x} - \psi_1\frac{\partial\psi_1^*}{\partial x}\right] \tag{235}$$

On substituting Eqs. (231), (232), (233) and (234), one obtains

$$(J_x)_1 = \frac{\hbar}{2im}\left[\left\{\left(Ae^{-ip_1x/\hbar} + B^*e^{ip_1x/\hbar}\right)\times\left(\frac{ip_1}{\hbar}\right)\left(Ae^{ip_1x/\hbar} - Ee^{-ip_1x/\hbar}\right)\right\}\right.$$

$$\left.-\left\{\left(Ae^{ip_1x/\hbar} + Be^{-ip_1x/\hbar}\right)\times\left(\frac{-ip_1}{\hbar}\right)\right\}\left\{\left(A^*e^{-ip_1x/\hbar} - B^{*\ ip_1x/\hbar}\right)\right\}\right]$$

$$= p_1\frac{(AA^* - B^*B)}{m} = \frac{p_1}{m}\left[|A|^2 - |B|^2\right] \tag{236}$$

From Eq. (236), it is evident that the current in the region I is equal to the difference between two terms, of which the first one is proportional to P_1A^2 represents the incident beam travelling from left to right, whereas the second one which is proportional to $p_1|B^2|$ represents the reflected beam travelling from right to left.

∴ The probability current of the incident beam

$$= |A|^2\frac{p_1}{m} \tag{237}$$

and the probability current of the reflected beam

$$= |B|^2\frac{p_1}{m} \tag{238}$$

In region II, we have

$$\psi_2 = Ce^{ip_2x/\hbar} \tag{239a}$$

Its complex conjugate is given by

$$\psi_2^* = Ce^{-ip_2x/\hbar} \tag{239b}$$

On differentiating Eqs. (239*a*) and (239*b*) with respect to x, one obtains

$$\frac{\partial\psi_2}{\partial x} = \frac{ip_2}{\hbar}Ce^{ip_2x/\hbar} \tag{239c}$$

and

$$\frac{\partial\psi_2^*}{\partial\psi} = \frac{-ip_2}{\hbar}C^*e^{-ip_2x/\hbar} \tag{239d}$$

The probability is regiono II is

$$(J_x)_{11} = \frac{\hbar}{2im}\left[\psi_2^*\frac{\partial\psi_2}{\partial x} - \psi_2\frac{\partial\psi_2^*}{\partial x}\right] \tag{239e}$$

On substituting Eqs. (239*a*) to (239*d*) in Eq. (239*e*), one obtains

$$(J_x)_{11} = \frac{\hbar}{2im}\left[\left\{C^* e^{-ip_2x/\hbar}\left(\frac{ip_2}{\hbar}\right)Ce^{ip_2x/\hbar}\right\} - \left\{Ce^{ip_1x/\hbar}\left(\frac{-ip_2}{\hbar}\right)Ce^{-ip_2x/\hbar}\right\}\right]$$

$$= \frac{p_2}{2m}\left[CC^* + CC^*\right] = \frac{p_2}{m}(CC^*) = \frac{|C|^2 p_2}{m} \tag{240}$$

It is obvious from Eq. (240) that there is only transmitted wave in region II, and hence Eq. (240) represents the transmitted current.

Now, we find the expressions for reflectance and transmittance for the case when $E > V_0$ or p_1 is real.

From Eq. (229), we have

$$\text{Reflectance (R)} = \frac{\text{Magnitude of reflected current}}{\text{Magnitude of incident current}}$$

On substituting Eqs. (237) and (238) in the above relation, one obtains

$$R = \frac{|B|^2}{|A|^2}\frac{p_1/m}{p_1/m} \tag{240a}$$

From Eq. (227), one obtains

$$B = \frac{p_1 - p_2}{p_1 + p_2}A$$

or

$$\frac{B}{A} = \frac{p_1 - p_2}{p_1 + p_2}$$

On substituting the above in Eq. (240*a*), one obtains

$$R = \frac{(p_1 - p_2)^2}{(p_1 + p_2)^2} \tag{241}$$

From Eq. (230), one obtains

$$\text{Transmittance } (T) = \frac{\text{Magnitude of transmitted current}}{\text{Magnitude of incident current}}$$

Substituting Eqs. (237) and (240) in the above, one obtains

$$T = \frac{|C|^2}{|A|^2}\frac{p_2/m}{p_1/m}$$

Using Eq. (227), one obtains

$$T = \left(\frac{2p_1}{p_1 + p_2}\right)^2 \frac{p_2}{p_1} = \frac{4p_1p_2}{(p_1 + p_2)^2} \tag{242}$$

From the definitions of reflectance and transmittance it follows that the sum of reflectance and transmittance must be equal to unity. Using Eqs. (241) and (242), one can easily verify it as follows:

$$R + T = \frac{(p_1 - p_2)^2}{(p_1 + p_2)^2} + \frac{4p_1p_2}{(p_1 + p_2)^2} = \frac{(p_1 + p_2)^2}{(p_1 + p_2)^2} = 1 \tag{243}$$

From Eq. (241) it is obvious that reflectance (R) approaches zero as p_2 approaches p_1 and unity as p_2 approaches zero.

But $$p_2 = \sqrt{2m(E - V_0)}$$

and $$p_1 = \sqrt{2mE}$$

From the above expressions it follows that p_2 would approach zero when V_0 is zero, *i.e.*, R would be zero when V_0 becomes equal to zero. This means that some reflection must take place even if $E \geq V_0$. The reflectance would be large only when V_0 is almost comparable in size with E. We must note that it is the wave nature of matter which gives rise to the property of reflection from a sudden change in potential. However, the classical theory does not give rise to this property, when $E > V_0$. Obviously, the property of reflection from a sudden change in potential is purely a quantum mechanical effect.

Case II: $E < V_0$

When $E < V_0$ is imaginary, *i.e.*

$$p_2 = i\sqrt{2m\{(V_0 - E)\}} \tag{244}$$

Its complex conjugate would be

$$p_2^* = i\sqrt{2m(V_0 - E)} = -p_2 \tag{245}$$

Since p_2 is imaginary, it is possible to calculate the probability current associated with wave function ψ_2 in the following way

$$\psi_2 = Ce^{ip_2x/\hbar} \tag{246}$$

and $$\psi_2^* = C^* e^{-ip_2^*x/\hbar} \qquad (\because p_2 \text{ is imaginary}) \tag{247}$$

On differentiating Eqs. (246) and (247), one obtains

$$\frac{\partial\psi_2}{\partial x} = \frac{ip_2}{\hbar} Ce^{ip_2x/\hbar} \tag{248}$$

$$\frac{\partial\psi_2^*}{\partial x} = \frac{ip_2^*}{\hbar} Ce^{ip_2^*x/\hbar} \tag{249}$$

In this case, the probability current is given by

$$J_2 = \frac{\hbar}{2im}\left[\psi_2^* \frac{\partial \psi_2}{\partial x} - \psi_2 \frac{\partial \psi_2^*}{\partial x}\right] \tag{250}$$

or $$J_2 = \frac{\hbar}{2im}\left[C^* e^{-ip_2^* x/\hbar}(ip_2/\hbar) C e^{ip_2 x} - C e^{ip_2 x/\hbar}(-ip_2^*/\hbar) C^* e^{-ip_2^* x/\hbar}\right] \tag{251}$$

But $$p_2^* = p_2$$

$$\therefore \quad J_x = \frac{\hbar}{2im}\left[C^* e^{ip_2 x/\hbar}\left(\frac{ip_2}{\hbar}\right) C e^{ip_2 x/\hbar}\right.$$

$$\left. CC^*(ip_2/\hbar)\, e^{ip_2 x/\hbar}\, e^{ip_2 x/\hbar}\right] = 0 \tag{252}$$

This means that the transmittance current is zero. Thus

$$\text{Transmittance } (T) = \frac{\text{Magnitude of transmitted current}}{\text{Magnitude of incident current}}$$

$$= \frac{0}{\text{Magnitude of incident current}} = 0$$

$$\therefore \quad T = 0 \tag{253}$$

Now,

$$\text{Reflectance } (R) = \frac{\text{Magnitude of reflected current}}{\text{Magnitude of incident current}}$$

$$= \frac{(BB^*)\, p_1/m}{(AA^*)\, p_1/m} \qquad \text{(Using Eq. 236)}$$

$$= \frac{\left(\frac{p_1 - p_2}{p_1 + p_2}\right) A \left(\frac{p_1 - p_2}{p_1 + p_2}\right)^* A^*}{AA^*}$$

$$= \frac{(p_1 - p_2)(p_1 - p_2)}{(p_1 + p_2)(p_1 + p_2)}$$

[By using Eq. (228) and its complex conjugate]

$$R = \left(\frac{p_1 - p_2}{p_1 + p_2}\right)\left(\frac{p_1 - p_2^*}{p_1 + p_2^*}\right) = \frac{(p_1 - p_2)(p_1 + p_2)}{(p_1 + p_2)(p_1 + p_2)} \qquad [\text{Since } p_2^* = -p_2]$$

$$R = 1 \tag{254}$$

From Eqs. (253) and (254), it is evident that the entire wave get reflected when $E < V_0$, *i.e.*, no electrons are transmitted but all are reflected.

The Rectangular Potential Barrier: Penetration of a Barrier: The Tunnel Effect

A rectangular one dimensional potential barrier of height V_0 and width of 'a' for a particle is shown in Fig. 8.28. It extends over the region II from $x = 0$ to $x =$ a in which the potential energy $V(x)$ of the particle is constant and is equal to V_0. On both the sides of the barrier, *i.e.* in region I and region III, $V(x) = 0$. Obviously, particle in regions I and III experiences no force on it, *i.e.*, it behaves as a free particle. Potential field can be represented as

$V(x) = 0$	$x < 0$	Region I
$V(x) = 0$	$0 < x < a$	Region II
$V(x) = 0$	$x > 0$	Region III

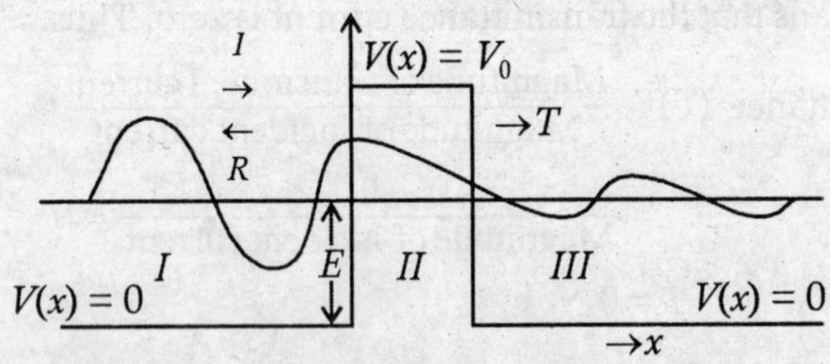

Fig. 8.28

Suppose a beam of particle (each of mass m) travelling parallel to the X-axis from left to right is incident on the potential barrier. In the region I and III, the energy E of a particle is wholly kinetic and in the region II it is partly kinetic and partly potential. If the kinetic energy of the particle $E > V_0$, then according to classical mechanics, the probability of any particle reaching the region III after crossing the region II is zero. However, according to quantum mechanics, the transmissions probability has a small but definite value. This behaviour is called tunnel effect. The phenomenon is known as tunnelling.

The situation shown in Fig. 8.28 represents the encounter between a free particle and a potential barrier. The incident particle from region I can be turned back by the barrier or it can penetrate through it. Obviously, in the region I, the desired wave function has to represent a particle moving to the right (incident particle) and a particle moving to the left (reflected particle). In region III the wave function has to represent only a particle moving to the right, *i.e.*, transmitted particle.

Let $\psi_1(x)$, $\psi_2(x)$ and $\psi_3(x)$ be the wave functions for the motion of particle in region I, II and III respectively. In these regions the time-independent Schrodinger wave equation are as follows:

Region I: $-\infty < x < 0$

In this region, particle is free, $\therefore$ $V = 0$. Schrodinger equation can be written as

$$\frac{d^2\psi_1}{dx^2} + \frac{2mE}{\hbar^2}\psi_1 = 0$$

or
$$\frac{d^2\psi_1}{dx^2} + K^2\psi_1 = 0 \tag{255}$$

where
$$K = \sqrt{\frac{2mE}{\hbar^2}}\,.$$

Region II: $0 \le x \le a$

If $E < V_0$, then Schrodinger equation is

$$\frac{d^2\psi_2}{dx^2} + \frac{2m}{\hbar^2}(E - V_0)\,\psi_2 = 0$$

or
$$\frac{d^2\psi_2}{dx^2} - \frac{2m}{\hbar^2}(V_0 - E)\,\psi_2 = 0 \tag{256}$$

where
$$\beta = \sqrt{\frac{2m(V_0 - E)}{\hbar^2}}$$

Region III: $a \le x \le \infty$

$$\frac{d^2\psi_3}{dx^2} + \frac{2mE}{\hbar^2}\psi_3 = 0$$

or
$$\frac{d^2\psi_3}{dx^2} + K^2\psi_3 = 0 \tag{257}$$

where
$$K = \sqrt{\frac{2mE}{\hbar^2}}\,.$$

The general solutions of Eqs. (255), (256) and (257) are

$$\psi_1 = Ae^{ikx} + Be^{-ikx} \tag{258}$$

$$\psi_2 = Ce^{\beta x} + De^{-\beta x} \tag{259}$$

$$\psi_3 = Ge^{ikx} + He^{-ikx} \tag{260}$$

Since there is no particle coming from left in the region III, we must have $H = 0$. Therefore

$$\psi_3 = Ge^{ikx} \tag{260a}$$

We can interpret the terms in the above equations as follows:

In Eq. (258), the term $A\exp(ikx)$ is a wave of amplitude A travelling in the positive x-direction and the term $B\exp(-ikx)$ is the wave of amplitude B reflected in the negative x-direction from the potential barrier, when the incident wave falls on the barrier. In Eq. (259), the term $D\exp(-\beta x)$ is an exponentially decreasing wave function representing a non-oscillatory disturbance which moves through the barrier in the positive x-direction, and the term $C\exp(\beta x)$ is the reflected disturbance within the barrier, it is an exponentially decreasing wave function. Eq. (260) represents the transmitted wave in the region III. Obviously this wave travels in the positive x-direction.

Expressions for the Coefficients A and B in Terms of the Coefficient G

The boundary conditions require the continuity of $\psi(x)$ and $\partial\psi/\partial x$ so that

(*i*) At $x = 0$, we have

$$\psi_1(0) = \psi_2(0)$$

∴ From Eqs. (258) and (259), we get

$$A + B = C + D \tag{261}$$

We also have

$$\left(\frac{\partial\psi_1}{\partial x}\right)_{x=0} = \left(\frac{\partial\psi_2}{\partial x}\right)_{x=0}$$

$$\therefore \quad Aik - Bik = C\beta - D\beta$$

or

$$(A - B) = \frac{\beta}{ik}(C - D) \tag{262}$$

From Eqs. (261) and (262), one obtains

$$A = \frac{1}{2}\left[\left(1 + \frac{\beta}{ik}\right)C + \left(1 - \frac{\beta}{ik}\right)D\right] \tag{263}$$

and

$$B = \frac{1}{2}\left[\left(1 - \frac{\beta}{ik}\right)C + \left(1 + \frac{\beta}{ik}\right)D\right] \tag{264}$$

(*ii*) At $x = a$, we have

$$\psi_2(a) = \psi_3(a)$$

∴ From Eqs, (259) and (260) one obtains

$$Ce^{\beta a} + De^{-\beta a} = Ge^{ika} \tag{265}$$

We also have

$$\left(\frac{\partial\psi_2}{\partial x}\right)_{x=a} = \left(\frac{\partial\psi_3}{\partial x}\right)_{x=a}$$

$$\therefore \quad C\beta e^{\beta a} - D\beta e^{-\beta a} = Gike^{ika}$$

or $$Ce^{\beta a} - De^{-\beta a} = \frac{ik}{\beta} Ge^{ika} \tag{266}$$

From Eqs. (265) and (266), one obtains

$$C = \frac{1}{2}\left(1 + \frac{ik}{\beta}\right) e^{-\beta a} Ge^{ika} \tag{267}$$

$$D = \frac{1}{2}\left(1 + \frac{ik}{\beta}\right) e^{\beta a} Ge^{ika} \tag{268}$$

Substituting the values of C and D in Eq. (263), one obtains

$$A = \frac{1}{2}\left[\left(1 + \frac{\beta}{ik}\right)\frac{1}{2}\left(1 + \frac{ik}{\beta}\right) e^{-\beta a} + \left(1 + \frac{\beta}{ik}\right)\frac{1}{2}\left(1 - \frac{ik}{\beta}\right) e^{\beta a}\right] Ge^{ika}$$

On simplification, one obtains

$$A = \left[\left(\frac{e^{\beta a} + e^{-\beta a}}{2}\right) - \frac{1}{2}\left(\frac{\beta}{ik} + \frac{ik}{\beta}\right)\left(\frac{e^{\beta a} - e^{-\beta a}}{2}\right)\right] Ge^{ika}$$

Using the following trigonometric relations:

$$\frac{e^{\beta a} + e^{-\beta a}}{2} = \cosh \beta a \text{ and } \frac{e^{\beta a} - e^{-\beta a}}{2} = \sin h\, \beta a,$$

one obtains

$$A = \left[\cosh \beta a - \frac{1}{2}\left(\frac{\beta}{ik} + \frac{ik}{\beta}\right) \sinh \beta a\right] Ge^{ika}$$

or $$A = \left[\cosh \beta a + \frac{i}{2}\left(\frac{\beta}{k} - \frac{k}{\beta}\right) \sinh \beta a\right] Ge^{ika} \tag{269}$$

Similarly, substituting the values of C and D in Eq. (264), one obtains

$$B = \frac{1}{2}\left[\left(1 - \frac{\beta}{ik}\right)\frac{1}{2}\left(1 + \frac{ik}{\beta}\right) e^{-\beta a} + \left(1 + \frac{\beta}{ik}\right)\frac{1}{2}\left(1 - \frac{ik}{\beta}\right) e^{\beta a}\right] Ge^{ika}$$

$$= \frac{1}{2}\left[-\frac{1}{2}\left(\frac{\beta}{ik} - \frac{ik}{\beta}\right) e^{-\beta a} + \frac{1}{2}\left(\frac{\beta}{ik} - \frac{ik}{\beta}\right) e^{\beta a}\right] Ge^{ika}$$

$$= \frac{1}{2}\left[\left(\frac{\beta}{ik} - \frac{ik}{\beta}\right)\left\{\frac{e^{\beta a} - e^{-\beta a}}{2}\right\}\right] Ge^{ika}$$

$$= -\frac{i}{2}\left(\frac{\beta}{k} + \frac{k}{\beta}\right) \sinh \beta a\, Ge^{ika} \tag{270}$$

Transmission Probability

The flux of particles in the incident beam is given by

$$S_1 = \{\text{Probability density of particles in the incident beam}\} \times \{\text{Particle velocity}\}$$

$$= \psi_1(\ell)\,\psi_1(\ell)^* v_1$$

$$= \left(Ae^{ikx}\right).\left(Ae^{ikx}\right)^* v_1 = Ae^{ikx}A^*e^{-ikx}v_1$$

$$= AA^*v_1$$

where A^* is the complex conjugate of A and v_1 is the particle velocity in the incident beam. Similarly the flux of particles in the transmitted beam is given by

$$S_2 = \left(Ge^{ikx}\right)\left(Ge^{ikx}\right)^* v_1 = GG^*v_1$$

The transmission probability T is defined as

$$T = \frac{GG^*v_1}{AA^*v_1} = \left(\frac{G}{A}\right)\left(\frac{G}{A}\right)^* = \left|\frac{G}{A}\right|^2 \tag{271}$$

Here $\left(\frac{G}{A}\right)^*$ is the complex conjugate of $\frac{G}{A}$. The quantity T is also referred as the transmission coefficient. From Eq. (269), one obtains

$$\frac{A}{G} = \left[\cosh\beta a + \frac{i}{2}\left(\frac{\beta}{k} - \frac{k}{\beta}\right)\sinh\beta a\right]e^{ika}$$

and

$$\left(\frac{A}{G}\right)^* = \left[\cosh\beta a - \frac{i}{2}\left(\frac{\beta}{k} - \frac{k}{\beta}\right)\sinh\beta a\right]e^{-ika}$$

Therefore

$$\left(\frac{A}{G}\right)\left(\frac{A}{G}\right)^* = \cosh^2\beta a + \frac{1}{4}\left(\frac{\beta}{k} - \frac{k}{\beta}\right)^2 \sinh^2\beta a$$

or

$$\left|\frac{A}{G}\right|^* = \cosh^2\beta a + \frac{1}{4}\left(\frac{\beta}{k} - \frac{k}{\beta}\right)^2 \sinh^2\beta a$$

Using $\cosh^2\beta a = 1 + \sinh^2\beta a$, one obtains

$$\left|\frac{A}{G}\right|^2 = 1 + \left[\frac{1}{4}\left(\frac{\beta}{k} - \frac{k}{\beta}\right)^2 + 1\right]\sinh^2\beta a$$

$$= 1 + \frac{1}{4}\left[\left(\frac{\beta}{k} - \frac{k}{\beta}\right)^2 + 4\right]\sinh^2\beta a$$

or
$$\frac{1}{T} = 1 + \frac{1}{4}\left(\frac{\beta}{k} - \frac{k}{\beta}\right)^2 \sinh^2 \beta a \tag{272}$$

Using the values of $k = \sqrt{\dfrac{2mE}{\hbar^2}}$ and $\beta = \sqrt{\dfrac{2m(V_0 - E)}{\hbar^2}}$, one obtains

$$\frac{\beta}{k} = \sqrt{\frac{V_0 - E}{E}} \text{ and } \frac{k}{\beta} = \sqrt{\frac{E}{V_0 - E}}$$

Substituting these values in Eq. (272), one obtains on simplification,

$$\frac{1}{T} = 1 + \frac{V_0^2}{4E(V_0 - E)} \sinh^2 \beta a \tag{273}$$

When $E < V_0$, the transmission probability is given by

$$T = \frac{1}{1 + \dfrac{V_0^2}{4E(V_0 - E)} \sinh^2 \beta a} \tag{274}$$

Expression (274) is the expression for transmission probability or the transmission coefficient in the case when $E < V_0$. This expression leads to the following conclusions:

Case I: $E < V_0$

(*i*) Classically, a particle with an energy $E < V_0$ cannot penetrate the potential barrier and it must rebound from the barrier. Expression (274) shows that quantum mechanically there is a finite probability for a particle to tunnel through a potential barrier of height V_0 even when the initial kinetic energy E of the particle is less then V_0.

(*ii*) When $\beta a >> 1$, then $e^{\beta a}$ is large, but $e^{-\beta a} = 1/e^{-\beta a}$ is very small compared to 1. Obviously, one can neglect $e^{-\beta a}$ compared to $e^{\beta a}$.

$$\therefore \quad \sinh^2 \beta a = \left(\frac{e^{\beta a} - e^{-\beta a}}{2}\right) \approx \frac{1}{4} \exp(2\beta a)$$

Then
$$\frac{1}{T} \approx 1 + \frac{V_0^2}{4E(V_0 - E)} \frac{1}{4} (2\beta a)$$

$$\approx \frac{V_0^2 \exp(2\beta a)}{16E(V_0 - E)}$$

$$\therefore \quad T \approx \frac{16E}{V_0^2}(V_0 - E) \frac{1}{4} \exp(2\beta a)$$

$$\approx \frac{16E}{V_0}\left(1 - \frac{E}{V_0}\right) \exp(-2\beta a) \tag{275}$$

Eq. (275) is true only when $E < V_0$.

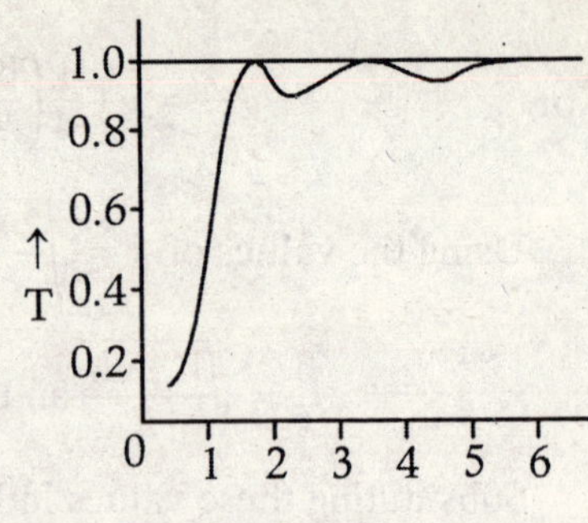

Fig. 8.29

Obviously, when $\beta a >> 1$, T becomes very small and it decreases exponentially with increase in the thickness of the barrier Eq. (275) also shows that if a is constant and if E/V_0 decreases, then T decreases exponentially. This variation is shown by the part AO of the curve T in Fig. 8.29.

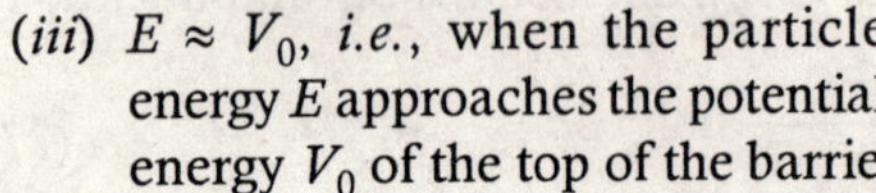

(*iii*) $E \approx V_0$, *i.e.*, when the particle energy E approaches the potential energy V_0 of the top of the barrier, then

$$\beta a = a\sqrt{\frac{2m(V_0 - E)}{\hbar^2}} << 1$$

$$\sinh^2 \beta a \approx a\sqrt{\frac{2m(V_0 - E)}{\hbar^2}}$$

and $$\sinh^2 \beta a \approx \frac{2m(V_0 - E)}{\hbar^2} a^2$$

Hence $$\frac{1}{T} \approx 1 + \frac{V_0^2}{4E(V_0 - E)} \frac{2m(V_0 - E)}{\hbar^2} a^2$$

$$\approx 1 + \frac{2m V_0^2 a^2}{4E \hbar^2}$$

Since $E \to V_0$, one obtains

$$\frac{1}{T} \approx 1 + \frac{mV_0 a^2}{2\hbar^2}$$

or $$T = \frac{1}{\frac{mV_0 a^2}{2\hbar^2}} \tag{276}$$

Obviously $T < 1$. Thus when E approaches V_0, T increases and when $E \approx V_0$ the transmission probability is given by Eq. (276) and it is less then 1, *i.e.*, $T < 1$.

Case II: $E > V_0$

When $E > V_0$ then $\beta = \sqrt{\frac{2m(V_0 - E)}{\hbar^2}} = i\sqrt{\frac{2m(E - V_0)}{\hbar^2}} = i\alpha$

where $$\alpha = \sqrt{\frac{2m(E - V_0)}{\hbar^2}}$$

$$\sinh\beta a = \frac{e^{\beta a} - e^{-\beta a}}{2} = \frac{e^{i\alpha a} - e^{-i\alpha a}}{2}$$

$$\beta = i\left(\frac{e^{i\alpha a} - e^{-i\alpha a}}{2i}\right) = i \sin\alpha\, a.$$

sparing both the side, one obtains

$$\sinh^2 a = -\sin^2 \alpha a.$$

Substituting this value in Eq. (273), one obtains

$$\frac{1}{T} = 1 - \frac{V^2}{4E(V_0 - E)} \sin^2 \alpha\, a$$

$$= 1 + \frac{V_0^2}{4E(E - V_0)} \sin^2 \alpha\, a \qquad (277)$$

$$\therefore \quad T = \frac{1}{1 + \dfrac{V^2}{4E(E - V_0)} \sin^2 \alpha\, a} \qquad (278)$$

The above expression for the transmission probability or the transmission coefficient in the case when $E > V_0$ leads to the following conclusions:

(*i*) When E approaches V_0, then

$$\alpha\, a = a\sqrt{\frac{2m(E - V_0)}{\hbar^2}} \text{ will be } << 1$$

$$\therefore \quad \sin\alpha\, a \approx a\sqrt{\frac{2m(E - V_0)}{\hbar^2}}$$

and $$\sin^2 \alpha\, a = \frac{2m(E - V_0)a^2}{\hbar^2}$$

Hence $$\frac{1}{T} \approx 1 + \frac{V_0^2}{4E(E - V_0)} \frac{2m(E - V_0)a^2}{\hbar^2}$$

$$\approx 1 + \frac{2m V_0^2 a^2}{4E\hbar^2}$$

Since $E \to V_0$, one obtains

$$\frac{1}{T} \approx 1 + \frac{m V_0^2 a^2}{2\hbar^2}$$

or $$T \approx \frac{1}{1 + \dfrac{m V_0^2 a^2}{2\hbar^2}}$$

We note that the above expression is the same as that obtained in the previous case for $E < V_0$.

(*iii*) Eq. (278) shows that for $E < V_0$, $T = 1$ when

$$\alpha a = n\pi, \text{ where } n = 1, 2, 3, \ldots$$

or
$$a = \frac{n\pi}{\alpha} = \frac{n\pi}{\sqrt{2m(E - V_0)/\hbar^2}} = \frac{n\pi\hbar}{\sqrt{2m(E - V_0)}} \tag{279}$$

or
$$a = \frac{nh}{2\sqrt{2m(E - V_0)}}$$

But $h/2\sqrt{2m(E - V_0)}$ is the de Broglie wavelength λ of the particle in the region of kinetic energy $(E - V_0)$. Hence $T = 1$, when

$$a = n(\lambda/2) \tag{280}$$

This oscillation is the analog of such phenomenon as the optical transmission through a non-reflecting thin film or the use of matching stubs in wave guides, or the Ramsauer-Townsend effect in the scattering of low energy electrons from noble gas atoms.

This shows that if the width of the barrier is an integral multiple of half wavelength, there is perfect transmission of the incident beam of the particles.

Eq. (278) also shows that as E increases T oscillates between the maximum value I and a value less then 1 as shown in Fig. 8.29.

Reflection Probability

The reflection probability is given by

$$R = \left(\frac{B}{A}\right)\left(\frac{B}{A}\right)^* = \left|\frac{B}{A}\right|^2 \tag{281}$$

where $(B/A)^*$ is the complex conjugate of (B/A). The quantity R is also termed as the reflection coefficient. Dividing Eq. (270) by Eq. (269), one obtains

$$\frac{B}{A} = \frac{-\frac{i}{2}\left(\frac{\beta}{k} + \frac{k}{\beta}\right)\sinh \beta a}{\cosh \beta a + \frac{i}{2}\left(\frac{\beta}{k} + \frac{k}{\beta}\right)\sinh \beta a}$$

and
$$\left(\frac{B}{A}\right)^* = \frac{-\frac{i}{2}\left(\frac{\beta}{k} + \frac{k}{\beta}\right)\sinh \beta a}{\cosh \beta a - \frac{i}{2}\left(\frac{\beta}{k} - \frac{k}{\beta}\right)\sinh \beta a}$$

$$\therefore \quad R = \left(\frac{B}{A}\right)\left(\frac{B}{A}\right)^*$$

$$= \frac{-\frac{1}{4}\left(\frac{\beta}{k}+\frac{k}{\beta}\right)^2 \sinh^2 \beta a}{\cosh^2 \beta a + \frac{1}{4}\left(\frac{\beta}{k}-\frac{k}{\beta}\right)^2 \sinh^2 \beta a}$$

$$= \frac{-\frac{1}{4}\left(\frac{\beta}{k}+\frac{k}{\beta}\right)^2 \sinh^2 \beta a}{1 + \frac{1}{4}\left(\frac{\beta}{k}-\frac{k}{\beta}\right)^2 \sinh^2 \beta a} \tag{282}$$

The above is the expression for the reflection coefficient. From Eq. (272), we have the expression for the transmission coefficient

$$T = \frac{1}{1 + \frac{1}{4}\left(\frac{\beta}{k}-\frac{k}{\beta}\right)^2 \sinh^2 \beta a} \tag{283}$$

Adding Eqs. (282) and (283), one obtains

$$R + T = 1 \tag{284}$$

The quantum mechanical tunnel effect or the transmission coefficient depends, as is clear from Eq. (276), on the width a and height V_0 of the potential barrier in addition to the mass of the particle. The tunnel effect has a great significance in physics and chemistry. It provides explanations for the following phenomena:

(*i*) the emission of α particles from a radioactive nucleus, *e.g.*

$$Ra \rightarrow Rn + \alpha$$

(*ii*) the field emission of electrons from a cold metallic surface,

(*iii*) the electrical breakdown of insulators,

(*iv*) the reverse breakdown of semiconductor diodes, and

(*v*) the switching action of a tunnel diode.

ILLUSTRATIVE EXAMPLES

Example 1

Calculate the value of the retarding potential needed to stop the photoelectrons ejected from a metal surface of work function 1.2 eV with light of frequency 5.5×10^{14} sec^{-1}.

Solution

From Einstein photoelectric equation, we have

$$eV_0 = h\nu - W$$

Here $e = 1.6 \times 10^{-19}\,\text{C}, \nu = 5.5 \times 10^{14}\,\text{sec}^{-1}$

$$W = 1.2\text{ eV} = 1.2 \times 1.6 \times 10^{-19}\text{ J} = 1.92 \times 10^{-19}\text{ J}$$

$$h = 6.6 \times 10^{-34}\text{ Js}$$

$$\therefore \quad 1.6 \times 10^{-19} V_0 = 6.6 \times 10^{-34} \times 5.5 \times 10^{14} - 1.92 \times 10^{-19}$$

$$= 1.71 \times 10^{-19}$$

$$V_0 = \frac{1.71 \times 10^{-19}}{1.6 \times 10^{-19}} = 1.06\text{ Volt}$$

Example 2

In an experiment tungsten cathode which as a threshold 2300 Å is irradiated by ultraviolet light of wavelength 1800 Å, calculate (*i*) the maximum energy of emitted photoelectrons and (*ii*) the work function for tungsten. Give your results in electron-volts. Given $1\text{ eV} = 1.6 \times 10^{-19}$ J and velocity of light $c = 3 \times 10^8$ m/sec.

Solution

We have from Einstein's photoelectric equation

$$\frac{1}{2} m v_{max}^2 = h\nu - hc = h\left(\frac{c}{\lambda} - \frac{c}{\lambda_0}\right)$$

Here $\lambda_0 \rightarrow$ is the threshold wavelength

$$\therefore \quad \frac{1}{2} m v_{max}^2 = hc\left(\frac{\lambda_0 - \lambda}{\lambda\lambda_0}\right)$$

$$\therefore \quad \frac{1}{2} m v_{max}^2 = 6.6 \times 10^{-34} \times 3 \times 10^8 \times \left[\frac{2300 \times 10^{-10} - 1800 \times 10^{-10}}{2300 \times 10^{-10} \times 1800 \times 10^{-10}}\right]$$

$$= \frac{55 \times 10^{-19}}{23 \times 1.6 \times 10^{-19}} = 1.485\text{ eV}$$

$$\therefore \text{ Work function } W = h\nu_0 = \frac{hc}{\lambda_0} = \frac{6.6 \times 10^{-34} \times 3 \times 10^8}{2300 \times 10^{-10}}$$

or

$$W = 8.608 \times 10^{19} = \frac{8.608 \times 10^{-19}}{1.6 \times 10^{-19}}\text{ eV} = 5.38\text{ eV}$$

Example 3

Find the maximum wavelength of light that can liberate electron from potassium. Work function of potassium is 2.24 eV.

Solution

$$\lambda_0 = \frac{hc}{W}$$

where $$W = 2.24 \text{ eV} = 2.24 \times 1.6 \times 10^{-19} \text{ J}$$

$$\therefore \quad \lambda_0 = \frac{6.6 \times 10^{-34} \times 3 \times 10^{8}}{2.24 \times 6 \times 10^{-19}} \text{ m}$$

$$= 5524 \times 10^{-10} \text{ m} = 5524 \text{ Å}.$$

Example 4

A metallic surface, when illuminated with light of wavelength λ_1, emits electrons with energies upto a maximum value E_1, and when illuminated with light of wavelength λ_2, where $\lambda_2 < \lambda_1$, it emits electrons with energies upto a maximum value E_2. Prove that Planck's constant h and the work function ϕ of the metal are given by

$$h = \frac{(E_2 - E_1)}{C(\lambda_1 - \lambda_2)} \quad \text{and} \quad \phi = \frac{E_2\lambda_2 - E_1\lambda_1}{\lambda_1 - \lambda_2}$$

Solution

The maximum kinetic energy E_1 is given by

$$E_1 = h\nu_1 - \phi$$

$$E_1 = \frac{hc}{\lambda_1} - \phi \qquad (i)$$

Similarly, the maximum kinetic energy E_2 is given by

$$E_2 = \frac{hc}{\lambda_2} - \phi \qquad (ii)$$

Subtracting (*i*) from (*ii*), one obtains

$$E_2 - E_1 = hc\left(\frac{1}{\lambda_2} - \frac{1}{\lambda_1}\right) = hc\left(\frac{\lambda_1 - \lambda_2}{\lambda_1\lambda_2}\right)$$

$$\therefore \quad h = \frac{(E_2 - E_1)\lambda_1\lambda_2}{c(\lambda_1 - \lambda_2)} \qquad (iii)$$

To obtain the expression for work function ϕ, one obtains from (*i*) and (*ii*)

$$E_1\lambda_1 = hc - \phi\lambda_1 \qquad (iv)$$

$$E_2\lambda_2 = hc - \phi\lambda_2 \qquad (v)$$

Subtracting (*iv*) from (*v*), one obtains

$$E_2\lambda_2 = E_1\lambda_1 = \phi(\lambda_1 - \lambda_2)$$

$$\therefore \quad \phi = \frac{E_2\lambda_2 - E_1\lambda_1}{\lambda_1 - \lambda_2} \qquad (vi)$$

Example 5

A photon of wavelength 3 Å suffers Compton scattering by a free electron originally at rest. If the angle of scattering is 90°. What is the kinetic energy of the recoil electron ?

Solution

$$\Delta\lambda = \frac{\lambda}{m_0 c}(1 - \cos\phi)$$

$$= 0.0242(1 - 0) = 0.0242 \text{ Å}$$

∴ Compton wavelength $\lambda_0 = \dfrac{\lambda}{m_0 c} = 0.0242$ Å

By the law of conservation of energy

$$h\nu + m_0 c^2 = h\nu + mc^2$$

But $\quad mc^2 = m_0 c^2 + \dfrac{1}{2} m_o v^2 \quad$ (for $v = c$)

K.E. of recoil electron = $h(\nu - \nu') = 1/2 m_0 v^2$

$$= hc\left(\frac{\lambda' - \lambda}{\lambda\lambda'}\right) \approx \frac{hc\Delta\lambda}{\lambda(\lambda + \Delta\lambda)}$$

Substituting the proper values, one obtains

$$K.E. = \frac{6.6 \times 10^{34} \times 3 \times 10^{8} \times 0.0242 \times 10^{-10}}{3 \times 10^{-10} \times 3.0242 \times 10^{-19}} \text{ J}$$

$$= \frac{6.6 \times 3 \times 0.0242 \times 10^{-16}}{3 \times 3.0242 \times 1.6 \times 10^{-19}} = 33.28 \text{ eV}.$$

Example 6

In an experiment of Compton scattering the incident radiation has wavelength 2 Å, while wavelength of radiation scattered through 180° is 2.048 Å. Calculate (*i*) the wavelength of scattered radiation if they are viewed at an angle of 60° to the direction of incidence, (*ii*) the energy of the recoil electron which scatter radiation through 60°.

Solution

We have $\quad \lambda' - \lambda = \dfrac{h}{m_0 c}(1 - \cos\phi)$

when $\quad \phi = 180°$

∴ $\quad \cos\phi = \cos 180° = -1$

$$\lambda = 2 \text{ Å} = 2 \times 10^{-10} \text{ m}$$

$$\lambda' = 2.048 \text{ Å} = 2.048 \times 10^{-10} \text{ m}$$

$$\therefore \quad (2.048 - 2) \times 10^{-10} = \frac{h}{m_0 c}(1 + 1)$$

or
$$\frac{h}{m_0 c} = \frac{0.048 \times 10^{-10}}{2}$$

$$= 0.024 \times 10^{-10} \text{ m} = 0.024 \text{ Å}$$

(*i*)
$$\phi = 60°, \cos\phi = \cos 60° = 1/2$$

$$\therefore \quad \lambda' = \lambda + \frac{h}{m_0 c}(1 - \cos 60°)$$

$$= 2 \times 10^{-10} + 0.024 \times 10^{-10} \times \left(1 - \frac{1}{2}\right)$$

$$= 2 \times 10^{-10} + 0.012 \times 10^{-10} \text{ m}$$

$$= 2.012 \text{ Å}$$

(*ii*) Energy of recoil electron which scatter photon through 60° is

$$(\text{K.E.})_{\text{recoil}} = (m - m_0)c^2 = h(\nu - \nu') = hc\left(\frac{1}{\lambda} - \frac{1}{\lambda'}\right)$$

$$= \frac{hc(\lambda' - \lambda)}{\lambda\lambda'}$$

$$= \frac{6.6 \times 10^{-34} \times 3 \times 10^{8} \times (2.012 - 2) \times 10^{-10}}{2 \times 10^{-10} \times 2.012 \times 10^{-10}}$$

$$= 5.91 \times 10^{-18} \text{ J.}$$

Example 7

Calculate the maximum percentage change in wavelength due to Compton scattering for incident photons of wavelengths 1 Å and 10 Å. What inference do you draw from the result?

Solution

The wavelength shift by Compton scattering is given by

$$\Delta\lambda = \frac{h}{m_0 c}(1 - \cos\phi)$$

For maximum value of $\Delta\lambda$, $\phi = \pi$

$$\therefore \quad (\Delta\lambda)_{\max} = \frac{h}{m_0 c}(1 + 1) = 2 \times 0.0242 = 0.0484 \text{ Å}$$

Maximum percentage change in wavelength

$$= \frac{(\Delta\lambda)_{\max}}{\lambda} \times 100 = \frac{0.0484}{1} \times 100 \quad (\text{for } \lambda = 1 \text{ Å})$$

$$= 4.8\%$$

And for $\lambda = 10$ Å, maximum change in wavelength

$$= \frac{0.0484}{10} \times 100 = 0.48\%$$

Obviously, $(\Delta\lambda)_{max}/\lambda$ is very small for larger wavelengths ~10 Å and hence it is very difficult to detect this change in wavelength, *i.e.*, Compton effect can be observed only for photons whose wavelength is sufficiently small.

Example 8

Show that

(*a*) the maximum recoil energy of a free electron of rest mass m_0, when struck by a photon of frequency ν is given by

$$K_{max} = \frac{(h\nu)^2}{h\nu + (1/2)m_0c^2}$$

(*b*) If λ is the wavelength of the photon and λ_c the Compton wavelength of the electron, then show that

$$K_{max} = \frac{2m_oc^2\lambda^2}{\lambda^2 + 2\lambda_c\lambda}$$

Solution

(*a*) The kinetic energy of the recoil electron

$$K = h\nu \frac{\frac{h\nu}{m_0c^2}(1 - \cos\phi)}{1 + \frac{h\nu}{m_0c^2}(1 - \cos\phi)}$$

$$= \frac{(h\nu^2)\, 2\sin^2\phi/2}{m_0c^2 + h\nu 2\sin^2\phi/2}$$

$$= \frac{2(h\nu^2)}{2h\nu + m_0c^2 \operatorname{cosec}^2\phi/2}$$

K will be maximum, when $\operatorname{cosec}^2\phi/2$ is minimum. The minimum value of $\operatorname{cosec}^2\phi/2$ is 1 and this value occurs when $\phi/2 = 90°$, *i.e.* when $\phi = 180°$

$\therefore$ The maximum value of K is given by

$$K_{max} = \frac{2(h\nu)^2}{2h\nu + m_0c^2} = \frac{(h\nu)^2}{h\nu + \frac{1}{2}m_0c^2} \qquad (i)$$

(*b*) From (*i*), one obtains

$$K_{\max} = \frac{2(m_0c)^2(h/m_0c)^2\nu^2}{2(m_0c)(h/m_0c)\nu + m_0c^2}$$

Substituting $\frac{h}{m_0c} = \lambda_0$ and $\nu = \frac{c}{\lambda}$ and simplifying, one obtains

$$K_{\max} = \frac{2(m_0c)^2\lambda_c^2\frac{c^2}{\lambda^2}}{2\lambda_c\frac{c}{\lambda + c}}$$

$$= \frac{2m_0\lambda_c^2c^2}{2\lambda\lambda_c + \lambda^2} = \frac{2m_0c^2\lambda^2}{\lambda^2 + 2\lambda_0\lambda}.$$

Example 9

Show that the de Broglie wavelength associated with an electron of energy V electron volts is approximately $\frac{12.27}{\sqrt{V}}$ Å.

Solution

We have $$\lambda = \frac{h}{\sqrt{2meV}}$$

Substituting the values of h, m and e, one obtains

$$\lambda = \frac{6.6 \times 10^{-34}}{\sqrt{2 \times 9.1 \times 10^{-34} \times 1.6 \times 10^{-19} V}}$$

$$= \frac{12.27 \times 10^{-10}}{\sqrt{V}} = \frac{12.27}{\sqrt{V}} \text{ Å}.$$

Example 10

A particle of rest mass m_0 has a kinetic energy K. Show that its de Broglie wavelength is given by

$$\lambda = \frac{hc}{\sqrt{[K(K + 2m_0c^2)]}}$$

Hence calculate the wavelength of an electron of kinetic energy 1 MeV. What will be the value of λ if $K << m_0c^2$?

Solution

The de Broglie wavelength associated with a particle moving with velocity v is given by

$$\lambda = \frac{h}{mv} \qquad (i)$$

where m is relativistic mass of the particle given by

$$m = \frac{m_0}{\sqrt{1 - v^2 / c^2}} \qquad (ii)$$

$$\therefore \qquad 1 - \frac{v^2}{c^2} = \frac{m_0^2}{m^2}$$

or
$$\frac{v^2}{c^2} = 1 - \frac{m_0^2}{m^2} = \frac{(m^2 - m_0^2)}{m^2}$$

or
$$mv = c\sqrt{m^2 - m_0^2} \qquad (iii)$$

From (*i*) and (*iii*), we obtain

$$\lambda = \frac{h}{c\sqrt{m^2 - m_0^2}} = \frac{hc}{c^2\sqrt{m^2 - m_0^2}} \qquad (iv)$$

But
$$c^2\sqrt{m^2 - m_0^2} = \sqrt{c^4(m - m_0)(m + m_0)}$$

$$= \sqrt{\{(m - m_0)c^2\}\,\{c^4(m - m_0)\,c^2 + 2m_0c^2)\}}$$

$$= \sqrt{K(K + 2m_0c^2)}$$

$$\because \qquad K = (m - m_0)c^2$$

$$\therefore \qquad \lambda = \frac{hc}{\sqrt{K(K + 2m_0c^2)}}$$

Given $m_0c^2 = 0.5$ MeV, so that when $K = 1$ MeV, we have

$$\therefore \qquad \lambda = \frac{6.62 \times 10^{-34} \times 3 \times 10^8}{1.414 \times 1.6 \times 10^{-13}}$$

$$= 8.78 \times 10^{-13} \text{ m} = 8.78 \times 10^{-3} \text{ Å}$$

If $K << m_0c^2$, one can replace $K + m_0c^2$ by m_0c^2 and hence one obtains

$$\lambda = \frac{hc}{\sqrt{2m_0Kc^2}} = \frac{h}{\sqrt{2m_0K}}.$$

Example 11

Show that the de Broglie wavelength for a material particle of rest mass m_0 and charge q, accelerated from rest through a potential difference of V volts relativistically is given by

$$\lambda = \frac{h}{\sqrt{\left[2m_0qV\left(1 + \frac{qV}{2m_0c^2}\right)\right]}}$$

Solution

In the given case the kinetic energy $T = qV$ as in non-relativistic case but $T \neq \frac{1}{2}mv^2$. Obviously, one cannot find momentum directly from T. Instead, one has to use the relativistic formula.

$$E^2 = p^2c^2 + m_0^2c^4$$

$$E = T + m_0c^2 = qV + m_0c^2$$

where m_0 = rest mass of electron, and T = kinetic energy

$$\therefore \quad p^2c^2 = E^2 - m_0^2c^4 = \left(qV + m_0c^2\right)^2 - m_0^2c^4$$

$$= q^2V^2 + 2m_0c^2qV$$

$$\because \quad p^2 = 2m_0qV\left(1 + \frac{qV}{2m_0c^2}\right)$$

or

$$p = \sqrt{\left[2m_0qV\left(1 + \frac{qV}{2m_0c^2}\right)\right]}$$

$\therefore$ de Broglie wavelength will be

$$\lambda = \frac{h}{p} = \frac{h}{\sqrt{\left[2m_0qV\left(1 + \frac{qV}{2m_0c^2}\right)\right]}}.$$

Example 12

Assume the uncertainty in position of a hydrogen molecule of mass about 2×10^{-27} kg is of the order of the diameter about 10^{-10} m. Determine the uncertainty in momentum.

Solution

The uncertainty in moment is given by

$$\Delta p_x \geq \frac{\hbar}{\Delta x} \geq \frac{1.054 \times 10^{-34}}{10^{-10}} \geq 1.054 \times 10^{-24} \text{ kg-m/sec.}$$

Example 13

The average period that elapes between the excitation of an atom and the time it emits radiation is 10^{-8} sec. Find the uncertainty in the energy emitted and the uncertainty in the frequency of light emitted. Given $\hbar = 1.054 \times 10^{-34}$ J-s and $h = 6.63 \times 10^{-34}$ J-s.

Solution

The uncertainty in the energy is given by

$$\Delta E = \frac{\hbar}{\Delta t} = \frac{1.054 \times 10^{-34}}{10^{-8}} = 1.054 \times 10^{-26} \text{ J}$$

The uncertainty in the frequency of light emitted is given by

$$\Delta \nu = \frac{\Delta E}{h} = \frac{1.054 \times 10^{-26}}{6.63 \times 10^{-34}} = 1.59 \times 10^{7} \text{ Hz}.$$

Example 14

A nucleon is confined to a nucleus of radius 5×10^{-15} m. Calculate the minimum uncertainty in the momentum of the nucleon. Also calculate the minimum kinetic energy of the nucleon. Given, mass of nucleon, 5×10^{-15} m and Planck's constant $h = 6.6 \times 10^{-34}$ J-s.

Solution

We have $(\Delta p)_{min} (\Delta x)_{max} \approx h$

Given $(\Delta x)_{max} = 2 \times 5 \times 10^{-15} \text{ m} = 10^{-14} \text{ m}$

$$(\Delta p)_{min} = \frac{h}{(\Delta x)_{max}} = \frac{6.6 \times 10^{-34}}{10^{-14}}$$

$$= 6.6 \times 10^{-20} \text{ kg-m/sec}$$

Since p can not be less than $(\Delta p)_{min}$, so we have

$$p_{min} = (\Delta p)_{min}$$

$$E_{min} = \frac{p_{min}^2}{2m} = \frac{(6.6 \times 10^{-20})^2}{2 \times 1.67 \times 10^{-27}} = 1.32 \times 10^{-12} \text{ J}$$

$$= 8.25 \times 10^{6} \text{ eV} = 8.25 \text{ MeV}.$$

Example 15

Find the lowest energy of an electron confined to move in one-dimensional potential box of length 1 Å.

Solution

Given $E_n = \dfrac{n^2 \pi^2 \hbar^2}{2ma^2} \quad \therefore \quad E_1 = \dfrac{1^2 \pi^2 \hbar^2}{2ma^2}$

For $E_1 = \dfrac{\pi^2 \times (1.054 \times 10^{-34})^2}{2 \times 9.1 \times 10^{-31} \times (10^{-10})^2}$

$$\frac{(1.054)^2 \times (3.14)^2 \times 10^{-17}}{2 \times 9.1} = 6 \times 10^{-18} \text{ J}$$

$$\frac{6 \times 10^{-18}}{1.6 \times 10^{-19}} \text{ eV} = 35.5 \text{ eV}.$$

Example 16

The size of the nucleus is 10^{-14} m. Treating it as a one dimensional box show why electron does not exist inside the nucleus.

Solution

Zero point energy $E_1 = \dfrac{\pi^2\hbar^2}{2ma^2}$

$$= 5.44 \times 10^{-10} \text{ J}$$

$$(\hbar = 1.054 \times 10^{-34} \text{ Js}, m = 9.1 \times 10^{-31} \text{ kg}, a = 10^{-14} \text{ m})$$

The Coulombic attraction energy is of the other hand is

$$V = \frac{e^2}{r} = -2.3 \times 10^{-14} \text{ J}$$

The Coulombic attraction energy is too weak for the nucleus to hold an electron.

Example 17

Calculate the length of a one dimensional box for which the difference between the lowest energy levels of a molecule becomes comparable to its average kinetic energy at a given temperature.

Solution

$$E_{n+1} - E_n = (2n+1)\frac{\pi^2\hbar^2}{2ma^2} = \frac{1}{2}k_B T$$

where k_B is Boltzmann's constant

$$\therefore \qquad a = \left[\frac{(2n+1)\pi^2\hbar^2}{mk_B / T}\right].$$

Example 18

For a particle in the states $n = 1, 2$ and 3 of a one dimensional box of length L, find the probability that the particle is in the region $0 < x < a/4$.

Solution

Probability $\qquad P = \dfrac{2}{a}\int_0^{1/4} \sin\dfrac{2n\pi x}{a}\,dx \qquad$ (using $\sin^2 x = 1 - \cos 2x/2$)

$$= \frac{1}{4} - \frac{\sin n\pi / 2}{2n\pi}$$

For $n = 1,\ P = \frac{1}{4} - \frac{1}{2}\pi = 0.0908$

For $n = 2,\ P = 0.25$ and for $n = 3,\ P = 0.3031$

Example 19

A particle of mass $m = 10^{-29}$ kg is confined to move in a box of length 2 Å whose ends are at $x = 0$ and $x = 2$ Å. What is the probability of finding the particle between 1.6000 Å and 1.6001 Å. Calculate for $n = 1$ and $n = 2$.

Solution

$$P = \frac{2}{a}\int_{1.6}^{1.6001} \sin\frac{2n\pi x}{a}\, dx$$

$$= \frac{1}{a}\left[x - \left\{a\left(\sin\frac{2n\pi x}{a}\right)\Big/ 2n\pi\right\}\right]_{1.6000}^{1.6001}$$

For $n = 1$,

$$P = \frac{1}{2.0}\left[0.0001 - \frac{2 \times \sin(2\pi \times 1.6001/2) - \sin(2\pi \times 1.6001/2)}{2\pi}\right]$$

$$= 3.45 \times 10^{-5}$$

Similarly, for $n = 2$, one obtains $P = 9.05 \times 10^{-5}$.

Example 20

(*a*) Calculate the lowest energy of an electron confined in a cubical box of each side 1 Å.

(*b*) Find the temperature of which the average energy of the molecules of a perfect gas would be equal to the lowest energy of the electron, $k_B = 1.38 \times 10^{-23}$ J/K.

Solution

(*a*) The possible energies of a particle in a cubical box of each side *a* are given by

$$E = \frac{3\pi^2\hbar^2}{2ma^2}\left(n_x^2 + n_y^2 + n_z^2\right)$$

For the lowest energy

$$n_x = n_y = n_z = 1$$

$$\therefore \quad E_1 = \frac{3\pi^2\hbar^2}{2ma^2} = \frac{3\pi^2 \times (1.054 \times 10^{-34})^2}{2 \times 9.11 \times 10^{-31} \times (10^{-10})^2}$$

$$= \frac{3 \times (1.054)^2 \times (3.14)^2 \times 10^{-17}}{2 \times 9.11} = 1.803 \times 10^{-17} \text{ J}$$

$$= 108 \text{ eV}$$

(*b*) Let T K be the required temperature, then

$$\frac{3}{2} k_B T = \frac{3\pi^2\hbar^2}{2ma^2} = 18.03 \times 10^{-18} \text{ J}$$

$$\therefore \quad T = \frac{2}{3} \times \frac{18.03 \times 10^{-18}}{1.38 \times 10^{-23}} = 8.71 \times 10^5 \text{ K.}$$

Note: The above model is not acceptable for H atom as the experimental value of ground state energy E_{111} is –13.6 eV.

Example 21

A beam of mono-energetic electrons strikes the surface of a metal at normal incidence. Calculate the reflection probability of these electrons if $E = 0.1$ eV and $V_0 = 8$ eV.

Solution

Since the electrons with energy 0.1 eV encounters a potential drop –8 eV at the metal surface, they will all enter the metal according to classical mechanics, and because of the law of conservation of energy they will acquire a final momentum $p_1 = \sqrt{2m(E - V_0)} / \hbar$ after doing so. From potential step expressions, we have

$$R = \frac{(p_2 - p_1)^2}{(p_2 + p_1)^2} \text{ and } T = \frac{4p_1p_2}{(p_2 + p_1)^2}$$

Thus for $E = 0.1$ eV, $V_0 = 8$ eV

$$R = \left[\frac{\sqrt{E} - \sqrt{(E - V_0)}}{\sqrt{E} + \sqrt{(E - V_0)}}\right]^2 = \left[\frac{1 - \sqrt{(1 - V_0 / E)}}{1 + \sqrt{(1 - + V_0 / E)}}\right]^2$$

$$= \left(\frac{8}{10}\right)^2 = 0.64.$$

Example 22

Find the width of the potential barrier for an α-particle emitted with kinetic energy 5.5 MeV from $^{222}_{86}\text{Rn}$.

Solution

The width of the potential barrier due to the nucleus is given by $a = r_1 - r_0$ where r_0 is the effective nuclear radius and r_1 is the radius from the centre of the nucleus, at which the potential energy of the α-particle is equal to the kinetic energy of the emitted α-particle.

Here $$r_0 = 1.5 \times 10^{-15}\, A^{1/3} = 1.4 \times 10^{-15} \times (222)^{1/3}$$
$$= 9.08 \times 10^{-15} \text{ m.}$$

r_1 is obtained from

$$E = \frac{2(Z-2)e^2}{4\pi\varepsilon_0 r_1} \qquad \therefore\ r_1 = \frac{2(Z-2)e^2}{4\pi\varepsilon_0 E}$$

$$= \frac{2 \times (86-2) \times (1.6 \times 10^{-19})^2}{4\pi \times 8.85 \times 10^{-12} \times 5.5 \times 1.6 \times 10^{-13}}$$

$$= 34.97 \times 10^{-15} \text{ m}$$

Now $$a = r_1 - r_0 = (43.97 \times 10^{-15} - 9.08 \times 10^{-15}) \text{ m}$$
$$= 43.98 \times 10^{-15} \text{ m.}$$

Example 23

A stream of electrons, each of energy E = 3 eV is incident on a potential barrier of height V_0 = 4 eV. The width of the barrier is 20 Å. Calculate the percentage transmission of the beam through this barrier.

Solution

The probability of transmission through a potential barrier is given by

$$T = \frac{16E}{V_0}\left(1 - \frac{E}{V_0}\right)e^{-2\beta a}$$

where $$2\beta a = 2a\sqrt{\frac{2m(V_0 - E)}{\hbar^2}} = \frac{2a}{\hbar^2}\sqrt{2m(V_0 - E)}$$

Given $$E = 3 \text{ eV} = 3 \times 1.6 \times 10^{-19} \text{ J}$$
$$V_0 = 4 \text{ eV} = 4 \times 1.6 \times 10^{-19} \text{ J}$$
$$a = 20 \text{ Å} = 20 \times 10^{-10} = 2 \times 10^{-9} \text{ m}$$

$$\therefore\quad 2\beta a = \frac{2a}{\hbar}\sqrt{2m(V_0 - E)} = 20.49$$

$$\therefore\quad T = \frac{16E}{V_0}\left(1 - \frac{E}{V_0}\right)e^{-2\beta a}$$

$$= \frac{16 \times 3}{4}\left(1 - \frac{3}{4}\right)e^{20.49}$$

$$= 12 \times \frac{1}{4} \times \frac{1}{(2.718)^{20.49}} = 3.797 \times 10^{-9}$$

∴ Percentage transmission

$$= 3.797 \times 10^{-7}\%.$$

Example 24

An α-particle having energy 10 MeV approaches a potential barrier of height equal to 30 MeV. Determine the width of the potential barrier if the transmission coefficient is 2×10^{-3}. Given mass of a α-particle = $4 \times 1.67 \times 10^{-27}$ kg and $\hbar = 1.054 \times 10^{-34}$ J-s.

Solution

$$T = \frac{16E}{V_0}\left(1 - \frac{E}{V_0}\right)e^{-2\beta a}$$

and

$$e^{2\beta a} = \frac{16E}{TV_0}\left(1 - \frac{E}{V_0}\right)$$

∴

$$2\beta a = 2.303 \log_{10} \frac{16E}{TV_0}\left(1 - \frac{E}{V_0}\right)$$

or

$$a = \frac{1}{2\beta} \times 2.303 \log_{10} \frac{16E}{TV_0}\left(1 - \frac{E}{V_0}\right)$$

$$\beta = \frac{1}{\hbar}\sqrt{2(E - V_0)}$$

$$= \frac{1}{1.054 \times 10^{-34}} \times \sqrt{2 \times 4 \times 1.67 \times 10^{-27} \times (30 - 10) \times 10^6 \times 1.6 \times 10^{-19}}$$

Hence

$$\beta = \frac{10^{14}}{1.054 \times 10^{-34}} \times \sqrt{2 \times 4 \times 1.67 \times 20 \times 1.6}$$

$$= \frac{20.68 \times 10^{14}}{1.054} = 19.62 \times 10^{14}$$

$$a = \frac{10^{-14}}{2 \times 19.62} \times 2.3031 \log_{10} \frac{16 \times 10}{2 \times 10^{-3} \times 30}\left(1 - \frac{10}{30}\right)$$

$$= \frac{10^{-14}}{2 \times 19.62} \times 2.3031 \log_{10} \frac{16}{9} \times 10^3$$

$$= \frac{10^{-14} \times 2.303 \times 3.25}{39.24} = 1.91 \times 10^{-15} \text{ m}$$

PROBLEMS

1. Explain the distribution of energy in blackbody spectrum. How far the classical theories explain the distribution? Write a note on Planck's law of distribution.
2. State Wien's displacement law and Rayleigh-Jean's law of radiation and indicate how far the two laws could explain the curves.
3. Discuss Planck's hypothesis for a blackbody radiation. Show that the average energy of Planck's oscillator of frequency ν in thermal equilibrium with a heat reservoir of temperature T is given by

$$\bar{\varepsilon} = \frac{h\nu}{[\exp(h\nu/k_BT) - 1]}$$

4. Discuss the distribution of energy in blackbody radiation and explain the success and limitations of the classical theory in explaining it. Derive Planck's radiation formula, explaining how the shortcomings of classical theory were overcome.
5. What is photoelectric effect? Give an account of the photoelectric emission of the electrons. Give Einstein's interpretation for the same.
6. How Einstein's photoelectric equation explains the phenomenon of photoelectric effect? What information is derived for the nature of light?
7. What are the laws of photoelectric effect? How has photoelectric effect been explained by Einstein?
8. What is Compton effect? Derive an expression for the change in wavelength of a photon when it is scattered by an electron. Calculate the maximum change in wavelength in a Compton scattering experiment.
9. Describe Compton effect. Derive the expression for the frequency of the scattered photon in terms of the frequency of incident radiation and scattering angle.
10. (*a*) Explain why Compton effect is not observed experimentally for visible rays?

    (*b*) Both photoelectric effect and Compton effect arise due to action of photons on electrons but the two effects are not the same. Explain.

    (*c*) Explain the presence of unmodified radiation in Compton scattering.
11. Explain the difference between photoelectric effect and Compton effect.
12. Explain Compton scattering. Derive an expression for the Compton scattering.
13. What is de Broglie hypothesis? Show that the wavelength λ associated with a particle of mass m and kinetic energy E is given by $\lambda = h/\sqrt{2meE}$, where h is Planck's constant.
14. Describe Davisson and Germer's experiment and indicate the important conclusion.
15. (*a*) Draw the schematic diagram of Davisson and Germer apparatus to study the electron diffraction. Explain how the experiment was carried out.

    (*b*) Show that the results of this experiment are closely in agreement with the electron wavelength calculated by the de Broglie relation.
16. State Heisenberg's uncertainty principle. Is this principle is the outcome of the wave description of a particle ? Give atleast one experiment to prove its validity.
17. (*a*) What is correct statement of Heisenberg's uncertainty principle? Discuss its significance and importance.

(*b*) From the uncertainty principle, show that the electrons can note reside in the nucleus of the atom. Given that nuclear radius = 10^{14} m, and the maximum possible kinetic energy of an electron in at atom = 4 MeV.

18. Show that the photon in a 1240 nm infrared light beam have energies of 1 eV.

$$\left[\text{Hint: } h\nu = \frac{hc}{\lambda} = \frac{6.63\times10^{-34}\times3\times10^{8}}{1240\times10^{-9}} = 1.6\times10^{-9}\text{ J} = 1\text{ eV}\right]$$

19. Calculate the energy of a photon of blue light of wavelength 450 nm. [**Ans.** 2.76 eV]

$$\left[\text{Hint: } E = \frac{hc}{\lambda} = \frac{6.63\times10^{-34}\times3\times10^{8}}{450\times10^{-9}} = 4.42\times10^{-9}\text{ J} = 2.76\text{ eV}\right]$$

20. In order to break a chemical bond in the molecules of human skin, causing sunburn, a photon energy of about 3.5 eV is required. To what wavelength does this correspond?

21. Calculate the average energy of an oscillator with frequency 10 Hz (s^{-1}) at a temperature 1000 K. [**Ans.** 5.5×10^{-20} J]

22. Planck's law of distribution of energy in the blackbody radiation is given by the equation $E(\nu)d\nu = \dfrac{8\pi h\nu^3}{c^3}\dfrac{d\nu}{\exp(h\nu/k_BT)-1}$

where $E(\nu)$ is the energy associated with the radiation of frequency ν. Show that (*a*) the total energy of all radiations from the blackbody is proportional to the fourth power of the temperature of the body and (*b*) the wavelength corresponding to the maximum energy density is inversely proportional to the temperature.

23. The work function of sodium metal is 2.3 eV. What is the longest wavelength of light that can cause photoelectron emission from sodium? [**Ans.** λ =540nm]

24. What potential difference must be applied to stop the fastest photoelectrons emitted by a nickel surface under the action of ultraviolet light of wavelength 2000 Å? The work function of nickel is 5.01 eV.
[Hint: Energy of photon

$$= \frac{hc}{\lambda} = \frac{6.6\times10^{-34}\times3\times10^{8}}{2000\times10^{-10}} = 9.95\times10^{-19}\text{ J} = 6.21\text{ eV}$$

Then, from the photoelectric equation, the energy of the fastest emitted electron is 6.21 eV – 5.01 eV = 1.20 eV. Obviously, a negative retarding potential of 1.2 eV is required]

25. Will photoelectrons be emitted by a copper surface, of work function 4.4 eV, when illuminated by visible light? [**Ans.** No]

[**Hint** : Threshold $\lambda = \dfrac{hc}{W_{\min}} = \dfrac{6.63\times10^{-34}\times3\times10^{8}}{4.4\times1.6\times10^{-19}} = 282$ nm. Obviously, visible light (400 nm to 700 nm) cannot eject photoelectrons from copper]

26. Suppose that a 3.64 nm photon going in the +*x* direction collides head on with a 2×10^5 m/s electron moving in the $-x$ direction. If the collision is perfectly elastic, find the conditions after collision.

[Hint: From the law of conservation of momentum,

Momentum before = Momentum after

$$\frac{h}{\lambda_0} - mv_0 = \frac{h}{\lambda} - mv$$

But $h/\lambda_0 = mv_0$ in this case and hence $h/\lambda = mv$. Also for a perfectly eleastic collision $(\text{K.E.})_{\text{before}} = (\text{K.E.})_{\text{after}}$

$$\frac{hc}{\lambda_0} + \frac{1}{2}mv_0^2 = \frac{hc}{\lambda} + \frac{1}{2}mv^2$$

Using $h/\lambda_0 = mv = mv_0$, the photon also rebounds and its original wavelength].

27. A photon ($\lambda = 0.400$ nm) strikes an electron at rest and rebounds at an angle of 150° to its original direction. Find the speed and wavelength of the photon after the collision. **[Ans.** 0.4045 nm]

28. An electron falls from rest through a potential difference of 100 V. What is its de Broglie wavelength? **[Ans.** 12.3 nm]

29. A photon of frequency ν is scattered by an electron of mass m_0. The scattered photon of frequency ν' travels in a direction inclined at 90° with the direction. Prove that the de Broglie wavelength of the recoil electron is given by

$$\lambda_r = \frac{C}{\sqrt{\nu^2 + \nu'^2}}$$

30. Show that the wavelength λ associated with an electron of mass m_0 and kinetic energy k is given by

$$\lambda = \frac{h}{\sqrt{2m_0 k}}$$

31. What potential difference is required in an electron microscope to give electrons a wavelength of 0.5 Å?

[Hint: K.E. of electron $= \frac{1}{2}mv^2 = \frac{1}{2}m\left(\frac{h}{m\lambda}\right)^2 = \frac{h^2}{2m\lambda^2}$

$$= \frac{(6.63 \times 10^{-34})^2}{2 \times 9.1 \times 10^{-31} \times (0.5 \times 10^{-10})^2}$$

$$= 9.66 \times 10^{-17} \text{ J}$$

But K.E. $= qV$

$$\therefore \quad V = \frac{K.E.}{q} = \frac{9.66 \times 10^{-17}}{1.6 \times 10^{-19}} \frac{\text{J}}{\text{C}} = 600 \text{ V}$$]

32. A beam of X-rays is scattered by loosely bound electrons at 45° from the diffraction of the beam. The wavelength of the scattered X-rays is 0.22 Å. What is the wavelength of X-rays in the direct beam. **[Ans.** 0.2129 Å]

$$\left[\text{Hint}: \lambda = \lambda' - \frac{h}{m_0 c}(1 - \cos\phi) = 0.22 - 0.02426\left(1 - 1/\sqrt{2}\right)\right]$$

33. X-rays with $\lambda = 2.00$ Å are scattered from a carbon block and the scattered radiation is viewed at 90° to the incident beam. Calculate the Compton shift.
[Ans. 0.02426 Å]

34. An X-ray photon of frequency 1.5×10^{19} Hz is scattered by a free electron and the frequency of the scattered photon is 1.2×10^{19} Hz. Find the kinetic energy imparted to the electron. [Hint : $K = h\nu - h\nu'$] **[Ans.** 1.989×10^{-15} J]

35. Assume that an electron is inside a nucleus of radius 10^{-15} m. From the uncertainty principle, estimate the kinetic energy of the electron.
[Ans. 0.95×10^{10} eV]

$$\left[\text{Hint}: \Delta x \sim 10^{-15} m \Delta x \Delta p \sim \frac{h}{4\pi\Delta x} = \frac{6.6 \times 10^{-34}}{4\pi \times 10^{-15}} \frac{\text{kgm}}{\text{sec}}\right.$$

Since the magnitude of the momentum cannot be less than that of uncertainty, the momentum of the electron will at least be equal to Δp. Thus the minimum K.E. of the electron will be

$$\left.\frac{(\Delta p)^2}{2m} = \frac{\left(6.63 \times 10^{-19}\right)^2}{2 \times \left(9.1 \times 10^{-31}\right)(4\pi)^2} \text{ J} \sim 0.95 \times 10^{10} \text{ eV}\right]$$

36. Deduce the time-independent Schrodinger equation. Explain the physical significance of wave function.

37. (*a*) Derive the time-independent Schrodinger wave equation for a particle.
(*b*) Give the physical meaning of the wave function. What do you understand by normalised wave function?

38. (*a*) Obtain the time-independent Schrodinger wave equation for a particle.
(*b*) Separate the wave function into time-dependent and time-independent parts and obtain the steady state Schrodinger equation.

39. Show that one-dimensional Schrodinger equation has solution of form $\phi(x)\ \psi(t)$ if the potential energy of the particle associated with the wave function is independent of time. Show that the most general solution of Schrodinger equation for such a particle can be written as

$$\phi = \sum_{n=1}^{\infty} a_n \psi_n(x)\, e^{-iEt/\hbar}$$

40. (*a*) Why should any wave function be finite, continuous and differentiable at every point?
(*b*) When and why do we call a solution of Schrodinger's equation a stationary one? Show that they occur only if the potential energy does not depend explicitly on the time.

41. Obtain Schrodinger's time dependent wave equation and separate it into space and time dependent parts. Given the probability interpretation of the wave function and show that the probability density ρ and the probability current density J satisfy the continuity equation. Explain the physical significance of this equation.

$$\frac{d\rho}{dt} + \nabla \cdot \boldsymbol{J} = 0$$

42. Explain why it is possible to choose as probability current-density in quantum mechanics by the following expression.

$$J = \frac{e\hbar}{2m}\left(\psi^{*}\nabla\psi - \psi\nabla\psi^{*}\right)$$

The symbols have their usual meanings.

43. Distinguish between phase velocity and group velocity. Show that for a non-relativistic free particle the phase velocity is half of the group velocity.

44. Write down Schrodinger's equation for a particle in a one dimensional box. Solve it to obtain eigen function and show that the eigen values are discrete.

45. Write one dimensional Schrodinger's time independent equation for a particle moving in a potential $V(x)$. Explain its significance. What are stationary state solutions.

46. Write down Schrodinger's equation for a particle in a three dimensional box. Solve it to obtain eigen functions and show that the eigen values are discrete. Explain clearly the meaning of degeneracy of levels.

47. Explain clearly, what do you understand by degeneracy of levels. Explain the statement "for a degenerate energy level there can be more than one different wave functions". Illustrate it by taking the example of particle in a box.

48. Formulate the problem of square well potential by writting the appropriate potential and Schrodinger equations. Obtain the solutions and an expression for the transmissibility (transmission coefficient). Discuss the expression with special emphasis on resonance.

49. A particle travelling with energy E along X-axis has in its path a rectangular potential barrier of height $V > E$ and width a. Calculate the transmission coefficient of the particle and discuss briefly its application to the observed phenomenon of α-decay in nuclei.

50. Determine the transmission coefficient for a particle of energy $E < V_0$ for a rectangular one dimensional potential barrier given by

$$V = 0 \text{ for } x < -a \text{ and } x > a$$
$$V = V_0 \text{ for } -a < x < a$$

Discuss briefly its application to the observed phenomenon of α-decay.

51. (*a*) Why would any wave function be finite, continuous and differentiable at every point?

(b) When and why do we call a solution of the Schrodinger's equation a stationary one? Show that they occur only if the potential energy does not depend explicitly on time.

52. Explain the problem of leaking of a particle through a rectangular potential barrier of finite width and explain the theory of α-decay.

53. (*a*) What do you mean by tunnelling through a barrier? A particle travelling with energy E along X-axis has a potential barrier defined as

$$V(x) = \begin{cases} 0 \text{ for } x < 0 \\ V_0 \text{ for } 0 < x < a \\ 0 \text{ for } x > a \end{cases}$$

Derive the expression for the equation and transmission coefficients of the particle.

(*b*) Show that coefficient of reflection of particles incident on barrier is same whether they approach from right or left.

54. A beam of particles with energy E is incident on a potential barrier with potential function?

$$V(x) = \begin{cases} 0 \text{ for } x < 0 \\ V_0 \text{ for } 0 < x < a \\ 0 \text{ for } x > a \end{cases}$$

Show that there is a finite probability of transmission even if $E < V_0$.

55. Describe the tunnelling phenomenon in quantum mechanics. Obtain the transmission coefficient for a particle of energy E crossing the barrier defined by $V(x) = 0$ for $x = 0$, $V(x) = V_0 \leq x \leq a$ and $V(x) = 0$ for $x = a$. Comment on the solution when the thickness is a multiple of the de Broglie wavelength.

56. Describe the tunnelling phenomenon in quantum mechanics. Show that tunnelling increases if the height and width of the barrier are reduced.

57. The width and depth of a finite square well potential are a and V_0 respectively. Derive an expression for the transmission coefficient of a particle having positive energy. What are transmission resonances?

58. Show that the wave functions for a particle in one dimensional box, $\psi = A\sin n\pi x/L$, are not eigen functions of the linear momentum operator P_x, but they can be written as combinations of the eigen functions of p_x.

[**Hint :** $\sin n\pi x/L = 1/2i(\exp)(in\pi x/\ell) - \exp(-in\pi x/\ell)$

Both exponential terms in the right hand side can be easily shown to be eigen functions of p_x with eigen values $\pm 2\pi n\hbar/2L)$].

59. Varify that $\psi = A\sin kx + B\cos kx$ is a general solution for a particle in a one dimensional box having infinite potential walls.

60. Find the energies of the six lowest energy levels of a particle in a cubical box. Which of the levels are degenerate?

[**Ans.** 3, 6, 9, 11, 12 and $14 \times \pi^2\hbar^2/2mL^2$, 1st and 5th are non-degenerate; 2nd, 3rd and 4th are threefold degenerate and the 6th is sixfold degenerate.]

61. Determine the degree of degeneracy of the energy-level $\dfrac{38\pi^2\hbar^2}{2mL^2}$ of a particle in a cubical potential box.

[**Ans.** The quantum numbers (611), (161), (116), (532), (523), (352), (325), (253), and (235) give the same energy levels therefore the level is ninefold degenerate.]

62. An electron is confined to one dimensional box of side 1 Å. Obtain the first four eigen values of eV of the electron. [**Ans.** 38 eV, 152 eV, 342 eV, 608 eV]

63. An electron is confined to move between two rigid walls separated by 10^{-9} m. Find the de Broglie wavelengths representing the first three allowed energy states of the electron and the corresponding energies.

64. Calculate the de Broglie wavelength associated with an electron associated by 5000 volts. Given mass of the electron = 9.1×10^{-28} gms, Planck's constant = 6.6×10^{-34} J-s. [Raj. BE, 90]

65. (*a*) What is Compton scattering? Obtain an expression for shift in wavelength of the scattered photon by Compton scattering? [Raj. BE, 94. 97, 98] (*b*) Calculate maximum percentage change in wavelength due to Compton scattering for incident photons of wavelengths 1 Å, 10 Å and 100 Å. What interference do you draw from this calculation? [BE]

66. (*a*) Explain Heisenberg's uncertainty principle for (*i*) Momentum and position (*ii*) Energy and time. Illustrate its validity with the help of one example of each.

(*b*) Derive Schrodinger's time independent wave equation. What is the physical significance of wave function ψ used in this equation? [Raj. BE, 94]

(*c*) Explain why Compton effect is not observed experimentally for visible rays. Explain Compton scattering.

(*d*) State Heisenberg uncertainty principle. Explain the validity by any thought experiment.

(*e*) What is tunnel effect? [Raj. BE, 96]

67. Light described at a place by the equation:

$$E = (100 \text{ volt/m} \left[\sin\left(5 \times 10^{15}\right)t + \sin\left(8 \times 10^{15}\right)t\right]$$

falls on a metal surface having work function 2.0 eV. Calculate the maximum kinetic energy of the photo electrons and the stopping potential (Planck's constant $h = 6.63 \times 10^{-34}$ J-s). [Raj. BE, 97]

68. Describe Schrodinger's time independent wave equation and eigen physical interpretation of wavefunction ψ. [Raj. BE, 97]

69. Write down Schrodinger equation and solve it to determine the eigen values and eigen functions for a particle in a box. [Raj. BE, 99]

70. Explain the meaning of the terms degeneracy and tunnelling. [Raj. BE, 98]

71. Write short notes on:

(*a*) Heisenberg uncertainty principle. [Raj. BE, 98]

(*b*) Photoelectric effect. [Raj. BE, 99]

72. (*a*) Derive Schrodinger's time independent wave equation. Give physical interpretation and essential requirements of wave function ψ used in this equation.

(*b*) Consider an electron whose total energy 5 eV approaching a barrier whose height is 6 eV and width is 7 Å. Find out de Broglie wavelength of incident electron and probability of transmission through the barrier. (Mass of electron $= 9.1 \times 10^{-31}$ kg, Planck's constant $= 6.6 \times 10^{-34}$ J). [Raj. BE, 2000]

73. A measurement establishes the position of a proton with an accuracy $\pm 10^{-11}$ m. Find the uncertainty in the proton's position one second later. Assume proton's speed $v << c$. (Mass of proton $= 1.67 \times 10^{-27}$ kg) [Raj. BE, 2000]

74. Write short note on Compton effect and its significance. [Raj. BE, 2000]

75. (*a*) What is Compton effect? Derive an expression for wavelength shift in Compton effect. [Raj. BE, 2002]

(*b*) Write short note on: (a) Tunnel effect. [Raj. BE, 2002]

76. (*a*) What is 'Compton scattering'? Obtain an expression for shift in wavelength of the scattered photon by compton scattering.

(*b*) In Compton scattering, the energy of an incident *X*-ray photon is 150 keV and that of scattered photon is 130 keV. Determine the angle of scattering.

(*c*) State Heisenberg's uncertainty principle. [Raj. BE, 2002]

77. Derive the Schrodinger time dependent equation. Explain the following:

(*i*) Hamiltonian of a free particle

(*ii*) Physical significance of wavefunction ψ.

(*iii*) Normalized and orthogonalized wavefunctions. [Raj. BE 2002]

78. (*a*) Describe Heisenberg's uncertainty principle and apply it to explain non-existence of electron in nucleus?

(*b*) Derive Schrodinger's time independent wave equation. What is the physical significance of wave function ψ used in this equation?

(*c*) Find the lowest energy of an electron confined to move in a one dimensional potential box of length 1 Å. [Raj. BE, 2000, 2001]

79. (*a*) Obtain an expression for the density of states for free electron gas in metal and hence find expression for the Fermi energy.

(*b*) Calculate Fermi energy in copper assuming that each copper atom contributes one free electron to the electron gas. Given density of copper 8.94×10^3 kg/m^3 and atomic mass is 63.5 μ.

(*c*) Write down Schrodinger's equation for a particle of mass *m* trapped in one-dimensional box of size *a*. Solve it for energy eigen values and eigen functions. How does solutions modify if particle were in three-dimensional cubical box of side *a*? Find lowest energy of the following states; (*i*) non-degenerate, (*ii*) doubly degenerate, (*iii*) triply degenerate and (*iv*) sixfold degenerate for 3-D cubical box. [Raj. BE, 2001]

80. (*a*) What is tunnel effect. Write down Schrodinger equation for potential barrier problem and steps to find out the transmission coefficient of a particle having less energy than the height of potential barrier.

(*b*) Show that the value of energy which a photon must have so that it may transfer half of its energy to an electron at rest is about 256 keV in a Compton scattering experiment. [Raj. BE, 2003]

81. (*a*) A measurement establishes the position of a proton with an accuracy of $\pm\ 10^{-11}$ meter. Find the uncertainty in the proton's position one second later. Assume proton's speed $v < c(m_p = 1.67 \times 10^{-27}$ kg). [Raj. BE, 2000]

SHORT ANSWER QUESTIONS

1. What will be the de Broglie wavelength of an electron with a velocity of 10^7 m/s.

Ans. $$\lambda = \frac{h}{mv} = \frac{6.63 \times 10^{-34} \text{ Js}}{(9.1 \times 10^{-31} \text{ kg}) \times (10^7 \text{ m/s})} = 7.3 \times 10^{-11} \text{ m}$$

2. What is a wavefunction?

Ans. The quantity whose variations make up matter waves is called the wavefunction (ψ). The value of the wave function associated with a moving body at the particular point (x, y, z) in space at the time *t* is related to the likelihood of finding the body at the time.

3. Why the wave function ψ itself has no direct physical significance?

Ans. ψ cannot be interpreted in terms of an experiment. The probability that something be in a certain place at a given time must lie between 0 (the object is definitely there) and 1 (the object is definitely there). An intermediate probability, say 0.2 means that there is a 20% chance of finding the object. But the amplitude of a wave can be negative as well as positive and negative probability is meaningless. Hence ψ by itself cannot be an observable quantity.

4. What $|\psi|^2$ signifies?

Ans. The probability of experimentally finding the body described by the wave function ψ at the point x, y, z at the time t is proportional to the value of $|\psi|^2$ there at t. A large value of $|\psi|^2$ means the strong possibility of the body's presence, while a small value of $|\psi|^2$ means the slight possibility of its presence. As long as $|\psi|^2$ is not actually zero (0) somewhere, however, there is a definite chance however small, of detecting it there.

5. A 10 g marble is in a box 10 cm. across. What will be its permitted energies?

Ans. $m = 10\text{ g} = 1 \times 10^{-2}\text{ kg}$, $L = 10\text{ cm} = 1.0 \times 10^{-1}\text{ m}$

$$E_n = \frac{(n^2)(6.63 \times 10^{-34}\text{ J - s})^2}{(8)(1.0 \times 10^{-2}\text{ kg})(1.0 \times 10^{-1}\text{ m})^2}$$

$$= 5.5 \times 10^{-64} n^2\text{ J}$$

The minimum energy the marble can have is, corresponding to $n = 1$. A marble with this kinetic energy has a speed of only 3.3×10^{-31} m/s and therefore one cannot distinguish it experimentally from a stationary marble.

6. Justify the statement; 'Classical mechanics is an approximation of quantum mechanics.'

Ans. The fundamental difference between classical mechanics and quantum mechanics lies in what they describe. In classical mechanics, the future history of a particle is completely determined by its initial position and momentum together with the forces that act upon it. In the everyday world one can determine these well enough for the predictions of Newtonian mechanics to agree with what we find.

Quantum mechanics also arrives at relationship between observable quantities, but the uncertainty principle suggests that the nature of an observable quantity is different in the atomic realm. Cause and effect are still related in quantum mechanics, but what they concern needs careful interpretation. In quantum mechanics the kind of certainty about the future characteristic of classical mechanics is impossible because the initial state of a particle cannot established with sufficient accuracy.

We may note that the quantities whose relationships quantum mechanics explores are probabilities. Classical mechanics turns out just an approximate version of quantum mechanics.

7. Write the condition for a normalized wave function ψ.

Ans. $\int |\psi|^2\, dv = 1$

8. Can Schrodinger's equation be derived from other basic principles of physics.

Ans. No, Schrodinger's equation is a basic principle in itself.

9. Is Schrodinger's equation is linear in the wave function ψ? What is the meaning of it?

Ans. Yes, Schrodinger's equation is linear in the wave function ψ. By this we mean that the equation has terms that contain ψ and its derivatives but no terms independent of ψ or that involves higher power of ψ or its derivatives.

10. Is Schrodinger's equation is linear in probabilities?

Ans. No.

11. What an operator in quantum mechanic signify?

Ans. An operator tells us what operation to carry out on the quantity that follows it.

12. What is an important property of Schrodinger's steady state equation?

Ans. If Schrodinger's steady state equation

$$\frac{\partial^2 \psi}{\partial x^2} + \frac{2m}{\hbar^2}(E - V)\psi = 0$$

has one or more solutions for a given system, each of these wave functions corresponds to a specific value of energy E. Thus energy quantization appears in wave mechanics as a natural element of theory, and energy quantization in the physical world is revealed as a universal phenomenon characteristic of all stable systems.

OBJECTIVE QUESTIONS

1. The average energy of Planck's oscillator of frequency ν is __________.

[**Ans.** $h\nu/(e^{h\nu/kT} - 1)$]

2. For the study of Compton effect, the best suited rays are __________.

[**Ans.** X-rays]

3. Bohr's quantum condition is __________. [**Ans.** $mvr = nh/2\pi$]

4. Quantum mechanical operator of momentum p is __________. $\left[\textbf{Ans. } \frac{\hbar}{i}\nabla\right]$

5. Quantum mechanical operator of kinetic energy T is __________.

$\left[\textbf{Ans. } -\frac{\hbar^2}{2m}\nabla^2\right]$

6. If A is an operator, operating on a state function ψ, then the possible values of any physical quantity of a system (e.g. energy, angular momentum, etc.) in operator equation $A\psi = a\psi$ are given by __________. [**Ans.** eigen value a]

7. If $A = 3x^2$ and $B = \frac{d}{dx}$, then __________. [**Ans.** $AB \neq BA$]

8. The energy of a particle of mass m in a one dimensional box of length $2L$ with ends at $\pm 2L$ *i.e.* the origin at the middle point is __________.

$\left[\textbf{Ans. } E_n = \frac{n^2h^2}{16\mu L^2}\right]$

9. The average value of P_x^2 for a particle in a box is __________. [**Ans.** $2\,mE$]

10. For a particle in a box, the fractional difference in the energy between adjacent eigen values is __________

$\left[\textbf{Ans. } \frac{\Delta E}{E_n} = \frac{2n+1}{n^2}\right]$

Energy Bands in Solids

9.1 INTRODUCTION

In a single atom, electrons exist in discrete energy levels. In a crystal, in which large numbers of atoms are held closely together in a lattice, electrons are influenced by a number of adjacent nuclei and sharply defined levels of the atoms become bands of allowed energy; this approach to energy levels in solids is often known as the *band theory.* Each band represents a large number of allowed quantum states. Between the bands are *forbidden bands.* The outermost electrons of the atoms (i.e. the ones responsible for chemical bounding) form the valence bond of the solid. This is the band, of those occupied, that has the highest energy.

The band structure of solids accounts for their electrical properties. In order to move through the solid, the electrons have to change from one quantum state to another. This can only occur if there are empty quantum states close to electrons. In general, if the valence band is full, electrons cannot change to new quantum states in the same band. For conduction to occur, the electrons have to be in an unfilled band—the conduction band. In this chapter band theory and free electron theory is presented.

9.2 ATOMIC STRUCTURE AND ENERGY LEVELS

To understand how materials work, one must have a good knowledge of atomic structure. We know that matter is composed of compounds and elements. The elements are the basic materials found in nature. When element are combined to form a new material, we have a compound. The smallest particle that an element can be reduced to and still retain its properties is called an atom.

Although the atoms of different elements have different properties, they all contain the same subatomic particles. There are a number of different subatomic particles, but only three of these are of interest in basic electronics—the proton, the neutron and the electron.

The proton and the neutrons are contained in the nucleus of the atom, and the electrons orbit around the nucleus. The electrons and the protons are the particles that have the electrical properties. Neutrons have no electrical charge. An, atom has the same number of electrons and protons, and so it is electrically neutral. If an atom does have more electrons, it is called a *negative* ion. If it has more protons, it is called a *positive* ion.

To understand the location and energy of each electron in an atom, one must have the knowledge of following four **quantum numbers.**

(*i*) Principal Quantum Number (*n*): This characterises the average distance of an electron from the nucleus and corresponds to the principal energy level in which electron resides. Obviously, n, gives some idea about the position of the electron around the nucleus. n can have positive integer values starting from 1, *i.e.* $n = 1, 2, 3, \ldots$. The principal energy levels or shells having different values of n are designated by the letters K, L, M, N and so on. The maximum numbers of the electrons that can be accommodated in a shell corresponding to n is equal to $2n^2$.

(*ii*) Azimuthal Quantum Number (*l*): This is also called as **orbital angular momentum quantum number** and gives a measure of the angular momentum of an electron in the orbit. Physically, l, indicates the shape of the classical orbit. For a given value of n, l can take all positive integer values from 0 to $(n - l)$. The particular l value defines the subshell, and the subshells with $l = 0, 1, 2, 3, \ldots$ are designated as $s, p, d, f, g, \ldots$ respectively.

For a given principal quantum number, the energies of various subshells are in the order of $s < p < d < f < \ldots$. Obviously, an electron in the *s*-subshell has lower energy than in the *p*-subshell with same n. The magnitude of this angular momentum if $\hbar = \sqrt{l(l+1)}$.

(*iii*) Magnetic Quantum Number (m_l): This is also called as **orbital magnetic number** and this determines the preferred orientation of the orbitals in space with respect to an applied magnetic field. We know that the magnetic moment of an electron due to its orbital motion gives rise to a magnetic field which can interact with an external magnetic field. The electrons orient themselves in certain preferred region of space around the nucleus under the influence of the external field.

For a given value of l, m_l can take integer values between -1 to $+1$ including 0, *i.e.* total allowed values of m_l are $(2l + 1)$. The magnitude of the component of angular momentum along the direction of the magnetic field is $m_l \hbar$.

(iv) Spin Quantum Number (m_s): Electron is spinning about its own axis in the atom. The spin of the electron produces a spin magnetic moment which can either parallel or antiparallel to the surrounding magnetic field. For an electron there are two spin states. Spin quantum number is concerned with the spinning of the electron about its own axis. Thus m_s can take only two possible states, $+\frac{1}{2}$ or $-\frac{1}{2}$.

Pauli Exclusion Principle

This states that in an atom no two electrons can exist in the same quantum state, *i.e.*, in an atom there cannot be two electrons with the same value of all the four quantum numbers. With the help of this principle, one can write the configuration of electrons. We may note that all the electrons with the same value of n constitute a shell and a shell can have a maximum of $2n^2$ electrons.

The energy state capacity of different shells and subshells are as follows:

K-shell: This corresponds to $n = 1$ and this has one subshell, namely, $1s$. This corresponds to $n = 1$, $l = 0$, $m_l = 0$, $m_s \pm\frac{1}{2}$ and has 2 states.

L-shell: This corresponds to $n = 2$ and this has two subshells, namely, $2s$ and $2p$.$2s$ subshell corresponds to $n = 2$, $l = 0$, $m_l = 0$ and has 2 states and $2p$ subshell correspond to $n = 2$, $l = 1$, $l = 2$, $m_l = 0, \pm 1$ and has 6 states.

M-shell: This corresponds to $n = 3$ and has $3s$, $3p$ and $3d$ subshells. The $3s$ subshell corresponds to $n = 3$, $l = 1$, and $m_l = 0, \pm 1$ and has 6 states. The $3d$ subshell corresponds to $n = 3$, $l = 2$, $m_l = 0, \pm 1, \pm 2$ and has 10 states.

Similarly, one can find subshells in other shells, e.g., N, etc. According to Pauli exclusion principle, the electronic distribution in first four shells are shown in Table 9.1. We may note that in all seven shells are required for all the chemical elements in the Periodic Table.

TABLE 9.1: ELECTRON DISTRIBUTION IN SUBSHELLS

Shell	*K*	*L*		*M*			*N*			
n	*1*	*2*		*3*			*4*			
l	0	0	1	0	1	2	0	1	2	3
Subshell	*s*	*s*	*p*	*s*	*p*	*d*	*s*	*p*	*d*	*f*
No. of electrons	2	2	6	2	6	10	2	6	10	14

The electrons in the inner most shells are very strongly attached to the atom and hence they cannot be easily separated. In other words, the electrons that are very near to the nucleus are most tightly bound and hence have the lowest energy. Clearly, that atoms that have electrons in the closed shells will have very stable configurations.

The atomic number Z gives the numbers of electrons orbiting about the nucleus and the superscripts are used to designate the number of electrons in a particular subshell, e.g. Z for silicon is 14 and its electronic configuration is given as $1s^2 2s^2 2p^6 3s^2 3p^2$. We may note that silicon has only two electrons in the outer most subshell. Electronic configurations of C, Si, Ge, Sn and Pb belonging to same subgroup (IV A) in the Periodic Table is given in Table 9.2.

TABLE 9.2: ELECTRONIC CONFIGURATION OF SOME ELEMENTS OF GROUP IV A

Element	*Atomic Number*	*Electronic Configuration*
C	6	$1s^2 2s^2 2p^2$
Si	14	$1s^2 2s^2 2p^6 3s^2 3p^2$
Ge	32	$1s^2 2s^2 2p^6 3s^2 3p^6 3d^{10} 4s^2 4p^2$
Sn	50	$1s^2 2s^2 2p^6 3s^2 3p^6 3d^{10} 4s^2 4p^6 4d^{10} 5s^2 5p^2$
Pb	82	$1s^2 2s^2 2p^6 3s^2 3p^6 3d^{10} 4s^2 4p^6 4d^{10} 4f^{14} 5s^2 5p^6 5d^{10} 6s^2 6p^2$

Figure 9.1 shows the representation of the atomic structures of H, B, Si, P and Ge atoms. Figure 9.1 (*a*) represents the hydrogen atom. It contains one electron revolving around one proton which is the nucleus. The nucleus of H atom contains no neutrons.

Figure 9.1 (*b*) represents the structure of a *Boron* atom. Its nucleus contains 5 protons (p) and 5 neutrons (n). There are 5 electrons revolving around the nucleus in different shells. There are two electrons in the first shell, 2 electrons in the second shell and only 1 electron in the third outermost or valence shell.

Figure 9.1 (*c*) represents the structure of a silicon atom. It contains 14 protons and 14 neutrons in the nucleus. There are 14 electrons revolving around the nucleus in different shells. There are 2 electrons in the first shell, 8 electrons in the second shell and 4 electrons in the outermost or valence shell.

Figures 9.1 (*d*) and 9.1 (*e*) represent the structure of Phosphorus and Germanium atoms respectively. We note that phosphorus contains 5 electrons in the outermost shell called valence electrons whereas Ge atom contains 4 electrons in the outermost shell.

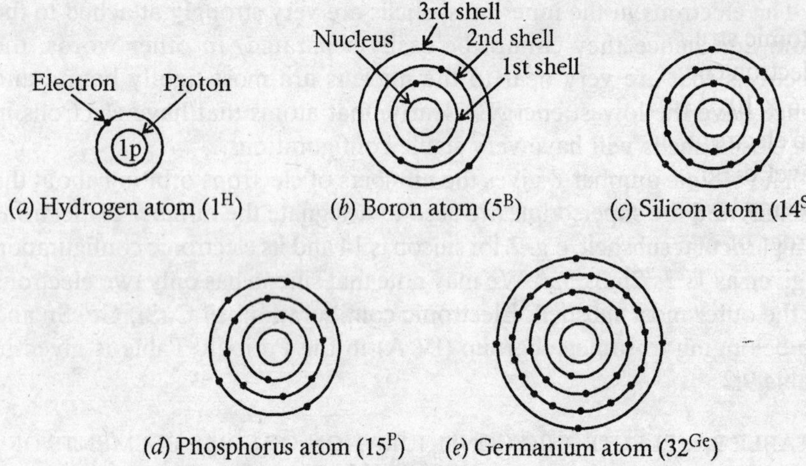

(*a*) Hydrogen atom (1^H) (*b*) Boron atom (5^B) (*c*) Silicon atom (14^{Si})

(*d*) Phosphorus atom (15^P) (*e*) Germanium atom (32^{Ge})

Fig. 9.1: Atomic structure of a few atoms

The electrons in the inner shells of an atom do not normally leave the atom. But the electrons which are in the outermost shell, so called valence shell do not always remain confined to the same atom. Some of these valence electrons move in a random manner and may travel from one atom to another in a crystal lattice. These electrons are called as *free electrons.* It is due to the presence of these free electrons in a material, that electrical conduction is possible. The electrons in the inner orbits of the atom remain bound to the nucleus and are, therefore, called *bound electrons.*

The tendency of an atom to give up its valence electrons depends on *chemical stability.* When an atom is stable, it resists giving up electrons, and when it is unstable, it tends to give up electrons. The level of stability is determined by the number of valence electrons, because the atom strives to have its outermost or valence shell completely filled.

If an atom's valence shell is more than half filled, then atom tends to fill its shell. So, since 8 is the maximum electrons that can be held in the valence shell, elements with 5 or more valence electrons make good *insulators,* since they tend to accept rather than give up electrons. On the other hand, atoms with less than 4 valence electrons tend to give up their electrons, thereby the valence shell is empty, this would allow the next shell, which is already filled, to be the outermost shell. These atoms make the best *electrical conductors.* The elements Si(14) and Ge(32) have 4 valence electrons, and are neither good conductors and nor good insulators. These are called semiconductors.

Most characteristics of semiconductors can be easily explained by means of an energy level diagram. We are familiar that each isolated atom has only a certain number of orbits available. These available orbits represent energy levels for the electrons in the atom. According to Boh'r theory of atomic structure only discrete values of electron energies are possible. An electron energy is usually expressed in electron volt (1 eV = 1.6×10^{-19} J = 1.6×10^{-12} erg). An electron can have only certain permissible values, i.e., no electron can exist at an energy level other than a permissible one. Energy level diagram for hydrogen atom is shown in Fig. 9.2. The permissible energy levels for hydrogen atom are numbered $n = 1, 2, 3, \ldots$ in increasing order of energy.

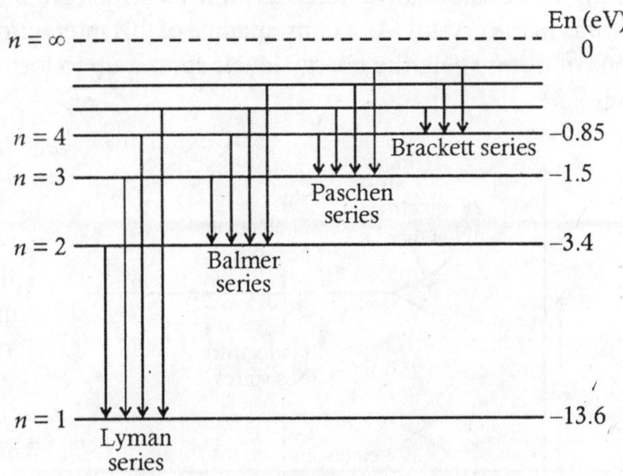

Fig. 9.2: Energy level diagram of hydrogen

In any atom, an electron orbiting very close to the nucleus in the first shell is tightly bound to the nucleus and possesses only a small amount of energy. The greater the distance of an electron from the nucleus, the greater is its total energy. The total energy of an electron includes Kinetic and Potential energies. Obviously, an electron orbiting far from the nucleus would have a greater energy, and hence it can be easily knocked out of its orbit. This makes it clear that why the valence electrons having maximum energy take part in chemical reactions and in bonding the atoms together to form solids.

When radiations impinge on an atom, the energy of the electrons increases. As a result, electrons are excited to higher energy levels. The excited state does not last long. Very soon, the electron after emitting out energy in the form of heat, light or other radiations, fall back to the original energy level.

9.3 ENERGY BAND THEORY

Most of the metals and semiconductors have crystalline structure. A crystal is made of a space array of atoms of molecules in regular repetition in three dimensions of some fundamental building block. The electronic energy levels for gaseous element are the same as for a single free atom because the individual atoms in a gas are well apart and has negligible influence on each other. However, in a crystal the individual atoms are so closely packed that the resulting energy levels gets modified due to interaction between the atoms in crystal lattice. We may note that when atoms form crystals, the energy, levels of the inner shell electrons are not affected appreciably. On the other hand, the energy levels of the outer-shell electrons are considerably altered as these electrons are shared by the adjacent atoms in the crystal. As a consequence of this interaction between the electrons of outer-shell, the energy levels spread up to form a band of energy (Fig. 9.3).

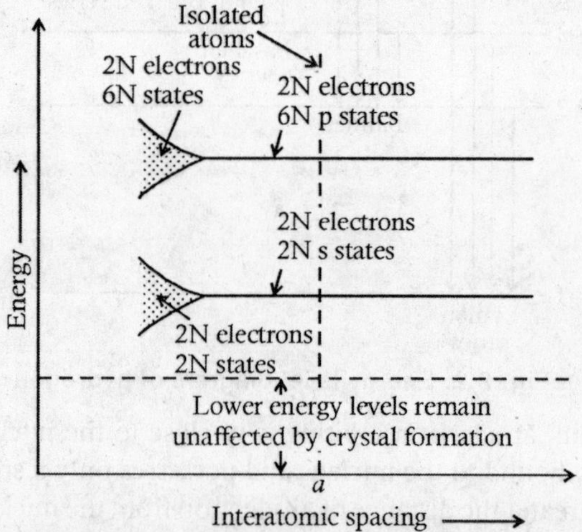

Fig. 9.3: Energy band formation in a crystal

To make more clear the formation of energy band in a crystal, we consider an example of silicon crystal made up of N atoms. Let us assume that the interatomic spacing can be varied without affecting the fundamental crystal structure. Let, the interaction between adjacent atoms is negligible for very large interatomic spacing, say a (Fig. 9.3). Silicon has 14 electrons and the outermost subshells, namely, $3s$ and $3p$ contains 2 electrons each. Thus in a silicon crystal consisting of N atoms, the outermost subshells $3s$ and $3p$ consist of $2N$ electrons each. Obviously, the 3s subshell has $2N$ electrons completely occupying the available $2N$ states, and the $3p$ subshell has only $2N$ electrons partially occupying the available $6N$ states, all at the same energy level.

Now, we consider that the interatomic spacing is gradually decreased, *i.e.* we are moving from right to left (Fig. 9.3). This means that there will be a gradual increase in the interaction between the neighbouring atoms. The atomic wave functions overlap due to this interaction. This makes the crystal an electronic system which obeys the Pauli exclusion principle. Clearly, $2Ns$ states spread out to form a band of energy. Since N is very large ($\sim 10^{23}/cm^3$) and separation between energy levels is small, the total spread between the minimum and maximum energy levels become large. This spread will have several volts of energy and termed as **energy band.** The lower shaded region in Fig. 9.3 shows this. In this band the $2N$ states are completely filled with $2N$ electrons. Similarly, we see that at the same value of interatomic spacing at *p*-level, $6Np$ states spread up to form a band. The upper shaded region in Fig. 9.3 shows this band we may note that there are $6N$ states available in this band but only $2N$ states are occupied and remaining $4N$ states remain unoccupied.

There exists an **energy gap (E_g)** between two energy bands, called as **forbidden energy gap**, as no electrons can occupy states in this energy gap. With the decrease in atomic spacing this forbidden energy gap decreases and becomes zero with further reduction in the interatomic spacing, say at '*b*' (Fig. 9.4). When the interatomic spacing is small enough, the two energy bands will overlap. Under these circumstances, the 6 Np states in the upper band merge with the $2Ns$ states in the lower band, giving a total of $8N$ states. Half of these $8N$ states, *i.e.*, $4N$ states are occupied by the $4N$ available electrons. Now, these $4N$ electrons no longer belong to either *p*-subshell or *s*-subshell but belong to the entire crystal. Clearly, at this interatomic spacing, each atom in the crystal can contribute 4 electrons to the crystal. The band occupied by these contributed electrons is known as **valence band.**

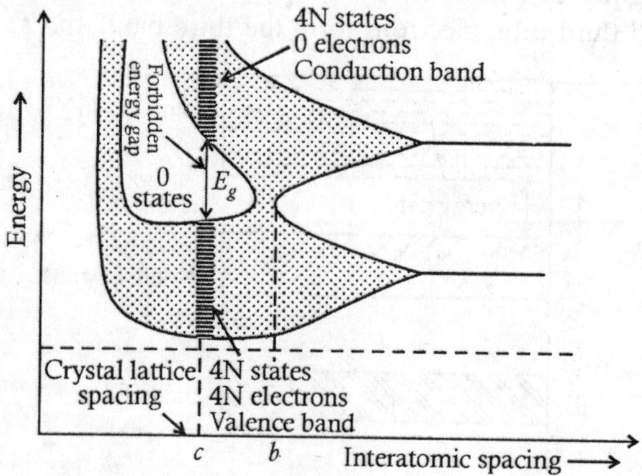

Fig. 9.4: Energy band in a crystal as a function of interatomic spacing

Now, if the interatomic spacing is further reduced, say '*c*' (Fig. 9.4), the interaction between the atoms becomes very large, and the energy band structure becomes as shown in Fig. 9.4. The exact shape of the energy band structure depends on: (*i*) the orientation of the atoms relative to each other in space, and (*ii*) the atomic number of the atom, which may be obtained from the solution of Schrödinger's wave equation. For spacing '*c*', 4*N* states in the valence band are completely filled by 4*N* electrons. This valence band is separated is separated from the upper unfilled or empty band by a forbidden energy gap (E_g). However, the forbidden energy gap contains no allowed energy states. The upper energy band has 4*N* unfilled energy states and this band is called as the **conduction band**.

9.4 ENERGY BAND DIAGRAMS IN SOLIDS

The simple energy level diagram of Fig. 9.2 for electron energies is no longer applicable when one discusses a solid. A solid is formed when atoms bond together. In a solid, the orbit of an electron is influenced not only by the charges in its own atom but by electrons and nuclei of every atom in the solid. Each electron in a solid occupies a different position inside the solid and hence no two electrons can have exactly the same pattern of surrounding charges. Obviously, the orbits of electrons in a solid are different.

When one is considering a solid in bulk, then the simple energy level diagram in Fig. 9.2 modifies to that shown in Fig. 9.5. All electrons belonging to the first orbits have slightly different energy levels because no two electrons see exactly the same charge environment. Since there are billions of first-orbit electrons in a bulk mass, the slightly different energy levels form a *group* or *band*. Similarly, the billions of second orbit electrons, all with slightly different energy levels, form the second energy band. Similarly all third orbit electrons form the third band and so on.

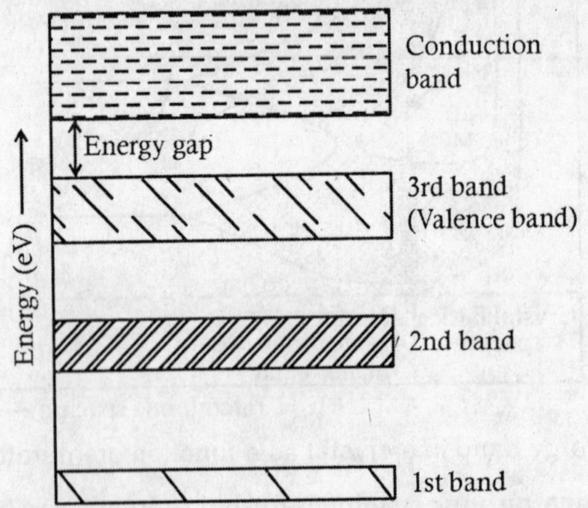

Fig. 9.5: Energy bands of silicon at absolute zero

Silicon is a commonly used semiconductor having atomic number 14. It has 4 electrons in its outermost or valence orbit. Here, the third band becomes the valence band (Fig. 9.5). In Fig. 9.5, all these three bands are shown completely filled. Although the third shell of an isolated atom of silicon is not completely filled as it has only 4 electrons whereas it could accommodate a maximum of 8 electrons, the third energy band or valence band of a bulk silicon material behaves as if completely filled. It is so, because in solid Silicon each atom positions itself between four other silicon atoms, and each of these neighbours share an electron with the central atom. In this manner, each atom now has 8 electrons filling the valence band completely, i.e. all the permissible energy levels in the band are occupied by electrons. Obviously, no electron in a filled band can move and hence an electron in a completely filled band cannot contribute to electric conduction current. At absolute zero temperature, electrons cannot move through the solid Silicon material and hence it acts like a perfect insulator. Beyond the valence band there is a conduction band (Fig. 9.6). At absolute zero temperature, the conduction band is empty, i.e., no electron has enough energy to travel in a conduction band. Definite amount of energy is needed to shift the electron from the valence band to the conduction band. This amount differs from one substance to another substance and helps to classify them as conductors, insulators and semiconductors.

9.5 CLASSIFICATION OF SOLIDS INTO CONDUCTORS (METALS), SEMICONDUCTORS AND INSULATORS

The band structure in a solid determines whether the solid is an insulator or a conductor or a semiconductor,. The bands are filled upto a certain level by the electrons within each atom. The highest band in which electrons are still predominantly attached to their atoms are found is called the *valence band*. This is the band in which the valence (outermost) electrons from each atom will be located. These are the electrons that are the possible carriers of electricity. However, in order for an electron to conduct, it must get up to a slightly higher energy so that is free of the grip of its atom.

Insulators: Let us consider the case when the valence band in full, *i.e.*, when there are no more available energy levels. Then, a valence electron must jump (increase it energy) into the next higher band to be free. If the energy gap between the valence band and the conduction band is too large, then the electron will not be able to make that jump. Such a material will not be a good conductor of electricity, and is called an *insulator*. Forbidden energy gap in an insulator is about 5 eV or even more (Fig. 9.6 (*a*)). The band theory of solids tells us that an insulator is a material in which the valence bands are filled and the forbidden energy gap between them is large and conduction band is at very high energy level for the valence electrons to jump at normal temperatures to it. An insulator does not conduct at room temperature because there are no conduction electrons in it. However, an insulator may conduct if its temperature is very high or if a high voltage is applied across it. This is known as the *breakdown* of the insulator.

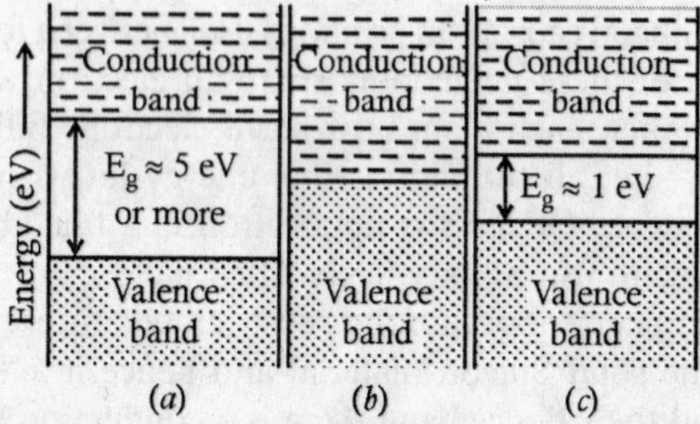

Fig. 9.6: Energy band diagram exhibiting energy band gap in (a) insulators, (*b*) conductors, (*c*) semiconductors

Conductors: There are actually two possibilities for a substance to be a conductor. One is that the valence band is not completely filled. Then an electron in the valence band can get free of its atom by simply jumping to a higher energy level, *i.e.* conduction band within the same band. This jump requires a very small amount of energy, and many electrons can, therefore, make that jumps (Fig. 9.6 (*b*)).

Another situation in which a conductor results stems from the fact that the size of the energy gaps between the valence band and the conduction band are different for different materials. In some materials this gap can be quite small. It can even disappear, when the valence band and conduction band overlap (Fig. 9.6 (*c*)). Therefore, the band theory tells us that we have a *conductor* when

(*i*) The valence band is not filled, so electrons can move to higher states in the valence band and be free, or
(*ii*) When there is no energy gap between the valence band and the conduction band, so electrons can easily make the transitions from the valence to the conduction band e.g., metals.

Semiconductors: There is a case in between 'conductor' and 'insulator'. A material with intermediate properties is called a *semiconductor e.g.* Ge, Si, Tellurium, Selenium, Graphite, etc. In a semiconductor

(*i*) the valence band is filled, and
(*ii*) although there is an energy gap between the valence band and the conduction band yet the energy gap is not very large.

The energy-band diagram for a semiconductor is shown in Fig. 9.6 (*c*). In this case, the forbidden energy gap is of the order of 1 eV (for Silicon,

E_g = 1.12 eV and for Ge, E_g = 0.72 eV). Silicon and Germanium are very good examples of semiconductors. The gaps between their valence and conduction bands are sufficiently small that the normal thermal energy of the solid at room temperature is enough to knock a few electrons into the conduction band. These electrons conduct a current, but, as the name 'semiconductor' implies, a semiconductor does not conduct as best as a real conductor. We will discuss it later on. The conductivity of these substances have a number of peculiarities:

(*i*) The dependence of conductivity on temperature is opposite to that of metals. The conductivity of semiconductors, in contradiction to that of conductors (metals), may decrease rapidly with temperature. At low temperature, a semiconductor may become an insulator. Hence, the distinction between semiconductors and insulators is purely quantitative and, to a large extent, conventional. The resistance of most semiconductors is considerably more sensitive to changes in temperature than metals. Compact temperature meters of high sensitivity may be constructed using semiconducting thermal resistors (thermistors).

(*ii*) In a number of cases semiconductors may possess positive as well as negative conductivity.

Few important properties of semiconductors are:

(*i*) these are crystalline in structure

(*ii*) the resistivity of these materials is less than that of an insulator and more than that of a conductor

(*iii*) the resistivity of these materials decreases rapidly with the increase in temperature

(*iv*) at high temperatures these behave just like good conductors but at low temperatures, they behave as insulators

(*v*) the conductivity of these substances is generally sensitive to temperature, illumination, magnetic field and the impurity added

(*vi*) the distinctive feature of these substances is the type and the degree of binding between atoms.

The semiconductors may be classified into following two types: (*i*) intrinsic (pure) semiconductors, and (*ii*) extrinsic (doped) or impure semiconductors.

9.6 FERMI LEVEL

In the preceding article, we have seen that as far as solid-state theory is concerned only the upper energy bands (valence band) are of interest, since electrons at lower levels practically do not take part in interactions between atoms. How can the behaviour of upper band electrons be described? Since we are dealing with a very large number of electrons, it is natural to use statistical physics methods and consider an aggregate of such electrons as a kind of gas, usually called electron gas.

The state of each electron of such a electron gas may be represented by a point (p_x, p_y, p_z) in momentum space. The direction of motion of an electron is parallel to its radius vector $\boldsymbol{P}$ and the energy of an electron depends on its momentum. Let us consider as a crude approximation that the electrons in solid behave like free particles, *i.e.*, we neglect the potential energy of the field in which the electrons move and the interaction among electrons.

If the electrons are free, the relationship between their energy and momentum is given by $E = (1/2m)p^2$. This means that in momentum space a surface of total energy is a sphere. Such a sphere is usually called a Fermi sphere. One may call the Fermi surface a surface of maximum energy as the states of an electron gas are contained in a sphere of radius

$$P_{\max} = \sqrt{2m E_{\max}}$$

The energy distribution of electrons is shown in Fig. 9.7. The energy E is shown as abscissa, and the number of electrons, N, having a particular value of energy as ordinate. E_0 is the maximum energy that electrons in a metal have at zero degree absolute ($T = 0$ K) and is called the *Fermi level*. At 0 K, all the states below the Fermi level are completely filled by electrons and above Fermi level are completely empty. From reference to the plot for $T = 0$, we see that the number of electrons having zero energy is zero. As the value of energy increases, the number of electrons possessing it also goes up. The most numerous are electrons at the Fermi level, E_0. From the plot for $T > 0$ in Fig. 9.7(*b*), it is seen that some electrons have an energy in excess of E_0, but there are fewer electrons with energy E_0. Those having $E > E_0$ decrease in number with increase of energy. The value of $E_{\max}$ rises with temperature.

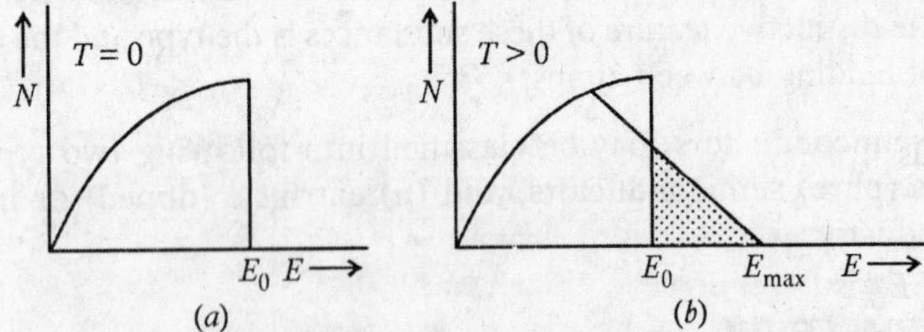

Fig. 9.7: Energy distribution of electrons in a metal

From the above discussion one can easily conclude that the concept of Fermi level serves the reference energy level from which all other energies are conveniently measured. The probability $F(E)$ of a state corresponding to energy E being occupied by an electron at temperature T K is given by

$$F(E) = \frac{1}{1+\exp\left(\frac{E-E_F}{kT}\right)} \qquad (1)$$

Here k is Boltzmann constant. Three cases of interest are:

(*i*) At $T = 0$ K, if $E > E_F$ then $F(E) = 0$, i.e., energy state is empty
(*ii*) At T = 0 K, if $E > E_F$ then $F(E) = 0$, i.e., energy state is occupied by an electron
(*iii*) At $T \neq 0$ K, and $E = E_F$ then $F(E) = 1/2$, i.e., energy state is 50% occupied.

Obviously, one finds that at $T = 0$ K, all the energy states above E_F are empty, whereas all those below Fermi energy (E_F) are filled with electrons. With the rise in temperature, states above Fermi energy level (E_F) no more remain empty. They are then occupied by the electrons to some extent. Figure 9.8 shows the position of Fermi level in energy band diagram in the case *pure* or *intrinsic* semiconductor. Figure 9.8 clearly depicts that Fermi level for an intrinsic semiconductor is situated in the middle of forbidden gap, *i.e.*, between conduction and valence band, and the position of Fermi level is independent of temperature.

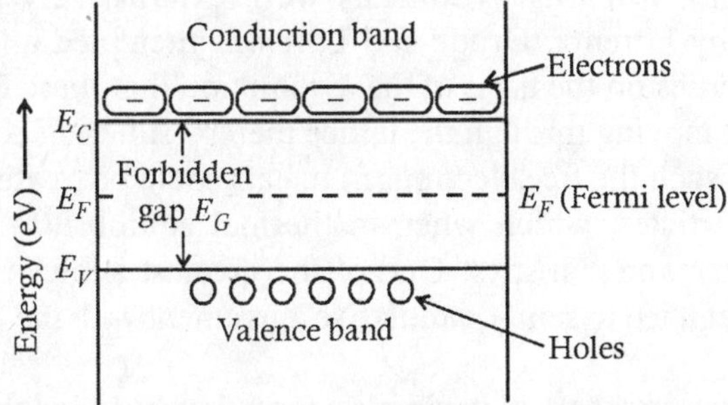

Fig. 9.8: Fermi level in intrinsic semiconductor at 0 K

Thus $$E_G = \frac{E_C + E_V}{2}$$

and $$E_G = 2(E_F - E_V)$$

where E_G is forbidden energy gap (≈ 1 eV)

This reveals that when the temperature is raised there is a greater possibility of electrons being found above the Fermi level with an equal possibility of finding an electron vacancy so called 'hole' below Fermi level.

9.7 FREE ELECTRON THEORY OF METALS

The outstanding properties of metals are:

1. High electrical and thermal conductivities,
2. **Ohms law:** Metallic conductors obey Ohms law, which states that the current density (j) in the steady state is proportional to the electric field strength (E) *i.e.*, $J = \sigma E$ where σ the electrical conductivity is independent of j or E.
3. **Temperature Dependence of Conductivity:** Resistivity of metals (ρ) above Debye temperature is proportional to the absolute temperature, *i.e.* $\rho \propto T$ and at low temperatures the resistivity of metals is proportional to the fifth power of absolute temperature, *i.e.* $\rho \propto T^6$.
4. **Wiedemann-Frantz Ratio:** The ratio of thermal to the electrical conductivity $\frac{k}{\rho}$ has the same value for nearly all metals at a give temperature and $\left(\frac{1}{T}\frac{k}{\sigma}\right)$ has the same value at any temperature.

The idea that the large electrical and thermal conductivity of metallic substances might be explained by the presence of large concentration of mobile free electrons in the metals was first proposed by Drude in 1900. The implications of this hypothesis were exhaustively investigated subsequently by Lorentz. Drude and Lorentz, attempted an explanation of these properties on the basis of the assumption that these free electrons are capable of moving through the lattice thereby suffering collisions with the atoms. As such the free electrons in metals could be treated as an ideal gas of free particles, which when in thermal equilibrium would obey Maxwell-Boltzmann statistics. One of the greatest achievements of the theory was that it led to semi-quantitative agreement with the Wiedemann-Frantz Law.

The free electron theory in its simplest form, however, led to a prediction of the electronic component of the specific heat which was in serious disagreement with experimental results. Another difficulty encountered in classical theory pertains to the magnetic properties of free electrons. Each electron has a magnetic moment associated with its spin: and classically, should therefore give rise a paramagnetic susceptibility inversely proportional to the temperature. Experimental results on the other hand, show that the paramagnetism of metals is nearly independent of temperature.

There is further difficulty: Using the classical theory we can not understand the occurrence of long electronic mean free paths. From many types of experiments it is abundantly clear that a conduction electron in a metal can move freely in a straight path over many atomic distances, undeflected by collision with other conduction electrons or by collision with the atom cores. In a very pure specimen at low temperatures the mean free path may be as long as 10^8 or 10^9 interatomic spacings (more than 1 cm). Vastly longer than we would expect from the known sizes of atoms. We must ask why condensed metallic matter is so transparent to conduction electrons. The conduction electrons act in this respect as a gas of non interacting particles.

There are two parts to the answer to our question.

(*a*) A conduction electron is not deflected by ion cores arranged on a periodic lattice, because matter waves propagate freely in a periodic structure

(*b*) A conduction electron is scattered only infrequently by other conduction electrons. This property is a consequence of the Pauli's exclusion principle.

These difficulties were resolved by Sommerfeld using Fermi-Dirac statistics rather than the classical Boltzmann statistics. The Fermi-Dirac free electron picture serves as a very simple and conceptually quite direct way of discussing and visualising transport effects in metals.

Classical Free Electron Theory of Metals

A model of metal seen as a combination of two parts: the ions, which are taken to be fixed positive charges, and the itinerant electrons which are treated as a gas that permeates the inter ionic space. The itinerant electrons in a metal are the valence electrons of the constituent atoms. They produce the electrical conductivity and so are called conduction electrons, as distinguished from the electrons localized in the cores of the ions. The conduction electrons are supposed to move freely throughout the metal, constrained only by the surface of the sample. In the early theory, these collisions were supposed to be with the fixed ions, but this view had to be abandoned with the advent of quantum mechanics. Whatever the source of scattering, in an electric field the electrons acquire a systematic drift superimposed on the random motion, leading to an electric current.

Drude Model

The first version introduced by P. Drude in the early 1900s, with improvements soon after by H. A. Lorentz. It treated the conduction electrons as a classical dilute gas that obeys Maxwell-Boltzmann statistics. The values of most parameters derived from this version were in serious disagreement with experiment, but the model did explain Ohm's law and fortuitously gave good agreement with the Wiedemann-Franz law. An outstanding difficulty was the prediction of a large specific heat and magnetic susceptibility for metals, contrary to observation. There are also properties of the Hall and magnetoresistance effects that could not be explained, and the electron mean free path deduced from experiment was vary large compared with the interionic spacing. This long mean free path was inconsistent with the Drude-Lorentz picture of electrons bouncing elastically off fixed, impenetrable ions. Now we present Drude's theory in detail.

According to Drude's theory the kinetic velocities of electrons are assumed to have a root mean square velocity $\bar{c}$ given by kinetic theory of gases in conjugation with the law of equipartion of energy, $\bar{c}$ can be obtained as follows:

We have for unit volume of the metal

$$p = \frac{1}{3}\rho\bar{c}^2 = \frac{1}{3}mn\,\bar{c}^2 \tag{2}$$

where n is the number of free electrons in unit volume, m is the mass of the electron and ρ is the density of the gas. For molar volume of the metal, we have

$$p = \frac{1}{3}\rho\bar{c}^2 = \frac{1}{3}\frac{mN_A\bar{c}^2}{V_m}$$

where V_m is molar volume and N_A is Avogadro's number. Now one finds

$$pV_m = \frac{1}{3}mN_A^2\bar{c}^2 = R_uT$$

Thus, one finds

$$m\bar{c}^2 = \left[\frac{3R_uT}{N_A}\right] = 3k_BT$$

where k_B and R_u are Boltzmann and universal gas constants respectively. Now, the kinetic energy of the electron is

$$\frac{1}{2}m\bar{c}^2 = \frac{3}{2}k_BT \tag{3}$$

and

$$\bar{c} = \sqrt{\frac{3k_BT}{m}} \tag{4}$$

Let us find the value of $\bar{c}$ at 20° C.

$$\bar{c} = \left(\frac{3 \times 1.38 \times 10^{-23} \times 293}{9.1 \times 10^{-31}}\right)^{1/2}$$

$$= 1.154 \times 10^5 \text{ m/s.}$$

From Eq. (4) it is obvious that the root mean square velocity of the electron is directly proportional to the square root of absolute temperature of the metal. We can see that at room temperature, the drift velocity imparted to the electrons by an applied electric filed is very much smaller than the average thermal velocity. However, the time τ taken by the electrons in traversing the distance λ will thus be determined not by the drift velocity due to the field but by much greater velocity $\bar{c}$ due to the kinetic thermal velocity. We have

$$\tau = \frac{\lambda}{\bar{c}} = \lambda\left[\sqrt{\frac{m}{3k_B T}}\right] \tag{5}$$

In order to test the validity of any theory of electrical conductivity, one will have to see whether they can account for Ohm's law ($V = IR$) or not. Ohm's law can be rewritten in terms of current density, as $J = I/A$, and field strength along the conductor,

$$E = -\frac{dV}{dx}. \quad \text{Now, we have}$$

$$\frac{V}{R} = I = AJ$$

or

$$\frac{I}{J} = \frac{AR}{V} = \frac{A\rho l}{AEl} = \frac{\rho}{E}$$

or

$$J = \frac{E}{\rho} = \sigma E \tag{6}$$

where σ is the conductivity and ρ is the resistivity of a conductor of length l and area of cross-section A. σ or ρ defined as

$$\rho = \frac{1}{\sigma} = \frac{RA}{l} \tag{7}$$

Now, we assume that there are n free electrons per cubic metre, and that in the absence of any applied field, these are moving about in all directions with no net velocity just similar to gas molecules in a container. When we apply a field E_x in x-direction, all electrons are accelerated in the x-direction, with the acceleration of the electrons a_{ix} as

$$a_{ix} = -\left(\frac{eE_x}{m}\right)$$

Alternatively, one can write this equation as

$$\left(\frac{dv_{ix}}{dt}\right) = E_x = -\left(\frac{eE_x}{m}\right) \tag{8}$$

where v_{ix} is the x component of the velocity of the ith electron and the subscript E_x shows that the acceleration arises from the applied field. We note that right hand side of this equation is the same for all electrons. One may write Eq. (8) in the another form as follows:

$$\left[\frac{d}{dt} < v_x >\right] E_x = -\left[\frac{eE_x}{m}\right] \tag{9}$$

where $< v_x >$ is the average velocity of all free electrons. $< v_x >$ is given by

$$< v_x > = \frac{1}{n}\left[\sum_{i=1}^{n} v_{ix}\right]$$

Here we have used $< v_x >$ to denote the average value of v. We know that the current density J is the charge flowing through unit area in unit time, so that J_x, *i.e.* x component must be the number of electrons times the charge on each in a volume of unit area and length v_x, *i.e.*

$$J_x = n < v_x >(-e) = -nev_x \tag{10}$$

The negative sign in Eq. (10), which is coupled to the charge on an electron, underlies the fact that the conventional direction of current, *i.e.*, of J_x is in the opposite direction to $< v_x >$. Differentiating Eq. (10), one obtains

$$\frac{d}{dt}[< v_x >] = -\frac{1}{ne}\left[\frac{dJ_x}{dt}\right] \tag{11}$$

On comparison Eq. (11) with Eq. (9), one obtains

$$\left(\frac{1}{ne}\right)\left[\frac{dJ_x}{dt}\right] = \left(\frac{e}{m}\right) E_x$$

or

$$\left[\frac{dJ_x}{dt}\right] = \text{constant} \times E_x \tag{12}$$

Obviously, for a constant E_x, the above free electron model leads to a current which increases linearly with time which is just a Ohm's law.

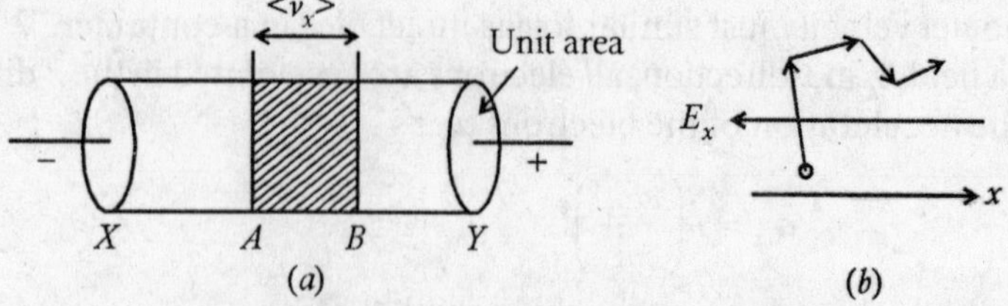

Fig. 9.9: Calculation of current density, J_x

From Eq. (12), we can see that the law can only be satisfied, i.e. $\frac{dJ_x}{dt} = 0,$

if $\frac{d}{dt} < v_x > = 0$. There is nothing wrong in just stated. However, the difficulty arises when we consider the omitted decelerating term, which when added to $\left[\frac{d}{dt} < v_x >\right]_{E_x}$, gives a net $\frac{d}{dt} < v_x > = 0$.

The above results clearly reveal that the electrons cannot be completely free. We know that a conductor always warms up when a current flows through it. This means that conduction electrons must collide and exchange energy with the ions. This process is usually referred as **electron-lattice scattering**. Obviously, the equation governing the sum of the accelerations should, therefore, contain two terms:

$$\left[\frac{d}{dt} < v_x >\right]_{E_x} + \left[\frac{d}{dt} < v_x >\right]_{\text{ele-lat}} = 0 \tag{13}$$

One can in principle find details of applied electric and magnetic fields as well as temperature gradient in a equation of type (13). One can solve it in order to obtain the resultant electric and thermal currents. However, this is an exceedingly difficult operation. One faces the main problem in obtaining the scattering term, which is possible only by applying quantum mechanical technique. Moreover, this method normally ends up with an integro-differential equation which is extremely difficult to solve. One can obtain an expression for $\left[\frac{d}{dt} < v_x >\right]_{\text{ele-lat}}$ by passing the above difficulties.

Let us suppose that at $t = 0$ the average velocity of the electron is $< v_{x_0} >_0$. Just at that instant, we switch the field off, so that $<v_x>$ subsequently tends to zero. In other words, one can say that the velocity will be randomized by electron-lattice collision. One can assume that this process follows the simplest law of decay.

$$<v_x> = <v_x>_0 e^{-t/\tau} \tag{14}$$

τ is a constant of the system, known as the **relaxation time.** Obviously, τ gives measure of the time that the system takes to relax when a constraint (the electric field) is removed [Fig. 9.9 (*a*)]. In fact, τ is the time taken for the drift velocity to decay to $1/e$ of its initial value. During this relaxation process, $E_x = 0$. One can use Eq. (14) to obtain electron-lattice term in Eq. (13) and this is done in such a manners that details of the collision process are avoided. One finally obtains

$$\left[\frac{d}{dt} < v_{x_0} >_0\right]_{\text{ele-lat}} = -\frac{< v_x >_0 e^{-t/\tau}}{\tau}$$

$$= -\frac{< v_x >}{\tau} \tag{15}$$

Let us assume that Eq. (15) remains unchanged by the presence of a field. We have from Eqs. (9), (13) and (15)

$$-\left(\frac{e}{m}\right)E_x - \frac{<v_x>}{\tau} = 0$$

or
$$<v_x> = -\left[\frac{e\tau E_x}{m}\right] \tag{16}$$

$<v_x>$ is the steady average velocity imposed by the field and proportional to it. This is called drift velocity of the electrons. The constant of proportionality $\mu = \left(\frac{e\tau}{m}\right)$ is called **mobility**. The drift velocity for the unit electric field is obtained as

$$\mu = \frac{<v_x>}{E_x} = \frac{e\tau}{m} \tag{17}$$

The unit of μ are $m^2V^{-1}s^{-1}$ (meter^2/volt/sec)

Now, combining Eqs. (6), (10) and (16), one obtains

$$\sigma = \frac{ne^2\tau}{m} \quad \text{or} \quad \rho = \frac{m}{ne^2\tau} \tag{18}$$

and
$$\sigma = ne\mu \quad \text{or} \quad \rho = \frac{1}{ne\mu} \tag{19}$$

We may note that Eq. (19) is of great importance in the theory of conductivity and embodies the difference between the electrical properties of metals and semi-conductors. We can that in metals, all the valence electrons are free at all temperature variation of σ is essentially the temperature variation of μ. However, in semiconductors, n may also vary with temperature and also it is usually the important factor.

Alternate Method

Taking the acceleration of an electron of mass m as eE/m, one can write the mean drift velocity as

$$\boldsymbol{v}_d = \left(\frac{-e\boldsymbol{E}}{m}\right)\tau \tag{20}$$

where τ is realization time. If there are a total of n electrons/m^3 in the metal, all with constant drift velocity $\boldsymbol{v}_d$, one obtains the following relation for the net electric current density $\boldsymbol{J}$, we have

$$\boldsymbol{J} = -ne\boldsymbol{v}_d = \left(\frac{ne^2\tau}{m}\right)\boldsymbol{E}$$

$$= \sigma\boldsymbol{E} \tag{21}$$

where σ is the positive scalar quantity

$$\sigma = \frac{ne^2\tau}{m} = ne\mu$$

where $$|\mu| = v_d/E \tag{22}$$

is called the electrical conductivity $\left(\sigma = \frac{1}{\rho}, \text{where } \rho \text{ is the electrical resistivity}\right)$.

We may note that form of the Eq. (22) remains the same in all models including those based on quantum physics. The difference lies only in the way *n*, *m* and τ are defined.

Drude put Eq. (22) in a useful form by exploiting the idea of kinetic theory such as to define τ in terms of thermal velocity *v* be expressed as

$$\tau = \frac{\lambda}{v_{\text{the}}} \tag{23}$$

$$\frac{3}{2}k_B T = \frac{1}{2}mv_{\text{rms}}^2 = \frac{1}{2}mcv_{\text{the}}^2 \tag{24}$$

where λ is the electron mean free path. Using these relations, one can write Eq. (22) as

$$\sigma = \frac{ne^2\lambda}{(3mk_B T)^{1/2}};\ \rho = \sqrt{\frac{3mk_B T}{ne^2\lambda}} \tag{25}$$

It was assumed by Drude and Lorentz that λ is independent of temperature and that is of the order of interatomic distance, *i.e.*, $\rho \propto \sqrt{T}$. This reveals that the specific resistance of an electric conductor is directly proportional to the square root of the absolute temperature. However, this is not in agreement with the experimental observation that $\sigma \propto T$. In addition to this discrepancy, it is also not correct to assume that the mean free path is independent of temperature. Clearly, Drude's classical theory is not acceptable. However, one can derive Ohm's law since the conductivity in Eq. (25) is independent of the field.

Drawbacks of Classical Theory

(*i*) Heat Capacity of the Electron Gas: According to classical theory of free electrons, all the valence electrons in a metal can absorbs thermal energy. According to the law of equipartition of energy, every free electron in a metal has an average kinetic energy $\frac{3}{2}k_B T$. Now, the energy associated with one *k* mol of such metal is expressed as

$$U = \frac{3}{2}N_A k_B T \tag{26}$$

where N_A is number of free electrons in one *k* mol of a metal.

If we supply heat to the metal, these free electrons also absorb part of the heat, and one obtains the **molar electronic specific heat as**

$$(C_v)_{el} = \left[\frac{dU}{dT}\right] = \frac{3}{2}k_B N_A = 1.3\, R_u \tag{27}$$

where R_u is universal gas constant, one obtains

$$(C_v)_{el} = 1.5 \times 1.38 \times 10^{-23} \times 6.02 \times 10^{26}$$
$$= 12.5 \times 10^3 \text{ J/k mol/K}$$
$$= 12.5 \text{ kJ/k mol/K.}$$

The above value (=1.5*R*) due to free electrons is nearly 100 times greater than the experimentally predicted value. The heat capacity of a solid due to atomic vibrations is $3/R_u$ and therefore free electrons should make a significant contribution to the total specific heat of a metal. We may note that at least at high temperatures, Dulong and Petit law regarding latent specific heat holds good and the total specific heat of a solid is 3 R_u. Obviously, this means that the free electrons do not contribute significantly to the heat capacity of a metal. One can conclude that the law of equipartition and hence Maxwell-Boltzmann classical statistics should not be applied to evaluate the electronic specific heat in metals.

(ii) **Mean Free Path:** We have the microscopic expression for the resistivity of a metal

$$\rho = \frac{m}{ne^2\tau}$$

we have the resistivity of copper at 20° C is 1.69×10^{-8} Ω-m and *n* (the concentration of free electrons in copper) = $8.5 \times 10^{28}/\text{m}^3$. Therefore,

$$\tau = \frac{m}{ne^2\rho} = \frac{9.1 \times 10^{-31}}{8.5 \times 10^{28} \times (1.6 \times 10^{-19})^2 \times 1.69 \times 10^{-8}}$$
$$= 2.47 \times 10^{-14} \text{ s}$$

Now, $$\tau = \frac{\lambda}{\bar{c}}$$

or $$\lambda = \tau\bar{c} = 2.47 \times 10^{-14} \times 1.154 \times 10^5$$
$$= 2.85 \text{ nm}$$

The experimentally observed value for λ is about 10 times above this theoretical value. Clearly, classical theory could not explain the large variation in λ values.

(*iii*) **Widemann-Franz Law (Relation between Electrical Conductivity and Thermal Conductivity):** We know that conduction in solids may take place either through the mechanism of the atoms or through that of free electrons. Moreover, in metals, electrons are the principal carriers of heat energy. Free electrons perform this by being excited by energetic scattering centres and carrying the extra energy to another scattering centre in the cooler part of the metal.

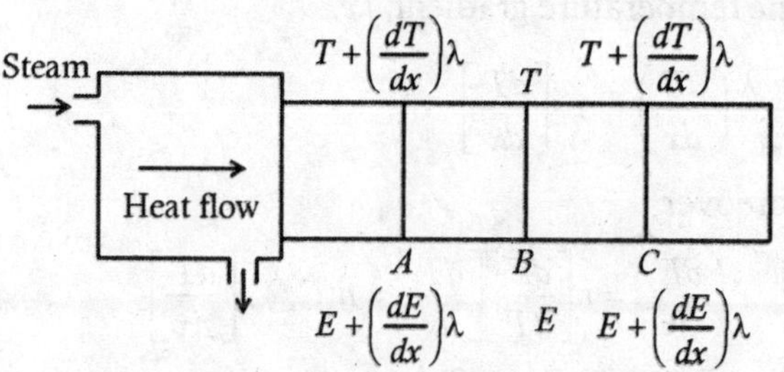

Fig. 9.10: Heat flow through a conductor

Let us consider a copper conductor in the form of a rod of uniform cross-section (say 1 sq.m). We assume that a thermal gradient be established. Normal to the direction of heat flow, consider three layers at *A*, *B* and *C*. Let us take that each layer is separated by a distance λ (the mean free path of the electron gas in the metal).

As shown in the Fig. 9.10 $\left[T+\left(\frac{dT}{dx}\right)\lambda\right]$, T *and* $\left[T-\left(\frac{dT}{dx}\right)\lambda\right]$ are the steady temperature of the layers. Thermal energy of an electron in these layers *A*, *B* and *C* be $\left[E+\left(\frac{dE}{dx}\right)\lambda\right]$, E *and* $\left[E-\left(\frac{dE}{dx}\right)\lambda\right]$ respectively.

Obviously, the excess energy carried by an electron from *A* to *B* is $\left(\frac{dE}{dx}\right)\lambda$. According to kinetic theory of gases, the number of electrons flowing in a given direction through unit area for unit time is $\frac{1}{6}n\bar{c}$, where *n* is the number of free electrons per cubic metre and $\bar{c}$ is the average thermal velocity of the electron. Now, the excess of energy transported by the process of conduction through unit time at the middle layer *B* is $\frac{1}{6}n\bar{c}\lambda\left(\frac{dE}{dx}\right)$. Similarly, one can see that the deficit of energy transported through *B* in opposite direction is $\frac{1}{6}n\bar{c}\lambda\left(\frac{dE}{dx}\right)$. Clearly the net heat energy transported through unit area in unit time from A to B

$$=\left[\frac{1}{6}n\bar{c}\lambda\left(\frac{dE}{dx}\right)\right]-\frac{1}{6}n\bar{c}\lambda\left(\frac{dE}{dx}\right)=\frac{n\bar{c}\lambda}{3}\left(\frac{dE}{dx}\right) \quad (28)$$

Now, this must be equal to the product of the thermal conductivity and the temperature gradient, *i.e.*

$$\frac{n\bar{c}\lambda}{3}\left(\frac{dE}{dx}\right)=\sigma_T\left[\frac{dT}{dx}\right] \quad (29)$$

Moreover

$$n\left(\frac{dE}{dx}\right)=n\left[\frac{dE}{dT}\right]\left[\frac{dT}{dx}\right]=n(C_v)_{el}\left[\frac{dT}{dx}\right] \quad (30)$$

Using Eq. (30), Eq. (29) becomes

$$\frac{n\bar{c}\lambda}{3}(C_v)_{el}\left[\frac{dT}{dx}\right]=\sigma_T\left[\frac{dT}{dx}\right]$$

or $$\sigma_T=\frac{1}{3}(C_v)_{el}\bar{c}n\lambda_{-4} \quad (31)$$

Now, the electronic specific heat associated with each electron is given by

$$(C_v)_{el}=\frac{3}{2}k_B \text{ and } \bar{c}=\sqrt{\frac{3k_BT}{m}} \quad (32)$$

$$\therefore \quad \sigma_T=\frac{1}{3}\lambda_n(1.5\,k_B)\sqrt{\frac{3k_BT}{m}}=\frac{n\lambda k_B}{2}\sqrt{\frac{3k_BT}{m}} \quad (33)$$

We have the expression for electrical conductivity as

$$\sigma=\frac{ne^2\lambda}{\sqrt{3mk_BT}} \quad (34)$$

Now, $$\frac{\sigma_T}{\sigma}=\left(\frac{n\lambda k_B}{2}\right)\sqrt{\frac{3k_BT}{m}}\left(\frac{\sqrt{3mk_BT}}{m}\right)$$

$$=\frac{3T}{2}\left(\frac{k_B}{e}\right)^2 \quad (35)$$

The above is **Wiedemann and Franz law,** which is classical one. According to relation (35) σ_T/σ is linearly dependent on absolute temperature. Further, the multiplying constant

$$L = \frac{3}{2}\left(\frac{k_B}{e}\right)^2 = 1.12 \times 10^{-8}\ \text{W}\Omega/\text{k}^2 \quad (36)$$

is called the **Lorentz number.**

One can check the validity of Eq. (35) by using the experimentally determined values of σ and σ_T. Eq. (35) is

$$\frac{\sigma_T}{\sigma(T)} = L$$

We have for copper at 20° C the electrical resistivity and thermal conductivity as $1.72 \times 10^{-8}\ \Omega$ and 386 $\text{Wm}^{-1}\text{K}^{-1}$ respectively. Now we have

$$L = \frac{\sigma_T}{\sigma(T)} = \frac{386}{581 \times 10^7 \times 293} = 2.26 \times 10^{-8}\ \text{W}\Omega/\text{k}^2$$

We can see that the value of Lorentz number does not agree with the calculated value from Eq. (36). This means that the classical assumption, *i.e.* all the free electrons of a metal particle in thermal conduction is not correct. We may note that according to quantum theory only the electrons near the Fermi level in accordance with the Fermi-Dirac statistics take part in thermal and electrical conduction.

(*iv*) Relaxation Time, Collision Time and Mean free path

(*a*) ***Relaxation Time***: Let us suppose that at a given time, $t = 0$, the average velocity of the electron is $<v_x>_0$. We switch the field off just at that instant, so the $< v_x >$ will be randomized by electron-lattice collisions. Suppose that this process follows the simplest law of decay

$$<v_x> = < v_x >_0 e^{-t/\tau} \quad (37)$$

where τ is a constant and called relaxation time. τ gives a measure of the time that the system takes to relax when a constraint (the electric field) is removed. For $t = \tau$, Eq. (37) takes the form

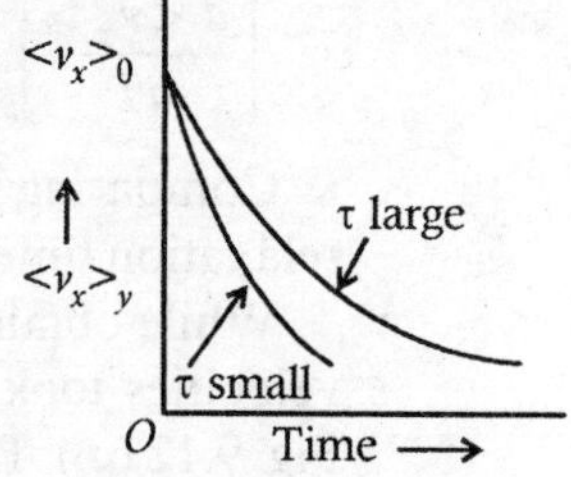

Fig. 9.11: Relaxation of a system of electrons to equilibrium after removing the electric field (Eq. 38)

$$<v_x> = \frac{< v_x >_0}{e}$$

Obviously, one can state the relaxation time as the time taken for the drift velocity to decay to $1/e$ of its initial value. Differentiating Eq. (37), one obtains

$$\left[\frac{d}{dt} < v_x >\right]_{\text{elec-latt}} = -\frac{< v_x >_0\ e^{-t/\tau}}{\tau}$$

or $$\left[\frac{d}{dt}<v_x>\right]_{\text{elec-latt}} = -\frac{<v_x>}{\tau} \quad (38)$$

Figure 9.11 shows the relaxation of a system of electrons to equilibrium after removing the electric field as represented by Eq. 38.

(*b*) ***Collision Time***: Let us suppose that the probability of an electron making a collision in time dt is dt/τ_T. This means that on an average there are $1/\tau_c$ collisions/sec. we will that $\tau = \tau_c$.

We assume that all electron collisions are elastic, *i.e.* energy is conserved. This means that the speed of an electron is the same before and after collisions. However, this assumption is not strictly correct, because a conductor gets warm up when a current is allowed to flow through it. This implies that energy is transferred to the lattice during electron-lattice collisions. However, at each collision, the actual change in energy is minute and the fractional change in velocity in a metal is typically 1 in 10^6. Clearly, the said assumption is reasonably correct. Let us assume that the velocities after collision are completely random. Thus, we have

$<v_x>_{\text{after}} = 0$, since $<v_x>_{\text{before}} = <v_x>$

This means that the change of $<v_x>$ on collision = $-<v_x>$. There are $1/\tau_c$ collisions/sec. Thus we have the rate of change of

$$<v_x> = \frac{<v_x>}{\tau_c} \quad (39)$$

The LHS of Eq. (39) can also be written as

$$\left[\frac{d<v_x>}{dT}\right]_{\text{elec-latt}}$$

Comparing Eq. (39) with Eq. (38), we find that $\tau = \tau_c$ and the relaxation time, there, equals the mean time between collisions.

While obtaining the above result, we have assumed that the velocities took up random directions after each collision (See Fig. 9.12 (*a*)). This is equivalent to saying that after each collision the electron had no memory of what went before. However, if the scattering is very weak, we can say that in such a case our assumption is unrealistic. If, on an average, the velocity direction changes by θ on collision (Fig. 9.12 (*b*)), then the change of $<v_x>$ is on an average, $-<v_x>(1 - \cos\theta)$. Thus, we have

$$\tau = \frac{\tau_c}{[1 - <\cos\theta>]} \quad (40)$$

where τ is the relaxation time and t_c is the mean time between collision.

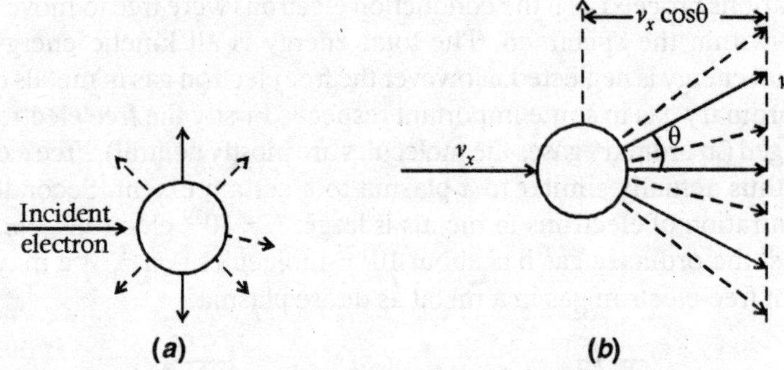

Fig. 9.12: (*a*) Scattering is equally probable in all directions, clearly the average angle is 90° (*b*) Scattering predominantly in the forward direction at an average angle θ. The change in any v_x is therefore, on an average $v_x(1 - \cos\theta)$.

(*v*) ***Mean Free Path (λ)***: One can define the mean free path λ of an electron as

$$\lambda = \bar{c}\,\tau_c \tag{41}$$

When the obstacles are hard spheres, the mean free path λ is determined by the concentration of these obstacles. In that case, for a given concentration of obstacles the collision time τ_c becomes inversely proportional to the velocity $\bar{c}$ of the electrons. Then, being $<\cos\theta>$ independent of velocity of electrons, the relaxation time τ becomes proportional to τ (in accordance with Eq. 40), *i.e.* inversely proportional to the the velocity $\bar{c}$. However, in the preceding section, we have considered relaxation time to have a single value, having no relevance to the distribution of electron velocities. However, we may note that for metals the relaxation time τ, which eventually determines the conductivity of metals, always pertains to a single velocity, *i.e.* one corresponding to the Fermi level E_f.

9.8 FREE ELECTRON MODEL

In the free electron model, the conductions electrons* are assumed to be free, except for a potential at the surface (Fig. 9.13), which has the effect of confining the electrons to the interior of the specimen. According to this model, the conduction electrons move about inside the specimen without any collision, except for an occasional reflection from the surface, much like the molecules in ideal gas. The forces between the conduction electrons and ion cores are neglected in the free electrons approximation: all

*The valence electrons are also called conduction electrons because they are responsible for the conduction of electricity in metals.

calculations proceed as if the conduction electrons were free to move every where within the specimen. The total energy is all kinetic energy, the potential energy is neglected. However the free electron gas in metals differs from ordinary gas in some important respects. Firstly the free electron gas is charged (in ordinary gases the molecules are mostly neutral). Free electron gas is thus actually similar to a plasma to a certain extent. Secondly the concentration of electrons in metals is large: $N \approx 10^{29}$ electrons – m^{-3}. By contrast the ordinary gas has about 10^{+25} molecules – m^{-3}. We may thus think of free-electron gas in a metal as dense plasma.

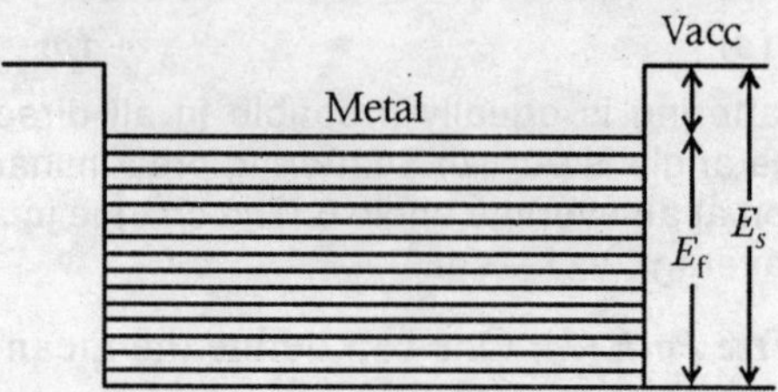

Fig. 9.13: The Sommerfeld model E_s is the energy difference between an electron at rest inside the metal and one at rest in vacuum. At $T = 0$ all energy levels up to E_f are filled; all higher ones are empty. Work function $\phi = E_s - E_f$.

Our model for metals, in short, is a box full of electrons and nothing else. Our problem is how to treat a box full of electrons quantum mechanically. First we have to apply the Pauli's exclusion principle, *i.e.* no two particles can have all the same quantum numbers, but we do not know the quantum numbers of particles in a box. The electrons in a box are free, the only constraint to which they are subjected is that they are in a box. We must find the allowed states in a box.

Free Electron Gas in a One-Dimensional Box

Consider an electron of mass m confined to a length L by infinite barriers Fig. 9.14. The wave function $\psi_n(x)$ of the electron is a solution of the Schrödinger equation

$$H\psi = E\psi \tag{42}$$

where Hamiltonian H is equal to the sum of kinetic energy and potential energy and E is a set of allowed energies or eigenvalues of the electron in the orbital.* Since the total energy is kinetic (The potential energy is neglected), we have

*We use the word orbital to denote a solution of the wave equation for a system of only one electron. The term allows us to distinguish an exact quantum state of the wave equation of a system of N electrons and approximate quantum state which we construct by assigning the N electrons to N different orbitals. Where each orbital is a solution of wave equation for one-electron. The orbital model is exact only if there are no interactions between electrons.

$$H = \frac{p^2}{2m}$$

where p is the momentum. In quantum mechanics p, *i.e.* the momentum is represented by operator $-i\hbar\frac{d}{dx}$, so that

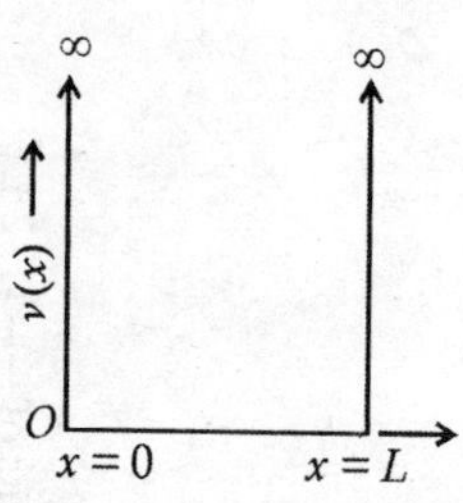

Fig. 9.14: An electron of mass m confined to alength L by infinite barriers

$$H\psi_n = \frac{\hbar^2}{2m}\frac{d^2\psi_n}{dx^2} = E_n\psi_n \quad (43)$$

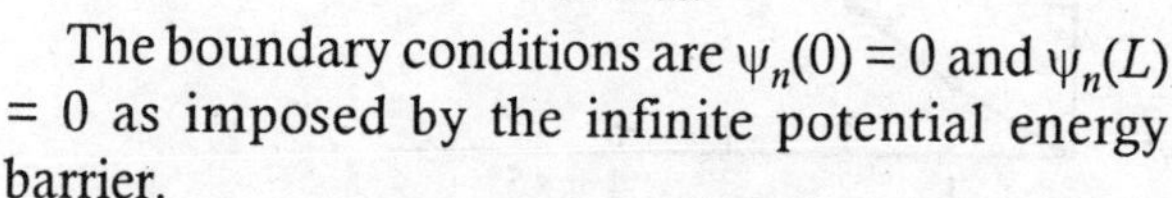

The boundary conditions are $\psi_n(0) = 0$ and $\psi_n(L) = 0$ as imposed by the infinite potential energy barrier.

The general solution of equation (43) is

$$\psi_n(x) = Ae^{ikx} + Be^{-ikx} \quad (44)$$

where $$k^2 = \left(\frac{2m}{\hbar^2}\right)E_n \quad (45)$$

and A and B are arbitrary contants.

The first boundary condition $\psi_n(0) = 0$ yields $A = -B$ leaving the wave function to be sine like. Applying the second boundary condition one singles out those solutions for which

$$\sin KL = 0$$

or $$\left.K_n = \frac{n\pi}{L} \text{ with } n = 1,2,3,\ldots\right\} \quad (46)$$

and the allowed energy eigen values E_n are given by

$$E_n = \frac{\hbar^2}{2m}\left(\frac{n\pi}{L}\right)^2 \quad (47)$$

and the corresponding wave functions are

$$\psi_n = A\sin\left(\frac{n\pi x}{L}\right) \quad (48)$$

Thus, only for integral values of n there are allowed wave functions $\psi_n(x)$ and the corresponding energy values E_n.

The number n is called quantum number. The plot of energy E_n versus n is given in Fig. (9.15). The energy spectrum evidently consists of discrete energy levels, with their separation depending on $\left(\frac{n^2}{L^2}\right)$. If L is large, the energy levels are thus spaced very closely together. For example if $L = 1$ cm, $E_n - E_{n+1} = 3.5 \times 10^{-19}$ eV. It is only when L is of atomic dimensions that the separation between the levels becomes appreciable.

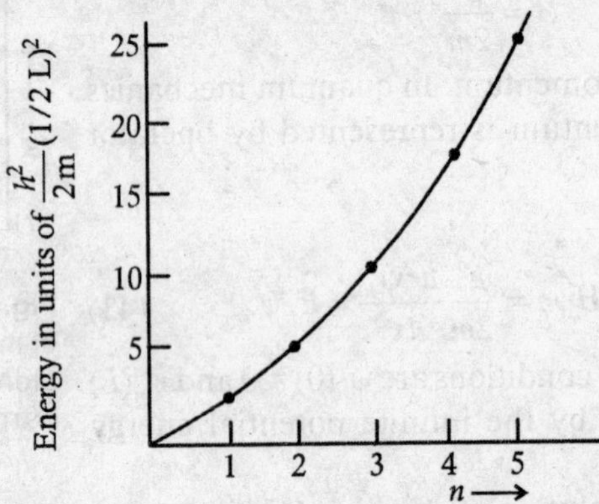

Fig. 9.15: The energy is a quadratic function of the quantum number *n*, for a free electron confined to a length *L* in one dimension

The value of the constant *A* in equation (47) is so chosen that there is unit probability of finding the electron somewhere on the line. Because $\psi_n^*(x)\psi_n(x)$ is the probability that the electron is in the line segment *dx* at *x*, we require that

$$\int_0^L \psi^*_n(x)\psi_n(x)dx = 1$$

or

$$\int_0^L A^2 \sin^2 \frac{n\pi x}{L} = 1$$

or

$$A^2 = \frac{2}{L}, \text{ i.e. } A = \sqrt{\frac{2}{L}} \quad \text{...(49)}$$

So the normalised wave functions are

$$\psi_n(x) = \frac{2}{L}\sin\frac{n\pi x}{L}$$

The first few lower energy state wave functions are represented in Fig. 9.16.

Now consider that we are interested in, to accomodate *N* electrons on this line. This can be done with the help of Pauli's exclusion principle. The elementary statement of the Pauli's exclusion principle is that no two electrons can have all their quantum numbers identical. That is each orbital* can be occupied by atmost one electron. This applies to electrons in atoms, molecules or solids. In a solid the quantum numbers of an electron in a

*More than one ortital may have the same energy. The number of ortitals with the same energy is called the degeneracy.

conduction electron orbital are n and m_S, while n is any positive integer in eq. (49) and $m_S = \pm\frac{1}{2}$ according to the spin orientation. Each energy level with quantum number n can accomodate two electrons. One with spin up and other with spin down. If there are eight electrons, then in the ground state of the system the filled orbitals are those given in the Table 9.3.

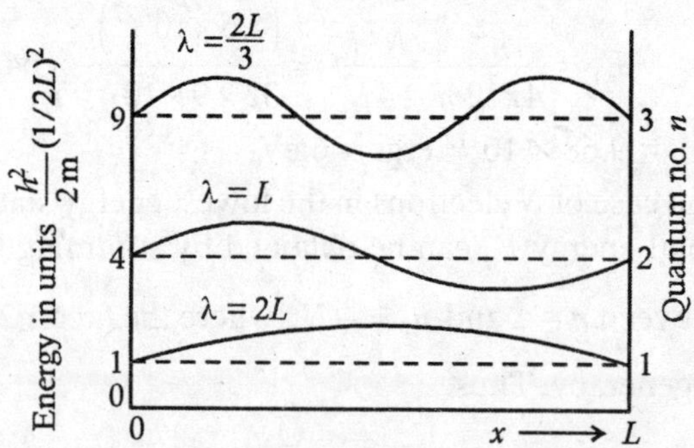

Fig. 9.16: First three energy levels and wave functions of a free electron of mass m confined to a line of length L. The energy levels are labelled according to the quantum no. n which gives the number of half-wave lengths in the wave functions. The wavelengths are indicated on the wave functions. The energy E_n of the level of quantum no. n is equal to $(h^2/2m)(n/2L)^2$

TABLE 9.3

n	m_s	*Electron Occupancy*	n	m_s	*electron Occupancy*
1	↑	1	4	↑	1
1	↓	1	4	↓	1
2	↑	1	5	↑	1
2	↓	1	5	↓	1
3	↑	1	6	↑	0
3	↓	1	6	↓	0

Let n_f denote the top most filled energy level, where we start filling the levels from bottom ($n = 1$) and continue filling the higher levels with electrons until all N electrons are accommodated. Considering N to be even we have

$$2n_f = N \tag{50}$$

and determines the value of n for the upper most filled level. The Fermi energy E_f is defined as the energy of the top-most filled level at 0 K. From equation (47) with $n = n_f$, we have

$$E_f = \frac{\hbar^2}{2m}\left(\frac{n_f\pi}{L}\right)^2 = \frac{\hbar^2}{2m}\left(\frac{N\pi}{2L}\right)^2 \tag{51}$$

We see that the energy of top electron depends on the number of electrons in the box and on the size of the box just as in an atom the ground state depends on the number of electrons (Z) and on the size of the central Coulomb field strength, *i.e.* the number of protons in the nucleus.

For example, if $N/L = 0.8$ electrons/angstrom $= 8 \times 10^7$ cm^{-1}, then from equation (51), we have

$$E_f = \frac{h^2}{4\pi^2 2m} \cdot \frac{N^2\pi^2}{4L^2} = \frac{\left(6.6\times 10^{-27}\right)^2}{32\times 9\times 10^{-28}} \times 64 \times 10^{14}$$

$$= 9.68 \times 10^{-12} \text{ ergs} = 6 \text{ eV}.$$

Consider the case of N electrons in the lowest energy state of the entire system, the total energy E_0 can be obtained by summing the individual energies E_n between $n = 1$ and $n_f = \frac{1}{2}N$; where the factor 2 is introduced due to spin degeneracy. Thus

$$E_0 = \sum_{n=1}^{N/2} E_n = \frac{2\hbar^2}{2m}\left(\frac{\pi}{L}\right)^2 \sum_{n=1}^{N/2} n^2$$

But

$$\sum_{n=1}^{\frac{1}{2}N} n^2 = 1^2 + 2^2 + 3^3 + \ldots + S^2 \quad \text{where } S = N/2$$

$$= \frac{1}{2}S\left(2S^2 + 3S + 1\right) \cong \frac{1}{3}S^3 \quad \text{for } S >> 1$$

$$\cong \frac{1}{3}\left(\frac{N}{2}\right)^3$$

Thus

$$E_0 = \frac{2\hbar^2}{2m}\left(\frac{\pi}{L}\right)^2 \frac{1}{3}\left(\frac{N}{2}\right)^3 = \frac{1}{3}\frac{\hbar^2}{2m}\left(\frac{N\pi}{2L}\right)^2$$

$$\cong \frac{1}{3} N E_f \tag{52}$$

Hence for one dimensional problem, the average K.E. in ground state is one third of Fermi energy.

Density of States

We now find the number of orbitals per unit energy, often called the density of states, equation (47) gives

$$E_n = \frac{\hbar^2}{2m}\left(\frac{n\pi}{L}\right)^2$$

Differentiating, we have

$$dE_n = \frac{\hbar^2}{2m}\left(\frac{\pi}{L}\right)^2 2n\,dn$$

Now $\frac{dn}{dE_n}$ denotes the energy levels per unit energy. Corresponding to two spin states, there are two quantum states and hence the density of states of a free electron gas given by

$$D(E) = \frac{2dn}{dE_n} = 2\cdot\frac{2m}{\hbar^2}\left(\frac{L}{\pi}\right)^2\frac{1}{2\pi}$$

$$= \frac{4L}{\hbar}\left(\frac{m}{2E_n}\right)^{1/2} \qquad (53)$$

This result is plotted in Fig. (9.17).

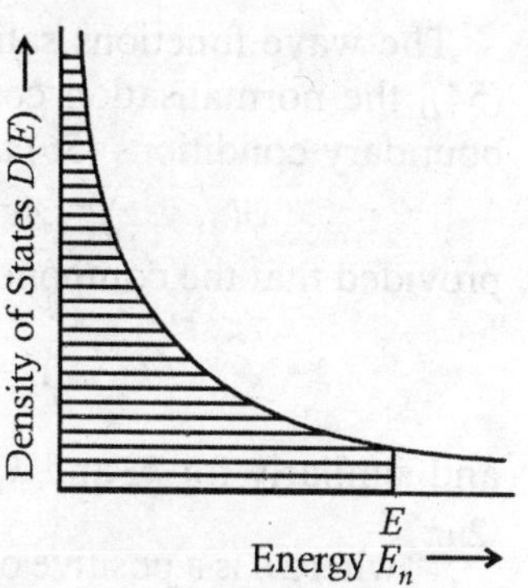

Fig. 9.17: Density of electronic states as a function of energy, for one dimensional line. At absolute zero, all the states are filled upto Fermi energy F_f.

Free Electron Gas in Three Dimensions

The free electron Schrödinger equation in three dimension is

$$-\frac{\hbar^2}{2m}\left(\frac{\partial^2}{\partial x^2}+\frac{\partial^2}{\partial y^2}+\frac{\partial^2}{\partial z^2}\right)\psi_k(r) = E_k\psi_k(r) \qquad (54)$$

where E_k is the total energy (in this case kinetic energy) of the electron in k-state.

If the electrons are confined to a cube of edge L the analogue to wave function (49) is

$$\psi_k(r) = \sqrt{\left(\frac{8}{L^2}\right)}\sin\left(\frac{\pi n_x x}{L}\right)\sin\left(\frac{\pi n_y y}{L}\right)\sin\left(\frac{\pi n_z z}{L}\right) \qquad (55)$$

where n_x, n_y and n_z are positive integers and $\sqrt{\left(\frac{8}{L^3}\right)}$ is normalising constant. This is a standing wave.

The solution of the wave equation is a consequence of particular boundary conditions imposed. It is convenient to introduce wave functions which satisfy periodic boundary conditions. We now require the wave functions to be periodic in x, y, z with period L. The appropriate boundary conditions are

$$\left.\begin{aligned}\psi(x,y,z) &= \psi(x+L,y,z)\\ \psi(x,y,z) &= \psi(x,y+L,z)\\ \psi(x,y,z) &= \psi(x,y,z+L)\end{aligned}\right] \qquad (56)$$

The wave functions satisfying the free particle Schrödinger equation (54), the normalisation condition over the volume L^3 and the periodic boundary conditions (56) are of the travelling plane wave form

$$\psi(x, y, z) = Ae^{ik \cdot r} = Ae^{i(k_x x + k_y y + k_z z)} \tag{57}$$

provided that the components of the wave vector $k(k_x, k_y, k_z)$ satisfy

$$k_x = 0, \pm\frac{2\pi}{L}, \pm\frac{4\pi}{L}, +\ldots \tag{58}$$

and similarly for k_y and k_z. That is, any component of k is of the form $\frac{2n\pi}{L}$ where n is a positive or negative integer. The components of k are the quantum numbers of the problem; along with the quantum number m_S for the spin direction. That is the state of an electron is specified completely when we are given the values of k_x, k_y, k_z and m_s.

On substituting (57) in (54), we have

$$E_k = \frac{\hbar^2}{2m}\left(k_x^2 + k_y^2 + k_z^2\right)$$

$$= \frac{\hbar^2}{2m}k^2 \tag{59}$$

for the energy E_k of the orbital with wave vector $\boldsymbol{k}$. The magnitude of wave vector is related to wavelength λ by

$$\boldsymbol{k} = \frac{2\pi}{\lambda} \tag{60}$$

Evidently the energy spectrum consists of discrete energy levels which lie usually very close together (~10^{-15} eV apart) and thus for most purposes may be taken as continuous; such energy levels are said to be quasicontinuous. In fact the separation between the energy levels depends on the size of the box. The above separation (~10^{-15} eV) is for the box having laboratory dimensions, on the other hand if the dimensions are atomic the levels are discrete as well as widely spaced (~3.6 eV for box of 5 Å side).

Using the normalising condition $\int\psi^*(r)\psi(r)d\tau = 1$. Constant A can be determined as

$$\int_0^L\int_0^L\int_0^L A^2 e^{-ik\cdot r} \cdot e^{ik\cdot r} dx\, dy\, dz = 1$$

or

$$A = \left(\frac{1}{V}\right)^{1/2}$$

Hence the normalised wave function is

$$\psi_k(r) = \left(\frac{1}{V}\right)^{1/2} e^{ik\cdot r} \tag{61}$$

Suppose there are N non-interacting electrons contained in the box at absolute zero temperature. Using the effect of Pauli's exclusion principle on the energy distribution of conduction electrons—as we did in case of one dimensional box, it can be shown that at absolute zero all the levels below a certain level will be filled with electrons and all the levels above it will be empty. The level which divides the filled and vacant-levels is called the Fermi level at absolute zero and is denoted by $E_{f(0)}$. We note that according to quantum theory, even at absolute zero not all the electrons are condensed into the state of zero energy as happens in the classical theory. They rather occupy states between the energy values 0 and $E_{f(0)}$. This feature modifies the parameters of the free electron gas in a significant manner.

Let us now discuss the *effect of temperature*. It is clear that at non zero temperatures, the distribution of filled states would not end abruptly rather some of the states below $E_{f(0)}$ would be empty and some above it would be filled. Enrico Fermi has shown that the probability that a particular quantum state at energy E is filled is given by the so called Fermi-Function

$$f(E) = \frac{1}{\exp\{(E - E_f)/k_B T\} + 1} \tag{62}$$

where E_f is the Fermi-energy. Figure 9.18 shows the plot of this function for several values of temperature including zero.

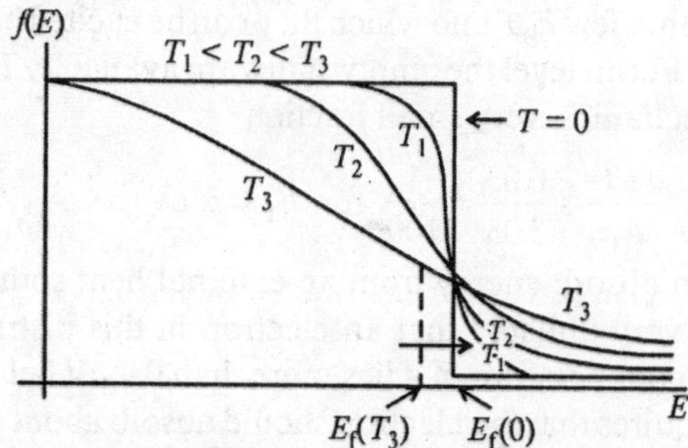

Fig. 9.18: Schematic representatiion of the Fermi distribution function for four different temperatures. Note the variation of the Fermi energy with temperature. The temperature dependence of the Fermi energy depicted here is, typical of a three-dimensional free-electron gas but the actual variation in any particular system will depend critically upon the density of states functions (or level degeneracies) for that system

From Eq. (62), it is clear that at absolute zero

$$\left.\begin{aligned} f(E) &= 1 \text{ for } E < E_{f(0)} \\ &= 0 \text{ for } E > E_{f(0)} \end{aligned}\right\} \tag{63}$$

i.e. the Fermi distribution function becomes simply the step function, implying that all the energy states below $E_{f(0)}$ are filled and all those above it are empty. As the temperature increases, the edges of the step are rounded off, and the distribution function varies rapidly from nearly unity to nearly zero over an energy range of a few times* $k_B T$ around the value $E = E_{f(0)}$. At the same time, the value E_f (energy of the highest filled level) itself changes as shown in Fig. (9.18). At very high temperatures, the distribution function loses its step like character and varies much slowly with energy. From (62) it is clear that the value of $f(E)$ at $E = E_f$ is just ½, that is

$$f(E_f) = \frac{1}{2}$$

hence a quantum state at the Fermi level has a probability of occupation of ½. This in turn means that E_f represents a level lying half way between the filled and empty states. But in case of metals since the allowed energy levels are very closely spaced (~10^{-19} eV) it is usually taken as referring to the highest filled (or partial filled) energy level.

It may therefore be concluded that if the electrons are heated up from absolute zero to a temperature T each electron would not increase its energy (by $k_B T$) as is expected classically, but those which are already within an energy range $k_B T$ below the Fermi level can do it of course. This is because absorption of energy by an electron requires its excitation into a higher energy state; for electrons lying deep into the energy there are no empty states within few $k_B T$ into which they can be excited but for electrons lying below the Fermi level the empty states are available. Thus according to quantum mechanics, very small fraction

$$\frac{k_B T}{E_f} \cong \frac{0.03}{3.03} = \frac{1}{100} \quad \text{for } E_f = 3 \text{ eV}$$

of electrons can absorb energy from an external heat source. Moreover, ordinarily it is very unlikely that an electron in this distribution will be excited at room temperature, if it lies more than 0.1 eV below Fermi level because this requires that the electron should absorb about $4k_B T$ (0.1 eV ≃ $4k_B T$) from external surrounding, this is not available. These modifications in the behaviour of the free electron gas have an important bearing on the problem of the heat capacity of metals.

Density of Available Electronic States $D(E)$

By $D(E)$ we mean the total number of available electronic states (number of orbitals) per unit energy range at E. This quantity is useful in the

---

*$k_B T$ is the order of the energy which each particle gains according to classical mechanics when temperature is increased by an amount T. At room temperature KT ~ 0.025 eV.

description of the behaviour of the free electron gas. To find the expression for $D(E)$ we consider the linear momentum P which in the quantum mechanics is represented by operator

$$P = -i\hbar\nabla$$

where as for the energy state (orbital)

$$P\psi_k(r) = -i\hbar\nabla\psi_k(r) = hK\psi_k(r) \tag{64}$$

So that the plane wave ψ_k is an eigen function of the linear momentum with eigen value $\hbar K$. The particle velocity is given by

$$v = \frac{\hbar K}{m} \tag{65}$$

In the ground state of a system of N free electrons the occupied orbitals may be represented as points inside a sphere in K-space. The energy at the surface of the sphere is the fermi energy, the wave vectors at the Fermi surface have a magnitude K_f Fig. (9.19) such that

$$E_{f(0)} = \frac{\hbar^2 K_f^2}{2m} \quad \text{...(66)}$$

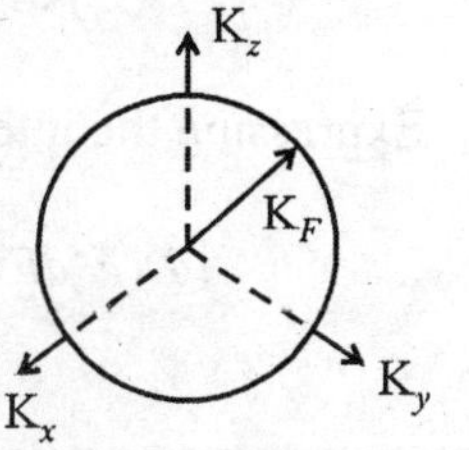

Fig. 9.19: In ground state of a system of N electrons the occupied orbitals of the system fill a sphere of radius K_f where $\frac{\hbar^2 K_f^2}{2m}$ is the energy of an electron having wave vector K_f on the surface of a sphere

From equation (58) and Fig. (9.19), we see that there is one allowed wave vector (one distinct triplet of quantum numbers K_x, K_y, K_z) for the volume element $\left(\frac{2\pi}{L}\right)^3$ of K-space. Thus in the sphere called Fermi sphere of volume $\frac{4\pi}{3}K_f^3$ the total number of orbitals is

$$\frac{2 \times 4\pi K_f^3/3}{(2\pi/L)^3} = \frac{V}{3\pi^2}K_f^3 = N \tag{67}$$

when the factor 2 on the left comes from the two allowed values of m_S, the spin quantum number for each allowed value of K. We have put the number of orbitals equal to N, the number of electrons. From (67), we have

$$K_f = \left(\frac{3\pi^2 N}{V}\right)^{1/3} \tag{68}$$

This depends only on the particle concentration and not on mass. Using (66), we have

$$E_{f(0)} = \frac{\hbar^2}{2m}\left(\frac{3\pi^2 N}{V}\right)^{2/3} \tag{69}$$

This relates the Fermi Energy to the electron concentration $\frac{N}{V}$ and the mass m. The electron velocity v_f at Fermi surface is

$$v_f = \frac{\hbar K_f}{m} = \frac{\hbar}{m}\left(\frac{3\pi^2 N}{V}\right)^{1/3} \tag{70}$$

The density of states functions $D(E)$ defined from the fact that all the energy states below $E_{f(0)}$ are occupied and this is equal to the total number of electrons, *i.e.*

$$\int_0^{E_{f(0)}} D(E)dE = N \tag{71}$$

Substituting the value of N from (68), we have

$$\int_0^{E_{f(0)}} D(E)dE = \frac{V}{3\pi^2}\left(\frac{2mE_{f(0)}}{\hbar^2}\right)^{3/2}$$

Expressing the integral as an indefinite integral, we have

$$\int D(E)dE = \frac{V}{3\pi^2}\left(\frac{2mE}{\hbar^2}\right)^{3/2}$$

$$D(E) = \frac{V}{2\pi^2}\left(\frac{2m}{\hbar^2}\right)^{3/2} E^{1/2} \tag{72}$$

or $$= CE^{1/2}$$

where $$C = \frac{V}{2\pi^2}\cdot\left(\frac{2m}{\hbar^2}\right)^{3/2}$$

This result may be obtained and expressed most simply by writing (69) as

$$\log N = \frac{3}{2}\log E_{f(0)} + \text{const.}$$

or $$\frac{dN}{N} = \frac{3}{2}\frac{dE_{f(0)}}{E_{f(0)}}$$

and we have

$$D(E_{f(0)}) = \frac{dN}{dE_{f(0)}} = \frac{3}{2}\frac{N}{E_{f(0)}} \quad \ldots(73)$$

within a factor of the order of unity, the number of orbitals per unit energy range at the Fermi-energy is just the total number of conduction electrons divided by the Fermi Energy.

These results apply to free electrons with energy (E) proportional to K^2. We can obtain a result for a general relation $E(K)$.

$$D(E) = \frac{2V}{(2\pi)^3} \int \frac{dS_E}{(\text{grad}E_K)} \tag{74}$$

where the factor 2 arises from the two spin orientations. V is the volume of the specimen and dS is the element of area in k-space of the surface of constant energy E.

It is seen from expression (72) that $D(E)$ is a parabolic function of energy. This is plotted in Fig. 9.20. Also it increases with increase in the crystal volume; this is in order to accommodate the total number of electrons present, which also increases with the size of the crystal.

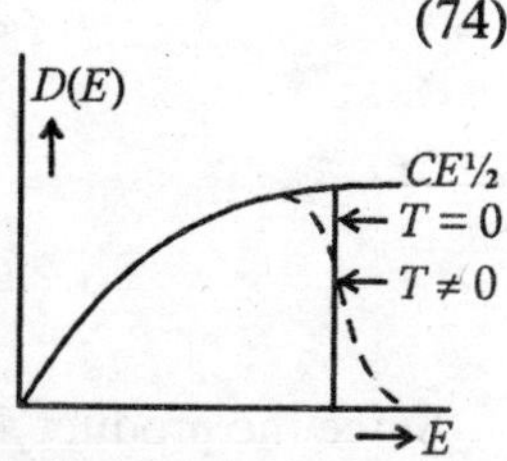

Fig. 9.20: The curve $CE^{1/2}$ represents $D(E)$ in accordance with (9.72). The energy distribution is obtained by multiplying $D(E)$ with $F(E.)$

Effect of Temperature on the Parameters of the Free Electron Gas

Till now we have assumed that the electron system is at zero K and in accordance with the Pauli's principle, we fill the levels until we have used up all electrons at the Fermi level of energy $E_{f(0)}$. But what will happen if we increase the temperature? Now the kinetic energy of the electron gas increases as the temperature is increased and some energy levels are occupied which were vacant at absolute zero and some levels are vacant which were previously occupied at absolute zero.

As an example the Fermi function at $T > 0$ is given in Fig. 9.18. For energies below E_f such that $E_f - E \gg k_B T$ (degenerate Fermi gas) the value of $f(E)$ is still practically unity, *i.e.* energy distribution in that region is the same as that for $T = 0$. It is only in the vicinity of E_f minus a few $k_B T$ the $f(E)$ begins to drop below the value at $T = 0$. For energies above E_f such that $E - E_f \gg k_B T$ one may neglect the term unity in the denominator of distribution function and obtain

$$f(E) = e^{-(E - E_f)/k_B T}$$

the Fermi distribution becomes identical with Boltzmann distribution.

Here it is of the prime importance to take into acount the effect of temperature on the Fermi energy and average kinetic energy. In order to do this, we shall have to evaluate the integral

$$I = \int_0^\infty f(E) \frac{\partial \phi(E)}{\partial E} dE \tag{75}$$

where $f(E)$ is the Fermi function and $\phi(E)$ is a function of energy (E) which has property that

$$\phi(0) = 0 \tag{76}$$

Integrating (75) by parts, we have

$$I = \left[f(E)\phi(E)\right]_0^{\infty} - \int_0^{\infty} \phi(E)\frac{\partial f(E)}{\partial E}dE$$

$$= -\int_0^{\infty} \phi(E)\frac{\partial f(E)}{\partial E}dE \tag{77}$$

Since the product $f(E)\ \phi(E)$ vanishes at both the limits. Expanding $\phi(E)$ s a Taylor's series about the point $E = E_f$ we have

$$\phi(E) = \phi(E_f) - (E - E_f)\left(\frac{\partial \phi}{\partial E}\right)_{E=E_f} + \frac{1}{2}(E - E_f)^2\left(\frac{\partial^2 \phi}{\partial E^2}\right)_{E=E_f} + \cdots \tag{78}$$

where by (77) can be expressed as

$$I = a_0\phi(E_f) + a_1\left(\frac{\partial \phi}{\partial E}\right)_{E=E_f} + a_2\left(\frac{\partial^2 \phi}{\partial E^2}\right)_{E=E_f} + \ldots \tag{79}$$

where
$$a_n = -\frac{1}{n!}\int_0^{\infty}(E - E_f)^n\frac{\partial f(E)}{\partial E}dE \tag{80}$$

But
$$\frac{\partial f(E)}{\partial E} = -\frac{e^{(E-E_f)/k_BT}}{\left\{1 + e^{(E-E_f)/k_BT}\right\}^2}\cdot\frac{1}{k_BT} = -\frac{1}{k_BT}f(1-f) \tag{81}$$

The plot of $\dfrac{\partial f(E)}{\partial E}$ versus E is given in Fig. 9.21

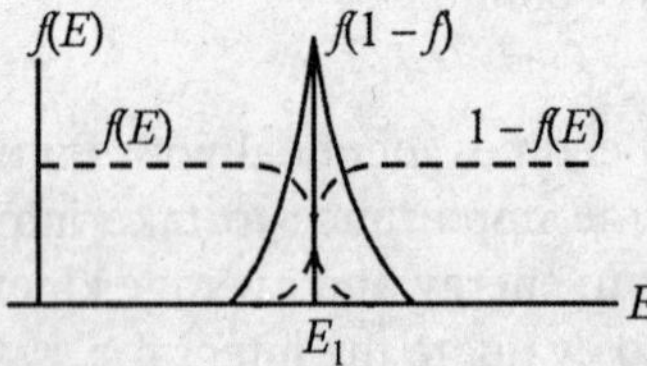

Fig. 9.21: The function $f(E)$, $1 - f(E)$ and $f(1 - f)$ for a Fermi system, plotted as a function of energy. The vertical scale factor for the $f(1 - f)$ curve is not the same scale as for the other curves

Now, if $k_BT \ll E_f\ \dfrac{\partial f(E)}{\partial E}$ is negligible for negative values of E and limits of integral (80) may be extended to $+\infty$ and $-\infty$. If this is done then, we have

$$a_0 = \int_{-\infty}^{+\infty} \frac{\partial f(E)}{\partial E} dE = -[f(E)]_{-\infty}^{+\infty} = 1 \quad (82)$$

$$a_1 = -\int_{-\infty}^{+\infty} \left(E - E_f\right) \frac{df}{\partial E} dE = 0 \quad (83)$$

The integral vanishes because $\dfrac{\partial f(E)}{\partial E}$ is an even function of $(E - E_f)$ and the integrand is then an odd function of that argument, which when integrated from $-\infty$ to $+\infty$ must yield zero.

$$a_2 = -\frac{1}{2} \int_{-\infty}^{+\infty} (E - E_f)^2 \frac{\partial f(E)}{\partial E} dE$$

and putting $x = \dfrac{E - E_f}{k_B T}$, we have

$$^*a_2 = \frac{(k_B T)^2}{2} \int_{-\infty}^{+\infty} \frac{x^2 e^x}{\left(1 + e^x\right)^2} dx = \frac{\pi^2}{6} (k_B T)^2 \quad (84)$$

Substituting these value in (79), we have

$$I = \int_0^{\infty} f(E) \frac{\partial \phi(E)}{\partial E} dE$$

$$= \phi(E_f) + \frac{\pi^2}{6} (k_B T)^2 \left\{ \frac{\partial^2 \phi(E)}{\partial E^2} \right\}_{E=E_f} + \ldots \quad (85)$$

This formula is convenient for working out approximation values of Fermi integrals. It is restricted by the condition (76) and by the condition that $k_B T \ll E_f$. The room temperature qualifies the latter condition. Suppose now that we chose

$$\phi(E) = \int_0^E D(E) dE$$

where $\dfrac{\partial \phi(E)}{\partial E} = D(E)$ and $\dfrac{\partial^2 \phi(E)}{\partial E^2} = \dfrac{\partial D(E)}{\partial E}$

*This form of integral can be found in most standard tables of integrals.

with $D(E)$ as given by (72) for free electron system, we have from (85)

$$N = \int_0^\infty f(E)D(E)dE$$

$$= \int_0^{E_f} D(E)dE + \frac{\pi^2}{6}(k_B T)^2 \left\{\frac{\partial D(E)}{\partial E}\right\}_{E=E_f} \tag{86}$$

But from (71), we have

$$\int_0^{E_{f(0)}} D(E)dE = N$$

Substituting the values of N in (86), we have

$$\int_0^{E_{f(0)}} D(E)dE = \int_0^{E_f} D(E)dE + \frac{\pi^2}{6}(k_B T)^2 \left\{\frac{\partial D(E)}{\partial E}\right\}_{E=E_f}$$

or

$$O = \int_{E_{f(0)}}^{E_f} D(E)dE + \frac{\pi^2}{6}(k_B T)^2 \left\{\frac{\partial D(E)}{\partial E}\right\}_{E=E_f}$$

$$\simeq D(E_f)\left\{E_f - E_{f(0)}\right\} + \frac{\pi^2}{6}(k_B T)^2 \left\{\frac{\partial D(E)}{\partial (E)}\right\}_{E=E_f} \tag{87}$$

In the above equation we have assumed that $D(E)$ does not vary much in the interval from $E_{f(0)}$ to E_f which for $k_B T \ll E_f$ will be only a small fraction of $E_{f(0)}$. Since $D(E)$ is of the form $CE^{1/2}$ (Eqn. 72) with C a constant, we have

$$\left\{\frac{\partial D(E)}{\partial E}\right\}_{E=E_f} = \frac{1}{2}CE_f^{-1/2} = \frac{1}{2}\frac{CE_f^{1/2}}{E_f} = \frac{1}{2}\frac{D(E_f)}{E_f}$$

and equation (87) can be written as

$$O = D(E_f)\left\{E_f - E_{f(0)}\right\} + \frac{\pi^2}{12}(k_B T)^2 \frac{D(E_f)}{E_f}$$

or

$$E_f = E_{f(0)} - \frac{\pi^2}{12}\frac{(k_B T)^2}{E_f} \tag{88}$$

For $k_B T \ll E_f$, the second term will be a small correction to be subtracted from relatively large quantity

$$E_{f(0)}\left\{\text{e.g. for } E_{f(0)} \sim 5 \text{ eV}, \ \frac{\pi^2}{12}\frac{(k_B T)^2}{E_f} \simeq \left\{10^{-4}\text{eV}\right\}\right.$$

and the difference between E_f and $E_{f(0)}$ will be small compared to $E_{f(0)}$. Under these circumstances not much error will be made if E_f in the correction term of (88) is replaced with $E_{f(0)}$, giving finally

$$E_f = E_{f(0)}\left[1 - \frac{\pi^2}{12}\left(\frac{k_B T}{E_{f(0)}}\right)^2\right] \tag{89}$$

Thus the Fermi energy is not constant, but decreases slowly as the temperature rises, the slowness of the variation following immediately from the occurrence of the factor $\frac{\pi^2}{12}\left(\frac{k_B T}{E_{f(0)}}\right)^2$ which is very small. The result (89) is quite satisfactory, but while using it we must remember that it is applicable only at temperatures such that $k_B T \ll E_f$, which incidentally concludes the whole range of temperatures for which the metals are solids (500° C to 3000° C).

Thermal Capacity of Free-electron System

The classical free electron theory proposed by Drude-Lorentz, in which the free electrons are assumed to obey Maxwell-Boltzmann statistics, leads of the free electrons should be $\frac{3}{2}Nk_B$. The total heat capacity would then be the sum of the lattice contribution given by Debye theory and electronic contribution $\frac{3}{2}Nk_B$. At temperatures large compared to Debye temperatures, the total specific heat of the metals can be expressed as

$$C_v = C_v\,(\text{lattice}) + C_v\,(\text{electronic})$$

$$= 3Nk_B + \frac{3}{2}Nk_B = \frac{9}{2}Nk_B \tag{90}$$

Dulong and Petit observed that the specific heat above the characteristic temperature is the same for all solids and is equal to $3Nk_B$; predicted by Debye theory for the lattice contribution alone. Since the large specific heat predicted by (90) is not observed in metals, it appears that the heat capacity of the free electrons is much smaller than the value predicted by Debye-Lorentz-Boltzmann model. This was one of the major short-comings of the original Drude-Lorentz model as first proposed.

With the advent of quantum mechanics, it became apparent that this short-coming is due to the incorrect assessment of the number of electrons that absorb thermal energy. When Fermi-Dirac statistics is used to describe the free electron energy distribution, the calculated electronic heat capacity is much smaller than the value $\frac{3}{2}Nk_B$ and in fact is negligible compared to the lattice contribution at moderate and high temperatures. The Fermi Free-electron model is thus in much better accord with the experiments observations, than the original Drude-Lorentz theory. The explanation of the electronic specific heat of metallic substances was one of the original triumph, of quantum statistics.

The reason for the much smaller heat capacity of the Fermi-Dirac free electron system is that only the electrons within a few $k_B T$ of the surface of the Fermi sphere can receive energy from an external heat source. Ordinarily a particle of an external heat bath at temperature T has an energy of only a few times $k_B T$ to give to an electron of the Fermi free-electron system. Electrons deep inside the Fermi sphere are incapable of interacting with such external exciting bodies because there are no unoccupied states within a few times $k_B T$ in energy into which they can be excited. As such these electrons do not contribute towards specific heat. Only electrons near the Fermi surface where there are unoccupied states available can participate in interactions with an external heat source and contribute towards the specific heat. The quantum mechanics, therefore, modifies the thermal behaviour of the free electrons in a simple and satisfying manner.

In order to make an accurate calculation of heat capacity it is necessary to find the average energy for the Fermi free electron gas. The average energy U is given by

$$U = \int_0^\infty ED(E)f(E)dE \tag{91}$$

In order to solve the integral, we choose

$$\phi(E) = \int_0^E ED(E)dE$$

Whence $$\frac{\partial\phi(E)}{\partial E} = ED(E)$$

and $$\frac{\partial^2\phi(E)}{\partial E^2} = \frac{\partial}{\partial E}\{ED(E)\}$$

Substituting in (85), we have

$$U = \int_0^\infty ED(E)f(E)\,dE = \int_0^\infty f(E)\frac{\partial\phi(E)}{\partial E}dE$$

and $$U = \int_0^{E_f} ED(E)dE + \frac{\pi^2}{6}(k_B T)^2\left[\frac{\partial}{\partial E}ED(E)\right]_{E=E_f}$$

Again using Eqn. (72), we have

$$D(E) = CE^{1/2}$$

$$ED(E) = CE^{3/2}$$

and $$\frac{\partial}{\partial E}\{ED(E)dE\}_{E=E_f} = \frac{3}{2}CE^{1/2} = \frac{3}{2}D(E_f) \tag{93}$$

Substituting in (92), we have

$$U = \int_0^{E_{f(0)}} ED(E)dE + \int_{E_{f(0)}}^{E_f} ED(E)dE + \frac{\pi^2}{6}(k_BT)^2\frac{3}{2}\cdot D(E_f)$$

$$\cong U_0 + \left[E_f - E_{f(0)}\right]E_{f(0)}D\left(E_{f(0)}\right) + \frac{\pi^2}{4}(k_BT)^2 D\left(E_{f(0)}\right) \tag{94}$$

Here U_0 is the absolute zero value of the internal energy as represented by the integral of $E\,D(E)$ from O to $E_{f(0)}$, and the integral of this quantity over the small range $E_{f(0)}$ to E_f has been approximated by the same procedure which was used in connection with (86). In the last term of (93) $D(E)$ is evaluated at $E_{f(0)}$ rather than at E_f; which; introduces only a small error. Substituting the value given by equation (88) for $E_f - E_{f(0)}$ into (93), we have

$$U = U_0 - \frac{\pi^2}{12}(k_BT)^2 D\left(E_{f(0)}\right) + \frac{\pi^2}{4}(k_BT)^2 D\left(E_{f(0)}\right)$$

$$= U_0 - \frac{\pi^2}{6}(k_BT)^2 D\left(E_{f(0)}\right) \tag{95}$$

Equation (72) gives

$$D\left(E_{f(0)}\right) = \frac{3}{2}\frac{N}{E_{f(0)}} = \frac{3}{2}\frac{N}{k_BT_F}$$

where T_F is Fermi temperature given by $E_{f(0)} = k_BT_F$, and $T_F = 30{,}000$ K a typical value for simple metals (alkali and noble) calculated from equation (88). Substituting the value of $D(E_{f(0)})$ in (94), we have

$$U = U_0 + \frac{\pi^2}{6}(k_BT)^2\cdot\frac{3}{2}\frac{N}{k_BT_F} = U_0\frac{\pi^2}{4}\frac{Nk_BT^2}{T_F} \tag{96}$$

The heat capacity is obtained in the usual way

$$C_V = \frac{\partial U}{\partial T} = \frac{\pi^2}{2}\frac{Nk_BT}{T_F} \tag{97}$$

The electronic component of specific heat as calculated using Fermi statistics, thus amounts to only a small fraction of the Drude-Lorentz value $\frac{3}{2}Nk_B$ and except at very low temperatures is small compared to the lattice contribution. At temperatures, small compared to the Debye temperature, the lattice contribution becomes quite small, approaching zero like T^3 as $T \to 0$. In this range the electronic contribution is often a significant factor and in certain temperature ranges may even be the dominant effect.

Comparison between Theory and Experiment

At temperatures much below the Debye temperature and very much below the Fermi temperature, the heat capacity of metals may be written as the sum of electronic and lattice contributions:

$$C_V = \gamma T + AT^3 \tag{98}$$

where γ and A are constant characteristics of the material and are given by (97). The electronic term is linear in T and is dominant at sufficiently low temperature. The measurements made in this limit determines primarily the electronic contributions and the direct comparison of the above theory with experiment is possible.

Kok and Keesom determined the value of γ for few metals—as a test of the above theory—by applying these considerations. They in fact represented the experimental results by plotting a curve of C_V/T versus T^2 and from the slope of the straight line curve thus obtained, determined the value of γ. Such a plot for Potassium is shown in Fig. (9.22) and corresponding value of γ is obtained. The observed values of the co-efficient γ are of the expected magnitude, but often do not agree very closely with the value calculated for free electrons of the mass m by use of the relation (97)

$$\gamma = \frac{\pi^2 N k_B^2 Z}{2E_f} \tag{99}$$

for a mole of material, where N is Avagadro's number and Z is the valency of the element. This is a difficulty which one encounters also for other metals and is a consequence of the over simplifying assumptions made in the free electron model. The departure involves three separate effects:

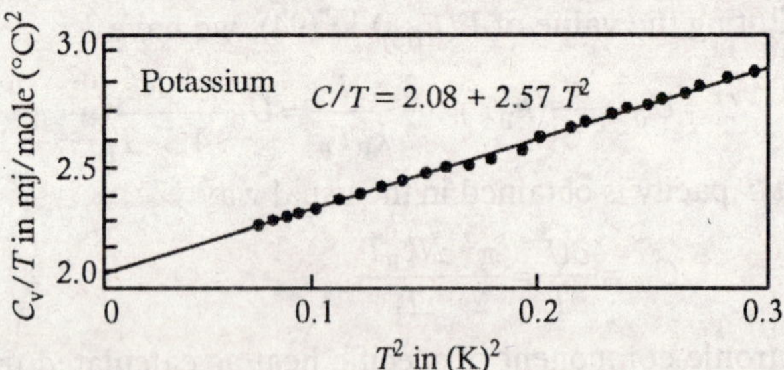

Fig. 9.22: Experimental heat capacity values for Potassium, plotted C_V/T versus T^2. The solid points were determined with an adiabatic demagnetisation of crystal

(*i*) The interaction of the conduction electrons with the periodic potential of the rigid crystal lattice. The effective mass of an electron in this potential is called the band effective mass.

(*ii*) The interaction of the conduction electrons with phonons. An electron tends to polarise or distort the lattice in its neighbourhood, so that the moving electron tries to drag nearby ions along, thereby increasing the effective mass of the electron. In ionic crystal the effect has a name, *i.e. the polaron effect.*

(*iii*) The interaction of conduction electrons with themselves. A moving electron causes an interaction reaction in the surrounding electron gas, thereby increasing the effective mass of electron. The effect of electron interactions are usually described within the framework of what is called the Landau theory of Fermiliquid.

9.9 SOMMERFELD THEORY OF ELECTRICAL CONDUCTIVITY

A theory of metallic conductivity based on average velocity, was developed by Drude in 1900. Lorentz in 1905 reinvestigated the problem using Boltzmann's transport equation and a simplified model for the collision between electrons and atoms in the lattice. However, the use of classical statistics led to serious difficulties. In 1928 Sommerfeld recalculated the conductivities along the lines of Lorentz theory, but recalculated the conductivities along the lines of Lorentz theory, but replacing classical statistics by Fermi-Dirac statistics. Sommerfeld did not investigate the actual mechanism of interaction between the electrons and the lattice any further, but assumed that a relaxation time can be defined which is a function of the energy of electrons only. As an application to the Boltzmann transport equation we shall discuss below Sommerfeld's theory of electrical conductivity based on the free electron approximation.

We shall consider first a uniform isotropic substance at constant temperature in a steady state condition under the influence of a constant applied electric field E_0. Since the system is in a steady state $\frac{df}{dt} = 0$ and since the material is uniform and at constant temperature, the distribution function must be the same in every part of the sample, whereby $\nabla f = 0$. Under these conditions the force upon an electron is $-eE_0$ and the Boltzmann equation in the relaxation time approximation may be written as

$$eE_0 \cdot \nabla_p f = \frac{f - f_0}{\tau}$$

If the co-ordinate system is chosen so that z-axis is in the direction of E_0, then E_0 has only a Z-component, and the above equation may be written as

$$\frac{eE_0\tau}{m}\frac{\partial f}{\partial v_z} = f - f_0 \tag{100}$$

In this equation the difference between f and f_0 is assumed to be quite small, so that $\frac{\partial f}{\partial v_z}$ can be approximated by $\frac{\partial f_0}{\partial v_z}$ · In addition, since the electrons obey the Pauli exclusion principle we should really use the Fermi-Dirac distribution function. The equilibrium Fermi-Dirac distribution function is

$$f_0(E) = \frac{1}{\exp\left(E - E_f\right) / k_B T + 1}$$

hence

$$\frac{\partial f_0}{\partial v_z} = \frac{\partial f_0}{\partial E} \cdot \frac{\partial E}{\partial v_z} = -\frac{m v_z e^{-(E-E_f)/k_B T}}{k_B T\left\{1 + e^{(E-E_f)/k_B T}\right\}^2}$$

$$= -\frac{m v_z}{k_B T} f_0\left(1 - f_0\right) \qquad (101)$$

Substituting this result into (100) and replacing $\frac{\partial f}{\partial v_z}$ by $\frac{\partial f_0}{\partial v_z}$ we have

$$f = f_0 - \frac{eE_0 \tau}{K_B T} f_0(1 - f_0) v \cos\theta \qquad (102)$$

Since $v_z = v\cos\theta$, where θ is the angle between the polar Z-axis and the direction v.

The electrical current density I_z is defined by

$$I_z = -ne\bar{v}_z \qquad (103)$$

n is the number of electrons per unit volume and $\bar{v}_z$ is expressed as

$$\bar{v}_z = \frac{\int v_z f(v) D(v) d^3 v}{\int f(v) D(v) d^3 v} \qquad (104)$$

where $D(v)$ is the density of states per unit volume of velocity space and $d^3 v$ is an appropriate volume element in velocity space. If we consider spherical co-ordinates (v, θ, ϕ) in velocity space we have volume element $d^3 v$.

$$dv^3 = v^2 \sin\theta \cdot d\theta \cdot d\phi \cdot dv$$

and knowing the fact that the density of states in velocity space is constant $(2m^3 V / h^3)$ we may write equation (104) in conjunction with (102) and noting that $v_z = v\cos\theta$ as

$$\bar{v}_z = \frac{\int_0^\infty \int_0^\pi \left\{ f_0 - \frac{eE_0 \tau}{k_B T} f_0(1 - f_0) v \cos\theta \right\} v^3 \sin\theta \cdot \cos\theta \, dv d\theta}{\int_0^\infty \int_0^\pi \left\{ f_0 - \frac{eE_0 \tau}{k_B T} f_0(1 - f_0) v \cos\theta \right\} v^2 \sin\theta \cdot dv \cdot d\theta} \qquad (105)$$

In Eq. (105) the integration over ϕ from 0 to 2π has been already performed. Since the integrands are both independent of ϕ, the factor 2π obtained in numerator and denominator cancels. Since $\int_0^\pi \sin\theta \cdot \cos\theta d\theta = 0$

hence the 1st term in the numerator and the second term in the denominator vanishes. Now integrating over θ from 0 to π, we have

$$\bar{v}_z = -\frac{eE_0}{3k_BT}\frac{\int_0^\infty f_0(1-f_0)\tau v^4 dv}{\int_0^\infty v^2 f_0 dv} \tag{106}$$

Although it is impossible to evaluate these integrals over the Fermi-Dirac distribution function analytically in general case, it is quite easy to obtain an approximate result which is very nearly correct for $T \ll T_F$, which of course, is usually the case for most metallic conductors under conditions which are encountered in practice. To accomplish this, it is most convenient to express the integral in (106) as integral over energy rather than velocity; whereby we obtain

$$\bar{v}_z = -\frac{2eE_0}{3mk_BT}\frac{\int_0^\infty E^{3/2}\tau(E)f_0\{1-f_0(E)\}dE}{\int_0^\infty E^{1/2}f_0(E)dE} \tag{107}$$

It must be now noted that for $T \ll T_F (k_BT \ll E_f)$ the Fermi function $f_0(E)$ is very small for E significantly in excess of E_f and the complimentary function $\{1 - f_0(E)\}$ is very small for values of E which are substantially smaller than E_f. It is easy to show analytically that $f_0(1 - f_0)$ is maximum at $E = E_f$ at which point $f_0(1-f_0) = \frac{1}{4}$, and that the width of the region over which $f_0(1 - f_0)$ is not small compared to this maximum value is of the order of k_BT. There will thus be no significant contribution to the integral in the numerator of (107) except where E takes on value with a few k_BT units of E_f. If $E^{3/2}\tau(E)$ is a function which does not vary much over this energy range we may regard $f_0(1 - f_0)$ as essentially a Dirac-δ-function of $(E - E_f)$ and write

$$f_0(E)\{1 - f_0(E)\} \cong A\delta(E - E_f) \tag{108}$$

where A is a normalising factor, which, on actual integration* comes out

---

* $\int_0^\infty f_0(E)[1-f_0(E)]dE = \int_0^\infty A\delta(E-E_f)dE = A$

L.H.S. is integrated by putting $= x = \frac{E-E_f}{k_BT}$ and remembering $f_0(E) = \frac{1}{e^{(E-E_f)/k_BT}+1}$

we have $\int_0^\infty f_0(E)\{1-f_0(E)\}dE = k_BT \int_{-E_f/k_BT}^{\infty} \frac{e^x dx}{(1+e^x)^2} = \frac{k_BT}{1+e^{-E_f/k_BT}} = k_BT$

to be equal to $k_B T$. Substituting (108) in (107) and in the denominator using $f_0(E) = 1$ for $0 < E < E_f$ and zero for $E > E_f$ which is a good approximation for $k_B T \ll E_f$, we obtain*

$$\bar{v}_z = -\frac{2eE_0}{3m}\frac{\int_0^\infty E^{3/2}\tau(E)\delta(E-E_f)dE}{\int_0^{E_f} E^{1/2}dE} = -\frac{eE_0\tau(E_f)}{m}$$

The current density $I_z = -ne\bar{v}_z$ is obtained by substituting $\bar{v}_z$

$$I_z = -\frac{ne^2E_0\tau(E_f)}{m}$$

and conductivity

$$\sigma = \frac{I_z}{E_0} = \frac{ne^2\tau(E_f)}{m} = ne\mu \qquad (109)$$

where μ is mobility and is defined as

$$\mu = \left|\frac{\bar{v}_z}{E_0}\right|$$

It is interesting to note that although all electrons take part in the conduction mechanism only the relaxation time of the electrons at the Fermi level occurs in the conductivity. The reason for this may be explained with reference to Fig. (9.23). The full circle represents the Fermi distribution for a two dimension case in the absence of an external field. In the presence of a field along the *Z*-direction, the velocity of all electrons is shifted by an amount Δv (the average drift velocity), leading to the dashed circle. It is evident that the distribution is changed only in the vicinity of the Fermi level, so that only the relaxation time of electrons near E_f is of importance.

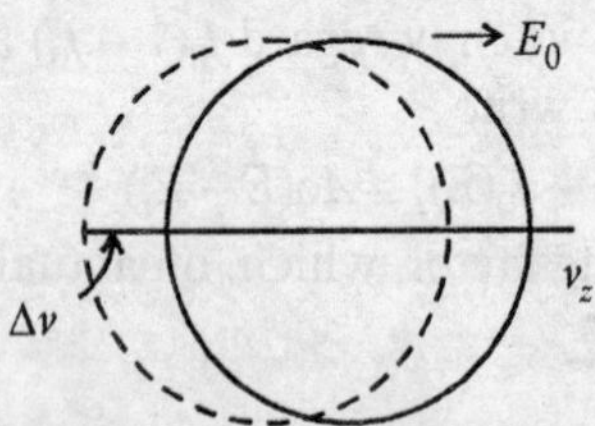

Fig. 9.23: Exaggerated representation of the influence of an electric field on the velocity distribution for a two-demensional crystal. The fully drawn circle corresponds to the Fermi distribution in the absence of a field, the field E_0 produces a shift Δv opposite to the field direction (dashed circle)

*In the integration of the numerator the Dirac-δ-function property

$\int F(x)\delta(x-a)dx = F(a)$, is used.

The result (109) obtained is essentially the same as obtained by Drude-Lorentz* using the Maxwell-Boltzmann distribution except that $\tau(v)$ has been replaced by $\tau(E_f)$. The statistics in this instance makes a difference only in the details of averaging, and does not affect the conductivity by orders of magnitude. This is understandable on physical grounds, since the effect of the field in either case is simply to superimpose a drift velocity of the order of $eE_0\tau/m$ upon velocity of all the electrons of the distribution as shown in Fig. 9.24 irrespective of the details of their distribution in energy. In any case the introduction of Fermi-Dirac statistics, while altering some of the details, preserves the essential features of the Drude-Lorentz-Boltzmann free electron picture of electrical conduction which is basically correct.

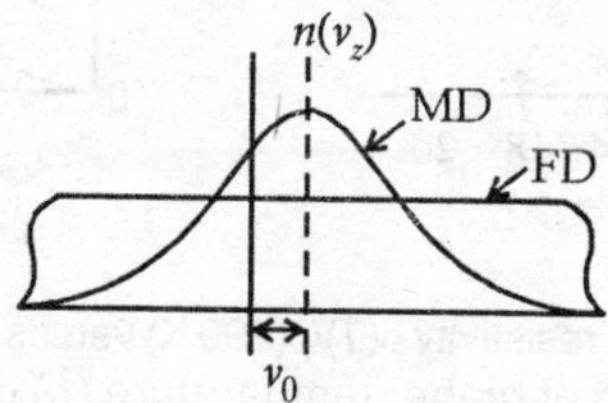

Fig. 9.24: The Maxwell-Boltzmann (MB) and Fermi-Dirac (FD) Z-velocity component distributions in the presence of an electric field

9.10 ELECTRICAL RESISTIVITY VERSUS TEMPERATURE

The electrical conductivity of a metal varies with the metal's temperature in a characteristic manner. The variation is usually discussed in terms of the behaviour of the resistivity ρ versus temperature T. Figure 9.25 shows the observed curve for Na. At $T = 0$ K, ρ has a small constant value, above that increases with temperature T, slowly at first, but afterwards ρ increases linearly with T. The linear behaviour continues essentially until the melting point is reached. This pattern is followed by most metals and usually at room temperature falls into the linear range. The linear behaviour is readily verified experimentally, as one may recall from elementary physics.

Recalling that $\rho = \sigma^{-1}$, we have from equation (109)

$$\rho = \frac{m}{ne^2} \cdot \frac{1}{\tau} \tag{110}$$

We note from the interpretation of τ that $\frac{1}{\tau}$ is actually equal to the probability of the electron suffering a scattering per unit time. Thus if $\tau = 10^{-14}$ s, then the electron undergoes 10^{+14} collisions in one second. But the electron undergoes a collision only because the lattice is not perfectly regular.

*Drude-Lorentz using classical Maxwell-Boltzmann distribution obtained the expression for electrical conductivity.

We group the deviation from a perfect lattice into two classes.

(*a*) Lattice vibrations (phonons) of the ions around their equilibrium position due to thermal excitation of the ions.

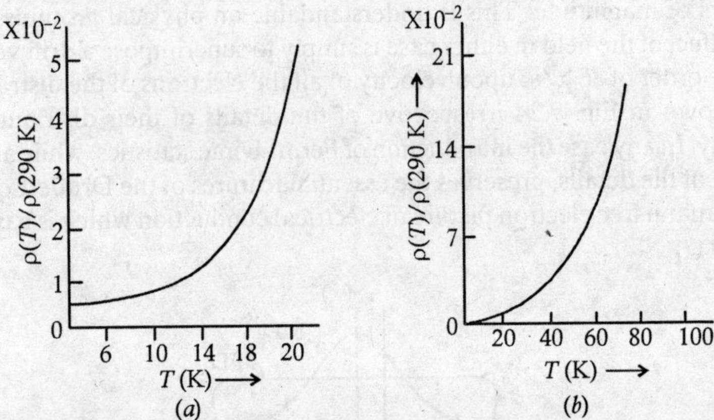

Fig. 9.25: The normalised resistivity ρ(*T*)/ρ(290 K) versus *T* for Na in low temperature region: (*a*) and at higher temperature (*b*) $\rho(290) \cong 2.10 \times 10^{-8}$ Ω-m.

(*b*) All static imperfections, such as foreign impurities or crystal defects. Of this latter group we shall take foreign impurities as an example. Now the probabilities of electrons being scattered by phonons and by impurities are additive, since these two mechanisms are assumed to act independently. Therefore, we write

$$\frac{1}{\tau} = \frac{1}{\tau_{ph}} + \frac{1}{\tau_i} \qquad (111)$$

where the first term on the right is due to phonons and the second is due to impurities. The former is expected to depend on temperature T and the latter on impurities but not on temperature T. Substituting (111) in (110) gives

$$\rho = \rho_i + \rho_{ph}(T) = \frac{m}{Ne^2}\cdot\frac{1}{\tau_1} + \frac{m}{Ne^2}\cdot\frac{1}{\tau_{ph}} \qquad (112)$$

We note that ρ has spilt into two terms, a term ρ_i due to scattering by impurities (independent of temperature) called the residual resistivity. Added to this is another term $\rho_{ph}(T)$ due to scattering by phonons, hence it is temperature dependent, and is called the *ideal resistivity*, in that it is the resistivity of a pure specimen.

At very low temperatures; scattering by phonons is negligible because the amplitudes of oscillations are very small, in that region $\tau_{ph} \to \infty_{ph} \to 0$ and hence $\rho = \rho_i$, a constant. This is in agreement with Fig. 9.25 (*a*). As T increases, scattering by phonons becomes more effective, and $\rho_{ph}(T)$

increases, this is why ρ increases. When T becomes sufficiently large, scattering by phonons dominates and $\rho \simeq \rho_{ph}(T)$. In high temperature region $\rho_{ph}(T)$ increases linearly with T, this is again in agreement with experimental result shown in Fig. 9.25 (*b*).

The statement that ρ can be split into two parts, one of which is independent of T is known as *Matthiessen rule.*

We except that ρ_i should increase with impurity concentrations and indeed it will be shown that for concentrations ρ_i is proportional to the impurity concentration N_i. We also remark that, for small impurity concentration $\rho_{ph} > \rho_i$ except at very low temperature. Let us now derive approximate expression for τ_i and τ_{ph} using arguments from the kinetic theory of gases. We shall assume for simplicity, that the collision is of the hard-spheres (billiard-ball) type.

Consider first the collision of electrons with impurities. We write

$$\tau_i = \frac{\lambda_i}{v_r} \tag{113}$$

where λ_i is the mean free path for collision with impurities and v_r random velocity. Given that the scattering cross-section of an impurity an is σ_i which is the area an impurity atom presents to the incident electron then, using an argument familiar from the kinetic theory of gases, one may write

$$\lambda_i \sigma_i N_i = 1 \text{ or } \lambda_i = \frac{1}{\sigma_i N_i} \tag{114}$$

It is expected that σ_i is of the same magnitude as actual geometrical area of the impurity atom, *i.e.* σ_i » 1 Å^2. By substituting τ_1 and τ_i from (113) and (114) in ρ_i is observed that

$$\rho_i = \frac{m}{Ne^2} . \sigma_i v_r N_i$$

i.e., ρ_i is proportional to N_i the concentration of impurities.

Calculating ρ_{ph} is more difficult; but equations similar to (113) and (114) still hold. In particular one may write

$$\lambda_{ph} = \frac{1}{N_{ion} \sigma_{ion}} \tag{115}$$

where N_{ion} is the concentration of metallic ions is lattice and σ_{ion} is the scattering cross-section per ion. We should note that σ_{ion} has no relation to the geometrical cross-section of the ion. Rather if is the area presented by the thermally fluctuating ion to the passing electron. Suppose that the distance of deviation from equilibrium is x, then the average scattering cross-section is about

$$\sigma_{ion} \approx \pi \langle x^2 \rangle \tag{116}$$

The value of $\langle x^2 \rangle$ can be estimated as follows. Since the ion is a harmonic oscillator, the average of its potential energy is equal to half the total energy.

Thus

$$\frac{1}{2}K < x^2 > = < E > = \frac{\hbar\omega}{e^{\hbar\omega/k_BT} - 1} \quad (117)$$

The frequency ω is either the Einstein or the Debye frequency, because in this rough argument we can neglect the difference between the two frequencies. We may introduce the Debye temperature θ_D so that $\hbar\omega = k_B\theta_D$. Making these substitutions, we can write

$$\rho_{ph}(T) = \frac{\pi\hbar}{k_B\theta_D M} \cdot \frac{1}{e^{\theta D/T} - 1} \quad (118)$$

where M is the mass of the ion. In the range $T >> \theta_D$ this can be written as

$$\rho_{ph}(T) = \frac{\pi\hbar^2}{k_B\theta_D M} \cdot \frac{T}{\theta_D} \quad (119)$$

which is linear in T, as depicted experimentally.

In low temperature range eqn. (118) gives $\sigma_{ph}(T)\alpha e^{-\theta_D T}$. However, the observed decreases is as T^5. The reason for this discrepancy is that we used the Einstein's model, in which the motion of the neighbouring ions was treated independently. When the correlation between ionic motion is taken into account, as in the Debye theory of lattice vibrations, one obtains the T^5 behaviour.

Deviations from Matthiessen's rule are often observed the best known being the Kondo effect. When some impurities of Fe, for example, are dissolved in Cu, ρ does not behave as in Fig. 9.25 (*a*) at low temperature *T*. Instead ρ has a minimum at low temperature *T*. This anomalous behaviour is due to an additional scattering of electrons by the magnetic moment on the impurity centres. Also deviations from Matthiessen's rule attributable to complications in the band structure of the conduction electrons have been reported. We see from these two examples that the behaviour of ρ versus *T* at very low temperature may be much more complex than that implied by the simple statement of Matthiessen's rule.

9.11 THERMAL CONDUCTIVITY

We can found an expression for thermal conductivity in terms of heat capacity C_V, mean free path λ and average thermal velocity *c*, as

$$K = \frac{1}{3}C_V\lambda c$$

The thermal conductivity of free electron gas can be obtained by using expression (97) for heat capacity and since only electrons at Fermi level are effective $E_f = \frac{1}{2}mv_f^2 = k_B T_F$. Thus, we have

$$K_{el} = \frac{1}{3}\frac{\pi_2 Nk_B T}{2T_F} = v_f \lambda$$

$$= \frac{1}{3}\frac{\pi^2 Nk_B^2 T \cdot v_f^2 \tau(E_f)}{2k_B T_F}$$

$$= \frac{\pi^2}{3}\frac{Nk_B^2 T}{m}\tau(E_f) \qquad (120)$$

Here $\lambda = v_f \tau(E_f)$, where $\tau(E_f)$ is the collision time or relaxation time and N is electron concentration.

Do the electrons or the phonons carry the greater part of the heat current in metals? At room temperatures normal pure metals tend to have values of the thermal conductivity one or two orders of magnitude higher than for dielectric solids, so that under these conditions the electrons must carry almost all the heat current. In pure metals the electronic contribution is dominant at all temperatures. In impure metals or in disordered alloys the phonons contribution may be comparable with electronic contribution.

Since $\tau(E_f)$ varies as $\frac{1}{T}$ for most metals. K is nearly temperature independent. However, it implies only to higher temperature because the relaxation time has its significance only at high temperatures. The theory of low temperature region is very complicated, but it can be said that at very low temperature $\tau(E_f)$ becomes constant and K_{el} thus varies directly with temperature and becomes zero as $T \rightarrow 0$. The variation of thermal conductivity of copper with temperature is shown in Fig. 9.26.

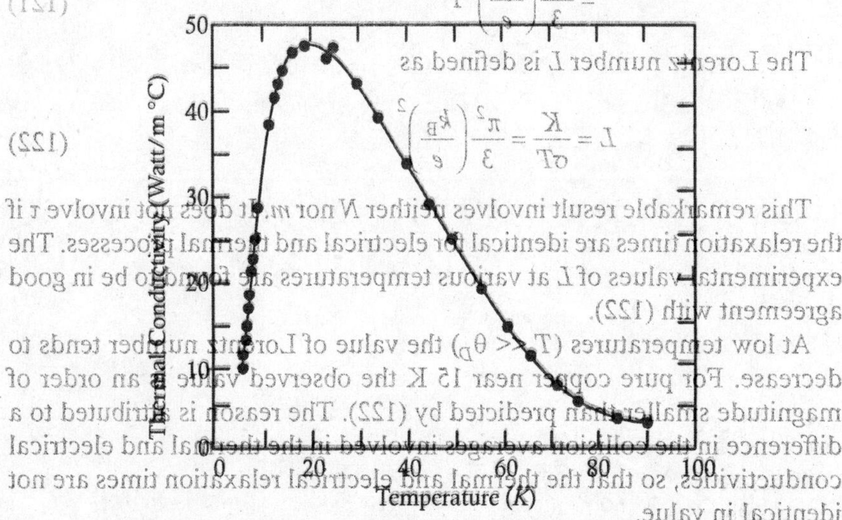

Fig. 9.26: Temperature variation of thermal conductivity in copper

It may also be mentioned in the last that if the metal contains impurities, then besides the electron-phonon scattering there will also be an electron-impurity scattering. In such a case if the two types of scattering are independent, we may write for the total conductivity as

$$\frac{1}{K} = \frac{1}{K_l} = \frac{1}{K_t}$$

where K_l is the contribution arising from electron-lattice scattering and K_t is the contribution from electron-impurity scattering. It is found that addition of impurities decreases the thermal conductivity.

Ratio of Thermal Conductivity to Electrical Conductivity (Wiedemann-Franz Law)

The Wiedemann-Franz law states that for metals at not too low temperatures the ratio of the thermal conductivity to the electrical conductivity is directly proportional to the temperature, with the value of the constant of proportionality independent of the particular metal. This result was most important in the history of the theory of metals, for it supported the picture of an electron gas. Using (109) for electrical conductivity and (120) for thermal conductivity, the Wiedemann Franz ratio can be written as

$$\frac{K}{\sigma} = \frac{n^2 K_B^2 T N \tau(E_f)/3m}{Ne^2 \tau(E_f)/m}$$

$$= \frac{\pi^2}{3}\left(\frac{k_B}{e}\right)^2 T \qquad (121)$$

The Lorentz number L is defined as

$$L = \frac{K}{\sigma T} = \frac{\pi^2}{3}\left(\frac{k_B}{e}\right)^2 \qquad (122)$$

This remarkable result involves neither N nor m. It does not involve τ if the relaxation times are identical for electrical and thermal processes. The experimental values of L at various temperatures are found to be in good agreement with (122).

At low temperatures ($T << \theta_D$) the value of Lorentz number tends to decrease. For pure copper near 15 K the observed value is an order of magnitude smaller than predicted by (122). The reason is attributed to a difference in the collision averages involved in the thermal and electrical conductivities, so that the thermal and electrical relaxation times are not identical in value.

9.12 CARRIER CONCENTRATION (n_C) IN METALS

One can obtain the actual number of electrons $N(E)dE$ by multiplying the actual number of available energy with f(E), the probability of occupation, *i.e.*,

$$N(E)dE = Z'(E)dE\,f(\mathrm{E})$$

or
$$N(E)dE = \frac{\pi}{2}\left(\frac{8ma^2}{h^2}\right)^{3/2} E^{1/2}dE\,f(E) \tag{123}$$

Thus the number of electrons per unit volume n_c (*i.e.*, density of electrons or carrier concentration) is

$$n_c = \frac{N(E)dE}{V} = \frac{N(E)dE}{a^3} = Z(E)dE \times f(E)$$

$$\left[\because \frac{Z'(E)dE}{a^3} = Z(E)dE\right] \text{(Density of states × Fermi function)}$$

$$\therefore \qquad n_c = \frac{\pi}{2}\left(\frac{8m}{h^2}\right)^{3/2} E^{1/2}dE\,f(E) \tag{124}$$

Thus, equation (124) represents the general expression for carrier concentration at any temperature in metals.

9.13 ENERGY DISTRIBUTION OF ELECTRONS IN METALS

The energy distribution gives, the energies possessed by electrons in a metal. The plot of $N(E)dE$ against the energy at $T = 0$ K is shown in Fig. 9.27.

$N(E)dE$ is proportional to $E^{1/2}$. Thus, with the increase of energy E, $N(E)dE$ also increases along a parabola from the origin. *i.e.*, $N(E)dE$ is a parabolic function of energy as shown in Fig. 9.27. At absolute zero, all the states or levels are filled upto the fermi energy E_F and those above E_F are empty.

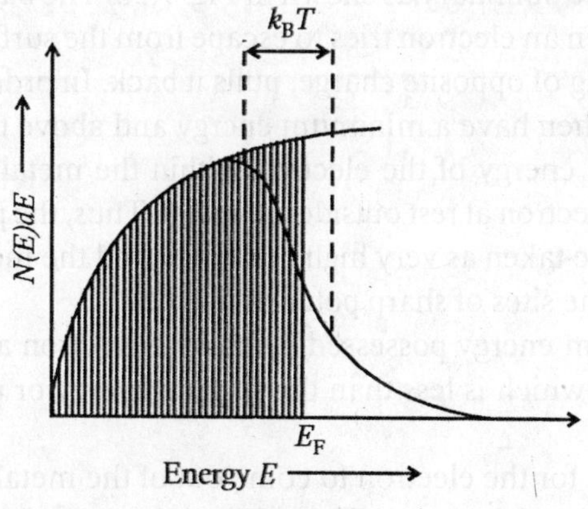

Fig. 9.27

From the Fig. 9.27, that, at high temperatures the energy states of only a small fraction of electrons gets disturbed, *i.e.* only those electrons near the Fermi level get disturbed due to increase of temperature, but not the electrons lying far below E_F.

At room temperature, electrons below the Fermi level will absorb thermal energy of the order $K_B T$ and will move to higher energy state.

9.14 WORK FUNCTION (ϕ)

The work function is defined as the minimum energy required to remove or liberate an electron from the metal and denoted usually by ϕ.

The work function can be explained on the basis of free electrons model of metals. According to the model, all the free electrons of the metal are assumed to move freely within the volume of the metal at room temperature, but the electrons are not free to come out of the metal surface.

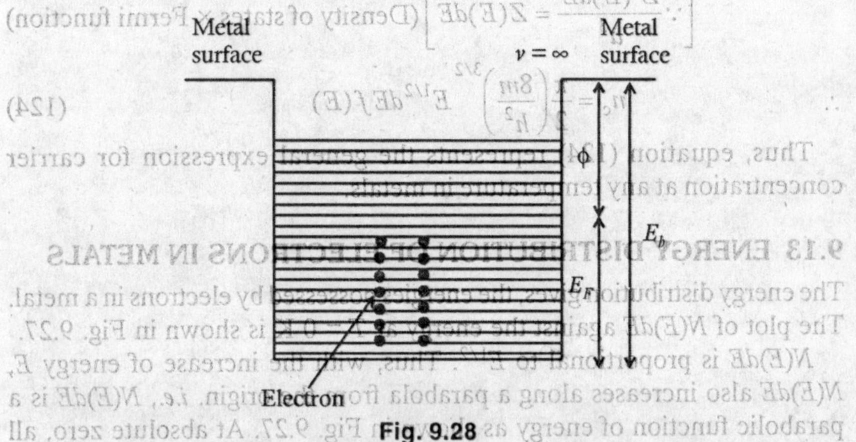

Fig. 9.28

The electrons in a metal may be regarded as particles in a potential well with barrier at the boundary as shown in Fig. 9.28. The barrier arises from the fact that when an electron tries to escape from the surface, its image in the surface, being of opposite charge, pulls it back. In order to escape, the electrons must then have a minimum energy and above the Fermi level.

The potential energy of the electron within the metal at rest is lower than that of an electron at rest outside the metal. Thus, the potential outside the metal may be taken as very high or infinity and the metal surfaces are assumed to be the sites of sharp potential barrier.

The maximum energy possessed by an by a electron at 0 K is only E_F (Fermi energy), which is less than the surface energy or potential barrier energy E_b.

Thus in order for the electron to come out of the metal surface it must possess an additional energy ϕ. Thus, the minimum energy $\phi = E_S - E_F$ is

called the work fraction. The work function is different for different metals. ϕ has a value of the order of a few eV, e.g. 2.3. eV for Na, about 4.5 eV for Cu, etc.

9.15 FAILURE OF FREE ELECTION THEORY

We have discussed the free-election model in great details to show how invaluable it is in accounting for observed metallic properties. Nevertheless, the model is only an approximation and as such has its limitations. Consider the following points:

1. The model suggest that other things being equal, electrical conductivity is proportional to electron concentration, equation (109). But it is surprising that the divalent metals (Be, Cd, Zn, etc.) and even trivalent metals (A1, In) are consistently low conductive than the monovalent metals (Cu, Ag and Au) despite the fact that the former have higher concentration of electrons.
2. A far more damaging testimony against the model is the fact that some metals exhibit positive Hall constant, for example Be, Zn, Cd. The free electron model always predicts a negative Hall constant.
3. Measurements of the Fermi surface indicate that it is often non-spherical in shape. this contradicts the model, which predicts a spherical Fermi surface.

These difficulties and others which need not be enumerated here, can be resolved by a more sophisticated theory which takes into account the interaction of the electrons with lattice.

ILLUSTRATIVE EXAMPLES

Example 1

Show that he kinetic energy of a three dimensional gas of *N* free electrons at 0 K is $E_0 = \frac{3}{5} NE_{f(0)}$.

Solution

Average kinetic energy of an electron $\overline{E}_0$ is given by

$$\overline{E}_0 = \frac{1}{N}\int_0^\infty ED(E)f(E)dE$$

To evaluate this integral, we substitute in it the value of $(D)E$ from (72) and use result $f(E = 1$ for $E < E_{f(0)}$ and $f(E) = 0$, $E > E_{f(0)}$ at $T = 0$ K. We thus have

$$\overline{E}_0 = \frac{1}{N}\int_0^{E_{f(0)}} E\cdot CE^{1/2}\cdot 1dE + \frac{1}{N}\int_{E_{f(0)}}^{x} E\cdot CE^{1/2} 0\cdot dE$$

$$= \frac{2}{5}\frac{C}{N} E_{f(0)}^{5/2}$$

Since (72) gives $CE_{f(0)}^{5/2} = D(E_{f(0)})$ and (73) gives

$D(E_{f(0)}) = \dfrac{3}{2}\dfrac{N}{E_{f(0)}}$. Combining the two together we have

$$CE_{f(0)}^{1/2} = \frac{3}{2}\frac{N}{E_{f(0)}}$$

or $$\frac{C}{N} = \frac{3}{2}\frac{1}{E_{f(0)}^{3/2}}$$

Substituting the value of C/N, we have

$$\overline{E}_0 = \frac{2}{5}\cdot\frac{3}{2}\frac{1}{E_{f(0)}^{3/2}}\cdot E_{f(0)}^{5/2}$$

$$\overline{E}_0 = \frac{3}{5}E_{f(0)}.$$

Example 2

(*i*) Derive an expression connecting the pressure and volume of an electron gas at 0 K.

(*ii*) Show that the bulk modules $B = -V(\partial P/\partial v)$ of an electron gas at 0 K is $\frac{5}{3}P$ or $10\overline{E}_0 / 9V$.

Solution

From thermodynamics we have, pressure P is negative derivative of free energy with restpect to volume

$$P = \frac{\partial \overline{E}_0}{\partial V} = -\frac{3}{5}\frac{\partial E_{f(0)}}{\partial V}$$

From (69) we have for $E_{f(0)}$

$$E_{f(0)} = \frac{h^2}{2m}\left(\frac{3\pi^2 N}{V}\right)^{2/3}$$

or $$\frac{\partial E_{f(0)}}{\partial V} = \frac{h^2}{2m}\left(3\pi^2 N\right)^{2/3}(-2/3)\left(\frac{1}{V}\right)^{5/3}$$

$$= -\frac{2}{3}\frac{h^2}{2m}(3\pi^2 N)^{2/3}\left(\frac{1}{V}\right)^{5/3}$$

$$\rho = -\frac{3}{5}\frac{\partial E_{f(0)}}{\partial V} = \frac{2}{5}\frac{\hbar^2}{2m}\left(3\pi^2 N\right)^{2/3}\left(\frac{1}{V}\right)^{5/3}$$

$$= A\left(\frac{1}{V}\right)^{5/3}$$

where $A = \frac{2}{5}\frac{\hbar^2}{2m}\left(3\pi^2 N\right)^{2/3}$ is a constant.

Now $\frac{dP}{dV} = -A \cdot \frac{5}{3}\left(\frac{1}{V}\right)^{8/3}$

and Bulk modulus $B = -V\frac{dP}{dv}$

$$\therefore \qquad -V\frac{dP}{dV} = B = A\frac{5}{3}\left(\frac{1}{V}\right)^{5/3} = \frac{5}{3}P.$$

Using result from kinetic theory of gases, we have

Average energy $\overline{E}_0$ of gas is $3/2\, Nk_B T = \frac{3}{2}PV$

or $P = \frac{2}{3}\frac{\overline{E}_0}{V}$

$$\therefore \qquad B = \frac{5}{3} \times \frac{2}{3}\frac{\overline{E}_0}{V} = \frac{10}{9}\frac{\overline{E}_0}{V}$$

Example 3

Estimate Bulk modules for potassium from following data:

$N = 1.40 \times 10^{22}/\text{cm}^2, E_{f(0)} = 2.12$ eV.

Compare it with the result obtained from elastic constants.

$C_{11} = 0.0416 \times 103$ dynes/cm², $C_{12} = 0.0341 \times 1012$ dynes/cm²

Solution

$$B = \frac{10}{9}\frac{\overline{E}}{V}$$

But average energy $\overline{E}_0$ per particle is

$$\overline{E}_0 = \frac{3}{5}E_{f(0)}$$

Hence $B = \frac{10}{9} \cdot \frac{3}{5}\frac{NE_{f(0)}}{V}$

$$\frac{N}{V} = 1.4 \times 10^{22}/\text{cm}^3 \text{ and } E_{f(0)} = 2.12 \text{ eV} = 2.12 \times 1.6 \times 10^{-12} \text{ ergs}$$

or
$$B = \frac{2}{3} \times 1.4 \times 10^{22} \times 2.12 \times 1.6 \times 10^{-12}$$

$$= 3.165 \times 10^{10} \text{ dynes/cm}^2$$

From elastic constants, we have

$$B = \frac{1}{3}(C_{11} + 2C_{12})$$

$$= \frac{1}{3}(0.0416 + 0.0682) \times 10^{22}$$

$$= \frac{0.1098}{3} \times 10^{12}$$

$$= 3.66 \times 10^{10} \text{ dynes/cm}^2.$$

Example 4

Copper has a mass density $\rho_m = 8.95$ gm/cm^3 and an electrical resistivity $\rho = 1.55 \times 10^{-8}$ ohm-m at room temperature. Calculate

(*a*) The concentration of the conduction electrons
(*b*) The mean free time τ
(*c*) The Fermi energy E_f
(*d*) The Fermi velocity v_f
(*e*) The mean free path λ_F at Fermi level.

Solution

1 gm mol of copper weighs 64 gms and its volume V is $64/\rho_m$ and it contains Avogadro's number 6.06×10^{23} atoms. Hence the concentration of conduction electron electrons (assuming the valency of copper to be 1) is

$$\frac{N}{V} = \frac{1 \times 6.06 \times 10^{23} \times 8.95}{64}$$

$$= 8.47 \times 10^{22}/\text{cm}^3 = 8.47 \times 10^{28} \text{ m}^{-3}$$

If K_F is Fermi wave vector, the Fermi energy $E_{f(0)}$ is given by

$$\frac{(\hbar K_F)^2}{2m} = E_{f(0)} = \frac{\hbar^2}{2m}\left(3\pi^2 \frac{N}{V}\right)^{2/3}$$

or
$$K_F = (3\pi^2 N/V)^{1/3}$$

$$= (29.609 \times 8.87 \times 10^{22})^{1/3}$$

$= 1.358 \times 10^8 \text{ cm}^{-1} = 1.358 \times 10^{10} \text{ m}^{-1}$

Fermi velocity v_f is given by

$$v_f = \frac{\hbar K_f}{m}$$

$= 1.157 \times K_f$

$= 1.157 \times 1.358 \times 10^8$

$= 1.566 \times 10^8 \text{ cm/sec} = 1.566 \times 10^6 \text{ m/s}$

Fermi energy $E_{f(0)} = \frac{1}{2} m v_f^2$

$= 0.284 \times 10\text{–}15\ v_f^2$ eV

$= 0.284 \times 10^{-15}(1.566 \times 10^8)^2$

$= 7.0$ eV

Fermi temperature $= T_F = E_f/k_B$ (k_B is Boltzmann const.)

The electrical resistivity is reciprocal of electrical conductivity. Thus, we have

$$f = \frac{1}{\sigma} = \frac{m}{ne^2\tau(E_f)}$$

Since $f = 1.55 \times 10^{-8}\ \Omega\text{-m}$, $m = 9 \times 10^{-81}$ kg

$$n = \frac{N}{V} = 8.47 \times 10^{22}/\text{cm}^2 = 8.47 \times 10^{28}/\text{m}^3 \text{ and}$$

$e = 1.6 \times 10^{-19}$ Coulomb. Thus

$$z(E_f) = \frac{m}{ne_f^2}$$

$$= \frac{9 \times 10^{-31}}{8.47 \times 10^{28}(1.6 \times 10^{-19})^2 \times 1.55 \times 10^{-8}}$$

$= 2.677 \times 10^{-14}$ sec.

The mean free path λ_F of a conduction electron is defined as

$\lambda_F = v_F t(E_t)$

$= 1.566 \times 10^8 \times (2.677 \times 10^{-14})$

$= 4.891 \times 10^{-6}$ cm.

Example 5

Show that if the mean free path is independent of the velocity, the electrical conductivity of Maxwell-Boltzmann free electron gas may be expressed as

$$\sigma = \frac{4}{3} ne^2\lambda / \sqrt{2\pi m k_B T}$$

Solution

When λ and τ is independent of velocity the Maxwell-Boltzmann distribution give average value of $\bar{\tau}$ as

$$\bar{\tau} = \frac{\lambda < v >}{< v^2 >}.$$

But kinetic theory and law of equipartition of energy gives

$$\frac{1}{2} m < v^2 > = \frac{3}{2} k_B T$$

or $$< v^2 > = \frac{3k_B T}{m}$$

Hence $$\bar{\tau} = \frac{\lambda \bar{c}}{3k_B T/m} = \frac{\lambda \bar{c}^2}{3k_B T/m \times \bar{c}}$$

But $$\bar{c}^2 = \frac{8k_B T}{\pi m}$$

$$\bar{\tau} = \frac{\lambda \cdot 8k_B T/\pi m}{3k_B T\bar{c}/m} = \frac{8\lambda}{3\pi \bar{c}}$$

Substituting the value of $\bar{\tau}$ in the expression for conductivity

$$\sigma = \frac{ne^2\bar{\tau}}{m}, \text{ we have}$$

$$\sigma = \frac{ne^2 \cdot 8\lambda}{3\pi m \bar{c}}$$

$$= \frac{ne^2 \cdot 8\lambda}{2\pi m \sqrt{8k_B T/\pi m}}$$

$$= \frac{8ne^2\lambda}{3 \times 2\sqrt{2\pi k_B T m}} = \frac{4}{3} \frac{ne^2\lambda}{\sqrt{2\pi k_B T m}}$$

Example 6

Calculate the number of energy states available for the electrons in a cubical box of side 1 cm lying below an energy of 1 electron volt.

Solution

Available energy states within energy range E and $E + dE$ are

$$D(E)dE = CE^{1/2}dE$$

where $$C = \frac{V}{2\pi^2} \cdot \left(\frac{2m}{\hbar^2}\right)^{3/2}$$

$$\therefore \qquad D(E)dE = \frac{V}{2\pi^2}\left(\frac{2m}{\hbar}\right)^{3/2} E^{1/2}dE$$

Hence total available states below energy level $E = 1$ eV are

$$\int D(E)dE = \frac{V}{2\pi^2}\left(\frac{2m}{\hbar}\right)^{3/2} \int_0^E E^{-1/2} \cdot dE$$

$$= \frac{V}{2\pi^2}\left(\frac{2m}{\hbar^2}\right)^{3/2} \cdot (2/3)E^{3/2} \text{ with } E = 1 \text{ eV} = 1.6 \times 10^{-19} \text{ J.}$$

We are given

$$V = 1 \text{ cm}^3 = 10^{-6} \text{ m}^3$$
$$m = 9.1 \times 10–31 \text{ kg}$$

$$\text{Available states are } = \frac{10^{-6}}{2\pi^2}\left(\frac{2 \times 9.1 \times 10^{-31}}{1.054 \times 10^{-34}}\right) \times \frac{2}{3} \times (1.6 \times 10^{-19})^{3/2}$$

$$= 4.5 \times 10^{21}.$$

Example 7

Evaluate the temperature at which there is one per cent probability that a state with an energy 0.5 eV above the Fermi energy will be occupied by an electron.

Solution

The probability that a particular quantum state at every E is filled given by

$$f(E) = \frac{1}{1 + \exp(E - E_f)/k_B T}$$

Here $\qquad E - E_f = 0.5$ eV and $f(E) = 1\% = 0.01$

$$\therefore \qquad 0.01 = \frac{1}{1 + e^x} \quad \text{where } x = \frac{E - E_f}{k_B T}$$

or $\qquad 1 + e^x = 100$ or $e^x = 99$

or $\qquad x = 2.303 \log 99$

or $\qquad \frac{0.5}{k_B T} = 2.303 \log 99$

or $\qquad k_B T = \frac{0.5}{2.303 \times \log 99} = 0.109 \text{ eV} = 0.10^9 \times 1.6 \times 10^{-19} \text{ J}$

or $\qquad T = \frac{0.109}{k_B T} = \frac{0.109 \times 1.6 \times 10^{-19}}{1.38 \times 10^{-23}} = 1264 \text{ K.}$

Example 8

The Fermi energy of silver is 5.51 eV. (*a*) What is the average energy of the electrons in silver at 0 K? (*b*) What temperature is necessary for the average molecular energy in an ideal gas to have this value? (*c*) What is the speed of the electron with this energy?

Solution

(*a*) The average electron energy at 0 K is given by

$$\overline{E}_0 = 3/5 E_F = 3/5 \times 5.51 = 3.306 \text{ eV}$$

$$= 3.306 \times 1.6 \times 10^{-19}\text{J} = 4.86 \times 10^{-19}\text{ J}.$$

(*b*) The average molecular energy in an ideal gas at temperature *T* K is $3/2 k_B T$.

or $$\overline{E}_0 = 3/2\, k_B T = 3.306 \times 1.6 \times 10^{-19}$$

or $$T = \frac{2}{3} \cdot \frac{3.306 \times 1.6 \times 10^{-19}}{k_B}$$

$$= \frac{2}{3} \times \frac{3.306 \times 1.6 \times 10^{-19}}{1.38 \times 10} = 2.560 \times 10^4 \text{ K.}$$

Example 9

(*a*) Show that the wavelength associated with an electron having an energy equal to Fermi energy is given by

$$\lambda_F = 2\left[\frac{\pi}{3N}\right]^{1/3}$$

(*b*) If this wavelength is 0.68 nm, then calculate the Fermi temperature.

Solution

(*a*) $$E_f = \frac{1}{2} m v_f^2$$

$$P_f = m v_f = \sqrt{2mE_f}$$

or $$\lambda_f = \frac{\hbar}{P_f} = \frac{\hbar}{\sqrt{2mE_f}}$$

But $$E_f = \frac{\hbar^2}{2m}\left(\frac{3\pi^2 N}{V}\right)^{2/3}$$

For unit volume $V = 1$ and

$$E_f = \frac{\hbar^2}{2m}(3\pi^2 N)^{2/3}$$

or $$\lambda_f = \frac{h}{\sqrt{2m}}\sqrt{\frac{2m}{h^2}\left(\frac{1}{3\pi^2 N}\right)^{1/3}}$$

$$= 2\pi\left(\frac{1}{3\pi^2 N}\right)^{1/3} = \left(\frac{8\pi}{3N}\right)^{1/3}$$

or $$\lambda_f = 2\left(\frac{\pi}{3N}\right)^{1/3}$$

(*b*) Given $\lambda_f = 0.68$ nm $= 0.68 \times 10^{-9}$ m.
and Fermi energy in terms of λ_f is

$$E_f = \frac{h^2}{2m}\left(\frac{3\pi^2 N}{8\pi^3}\right)^{2/3} = \frac{h^2}{2m}\left(\frac{3N}{8\pi}\right)^{2/3}$$

$$= \frac{h^2}{2m}\left(\frac{1}{\lambda_f}\right)^2$$

or $$E_f = k_B T_f = \frac{h^2}{2m}\left(\frac{1}{\lambda_f}\right)^2$$

or $$T_f = \frac{h^2}{2m}\cdot\frac{1}{\lambda_f^2}\cdot\frac{1}{k_B}$$

$$= \frac{(6.62 \times 10^{-34})^2}{2 \times 9.1 \times 10^{-31} \times (0.68 \times 10^{-9})^2 \times 1.38 \times 10^{-23}}$$

$$= 3.77 \times 10^4 \text{ K.}$$

Example 10

Calculate the number of states lying in an energy interval of 0.02 eV above the Fermi energy for sodium crystal of unit volume (E_f = 3.22 eV for sodium).

Solution

The number of energy states available in an energy interval E_1 and E_2 per units volume is given by

$$\int D(E)dE = \left[\frac{\pi}{2}\right]\left[\frac{8m}{h^2}\right]^{3/2}\int_{E_1}^{E_2} E^{1/2}dE$$

or number of states available $= \frac{\pi}{2}\left(\frac{8m}{h^2}\right)^{3/2}\cdot\left(\frac{2}{3}\right)[E_2^{3/2} - E_1^{3/2}]$

where $E_1 = E_f = 3.22$ eV and $E_2 = E_f + 0.02 = 3.24$ eV

Hence number of states available $= \frac{\pi}{2}\left(\frac{8m}{h^2}\right)^{3/2}\left(\frac{2}{3}\right)$

$$= \frac{\pi}{2}\left(\frac{8m}{h^2}\right)^{3/2}\left(\frac{2}{3}\right)\left\{(3.24)^{3/2} - (3.22)^{3/2}\right\}\left(1.6 \times 10^{-19}\right)^{3/2}$$

$$= \frac{\pi}{3}\left[\frac{8 \times 9.1 \times 10^{-31}}{(6.62 \times 10^{-34})^2}\right]^{3/2} [(3.24)^{3/2} - (3.22)^{3/2}] \times (1.6 \times 10^{-19})^{3/2}$$

$$= 2.47 \times 10^{26}.$$

Example 11

Calculate the extent of the energy range between $F(E) = 0.9$ and $F(E) = 0.1$ at a temperature of 200 K, and express it as a function of E_f which is 3 eV.

Solution

We have

$$F(E) = \frac{1}{1 + \exp(E - E_F)/k_B T} = \frac{1}{1 + \exp-(E_F - E)/k_B T}$$

when $F(E) = 0.9$, then

$$0.9 = \frac{1}{1 + e^{-x}} \quad \text{where} \quad x = \frac{E_F - E}{k_B T}$$

or $$1 + e^{-x} = \frac{1}{0.9}$$

or $$e^{-x} = \frac{1}{0.9} - 1 = \frac{1}{9}$$

or $$e^x = 9$$

or $$x = 2.303 \log 9 = 2.303 \times 0.954 = 2.198$$

or $$\frac{E_F - E}{k_B T} = 2.198$$

or $$E_F - E = 2.198\, k_B T = 2.198 \times 1.38 \times 10^{-23} \times 200 \text{ J}$$

$$= \frac{2.198 \times 1.38 \times 10^{-23} \times 200}{1.6 \times 10^{-19}} = 0.038 \text{ eV}$$

or $$E = E_F - 0.038 = 3 - 0.038 = 2.962 \text{ eV}.$$

when $F(E) = 0.1$ then

$$0.1 = \frac{1}{1+e^x} \quad \text{where} \quad x = \frac{E - E_f}{k_B T}$$

or $$1 + e^x = 10 \quad \text{or} \quad e^x = 9$$

or $$x = \frac{E - E_F}{k_B T} = 2.198$$

and $$E - E_F = 0.038 \text{ eV}$$

or $$E = E_F + 0.038 \text{ eV} = 3.038 \text{ eV}$$

or $$\Delta_E = 3.038 - 2.962 = 0.076 \text{ eV}$$

or $$\frac{\nabla E}{E_F} = \frac{0.076}{3} = 0.025 = 2.5\%$$

Example 12

Estimate the relative contribution of electrons and lattice to the specific heat of sodium (at constant volume) at 20 K. The Fermi temperature of sodium is 3.8×10^4 K and its Debye temperature is 150 K.

Solution

The electronic specific heat at constant volume is given by

$$C_{cl} = \frac{\pi^2}{2}\frac{T}{T_f} = Kn_0$$

where T_f is Fermi temperature and n_0 is Avogadro number. Substituting the values, we have

$$C_{cl} = \frac{\pi^2}{2} \times \frac{20}{3.8 \times 10^4} \times 1.38 \times 10^{-23} \times 6.02 \times 10^{23}$$

$$= 21.5 \times 10^{-3} \text{ J/mol(K)}$$

$$= 21.5 \text{ J/K mol(K)}$$

The lattice specific heat is given by

$$C_{la} = \frac{12\pi^4}{5} kn_0 (T / \theta_D)^3$$

where θ_D is Debye temperature. Substituting the values, we have

$$C_{la} = \frac{12\pi^4}{5} \times 1.38 \times 10^{-23} \times 6.02 \times 10^{23} \left(\frac{20}{150}\right)^3$$

$$= 4.6 \text{ J/mol K} = 4.6 \times 10^3 \text{ J/mol K}$$

Percentage contribution is

$$\frac{21.5\times10^{-3}}{4.6}\times100 = 0.467\%$$

Example 13

Estimate the paramagnetic contribution to the magnetic susceptibility per m^3 of potassium, for which the Fermi energy is 2.1 eV. Atomic weight of potassium is 39.1 and density of potassium is 0.86×10^3 kgm^{-3}.

Solution

Paramagnetic susceptibility is given by

$$\chi_P = \frac{\mu_0 n \mu_B^2}{k_B T_F}$$

where μ_B is Bohr magneton = 9.3×10^{-24} J/tesla

μ_0 is magnetic permeability of free space = $4\pi \times 10^{-7}$ henry/m.

Assuming one free electron per atom the number of atoms per cubic metre of potassium is $\dfrac{6.02\times10^{23}\times0.86\times10^3}{39.1\times10^{-3}}$

Hence number of free electrons per m^3 are $n = \dfrac{6.02\times10^{23}\times0.86\times10^3}{39.1\times10^{-3}}$

or $$n = 1.3 \times 10^{28}\ m^{-3}$$

$$E_F = k_B T_F \quad \text{or} \quad T_F = \frac{E_F}{k_B}$$

$$\therefore \quad T_F = \frac{2.1\times1.6\times10^{-19}}{1.38\times10^{-23}} = 2.43\times10^4\ \text{K}$$

Substituting the values, we have

$$\chi_P = \frac{4\pi\times10^{-7}\times1.3\times10^{28}\times(9.3\times10^{-24})^2}{2.1\times1.6\times10^{-19}} = 420.5\times10^{-8}.$$

REVIEW QUESTIONS

1. Show that if τ is independent of v, the distribution function

$$f(v) = Ae^{-m}\frac{(v-v_0)^2}{2k_BT} \qquad \left(\text{where } v_o = \frac{e\tau}{m}E_0\right)$$

is an approximate solution of Botzmann equation is the case where the Maxwell Boltzmann distribution represents the equilibrium distribution function and where $v_0 << \left(\dfrac{k_BT}{m}\right)^{1/2}$.

2. Estimate the fraction of electrons excited above the Fermi level at room temperature for Cu and Na.
3. Calculate the ratio of electrons to lattice heat capacity for Cu at $T = 0.3$ and 300K.
4. The atom He^2 has spin $\frac{1}{2}$ and is a Fermion. The density of liquid He^3 is 0.081 gm cm^{-3} near absolute zero. Calculate the Fermi energy E_f and the temperature T_F.
   [A Fermi gas is a system of non-interacting identical particles subject to the Pauli exclusion principle. The system with interactions is called a Fermi liquid. Conduction electrons in a metal form a Fermi liquid, and liquid He^3 is also a Fermi liquid].
5. If the average inter-atomic spacing in solids were increased ten fold how would the values of electronic specific heat and para magnetic susceptibility at room temperature be affected.
6. Derive an expression for the magnetic susceptibility of a free electron gas. Discuss its application to alkali metals.
7. Calculate specific heat of a free electron gas by using Maxwell Boltzmann and Fermi-Dirac distributions. Explain why experimental result agree with latter.
8. Derive an expression for Fermi energy E_f of a free electron gas in three dimension.
9. Derive an expression for Hall constant by free electron theory. Describe an experimental set up to observe the Hall effect. What important information can be obtained from the measuremental of this effect.
10. Show that the heat capacity of metal with N valence electrons at temperature T is given by $C_v = \dfrac{9Nk_B^2T}{4E_f}$ where E_f is the Fermi energy. How does this result change if thermally excited electrons are to obey the Pauli's exclusion principle?
11. Derive an expression for the electrical conductivity of a free electron gas using the collision time concept. Does this result explain the experimental values of resistivity of a normal metal like sodium.
12. How are the free electrons distributed among the allowed energy states at a finite temperature? Obtain the change in Fermi level with respect to its position at 0 K.
13. Set up linearised Boltzmann transport equation for electrons in crystals and establish the collision term. Assuming the collisions to be elastic and isotropic, derive an expression for the collision integral.
14. Write down Boltzmann transport equation for an electron gas under external elastic field. Solve it to obtain an expression for electrical conductivity in solids.
15. On the basis of free electrons theory derive an expression for the electrical and thermal conductivity of metal and hence establish Wiedmann Franz law.
16. Consider standard geometry for Hall effect where a rod-shaped specimen is placed in a magnetic field along Z-direction. A electric field is applied along the length of rod which is also taken as X-direction. Show that the resistance is independent of the magnetic field provided we assume a single relaxation time for all the electrons.

17. (*a*) Derive an expression for electrical conductivity of metal $\left(\sigma = \frac{Ne^2\tau}{m}\right)$.

    (*b*) Define Fermi-Dirac distribution function and discuss its behaviour with change in temperature.

18. What is Wiedmann-Franz law? Show that the Lorentz number is equal to $\frac{\pi^2}{3}\left(\frac{K}{e}\right)^2$, where the symbols have their usual meaning.

19. Consider a two dimensional square lattice. Take a value of electron concentration such that the Fermi surface extends to the fourth zone. Describe through Harrison construction or otherwise how we can draw Fermi surface in the reduced zone scheme in the first, the second and the third zone. Assume that the electrons behave as free electrons.

20. Obtain expression for the wave function and the energy eigen values for free electrons confined to a line of length *L*. Derive expressions for the Fermi energy, the total energy and the density of states of this system. Plot density of states as a function of energy for this system.

21. Calculate the electrical conductivity of free electron Fermi gas using Boltzmann equation and the relaxation time approximation.

22. Explain why pure metals show a constant resistivity at low temperature.

23. Derive Boltzmann equation. Solve the equation to obtain an expression for electrical conductivity of a metal.

24. Show that heat capacity of metal is given by $C = \gamma T + AT^3$. Explain the physical significance, of γ and A. Which of these constants define thermal effective mass of electron?

25. Write down the Boltzmann transport equation for electrons in a metal. Solve the equation to determine Hall Coefficient.

26. Prove that for a two-dimensional metallic system, the density of states per unit area is

$$\rho(E) = \frac{2\pi}{h^2} g_n m^*$$

    where m^* is the effective mass for electron and g_n for simple bands is the degeneracy factor for the atomic level.

27. By considering the variation of E_F with volume, show that there is a relation between pressure and volume of the free-gas $PV = \left(\frac{2}{3}\right) U$, where U is the internal energy of the gas at zero temperature. Hence show that $\beta_c = 10U/9V$.

30. Construct Fermi surface in two dimensional square lattice for free electron gas in first, second and third Brillouin zones.

SHORT ANSWER QUESTIONS

1. If E_1 is the energy of electron in the lowest energy state of a one dimensional potential box of side *a* and E_2 is the energy of the electron in energy state just

the lowest state of three dimensional cubical box of side $2a$. Then find E_1/E_2?

Ans. In one dimensional potential box $E = \frac{n^2h^2}{8ma^2}$ $\therefore E_1 = \frac{h^2}{8ma^2}$, $n = 1$. In three dimensional potential box $E = \frac{h^2}{8ma^2}(n_x^2 + n_y^2 + n_z^2)$ for E_2 $n_x = 2$, $n_y = 2$, $n_z = 1$, and side $= 2a$. $\therefore$ $E_2 = \frac{6h^2}{4 \times 8ma^2}$ or $E_1/E_2 = \frac{4}{6} = \frac{2}{3}$

2. Find the energy of free electron in the state (2, 1, 1) in rectangular box of side $a = b \neq c$?

Ans. Energy in rectangular box is $E = \left(\frac{h^2}{8m}\right)\left(\frac{n_x^2}{a^2} + \frac{n_y^2}{b^2} + \frac{n_z^2}{c^2}\right)$ Here $n_x = 2$, $n_y = 1$ $n_z = 1$. Hence $E = \frac{h^2}{8m}\left[\frac{2}{a^2} + \frac{1}{a^2} + \frac{1}{c^2}\right] = \frac{h^2}{8m}\left(\frac{5}{a^2} + \frac{1}{c^2}\right)$

3. If E_1 is the energy of the electron in six fold degenerate state (3, 2, 1) in a cubical box of side a and E_2 is the energy of a particle of mass twice that of electron in the second non degenerative energy of side $a/2$; then find $(E_2 - E_1)$?

Ans. $E_1 = \frac{h^2}{8m^2}(3^2 + 2^2 + 1^2) = \frac{14h^2}{8ma^2}$. For second non degenerative state $n_x = 2$, $n_y = 2$, $n_z = 2$. $E_2 = \frac{h^2}{8(2m)(a/2)^2}(2^2 + 2^2 + 2^2) = \frac{24h^2}{8ma^2}$. Hence $E_2 - E_1$ $= \frac{10h^2}{8ma^2}$.

4. The wave function for the motion of the particle in one dimensional potential box of length 'a' is given by $\psi_n = A\sin\frac{n\pi x}{a}$. Where A is normalization constant. Find the value of A?

Ans. For normalized wave function we have $\int_0^a |\psi_n|^2 dx = 1$ for $\psi_n = A\sin\frac{n\pi x}{a}$ we have $\int_0^a A^2 \frac{\sin^2 n\pi x}{a} = \frac{A^2 a}{2} = 1$ or $A = \sqrt{\frac{2}{a}}$.

5. What is the spacing between nth energy level and the next higher level in one dimensional potential box?

Ans. $E_n = \frac{n^2h^2}{8ma^2}$ and $E_{n+1} = \frac{(n+1)^2h^2}{8ma^2}$. $E_{n+1} - E_n = \frac{h^2}{8ma^2}[(n+1)^2 - n^2]$ $= \frac{h^2}{8ma^2}(2n+1)$, *i.e.* spacing is $(2n + 1)$ times the energy of the lowest state.

6. The collision time and root mean velocity of the electron at room temperature are 2.5×10^{-14} and 10^5 m/s respectively. Calculate the classical value of mean free path of electron ?

Ans. $\lambda = v\tau = 10^5 \times 2.5 \times 10^{-14} = 2.5 \times 10^{-9}$ m = 2.5 nm.

7. What is the mobility of electron? What are its dimensions?

Ans. The average drift velocity per unit electric field is called mobility $\mu = \frac{v}{E} = \frac{M^0 L^1 T^{-1}}{M^1 L^1 T^{-2} Q^{-1}} = M^{-1} L^0 T^1 Q^1 = M^{-1} L^0 T^2 I^1$ where Q is charge and I is current.

8. What is the value of Lorentz number L?

[**Ans.** $L = \frac{\pi^2}{3}\left[\frac{k_B}{e}\right]^2 = \frac{\pi^2}{3}\left[\frac{1.4 \times 10^{-23}}{1.6 \times 10^{-19}}\right]^2 = 2.5 \times 10^{-8}$ watt ohm (kelvin)$^{-2}$]

9. For copper the experimental value of Lorentz number is 2.5×10^{-8} WΩ(K)$^{-2}$ and classical theoretical value is 1.15×10^{-8} WΩ(K)$^{-2}$; what is the cause of this discrepancy?

Ans. The classical assumption that all the free electrons of a metal participate in thermal conduction is not correct. Quantum theory with Fermi-Dirac statistics reveals that only the electrons near the Fermi level take part in thermal and electrical conductivity.

10. If $E_{f(0)}$ and E_f are Fermi levels of sodium at 0 K and 10000 K, then what is the value of $E_f/E_{f(0)}$ if $E_{f(0)} = 3$ eV?

[**Ans.** $E_f = E_{f(0)}\left[1 - \frac{\pi^2}{12}\left(\frac{k_B T}{E_{f(0)}}\right)^2\right]$. Hence the ratio $E_f/E_{f(0)} = 1 - \frac{\pi^2}{12}\left(\frac{k_B T}{E_{f(0)}}\right)^2$

or the ratio $= 1 - \frac{\pi^2}{12}\left(\frac{1.4 \times 10^{-23} \times 10^2}{3 \times 1.6 \times 10^{-19}}\right) = 1 - 0.7 = 0.03$.

11. Find the state of a particle in which the energy of a particle in cubical box is $\frac{11h^2}{8ma^2}$? a is the side of the box.

Ans. $E_n = \frac{h^2}{8ma^2}(n_x^2 + n_y^2 + n_z^2) = \frac{11h^2}{8ma^2}$ or $n_x^2 +_y^2 + z_z^2 = 11$ then the possible values of n_x, n_y, n_z are (113). This state has 6 fold degeneracy.

12. Find the ratio of electronic specific heat calculated quantum mechanically and classically for silver of $E_f = 5$ eV at T K?

Ans. $C_V(\text{Q} - \text{mech}) = \frac{\pi^2}{2} R\left(\frac{k_B T}{E_f}\right)$ and $C_V(\text{classical}) = \frac{3}{2}R$.

$\therefore$ Ratio $= \frac{\pi^2}{3}\left(\frac{k_B T}{E_f}\right) = \frac{\pi^2}{3}\left(\frac{1.4 \times 10^{-23} T}{5 \times 1.6 \times 10^{-19}}\right) = 0.57 \times 10^{-4} T$

14. What do you mean by Fermi energy? In one dimensional problem what is the ratio of the average kinetic energy (E) in the ground state and Fermi energy E_f?

[**Ans.** The Fermi energy E_f is defined as the energy of the top most level at 0 K. It is also defined as the energy of the level whose probability of occupation is 1/2. E is $\frac{1}{3} E_f$ *i.e.*, $E/E_f = \frac{1}{3}$

15. What is Mathiessen rule?

Ans. This rule says that resistivity ρ can be split into two parts, one of which is independent of temperature T (= ρ_0) and other the temperature dependent $\sigma(T)$. ρ_0 is called residual resistivity and $\rho(T)$ is called ideal resistivity. Thus $\rho = \rho_0 + \rho(T)$.

MULTIPLE CHOICE QUESTIONS

1. If E_1 is the energy of lowest state of a one dimensional box of length a and E_2 is the energy of the lowest state when the length of the box is halved, then

(*a*) $E_2 = E_1$ (*b*) $E_2 = 2E_1$ (*c*) $E_2 = 3E_1$ (*d*) $E_2 = 4E_1$.

2. The relation between average drift-velocity (v_d) of electrons in a metal is related to the applied electric field E and collision time τ as

(*a*) $\frac{m}{eE\tau}$ (*b*) $\sqrt{\frac{m}{eE\tau}}$ (*c*) $\sqrt{\frac{eE\tau}{m}}$ (*d*) $\frac{eE\tau}{m}$.

3. Which of the following relation gives Wiedmann-Franz law?

(*a*) $\frac{\sigma_\tau}{\sigma_e} = LT$ (*b*) $\frac{\sigma_e}{\sigma_\tau} = LT$ (*c*) $\frac{\sigma_T}{\sigma_e} = \frac{L}{T}$ (*d*) $\frac{\sigma_\tau}{\sigma_e} = T/L$.

4. The classical expression for the electrical conductivity of a metal in terms of mass of the electron, charge of electron, concentration of electrons and collision time is given by

(*a*) $mne\tau$ (*b*) $\frac{me\tau}{n}$ (*c*) $\frac{ne^2\tau}{m}$ (*d*) $\frac{ne^2\tau^2}{m}$

5. The temperature dependence of the classical expression for electrical resistivity of metal is

(*a*) $\rho \propto T^2$ (*b*) $\rho \propto \frac{1}{T^2}$ (*c*) $\rho \propto \frac{1}{T}$ (*d*) $\rho \propto \frac{1}{\sqrt{T}}$.

6. Ohms law relates to the electric field E, conductivity σ and current density J as

(*a*) $J = \sigma E$ (*b*) $J = \sigma E^2$ (*c*) $J = \frac{E}{\sigma}$ (*d*) $J = \sigma/E$.

7. If e, μ and m respectively represents charge, mobility and concentration of electrons then the electrical conductivity of the material is given by

(*a*) $\frac{\mu e}{n}$ (*b*) $\frac{n}{\mu e}$ (*c*) $ne\mu$ (*d*) ne.

8. The classical expression for molar electronics specific heat is $1.5R$ (where R is gas constant). Its value with units is expressed as

(*a*) 12.5 J(k mol)$^{-1}$K^{-1} (*b*) 1250 KJ (mol)$^{-1}$K^{-1}

(*c*) 125 kJ (mol)$^{-1}$K^{-2} (*d*) 12.5 KJ(k mol)$^{-1}$K^{-1}

9. The energy of the lowest state in a one dimensional potential box of length a is

(*a*) zero (*b*) $\frac{2h^2}{8ma^2}$ (*c*) $\frac{h^2}{8ma^2}$ (*d*) $\frac{h}{8ma^2}$.

10. The energy of particle moving a three dimensional potential box of side a is given by $E = \frac{h^2}{8ma^2}(n_x^2 + n_y^2 + n_z^2)$. The degeneracy of state (1 2 3) is (*a*) 2 (*b*) 3 (*c*) 6 (*d*) 8.

11. The general expression for Fermi energy of a metal at 0 K is (n is concentration of electrons)

(*a*) $E_{f(0)} \propto n^{3/2}$ (*b*) $E_{f(0)} \propto n^{2/3}$ (*c*) $E_{f(0)} \propto n^{1/3}$ (*d*) $E_{f(0)} \propto n^3$

12. The quantum mechanical expression for electrical conductivity is

(*a*) $\sigma = \frac{ne^2\lambda_f}{m^* v_f}$ (*b*) $\sigma = \frac{m^* v_f}{\lambda_f ne^2}$ (*c*) $\sigma = \frac{ne^2\lambda_f}{m^*}$ (*d*) $\sigma = \frac{m^*}{ne^2\lambda_f}$

13. The quantum mechanical expression for Lorentz number is

(*a*) $\left(\frac{\pi}{3}\right)\left(\frac{k_B}{e}\right)^2$ (*b*) $\frac{\pi^2}{3}\left(\frac{k_B}{e}\right)^2$ (*c*) $\frac{\pi}{3}\left(\frac{e}{k_B}\right)^2$ (*b*) $\frac{\pi^2}{3}\left(\frac{e}{k_B}\right)^2$

14. The ratio of classical Lorentz number L_c and quantum mechanical Lorentz number L_q is (L_c/L_q)

(*a*) $\frac{9}{\pi^2}$ (*b*) $\frac{\pi^2}{9}$ (*c*) $\frac{2\pi^2}{9}$ (*d*) $\frac{9}{2\pi^2}$

15. If the Fermi energy of silver at 0 K is 5 eV. The mean energy of electron in silver at 0 K is

(*a*) 2.5 eV (*b*) 3 eV (*c*) 5 eV (*d*) 10 eV.

16. If the Fermi energy of metal is 2.8 eV the Fermi temperature of metal is

(*a*) 3.2×10^4 K (*b*) 1.6×10^4 K (*c*) 3.2×10^3 K (*d*) 1.6×10^3 K.

17. If E_1 is the energy of electron in lowest energy state in cubical potential box of side a and E_2 is the enery of the electron in the state (2, 2, 2) when the side of the box is $a/2$, then

(*a*) $E_2 = 16E_1$ (*b*) $E_2 = 4E_1$ (*c*) $E_2 = \frac{E_1}{2}$ (*d*) $E_2 = 8E_1$.

18. If E_1 is the lowest energy value of the electron in a one dimensional potential box of side a and E_2 is the lowest energy value of the electron in a cubical box of side $2a$ then E_2/E_1 is

(*a*) $\frac{3}{2}$ (*b*) $\frac{3}{4}$ (*c*) $\frac{2}{3}$ (*d*) $\frac{4}{3}$

19. The correct relation between Fermi energy at 0 K $(E_{f(0)}]$ and at T K (E_f) is

(*a*) $E_f = E_{f(0)}\left[1 + \frac{\pi^2}{12}\left(\frac{k_BT}{E_{f(0)}}\right)^2\right]$ (*b*) $E_f = E_{f(0)}\left[1 - \frac{\pi^2}{12}\left(\frac{k_BT}{E_{f(0)}}\right)^2\right]$

(*c*) $E_f = E_{f(0)}\left[1 + \frac{\pi^2}{12}\left(\frac{k_BT}{E_{f(0)}}\right)\right]$ (*d*) $E_f = E_{f(0)}\left[1 - \frac{\pi^2}{12}\left(\frac{k_BT}{E_{f(0)}}\right)\right]$

20. If σ and E are electric conductivity and applied field respectively for a current carrying conductor, the heat developed per unit volume is

(*a*) $\sigma^2 E$ (b) σ^2/E (*c*) σE^2 (*d*) E^2/e.

21. If E, J and B are the Hall field, current density and magnetic field strength, then Hall constant is given by

(*a*) $R_H = \frac{EJ}{B}$ (*b*) $R_H = \frac{JE_H}{B}$ (*c*) $\frac{B}{EJ}$ (*d*) none of these.

22. If ρ_0 and $\rho(T)$ are temperature independent and temperature dependent parts of resistivity of a metal then the total resistivity is

(*a*) $\rho = \rho_0 + \rho(T)$ (b) $\rho = \rho_0 - \rho(T)$

(*c*) $\rho = \rho_0\{1 + \rho(T)\}$ (*d*) $\rho = \rho_0\{(1 - \rho(T)\}$.

23. The normalised wave function for a particle in rectangular box of dimensions a, b, c is

(*a*) $\Psi_n = \sqrt{\frac{abc}{2}} \sin\frac{n_x \pi x}{a} \sin\frac{x_y \pi y}{b} \sin\frac{n_z \pi z}{c}$

(*b*) $\Psi_n = \sqrt{\frac{2}{abc}} \sin\frac{n_x \pi x}{a} \sin\frac{x_y \pi y}{b} \sin\frac{n_z \pi z}{c}$

(*c*) $\Psi_n = \sqrt{\frac{4}{abc}} \sin\frac{n_x \pi x}{a} \sin\frac{x_y \pi y}{b} \sin\frac{n_z \pi z}{c}$

(*d*) $\Psi_n = \sqrt{\frac{8}{abc}} \sin\frac{n_x \pi x}{a} \sin\frac{x_y \pi y}{b} \sin\frac{n_z \pi z}{c}$

24. The energy spectrum of one dimensional free electron gas is discrete, the energy level separation depends on: (n is an integer called quantum number and L is length)

(*a*) $\frac{n^2}{L^2}$ (*b*) $\frac{2n+1}{L^2}$ (*c*) $\frac{(2n+1)}{L}$ (*d*) $\frac{n^2}{L}$

25. Fermi energy is given by $E_f = 3.65 \times 10^{-19}(n)^{2/3}$ eV. Where n is the number of free electrons per unit volume. What is Fermi energy of electrons of Potassium. Atomic weight = 39 and density = 860 kgm^{-3}.

(*a*) 1.09 eV (*b*) 2.02 eV (*c*) 4.04 eV (*d*) 5.05 eV.

26. Silver is (f_∞) has an atomic radius of 1.44 Å. Assume expression for Fermi energy as given in example 4, the Fermi energy of silver is

(*a*) 1.1 eV (*b*) 2.2 eV (c) 4.4 eV (*d*) 5.5 eV.

ANSWERS

1. (*d*)	**2.** (*d*)	**3.** (*a*)	**4.** (c)	**5.** (*d*)
6. (*a*)	**7.** (*c*)	**8.** (*d*)	**9.** (*c*)	**10.** (*c*)
11. (*b*)	**12.** (*a*)	**13.** (*b*)	**14.** (*d*)	**15.** (*b*)
16. (*a*)	**17.** (*a*)	**18.** (*b*)	**19.** (*b*)	**20.** (*c*)
21. (*d*)	**22.** (*a*)	**23.** (*d*)	**24.** (*b*)	**25.** (*b*)
26. (*d*).				

Physics of Materials

10

Superconducting Materials

10.1 INTRODUCTION

When cooled to sufficiently low temperatures, a large number of metals and alloys can conduct electric current without resistance. Obviously, these specific materials undergo a phase transition to a new superconducting state characterized by the complete loss of DC resistance below a well defined critical temperature, T_C. Thus zero resistivity ($\rho = 0$), *i.e.* infinite conductivity is observed in a superconductor at all temperatures, below a critical temperature ($\rho = 0$ for all $T < T_C$). However, if we pass a current higher than the critical current density J_C, superconductivity disappears. This limits the maximum current the material can sustain and is an important problem for applications of superconducting material. For elements, the transition temperature, T_C lies generally below 10 K. Figure 10.1 shows resistance versus temperature for a low temperature superconductor. At the transition temperature T_C the resistance drops abruptly to an unmeasurably small value. The transition from normal to the superconducting phase is often sharp and occurs within 10^{-2} to 10^{-4} K. This behaviour is remarkably different from the steadily decreasing resistance of non-superconducting metals (Fig. 10.2) and suggests the existence of a physically different superconducting state. In pure metals the zero resistance state can be reached within a temperature range of 1 mK. In the case of impure metals the transition to the superconducting state may be considerably broadened. A transition width of ≈ 0.05 K was observed for impure tin.

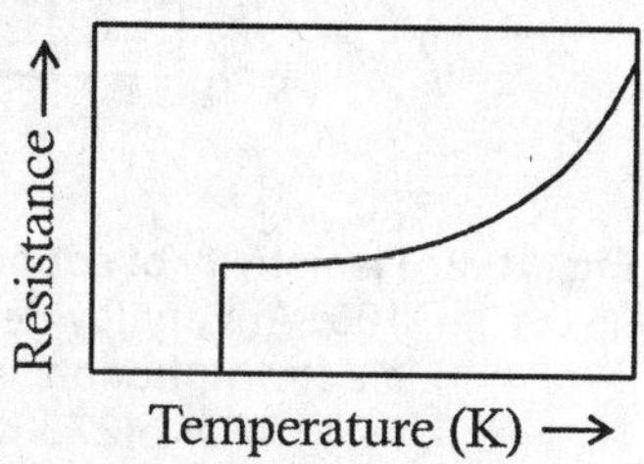

Fig. 10.1: Resistance versus temerature for a low temperature superconductor

The resistivity of a superconductor to direct current is zero as far as it can be measured. The estimates of the resistivity in the superconducting phase place it at less than 4×10^{-25} Ω-m, which is essentially zero for all practical purposes. A striking way to demonstrate zero resistivity is to induce a current around a close ring of a superconducting metal. Experiments have been performed in which a 'persistent current' has run for over two and a half years without any measurable decay. The time dependence of the current I in the loop is given by $I(t) = I_0 e^{-t/\tau}$, where I_0 is the initial value of the current and t is the time which has elapsed since the supercurrent has been induced. The ratio of resistance R and self inductance L of the superconducting loop determines the time constant τ for the decay of the current. Above the critical temperature T_C, the metal is in the normal state and resistance is proportional to T^5. In many metals the exponent is between 2 and 6, considerably different from the value of 5 predicted by Bloch theory.

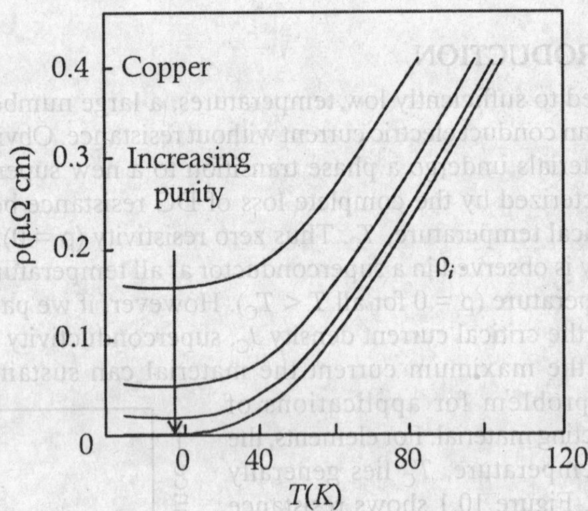

Fig. 10.2: Resistivity of copper for residual ratio (RRR) values of 10, 30 and 100. At very low temperatures the intrinsic resistivity ρ(*T*) due to electron-phonon interaction approaches zero. RRR is defined as: RRR = ρ(273 K)/ρ(4 K)

The critical temperature T_C varies from superconductor to superconductor but lies between less then 1 K and approximately 20 K for metals and metal alloys. Until 1986 the maximum T_C was observed in an alloy of niobium, aluminium and germanium. Recently it has been demonstrated that some complex cuprate oxide ceramics have critical temperatures in excess of 100 K. Today, the highest known T_C is 133 K for mercury based cuprate oxide, $HgBa_2Ca_2Cu_3O_{8+\delta}$. When this compound is subjected to high pressure ~30 GPa, the onset of T_C increases to ~164 K. The dramatic evolution of critical temperatures that have been observed since its discovery in 1911 by K. Onnes are illustrated in Fig. 10.3 where the maximum value of T_C is plotted versus date. The superconductors with $T_C < 25$ K are called *conventional* or low T_C superconductors, whereas cuprate oxides and some other recently discovered superconductors with $T_C > 25$ K are termed as high temperature superconductors (HTSC).

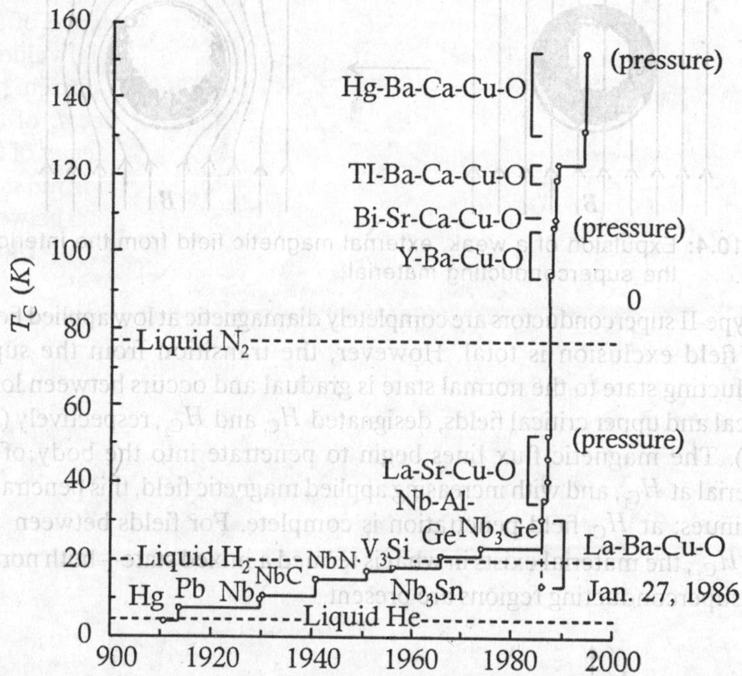

Fig. 10.3: The evolution of superconductivity critical temperature since its discovery

In addition to resistanceless current transport, the superconducting state is characterized by perfect diamagnetism, *i.e.* $B = 0$ inside the superconductor. The magnetic inductance becomes zero inside the superconductor when it is cooled below T_C in a weak external magnetic field: the magnetic flux is expelled from the interior of the superconductor (Fig. 10.4). This effect is called the *Meissner-Oschsenfeld effect* after its discoverers and it is the ultimate practical test in any new material. We must note that there always exists some critical field H_C above which superconductivity disappears. Supercondctivity disappears and the material returns to the normal state if one applies an external magnetic field of strength greater than H_C.

On the basis of magnetic response, superconducting materials may be divided into two classes designated as type-I an type -II. Type-I materials, while in the superconducting state, are completely diamagnetic, *i.e.* in an applied magnetic field, field will be excluded from the body of material

(Miessner effect). Several metallic elements including aluminium, lead, tin and mercury belong to type-I group.

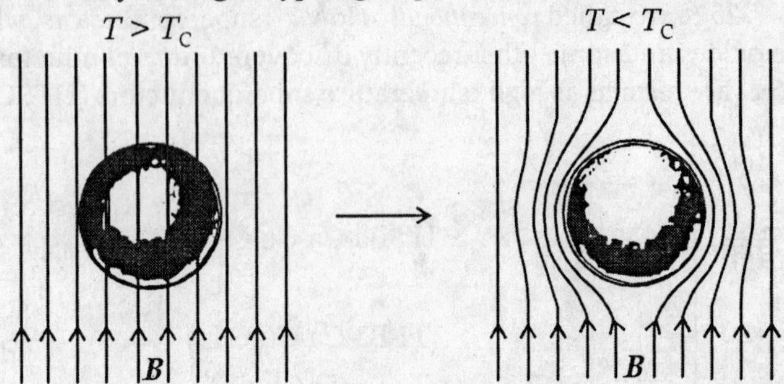

Fig. 10.4: Expulsion of a weak, external magnetic field from the interior of the superconducting material

Type-II superconductors are completely diamagnetic at low applied fields, and field exclusion is total. However, the transition from the superconducting state to the normal state is gradual and occurs between lower critical and upper critical fields, designated H_{C_1} and H_{C_2}, respectively (Fig. 10.5). The magnetic flux lines begin to penetrate into the body of the material at H_{C_1}, and with increasing applied magnetic field, this penetration continues; at H_{C_2} field penetration is complete. For fields between H_{C_1} and H_{C_2}, the material exists in what is termed a mixed state—both normal and superconducting regions are present.

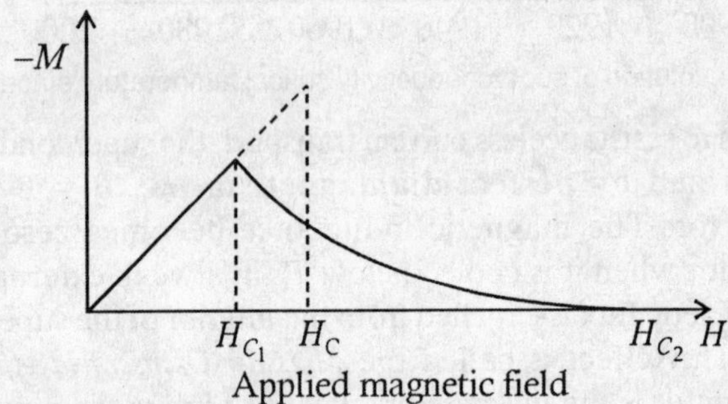

Fig. 10.5: Variation of magnetization as a function of the magnetic field for (*i*) type-I superconductor (dotted line) (*ii*) type-II ideal superconductor (solid line)

Type-II superconductors are preferred over type-I for most practical applications by virtue of their higher critical temperatures and critical magnetic fields.

The three material parameters, T_C, H_C and J_C are of extreme importance in the practical applications of superconductivity. Figure 10.6 shows schematically the boundary in temperature magnetic field-current density space separating normal and superconducting states. The position of this boundary will, of course, depend on the material. For temperature, magnetic field, and current density values lying between the origin and this boundary, the material will be superconductive; outside the boundary, conduction is normal.

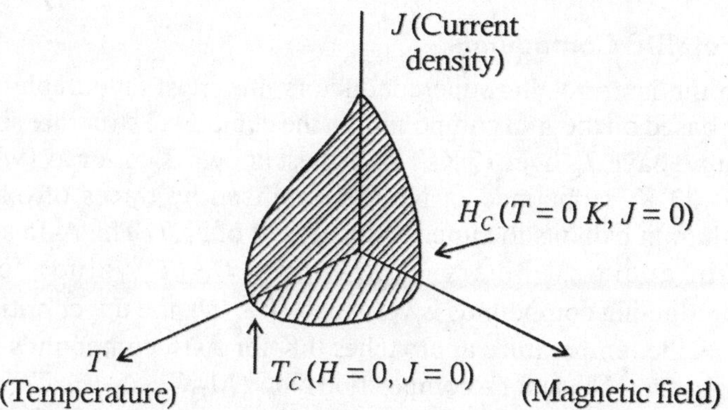

Fig. 10.6: Critical temperature (T_C), current density (J_C) and magnetic field boundary separating superconducting and normal conducting states

10.2 SUPERCONDUCTING MATERIALS

(*i*) **Superconducting Elements:** Metallic elements are mostly superconductors. Their T_C are typically of the order of a few Kelvin. Among metals, niobium exhibits the highest critical temperature of the pure elements, $T_C = 9.2$ K. Noble metals, copper, silver and gold are excellent conductors of electricity at ambient temperatures but are not superconductors down to very low temperatures at all. Magnetic metallic elements do not exhibit superconductivity.

The best known semiconductors, Si and Ge, become superconductors under a pressure of ~2 K bar with $T_C = 7$ and 5.3 K respectively. Other elements that become superconductors under pressure include P, As, Se, Y, Sb, Te, Ba, Bi, Ce, and U.

(*ii*) **Binary Alloys and Compounds:** In the most alloys and compounds, the critical temperatures are usually somewhat higher. Nb compounds, like Nb_3Sn, Nb_3Ge and in particular, Nb-Ti, are of technological interest. While the maximum current density one can pass through the standard water cooled copper wire at 300 K is about 2000 A-cm^2, one can pass very high current densities of upto 10^4 A-cm^2 in high magnetic fields of 10 Tesla at 4.2 K through a wire made of Nb-Ti alloy without destroying superconductivity. This enables the construction of powerful supermagnets which provide a basis for a range of large scale applications like energy storage or levitation of trains, etc.

Transition metals combined with other elements often, produce binary alloys or compounds with T_C's higher than those of starting elements. The intermetallic compounds and ordinary compounds usually exhibit the highest T_C's.

Intermetallic Compounds

Among the intermetallic superconductors, the most favourable group is the one based on the A_3B compound. In the cubic A-15 structure six binary compound have T_C over 17 K. The highest known T_C prior to 1986, close to T_C = 23 K is obtained in Nb_3Ge stabilized by traces of oxygen or aluminium; it exhibits the upper critical field of 38 *T*. The A-15 structure exists in about 70 binary compounds. $H_{C_2}(T)$ values for A-15 superconducting compounds is very high. $H_{C_2}(O)$, the upper critical field values as the temperature approaches 0 K for A-15 compounds are also very high, e.g. 44 tesla for a composition $Nb_{79}(Al_{73}Ge_{27})_{21}$, 32 *T* for Nb_3Al, 39 *T* for Nb_3Ge, 23 *T* for Nb_3Sn, 21 *T* for V_3Ga and 25 *T* for V_3Si

(*i*) **Chevrel Phases:** In 1971, Chevrel *et al.*, discovered a series of ternary molybdenum chalcogenides having the general formula $M_xMO_6X_8$ where M represents any of a large number (nearly 40) of metallic elements and rare earths (RE) throughout the Periodic Table; *x* has values between 1 and 4, depending on the M element; and X is a chalcogens; sulphur (S), selenium (Se) or tellurium (Te). The highest T_C = (15 K) in the series is obtained in $PbMO_6S_8$ with unusual high H_{C_2} value (= 60 *T* = 600 kilogauss). This was the highest value of H_{C_2} prior to the discovery of HTSC cuprates. The large values of H_{C_2} as compared with Nb_3Sn and NbTi make this material interesting for making superconducting wires. Critical currents (J_C) as high as ~3 × 10^{-5} A/cm^2 have been reported at 4.2 K and this provides an impetus for making wires out of these very brittle materials. It has been reported that antiferromagnetism of the rare earth can coexist with superconductivity like in Gd, Tb, Dy, Tr compounds where T_C is 1.4, 1.65, 2.1 and 1.85 K and T_N (Neel Temperature) is 0.84, 0.9, 0.4 and 0.15 K respectively. The compound $HoMo_6S_8$ exhibits coexistence of superconductivity and ferromagnetism. The material is superconducting only between two critical temperatures, H_{C_1} = 2 K and H_{C_2} below 0.65 K. Below 0.65 K the material is ferromagnetic.

(*ii*) **Tetragonal Rare Earth Rhodium Borides:** This series has general formula $RERh_4B_4$. The compounds for RE = Y, Er, Tm and Lu exhibit superconductivity. The compound $ErRh_4B_4$ is a typical reentrant ferromegnetic superconductor ($T_{C_1} = 8.67$ K, $T_{C_2} = 0.775$ warming and 0.710 cooling and Curie temperature 1.2 K).

(iii) **Organic Superconductors:** These are a novel group of materials. The first organic superconductor $[TMTSF]_2PF_6$, where TMTSF denotes tetramethyletetraselenafulvalene, had a T_C of 1 K. Subsequent developments in the field have led to higher T_C organic materials. These systems were characterized by their nearly one-dimensional properties and by low carrier concentration. Later, a new series of organic materials with a two dimensional character was discovered: $(BEDT\text{-}TTF)_2X$, where BEDT-TTF denotes bisethylenedi-thio-tetrathiafulvalene. The κ-modification of the X = $Cu(NCS)_2$ has T_C = 10.4 K and κ-$(BEDT\text{-}TTF)_2Cu[NCN]_2Br$ has T_C = 11.2 K, which is probably the highest known T_C so far observed in these materials.

(iv) **Rare-Earth Transition Metal Borocarbides:** Superconductivity was reported recently in a series of compounds with the formula RNi_2B_2C with a maximum T_C of 16.5 K for R = Lu. These materials display both superconductivity and magnetic order and effects associated with them. Recently, intermetallic compounds YPd_5B_3C (T_C = 23 K) and $ThPd_3B_3C$ (T_C = 21 K) have been discovered.

(v) **Fullerene Superconductors:** Fullerene are stable, cagelike molecules that constitute the third form of pure carbon; the other two are diamond and graphite. The archetype fullerene is C_{60}. When C_{60} is doped with alkaline metals, superconductivity is observed with T_C = 18 K, 30 K, 33 K and 40 K for K_3C_{60}, Rb_3C_{60}, $RbCs_2C_{60}$ and Cs_3C_{60} respectively. Hole doped C_{60} (for $C_{60}/CHBr_3$ with 3 to 3.5 holes per C_{60} molecule) was recently reported to be superconducting with a T_C as high as 117 K, although the nature of experiment meant that the supercurrents were confined to the surface of the C_{60} crystal, rather than probing the bulk.

(vi) **Heavy Fermion Superconductors:** A small number of compounds which include one Ce compound $CeCu_2Si_2$ and few uranium compounds: UBe_{13}, UPt_3, URu_2Si_2, UNi_2Al_3, UPd_2Al_3 exhibit superconductivity with $T_C < 1$ K. These compounds are characterized by enormous volumes of linear coefficient of the electronic specific heat ($C_e = \gamma T$) which can be as high as ~1 J mol^{-1}K^{-2}, and a corresponding large electron mass (m ~10^2-10^3 m_e, where m_e is free electron mass). Most of these systems exhibit the coexistence of superconductivity and antiferromagnetism.

(*vii*) **Itinerant Electron Ferromagnetic Superconductors:** The intermetallic compound Y_9Co_7 has been shown to exhibit an interesting interplay between very weak ferromagnetism and some from of superconductivity. The system shows ferromagnetism below 6-8 K and at a lower temperature (~3 K) superconductivity sets in. Recently, superconductivity in UGe_2 is reported below 1 K on the border of weak ferromagnetism.

(*viii*) **Quantum Spin Ladder Materials:** These materials consist of ladders made of AFM chains of $s = 1/2$ spins coupled by interchain AFM bonds. $SrCu_2O_3$ and $LaCuO_{2.5}$ 2-leg ladder materials, whereas $Sr_2Cu_2O_5$ is a 3-leg ladder material. Superconductivity has apparently been discovered in the ladder material $Sr_{0.4}Ca_{13.6}Cu_{24}O_{41.84}$ under pressure with $T_C \approx 12$ K at 3 GPa.

(*ix*) **Magnesium Diboride (MgB_2):** Very recently, superconductivity is reported at 39 K in the simple binary ceramic compound MgB_2. This is probably highest T_C yet determined for a non-copper oxide bulk superconductor.

(*x*) **Sr_2RuO_4:** This superconducting compound has the same structure as the $La_{2-x}M_xCuO_4$ (M = Ba, Sr, Ca, Na) HTSC cuprates. While the T_C of Sr_2RuO_4 is only ~1 K, this compound is of considerable interest because it is the only layered perovskite superconductor without copper.

(*xi*) **High Temperature Cuprate Superconductors:** Until 1986, most of the superconducting compounds were metals and alloys. However, some oxide superconductors were known for decades, but their transition temperatures were rather low. This was mainly due to a low number of carriers in the metallic state. Two known exceptions were $LiTi_2O_4$ and $BaPbBiO_3$ with T_C ~13 K. This was quite unusual as their densities of carriers were also very small. The major breakthrough came in 1986 when J.G. Bednorz and K.A. Muller discovered superconductivity with a T_C (onset) of 35 K a new record in a mixed phase copper-oxide ceramic containing $La_{2-x}Ba_xCuO_4$. The end of 1986 and the beginning of 1987 was marked by synthesis of rare-earth metal oxides with the discovery of Yttrium barium copper oxide, $YBa_2Cu_3O_7$, which has a critical temperature of about 92 K. This was a significant breakthrough as it meant that for the first time the world has witnessed the existence of a superconductor with a T_C above that a liquid nitrogen (boiling point 77 K). Nitrogen is much more abundant than helium, much less expensive, and liquid nitrogen cryogenic systems using helium refrigeration. One application which could benefit from nitrogen cooling is the development of hybrid microelectronic technology (semiconductor-superconductor devices) both gallium arsenide and silicon can be tailored to perform better at liquid nitrogen temperatures.

In early 1988, Bismuth (Bi) and Thallium (Tl) based cuprates $Bi_2Sr_2Ca_2Cu_3O_{10}$ (T_C = 110 K and $Tl_2Ba_2Ca_2Cu_3O_{10}$ (T_C = 125 K) were discovered. These new HTSC cuprates may have some advantages over ceramic superconductors containing rare earths. Since the critical current density increases as T/T_C decreases, a T_C far above the opening temperature of liquid nitrogen (77 K) is advantageous. Moreover, the new materials are more stable than the rare earth cuprate superconductors; they do not lose oxygen or react with water.

The maximum value of T_C has now increased to 133 K for mercury based cuprate, $HgBa_2Ca_2O_{8+\delta}$. When this compound is subjected to high pressure ~30 GPa, the onset of T_C increases to ~164 K (more than half way to room temperature). While $HgBa_2Ca_2O_{8+\delta}$ cannot be used in applications at such high pressures, this striking result suggests that values of T_C in the neighbourhood of 160 K, or even higher, are attainable in cuprate oxides. More than 100 different cuprate materials many of which are superconducting have been discovered since 1986. Several of the more important are listed in Table 10.1, along with the maximum values of T_C observed in each class of materials.

TABLE 10.1

(*i*) Some important classes of HTSC cuprates along with the maximum value of T_C observed in each class.

(*ii*) Examples of the abbreviated names (nicknames) used to denote superconducting cuprate materials.

(*i*) *Material*	*Maximum* T_C *(K)*
$La_{2-x}MCuO_4$; M = Ba, Sr, Ca, Na	~40
$Ln_{2-x}MCuO_4$; Ln = Pr, Nd, Sm, Eu; M = Ce, Th	~25
$YBa_2Cu_3O_{7-\delta}$	92
$LnBa_2Cu_3O_{7-\delta}$	~95
$RBa_2Cu_4O_8$	~80
$Bi_2Sr_2Ca_{n-1}Cu_nO_{2n+4}$ (n = 1, 2, 3, 4)	(n = 3) 110
$TlBa_2Ca_{n-1}Cu_nO_{2n+3}$ (n = 1, 2, 3, 4)	(n = 4) 122
$Tl_2Ba_2Ca_{n-1}Cu_nO_{2n+4}$ (n = 1, 2, 3, 4)	(n = 3) 122
$HgBa_2Ca_{n-1}Cu_nO_{2n+2}$ (n = 1, 2, 3, 4)	(n = 3) 133

(*i*) *Material*	*Nickname*	*Maximum* T_C *(K)*
$YBa_2Cu_3O_{7-\delta}$	YBCO; YBCO-123; Y-123	92
$Bi_2Sr_2Ca_2Cu_3O_{10}$	BSCCO; BSCCO-2223; Bi-2223	110
$Tl_2Ba_2Ca_2Cu_3O_{10}$	TBCCO; TBCCO-2223; Tl-2223	122
$HgBa_2Ca_2Cu_3O_8$	HBCCO; HBCCO-1223; Hg-1223	133

10.3 HTSC CUPRATE MATERIALS CHARACTERISTICS

(*i*) **Highly Anisotropic, Layered Structures:** Except for one material (isotropic, cubic $Ba_{1-x}K_xBiO_3$), all HTSC oxides are layered perovskites. One of the important characteristics of all cuprates is the presence of CuO_2 planes in their layered perovskite-like crystal structures (Fig. 10.7). The mobile charge carriers which can be electrons but are usually holes, are believed to reside primarily within the CuO_2 planes. Obviously, the presence of CuO_2 layers dominate most properties of these systems.

(*ii*) **Metallic Oxides:** While most oxides are insulating materials, HTSC cuprates exhibit metallic behaviour. Many of these cuprates can be doped with charge carriers and rendered superconducting by substitution of appropriate elements into an insulating parent compound. For example, substitution of divalent Sr for trivalent La in the antiferromagnetic insulator La_2CuO_4 dopes the CuO_2 planes with mobile holes and produces superconductivity in $La_{2-x}Sr_xCuO_4$ with a maximum T_C of ~40 K at $x \approx 0.17$. Similarly, substitution of tetravalent Ce for trivalent Nd in the antiferromagnetic insulating compound Nd_2CuO_4 apparently dopes CuO_2 planes with electrons, resulting in electron doped superconductivity in $Nd_{2-x}Ce_xCuO_{4-y}$ with a maximum T_C of ~25 K at $x \approx 0.15$ for $y \approx 0.02$. The temperature T versus x phase diagrams for both these systems are shown in Fig. 10.8.

(*iii*) **Ceramic Materials:** The original materials $La_{2-x}Sr_xCuO_4$ and $YBa_2Cu_3O_7$ were synthesized by their discoverers as ceramic pellets. One mixes the correct ratio of constituent oxides, grinds and sinters them, make a pallet, and following a calcining procedure (at ~950° C) cools it down in oxygen. Typical ceramics, HTSC oxides also contain grains, grain boundaries, twins, voids and other imperfections. Even some of the best thin films may consist of grains a few microns in diameter, all these are mostly detrimental to high critical current densities that are required for applications.

It is worthwhile to mention that even the best single crystals of HTSC cuprates often contain various defects and imperfections like oxygen vacancies, twins, impurities, etc. These imperfections are not only very relevant to their basic thermodynamic (meta) stability. We may note that various imperfections found in HTSC crystals are intrinsic to these materials.

We may note that the understanding of phase diagrams, crystal chemistry, preparation and stability of these oxides is still very much in progress.

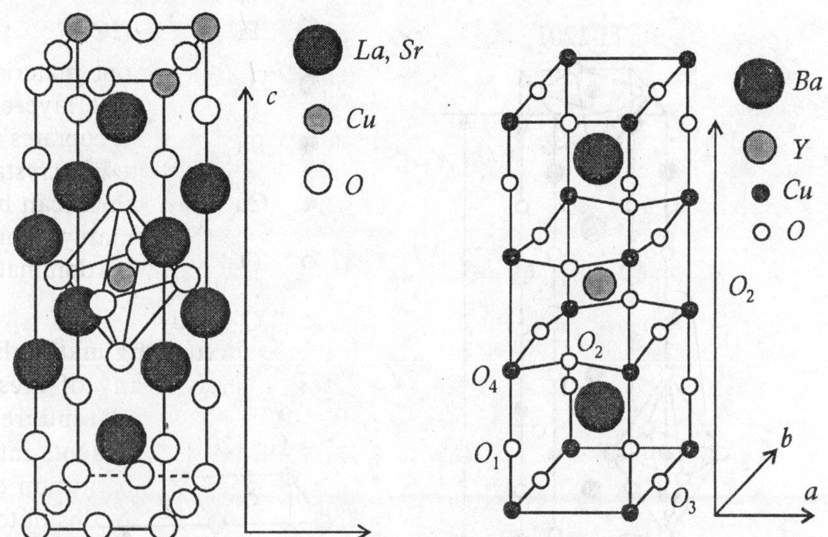

(*a*) Lattice structure of $La_{2-x}Sr_xCuO_4$ (*b*) Orthorhombic unit cell of $YBa_2Cu_3O_{7-\delta}$. A special feature of the Y-123 structure are CuO chains in *b* direction

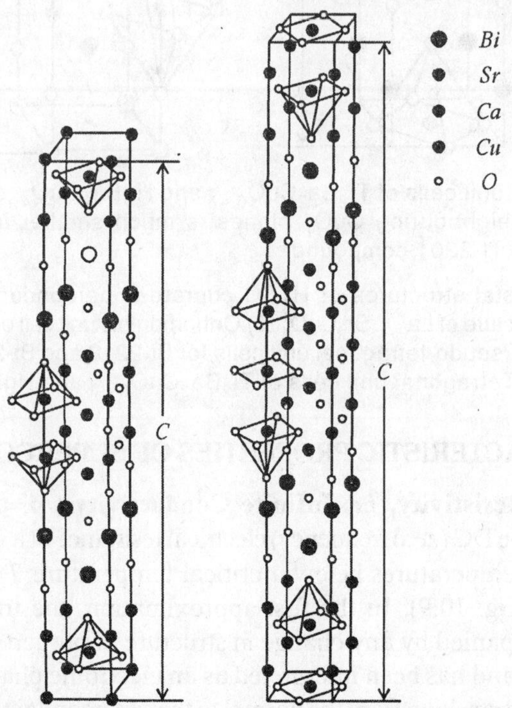

(*c*) Pseudo-tetragonal unit cells for the Bi-2212 and Bi-2223 structures

Fig. 10.7 **(Contd.)**

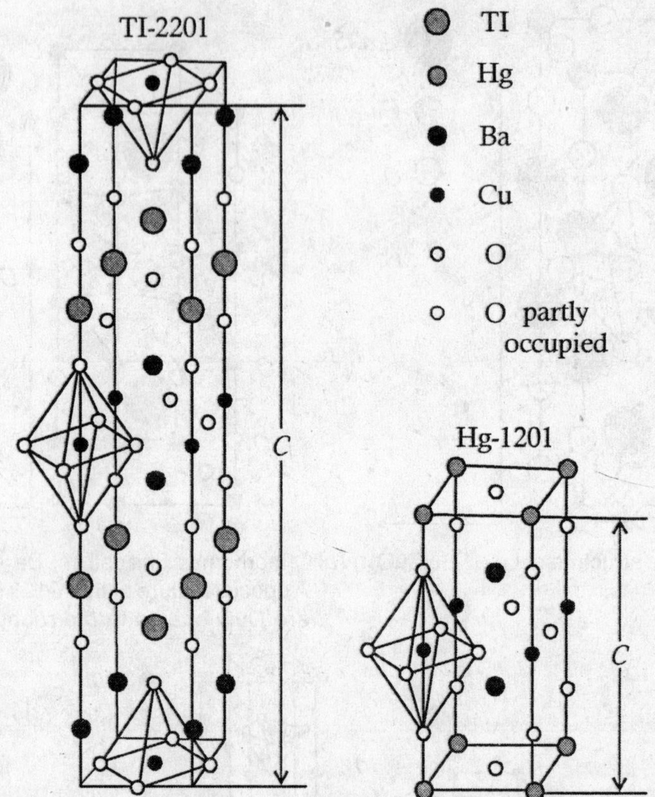

(d) Tetragonal unit cells of $Tl_2Ba_2CuO_{6+\delta}$ and $HgBa_2CuO_{4+\delta}$. The distance between neighbouring CuO_2 planes is much smaller in the Hg-1201 than in the Tl-2201 compound

Fig. 10.7: Crystal structures of HTSC cuprate superconductors (a) Lattice structure of $La_{2-x}Sr_xCuO_4$ (b) Orthorhombic unit cell of $YBa_2Cu_3O_{7-\delta}$. (c) Pseudo-tetragonal unit cells for Bi-2212 and Bi-2223 structures (d) Tetragonal unit cells of $Tl_2Ba_2CuO_{6+\delta}$ and $HgBa_2CuO_{4+\delta}$.

10.4 CHARACTERISTIC PROPERTIES OF SUPERCONDUCTORS

(*i*) Zero Resistivity, *i.e.* Infinite Conductivity ($\rho = 0$ for all $T < T_C$): The DC (zero frequency) electrical resistance of a superconductor at all temperatures below a critical temperature T_C is practically zero (Fig. 10.9). In the first approximation, the transition is not accompanied by any change in structure of property of the crystal lattice and has been interpreted as an electronic phase transition.

The transition from the normal to the superconducting state occurs sharply in pure metals (Fig. 10.1) but not so in some impure, deformed and HTSC oxides. Bi-, Tl-, and Hg-based cuprate super-

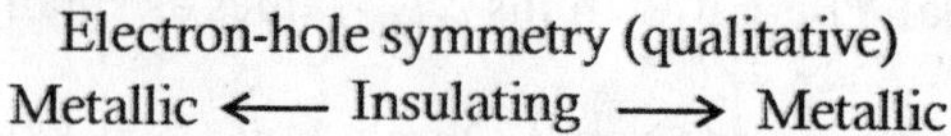

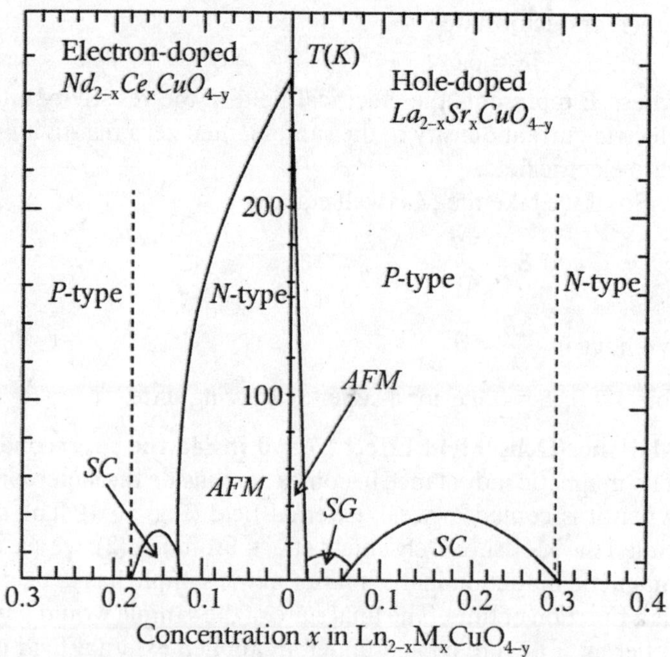

Fig. 10.8: Temperature-dopant concentration ($T-x$) phase diagram delineating the regions of superconductivity and antiferromagnetic-ordering of the Cu^{2+} ions for the hole-doped $La_{2-x}Sr_xCuO_4$ and electron doped $Nd_{2-x}Ce_xO_{4-y}$ systems. AFM→ Antiferromagnetism SG→ spin glass phase, and SC→ superconducting phase

conductors are chemically complex materials, in which there may exist several superconducting phases in one specimen. A two-step transition reflects the presence of atleast two superconducting phases. Figure 10.9 shows resistance versus temperature for a single and multi-phase high temperature superconductor. The transition width ΔT_C for single-phase high temperature superconductors is typically ≈ 1 K.

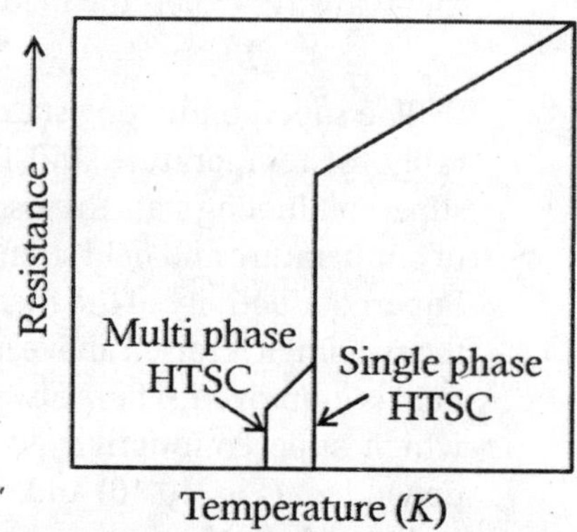

Fig. 10.9: Resistance versus temperature curves of a single- and a multi-phase high temperature superconductor

In epitaxial $YBa_2Cu_3O_7$ films ΔT_C-values as small 0.3 K has been achieved.

If we assume the usual Ohm's law ($V = RI$) describing the superconductor state

$$\boldsymbol{E} = \rho \boldsymbol{J} \qquad (1)$$

where $\boldsymbol{E}$ represents the electrical field, ρ the resistivity and $\boldsymbol{J}$ the electric current density in the sample, then zero resistivity implies zero electric field.

So, if we take the Maxwell equation

$$\text{curl}\, \boldsymbol{E} = \frac{\partial \boldsymbol{B}}{\partial t}$$

We have $$\frac{\partial \boldsymbol{B}}{\partial t} = 0 \qquad (2)$$

for $\rho = 0$, $\boldsymbol{E} = 0$, *i.e.* for a superconducting state.

(*ii*) Meissner-Ochsenfeld Effect (B = 0 inside the superconductor): The magnetic inductance becomes zero inside the superconductor when it is cooled in weak external field (Fig. 10.4). The effect is called the Meissner-Ochsenfeld effect. From Eq. (2), we see that the magnetic induction in the interior of the sample has to be constant as a function of time. The final state of the sample would have been different if it were cooled under an applied external field or if the field were applied after the sample has been cooled below T_C. In the former case the field would have remained within the sample, while in the latter it would have been zero. For the specimen to be in the same thermodynamic state, independent of the precise sequence that one uses in cooling or in applying the field, the superconducting metal always expels the fields from its interior, and has

$$\boldsymbol{B} = 0$$

The superconducting state of a metal exists only in a particular range of temperature and field strength. The condition for the superconducting state to exist in the metal is that some combination of temperature and field strength should be less than a critical value. Superconductivity of the metal will disappear if the temperature of the specimen is raised above its T_C or if a sufficiently strong magnetic field is employed. There always exists some critical field H_C, above which superconductivity disappears. This field is temperature dependent (Fig. 10.10) and the empirical relation which describes well this dependence is

$$H_C(T) = H_0[1 - (T/T_C)^2] \qquad (3)$$

where H_0 is the value of H_C at absolute zero.

We have already mentioned in section 1 that the superconducting materials that completely expel magnetic flux until they become completely normal are called *type-I superconductors*. For a *type-II superconductor* there are two critical fields: the lower H_{C_1}, and the upper H_{C_2} (Fig. 10.5). So in applied field smaller than H_{C_1}, the type-II superconductor behaves just like a type-I superconductor below H_C. Above H_{C_1} the flux partially penetrates into the material until the upper critical field, H_{C_2}, is reached. Above H_{C_2} the material returns to the normal state. Between H_{C_1} and H_{C_2} the superconductor is said to be in *mixed state*. For all applied fields $H_{C_1} < H < H_{C_2}$, magnetic flux partially penetrates the superconducting specimen in the form of tiny microscope filaments called *vortices*. The diameter of vortex in conventional superconductors is typically 100 nm.

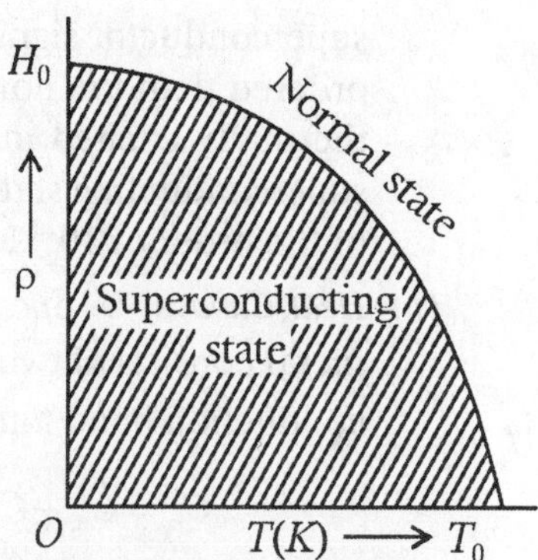

Fig. 10.10: The critical magnetic field at which superconductivity of a metal disappears

Type-II superconducting materials can withstand strong applied magnetic fields without returning to the normal state due to partial flux penetration. Superconductivity can and does persist in the mixed state upto the upper critical field, H_{C_2}, which is sometimes as high as 60 Tesla (Chevrel phases) or even ~150 Tesla in HTSC cuprates. At fields higher than H_{C_2} the superconductor returns to the normal state.

Practically all technologically interesting materials like Nb compounds. Chevrel phases or HTSC cuprates are all type-II superconductors. The reason for this is that the creation of vortices keeps the magnetic energy smaller than the condensation energy, so the over all free energy of the mixed superconducting state remains more favourable than the normal state even upto high magnetic fields. Since the supercurrent can flow in the mixed state through the superconducting regions between vortices, type-II superconductors allow one to construct wires needed for high field magnets.

(*iii*) Thermal Properties of Superconductors

(*a*) *Entropy:* Entropy of all superconductors decreases considerably upon cooling-below T_C. The entropy is a measure of the degree of disorder of a given system and hence this decrease in a superconductor signifies that the superconducting state is more ordered than the normal state. The fraction of electrons that is thermally excited in the normal state becomes ordered in the superconducting state. The entropy variation is relatively weak, of the order of 10^{-4} k_B/atom ($k_B \rightarrow$ Boltzmann constant).

(*b*) *Heat Capacity or Specific Heat:* The specific heat C_n is a normal metal consists of two contributions C_n^e from the electrons in the conduction band and C_n^l from the lattice. Thus

$$C_n = C_n^e + C_n^l = \gamma T + \beta T^3 \tag{4}$$

The first term in Eq. (4) is linearly proportional to T while the second term is proportional to T^3. The specific heat of the superconductor (C_{es}), changes at T_C in a characteristic way (Fig. 10.11). In zero magnetic field there appears a discontinuity at T_C. At temperatures well below T_C, the heat capacity fits an exponential form

$$C_{es}(T < T_C) = A\exp(-\Delta / k_B T) \tag{5}$$

where $E_g = 2\Delta$ is a constant of given material, called the *energy gap*. Such an exponential temperature dependence is a hallmark of a system with a gap Δ in the spectrum of allowed energy states, separating the excited states from the ground states by energy Δ. Although there are also materials with gapless superconductivity, most materials of interest do have an energy gap. The relation between Δ and T_C is given by Bardeen-Cooper-Schrieffer (BCS) theory of superconductivity.

$$2\Delta = 3.5 k_B T_C \tag{6}$$

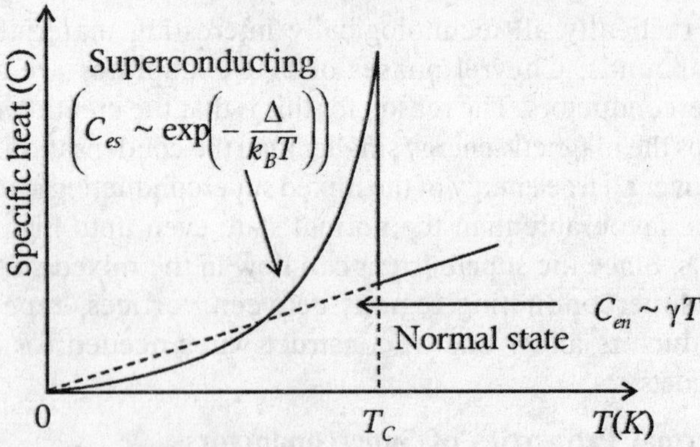

Fig. 10.11: Schematic representation of the specific heat of a metal in the normal and the superconducting phase. There is a characteristic jump of the transition point

Heat capacity measurements provided the first indications of such a gap in superconductors, and one of the key features of BCS theory is its prediction of such a gap in superconductors.

Estimates shows that the gap in conventional superconductors with $T_C < 20$ K is of the order of 1 meV, while in HTSC oxides with $T_C \sim 100$ K, $\Delta \sim$ several (1-10) meV. We may note that in superconductor 2Δ corresponds to the energy needed to break a Cooper pair, whereas in semiconductors the gap in the energy spectrum corresponds to the energy difference between the valence and the conduction band (~1 eV).

(*c*) *Thermal Conductivity:* The thermal conductivity of superconductors undergoes a continuous change between the two phases and usually lower in a superconducting phase and at very low temperatures approaches zero. This suggests that the electronic contribution drops, the superconducting electrons possibly plays no part in heat transfer. The thermal conductivity of tin (T_C = 3.73 K) at 2 K is 16 $\text{Wcm}^{-1}\text{K}^{-1}$ for the superconducting phase and 34 $\text{Wcm}^{-1}\text{K}^{-1}$ for the normal phase.

(*d*) *Thermoelectric Properties:* A combined thermal and electric effect of interest and practical importance is the Peltier effect, which is the basis of operation of thermocouples used for temperature measurement. If the two junction regions of a loop made of two different metals are maintained at different temperatures, an electrical current is driven around the loop. A greatly weakened form of this current occurs in the superconducting state, because some 'normal' electrons remain at any non-zero temperature. However, the presence of superconductivity 'shorts out' the usual thermoelectric voltage.

(*iv*) **Acoustic Attenuation:** When sound wave propagates through a metal, the microscopic electric fields due to the displacements of the ions can impart energy to the electrons near the Fermi level thereby removing energy from the wave. One can express this by the attenuation coefficient α, of acoustic waves. The ratio of α for superconducting and normal state is given by

$$\frac{\alpha_s}{\alpha_n} = \frac{2}{1 + \exp(\Delta / k_B T)} \tag{7}$$

At low temperatures, one finds

$$\frac{\alpha_s}{\alpha_n} = 2\left[\exp\left(\Delta / k_B T\right)\right] \tag{8}$$

(*v*) **High Frequency Electromagnetic Properties:** For all frequencies much higher than the frequency corresponding to the energy gap, $E_g = h\nu$ where ν is the frequency in Hz, the electromagnetic response in the superconducting state is identical to the response of the normal state. Remembering that 1 eV ~ 10^{14} Hz, one can easily understand that the change in the frequency response occurs and at $\nu \sim 10^{11}$ and $\sim 10^{12}$ Hz in the conventional and HTSC oxides respectively.

High frequency electromagnetic properties of superconductors differ from zero frequency (or low frequency) behaviour. In the radio-frequency ($<10^8$ Hz) and microwave frequency range ($\sim 10^8$-10^{11} Hz) the resistance of the superconductor to current flow is not zero. However, the resistance and the accompanying energy loss are still rather small.

In the optical region of the spectrum ($\sim 10^{15}$ Hz), one finds no difference in electromagnetic response between the normal and the superconducting state so one does not see any change in the appearance of the sample as it undergoes the transition. In the range $\sim 10^{11}$ to 10^{12} Hz (depending on the material), there is a sharp increase in the absorption of electromagnetic radiation by the superconductor. This is due to the existence of energy gap (E_g) in the electronic energy spectrum.

We may note that the sharp rise in the absorption occurs at the energy ($E = h\nu$) of a single photon which is just sufficient to produce an excitation (by depairing weakly coupled Cooper pairs).

HTSC oxides exhibit light excitations on the T_C values. These materials are ceramic in nature, black in colour and P-type (hole-type) in carriers. Though these materials are black in colour, they do not behave like black body. Their optical absorption spectra are continuous but consist of discrete bonds corresponding to different energy levels. These are responsible for additional *photoinduced* charge carriers and giving rise to change in T_C. One may call it as photodoping and superconductors as photoinduced superconductors.

(*vi*) **The Energy Gap:** We have seen that the heat capacity in the superconducting state well below T_C varies with temperature in an exponential manner in accordance with relation (5) with $\Delta = bk_B T_C$, where *b* is a constant. Near the transition temperature, the half width ($\Delta = E_g/2$) of the energy gap is approximately

$$\Delta(T) = 3.2k_B T_C (1 - T/T_C)^2 \tag{9}$$

Such a temperature dependence is characteristic of a system that has an energy gap in its spectrum of allowed energy states. One can determine the energy gap of a superconductor by knowing the absorption of electromagnetic waves. One can detect this gap by photo absorption, quantum tunnelling and other experiments. A plot of the temperature dependence of energy gap parameter $\Delta(T)$ is shown in Fig. 10.12. We note that the energy gap is zero at $T \to T_C$ and reaches a maximum value $\Delta_0(T)$ as the temperature lowered towards 0 K.

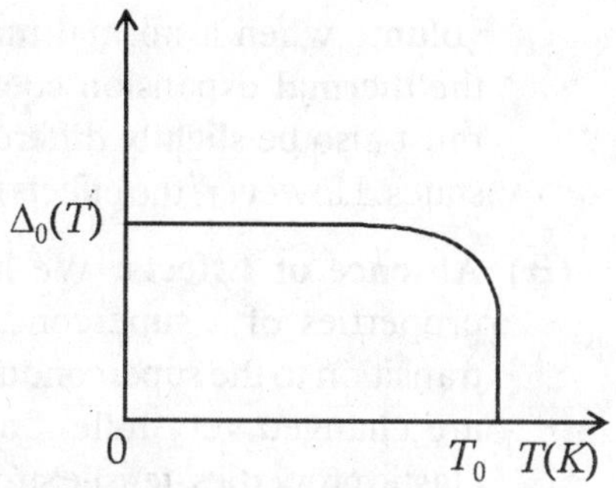

Fig. 10.12: The energy gap $\Delta(T)$ versus temperature. $\Delta(T)$ vanishes with infinite slope as $T \to T_C$ leading to the second order phase transition

(*vii*) Isotope Effect: It has been observed that the critical temperature (T_C) of superconductors varies with the ionic or isotopic mass. The relation valid for some simple metals in given by

$$T_C \propto M^{-\alpha}$$

or $$M^{\alpha} T_C = \text{constant} \tag{10}$$

where M is the atomic mass of the isotope and α is roughly 0.5. Obviously, the existence of the isotope effect indicated that, although superconductivity is an electronic phenomenon, it nevertheless depends in an important way on the vibrations of the crystal lattice in which the electrons move. Fortunately, not until after the development of the BCS theory was it discovered that the situation is more complicated than it had appeared. For some superconductors the exponent of M, *i.e.* α is not $-1/2$, but near zero, e.g. Ru and Zr, and for at least one it is positive.

The isotope effect, as studied by substitution of ^{18}O for ^{16}O in HTSC cuprates is very weak. The substitution of ^{18}O for ^{16}O in Eq. (10) gives $\alpha \approx 0.02$ in YBCO and $\alpha \approx 0.15$ in LBCO. This has prompted the exploration of non-phonon electronic coupling mechanisms responsible for superconductivity in these cuprate systems. However, a weak isotope effect is not conclusive.

(*viii*) Mechanical Effects: When a superconducting material is mechanically stressed, it is found experimentally that both T_C and H_C are slightly altered. One can easily see that many of the mechanical properties of the superconducting and normal states are thermodynamically related to the free energies of these states, and the critical field strength depends on the difference in the free energies of the two states. There is an extremely small change in volume when a normal material becomes superconducting, and the thermal expansion coefficient and bulk modulus of elasticity must also be slightly different in the superconducting and normal states. However, the effects are extremely small.

(*ix*) Absence of Effects: We have seen that most of the electronic properties of a superconductor are profoundly affected by the transition to the superconducting state, while many other properties are changed very little if all. These include the mechanical and elastic properties, tensile strength, sound velocity and density among others.

(*x*) Characteristics Phenomenological Parameters

(*a*) *Penetration Depth* (λ): While studying Meissner effect, we mentioned that the superconductor expels a (weak) magnetic field B from its interior, *i.e.* $\boldsymbol{B} = 0$ in the interior of a superconductor. The finer experiments reveal that the field B penetrates into the superconductor within a very thin surface layer. Consider the boundary of a semi-infinite slab. When the external field is applied parallel to the boundary, the applied field does not suddenly drop to zero at the surface of the superconductor, but decays exponentially according to the relation

$$H(x) = H(0)\exp(-x/\lambda) \qquad (11)$$

where $H(0)$ is the value of the magnetic field at the surface and λ is a characteristics length known as the penetration depth; λ is the distance for H to fall from $H(0)$ to $H(0)/e$. In most of the superconductors, λ is of the order of 500 Å. λ depends on the material and on the temperature, the latter variation being given approximately by

$$\lambda = \lambda_0\left[1-\left(\frac{T}{T_C}\right)^4\right]^{-1} \qquad (12)$$

where λ_0 is the penetration depth at zero temperature for the particular material and is typically of order 500 Å.

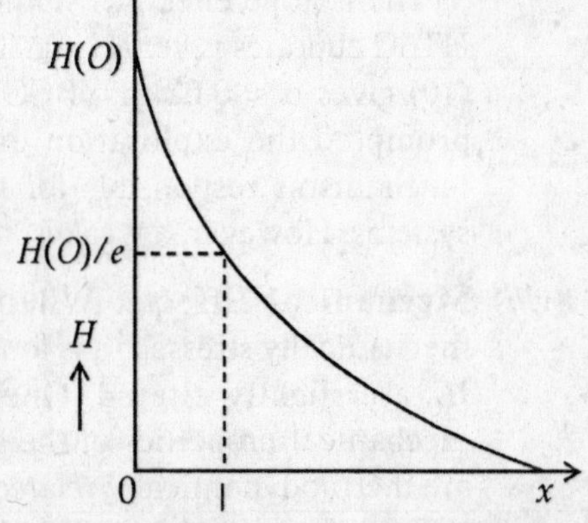

Fig. 10.13: Decay of the magnetic field in the interior of a superconductor

If a superconducting film or filament is thinner than λ, its properties are significantly different from those of the bulk material. In particular, the value of H_C increases as thickness decreases, and the special property of type II superconductor arise from this.

(*b*) *The Coherence Length* (ξ): It is a measure of the distance over which the gap parameter (Δ), can vary, for instance in a spatially varying magnetic field or near a superconductor-normal metal boundary. It is also referred as the distance between two electrons of the Cooper pair within the highly coherent superconducting state. The intrinsic or BCS coherence length ξ_0 is defined as

$$\xi_0 = \frac{\hbar v_F}{\pi \Delta} \qquad (13)$$

where v_F is the Fermi velocity and Δ is the energy gap.

Using order of magnitude values for v_F and Δ, one obtains ξ = 1600 Å in pure Al, ξ = 380 Å in pure Nb, but only about 10 Å in the new HTSC oxides. Table 10.2 provides values of λ and ξ for few selected superconductors.

When a small impurity is added to a metal λ increases very rapidly while ξ decreases.

TABLE 10.2: T_C, H_{C_2}, **λ AND ξ FOR FEW SELECTED SUPERCONDUCTORS**

Material	T_C (K)	E_{C_2} (*Tesla*)	λ (Å)	ξ(Å)
Al	1.1	0.02 (Also exhibit type- I superconductivity)	500	16000
Nb	9.2	0.2	400	380
Nb-Tl	9.5	14	600	450
Nb_3Sn	18.3	24	800	35
Rb_3C_{60}	29.3	~50	1600	~20
$La_{1-x}Sr_xCuO_4$	38	~65	2500	~15
$YBa_2Cu_3O_7$	92	~120	4000	~10

(*c*) *Ginzburg-Landau Parameter (κ)* : The ratio of two characteristic length λ and ξ is called the Ginzburg-Landau ratio κ

$$\kappa = \frac{\lambda}{\xi} \qquad (14)$$

Close to T_C, κ is independent of temperature and it allows one to distinguish between type-I and type-II superconductors.

If $\kappa < 0.7$ material is type-I superconductor, and if $\kappa > 0.7$, the material is type-II superconductor. The exact critical value of κ which separates type-I from type-II behaviours is $1/\sqrt{2} \approx 0.7$.

In the latter case the magnetic flux does penetrate the sample in the form of the cylindrical tubes called vortices. The vortices have a radius λ and destroy superconductivity locally within a cylinder of radius ξ. It is energetically favourable for type-II superconductors to let the flux penetrate partially in the form of vortices.

(*xi*) *Flux Quantization:* This refers to the fact that the magnetic flux threading a superconductor loop cannot have an arbitrary value; it has to be a multiple of $\phi_0 = h/2e$. The flux quantum is defined as

$$\phi = \frac{h}{2e} = 2.0678 \times 10^{-15} \text{ Weber} \tag{15}$$

Superconducting devices can measure this tiny variation of magnetic flux which is exceedingly important in metrology and advanced instrumentation.

10.5 JOSEPHSON EFFECTS

In 1962 B. D. Josephson predicted theoretically that Cooper pair tunneling through a very thin insulating layer (~2 nanometer) is possible. His theory predicted in addition to the Giaever current, the existence of supercurrent, arising from tunneling of the bound electron pairs. Such a junction is called *weak link.* The effects of pair tunneling include:

(*i*) **D.C. Josepheon Effect:** A DC current flows across the junction resulting Cooper pair-tunneling in the absence of any electric or magnetic field, *i.e.* at zero voltage across the insulating layer. One obtains the following expression for current through the contact for two identical superconductors as

$$I = I_C \sin\delta$$

where I_C is the critical current for the Josephson junction and δ is the phase difference of the two Cooper wave functions. Theoretical results for voltage across the junction yield

$$V_t = \frac{\pi\Delta}{2e} \tag{16}$$

For metals most commonly used as electrodes, lead and niobium, V_t is of the order of ~2.5 mV.

(*ii*) **A. C. Josephson Effect:** Josephson also predicted that if a constant non zero voltage V were maintained across the tunnel barrier, an alternating supercurrent could flow through the barrier in addition to the DC current produced by the tunneling of unpaired electrons.

This effect has been utilized in a precision determination of $\hbar/e$. Further, an r_f voltage applied with the DC voltage can then cause a DC current across the junction. The frequency of the A.C. Josephson current given by the Josephson relation

$$\omega = \frac{2eV}{\hbar} \quad (17)$$

where V is the DC voltage applied across the junction.

Josephson Junctions

These are sophisticated sandwich structures of superconducting films (usually of Nb) separated by extremely thin (~10 Å) insulating oxide layers. In a suitable submicroscopic circuit they act as the fastest switching elements available today.

Josephson junction have also the lowest power consumption of their operating temperature, so they might provide the basis for the architecture of some of the fastest computers of the future. They are already used in the fastest commercially available oscilloscopes that operates at 10 GHz.

10.6 PROPERTIES OF HTSC OXIDES

All HTSC oxides possess the main properties of elemental of conventional superconductors, including zero resistivity, Meissner effect, flux quantization. Josephson effect, etc. In addition HTSC oxides are characterized by (*i*) highly anisotropic layered structures (*ii*) short zero temperature coherence length (ξ = 10 Å). (*iii*) transport properties in the normal state significantly different from those of normal metals, with strongly anisotropic resistivity, anomalous magnetoresistance, pseudogap, etc. This is why that HTSC oxides are termed as *unconventional superconductors.*

T_C, λ, ξ, H_{C_2} and H_{C_1} for few selected HTSC cuprates are given in Table 10.3.

TABLE 10.3: CRITICAL TEMPERATURE (T_C), PENETRATION DEPTH (λ_i), COHERENCE LENGTH (ξ_i) AND THE CRITICAL FIELDS H_{C_1} AND H_{C_2} OF FEW HTSC CUPRATES ($i = ab$ OR c)

HTSC cuprates	T_C(K)	λ_{ab}(Å)	λ_C(Å)	ξ_{ab}(Å)	H_{C_2}/T	H_{C_1}/T
LBCO	38	800	4500	35	80	15
YBCO	94	1500	7000	15	150	40
BSCCO	110	2500	9000	13	260	32

10.7 THERMODYNAMICS OF A SUPERCONDUCTOR

The transition between the normal and superconducting state is thermodynamically reversible. This change occurs because the Gibbs free energy in the superconducting state, G_S, is lower than its value G_N in the normal state. Gibbs free energy (G) is defined as

$$G(T, p) = U - TS + PV \tag{18}$$

where U, T, S, P and V are internal energy, temperature, entropy, pressure and volume of the system respectively.

Let us consider that a superconductor is placed in a magnetic field, H_a. In analogy with Eq. (18), the Gibbs free energy in this case can be expressed by adding the magnetic contribution to the energy density of a magnetized specimen. If susceptibility, χ is independent of H, the magnetic energy is

$$U_M = -\int_0^{H_0} \mu_0 H dM = -\int_0^{H_0} \mu_0 H \chi dH = -\frac{1}{2}\mu_0 x H_a^2 \tag{19}$$

Now, neglecting the term PV in Eq. (18) for a magnetic solid and adding the magnetic energy term of Eq. (19), one obtains

$$G(T, H_a) = U - TS - \frac{1}{2}\mu_0 \chi H_a^2 \tag{20}$$

In the superconducting state $\chi = -1$, and hence

$$G_S(T, H_a) = G_S(T, 0) + \frac{1}{2}\mu_0 H_a^2 \tag{21}$$

In the normal state χ is very small and we may assume $\chi = 0$. Thus

$$G_N(T, H_a) = G_N(T, 0) \tag{22}$$

Normal and superconducting phases are at equilibrium at $H_a = H_C$. This means

$$G_S(T, H_C) = G_N(T, H_C) = G_N(T, 0)$$
$$= G_S(T, 0) + \frac{1}{2}\mu_0 H_C^2(0) \tag{23}$$

or

$$G_N(T, 0) - G_S(T, 0) = \frac{1}{2}\mu_0 H_C^2 \tag{24}$$

From Eqs. (21), (22 and (23), it is obvious that for $H_a > H_C$, $> G_S > G_N$ while for $H_a < H_C$, $G_S < G_N$, and we note that the superconducting state is stable for $H_a > H_C$. Fig. 10.14 shows free energy *vs* applied fields. We can see that the transition from the superconducting state to the normal state for $H_a > H_C$, occurs because G_S exceeds G_N at this magnetic field.

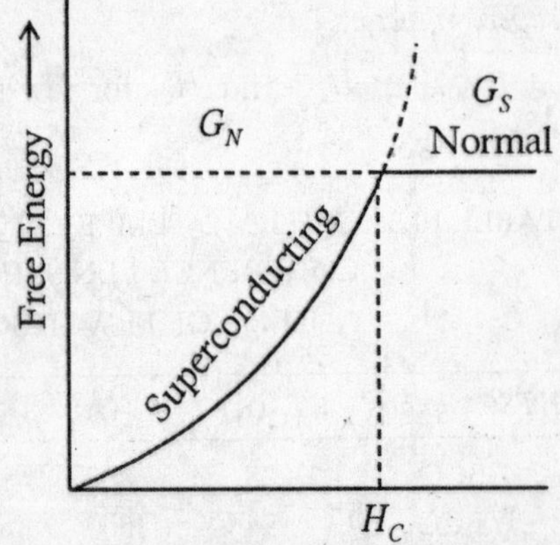

Fig. 10.14: Free energy vs applied field

From Fig. 10.14, we note that the normal and superconducting phases are in equilibrium along the critical field. For all the points along this curve, Eq. (23) is satisfied and hence for these points

$$dG_N(T, H_C) = dG_S(T, H_C)$$

or $$\left(\frac{\partial G_N}{dT}\right)_{H_C} dT + \left(\frac{\partial G_N}{\partial H_C}\right)_T dH_C = \left(\frac{\partial G_N}{\partial T}\right)_{H_C} dT + \left(\frac{\partial G_S}{\partial H_C}\right)_T dH_C \quad (25)$$

We have $G = U - TS + \frac{1}{2}\mu_0 H_a^2$. When H_a is constant.

$$dG = dU - TdS - SdT \text{ and } dU = TdS - C_v dT.$$

Thus $$S = -\left(\frac{\partial G}{\partial T}\right)_{H_C} \quad (26)$$

and $$C_v^e = T\left(\frac{\partial G}{dT}\right)_{H_C} \quad (27)$$

The relation between entropy in the normal state, S_N, and that in the superconductivity state, S_S is as

$$S_N - S_S = -\frac{1}{2}\mu_0 \frac{d}{dT} H_C^2$$

or $$S_N = S_S - \mu_0\left(H_C \frac{dH_C}{dT}\right) \quad (28)$$

From Eq. (3)

$$\frac{dH_C(T)}{dT} = -2H_C(0)\frac{T}{T_C^2} < 0 \quad (29)$$

Since H_C (dH_C/dT) is always negative and hence S_S is lower than S_N. Obviously, the superconducting state is more ordered than the normal state.

Using Eqs. (27) and (28), one obtains the difference in specific heats C_N and C_S in the normal ad superconducting states respectively as

$$C_N - C_S = -T\mu_0\left[\left(\frac{dH_C}{dT}\right)^2 + H_C \frac{d^2 H_C}{dT^2}\right]$$

$$= \frac{2H_C^2(0)\,\mu_0}{T_C}\left[\frac{T}{T_C} - 3\left(\frac{T}{T_C}\right)^3\right] \quad (30)$$

Specific heat in the superconducting state has no linear term, as such we may identify

$$C_N = \gamma T = \frac{2H_C^2(0)}{T_C^2}\mu_0 T \quad (34)$$

with $$\gamma = \frac{2H_C^2(0)}{T_C^2}\mu_0 \tag{35}$$

Thus $$C_S = \frac{6H_C^2(0)\,\mu_0}{T_C}\left(\frac{T}{T_C}\right)^3 = 3\gamma T_C\left(\frac{T}{T_C}\right)^3 \tag{36}$$

Relation (36) explains the experimental result on the specific heat of the conventional superconductors.

10.8 THEORY OF SUPERCONDUCTIVITY

One of the simplest idea which was put forward in 1934 for describing superconductivity was the *two-fluid model*. Some properties can be understood with the simple assumption that some electrons behave in the normal way as nearly free electrons, while other exhibit anomalous behaviour. Developing this idea F. and H. London were able to describe the electrodynamics of what is now called type-II superconductors.

One cannot describe the unique characterization of the superconducting state, *i.e.*, $\boldsymbol{B} = 0$ in the applied magnetic field by considering the superconductor as a medium of zero resistivity. This means that one will have to modify the Ohm's law drastically, in order to describe the zero resistivity and Meissener effect in the superconducting state, without modifying the Maxwell's equations. We have Ohm's law

$$\boldsymbol{J} = \sigma\,\boldsymbol{E},\ (\therefore\ \sigma = \frac{1}{\rho} = \text{conductivity},\ \boldsymbol{J} = \text{current density}) \tag{37}$$

In the superconductivity state electron behave differently, so it was assumed that in $\boldsymbol{J}$ (current density) is directly proportional to the vector potential $\boldsymbol{A}$ to local magnetic field, where $\boldsymbol{B} = \nabla \times \boldsymbol{A}$ ($\boldsymbol{B}$ magnetic induction, and $\nabla = \hat{i}\frac{\partial}{\partial x} + \hat{j}\frac{\partial}{\partial y} + \hat{k}\frac{\partial}{\partial z}$, *i.e.*, $\boldsymbol{B}$ = curl$\boldsymbol{A}$, and electric field intensity $\boldsymbol{E} = -\frac{\partial \boldsymbol{A}}{\partial t}$. We have

$$\boldsymbol{J} \propto \boldsymbol{A} \tag{37(a)}$$

where $$\boldsymbol{B} = \nabla \times \boldsymbol{A} = \text{curl}\boldsymbol{A} \tag{37(b)}$$

The constant of proportionality in Eq. (37*a*), thus has taken in SI unit as $-1/\mu_0\lambda_L^2$, where λ_L is a constant with dimensions of length and of value $\lambda_L^2 = m/\mu_0 ne^2$. Hence Eq. (37*a*) takes the form

$$\boldsymbol{J} = -\frac{\boldsymbol{A}}{\mu_0\lambda_L^2} = -\frac{ne^2}{m}\boldsymbol{A} \tag{38}$$

instead of $\boldsymbol{J} = \sigma\boldsymbol{E}$ as in Ohm's law [Eq. (37)] where n is the free electron density, e the electronic charge, m the electronic mass, Eq. (38) is called the *London Equation.*

One can also derive London equation in a superconducting state in the following manner: The resistivity $\rho = 0$ in a superconducting state and so the electrons can be freely accelerated. The acceleration is given by

$$\frac{dv}{dt} = \frac{eE}{m} \tag{39}$$

where v is the velocity of the free electrons. Multiplying both sides of Eq. (39) by ne, one obtains

$$\frac{d}{dt}(nev) = \frac{ne^2}{m}E = \frac{ne^2}{m}\frac{(-\partial A)}{\partial t}$$

We have $nev = J$ current density. Hence by integrating the above equation with respect to t, one obtains

$$J = -\frac{ne^2}{m}A = -\frac{1}{\mu_0\lambda_L^2}A \tag{39a}$$

which is same as Eq. (38), the London equation, where constant of integration is taken to be zero.

One can explain Meissner effect in superconducting state by London equation as follows:

Taking curl on both sides of Eq. (38), we have

$$\nabla \times J = -\frac{ne^2}{m}\nabla \times A = -\frac{\nabla \times A}{\mu_0\,\lambda_L^2} = -\frac{B}{\mu_0\,\lambda_L^2} \tag{39b}$$

Now considering curl of the Maxwell's equation

$\nabla \times B = \mu_0 J$, we have

$$\nabla \times (\nabla \times B) = \mu_0(\nabla \times J)$$

or $\quad \nabla \times (\nabla \times B) = \nabla(\nabla \cdot B) - \nabla^2 B = \mu_0(\nabla \times J)$

However, $\nabla \cdot B = 0$ from Maxwell's equation (2)

Thus, one finds

$$\nabla \times (\nabla \times B) = -\nabla^2 B = \mu_0(\nabla \times J) \tag{39c}$$

Comparing Eq. (39*b*) and Eq. (39*c*), one obtains

$$\nabla^2 B = \frac{B}{\lambda_L^2} \quad \text{constant} \tag{39d}$$

Eq. (39*d*) accounts for the *Meissner effect*, as it *does not allow a solution uniform in space*, which clearly reveals that a uniform magnetic field cannot exist in a superconductor. No doubt, uniform magnetic field B exists outside the superconductor, it gradually decreases exponentially to zero according to the solution of Eq. (39*c*) as

$$B(x) = B(o)\exp(-x/\lambda_L) \tag{39e}$$

where λ_L is the *depth of penetration* of the magnetic field and called as *'London Penetration Depth'* having dimension of length and of magnitude as

$$\lambda_L^2 = \frac{m}{\mu_0 n e^2} \tag{39f}$$

We may note that an applied external magnetic field H will penetrate a thin film, if the thickness is much less then λ_L, *i.e.* in a *thin film* Meissner effect is incomplete. For bulk specimen of superconductor, Meissner effect is complete. Obviously the shape of the specimen has very important effect, which are simple only when the specimen has the shape of a long cylinder whose axis is parallel to the applied magnetic field.

The London theory is not applicable to situations in which the number of superconducting electrons, n_S, varies; it does not link n_S with the applied voltage or current.

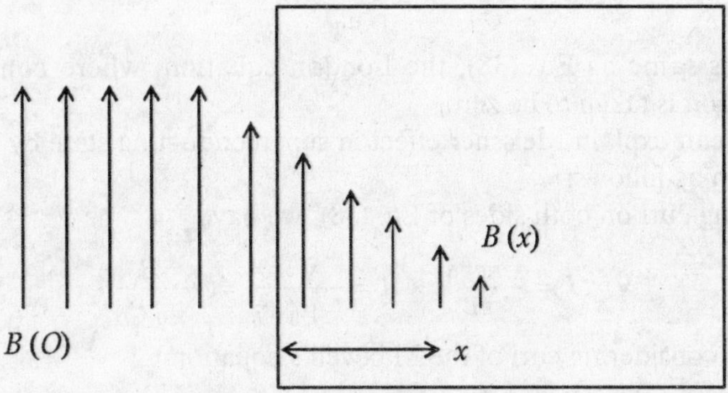

Fig. 10.15: Penetration depth

Ginzburg and Landau (GL) in 1950, using the Landu's theory of second order phase transitions, proposed a highly innovative phenomenological theory of superconductivity: *The Ginzburg-Landau Theory.* They first derive two equations which can be used to calculate both the distribution of fields and the variation of the number of superconducting electrons. *This theory* describes a superconductor close to T_C. The order parameter $\psi(r) = |\psi| \exp i\phi(r)$ is a complex scalar and $\phi(r)$ is small. They introduce two unknown parameters α and β which can be estimated from experimental measurements. The characteristic quantities of the materials λ, ξ or H_C, α and β can be eliminated and replaced by two of the measured characteristic quantities (λ, ξ or H_C). While both the characteristic lengths λ and ξ diverge at T_C, their ratio $\kappa = \lambda/\xi$, which is called the GL parameter, does not depend on temperature when close to T_C. If $\kappa < 1/\sqrt{2}$, one is dealing with type-I superconductor and if $\kappa > 1/\sqrt{2}$ one has a type-II superconductors. As flux expulsion is incomplete in type-II superconductor, the mixed state exists between H_{C_1} and H_{C_2}. One of the greatest success of GL theory was prediction of the existence of type-II superconductors. An interesting relation can be obtained for the H_{C_1} H_{C_2} as

$$H_{C_1} H_{C_2} = H_C^2 \log \kappa \tag{40}$$

GL theory, although originally phenomenological, proved to be exact and very powerful.

Microscopic BCS Theory

The microscopic theory of superconductivity was formulated by Bardeen, Cooper and Schrieffer (BCS theory) in 1957, almost five decades after the discovery of the phenomenon in 1911 by Kammerlingh Onnes. It is an elegant but mathematically complex theory. We therefore present a brief discussion of the results of this theory.

The first clue to the basic interaction between electrons which gives rise to superconductivity was provided by the isotope effect. Two different isotopes of the same metal exhibit different T_C's (Eq. 10). Why is the mass of an atom involved in a purely electronic property? Obviously, the motion of ions has something to do with superconductivity.

The second clue was found by Leon Cooper. He showed that a normal metal (with standard metallic properties) could not be formed if there was a small attraction between electrons. In such a case, two electrons would form pairs however small the attractive interaction. And, if electrons did form pairs, completely different properties for the whole ensemble of electrons would be observed.

Thus the central idea is that motion of ions can lead to attractive interaction between electrons. How can this happen? When an electron moves among the positive ions of lattice in a solid, it interacts with the lattice vibrations (called *phonons*) and attracts them. However, the motion of ions is slow so the electron advances a great deal while the ions somwhat converge towards each other. They build a region of positive charge, which before relaxing attracts another electron. So, due to this slow response of ions, there appears an effective attractive interaction between electrons, called the *electron-phonon interaction* (Fig. 10.16). Of course the repulsive Coulomb interaction is also present and there is a delicate 'competition'

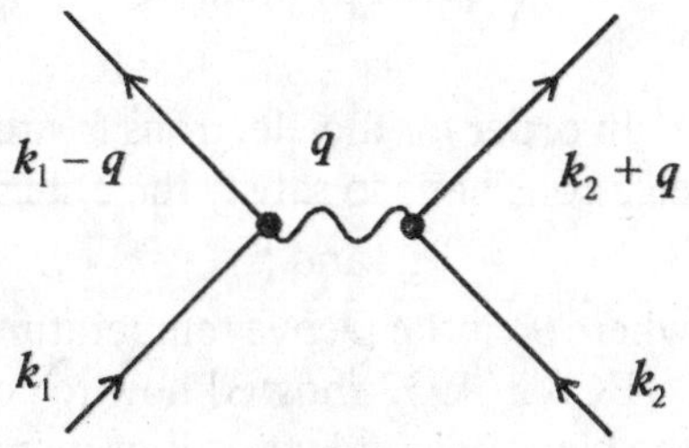

Fig. 10.16: Electron-phonon interaction. An electron of wave vector $\boldsymbol{k}_1$ emits a virtual phonon q which is absorbed by electron $\boldsymbol{k}_2$, $\boldsymbol{k}_1$ is thus scattered as $\boldsymbol{k}_1 - \boldsymbol{q}$ and $\boldsymbol{k}_2$ as $\boldsymbol{k}_2 + \boldsymbol{q}$. If this phonon energy exceeds electronic energy, the electron-phonon interaction is attractive

between the Coulomb interaction and electron-phonon interaction which makes some metals superconducting and other non-superconducting.

Cooper in 1956 showed that two electrons with an attractive interaction can bind together to form a 'bound pair' (often called a *Cooper pair*) if they are in the presence of a high density fluid of other electrons, no matter how weak the interaction is. The two partners of Cooper pair have opposite momenta (***K***) and spin angular momenta, *i.e.* Cooper pair composed of particle pair ***K***↑ and –***K***↓. This pairing is known as *s*-wave pairing. There are other forms of particle pairing possible, e.g. *d* wave, *d* + *s* wave, *p* wave, etc. with the BCS theory. It is now established that superconducting gap in HTSC oxides is of $d_{x^2-y^2}$ symmetry. These electrons due to their charge also repel each other with a Coulomb interaction, V_C but BCS assume that in the superconducting phase electron-phonon mediated interaction V_{ph} dominates over V_C and net interaction V is positive, *i.e.*,

$$V = -V_{ph} + V_C < 0 \tag{41}$$

The main assumptions of BCS theory are

(*i*) The superconducting ground state can be expressed in terms of Cooper pairs so that the states (***K***↑, –***K***↓) are simultaneously occupied or empty.

(*ii*) Various interactions in the normal and superconducting states are identical and only the effective screened Coulomb interaction has to be considered.

(*iii*) The effective interaction is zero, except when two electrons of wave vector ***K*** and ***K'*** have energies close to the Fermi energy. Then the attractive interaction is taken as a constant $-V$. If ξ_k is the energy measured from the Fermi energy, or more precisely from the chemical potential μ, then:

$$\xi_k = \varepsilon_k - \mu \tag{42}$$

In order for the electrons to attract each other, the energies of both the electrons have to satisfy the criterion

$$|\xi_k| \text{ and } |\xi_k| < k_B\,\theta_D \tag{43}$$

where θ_D is the Debye temperature.

BCS in 1957 showed how to construct a wave function in which all of the electrons (at least, all of important ones) are paired. Once this wave function is adjusted to minimize the free energy, it can be used as the basis for a complete microscopic theory of superconductivity.

At zero temperature, electrons in a superconductor are paired. In the superconducting state excitations are obtained by breaking up pairs which costs a minimum energy $2\Delta(T)$. The excitation spectrum is given by

$$E_k = \sqrt{\xi_k^2 + \Delta^2} \tag{44}$$

where Δ is a fundamental quantity introduced by BCS is called the gap or the superconducting order parameter. Δ depends on the temperature and obeys the relation, known as the *self-consistent BCS equation.*

$$\frac{1}{VN(E_F)} = \int_0^{\hbar\omega_D} d\xi \left(\xi^2 + \Delta^2\right)^{-1/2} \tan h \left[\frac{1}{2k_B T}\left(\xi^2 + \Delta^2\right)^{1/2}\right] \quad (45)$$

where $\hbar\omega_D$ is the characteristic energy of the attractive potential and $N(E_F)$ is the density of states as Fermi level.

When $\Delta(T) = 0$, *i.e.*, for $T = T_C$ the energy of the normal and the superconducting states are equal. An important BCS relation which relates $\Delta(0)$ and T_C is

$$2\Delta(0) = 3.52 k_B T_C \quad (46)$$

Close to T_C we have an approximate relation

$$\frac{\Delta(T)}{\Delta(0)} = 1.74\left(1 - \frac{T}{T_C}\right)^{1/2} \quad (47)$$

The critical temperature (T_C) of simple conventional superconductor is given by the well-known BCS formula

$$k_B T_C = 1.13\,\hbar\omega_D \exp\left[\frac{-1}{\lambda_{ep}}\right] \quad (48)$$

λ_{ep} is the dimensionless electron-phonon coupling parameter. Its value for conventional superconductors is very close to ~0.3. ω_D is Debye or characteristic frequency, varies from one metal to another but only over a small range of value. Instead of ω_D, one can use the Debye temperature θ_D ($\because k_B\theta_D \approx \hbar\omega_D$), θ_D ranges from 100 K to 500 K. Such a range of θ_D (and λ_{ep} ~ 0.3) implies a maximum 'BCS' value of T_C ~25 K.

The extension of BCS weak coupling theory to strong coupling was presented by Eliashberg in 1960. MacMillan in 1968 obtained an approximate solution of Eliashberg's equations, good for $\lambda \leq 2$. BCS theory alone cannot explain the anomalous behaviour of HTSC oxides in normal and superconducting states. Several theoretical proposals have been advanced, *e.g.* Anderson's Resonance Valence Bond (RVB) theory; *t-J* model, interlayer tunneling model, Spin-Fermion model, Varma's Marginal Fermi Liquid Model; SO(2) and SO(5) symmetry models, Boson Fermion model. None of these models provide satisfactory explanation of the anomalous features of HTSC oxides. However, the efforts to understand the HTSC oxides pairing mechanism, with enhancing future prospects for new HTSC materials and novel applications are in progress.

10.9 QUANTUM TUNNELING

This is a quantum mechanical process which permits electrons to penetrate from one side to the other through an extremely thin potential barrier to electron flow. Tunneling had been considered to be a possible electron transport mechanism between metal electrodes separated by either a narrow vacuum or a thin insulating film usually made of oxides. In 1960, Giaever demonstrates for the first time that, if one or both of the metals were in a superconducting state, the current-voltage curve in such metal tunnel junctions revealed many details of that state. Giaever's technique was sensitive enough to measure the most important feature of the BCS theory—the energy gap which forms when the electrons condense into corr bound pairs, called as Cooper pairs.

A thin film of a superconductor, is deposited on a smooth glass surface by evaporation. The surface of the layer is coated by an insulator layer of thickness 10 nm. In case of aluminium, this can be achieved by oxidizing the layer first deposited. Let us suppose that a second metal coating (normal metal) is made on the top of the insulating layer. Electrodes be connected to the metal states on either side of the insulating layer and then a gradually increasing potential is applied. This will raise the Fermi level on the right side. Until a certain critical voltage V_C is attained, there will be no current in circuit. The current begins to grow afterwards as shown in Fig. 10.18. One can explain this on the basis of quantum tunneling. We know that quantum mechanically an electron on one side of the potential barrier has finite probability of tunneling through it, if there is an allowed state of equal energy is available on the other side of the potential barrier. The density of states

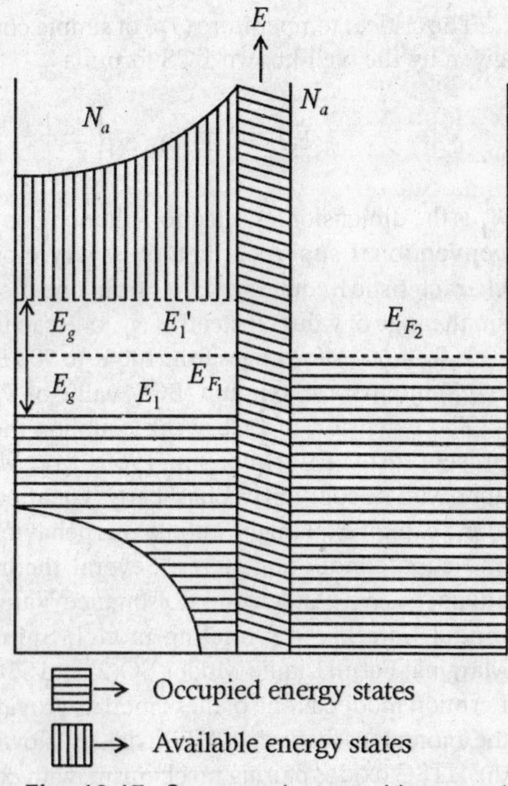

Fig. 10.17: Superconductor-oxide normal junction at 0 K

function in the energy space for a sandwich consisting of a superconductor, an insulator and a normal metal (all at absolute zero) is shown in Fig. 10.17. We know that in a normal conductor the electrons fill up to the Fermi level.

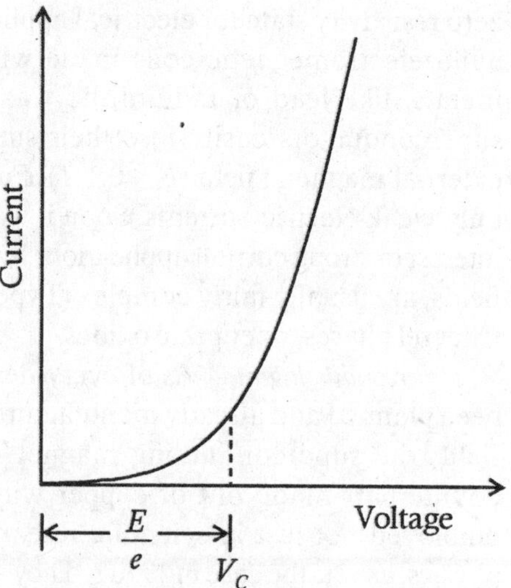

Fig. 10.18: Current (I)-Voltage (V) characteristic curve for Giaever tunneling

The Fermi level in the superconductor is in the middle of the band gap. When the sandwich is formed the Fermi levels are aligned (Fig. 10.17). Let us suppose that a voltage is applied across the barrier so as to raise the Fermi level on the right side of the junction. Until the Fermi level E_{F_2} on the right side of the junction raised upto the level E_1' electrons cannot move from right to the left as there are no available energy levels on the left side; since in that portion the band gap exist. Obviously, once the Fermi level E_{F_2} rises above the level of E_1' electrons can easily tunnel through the barrier from the right to the left since there are energy states available for the electrons on the left band side. Thus the current increases. However, there will be no current until the voltage becomes available equal to the band gap ($E_1' - E_2$). This has been confirmed experimentally.

The tunneling phenomenon has been exploited in many fields. For example, small area tunnel-junctions are used for mixing and synthesis of frequencies ranging from DC to the infrared region of the spectrum. This leads to absolute measurement of frequency in the infrared and provided the most accurate determination of the speed of light.

To study the non equilibrium superconducting properties, two tunnel junctions, one on the top of the other sharing the middle electrode, are used. Tunnel diodes are also used as a spectroscopic tool to study the phonon and plasmon spectra of the metals and the vibrational spectra of complex organic molecules introduced inside the insulating barriers.

10.10 APPLICATIONS OF SUPERCONDUCTIVITY

Though concerning its technological applications are as old as superconductivity itself, Onnes himself tried to explore the potential of zero resistivity state for electrical applications. However, when he tried to build electromegnetic coils made with wires of pure superconducting metals, like lead or indium, he was unable to do so. These metallic superconductors easily lose their superconducting properties in weak external magnetic field ($B < 0.1\ T$). Furthermore, they were able to carry only weak electric currents. Soon it was realized that superconductors of interest to strong current applications, which can sustain very high magnetic fields, are usually fairly complex (Type-II) materials, like Nb compounds, Chevrel phases, or cuprate oxides.

Superconducting magnets of every desired size and geometry have now been planned and already manufactured. For equal power and uppermost field, the superconducting magnet is much smaller than its normal counterpart made out of copper wires. Superconducting magnets are employed not just in scientific research, in high energy and solid state physics, as was the case in 1960's. They are already in planning for controlled thermonuclear fusion, for superconducting motors with output of several kilowatts, for energy storage facilities and for the magnetic suspension of trains (*Levitation trains*). Powerful superconducting magnets, *supermagnets*, are also used in medicine for diagnostics, for e.g. in nuclear spin tomographs. Superconducting magnets generate homogeneous, low noise and extremely high magnetic fields. A number of manufacturers now sell 20-T superconducting magnets with bore size in the range of 50 nm. Efforts are in process to design construct solenoidal magnets having fields approaching 25 *T* using new superconducting materials, including HTSC oxides.

Superconducting magnets seem more complicated than electromagnetic, especially because of their requirement of low temperature to keep the magnetic solenoid in its superconducting state.

Superconducting magnet technology allows users to produce extremely high magnetic fields without the kilowatt or megawatt power supplies needed for electromagnets. Once superconducting magnets are energized, user can easily disconnect them from their power source and they will remain energized which significantly reduces electricity costs.

However, the consideration of superconductivity for large scale applications is not limited to superconducting magnets. A great deal of effort is going into using superconductivity for the *generation and transmission of electric power*. Superconducting cables and storage elements will offer crucial advantages.

The application of superconductors in *measurement technology* is not revolutionary; here it opens up for us the opportunity to increase the sensitivity of many determinations by orders of magnitude over what could be achieved with normal conducting circuitry. Superconducting circuit and memory elements could bring decisive improvements to large computers, under certain conditions.

Thin film applications in electronics, together with superconducting magnets, are usually considered the most important area of superconductivity-based technology. Most of these applications are based on Josephson effect that enables the construction of the fastest nanoscopic switches, Josephson junctions, and related device structures, SQUIDs (Superconducting Quantum interference devices). SQUIDs explore subtle quantum interference effects: an analysis of superconducting loop shows that the magnetic flux that can thread the loop is quantized in quantum units of flux which have a value of 2×10^{-15} Weber. SQUIDs are suitably processed superconducting loops which detect minute changes in magnetic flux, *i.e.* they are high sensitivity magnetic flux detectors that can be used in the finest precision instruments at the forefront of meteorology.

The HTSC oxides provide us with superconductors which can even be employed at the temperature of liquid nitrogen ($T_B = 77$ K) or even higher temperature. While the widespread use of HTSC cuprate superconductors in technology has not yet been realized, steady and significant progress has been made towards this objective during the past 15 years. Recent developments indicate that HTSC cuprates will begin to have a significant impact on technology during the next 5 to 10 years. The applications in superconducting electronics that are likely to be realized on this time scale include SQUIDs, NMR (nuclear magnetic resonance) coils, wireless communications subsystems, MRI coils and NMR microscopes, and digital instruments. In the area of superconducting wires and tapes, applications that appear to be feasible within this same time period include: power transmission lines, motors and generators, transformers, current limiters, magnetic energy storage, magnetic separation, research magnetic systems and current leads.

An example of recent progress in the area of superconducting wires and tapes is the development of flexible superconducting ribbons consisting of deposits of YBCO on textured substrates which have critical-current densities $J_C \sim 10^6$ A/cm^2 in fields up to $8T$ at 64 K, a temperature that can be achieved by pumping on liquid nitrogen. The performance of these ptototype conductors in storing magnetic fields already surpasses that of NbTi and Nb_3Sn, which are currently used in commercial superconducting wires at liquid helium temperatures in a comparable field range.

It is certain that HTSC oxides have opened new opportunities for technological applications of superconductivity. It is also certain, however, that a great deal of development will be necessary before we have in our hands suitable conductors for use in say magnet construction or for microcircuitry. HTSC oxides will have some impact in electronics related applications: initially in passive microwave devices and SQUIDs. The mixtures of HTSC oxides with silver may offer somewhat improved properties and might provide the basis for future wires and cables. It is economic question that will play a crucial role here.

A brief account of potential applications of superconductivity in vital areas is given in Table 10.4.

TABLE 10.4: POTENTIAL APPLICATIONS OF SUPERCONDUCTIVITY IN VITAL AREAS

Field	*Applications*
• Magnets	• High field Magnets • Magnetic Levitation • Magnetic Shielding • NMR (Magnetic diagnostics and spectroscopy) • Large Machines (colliders, Magnetic fusion confinement, r.f. cavities • Ore refining
• Energy or Power-related	• Electric power storage and transmission • Magnetic energy storage • Power production by magnetic fusion and magnetohydrodynamics (MHD)
• Transportation	• High speed maglev trains • ship-drive system
• Electronic and other small device	• SQUIDs • Bolometer • Electromagnetic shielding • Josephson devices (square-law detector, parametric amplifier, mixer)
Computer and information processing	• Optoelectronics • Voltage standard • Active superconducting elements (FETs) • Malchel filters • Semiconductor-superconductor hybrids (A-D converters)

The requirements for critical current density (J_C) in a given magnetic field for few selected applications are presented in Table 10.5.

We may note that most high current applications of superconductors require $J_C > 10^5$ A/cm^2 and $T_{use} \sim T_C/2$ and for weak current and most applications in electronics.

$$T_{use} \sim \frac{2}{3} T_C$$

TABLE 10.5: REQUIREMENTS FOR J_C IN A GIVEN MAGNETIC FIELD FOR SEVERAL APPLICATIONS OF SUPERCONDUCTORS

Application	*H (Tesla)*	J_C (A/cm²)
Interconnects	0.1	5×10^6
AC transmission lines	0.2	10^5
DC transmission lines	0.2	2×10^4
SQUIDs	0.1	2×10^2
Motors and generators	~4	~10^4
Fault current limiters	>5	>10^5
Power generation	5-7	2×10^4
Power storage	5-10	10^3-10^4
MRI scanners	2	10^4
Mineral separation	2-5	10^4

T_C, upper critical field (H_{C_2}), λ and ξ of some technologically important superconductors are given in Table 10.6.

TABLE 10.6

Material	T_C *(K)*	*Penetration depth λ (nm)*	*Coherence length ξ (nm)*	H_{C_2} *(T)*
Nb	9.2	40	38	0.2
Nb-Ti	9.2	60	40	14
NbN	16	250	4	16
Nb_3Sn	18	80	3	24
$YBa_2Cu_3O_7$	92	150/600	1.5/0.4	150/40
$Bi_2Sr_2Ca_2Cu_3O_{10}$	110	200/1000	1.4/0.2	250/30

ILLUSTRATIVE EXAMPLES

Example 1

The critical temperature of mercury is 4.2 K.

(*a*) Calculate the energy gap in electron volts at $T = 0$?

(*b*) Calculate the wavelength of a photon whose energy is just sufficient to break up Cooper pairs in mercury at $T = 0$. In what region of the electromagnetic spectrum are such photons found?

Solution

(*a*) The Cooper pair binding energy, or gap energy, is

$$E_g = 3kT_C$$

$$= 3 \times 1.4 \times 10^{-23}\ \text{J/K} \times 4.2\ \text{K}$$

$$= 1.8 \times 10^{-22}\ \text{J} \simeq 1.1 \times 10^{-3}\ \text{eV}$$

(*b*) $E_g = h\nu = hc/\lambda$

$$\therefore \quad \lambda = \frac{hc}{E_g} = \frac{6.6 \times 10^{-34}\text{Js} \times 3 \times 10^{8}\ \text{m/s}}{1.8 \times 10^{-22}\ \text{J}} = 1.1 \times 10^{-3}\ \text{m}$$

Obviously, these photons are in the very short wavelength part of the microwave region.

Example 2

Does the metal look like a superconductor to electromagnetic waves having wavelengths shorter than that found in example 1(*b*)?

Solution

No, since the energy content of shorter wavelength photons is sufficiently high to break up the Cooper pairs, or excite the conduction electrons through the energy gap into the non-superconducting states above the gap.

Example 3

How much current can a lead wire, 1 mm in diameter, carry in its superconducting state at 4.2 K? (Given T_C for Pb is 7.2 K and). $B_C(0) = 0.0803$ Wb/m²)

Solution

We have $$B_C(T) = B_C(0)\left[1 - \left(\frac{T}{T_C}\right)^2\right].$$

Given $$B_C(0) = 0.0803\ \text{Wb/m}^3$$

$$T = 4.2\ \text{K},\ T_C = 7.2\ \text{K},\ 2r = 10^{-3}\ \text{m}$$

$$\therefore \quad B_C(4.2) = 0.0803\left[1 - \left(\frac{4.2}{7.2}\right)^2\right] = 0.0548T$$

Now $$I_C = \frac{2\pi r B_C(T)}{\mu_0} = \frac{\pi(2r)B_C(T)}{\mu_0}$$

$$= \frac{\pi(10^{-3})0.0548}{4\pi \times 10^{-7}} = \frac{548}{4} = 137\ \text{Å}.$$

Example 4

For a superconductor $T_C = 3$ K and $n_S = 10^{28}/\text{m}^3$. Find the penetration depth at 0 K and 1 K.

Solution

$$\lambda_0 = \left(\frac{m}{\mu_0 n_S e^2}\right)^{\frac{1}{2}} = \left(\frac{m}{\mu_0 n_S}\right)^{\frac{1}{2}} \cdot \frac{1}{e}$$

We have $m = 9.1 \times 10^{-31}$ kg, $\mu_0 = 12.56 \times 10^{-7}$ SI units, $n_S = 10^{28}$ m^{-3} and $e = 1.6 \times 10^{-19}$ C

$$\therefore \quad \lambda_0 = \left(\frac{9.1\times10^{-31}}{12.56\times10^{-7}\times10^{28}}\right)^{\frac{1}{2}} \times \frac{1}{1.6\times10^{-19}}$$

$$= \frac{0.85\times10^{-26}}{1.6\times10^{-19}} = 0.53\times10^{-7}\ \text{m} = 530\ \text{Å}$$

$$\lambda_{T=3K} = \frac{\lambda_0}{\left[1-\left(\frac{I}{T_C}\right)^4\right]^{\frac{1}{2}}} = \frac{530\times10^{-10}}{\left[1-\left(\frac{1}{3}\right)^4\right]^{\frac{1}{2}}} = \frac{530\times10^{-8}\times9}{8.95}$$

$$= 533 \times 10^{-10}\ \text{m}$$

$$= 533\ \text{Å}.$$

REVIEW QUESTIONS

1. What is superconductivity? Mention some important property changes that occur in materials when they undergo phase change from normal to the superconducting state.
2. What are type-I and type-II superconducting materials? Give three examples of each why type-II materials are preferred for applications of superconductivity. [BE]
3. Define critical temperature, critical field and critical current density for a superconductor. [AMIE]
4. Explain the behaviour of a superconductor in a magnetic field. What is Meissner effect? How does the critical magnetic field varies with temperature?
5. What is isotope effect in superconductors? How it provided a clue to the microscopic BCS theory of superconductivity? [BE]
6. Describe the effects of (*i*) magnetic field (*ii*) frequency and (*iii*) isotopes on superconductors.
7. What are the characteristic lengths of a superconductor? How penetration depth varies with temperature? [BE]
8. Give a brief account of thermodynamics of a superconductor.
9. What is BCS theory of superconductivity? Write its salient features.
10. Write a short note on potential applications of superconductors. [BE]
11. What are DC and AC Josephson effects? Explain their importance.
12. What do you understand by quantum tunneling? Explain its significance. [BE]

13. What are HTSC oxides? How they are different from conventional superconductors? What is their future prospects from applications of superconductivity point of view?

PROBLEMS

1. (*a*) Show from Maxwell's equations that resistivity $\rho = 0$ (a perfect conductor) implies that $\boldsymbol{B} = 0$ inside the material (*b*) Show, from Maxwell's equations, that $\boldsymbol{B} = 0$ inside a material (a superconductor) implies that the resistivity (ρ) of the material is zero, *i.e.*, $\rho = 0$.
2. Show that Meissner effect implies perfect conductivity, but *vice versa* is not true.
3. The resistivities of Cu, Ag and Au at room temperature in units of 10^{-8} Ω-m are 1.6, 1.5 and 2.4 and while those of and Ti, Zr and Hf are 89, 45 and 32 respectively. Explain, why the former do not show superconductivity, whereas the latter become superconductors?
4. The critical temperature (T_C) for Hg with isotopic mass 199.5 is 4.185 K. Calculate its critical temperature when its isotope mass changes to 203.4.

   [Ans. 4.139 K]
5. Show that the frequency of the electromagnetic waves radiated by a Josephson junction having a voltage of 650 μV across its terminals is 3.15×10^{11} Hz.

SHORT ANSWER QUESTIONS

1. What is the relaxation time for a superconductor?

Ans. Infinite

2. What is the susceptibility of a superconductor?

Ans. Negative and unity.

3. What is superconducting electron density at absolute zero?

Ans. Finite.

4. What is critical field for a superconductor?

Ans. The magnetic field above which superconductivity disappears is called critical field B_C, given by

$$B_C(T) = B_C(0)\left[1 - \frac{T^2}{T_C^2}\right]$$

For type II superconductor there are two critical fields e.g. B_{C_1} (lower critical field and B_{C_2} (upper critical field).

5. For a material to be superconductor what should be its two distinctive properties?

Ans. Zero resistivity, *i.e.* infinite conductivity and $\boldsymbol{B} = 0$ inside the superconductor.

6. The Ginzburg-Landau parameter, $\kappa = \lambda/\xi$ for a type-II superconductor should be?

Ans. $\kappa > 1$

7. What is the BCS relation for critical temperature (T_C) for simple conventional weak coupling superconductors?

Ans. $k_B T_C = 1.13\hbar\omega_D \exp[-1/\lambda N(0)]$

8. What is the BCS relation between zero temperature gap $\Delta(0)$ and critical temperature, T_C?

Ans. $2\Delta(0) = 3.552\, k_B T_C$ and close to T_C.

$$\Delta(T) = 1.74\, \Delta(0) \left[1 - \frac{T}{T_C}\right]^{\frac{1}{2}}$$

OBJECTIVE QUESTIONS

1. A superconductor has no --------- for all $T < T_C$. [resistivity]

2. Meissner-Ochsenfeld effect shows that --------- inside the superconductor. [$B = 0$]

3. BCS theory assumes a spherical --------- and isotropic energy gap. [Fermi surface]

4. Critical magnetic fields of a superconductor --------- if temperature decreases. [increases]

5. Mercury is a type --------- superconductor. [I]

6. Superconducting electron density is --------- at absolute zero. [finite]

7. Nb_3Sn is a type --------- superconductor. [II]

8. The surface energy for a type-I is always ---------. [positive]

9. Flux quantum in a superconductor is ---------. [$h/2e$]

10. The total charge on a Cooper pair in a superconductor is ---------. [$-2e$]

11. A Cooper pair is a system of two electrons bound by exchange of --------- between them. [phonon]

Dielectrics

11.1 INTRODUCTION

These are the materials or insulators which have the unique characteristic of being able to store electric charge. The electrons in these substances are localized in the process of bonding the atoms together. Obviously, covalent or ionic bonds, a mixture of both, or Vander Waals bonding between closed shell atoms give rise to solids or gases which exhibit dielectric or insulating properties. Dielectric materials may be gases, liquids or solids with the exception of air which is the insulating material between the bare conductors of the overhead electric grid system. Liquid dielectrics are used mainly as impregnants for high voltage paper, insulated cables and capacitors as filling and cooling media for transformers and circuit breakers. Most common properties of dielectric materials are: (*i*) dielectric constant (*ii*) dielectric strength (*iii*) insulation resistance (*iv*) surface resistivity (*v*) loss factor (*vi*) tangent of loss factor in terms of a capacitor or phase difference (*vii*) polar and non-polar materials.

Materials, which are capable of separating electrical conductors circuit breakers, e.g. silicon, oils, liquid dielectrics have high dielectric constant, high resistance, high dielectric strength when moisture and impurities are removed from them. They have high temperature dissipation capacity and least dielectric losses. Their dielectric constant is greater than one. Following materials are important from engineering point of view:

Mica: It is the widely used insulating material in switch gears armature windings, electrical heating devices like iron, hot plates, etc. It is also used in capacitors for high frequency application. Mica is an inorganic compound of silicates of aluminium, soda potash and magnesia. It is crystalline in nature and can be easily split into very thin falt sheets. The two important types of mica are: (*i*) *muscovite* and (*ii*) *phlogopite*. Mica has a good dielectric strength and mechanical strength. Its dielectric constant varies between 5 and 7.5, loss tangent between 0.0003 and 0.015 and dielectric strength between 700 and 1000 kV/mm.

Asbestos: It is also used as an insulator in the form of paper, tape, cloth and board. Asbestos is widely used in panel boards, insulating tubes and cylinders in the construction of air cooled transformers. Asbestos is an inorganic material, which is used to designate a group of naturally occurring fibre materials. Asbestos has good dielectric and mechanical properties.

Ceramics: These are generally non-metallic inorganic compounds, e.g. silicates, aluminates, oxides, carbides, borides, nitrides and hydroxides. Ceramics used as dielectrics may be broadly described as alumina, porcelains, ceramics, titanates, etc. These have excellent dielectric and mechanical properties. The dielectric constant of most commonly used ceramics varies between 4 and 10. These are used in switches in plug holders, thermocouples, cathode heaters, vacuum type ceramic metal seals, etc. Ceramic capacitors may be operated at high temperatures and can be moulded into any shape and size.

Electric grade ceramics are used for the manufacturing of insulators, terminal blocks, plates, frames, coils, etc. They must have low losses, good insulating properties and high strength.

Insulators for operation at low frequencies are made of electric-grade porcelain which possesses fairly good electric properties. A drawback is that it has high losses which increase sharply on heating above 200° C, and a low strength.

Insulating parts for operation at high frequencies are mostly made of steatite, a talc bace material. Steatites contain no harmful impurities and retain their properties at temperatures upto 100° C. They can be pressed easily, shrink on burning by only 1-2%, and are suitable for making parts in which dense but porous structure and accurate dimensions are essential. In contrast to other kinds of ceramics, steatite can be cut quite easily (after burning). Among their drawbacks are cracking under the action of rapidly varying temperatures and some difficulties in the burning involves.

Glass: It is an inorganic insulating material, which comprises of complex system of oxides. Silica (SiO_2) is the most essential constituent of many commercially used glasses. It is fused with alkali (lide potash, soda, etc.) and some base (like lime, lead oxide, etc.). The silica glass (having 100% SiO_2) is the best insulating material. The dielectric constant of glass varies between 3.7 and 10 and loss tangent between 0.0003 and 0.01 and dielectric strength between 2.5 and 50 kV/mm. Glass is used in electric bulbs, *X*-ray tubes, mercury switches, electronic valves as insulating material. It is also used in capacitors as dielectric material.

Resins: These are organic polymers and may be natural or synthetic. The synthetic resins are produced artificially. The commonly used synthetic resins are polyethylene, polystyrene, polyvinyl chloride, acrylic resins, teflon, nylon, etc. These have good dielectric and mechanical properties. The dielectric constant of resins varies between 2 and 4.5, the loss tangent between 0.0002 to 0.04 and dielectric strength is quite high. These are used in transformers, high frequency capacitors. These are also used as a dielectric material in DC capacitors.

Rubber: These are organic polymers and may be natural or synthetic. The natural rubber obtained from rubber tree has limited applications due to its resistance to low and high temperatures. The synthetic rubbers are produced artificially by *copolymerisation* of isobutylene and isoprene. These have good electrical and thermal properties. The dielectric constant of rubber varies between 2.5 and 5, and loss tangent between 0.01 and 0.03. Rubber's are used as an insulating material for electric wires, tapes, cables, coatings, motor windings, transformers, etc.

Gaseous dielectrics, e.g. nitrogen, hydrogen, etc. have dielectric constant 1.0.

The best dielectrics are polystyrene, polyethylene, polymide, mineral oil, pure alumina and pure silica.

11.2 DIELECTRIC CONSTANT

In an electric field, the charge density D is directly proportional to the applied field, *i.e.*

$$D = \varepsilon E \tag{1}$$

where ε is the dielectric constant or permittivity of the material placed between the electrodes. For vacuum,

$$D = \varepsilon_0 E \tag{1a}$$

where $\varepsilon_0 = 8.854 \times 10^{-12}$ farad/m is called the absolute permittivity.

One can define the relative permittivity (ε_r) of a dielectric as:

(*i*) The ratio of electric field density produced in the dielectric medium (E_0) to that produced in vacuum (E) by the same electric field strength, *i.e.*

$$\varepsilon_r = \frac{E_0}{E}$$

ε_r is also called as macroscopic dielectric constant.

(*ii*) The ratio of capacitance of a condenser containing a given dielectric to the same condenser with vacuum as dielectric.

Thus relative permittivity can be expressed as

$$\varepsilon_r = \frac{\varepsilon}{\varepsilon_0} \tag{2}$$

Now, charge density = $\varepsilon_0\varepsilon_r E = \varepsilon_0 E + \varepsilon_0(\varepsilon_r + 1)E$ where $\varepsilon_r - 1 = \chi$ is called *electric susceptibility*. χ = Bound charge density/free charge density.

Permittivity is influenced by permeability (μ), stress distribution, temperature and frequency.

The basic phenomenon that takes place in dielectric materials can be easily understood if we consider a solid dielectric located in the region between the two plates of a *parallel plate condenser (capacitor).*

Let us consider a parallel plate capacitor connected to a DC voltage source, V_0 as shown in Fig. 11.1(*a*). The DC source charges the plates of capacitor till the potential difference across the plates acquire the voltage V_0. Let the charges on the two plates are $+Q_0$ and $-Q_0$, when the potential difference between two plates is V_0.

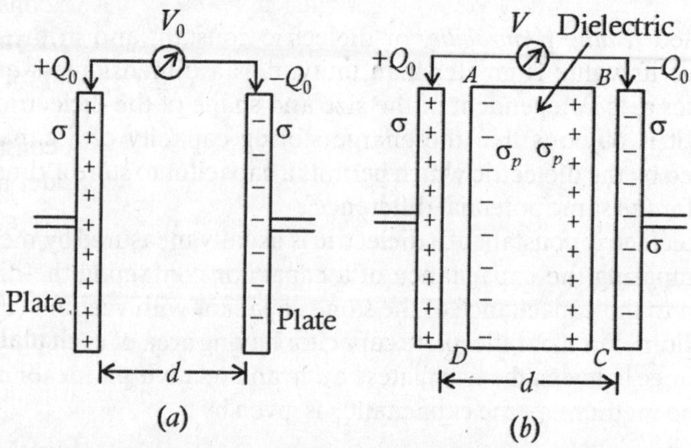

Fig. 11.1: Parallel plate capacitor (*a*) when there is no dielectric present, *i.e.* there is vacuum between the plates and plates are charged to charge Q_0 and a DC potential difference V_0 is established (*b*) when dielectric (*ABCD*) is introduced, Q_0 remains constant whereas V_0 falls to V

The *capacitance* C_0 is related to the quantity of charge stored on either plate of the capacitor is given by

$$C_0 = \frac{Q_0}{V_0} \tag{3}$$

Next, we disconnect the plates of the capacitor from the voltage source V_0. Obviously, the circuit is open and therefore the magnitude of the charge Q_0 on either plate must remain constant. Let us suppose that we now introduce a dielectric *ABCD* in the form of a rectangular parallelopiped between the two plates (Fig. 11.1(*b*)) and dielectric fills the space between the plates completely. Now, the potential difference between the plates falls to V. When we pull out the dielectric slab, the potential difference returns to the original value V_0. This experimental observation clearly reveals that the original charges on the plates have not been affected by the introduction of the dielectric between the plates. One can therefore conclude that the value of the capacitance increased from C_0 to C in the presence of the dielectric between the plates, *i.e.*

$$C = \frac{Q_0}{V} \tag{4}$$

Let us write $V = \dfrac{V_0}{\varepsilon_r}$ (5)

$$\therefore \quad C = \frac{Q_0}{V_0/\varepsilon_r} = \varepsilon_r C_0 \tag{6}$$

or

$$\varepsilon_r = \frac{C}{C_0} \tag{7}$$

ε_r is called *relative permeability* or dielectric constant and in the case of dielectics its value is greater than unity. ε_r is a dimensionless quantity. Moreover ε_r is independent of the size and shape of the dielectric. From Eq. (6) it is obvious that the charge storing capacity of a capacitor is enhanced by the dielectric which permits a capacitor to store ε_r times more charge for the same potential difference.

The dielectric constant of a dielectric is usually measured by measuring and comparing the capacitance of a capacitor containing the dielectric medium to the capacitance of the same capacitor with vacuum (or air) as the medium. For a parallel plate capacitor having area of each plate A m^2, the distance between the two plates be d m and relative permeability of the dielectric medium ε_r, the capacitance is given by

$$C_0 = \frac{\varepsilon_0 A}{d} \tag{8}$$

(when the medium between the two plates is vacuum or air). where $\varepsilon_0 = 8.85 \times 10^{-12}$ F/m is the permittivity of free space. When the capacitor is completely filled with a die!ectric, the capacitance is

$$C = \frac{\varepsilon_0 \varepsilon_r A}{d} \tag{9}$$

Energy of a Charged Capacitor

We know that work is done when two equal and opposite charges are separated and the work done is stored in the form of electric potential energy. We can recover this energy if the charges are allowed to assume their original configuration. A capacitor gets charged by a voltage source (say by a battery) at the expense of energy stored in the battery, *i.e.* chemical energy. The work done by the voltage source is given by

$$W = \frac{1}{2}\frac{Q^2}{C} = \frac{1}{2} C_0 V_0^2 = U \tag{10}$$

The work W is stored in the form of potential energy (U) in the electric field between the plates of the capacitor. If the capacitor is discharged, the energy can be recovered from the electric field, e.g. if we connect an electric motor to a charged capacitor, current will flow through the motor, the shaft of the motor will turn and mechanical work will be done. The total amount of mechanical work equal to $\frac{1}{2}C_0V_0^2$, when the capacitor is completely discharged.

Since the electric field $E_0 (= V_0/d)$ is constant between the plates of the capacitor, the *energy density* U_0, which is stored energy per unit volume, will be constant. Thus

$$U_0 = \frac{U}{Ad} = \frac{\frac{1}{2}C_0V_0^2}{Ad} = \frac{1}{2}\varepsilon_0\left(\frac{V_0}{d}\right)^2 \tag{11}$$

$$= \frac{1}{2}\varepsilon_0 E_0^2 \tag{12}$$

When a dielectric medium is present between the plates of the capacitor, the energy density is given by

$$U_0 = \frac{1}{2}\varepsilon_0\varepsilon_r E_0^2 \tag{13}$$

Induced Charges

According to Gauss's theorem, the flux of the electric field E_0 through a closed surface is equal to the algebraic sum of charges enclosed by this surface divided by ε_0. We have

$$E_0 = \frac{\sigma}{\varepsilon_0} \tag{14}$$

where σ is the magnitude of *surface charge density* on either plate of the capacitor.

When a dielectric is introduced between the plates of a capacitor, there occurs a reduction in potential difference across the plates, *i.e.* it implies a decrease in the electric field strength E_0 in the region. The decrease in the electric field strength E_0 can occur only when there is a reduction in the charge on the plates of the capacitor. Since the charges cannot leak off the plates, and hence the reduction in electric field strength suggests that the effect of some of the original charges is annulled by the charges of opposite sign appearing on the two surfaces of the dielectric between the two plates. Obviously, negative charge is induced by the electric field on the dielectric surface adjacent to the positive plate while a positive charge of equal magnitude is induced on the dielectric surface adjacent to the negative plate. Clearly, the electric field is inducing charges on the surfaces of dielectric. When charges of opposite polarity are induced on the surface of dielectric, we say that dielectric is *polarized*. Due to induced surface charges on the dielectric, an induced electric field E_i is developed, which opposes the external applied field E_0 (Fig. 11.2). Thus the net electric field E_0 in the dielectric has a magnitude given by

$$E = E_0 - E_i \tag{15}$$

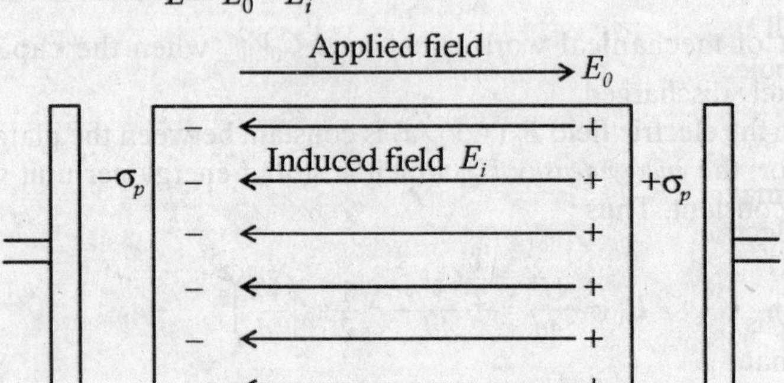

Fig. 11.2: Polarization of a dielectric. Induced charges appear on the opposite side of the dielectric. An induced (internal) field E_i directed opposite to the applied field E_0 appears within the dielectric

Let σ_p be the magnitude of the induced surface charge density (Fig. 11.1(*b*)). Obviously, this charge density nullifies the effect of part of the free charge density σ present on the plates of the capacitor.

The relation between the induced electric field E_i and induced charge σ_p is

$$E_i = \frac{\sigma_p}{\varepsilon_0} \tag{16}$$

Now, the net surface charge density that contributes to the electric field E within the dielectric is $(\sigma - \sigma_p)$. On applying Gauss's theorem, one obtains

$$E = \frac{\sigma - \sigma_p}{\varepsilon_0} \tag{17}$$

Using $C/C_0 = \sigma_p$, and $\frac{E_0}{E} = \varepsilon_r$, one finds

$$E = \frac{E_0}{\varepsilon_r} = \frac{\sigma}{\varepsilon_0 \varepsilon_r} = \frac{\sigma}{\varepsilon} \tag{18}$$

where $\varepsilon = \varepsilon_0 \varepsilon_r$ (19)

Now, equating Eqs. (17) and (19) , one finds

$$\frac{\sigma - \sigma_p}{\varepsilon_0} = \frac{\sigma}{\varepsilon_0 \varepsilon_r}$$

or $$\sigma_p = \sigma\left(1 - \frac{1}{\varepsilon_r}\right) \tag{20}$$

Since $\varepsilon_r > 1$, we have $\sigma_p < \sigma$. We may note that for moderately large values of ε_r, σ_p is nearly equal to σ. The presence of polarization charge diminishes the field E within the dielectric, causing $E >> E_0$.

11.3 POLARIZATION

We consider an electrically neutral slab of an isotropic dielectric inserted between the plates of a parallel plate capacitor. Let us imagine that the dielectric is divided into a large number of identical cells of volume dV. Now, under the action of applied external field, charges are induced in each cell and each cell acquires a charge dq. We define the intensity of polarization as the total dipole moment per unit volume of the material. Thus, one can write

$$P = \sum \frac{d\mu}{dV} = \frac{\Sigma d\mu}{V} \tag{21}$$

The magnitude of polarization is directly proportional to the intensity of the electric field, E. Thus, we have

$$P = \chi \varepsilon_0 E \tag{22}$$

where χ is the proportionality constant and is called *dielectric susceptibility* of the material. χ characterizes the ease with which a dielectric material can be influenced by an external field.

In the dielectric within each cell, the induced charge appear on the front and rear faces of each cell. These are called *bound* charges. We may note that the bound charges are not free to migrate through the dielectric. The common wall between two adjacent cells carries a charge $+q$ from one cell and $-q$ from the other cell. Obviously, in the interior of the dielectric these charges cancel. However, the charges on the opposite faces of the dielectric do not cancel. Fig. 11.3 depicts this situation and shows that the layer of charge nearest to the boundary of the slab remains and constitutes the polarization charge of surface density σ_p. We can say that the polarized

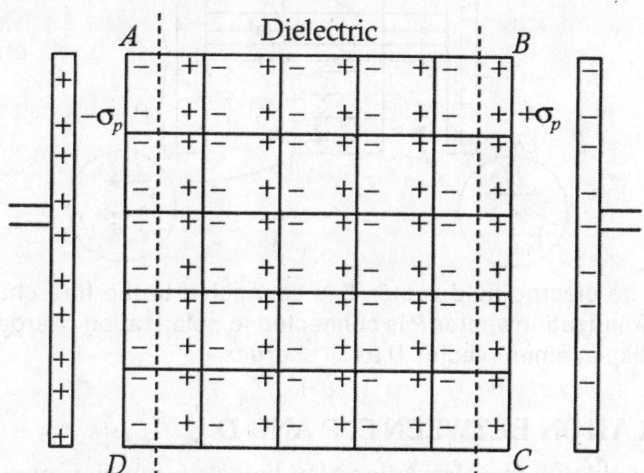

Fig. 11.3: A dielectric slab may be considered to be made up of a large number of identical and infinitely small cells with a dipole moment μ. The bound charges on the walls of the small cells cancel throughout the cell and leaving surface charges only. The magnitude of bound charges left on the sides of the slab is σ_p

dielectric is equivalent to a big *dipole* consisting of polarization charges separated by a distance d, which is also the thickness of the slab. The dipole of moment of entire slab $ABCD$ is

$$\mu = (A\sigma_p)d = \sigma_p V \tag{23}$$

where A is the surface area of the slab and V its volume. We have from (21)

$$P = \frac{\mu}{V}$$

$$\therefore \quad P = \sigma_p \tag{24}$$

From Eq. (24), it is evident that polarization is equal to the surface density of the induced charges in the dielectric. We may note that polarization is vector quantity and its direction is from negative charges to positive charges. We may see that (Fig. 11.4) direction of $\boldsymbol{P}$ is the same as the electric field. Due to polarization, an internal field $\boldsymbol{E}_i$ is produced in the dielectric, which opposes the external field and therefore weakens it. We may note that the absolute value of P equals the sum of the projections of the dipole moments of all the elementary volumes on the field direction

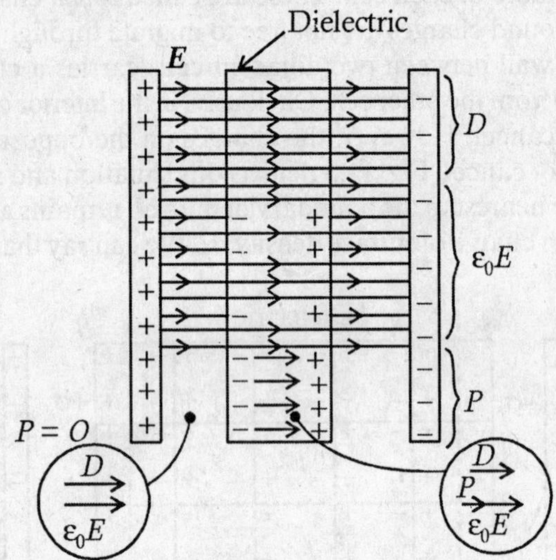

Fig. 11.4: The electric field vector $\boldsymbol{E}$ is connected to the free charges, the polarization vector $\boldsymbol{P}$ is connected to polarization charges and the displacement vector $\boldsymbol{D}$ to all charges

11.4 RELATION BETWEEN *E*, *P* AND *D*

If $\boldsymbol{E}_i$ is the field due to polarization of a dielectric, which is opposite in its direction to the external field, then the effective field $\boldsymbol{E}$ in the dielectric is given by

$$E = E_0 - E_i = \frac{\sigma}{\varepsilon_0} - \frac{\sigma_p}{\varepsilon_0}$$

Using Eq. (24), we obtain

or $$\varepsilon_0 E = \sigma - \sigma_p \quad (25)$$

Let us introduce an auxiliary vector **D**, called the *displacement vector*. Its magnitude is equal to the surface density of charges, *i.e.*

$$D = \sigma \quad (26)$$

$$\therefore \quad \varepsilon_0 E = D - P$$

$$\therefore \quad D = \varepsilon_0 E + P \quad (27)$$

Relation between ε_r and χ

We have $$P = \sigma - \varepsilon_0 E = \varepsilon_0 \varepsilon_r E - \varepsilon_0 E$$

or $$P = \varepsilon_0(\varepsilon_r - 1)E \quad (28)$$

Comparing (22) and (28), we find

$$\chi = \varepsilon_r - 1$$

or $$\varepsilon_r = 1 + \chi \quad (29)$$

Thus like susceptibility, the dielectric constant is also a measure of polarization of the material. Larger the polarization per unit resultant field, greater will be the dielectric constant of the dielectric.

11.5 INDUCED DIPOLES

When a dielectric is placed in an electric field, the question arises, what will be its effect on an atom?

The nucleus of an atom is about 10^{-15} m in diameter and one can regard it as a point. The electron cloud is about 10^{-10} m in diameter and we may regard that its negative charge is concentrated at its centre. Obviously the centres of gravity of positive and negative charges in an atom coincide (Fig. 11.5(*a*)). This means that, such an atom does not produce any electric field at its own.

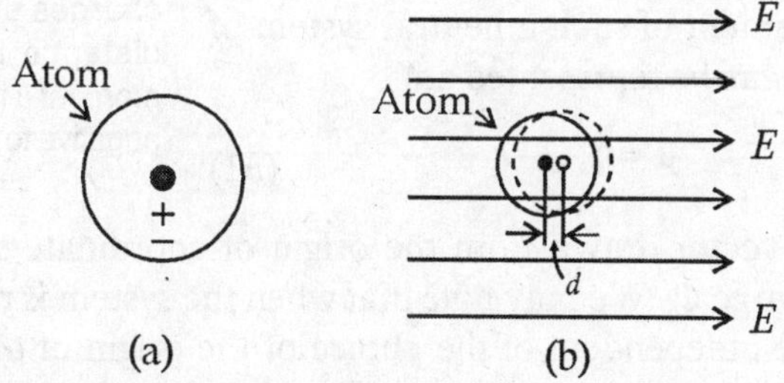

Fig. 11.5: (*a*) The centres of positive and negative charges in an atom coincide and atom does not possess electric dipole moment. (*b*) Atom placed in an electric field of strength *E*. Electron cloud is displaced with respect to the nucleus. The centres of negative and positive charges no longer coincide and separated. An electric dipole is induced

Let the atom is placed in an electric field of field strength *E*, the electron cloud will be displaced in the direction of *E*. Let this distance is '*d*' with respect to the nucleus (Fig. 11.5(*b*)). Obviously, the centres of gravity of positive and negative charges in the atom no more coincide (Fig. 11.5(*a*)). Clearly, the atom is now equivalent to a system of two charges, $q = Ze$, of equal magnitude but opposite in sign separated by a distance d (Fig. 11.5(*b*)). Such a system is usually called an electric dipole or simply a dipole.

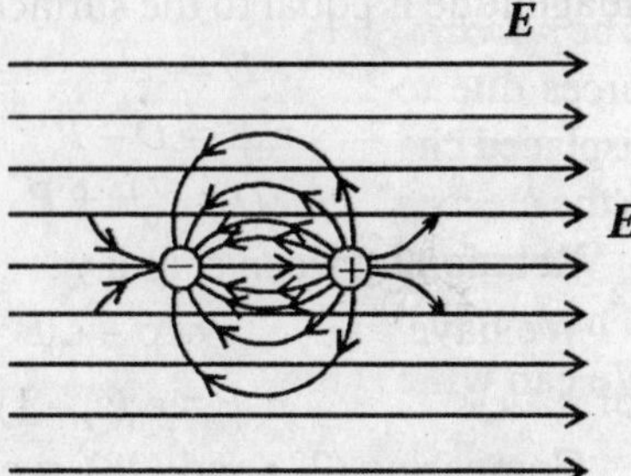

Fig. 11.6: The electric dipole as a whole is electrically neutral but it sets up its own electric field in space-which is opposite in direction to the external field

We may note that dipole induced in the atom due to the action of external field is electrically neutral as a whole. However, the induced dipole sets up its own electric field which is opposite in direction to the electric field (Fig. 11.6). The product of the magnitude of the charge (q) and separation distance between any two charges (d) is called the dipole moment (μ), *i.e.*

$$\mu = qd \tag{30}$$

The dipole moment (μ) is a vector directed along the axis of the dipole from the negative charge to the positive charge (Fig. 11.7).

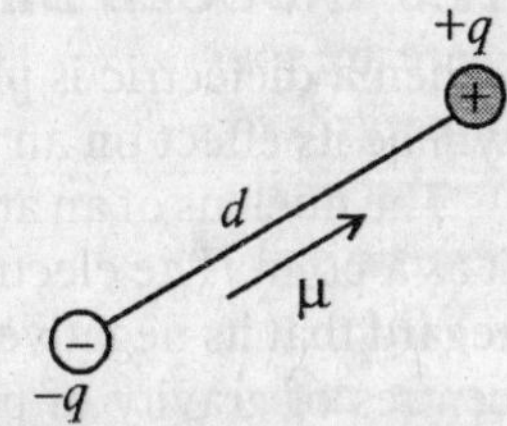

Fig. 11.7: An electric dipole. This consists of two equal and opposite charges separated by a distance *d*. The dipole moment μ is directed from negative to positive charge

Generalizing the above, we can say that any neutral system of *N* point charges $Q_1, Q_2, \ldots, Q_N$ occupying a volume having linear dimensions '*d*' acts as a dipole. However, the sum of the charges ΣQ_i in the volume should be equal to zero so that the system remain neutral. Now, the dipole moment of such a neutral system of point charges can be represented as

$$\mu = \sum_i Q_i \cdot \boldsymbol{r}_i \tag{31}$$

where $\boldsymbol{r}_i$ is a vector drawn from the origin of coordinate system to the position of charge Q_i. We may note that when the system is really neutral, then μ will be independent of the choice of the origin of the coordinate system.

We have already said that dielectric materials are made up of atoms and molecules which are neutral systems. Whenever a atom or molecule is subjected to an external field, the electric field tends to displace the equilibrium positions of bound charges, as a result of which dipole moment is induced in the atom or molecule. The amount of induced dipole moment will be proportional to the electric field strength and larger the field, the greater will be the displacement of charges and hence the larger the induced dipole moment.

Since the charges are displaced along the the field direction and hence the dipole moment is induced in the same direction. The molecule is said to be *polarized* by the electric field. When a molecule gets polarized, restoring forces due to Coulomb attraction come into play and tend to pull the displaced charges together. Until the restoring force balances the force due to the electric field, the charges separate.

We may note that the restoring forces vary in magnitude from one kind of molecule to another and therefore, the extent of dipole moment differs. We can write the induced dipole moment as

$$\mu = \alpha \boldsymbol{E} \tag{32}$$

where α is the proportionality constant and called as the *polarizability* of the molecule. The polarizability (α) characterizes the capacity of the electric charges in the molecule to suffer displacement in external electric field. α has the dimensions of volume. As soon as the electric field is switched off, the induced dipole moment vanishes.

11.6 PERMANENT DIPOLES

There are some molecules, known as polar molecules. In these molecules the centres of gravity of the charges of opposite sign are separated even in the absence of an external electric field. Such molecules are said to have intrinsic dipole moment and carry permanent dipoles.

When a molecule having intrinsic dipole moment is placed in a uniform electric field E as shown in Fig. 11.8, the electric field E exerts a force $+qE$ on the charge $+q$ and $-qE$ on $-q$. Since the two forces acting on the dipole are equal and opposite to each other and hence the net force on it is zero. Thus, there is no translational force on the dipole in a uniform field. However, the two forces are antiparallel and constitute a couple which tends to rotate the dipole. The torque on the dipole is

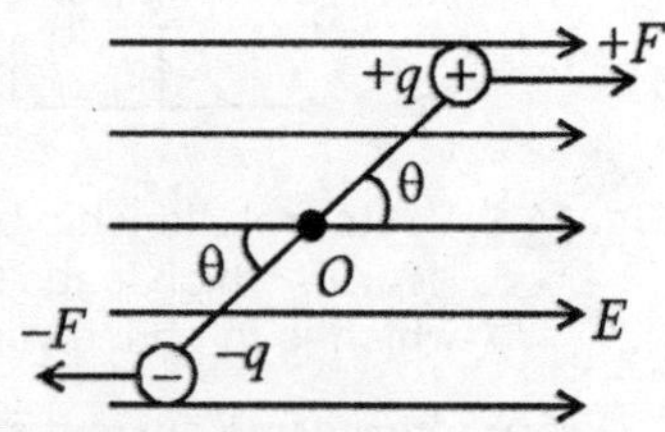

Fig. 11.8: An electric dipole in uniform electric field experiences a torque equal to $\mu E \sin\theta$, where θ is the angle between μ and E

$$\tau = qEd \sin\theta = \mu E \sin\theta$$

or

$$\tau = \mu \times \boldsymbol{E} \tag{33}$$

Summarizing, we can say that a dipole in a uniform electric field does not undergo translational motion but rotates in an attempt to align with the field direction. Clearly, a free dipole aligns its axis with the field direction.

We may note that the electric field can also induce a dipole moment in the molecule. Thus the total dipole moment of the molecule is a sum of the induced and permanent dipole moments. Thus

$$\mu_{Total} = \mu_{ind} + \mu_{per}$$

In case of polar molecules, $\mu_{ind} << \mu_{per}$ and

$$\mu_{Total} \approx \mu_{per}$$

11.7 PHASE DIFFERENCE AND ELECTRIC LOSS

A dipole tends to align itself along the direction of applied electric field and for *AC fields* tends to follow the field and be in a phase with it. However, the interaction of this dipole with other dipoles in the medium prevents this and this leads to dielectric loss. This loss appears as heat. This dielectric loss is connected with ε_r'', the imaginary part of the dielectric constant, The rate of loss of energy in unit volume of the material is obtained as

$$W = \frac{1}{2}\varepsilon_0\varepsilon_r''\omega E_0^2 \tag{34}$$

Obviously, the energy loss is proportional to ε_r''. Usually the dielectric loss is expressed in terms of a quantity called 'the loss tangent', tanδ. This is defined as $\tan\delta = \varepsilon_r''/\varepsilon_r'$. The angle δ is the compliment of angle φ, *i.e.* $90 - \phi = \delta$, where φ is the angle between the applied field and the resultant current vector (Fig. 11.9)

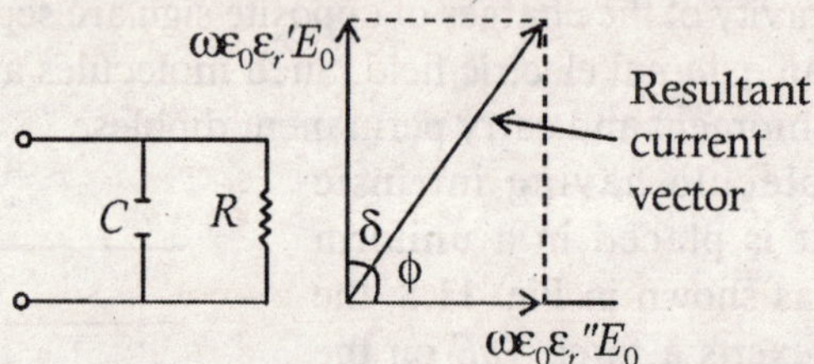

Fig. 11.9: Vector relationship between the field vector E_0, the current vector $\omega\varepsilon_0\varepsilon_r'E_0$ which leads the field by 90° and the current vector $\omega\varepsilon_0\varepsilon_r''E_0$ which is in phase with the field. The angle δ is shown

From alternating current studies, we have

$$\text{Power} = V_R I_R \cos\phi$$

where V_R and I_R are rms values of voltage and current, and for a circuit containing capacitance C and resistance R, we have

$$\tan\delta = \frac{1}{\omega CR} \tag{35}$$

Using Eq. (2), one obtains for a dielectric

$$W = \frac{1}{2}\varepsilon_0\varepsilon' \tan\delta\, \omega E_0^2 = \frac{1}{2}\frac{E_0^2}{\rho} \tag{36}$$

where ρ is the resistivity. Thus

$$\tan\delta = \frac{1}{\omega\varepsilon_0\varepsilon'\rho} \tag{37}$$

The energy losses in a dielectric material are due to: (*i*) *ionization*, (*ii*) *leakage current*, (*iii*) *polarization*, and (*iv*) *structural inhomogeneity*.

Ionization losses occurs in gases and solids having pores with entrapped gases. With the rise in field strength applied to a gas, a stage is reached when the gas molecules gets ionized due to collisions. This lead to enhanced conduction leading to dielectric losses. At low or ordinary electric fields, the conductivity is low and hence the loss is also small (Eq. 36). For example, for $\rho \approx 10^{16}$ Ω-m, $\varepsilon_r' \approx 1$ and at 100 Hz, $\tan\delta < 2 \times 10^{-8}$. In case of solids with gas inclusions the loss tangent increases with voltage from the voltage required for the ionization of gas molecules to the value of voltage at which ionization is complete. When such inclusions are high, there are chances of the failure of the dielectric.

Due to the leakage current, the losses in highly conducting liquids and solids are high (Eq. 35). In case of neutral liquids the losses will be small as their conductivity is low. Purity of oil is quite important as otherwise dielectric losses increase due to conduction. Pure transformer oil has $\tan\delta \approx 0.0005$.

Dipole losses in the radiofrequency region are usually due to dipole rotation or ions jumping from one equilibrium position to another. The dielectric losses associated with ions, the frequencies of which fall in the infrared region, are usually referred to as *infrared absorption*. Similarly, the losses in the optical region, associated with the electrons, are referred to as *optical absorption*.

10.8 DIELECTRIC STRENGTH

As the voltage applied across a dielectric is increased, dielectric losses its insulation property at a certain voltage called the *breakdown voltage* (V_{br}). The corresponding field strength (E_{br}) is given by

$$E_{br} = \frac{V_{br}}{t} \tag{38}$$

Here t is the thickness of the material. This voltage, V_{br} per unit thickness of the material is called the *dielectric strength* and usually expressed in volts/millimetre or kV/millimetre or mega volt/m. The breakdown of a dielectric is due to the collision of accelerated electrons or ions with molecules. At relatively high fields, the electrons in the dielectric gain enough energy to knock other charged particles and make them available for conduction.

Dielectric failure involves deterioration and electric breakdown or cascading. Resistance to electronic breakdown in a dielectric is *intrinsic dielectric strength* of the material. Intrinsic breakdown in a dielectric is caused by imperfection. When the applied potential across the dielectric becomes high enough, a few electrons are broken loose at points where their bonds are strained by the presence of imperfection. Upon being freed, these electrons are accelerated rapidly through the material. The dielectric material loses completely its insulating capacity as a result of this process.

Obviously, dielectric strength is structure sensitive property of the material. We may note that the actual thickness of the material affects the breakdown potential (V_{br}) per unit thickness, *i.e.* dielectric strength. Usually thicker materials possess lower dielectric strength than thinner materials.

The fundamental breakdown mechanisms in a solid are: (*i*) *intrinsic breakdown* (*ii*) *thermal breakdown* (*iii*) *discharge breakdown* and (*iv*) *electro-chemical breakdown.*

11.9 POWER FACTOR

The ratio of power loss in a material to the product of applied voltage and current is called power factor. The power factor and loss angle depend upon the nature of the material, applied voltage, humidity, temperature and frequency.

11.10 POLAR AND NON-POLAR MATERIALS

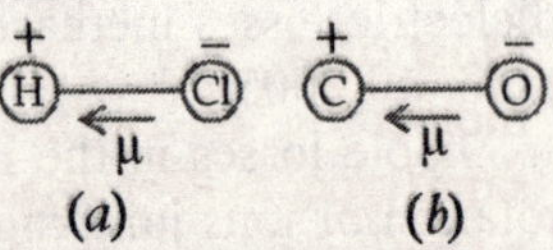

Fig. 11.10: Polar diatomic molecules

Permanent dipole moment exist in some molecules by virtue of an asymmetrical arrangement of positively and negatively charged regions. Such molecules are termed polar molecules. For example, in the HCl molecule (Fig. 11.10), the electron of hydrogen atom spends more time moving around the Cl atom than around the *H* atom. Therefore the centre of negative charges does not coincide with that of positive charges, and the molecule has a dipole moment directed from the Cl atom to the H atom. That is, we may write H^+Cl^-. The electric dipole of HCl molecule is $p = 3.43 \times 10^{-30}$ C-m. In the CO molecule [Fig. 11.10 (*b*)], the charge distribution is only slightly asymmetric and the electric dipole moment is relatively small, about 0.4×10^{-30} C-m, with the carbon atom corresponding to the positive and the oxygen atom to the negative end of the molecule.

Polar molecules can also induce dipoles in adjacent non-polar molecules, and a bond will form as a result of attractive forces between the two molecules. Moreover, the magnitude of this bond will be greater than for fluctuating induced dipoles.

Vander Waals forces will also exist between adjacent polar molecules. We may note that the associated bonding energies are significantly greater than for bonds involving induced dipoles.

The strongest secondary bonding type, the hydrogen bond, is a special case of polar molecule bonding. We find that it occurs between molecules in which hydrogen is covalently bonded to fluorine (as in HF molecule), oxygen (as in H_2O molecule), and nitrogen (as in NH_3 molecule). For each H—F, H—O or H—N bond, the single hydrogen electron is shared with the other atom. Obviously, the hydrogen end of the bond is essentially a positively charged bare proton that is unscreened by any electrons. This highly positively charged end of the molecule is capable of a strong attractive force with the negative end of an adjacent molecule (Fig. 11.11). Clearly, this single proton forms a bridge between two negatively charged atoms. We may note that the magnitude of the hydrogen bond is generally greater than that of the other types of secondary bonds and may be as high as 0.52 eV/molecule (= 51 kJ/mol). We find that the melting and boiling temperatures for hydrogen fluoride and H_2O are abnormally high in light of their low molecular weights, as a consequence of hydrogen bonding.

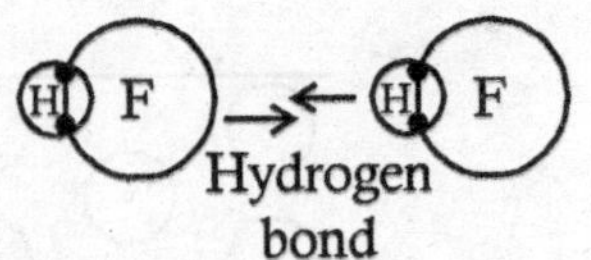

Fig. 11.11: Representation of hydrogen bonding in HF molecule

Non–Polar materials

Most of the hydrocarbons are non-polar. In non-polar materials the molecules, which are usually diatomic and composed of two atoms of the same type, may be represented as positive nuclei of charge q surrounded by a symmetrically distributed negative electron cloud of charge $-q$. In the absence of an applied electric field the centres of gravity of the positive and negative charge distributions coincide. When the molecules are subjected to an external electric field the positive and negative charges experience electric force tending to move them apart in the direction of external electric field. The distance moved is very small ($\sim 10^{-10}$ m) since the displacement is limited by restoring forces which increases with increasing displacement. The centres of positive and negative charges no longer coincide and the molecules are said to be polarized.

Polarization

When a material is placed in an electric field (Fig. 11.12), e.g. between the plates of a condenser, the field strength of charged particles within the material interact with the electric field. If the material is a conductor, some of the free electrons simply move to the side nearest the positive electrode until they totally counteract the applied electric field. Obviously, no field is left within the material. The displacement of charged particles occurs simultaneously, bringing about equilibrium.

If the material is non-conducting or insulator (dielectric), electrons can only be displaced locally as they are bound to the individual atoms. This local displacement of electrons, however, is sufficient to polarize the material. The negative electron cloud is displaced in each atom relative to the positive nucleus, thereby creating a small induced dipole whose negative pole is toward the positive side of the electric field. All dielectric materials are subjected to such *electronic polarization* (Fig. 11.12).

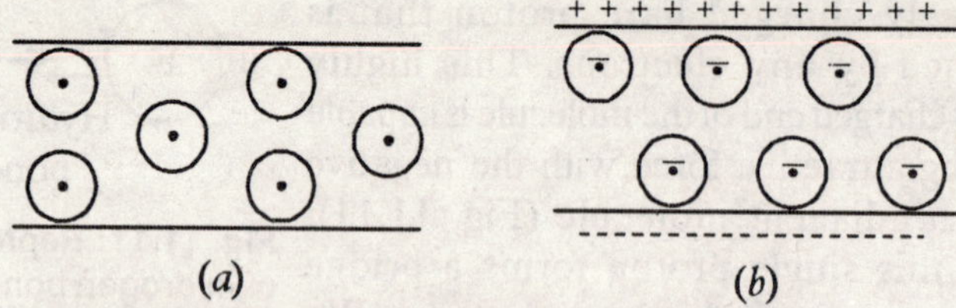

Fig. 11.12: Electronic polarization (*a*) In the absence of field (*b*) In the presence of electric field, the formation of induced dipole

Induced polarization also occurs in ionic materials. In an ionic crystal, for e.g. the negative ions are attracted towards the positive side and *vice versa* as illustrated in Fig. 11.13. NaCl is a good example of ionic polarization. Its interionic separation is 0.1 nm and dipole moment is of the order of 10^{-29} C-m.

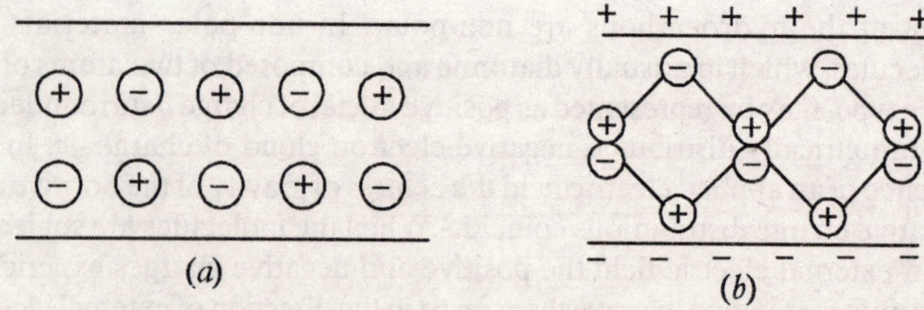

Fig. 11.13: Ionic polarization (*a*) In the absence of external field (*b*) Ions displacement toward opposite charged plates in an external field

There are many molecular structures which have permanent electronic dipoles due to asymmetric nature of their atomic structures. Few examples of such molecules are: H_2O, HCl, polyvinylchloride, etc. Some molecules also contain permanent charges which produce dipole moments in the individual molecules. There are many polymers of this type (polymers) and glasses contain dipoles. The permanent molecular dipoles in such materials can rotate about their axis of symmetry to align with an applied field which exert a torque in them. This is known as *orientation polarization.* This type of polarization is rarely perfect as thermal fluctuations do not permit all the dipoles to align with the applied field. With the rise in temperature the degree of polarization decreases. In the process of orientation polarization carriers collect at the boundaries between the constituent materials resulting in local distortion of the applied field. This is called as *Space-Charge polarization.*

The dipole moment of a single pair of the charges is given by

$$\mu = qd \qquad (39)$$

where q is one of the charges and d is the distance between charges.

Atomic View of Polarization

Let us consider that a slab of a dielectric located between the two plates of a parallel plate capacitor. When there is no external field, each volume of the dielectric has no dipole moment. When the dipolar is a non-polar material, the constituent molecules of the dielectric do not possess intrinsic dipole moment and if the material is a polar material, the individual molecular dipoles are randomly oriented so that in an elementary volume $\Sigma\mu = 0$. Clearly, the polarization is zero. When we apply electric field, *i.e.*, field is switched on. dipoles are induced in nonpolar molecules which form chains along the field lines (Fig. 11.14).

We have the polarization

$$P = N\mu = N \propto E \text{ (for nonpoler molecules)} \tag{40}$$

where N is the number of molecules per unit volume.

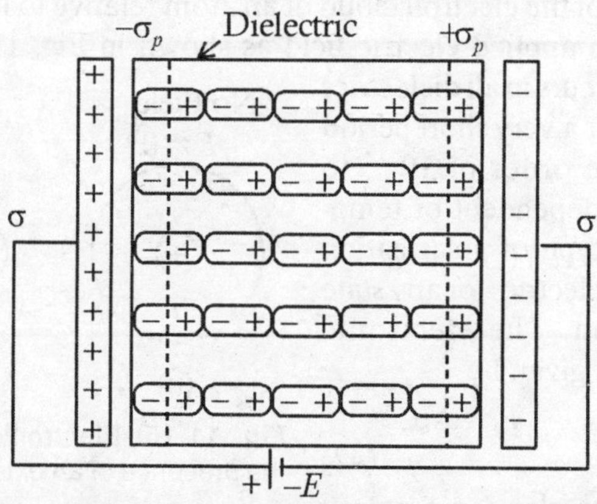

Fig. 11.14: Atomic view point of a polarization of a dielectric. The induced charge density (σ_P) on the dieletric is less than the free charge density (σ) on the plates

In case of polar dielectrics, the molecular dipoles experience a torque that tends to align them with a electric field direction. Total alignment is not achieved due to the disordering effects of the thermal agitation. In the direction of the field, and average alignment $<\mu>$ is achieved. The polarization is thus given by

$$P = N<\mu> \text{ (for polar molecules)} \tag{41}$$

Obviously, the action of electric field brings the dipoles into a certain ordered arrangement in space. One finds that the ends of adjacent dipoles carrying opposite charges neutralise each other. However, the charges of dipole ends terminating on the opposite face of the slab remain uncompensated. This is why the application of an electric field to the dielectric produces a displacement of charges within the dielectric material through a progressive orientation of intrinsic or induced dipoles. This is termed as dielectric polarization.

Sources of Polarizability

As stated earlier, the net polarizability of a dielectric material results mainly from the following three types of contributions:

(*i*) Electronic polarizability, (*ii*) Ionic polarizability, and (*iii*) Dipolar or orientational polarizability

The extent to which a particular polarizability contributes depends on the nature of dielectric and the frequency of the applied electric field. One or two of the said polarizations are always present and at a particular temperature or frequency of the applied field. However, one or another type of source of polarization may contribute in a large measure to the total polarization.

Electronic Polarization: This type of polarization arises due to the displacement of the electron cloud of an atom relative to its nucleus in the presence of an applied electric field as shown in Fig. 11.15. Electronic polarization occurs in all dielectrics and sets in over a very short period of time of the order of 10^{-14} to 10^{-15} s and independent of temperature. This type of polarization occurs in all dielectrics for any state of aggregation. The electronic polarization is given by

$$P_e = N\alpha_e E \tag{42}$$

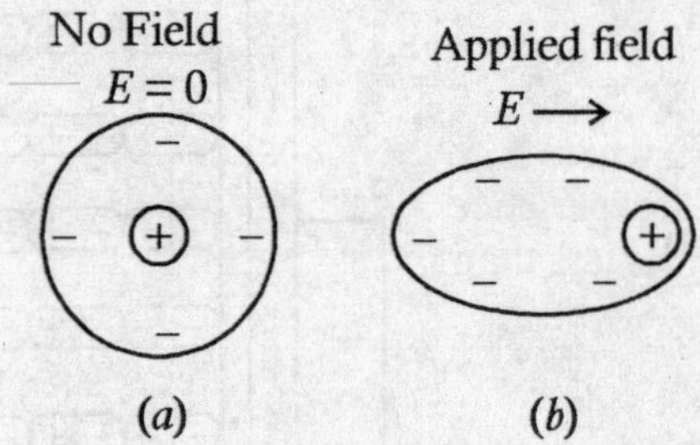

No Field $E = 0$ — Applied field $E \longrightarrow$

(*a*) (*b*)

Fig. 11.15: Electronic polarization in the presence of an external electric field

where α_e is the electronic polarizability. One can obtain the contribution of P_e to the dielectric constant as follows:

$$\varepsilon_r = 1 + \chi = 1 + \frac{P_e}{\varepsilon_0 E} = 1 + \frac{N\alpha_e E}{\varepsilon_0 E}$$

or
$$\varepsilon_r = 1 + \frac{N\alpha_e}{\varepsilon_0} \tag{43}$$

The expression (43) indicates the dielectric constant due to electronic polarization alone and thus gives the dielectric of a nonpolar gaseous dielectric. This reveals that the dielectric constant of a nonpolar gas depends on the polarizability of a molecule and the number of molecules in unit volume of the dielectric. For a monoatomic gas, the electronic polarization is given by

$$\alpha_e = 4\pi\varepsilon_0 R^3 \tag{44}$$

$$\therefore \quad \varepsilon_r = 1 + 4\pi N R^3 \tag{45}$$

where R denotes the radius of the atom.

Ionic Polarization: The ionic polarizability arises due to displacement of a charged ion relative to other ions in a solid or ionic crystals. In other words, we can say that ionic polarization occurs by the elastic displacement of positive and negative ions from their equilibrium positions, e.g. NaCl. NaCl molecule consists of Na^+ ions bound to Cl^- ions through ionic bond. If the interatomic distance is d, the molecule exhibits an intrinsic dipole moment equal to 'ed'. When an external electric field is applied to NaCl molecule, the Na^+ and Cl^- are displaced in opposite direction until ionic bonding forces stop the process.

The dipole moment of the molecule increases consequently. When we reverse the field direction, the ions move closer and again the dipole moment undergoes a change. Clearly, dipoles are induced. Assuming the forces near equilibrium as simple harmonic, the displacement Δx in the presence of electric field E is given by

$$\beta \Delta x \cong eE \tag{46}$$

where β is force constant, The induced dipole moment is proportional to the applied field, *i.e.*

$$\mu_i = \alpha_i E \tag{47}$$

where α_i is known as *ionic polarizability*. We may note that the ions also experience electronic polarization. For most materials ionic polarization is less than the electronic polarizability. Typically,

$$\alpha_i = \alpha_e / 10$$

We know that ionic polarization is given by

$$P_i = N\alpha_i E \tag{48}$$

For $\beta \cong 20$ Nm^{-1}, $\alpha_i = 10^{-39}$ Fm^2. The ionic contribution is important at low frequencies. NaCl has $\varepsilon_r \cong 5.6$ at low frequencies, whereas the value reduces to about 2.25 at optical frequencies. Ionic polarizability takes about 10^{-11} to 10^{-14} s to build up, and almost independent by temperature variations. Figure 11.16 illustrates the phenomenon of ionic polarizability.

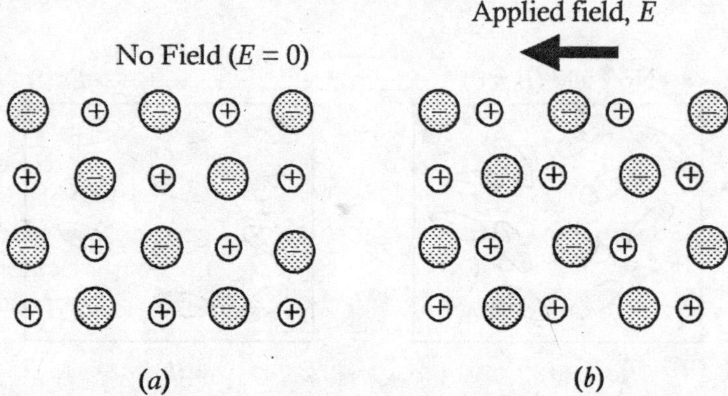

Fig. 11.16: Ionic polarization: (*a*) $E = 0$, unpolarized dielectric, (*b*) In an applied field, E, ionic polarization takes place due to relative motion of electrically charged positive (+) and negative (–) ions

Orientation or Dipolar Polarizability: This is the characteristic of polar dielectrics which consist of molecules having permanent dipole moment, e.g. water molecule, H_2O. In the absence of an external electric field, the dipoles have random orientations and there is no net polarization [Fig. 11.17 (*a*)]. However, when the electric field is applied, the dipoles orient themselves and the field produce orientational or dipolar polarizability. In the case of the electronic and ionic polarizations, external field force is balanced by a restoring force due to Coulomb attraction, but for orientation polarization, restoring forces do not exist. However, the dipolar alignment is counteracted by thermal agitation and an equilibrium state is reached where in the different dipoles make all possible angles varying from 0 to π radiations with the field direction [Fig. 11.17 (*b*)]. Even in the case of liquids or gases, where all molecules are free to rotate, a complete alignment cannot be achieved due to the randomizing effect of the temperature. However, one finds that if one molecular dipole is 10^5 completely aligns with the field to produce orientation polarization of the order of electronic polarization.

Clearly, orientation polarization is strongly temperature dependent. The rotation of polar molecules in the case of solids may be highly restricted by the lattice forces and leads to a great reduction in their contribution to orientation polarization. Due to this reason, the dielectric constant of water is about 80, whereas that of ice is about 10 only.

The process of orientation polarization takes relatively longer time than the other two polarizations due to involvement of rotation of molecules. The build-up time is of the order of ~10^{-10} s or more.

One can calculate the orientational polarization α_0 form

$$\alpha_0 = \frac{\mu^2}{3kT} \tag{49}$$

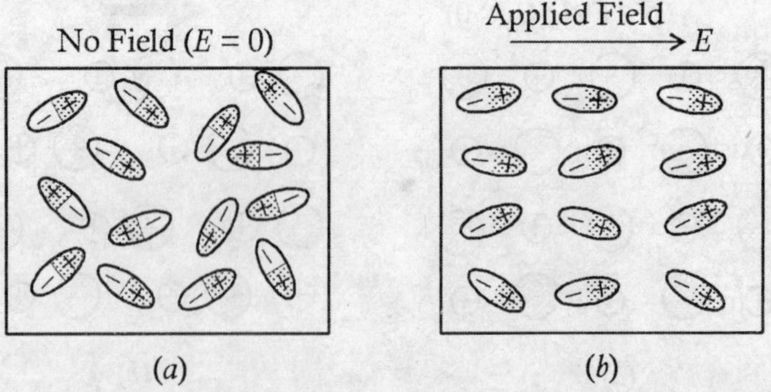

Fig. 11.17: Dipolar polarization: (*a*) No field ($E = 0$), molecular dipoles are randomly oriented, (*b*) Dipoles are partially aligned in the applied field

Orientational polarization is given by Debye's law

$$P_0 = \frac{N\mu^2 E}{3kT} \tag{50}$$

Obviously, the orientational polarization is inversely proportional to the temperature and proportional to the square of the permanent dipole moment.

The existence of dipolar polarizability depends on whether the molecules possess a permanent dipole moment. The dipolar polarizability also depends on temperature. At room temperature, $\alpha_d \sim 10^{-39}$ Fm^2 which is comparable to electronic polarizability.

Total Polarization

The total polarizability of a dielectric is given by

$$\alpha = \alpha_e + \alpha_i + \alpha_o \tag{51}$$

or

$$\alpha = \alpha_e + \alpha_i + \frac{\mu^2}{3kT} \tag{52}$$

The total polarization P may be expressed as

$$P = P_e + P_i + P_o$$

$$= N\left[\alpha_e + \alpha_i + \frac{\mu^2}{3kT}\right]E \tag{53}$$

One can express the dielectric constant of a polar gaseous dielectric as

$$\varepsilon_0(\varepsilon_r - 1) = N\left(\alpha_e + \alpha_i + \frac{\mu^2}{3kT}\right) \tag{54}$$

A plot of ε_r or P with $1/T$ is a straight line (Fig. 11.18)

As stated earlier, it is possible for one or more of the contributions to the polarization to be either negligible in magnitude relative to the others or absent. For instance, the orientation polarization does not exist in non-polar dielectrics and ionic polarization in covalently bonded materials. In polar dielectrics, the electronic polarization is negligible as compared to dipolar polarization.

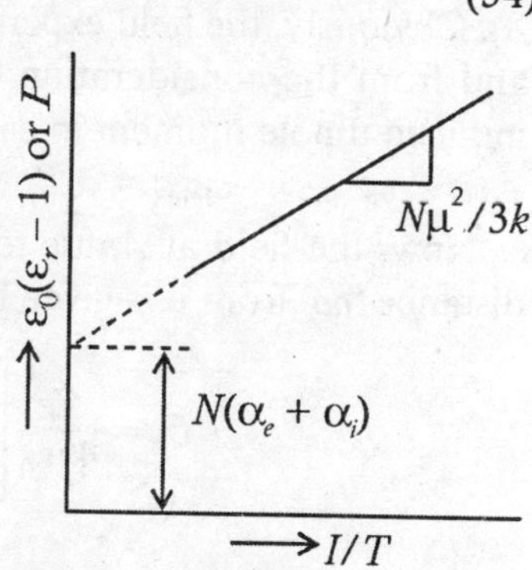

Fig. 11.18: Variation of P or ε_r with $1/T$. The permanent dipole moment can be determined from the slope of the curve

Internal Field in Solids

The internal field (E_i) is defined as the electric field acting at the location of given atom, is equal to the sum of the electric fields created by the neighbouring atoms and the applied field. We know that the atoms in gases are in constant random motion and are separated by sufficiently large distances. Therefore, one can neglect the interaction between atoms. Obviously, when an external field E is applied, the intensity of the electric field experienced by a given atom in the gas equals to the applied field E. In solids and liquids, the atoms are so close that they touch each other leading to a strong interaction between them. Since the atoms are surrounded on all sides by other polarized atoms, the internal intensity of the field at a given point of the material is

$$E_i = E + E' \tag{55}$$

One can evaluate the value of E' by the summation of all the effects of the surrounding atoms. Let us consider a one dimensional solid consisting of a string of equidistant identicals, each of polarizability α_e (Fig. 11.19)

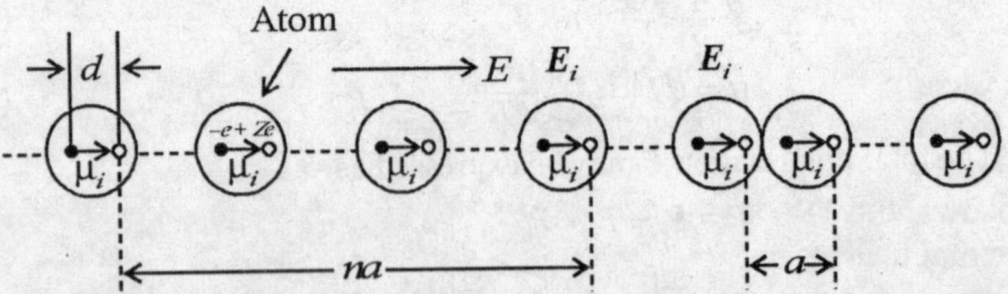

Fig. 11.19: A one dimensional solid. The internal field E_i is different from the external field, $E\,(=V/l\,)$.

Let the external field E is applied in a direction parallel to the string. Let us determine the internal field E_i experienced by any one of the atom (say A). Obviously, the field experienced by all other atoms will also be same, and from the consideration of symmetry E_i will be parallel to E. The induced dipole moment in each of the atoms of the string is as

$$\mu_{\text{ind}} = \alpha_e E_i \tag{56}$$

Now, the field at A due to the dipole induced in an atom located at distance 'na' from it is given by

$$E_n = \frac{Ze}{4\pi\varepsilon_0}\left[\frac{1}{(na)^2} - \frac{1}{(na+d)^2}\right]$$

$$= \frac{Ze}{4\pi\varepsilon_0}\left[\frac{(na+d)^2 - (na)^2}{(na+d)^2\,(na)^2}\right]$$

$$= \frac{2Zed}{4\pi\varepsilon_0\,(na)^3} \qquad (\because d << na)$$

$$= \frac{\mu_i}{2\pi\varepsilon_0 (na)^2} \qquad (\because \mu_i = Zed)$$

$\therefore$ The total field E_i at A is obtained as

$$E_i = E + \frac{\mu_i}{2\pi\varepsilon_0}\left[2\sum_{n=1}^{\infty}\frac{1}{(na)^3}\right]$$

In order to take into account of the atoms to the left and right of A, we have multiplied by a factor of 2. Now,

$$E_i = E + \frac{\mu_i}{\pi\varepsilon_0 \alpha^3}\sum_{n=1}^{\infty}\frac{1}{n^3}$$

$$= E + \frac{1.2\,\mu_i}{\pi\varepsilon_0 \alpha^3} \tag{57}$$

Clearly, the combined effect of induced dipoles of neighbouring atoms is to produce a net field at the location of a given atom which is larger than the applied field. From Eq. (57) we can see that the greater the polarizability α_e or smaller the intermolecular spacing 'a', the larger is the internal field.

One can determine the local field in three dimensional solid with the help of the structure of the solid. An accurate calculation of the internal field in solids and liquids is in general very complicated. One finds the general expression for E_i very much similar to the expression (57) where the number density N of atoms replaces $1/a^3$. As $P = N\mu_i$, and hence we can write the general expression as

$$E_i = E + \frac{\gamma P}{\varepsilon_0} \tag{58}$$

where γ is the proportionality constant and termed as the *internal field constant*.

We may note that the value of γ is independent on the internal arrangement of atoms in dielectric. In general $\gamma \approx 1$. In the case of an infinite chain of molecules, $\gamma = 1.2/\pi$. One can evaluate γ in other simple cases also. In case of crystals having cubic symmetry, $\gamma = 1/3$ and the internal field is given by

$$E_i = E + \frac{P}{3\varepsilon_0} \tag{59}$$

This is called as *Lorentz field*.

Clausius–Mossotti Relation

We now consider the simple case of an elemental solid dielectric which exhibit only electronic polarizability. Silicon, germanium and diamond are such solids which are made up of single type of atoms. The electronic polarizability per atom, α_e is related to the bulk polarization P as

$$\alpha_e = \frac{P}{NE_i} \tag{60}$$

where N is the number of atoms/m^3 and E_i is the local field, Using Eq. (58), Eq. (60) becomes

$$\alpha_e = \frac{P}{N\left(E + \frac{\gamma P}{\varepsilon_0}\right)} \tag{61}$$

But $$E = \frac{P}{\varepsilon_0(\varepsilon_r - 1)} \tag{61a}$$

Now, if we assume that internal field is Lorentz field, $\gamma = 1/3$ and Eq. (61) becomes

$$\alpha_e = \frac{P}{N\left(E + \frac{P}{3\varepsilon_0}\right)} \tag{62}$$

Using Eq. (61*a*), Eq. (62) becomes

$$\alpha_e = \frac{P}{N\left[\frac{P}{\varepsilon_0(\varepsilon_r - 1)} + \frac{P}{3\varepsilon_0}\right]}$$

or $$\frac{N\alpha_e}{E_0} = \frac{3(\varepsilon_r - 1)}{(\varepsilon_r + 2)}$$

or $$\frac{\varepsilon_r - 1}{\varepsilon_r + 1} = \frac{N\alpha_e}{3\varepsilon_0} \tag{63}$$

This is known as the *Clausius-Mossotti relation*. It relates the dielectric constant to the atomic polarizability for nonpolar solids provided the condition of cubic symmetry holds. In a more general form, it is expressed as

$$\frac{\varepsilon_r - 1}{\varepsilon_r + 1} = \frac{1}{3\varepsilon_0}\sum_j N_j \alpha_j \tag{64}$$

The measured values of ε_r for diamond, silicon and germanium are 5.68,12 and 16 respectively.

Frequency Dependence of Total Polarizability

The total polarizability of a dielectric is given by

$$\alpha = \alpha_e + \alpha_i + \alpha_o \tag{65}$$

It decreases with increase in frequency as shown in Fig. 11.20. This type of variation of polarizability can be explained on the basis of the relaxation times of the various contributing polarization processes. When the frequency of the applied field is quite large as compared to the inverse of the relaxation time for a particular polarization process, the contribution of that process to polarizability is negligible. As relaxation time is maximum for the dipolar process and minimum for the electronic process, the dipolar contribution disappears first followed by ionic electronic contributions.

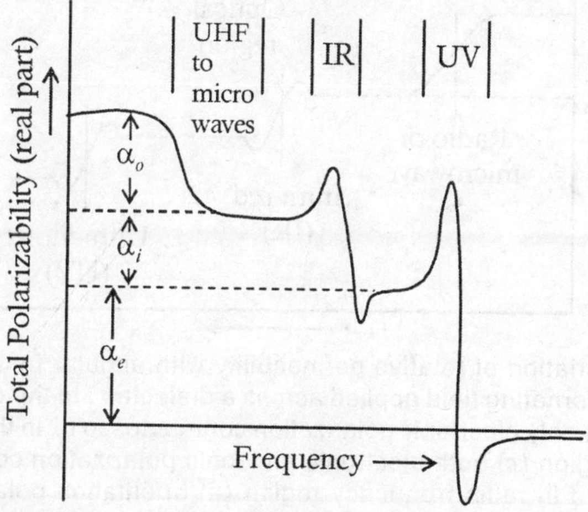

Fig. 11.20: Frequency dependence of total polarizability. The curve also shows the various contributions

Frequency Dependence of the Dielectric Constant

In several practical situations, a dielectric is subjected to an alternating field. As an AC changes its direction with time and with each direction reversal, the polarization components are required to follow the field reversals in order to contribute to the total polarization of the dielectric. Obviously, the total polarization depends on the ability of dipoles to orient themselves in the direction of field during each alteration of the field. The relative permeability which is a measure of the polarization shows marked differences in the behaviour at different frequencies of the electric field for a polar dielectric as shown in Fig. 11.21. We can see that in audiofrequency region, all the three types of polarization are possible and the dielectric is characterized by a polarizability $\alpha = \alpha_e + \alpha_i + \alpha_0$ and the polarization $P = P_e + P_i + P_0$. At low frequencies, one finds that the dipoles will get sufficient time to orient themselves completely along the instantaneous direction of the field. This orientation of the dipoles occurs first in one direction and then in the other, following the changes in the direction of the applied electric field E (Fig. 11.22). The average time taken by the dipoles to orient in the electric field direction is known as *relaxation time* (τ). The reciprocal of τ, *i.e.* τ^{-1} is called the *relaxation frequency* (τ^{-1}). When the frequency of the applied electric field is much higher than the relaxation frequency of the dipoles, the dipoles cannot reverse fast enough. However, when the dipole relaxation time (τ) is less than half the period of electric field T, *i.e.* $T/2$, the dipole can follow alterations in AC electric field and contribute to orientation polarization. Consequently, the orientation polarization,

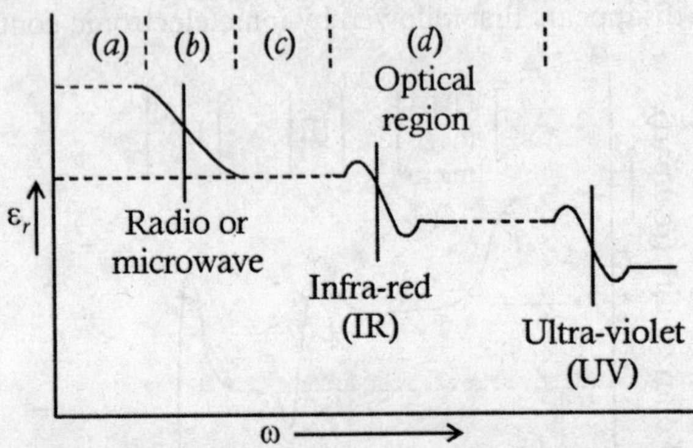

Fig. 11.21: Variation of relative permeability with angular frequency of the alternating field applied across a dielectric . In the optical region (*d*) only electronic polarization contributes to ε_r, in the microwave region (*c*) both electronic and ionic polarization contribute to ε_r and in radio frequency region (*a*) orientation polarization also contributes to ε_r

which is effective at low frequencies, gets damped out for higher AC field frequencies. ($f_{\text{field}} > f_{\text{relax}}$). Normally in the radiofrequency or microwave band region, the permanent dipoles fails to follow the field reversal and the polarization falls to $P = P_i + P_e$, only. As a result, ε_r decreases considerably.

In the infrared region, the ionic polarization fails to follow the AC field reversals due to the inertia of the system and hence the contribution of ionic polarizability becomes almost negligible and total polarization $P = P_e$ only. In the optical region, the electron cloud follows the AC electric field variations and the dielectric material exhibits an electronic polarizability α_e and the relative permeability,

$$(\varepsilon_r)_{\text{opt.reg.}} = n^2 \tag{66}$$

i.e. $(\varepsilon_r)_{\text{opt. reg.}}$ is equal to the square of the refractive index of the dielectric 'n'. In the ultraviolet region, the electron cloud too fails to follow the electric field alterations and electronic field contribution to the polarization ceases and total polarization, $P = 0$. Above the ultraviolet range, ε_r approaches unity, *i.e.*

$$(\varepsilon_r)_{\text{X-ray}} = 1 \tag{67}$$

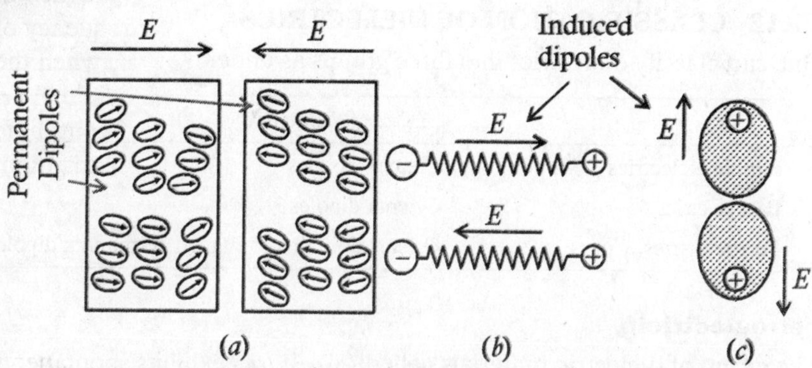

Fig. 11.22: The behaviour of (*a*) permanent, and (*b*) and (*c*) induced dipoles in an applied AC electric field. The orientation of dipoles first takes place in one direction and then in other in accordance with the changes in the direction of the electric field

We take the example of water. The dielectric constant of water at low frequency is generally referred to as *static dielectric constant*, at room temperature its value is about 80. In the optical region it falls to about 1.8.

11.11 INSULATION RESISTANCE

This separates a number of conductors at different electrical potentials and does not allow a large flow of electric current between them. The difference of potential may cause leakage of current along two paths, *i.e.* (*i*) over the surface of insulation, and (*ii*) through the solid material of the insulators.

We may note that the resistance offered along the said two paths is not the same due to the nature of the material. The former is due to the surface resistivity of the material, whereas the latter is the result of the volume resistivity of the material. The combined effect of two is called the insulation resistance of the material.

We may define the surface resistivity as the resistance between the two opposite edges of a square of unit area of insulation surface. The volume resistivity is defined as the resistance presented to the flow of an electric current by a material of unit cross section and of unit length at 0° C.

The resistive properties of real dielectrics are:

(*i*) Under the action of a potential, mobility with which any conducting species moves.

(*ii*) Temperature variation of conductivity to state why the conductivity of a dielectric varies in exponential manner with temperature.

(*iii*) To analyse the time response of dielectrics, frequency or time variation of conductivity.

(*iv*) While designing an equipment, the knowledge of breakdown or the highest voltage a dielectric will withstand is of interest. Usually insulators breakdown at 10^6 V/m to 10^9 V/m to DC at 20° C.

11.12 CLASSIFICATION OF DIELECTRICS

One can classify dielectrics into three groups as under

Class	*Effect of electric field*
Simple dielectrics	Creates dipoles
Paraelectrics	Orients dipoles
Ferroelectrics	Orients domains of aligned permanent dipoles

Ferroelectricity

The group of dielectric materials called *ferroelectrics* exhibits spontaneous polarization *i.e.*, polarization in the absence of an electric field. In a sense, ferroelectrics are the electric analog of the ferromagnets, which may display permanent magnetic behaviour. In ferroelectrics, the polarization can be changed and even reversed by an external electric field. The reversibility of the spontaneous polarization is due to the fact that the structure of a ferroelectric crystal can be derived from a non-polarized structure by small displacement of ions. In most ferroelectric crystals, this non-polarized structure becomes stable if the crystal is heated above a critical temperature, the *ferroelectric Curie temperature* (T_C); *i.e.* the crystal undergoes a phase transition from the polarized phase (ferroelectric phase) into an unpolarized phase (paraelectric phase). The change of the spontaneous polarization at T_C can be continuous or discontinuous. The T_C of different types of ferroelectric crystals range from a few degrees absolute to a few hundred degrees absolute.

From a practical standpoint ferroelectric can be divided into two classes: (*i*) In this first class of ferroelectrics, polarization can occur only one crystal axis, e.g. Rochelle salt, KH_2PO_4, $(NH_4)_2SO_4$, guanidine aluminium sulphate hexahydrate, glycine sulphate, colemanite and thiourea, (*ii*) In this *second* class of ferroelectrics, spontaneous polarization can occur along several axes that are equivalent in the paraelectric phase, e.g. $BaTiO_3$,- type (or perovskite type) ferroelectrics, $Cd_2Nb_2O_7$, PbN_2O_6, certain alums, such as methyl ammonium alum and $(NH_4)_2 \cdot Cd_3(SO_4)_3$.

One can also place ferroelectrics as *proper ferroelectrics and improper ferroelectrics.* In proper ferroelectrics, for e.g. $BaTiO_3$, KH_3PO_4, and Rochelle salt, the spontaneous polarization is the order parameter. In improper ferroelectrics the spontaneous polarization can be considered a by-product of another structural phase transition. The properties of some representative ferroelectrics are given in Table 11.1

The spontaneous polarization in ferroelectrics can occur in at least two equivalent crystal directions, thus a ferroelectric crystal consists in general of regions of homogeneous domains. Ferroelectrics of the first class consist of domains with parallel and antiparallel polarization (Fig. 11.23*a*), whereas ferroelectrics of the second class can assume much more complicated domain configurations (Fig. 11.23*b*). The region between two adjacent domains is called a *domain wall.* Within this wall, the spontaneous polarization changes its direction.

TABLE 11.1: PROPERTIES OF SOME REPRESENTATIVE FERROELECTRIC CRYSTALS

Group	*Crystal*	*Curie temperature* T_C (K)	P_s $C/m^{-2} \times 10^{-2}$	*At T* (K)
Ilmenites and	GeTe	670	–	–
Perovskites	$LiNbO_3$	1480	71	296
	$KNbO_3$	710	30	600
	$BaTiO_3$	393	26	300
	$SrTiO_3$	32	3	4.2
KDP type	KH_2PO_4 (KDP)	123	4.7	100
	KD_2PO_4	213	5.5	100
	$Rb_2H_2PO_4$	147	5.6	90
	KH_2AsO_4	97	5.0	78
TGS type	$(NH_2CH_2COOH)_3$ H_2SO_4 (Triglycine sulphate)	322	2.8	275
Rochelle salt type	$NaKC_4H_4O_6 \cdot 4H_2O$ (Rochelle salt)	296 (upper) 255 (lower)	0.25	275

When an electric field is applied to specimen of a ferroelectric crystal, the polarization first rises rapidly with applied field to a value above which the dependence is linear. Linear extrapolation to zero field gives the *saturation of spontaneous polarization.* On subsequently reducing the field to zero, remanent (residual) polarization remains. The negative field to reduce the polarization to zero is called the coercive field. Obviously, a ferroelectric crystal can be switched by the application of an electric field and a *hysteresis loop* is associated with the switching. The existence of a dielectric hysteresis

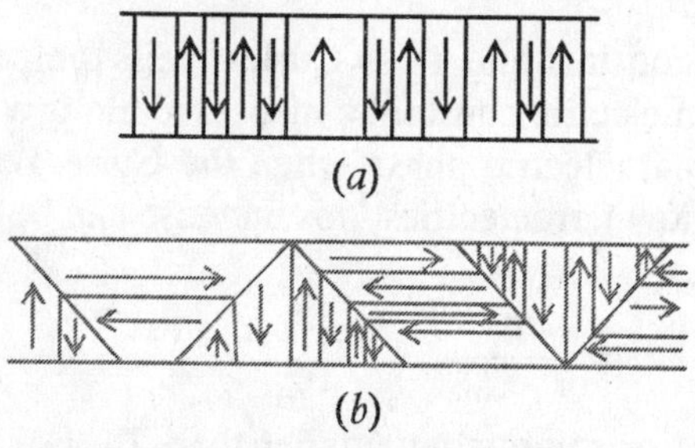

Fig. 11.23: Domain configurations (simplified) encountered in ferroelectric. (*a*) First class (*b*) Second class

loop in a dielectric material implies that the substance possesses a spontaneous polarization, P_s and the value of spontaneous polarization (depending upon the shape of hysteresis loop) depends upon a number of factors such as the dimension of the specimen, the temperature, the texture of the crystal, the thermal and electrical properties of the crystal. A typical hysteresis loop is shown in Fig. 11.24. For most ferroelectrics, the values of P_s are between 10^{-7} and 10^{-4} C/cm^2. In non-ferroelectric dielectrics, electric field between 10^5 and 10^8 V/cm would be necessary in order to achieve such large polarizations.

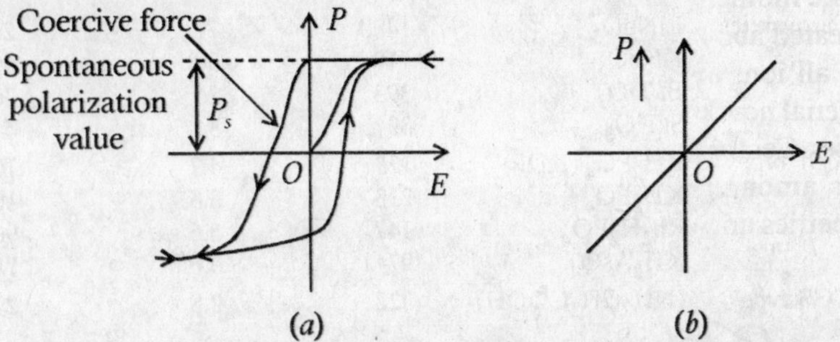

Fig. 11.24: Net polarization P of a ferroelectric crystal versus externally applied electric field. (*a*) Hysteresis (*b*) $P-E$ relation above Curie temperature

The hysteresis loop of a ferroelectric material changes its shape as the temperature is increased. The height and width decrease with increase of temperature. At a certain temperature known as *ferroelectric Curie temperature* (T_C), the loop merges to a straight line and the ferroelectric behaviour of the material disappears. The electric susceptibility χ_E in the paraelectric phase (above Curie temperature, T_C) is related to the temperature by the Curie-Weiss law

$$\chi_E = \frac{C}{T - T_C} \tag{68}$$

where C is Curie constant and $T_C \rightarrow$ Curie–Weiss temperature.

As a rule, the dielectric constant ε measured along a ferroelectric axis increases in the paraelectric phase when the Curie Temperature, T_C is approached in many ferroelectrics, this increase can be approximated by Curie-Weiss law,

$$\varepsilon = \frac{T}{T - T_0} \tag{68a}$$

where T_0 is equal or some what smaller than T_C. For $BaTiO_3$, this law holds unaltered upto frequencies of 24×10^{10} Hz. Ferroelectric crystals will have an extremely large dielectric constant.

One of the most common ferroelectrics is the barium titanate ($BaTiO_3$). The spontaneous polarization of this crystal is a consequence of the positioning of the Ba^{2+}, Ti^{4+}, and O^{2-} ions within the unit cell (Fig. 11.25). The Ba^{2+}, ions are located at the corners of the unit cell, which is of tetragonal symmetry (a cube that has been elongated slightly in one direction). The dipole moment results from the relative displacements of the O^{2-} and Ti^{4+} ions from their symmetrical positions as shown in the side view of the unit cell. We can see that O^{2-} ions are located near, but slightly below the centres of each of six faces, whereas the Ti^{4+} ion is displaced upward from the unit cell corner. Obviously, a permanent ionic dipole moment is associated with each unit cell. However, when $BaTiO_3$ is heated above its *ferroelectric* T_C (=120° C), the unit cell becomes cubic, and all ions assume symmetric positions within the cubic unit cell; the material now has a perovskite crystal structure, and ferroelectric behaviour ceases. Ceramic materials and especially ferroelectrics are the most important ones among dielectrics. Ceramics may have very diversified electric properties and are almost insusceptible to ageing and heating.

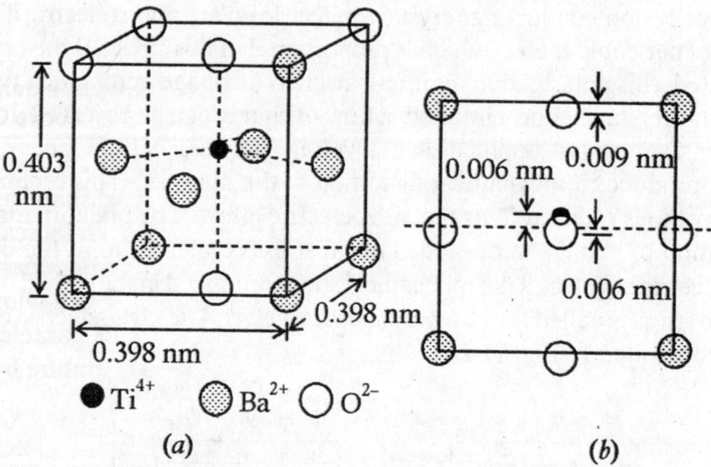

Fig. 11.25: $BaTiO_3$ unit cell (*a*) in an isomeric projection, and (*b*) looking at one face, exhibiting the displacement of Ti^{4+} and O^{2-} ions from the center of face

Spontaneous polarization of this group of materials results as a consequence of interactions between adjacent permanent dipoles wherein they mutually align, all in the same direction. For e.g., with $BaTiO_3$, the relative displacements of O^{2-} and Ti^{4+} ions are in the same direction for all the unit cells within some volume region of the specimen. At room temperature, ε for $BaTiO_3$ may be as high as 5000. Consequently, capacitors made from these ferroelectric materials can be significantly smaller than capacitors made from other dielectric materials. This is why that these materials are widely used in the manufacture of miniatured capacitors.

11.13 ELECTROSTRICTION

This is mechanical deformation which always accompanies polarization in dielectric. An electric field polarizes any material by inducing dipole moments. This displacement of charges from their equilibrium positions alters the mechanical dimensions of a solid; it causes electrostriction. However, mechanical stress applied to a neutral material cannot induce dipole moment; *i.e.*, *electrostriction has no inverse*. If a mechanical distortion creates a voltage, the effect must be caused by permanent dipole moment anchored in the structure without a centre of symmetry.

11.14 PIEZOELECTRICITY OR PRESSURE ELECTRICITY

Electricity, or electric polarity, resulting from the application of mechanical pressure on a dielectric crystal is called piezoelectricity or pressure electricity. The application of a mechanical stress produces in certain dielectric (electrically non-conducting) crystal an electric polarization (electric dipole moment per cubic meter) which is proportional to this stress. If the crystal is isolated, this polarization manifests itself as a voltage across the crystal, and if the crystal is short-circuited, a flow of charge can be observed during loading. Conversely, application of a voltage between certain faces of the crystal produces a mechanical distortion of the material. This reciprocal relationship is referred to as the piezoelectric effect. The phenomenon of generation of voltage under mechanical stress is referred to as the direct piezoelectric effect, and the mechanical strain produced in the crystal under electric stress is called the *converse piezoelectric effect*. The piezoelectric effect is demonstrated in Fig. 11.26

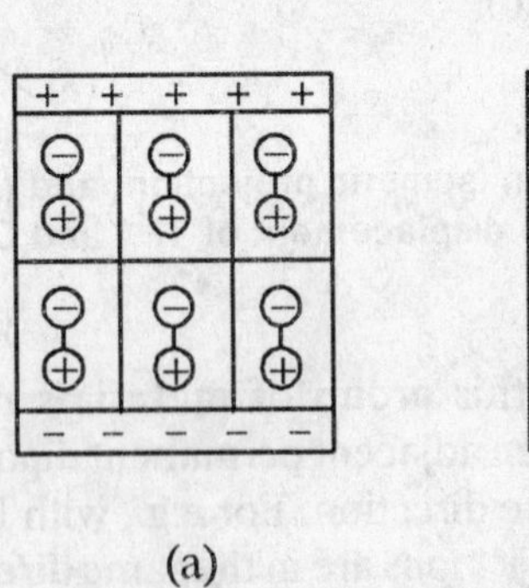

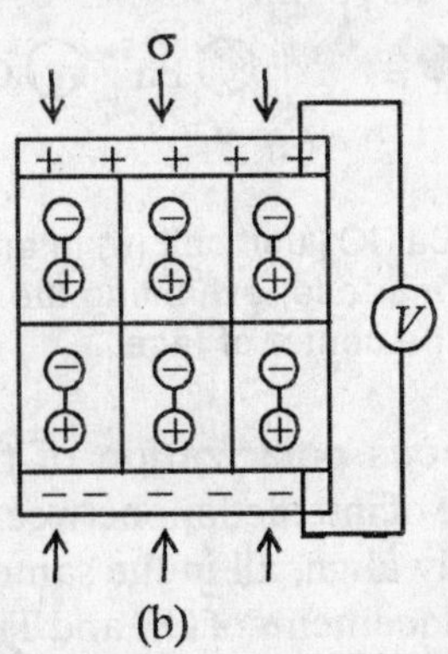

Fig. 11.26: Piezoelectric effect (*a*) dipoles within a piezoelectric material (*b*) a voltage is generated when the material is subjected to a compressive stress

Piezoelectricity occur only in insulating materials. Only few ceramic materials exhibit this property. Piezoelectric materials are used extensively in *transducers* for converting a mechanical strain into an electrical signal. Such devices include microphones, phonograph pick-ups, vibration-sensing element, and the like. The converse effect, in which a mechanical output is derived from an electrical signal input, is also widely used in such devices as sonic and ultrasonic transducers, headphones, loudspeakers, and cutting heads for disk recording. Both the direct and converse effects are employed in device in which the mechanical resonance frequency of the crystal is of importance. Such devices include electric wave filters and frequency control elements in electronic oscillator circuits.

The necessary condition for the piezoelectric effect is the absence of symmetry in the crystal structure. Of the 32 crystal classes, 21 lack a centre of symmetry, and with the exception of one class, all of these are piezoelectric. Piezoelectric materials include titanates of barium and lead, zirconate ($PbZrO_3$), ammonium dihydrogen phosphate ($NH_4H_2PO_4$), natural quartz. As stated earlier, this property is a characteristic of materials having complicated crystal structure with a low degree of symmetry. One may improve the piezoelectric behaviour of a polycrystalline specimen by heating above its Curie temperature and then cooling to room temperature in strong electric field.

Piezoelectrics are required to have high piezoelectric modulus and low losses. Ferroelectric ceramics have the piezoelectric modulus around 6×10^{-10} C/N, which is one or two order of magnitude higher than that of quartz. The structure of piezoelectric ceramics is a solid solution of barium titanate, barium niobate, lead niobate of lead titanate.

Piezoelectric materials have the following important applications: (*i*) frequency resonators (*ii*) gramophone pick-ups (*iii*) filters (*iv*) ultrasonic flaw detectors (*v*) underwater sonar transducers and (*vi*) air transducers (ear-phones, micro phones, hearing aids, etc.).

11.15 USES OF DIELECTRICS

We have seen that dielectric materials are electrically insulative, yet susceptible to polarization in the presence of an electric field. This polarization phenomenon accounts for the ability of the dielectrics to increase the charge storing capacity of capacitors. Now, we can summarize the main uses of dielectrics as follows: (*i*) piezoelectric and electro-optic devices (*ii*) in capacitors, resistors and strain gauges (*iii*) thermionic valves, radiation detectors, electric devices, dielectric amplifier (*iv*) dielectrics are usually used as ordinary insulators in power cables, signal , electric motors, etc. (*v*) dielectics are used in transformers and various form of switchgear and generators where the dissipation problem of heat is active, and a common way of getting rid of it is to insulate with a transformer oil, *i.e.* mineral oil.

ILLUSTRATIVE EXAMPLES

Example 1

Show that the relative dielectric constant of a barium titanate crystal, which, when inserted in a parallel plate condenser of area 10 mm × 10 mm and distance of separation of 2 mm, gives a capacitance of 10^{-9} F is 2259.

Solution

We have $C = 10^{-9}$ F, $d = 2 \times 10^{-3}$ m, $\varepsilon_0 = 8.854 \times 10^{-12}$

$$A = 10 \times 10^{-3} \times 10 \times 10^{-3} = 10^{-3} \text{ m}^2$$

$$C = \frac{\varepsilon_r \varepsilon_0 A}{d}$$

or

$$\varepsilon_r = \frac{Cd}{\varepsilon_0 A}$$

$$= \frac{10^{-19} \times 0 \times 10^{-3}}{8.854 \times 10^{-12} \times 10^{-4}} = 2259$$

Example 2

Using the given data foe $BaTiO_3$ crystal, calculate the polarization. The shift of the titanium ion from the body centre is 0.06 Å. The oxygen anions of the side faces shift by 0.06 Å, while the oxygen anions of the top and bottom faces shift by 0.08 Å, all in a direction opposite to that of the titanium ion.

Solution

Let us calculate the dipole moments due to the effective number of each type of ion in the unit cell of $BaTiO_3$ as:

(*i*) dipole moment due to two O^{2-} ions on the four side faces

$$\mu_1 = Qd \times 2 \times 2 \times 1.6 \times 10^{-19} \times 0.06 \times 10^{-10} \text{ C-m}$$

(*ii*) dipole moment due to one O^{2-} on top and bottom

$$\mu_2 = Qd = 1 \times 2 \times 1.6 \times 10^{-19} \times 0.08 \times 10^{-10} \text{ C-m}$$

(*iii*) dipole moment due to one Ti^{4+} ion at body centre

$$\mu_3 = Qd = 1 \times 4 \times 1.6 \times 10^{-19} \times 0.06 \times 10^{-10} \text{ C-m}$$

∴ Total dipole moment = 1.02×10^{-29} C-m

Now, the polarization is the total dipole moments per unit volume, Now, ignoring the effect due to barium ions, one obtains

$$P = \frac{\mu}{A} = \frac{1.02 \times 10^{-29}}{4.03 \times (3.98)^2 \times 10^{-30}}$$

$$= 0.16 \text{ C-m}^2$$

Example 3

The atomic weight and density of sulphur are 32 and 2.08 gm/cm^3 respectively. The electronic polarizability of the atom is 3.28×10^{-40} Fm^2. If sulphur solid has cubical symmetry, what will be its relative dielectric constant?

Solution

$$\frac{\varepsilon_r - 1}{\varepsilon_r + 2} = \frac{N\alpha_e}{3\varepsilon_0}$$

$$\therefore \qquad N = \frac{N_A P}{M}, \frac{\varepsilon_r - 1}{\varepsilon_r + 2} = \frac{N_A \alpha_e}{3M\varepsilon_0}$$

$$\therefore \qquad \frac{\varepsilon_r - 1}{\varepsilon_r + 2} = \frac{(6.023 \times 10^{26})(2.08 \times 10^3 \text{ kg / m}^3)(3.28 \times 10^{-40} \text{ F/m}^2)}{3 \times 32(8.854 \times 10^{-12} \text{ F/m})} = 0.483$$

$$\varepsilon_r = \frac{1.966}{0.517} = 3.8$$

Example 4

Calculate the electronic polarizability of argon atom. Given $\varepsilon_r = 1.0024$ at NTP and $N = 2.7 \times 10^{25}$ atoms/m^3 [B.Tech, BE]

Solution

$$P = \varepsilon_0(\varepsilon_r - 1)E$$

$$P = N\alpha_e E$$

$$\therefore \qquad \alpha_e = \frac{\varepsilon_0(\varepsilon_r - 1)}{N} = \frac{(8.85 \times 10^{-12} \text{ F/m})(1.0024 - 1)}{2.7 \times 10^{25}/\text{m}^3}$$

$$= 7.9\ 10^{-40} \text{ F-m}^2$$

Example 5

For ice, relaxation time is given as 18×10^{-6} at 22° C. Calculate the frequency when the real and imaginary parts of the complex dielectric constant will become equal. What will be the phase difference between the current and voltage at this frequency? Which of the two will be leading in phase?

Solution

The real and imaginary parts of dielectric constant will be equal when $\omega\tau = 1$.

$$\omega = \frac{1}{\tau}$$

or
$$f = \frac{1}{2\pi\tau} = \frac{1}{2 \times 3.142 \times 18 \times 10^{-6}\ \text{s}}$$

$$\therefore \quad f = 8.8 \text{ kHz}$$

The phase difference between the current and voltage at 8.8 kHz is given by $\phi = (90 - \delta)$

$$\delta = \tan^{-1}(\varepsilon_r''/\varepsilon_r') = \tan 1 = 45°$$

$$\phi = 45°$$

Example 6

The number of atoms in volume of one cubic metre of hydrogen gas is 9.8×10^{26}. The radius of the hydrogen atom is 0.53. Calculate the polarizability and relative permittivity. [BE]

Solution

$$\alpha_e = 4\pi\varepsilon_0 R^3$$

$$= 4(3.142)\ (8.85 \times 10^{-12}\ \text{F/m})\ \ (0.53 \times 10^{-10}\ \text{m})^3$$

$$= 1.66 \times 10^{-41}\ \text{F-m}^2$$

$$\varepsilon_r = 1 + 4\pi N R^3$$

$$= 1 + 4(3.142)\ (9.8 \times 10^{26}/\text{m}^3)\ (0.53 \times 10^{-10}\ \text{m})^3$$

$$= 1.0018.$$

Example 7

The value of ε_r for glass is 6.75 at frequencies 10^9 Hz. What mechanisms are contributing towards dielectric constant? What percentage may be attributed to ionic polarizability? Given refractive index of glass is 1.5. [B.Tech, BE]

Solution

We have the polarization at 10^9 Hz

$$P = P_i + P_e = \varepsilon_0(\varepsilon_r - 1)E$$

At optical frequencies, we have $\varepsilon_r = n^2$, $P_e = \varepsilon_0(n^2 - 1)E$

and
$$P_i = \varepsilon_0(\varepsilon_r - 1)E - \varepsilon_0(n^2 - 1)E$$

$$= \varepsilon_0(\varepsilon_r - n^2)E$$

% of ionic polarizability

$$= \frac{P_i}{P_e + P_i} \times 100 = \frac{\varepsilon_0(\varepsilon_r - n^2)E}{\varepsilon_0(\varepsilon_r - 1)E} \times 100$$

$$= \frac{\varepsilon_r - n^2}{\varepsilon_r - 1} \times 100 = \frac{6.75 \times 2.25}{6.75 - 1}$$

$$= 78.3\ \%$$

Example 8

A parallel plate capacitor has an area of 7.45×10^{-4} m^2 and the plates are separated by a distance of 2.45×10^{-3} m, across which a potential of 10 V is applied. If a material with delectric constant 6 is introduced between the plates, determine the capacitance, the charge stored on each plate, the dielectric displacement D and polarization. [BE]

Solution

$$C = \frac{\varepsilon_r \varepsilon_0 A}{d}$$

$$A = 7.45 \times 10^{-4}\ \text{m}^2$$

$$d = 2.45 \times 10^{-3}\ \text{m}$$

$$V = 10\ \text{volt}$$

$$\varepsilon_r = 6,\ \varepsilon_0 = 8.85 \times 10^{-12}$$

$$\therefore \quad C = \frac{6 \times 8.85 \times 10^{-12} \times 7.45 \times 10^{-4}}{2.45 \times 10^{-3}}$$

$$= 1.61 \times 10^{-11}\ \text{F}$$

$$Q = CV$$

$$= 1.61 \times 10^{-11} \times 10 = 1.61 \times 10^{-10}\ \text{C}$$

Polarization

$$P = \varepsilon_0(\varepsilon_r - 1)E = \varepsilon_0(\varepsilon_r - 1)\frac{V}{d}$$

$$= 8.85 \times 10^{-12}(6 - 1) \times \frac{10}{2.45 \times 10^3}$$

$$= 1.8 \times 10^{-7}\ \text{C/m}^2$$

Example 9

A parallel plate capacitor consists of two plates each of area 5×10^{-4} m^2 They are separated by a distance 1.5×10^{-3} m and filled with dielectric of relative permeability 6. Calculate the charge on the capacitor if it is connected to a 100 V DC supply. [BE]

Solution

$$Q = CV \quad (i)$$

and
$$C = \frac{K\varepsilon_0 A}{d} \quad (ii)$$

From Eqs. (*i*) and (*ii*)

$$Q = \frac{K\varepsilon_0 AV}{d}$$

$$\varepsilon_0 = 8.85 \times 10^{-12}$$
$$K = 6$$
$$A = 5 \times 10^{-4} \text{ m}^2$$
$$d = 1.5 \times 10^{-3} \text{ m}$$
$$V = 100 \text{ volt}$$

$$\therefore \quad Q = \frac{8.85 \times 10^{-12} \times 6 \times 5 \times 10^{-4} \times 100}{1.5 \times 10^{-3}}$$
$$= 1.77 \times 10^{-9} \text{ C.}$$

REVIEW QUESTIONS

1. What are dielectrics? Derive an expression for internal field for one dimensional dielectric solid. [Visveswaraiah Tech. Univ 2004]
2. Define dielectric polarization. Derive an expression for internal field in case of solid and liquid dielectric by considering one dimensional array of atoms. [Visveswaraiah Tech. Uinv, 2003]
3. (*a*) Discuss different polarization mechanisms. (*b*) Deduce an expression for internal field in case of one dimensional array of atoms in solids. [Visveswaraiah, 2003]
4. Explain the behaviour of dielectrics under static electric fields. Derive the relation between the polarization P and the external field E. [BE]
5. Explain the significance of $\boldsymbol{D}$, $\boldsymbol{E}$ and $\boldsymbol{P}$ in $\boldsymbol{D} = \varepsilon_0\boldsymbol{E} + \boldsymbol{P}$ and derive the relation. [BE]
6. Define dielectric susceptibility and polarizability of a dielectric. Derive the relation connecting the two. [B.Tech]
7. Explain electronic and ionic polarizability and show that the electronic polarizability for a monoatomic gas increases as the size of the atoms becomes larger. [BE]

8. What is ionic polarizability? Explain, why it is found to be rather insensitive to temperature. [BE, B.Tech]
9. Explain, why certain molecules possess permanent electric dipole moment even in the absence of an applied electric field. [B.E.]
10. What is meant by a permanent dipole moment for a polyatomic gas? Show how the potential energy varies with its orientation in the electric field. Write an expression for the orientation polarizability. [BE]
11. Write the expression for total polarization P for a polyatomic gas. How you can determine permanent dipole moment of molecules if P is known at different temperatures? [B. Tech]
12. Define dipole moment and classify dielectric materials on its basis. Give suitable examples in support of your answer. [B.E.]
13. What is dipolar polarizability? Obtain an expression for dipolar polarizability of a dielectric at moderate temperature. [B.E.]
14. What are different mechanisms of polarization in a dielectric? Explain in brief with examples. [BE]
15. What are polar and non-polar molecules? Explain electrical breakdown in solid dielectrics?
16. What is dielectric loss? Show that dielectric loss is given by $\tan\delta = \varepsilon_r''/\varepsilon_r'$. What do you understand by loss tangent as referred to polar dielectrics? [B.E., B.Tech.]
17. What are the various mechanisms that contribute to dielectric loss? Sketch the loss spectrum for a polar material. [BE]
18. Explain the behaviour of a dielectric in an alternating field and write down the expression for energy absorbed per second in dielectric material when it is subjected to an alternating electric field of amplitude E. [BE]
19. What are the main causes for the dielectric losses occurring in rf, IR and visible region of the electromagnetic spectrum. [BE]
20. Describe the characteristic properties of ferroelectric materials. What do you understand by polarization catastrophe? [BE]
21. Explain the meaning and origin of piezoelectricity. Justify the statement. "*All ferroelectric crystals are piezoelectric, but all piezoelectric crystals are not necessarily ferroelectric*".
22. What are ferroelectric materials? How do dielectric constant and polarization change with temperature in a ferroelectric material? [BE]
23. In what respects ferroelectrics differ from ordinary dielectrics? Explain the phenomenon of spontaneous polarization in ferroelectrics. [B. Tech]
24. What are ferroelectric materials? Sketch the variation of spontaneous polarization with temperature in ferroelectrics. Do ferroelectrics exhibit domains? [BE]
25. What are direct and inverse piezoelectric effects? Explain their importance and practical applications: [BE]
26. Explain: (*a*) How is dielectric loss utilized in cooking food in microwave oven? (*b*) How do a gas lighter and quartz watch utilize piezoelectric phenomenon in their functioning. [BE]

27. Write short notes on

(*i*) Dielectric materials
(*ii*) Polar and non-polar dielectrics
(*iii*) Ferroelectric materials
(*iv*) Characteristics of ferroelectric $BaTiO_3$

PROBLEMS

1. A parallel plate capacitor of area 650 mm^2 and a plate separation of 4 mm has a charge of 2×10^{-10} C. Show that the resultant voltage across the capacitor when a material of dielectric constant 3.5 is introduced between the plates is 39.7 volt.

$$\left[\text{Hint: } C = \frac{\varepsilon_0 \varepsilon_r A}{d} \text{ and } C = Q/V,\ V = \frac{Qd}{\varepsilon_0 \varepsilon_r A}\right.$$

$$\left. = \frac{2 \times 10^{-10} \times 4 \times 10^{-3}}{8.85 \times 10^{-12} \times 3.5 \times 650 \times 10^{-6}} = 39.7 \text{ V}\right]$$

2. A parallel plate capacitor has circular plates of radius 8 cm and plate separation 1 mm. If the potential difference between the plates is 100 volts, show that the charges on the plates are $+1.78 \times 10^{-8}$ C and -1.78×10^{-8} C respectively.

$$\left[\text{Hint: } C = \frac{\varepsilon_0 A}{d} = \frac{8.85 \times 10^{-12} \times \pi \times (0.08)^2}{10^{-3}} \quad Q = CV\right.$$

$$\left. = \frac{8.85 \times 10^{-12} \times \pi \times (0.08)^2}{10^{-3}} \times 100 = 1.78 \times 10^{-8} \text{ C}\right]$$

Thus the charges on the plates are $+1.78 \times 10^{-8}$ C and -1.78×10^{-8} C respectively.

3. A parallel plate capacitor consists of two plates each of area 5×10^{-4} m^2 separated by a distance of 1.5×10^{-3} m and filled with a dielectric of relative permeability 6. The capacitor is connected to a 100 V DC source. Show that the charge on the capacitor is 1.77×10^{-9} C.

4. A parallel plate capacitor of area 2 sq.m. with a medium of relative permeability 7 is charged to a potential of 100 volts. Calculate the capacitance and the energy stored in the condenser. Given the distance between the plates = 10^{-4} m.

$$\left[\text{Hint: } C = \frac{\varepsilon_0 \varepsilon_r A}{d} = \frac{8.85 \times 10^{-12} \times 7 \times 2}{10^{-4}} = 1.24 \times 10^{-6} \text{ F} = 1.24\ \mu\text{F}\right.$$

$$\left. E = \frac{1}{2} CV^2 = \frac{1}{2} \times 123.9 \times 10^{-8} \times (100)^2 = 61.95 \times 10^{-4} \text{ J}\right]$$

SHORT ANSWER QUESTIONS

1. What are dielectric materials?

Ans. These are insulators which possess high electrical resistivity.

2. What is the significance of relative permittivity, ε_r?

Ans. ε_r is an important characteristic of a of dielectric and it is a measure of polarization of dielectric subjected to electric field.

3. What polarization of dielectrics signifies?

Ans. Polarization results in a dielectric due to inducement of charges and accounts for the ability of dielectrics to increase the charge storing capability of capacitors.

4. What is dielectric susceptibility?

Ans. The ease with which a dielectric is polarized in the presence of an electric field is called dielectric susceptibility.

5. How will you define polarizability?

Ans. The polarizability ***P*** is defined as the dipole moment per unit volume of the material and is related to the electric susceptibility, χ_e, as

$$\boldsymbol{P} = \varepsilon_0 \chi_e \boldsymbol{E}$$

6. What do you understand by electric dipole and dipole moment?

Ans. An electric dipole is said to exist when there is a net spatial separation of positively and negatively charged entities. In a molecule, an electric dipole is said to exist due to inherent asymmetry of molecule or may be induced in it due to the action of an external field.

7. What is dipole moment?

Ans. A dipole moment is the product of either charge (q) on the dipole and the distance of separation (d) between the charges, *i.e.*

$$\mu = q\,d.$$

8. What is the effect of induced dipoles?

Ans. Bulk polarization of the dielectric.

9. What is atomic polarizability?

Ans. The atomic polarizability (α) is the dipole moment (P) per unit electric field at an atom and is given by the relation

$$P = \alpha E_{\text{loc}}$$

10. What does polarizability and dielectric constant of the material signifies? How these two parameters are related with each other?

Ans. The polarizability is an atomic property whereas the dielectric constant is a macroscopic property of the material. The two parameters are related to each other by the Clausius-Mossotti relation given as

$$\frac{\varepsilon_r - 1}{\varepsilon_r + 2} = \frac{N\alpha}{3\varepsilon_0}$$

11. What are the main contributions to the net polarizability of dielectric material?

Ans.

(*i*) *Electronic polarization*: This arises due to shift of electron cloud with respect to the nucleus in the atom subjected to an electric field:

$$P_e = N\mu_{\text{ind}} = N\alpha_e E$$

(*ii*) *Ionic Polarization*: This occurs due to shift of positive and negative ions from their equilibrium positions

$$P_i = N\alpha_i E$$

(*iii*) *Dipolar or orientation polarization*: This consists in partial alignment of molecular dipoles with the applied field E.

$$P_0 = \frac{N\mu^2}{3kT}E$$

At low frequencies of the applied electric field, all the three contributions are present. As the frequency increases, the dipolar contribution disappears first followed by ionic and electronic contributions at higher frequencies. At optical frequencies, the polarization is mainly due to electronic contribution.

12. Write the local field within a dielectric.

Ans. $$E_i = E + \frac{\gamma P}{\varepsilon_0}$$

In case of cubic symmetry

$$E_i = E + \frac{P}{3\varepsilon_0} \text{ (Lorentz field)}$$

13. What is dielectric loss?

Ans. Electric power is absorbed by the dielectric from the source whenever a particular polarization mechanism tends to become ineffective, *i.e.* disappears. The maximum power absorption will occur at the relaxation frequency and known as *dielectric loss.* Power dissipation per unit volume of dielectric is given by

$$P_L = \omega\, \varepsilon_0\, \varepsilon_r E^2$$

and dielectric loss in terms of loss angle δ as

$$\tan\delta = \frac{\varepsilon''_r}{\varepsilon'_r}$$

14. What is dielectric strength?

Ans. This is defined as the maximum electric field which can be applied across a dielectric without causing an electrical conduction in it.

15. What is electrostriction?

Ans. This refers to the change in dimensions of a dielectric subjected to an electric field.

16. What are ferroelectric materials?

Ans. These are the materials which exhibit spontaneous polarization, *i.e.* polarization in the absence of an applied field. These are analogous to ferromagnetic materials. These are basically *non-linear* dielectrics, in which polarization does not vary linearly with the applied field. These substances, exhibit Curie temperature, domains and hysteresis. The examples are barium titanate, the Rochelle salt, etc.

17. What is piezoelectric effect?

Ans. The polarization of a dielectric in response to mechanical deformation is called direct piezoelectric effect. Mechanical deformation of a dielectric caused by an external voltage is called the inverse piezoelectric effect. The absence of the centre of inversion is the pre-requisite for the occurrence of piezoelectricity. All ferroelectric crystals are piezoelectric, but all piezoelectric crystals are not necessarily ferroelectric. Example of piezoelectric crystal is quartz.

OBJECTIVE QUESTIONS

1. A forbidden energy gap, E_g is very________in dielectrics and excitation of electrons from the normally full valence band to empty conduction band is________under ordinary conditions. [large, not possible]
2. The relative permeability, ε_r is a ________quantity which is always greater than ________in case of dielectrics. [dimensionless, unity]
3. Basing on dipole moment, dielectrics can be broadly classified into two major groups________and________dielectrics. [polar, non-polar]
4. Ionic polarization in ionic crystals is brought about by the elastic displacement of________and negative ions from their equilibrium positions. [positive]
5. The orientation polarization is characteristic of ________dielectrics which consists of molecules having permanent________. [polar, dipole moment]
6. The average time taken by the dipoles to reorient in the field direction is known as the________. [relaxation time]
7. The mechanical deformation of a dielectric induced by an electric field is called ________. [electrostriction]
8. Ferroelectrics are anisotropic crystals which exhibit________ polarization. [spontaneous]

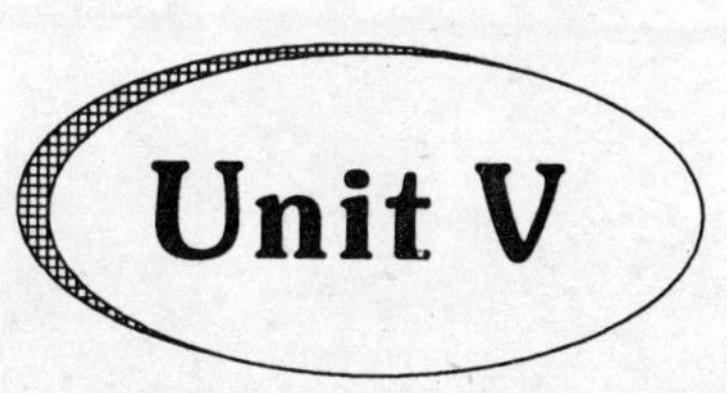

Modern Engineering Materials

12

Modern Engineering Materials

12.1 INTRODUCTION

Today, we are living in the materials age having numerous varieties of materials. In the past few years, several new engineering materials have been discovered which find major application in industries. In general, we can classify the materials as metals and alloys, ceramics, glasses, glass ceramics, composites, polymers and electronic (semiconductor) materials. Today the old statements that 'metals are crystalline', 'polymers are insulators' and ceramics are brittle have lost their significance noncrystalline or amorphous metals also called as *metallic glasses* have been developed and promise bright technological potential. Extensive research work is in progress to produce *bulk metallic glasses*. Today we have *conductive polymers* which exhibit unique properties and applications. Smart or *intelligent materials* form a group of new and state-of-the-art materials now being developed that will have a significant influence on many of our technologies. Nanostructured materials is a new group of materials whose structural elements clusters, crystallites or molecules have dimensions in the range of 1-100 nm. These small groups of atoms, in general, go by different names such as nanoparticles, nanocrystals, quantum dots and quantum boxes. Quantum dot represent the ultimate reduction in the dimensionality of semiconductor devices. A revolutionary new class of semiconductor electronics based on the spin degree of freedom called as *spintronics* could be developed. Very recently a new form of matter called fermionic condensate matter have been created which could lead to the next generation of superconductors for use in electricity generation, more efficient trains and countless other applications.

In spite of the tremendous progress that has been made in the field of material science within the past few years, there still remain technological challenges, including the development of more sophisticated and specialized materials as well as the impact of material production on environment.

12.2 METALLIC GLASSES OR METGLASSES

These are the newly developed metallic materials which exhibit a disordered atomic structure and closely resemble that of glass rather than the conventional crystalline structure of metals. These are formed from reasonably inexpensive metals like, Fe, Ni, Co and have number of superior properties for technical applications compared to those progressed by the crystalline solids, e.g. the filaments of metglasses are stronger and stiffer and hence offer potential application as low-cost high performance structural reinforcement elements. Unlike the covalent and/or ionic bonds of ceramic glasses, the bonding of metglasses is metallic in nature which confers on material ductility and cohesiveness similar to crystalline materials.

The glassy structure of metglasses is highly metastable. These are metal alloys that are amorphous. That is why metglasses do not have a *long range aromic order*. The major advantages of these metglasses are generally, homogeneous in composition, strong and superior corrosion resistance. Hence met glasses offer an economical and more efficient alternative to the crystalline solids. Most of these possibilities are still under development stages and it is likely that successful commercial exploitation would increase the production of amorphous materials. Metglasses are produced when molten metals or alloys are cooled rapidly (at a cooling rate of $10^6 - 10^8$ Ks^{-1}). During this process of solidification, the atoms do not find enough time or energy to rearrange for crystal nucleation. Clearly, the liquid upon attaining the glass transition temperature (T_g) solidifies as metallic glass. This technique is called as *rapid solidification technique*. We may note that upon heating, metallic glasses show a reversible glass-liquid at glass transition temperature (T_g).

Metallic glasses show very high strength levels (approaching that of dislocation free single crystals or hard drown piano wire, e.g. $Fe_{80}B_{20}$ metglass has a tensile yield strength of 3.63 GPa), good ductility in bending or compression (but not in tension), excellent corrosion resistance (due to absence of grain boundaries and formation of resistance surface film), high electrical resistivity with a small variation with temperature, superconductivity and radiation stability. However, it is not necessary that these attributes are present simultaneously in a given metglass. Metglasses suffer from the major disadvantage of poor fracture and fatigue behaviour.

Types of Metallic Glasses

Based upon their base material used for preparation, metglasses are placed under two categories:

(*i*) **Metal-Metal Glasses:** Ni-Nb, Mg-Zn and Cu-Zr are few examples of metal-metal glasses.

(*ii*) **Metal-Metalloid Glasses:** Transition metals like Fe, Co, Ni, and metalloids like B, Si, C and P are uses for metal-metalloid glasses.

A comparison of traditional metals, traditional glasses and metallic glasses is presented in Table 12.1.

TABLE 12.1: A COMPARISON OF PROPERTIES OF TRADITIONAL METALS, TRADITIONAL GLASSES AND METGLASSES

Properties	*Traditional Metals*	*Traditional Glasses*	*Metglasses*
Structure	Crystalline	Amorphous	Amorphous
Bonding	Metallic	Ionic/covalent	Metallic
Hardness	Low to high	Very high	Very high
Tensile strength	Low to high	Low	High to very high
Workability	Good, ductile	Poor, Brittle	Good, ductile
Thermal conducting	Very good	Poor	Very good
Electrical resistivity	Very Low	High	High
Optical Property	Opaque	Transparent to opaque	Opaque
Corrosion resistance	Poor to good	very good	very good

Preparation

For the production of metallic glasses or metglasses, various rapid cooling techniques e.g. spraying, spinning and laser deposition are used.

Figure 12.1 shows the melt spinning process technique used in the preparation of metglasses.

In the melt spinning process, there is a spinning disc made of copper. To prepare a metglass of a particular type, one takes a suitable combination of metal-metalloid alloy in their stoichiometric ratio in a refractory tube having fine nozzle at its bottom. The nozzle side of the tube is placed just over the spinning disc.

The alloy gets melted with the help of an induction heater attached to the refractory tube and this melt is kept above its melting point till it gets transformed into a homogeneous mixture. An inert gas, e.g. helium is allowed to show through the tube containing the homogeneous mixture.

As a consequence, the melt gets ejected through the nozzle and the ejected melt is cooled at a faster rate with the help of the spinning cooled copper disc. One can increase the ejection rate by increasing the pressure of the inert gas. Thus, a glass alloy ribbon starts getting formed over the spinning disc. One may very the thickness of the glass ribbon by increasing or decreasing the speed of the spinning disc. One may vary the thickness of glassy ribbon by increasing or decreasing the speed of the spinning disc.

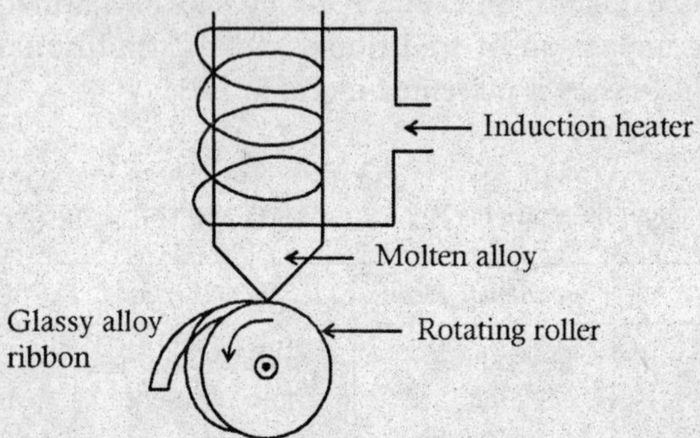

Fig 12.1: Melt spinning process technique involved in the preparation of metglasses

The following other techniques are also used for the production of metallic glasses.

(*i*) **Twin Roller System:** To produce metallic glasses by this technique, a molten alloy is passed through two rollers rotating in opposite directions.

(*ii*) **Melt Extraction System:** In this technique to produce met glasses, a fast moving roller sweeps of molten droplet into a strip from a solid rod.

Properties of Metglasses

Metallic glasses exhibit very high strength levels (nearly twice that of stainless steel) approaching that of dislocation free single crystals or hard drawn plano wire, e.g. $Fe_{80}B_{20}$ metallic glass has a tensile yield strength of 3.63 G Pa.

Metglasses have good ductility in bending or compression (but not in tension). They have excellent corrosion resistance (due to absence of grain boundaries and formation of a resistant surface film). They are malleable, brittle and opaque. The hardness of metglasses is very high. Metglasses have high electrical resistivity with a small variation with temperature, superconductivity and radiation stability. Metglasses have high elasticity, *i.e.* the yield strength is high. Metglasses are free from crystalline defects, e.g. point defects, dislocation, stacking faults, etc.

Metglasses are soft magnetic materials. This is why, easy magnetization and demagnetization is possible. Magnetically soft metallic glasses have very narrow hysteresis loop (Fig. 12.2).

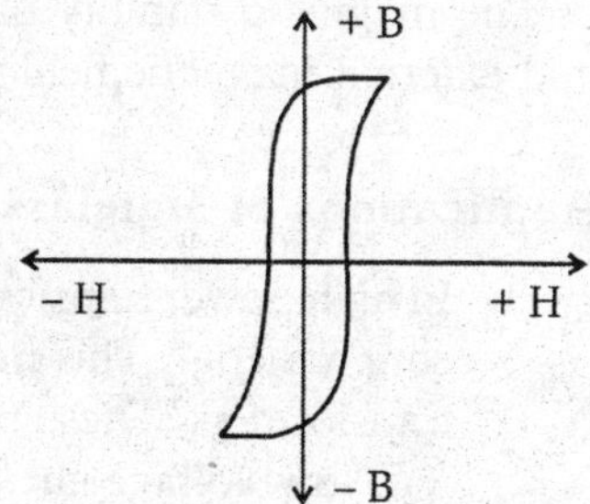

Fig. 12.2: Hysteresis loop of a ferromagnetic metallic glass

Metglasses are extremely strong solids. They have E/σ_y (where E is Young's modulus and σ_y is the yield strength) ratio close to 50 which is matched by very few material, e.g. whiskers of Fe and SiC have values of 15.5 and 33.3 respectively.

Metglasses based on the first transition series of elements have significantly higher strengths than those from second and third transition series. The introduction of early transition elements like Cr in Fe-based alloys improves the strength while late transition series elements like Ni do not have any perceptible effect.

The correlation between the electrical resistance ρ and the temperature coefficient of resistivity $\alpha = \rho^{-1}(\partial\rho/\partial T)$ has been studied in detail for amorphous alloys. The following conclusions have been drawn.

(*i*) At high resistivity ($\rho > 100$ μΩ-cm) the resistance is insensitive to the atomic and electronic structure of the alloy.

(*ii*) It $\rho > 150$ μΩ-cm the alloy exhibits negative temperature coefficient (NTC) of resistance ($\alpha < 0$).

(*iii*) Irrespective of the magnitude of residual resistivity of the alloy at low temperature all alloys tend towards a temperature-independent resistivity of 150 to 200 μΩ-cm at high temperatures. The limiting resistivity at high temperature is called the 'saturation resistivity' (ρ^*). The explanation of the electrical resistivity of metallic glasses has not been very satisfactory.

The dynamic and static magnetic properties of amorphous metglass ribbons have been extensively studied. These properties are very similar to those observed for these materials in the crystalline form and hence are easily interpreted in turms of very simple models based on domain configurations. The energy of domain wall per unit area (σ_w) is given by

$$\rho_w = 2\pi(JKS^2/a)^{1/2}$$

where J is exchange constant, K is the anisotropy constant, S is the spin and a is the distance between neighbouring spins.

The amorphous metglasses have somewhat superior magnetic properties compared to their crystalline counterparts.

The magnetic stability of the amorphous alloys against temperature and external magnetic field fluctuating is still not well known.

Applications of Metglasses

1. In high power transformers, metallic glasses are used as transformer core material. This improve the efficiency of power distribution in transformers. These transformers can convert high voltage current into low voltage current and can be used for domestic appliances (120 V and 240 V).

   In comparison to conventional transformers, the power transformers with metallic glasses as core is useful in reducing the core loss. Moreover, the size and weight is also very less. In the later case the loss is less due to the fact that its ferromagnetic property with hysteresis loss and high electrical resistivity property.
2. Metglasses are used in making cryothermometers, magnetoresistance sensors and computer memories due to their high electrical resistivity and nearly zero temperature coefficient of resistance.
3. The mechanical properties of metglasses makes them suitable for high strength control cables, pressure vessels, inertial energy storage devices such as flywheels, cutting tools, spring components, extrusion dies, casting, dies, abrasive wheels mechanical transmission belts, torque transmission tubes and reinforcing belts in rubber tyre. Metglasses also find use as reinforcement in concrete and refractories.
4. Magnetic properties of metglasses are not affected by irradiation and therefore they are used in making container for nuclear waste disposal. Based on the magnetic properties metglasses finds major applications perhaps in power devices, particularly distribution transformers. The other area is in information handling equipment such as tape recorder heads, where the initial permeability is the most important characteristic requirement. Another area of application includes transducers and magnetic delay lines. The other applications are in electronic devices, material for magnetic shielding, transformers, leakage current, over current alarm, small power watt meter, phase meters, speaker diaphragms, magnetic amplifiers, chokes and electrical motors.
5. Metglasses are used for magnets for confinement of hot plasma for fusion reactors, magnetic levitation trains.
6. The excellent corrosion resistance property of metglasses make these materials suitable.
7. Due to their high electrical resistivity and nearly zero temperature coefficient of resistance, metglasses are used in making cryothermometers, magnetoresistance sensors and computer memories.

8. Metglasses can also be used for making watch cases to replace nickel (Ni) and other metals which may cause allergic reactions.
9. These materials with a very small and often zero value of temperature coefficient of resistivity can be used as resistance standards and also as magnetoresistance sensors for detecting the presence of magnetic fields, notably in bubble memory devices. Other uses include heating elements (low temperature heating element instead of high temperature Nichrome) required for products such as TVs, radios, refrigerators, inductance, thermometers, toasters, resistance thermometers, electron sources, electrical fuses, temperature alarms, batteries, sensors, superconductors, timing devices, delay lines, flash bulbs, etc.
10. The excellent corrosion resistance property of metglasses makes these materials ideal for cutting and in surgical instruments. They can also be used as prosthetic material for implantation in the human body.

12.3 SHAPE MEMORY ALLOYS (SMA's)

Smart or Intelligent Materials

Smart or intelligent materials form a group of new and state-of-art materials now being developed that will have a significant influence on many of our technologies. The adjective 'smart' implies that these materials are able to sense changes in their environments and then respond to these changes in predetermined manners—traits that are also found in living organisms. In addition, the concept of smart materials is being extended to rather sophisticated systems that consist of both smart and traditional materials.

The field of smart materials attempts to combine the sensor (that detects an input signal), actuator (that performs a responsive and adaptive function) and the control circuit on as one integrated unit. Actuators may be called upon to change shape, position, natural frequency, or mechanical characteristics in response to changes in temperature, electric fields, and/or magnetic fields.

Usually, four types of materials are commonly used for actuators: shape memory alloys, piezoelectric ceramics, magnetostrictive materials, and electrorheological/magnetorheological fluids. Shape memory alloys are metals that, after having been deformed, revert back to their original shapes when temperature is changed. Piezoelectric ceramics expand and contract in response to an applied electric field (or voltage); conversely, these materials also generate an electric field when their dimensions are altered. The behaviour of magnetostrictive materials is analogous to that of piezoelectric ceramic materials, except that they are responsive to magnetic field. Also, electrorheological and magnetorheological fluids are liquids that experience dramatic changes in viscosity upon the application of electric and magnetic field, respectively.

The combined system of sensor, actuator and control circuit on as one IC unit, emulates a biological system (Fig. 12.3)

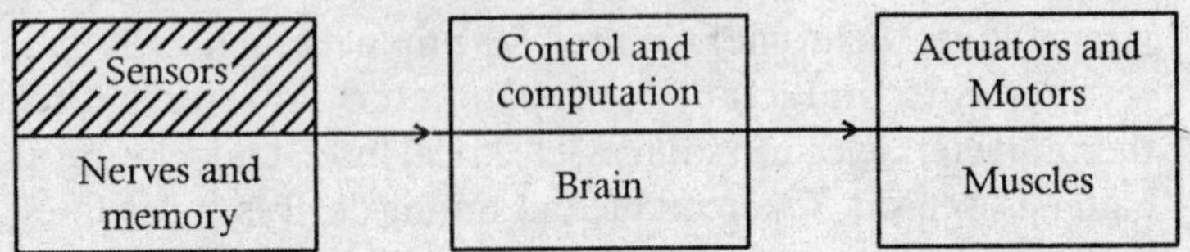

Fig. 12.3: Integrated sensor-actuator systems with controller are analogous to biological systems

These are known as smart sensors, microsystem technology (MST) or microelectromechanical systems (MEMS). Materials/devices employed as sensors include optical fibres, piezoelectric materials (including some polymers), and MEMS

For example, one type of smart system is used in helicopters to reduce aero-dynamic cockpit noise that is created by the rotating rotor blades. Piezoelectric sensors inserted into the blades, monitor blade stresses and deformations; feedback signals from these sensors are fed into a computer controlled adaptive device, which generates noise cancelling antidose.

MEMS devices are small in size, light weight, low cost, reliable with large batch fabrication technology. They generally consist of sensors that gather environmental information such as pressure, temperature, acceleration, etc., integrated electronics to process the data collected and actuators to influence and control the environment in the desired manner.

The MEMS technology involves a large number of materials. Silicon forms the backbone of these systems also due to its excellent mechanical properties as well as mature micro-fabrication technology including lithography, etching, and bonding. Other materials having piezoelectric, piezoresistive, ferroelectric and other properties are widely used for sensing and actuating functions in conjunction with silicon. The field of MEMS is expected to touch all aspects of our lives during this decade with revolution in aviation, pollution control, and industrial processes.

Smart materials can respond to external stimuli, e.g. pressure, temperature, optical, electric field, magnetic field, moisture, etc.

Smart materials offer numerous possible applications in medical engineering, automobile engineering, civil engineering, aerospace and consumer electronics.

Among several smart materials, shape memory alloys, fibre optic sensors and piezoelectric materials are currently receiving wide attention.

Shape Memory Alloys (SMAs)

These are metals which exhibit two unique properties: (*i*) *Shape Memory Effect* (SME) and (*ii*) *Pseudo Elasticity* or *Super Elasticity* (SE).

Sharp memory alloys are a unique class of alloys or materials, which remember their shape even after severe deformation, i.e. when a sharp memory alloy is once deformed in the cold shape (martensite) these materials will stay deformed until heated, where, upon heating they will spontaneously return to their original predetermined hot shape (austenite).

It is reported that the structural changes at the atomic level contributes to their unique properties of the materials.

Type of Shape Memory Alloys

Shape memory alloys can be placed under two categories:

(*i*) **One Way Shape Memory Alloys:** The materials which exhibit shape memory effect, (i.e. taking their own shape) only upon heating are termed as one way shape memory alloys.

(*ii*) **Two Way Shape Memory Alloys:** There are some materials which exhibit shape memory effect both during heating and cooling. Such materials are termed as two way shape memory alloys.

Crystal Structure of Shape Memory Alloys

These materials have two distinct types of crystal structure. At any particular instant the crystal structure or the phase of the SMA is determined by the temperature and internal stresses.

At high temperature, the phase of SMA is called *austenite* phase (microscopically they possess small *plate let structure*) and at low temperature the phase is called the martensite (microscopically they possess needle like structure). In both the cases the material remain in solid form.

It possesses a highly symmetric structure such as the one in cubic system in austenite phase. However, in the martensite phase the structure is of low symmetry such as monoclinic. The basis for shape memory effect is that, these materials can very easily transform to and from martensite.

One can broadly classify the shape memory alloys into : *Thermal induced shape memory*, *Stress induced shape memory* and ferromagnetic shape memory materials.

Temperature Induced Transformation

The phase transformation in smart material alloys do not take place at a particular temperature but over a range of temperature (Fig. 12.4). This transformation is named as the *temperature induced transformation.*

The temperature induced transformation in SMA's is characterised by four temperatures, M_s and M_f during cooling and A_s and A_f during heating (Fig. 12.4). M_s and M_f in Fig. 12.4 indicates the temperatures at which the transformation from the parent phase austenite into martensite starts and ends respectively. Similarly, A_s and A_f in Fig. 12.4 indicates the temperatures at which the reverse transformation from martensite to austenite starts and ends upon heating. The overall transformation has a temperature range of 10-15° C depending on the chemical composition of the alloy. It is observed that this overall transformation exhibit hysteresis in which, the transformations on heating and cooling do not develop. However, the hysteresis H depends on the composition of the alloy system.

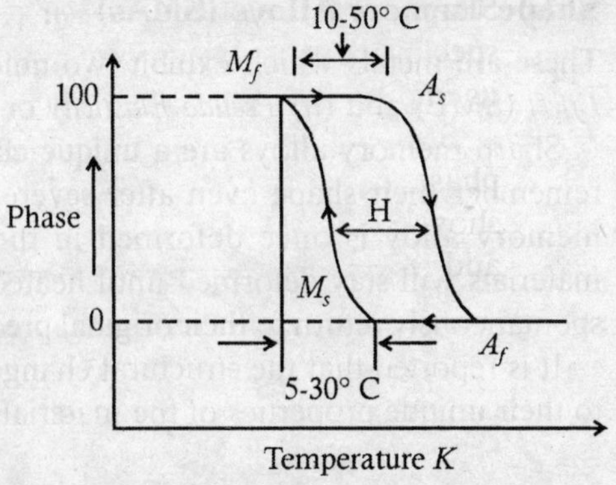

Fig. 12.4: Temperature inducted transformation in SMAs. Temperature M_s and M_f during cooling and A_s and A_f during heating

Stress Induced Transformation

When a stress is applied over the shape memory alloys, there is also a possibility for phase trans-formation.

The stress induced transformation in shape memory alloys takes place at a constant temperature and at a temperature above the A_f temperature, the martensite phase can be induced by applying stress over the austenite phase (Fig. 12.5).

The austenite phase undergoes a large elastic deformation when stressed above a certain value. Stressed beyond the elastic limit causes permanent plastic strains. When the stress is removed, the material almost completely recovers to the parent austenite phase at a much lower value of stress (Fig. 12.5).

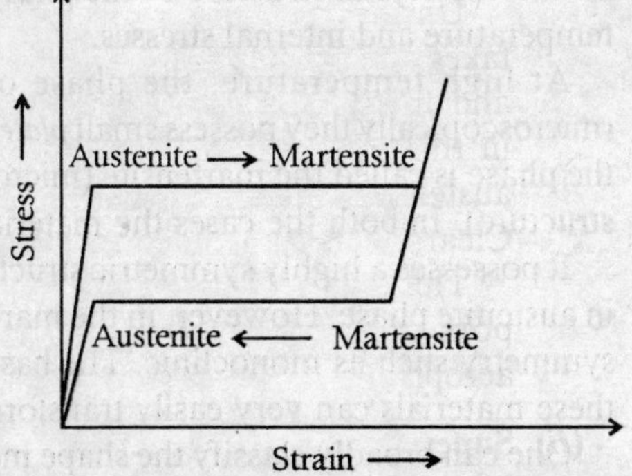

Fig. 12.5: Stress induced transformation in SMA

Functional Properties of SMA

There are two important characteristic properties of SMAs. These are:

(i) Shape Memory Effect (SME): This is the phenomenon in which a specimen apparently deformed at lower temperature and regains to its undeformed original shape when heated to higher temperature.

SME is a consequence of a crystallographic reversible martensitic phase transformation taking place in the solid state. Figure 12.6 shows schematically the crystallographic formation of martensite and reversion to austenite on heating.

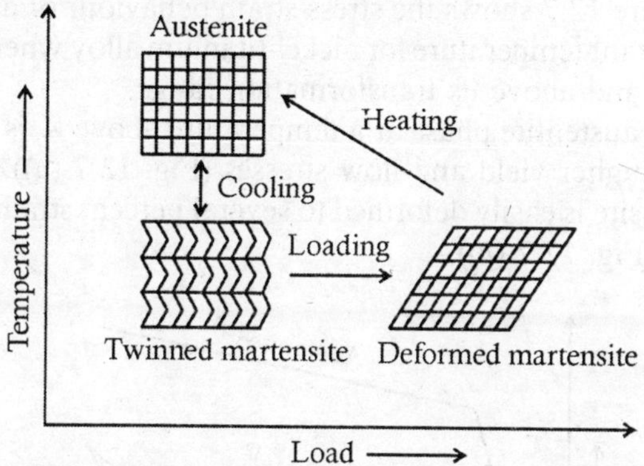

Fig. 12.6: Shape memory effect in SMAs

From the Fig. 12.6, we note that the high temperature austenitic structure undergoes twinning as the temperature is lowered. This twinned structure having microscopically needle like structure is termed *martensite*. This phase of SMA is relatively soft and easily deformed phase which exists at lower temperature.

Under deformation, *i.e.* on applying external stress, this phase takes a particular shape termed as deformed (twinned) martensite and undergoes a large elastic strain during this process. When heated in this condition, the deformed martensite returns to the stable austenite structure and in the process recovers the elastic strain. Clearly, the shape memory phenomenon is observed.

This shape memory effect is being implemented in making coffee pots, thermostats, vascular stents and hydraulic fittings for aeroplanes.

(ii) Superelasticity (SE) or Pseudoelasticity: This refers to the ability of shape memory alloys to return to its original shape upon unloading after a substantial deformation. This superelastic effect is based on stress induced martensitic (SIM) transformation. This superelasticity in shape memory alloys occurs at a constant temperature when the alloy is completely composed of austenite phase (temperature is greater than A_f temperature).

The stress, *i.e.* load on shape memory alloy is increased till the austenite phase gets transformed into a martensite phase simply due to loading. However, the martensite begins to transform back to austentite as soon as the load is decreased, since the temperature is still above A_f. As a result, it regains its original shape.

This effect, which causes material to be extremely elastic is called superelastic effect or pseudoelasticity. This effect is nonlinear and also temperature and strain dependent. Superelasticity is a mechanical type of behaviour of SMAs.

Figure 12.7 shows the stress-strain behaviour of different phases at constant temperature for nickel-titanium alloy when tested below, within and above its transformation range.

The austentite phase at a temperature above A_f is found to have much higher yield and flow stresses (Fig. 12.7 (*a*)). However, the martensite is easily deformed to several percent strain at quite a low stress (Fig. 12.7 (*b*)).

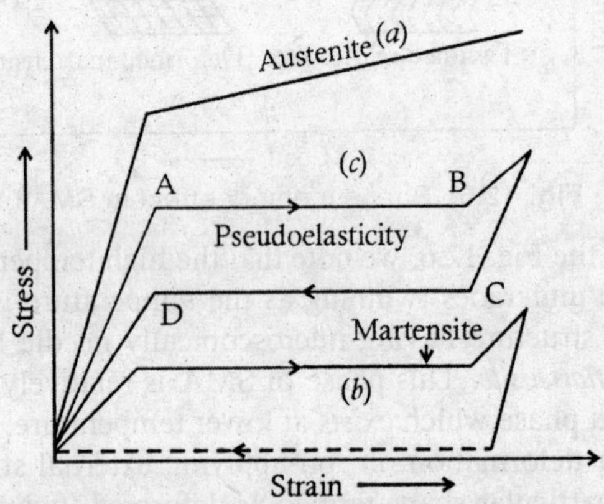

Fig. 12.7: Stress-strain behaviour of different shapes of shape memory alloys at constant temperature

After removing the stress, this martensite phase return back to the austenite phase upon heating (shown in Fig. 12.7 by dashed line). However, no such shape recovering is found in the austenite phase upon straining and heating, because no phase change takes place.

The above mentioned two behaviours of SMAs are called as thermomechanical behaviour. An interesting stress-strain behaviour is depicted by Fig. 12.7 (c) where the material is completely composed of austenite phase ($t > A_f$). The martensite phase at this temperature can be stressed induced (curve *AB* in Fig. 12.7). Again, upon unloading, the material return to its original shape (shown by curve *CD* in Fig. 12.7) austenite without applying heat. This is a pure mechanical behaviour and is termed as *superelasticity* (*SE*) or *pseudoelasticity*.

SE behaviour of SMAs is applied in making eye glass frames, cellular phone antennae, medical tools and orthodontic arches.

Important Smart Material Alloys Used Commercially

There are wide variety of alloys known to exhibit SME, but there are only few alloys which find commercial alloys. The most common smart material alloys is nickel titanium (Ni-Ti). This smart material alloy of Ni-Ti has been reported to be the most useful of all SMAs. The generic name for Ni-Ti alloy family is *Nitinol*, which stands for Nickel-Titanium Naval Ordnance Laboratory.

Cu-As-Ni, Cu-Zn-Al, and Co-Ni-Al are other shape memory alloys. In addition, there is also a subfamily of SMAs called as ferromagnetic shape memory alloys (FSMAs) which are magnetically driven. In FSMAs, the frequency of phase transformation is high as compared to the cycles of heating and cooling. Only Ni-Mn-Ga ternary alloy has shown satisfactory performance in this category.

Mn-Cu, Ni-Ti-Hf, Ti-Pd-Ni, and In-Tl, etc. are other examples of such SMAs.

Applications of SMAs

The crystal properties of SMAs makes them suitable for use in a variety of fields. The important applications of SMAs are particularly in the following fields:

(*i*) **Aircraft and Space Industry:** SMAs can be used in this industry as fine tuned helicopter blades and as lock ring electrical connectors. In addition, they also find applications in antenna opening high damping parts, triggering devices and hubble telescope.

(*ii*) **Automobile Industry:** SMAs are used in this industry in making spring actuators, clutch systems, thermostats, oil pressure control unit and high pressure sealing plugs.

(*iii*) **Medical Field:** SMAs find largest commercial applications in the field of bioengineering and medical science. The potential applications in these fields are:

(*a*) SMAs are used as dental arch wires. These wires helps the misaligned teeth to return gradually to their original shape exerting a small and almost constant force on the misaligned teeth. The use of this wire is more efficient in faster movement of the teeth and less discomfort.

(*b*) SMAs also find use as blood clot filter

(*c*) SMAs make tweezers to remove foreign objects through small incisions.

(*d*) Nitinol needle wire localizers are used to locate and mark breast tumours. This helps subsequent more exact and less invasive.

(*e*) SMAs find use in designing microsurgical instruments and microgrippers, etc.

(*f*) SMAs find use as guide wires for catheters through blood vessels.

(*iv*) Consumer Products

(*a*) These are used in making frames for eye glass, which offers improved comfort and flexibility. They also find use in phone antenna.

(*b*) Ni-Ti springs in coffee pots can be trained to open a valve and helps to release hot water at proper temperature.

(*c*) Nitinol find use in robotic actuators and micromanipulators to stimulate human muscle motion.

(*d*) Using SMAs shape memory based toys and ornamental goods have been fabricated.

(*e*) SMAs can be used as couplers and fasteners.

(*f*) For industrial facilities, SMAs based fixed safety valves have been developed.

(*g*) Temperature sensitive SMA valves are widely used to instantly restrict water flow in sinks or showers.

(*h*) SMAs can be used to design safety valves. These valves provide emergency shutdown of process control lines that handle flammable and toxic fluid and gases.

Merits and Demerits of SMAs

SMAs are advantageous biocompatibility and finds application in diverse fields. They have good mechanical properties.

The major disadvantage of SMAs is that they are highly expensive to manufacture and machine it.

12.4 NON LINEAR MATERIALS

Active Materials

Today, the smallest feature size in production systems is about 250 nanometers—the smallest feature size in computer chips. Since atoms are an angstrom or so across and carbon nanotubes have a diameter as small as 0.7 nanometers, atomically precise molecular machines can be smaller than current MEMS devices by two to three orders of magnitude in each dimension, or six to nine orders of magnitude smaller in volume (and mass). For example, the size of the kinesin motor, which transports material in cells, is 12 nm. Han computationally demonstrated that molecular gears fashioned from single-walled carbon nanotubes with benzyne teeth should operate well at 50-100 gigahertz. These gears are about two nanometers across. Han computationally demonstrated cooling the gears with an inert atmosphere. Srivastava simulated powering the gears using alternating electric fields generated by a single simulated laser. In this case, charges were added to opposite sides of the tube to form a dipole.

To make active materials, a material might be filled with nano-scale sensors, computers, and actuators so the material can probe its environment, compute a response, and act. Although this document is concerned with relatively simple artificial systems, living tissue may be thought of as an active material. Living tissue is filled with protein machines which gives living tissue properties (adaptability, growth, self-repair, etc.) unimaginable in conventional materials.

Nonlinear Optical Materials (NLO)

Photons can be used to generate and transmit data. Though it is beyond the scope of this text, there are also devices for storing data in an optical fashion. This emerging area of using photons to replace what was previously performed by electrons is called *photonics*. The integration of photonic and electronic devices is called *optoelectronics*. In this section, we describe how the optical properties of materials can be used in yet another manner to create devices that can act as switches, multiplexers, and mirrors and even change the wavelength of light. The materials that exhibit these effects and that are used in these photonic and optoelectronic devices are generally referred to as *nonlinear optical materials* (NLO), because their outputs are not a linear response to some input. We are most familiar with linear responses of materials; for example, the intensity of photons emitted from a tungsten filament in a light bulb increases (more or less) linearly with the power we put into it. But even in this example, nonlinear behaviour can occur in the form of *saturation*, where further increases in power do not result in more output, or in the form of *breakdown*, when the filament buns out. The diodes described in the preceding sections are nonlinear electronic components—their output currents remain negligibly low until the driving voltages reach a threshold value whereupon the output current increases sharply.

The best way to describe a nonlinear optical response is first through some examples then through some mathematics. Shortly after the invention of lasers, it was observed that a ruby laser beam passing through a quartz crystal produced a faint beam at the laser's second harmonic—that is, at twice the fundamental frequency of the ruby laser. A much simpler-to-perform example of nonlinear optical behaviour is found when the polarization of light passing through a crystal is modified upon application of an electric field to the crystal. The origin of both of these NLO effects is related to the change in refractive indices that result by an applied electric field and the modulation of light beams by these field-dependent indices. The relationship between refractive index and polarizability is well established, and this is where our mathematical description of NLO effects begins.

We can examine how induced polarization behaves as a function of an applied electric field, ξ, by considering the induced electric dipole moment p_e, as a Taylor series expansion in ξ:

$$p_e = p_{e,0} + \xi(\partial p_e/\partial \xi)_{\xi \to 0} + 1/2\xi^2(\partial^2 p_e/\partial \xi^2)_{\xi \to 0} + 1/6\xi^3(\partial^3 p_e/\partial \xi^3)_{\xi \to 0} + \ldots \quad (1)$$

or, more simply,

$$p_e = p_{e,0} + \alpha\xi + (\beta/2)\xi^2 + (\gamma/6)\xi^3 + \ldots \quad (2)$$

where $P_{e,0}$ is the static dipole of the molecule, α is the linear polarizability previously called just "polarizability" and the higher-order terms β and γ are called the *first* and *second hyperpolarizabilities*, respectively. Note that once again, we consider only the magnitude of the electric field, which is actually a vector quantity.

The terms beyond $\alpha\xi$ in Eq. (2) are not linear in ξ, so they are referred to as the *nonlinear polarization* and give rise to NLO effects. You should now see why nonlinear polarization becomes more important at higher electric field strengths, since it scales with higher powers of the field. For most materials, $\alpha\xi > (\beta/2)\xi^2 > (\gamma/6)\xi^3$, so the first few observations of NLO effects were made prior to the invention of lasers and their associated large electric fields. The bulk polarization, P, which you will recall is the dipole moment per unit volume, can then be expressed as

$$P = P_0 + \chi^{(1)}\xi + \chi^{(2)}\xi^2 + \chi^{(3)}\xi^3 + \ldots \quad (3)$$

where the $\chi^{(n)}$ *susceptibility coefficients* (not to be confused with the magnetic susceptibility of the same symbol) arc tensors of order $n + 1$ and P_0 is the intrinsic static dipole moment density of the material.

We now consider the polarization induced by an oscillating electric field, such as that found in electromagnetic radiation (light), which can he expressed as

$$\xi = \xi_0\cos(\omega t) \quad (4)$$

where ω is the frequency of the electric field and t is time. Substitution of Eq. (4) into Eq. (3) yields

$$P = P_0 + \chi^{(1)}\xi_0\cos(\omega t) + \chi^{(2)}\xi_0^2\cos^2(\omega t) + \chi^{(3)}\xi_0^3\cos^3(\omega t) + \ldots \quad (5)$$

Because $\cos^2(\omega t) = 1/2 + 1/2\cos(2\omega t)$, the first three terms of Eq. (5) become

$$P = (P_0 + 1/2\chi^{(2)}\xi_0^2) + \chi^{(1)}\xi_0\cos(\omega t) + 1/2\chi^{(2)}\xi_0^2\cos(2\omega t) + \ldots \quad (6)$$

Equation (6) indicates that the polarization consists of a second-order, direct-current (not dependent upon ω) field contribution to the static polarization (the first term), a frequency component, ω, corresponding to the incident light frequency (the second term), and a new, frequency-doubled component, 2ω (the third term). Thus, if an intense light beam passes through a second-order NLO material, light at twice the input frequency will be produced, as well as a static electric field. The process of frequency-doubling, which explains the first example of NLO behaviour from the beginning of this section, is called *Second Harmonic Generation* (SHG). The oscillating dipole reemits at all of its polarization frequencies, so light is observed at both ω and 2ω .The process of static field production is called *optical rectification.*

The general condition of second-order NLO effects involves the interaction of two distinct waves of frequencies ω_1 and ω_2 in an NLO material. In this case, polarization occurs at sum $(\omega_1 + \omega_2)$ and difference $(\omega_1 - \omega_2)$ frequencies. This electronic polarization will therefore reemit radiation at these frequencies, with contributions that depend on $\chi^{(2)}$ which is itself frequency-dependent. The combination of frequencies is called *sum* (or difference) *frequency generation* (SFG). SHG is a special case of SFG, in which the two frequencies are equal.

An applied electric field can also change a material's linear susceptibility, and thus its refractive index. This effect is known as the *linear electro-optic* (LEO) or *Pocket's effect*, and it can be used to modulate light by changing the voltage applied to a second-order NLO material. The applied voltage anisotropically distorts the electron density within the material, and the optical beam "sees" a different polarizability and anisotropy of the polarizability. As a result, a beam of light can have its polarization state changed by an amount related to the strength of the orientation of the applied voltage and can travel at a different speed, and possibly in a different direction. The change in refractive index as a function of the applied field is approximated by the general expression

$$1/N_{ij}^2 = 1/n_{ij}^2 + r_{ijk}\xi_k + s_{ijkl}\xi_k\xi_l + \cdots \tag{7}$$

where N_{ij} are the induced refractive indices, n_{ij} are the refractive indices in the absence of the electric field, r_{ijk} are the linear or Pocket coefficients, s_{ijkl} are the quadratic or Kerr coefficients, and ξ is the applied field. Subscripts indicate the orientation of the field with respect to a material coordinate system. Equation (7) indicates that light travelling through an electro-optic material can he phase- or polarization-modulated by refractive index changes induced by an applied electric field. Devices exploiting this effect include optical switches, modulators, and wavelength filters.

An example of an electro-optic switch based on NLO materials is shown in Figure 12.8. The switch is comprised of two parallel waveguides made of NLO materials. The waveguide channels have a different refractive index from the surrounding material. The light can be switched back and forth between the channels by applying and removing a voltage across the bottleneck. In the absence of an electric field, the light travelling through the lower waveguide interacts with the upper waveguide in a nonlinear manner at the bottleneck, causing the light to switch channels. Switching *does not* occur when an electric field is applied. The electric field polarizes the NLO material and alters the refractive indices of the two channels, such that the nonlinear interaction at the bottleneck is modified and the light stays in the lower waveguide.

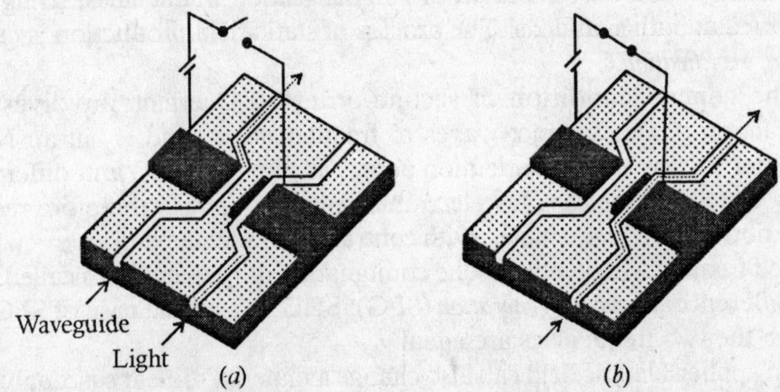

Fig.12.8: Schematic illustration of an electro-optic switch, (*a*) An open circuit causes light to switch channels, and (*b*) a closed switch keeps light in the same channel

Many current NLO devices are based upon crystalline materials (such as lithium niobate for the electro-optic switch) and nonlinear optical glasses, but there is intense interest in utility polymers low these applications. Due to their inherently low densities and potential ease of processing. Some of the polymers that have been used for photonic devices are based upon architectures that have highly polarizable electronic structures. Several polyimides fall into this category (Fig. 12.9 (*a*)), as do polyacrylates and polyurethanes with optically responsive side groups (Fig. 12.9 (*b*)).

Nonlinear optical (NLO) materials capable of generating the second harmonic frequency play an important role in the domain of optoelectronics and photonics. Nonlinear optical (NLO) crystals with high conversion efficiences for second harmonic generation (SHG) and transparent in visible and ultraviolet ranges are required for numerous device applications. Within the last decade much progress has been made in the development of these NLO organic materials having large nonlinear optical coefficients. However, most of the organic NLO crystals are constituted by weak Van der Waals and hydrogen bonds with conjugated π electrons. So they are soft in nature and difficult to polish and these materials also have intense absorption in UV region. In view of these problems, new types of hybrid

NLO materials have been explored from organic-inorganic complexes with stronger ionic bond such as L-arginine phosphate monohydrate (LAP) and others have been synthesized and single crystals have been grown. LAP crystals are used for second, third and fourth harmonic generation from the fundamental radiation 1.06 μm and also sum and difference frequencies generation in wide spectral range from UV to IR. LAP has three times the nonlinearity, two or three times high damage threshold of KDP, chemical stability, and also lower cost. A problem related with LAP growth is the presence of microorganisms arising during growth, which lead to deterioration in L-histidine tetrafluoroborate (L-HFB), L-arginine tetrafluoroborate (L-AFB) and others.

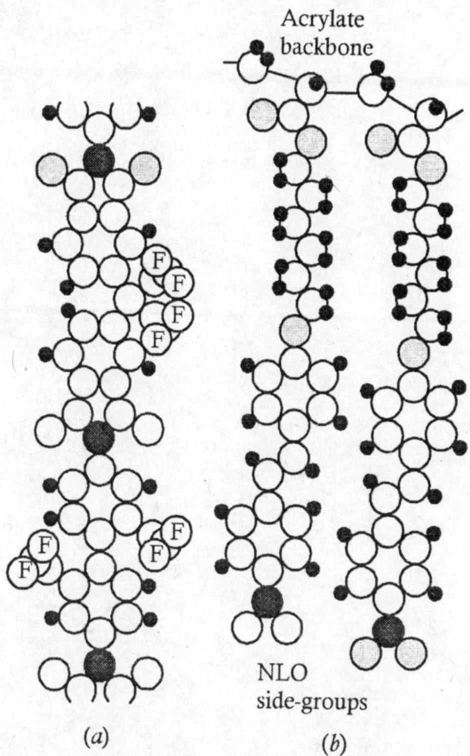

Fig. 12.9: Examples of nonlinear optical polymers (a) fluorinated polyimide and (*b*) polyacrylate with NLO side groups

Polymeric Materials

Polymers are very large molecular structures made by reacting together small and comparatively simple molecules. The reactions are termed polymerisation processes and may be of the simple addition type, addition copolymerisation, or condensation polymerisation. The structure of the final compound and hence, to a large extent, its properties will be determined by the structure of the original reactants and the type of polymerisation process.

On the basis of structure we may subdivide polymers into two major groups *linear* and *non-linear*. Each of these two groups may be further sub-divided as shown in Fig. 12.10. Many addition polymers are of the linear

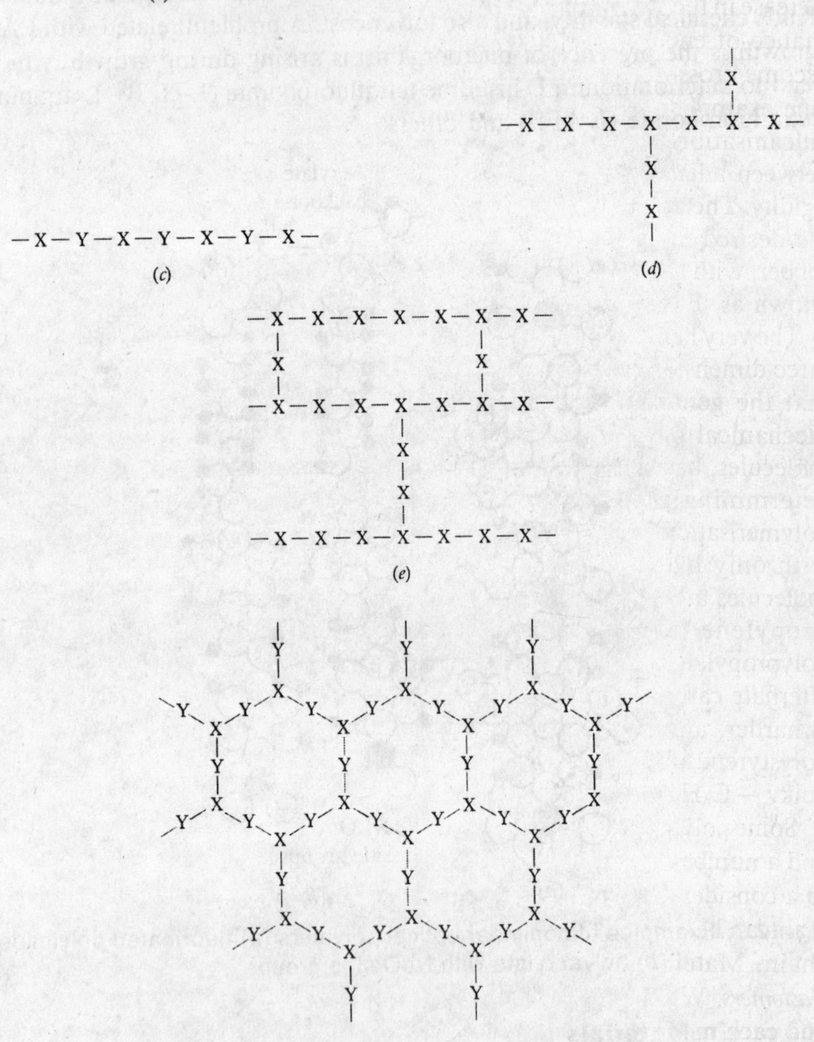

Fig. 12.10: Polymer structure types: (*a*) linear, (*b*) random linear copolymer, (*c*) ordered linear copolymer, (*d*) branched, (*e*) cross-linked, (*f*) network. (*X* and *Y* are manomers), (*d*), (*e*) and (*f*) are examples of *non-linear polymers*

type. The non-linear is really a misnomer, as the molecular chains are far from straight. The bond angle between adjacent carbon atoms is 109.5° and so the chain molecules become fibres of irregular shape. The use of the word linear to describe a polymer molecule signifies that there are no side chains or branches in the molecule. If a branch chain is formed the polymer becomes non-linear. Branch chains may be formed some polymers in the presence of a catalyst. The presence of branch chains causes an increase in the strength and rigidity of the material because there is a greater chance of entanglement between molecules. If the polymer molecules become cross-linked the hardness and rigidity are increased still further. One example of cross-linking is the vulcanisation of rubber. During the vulcanisation reaction short chains of sulphur atoms form cross-links between adjacent molecules of the rubber increasing the hardness and rigidity. The degree of vulcanisation can be carefully controlled to produce the desired degree of hardness in the rubber. Fully vulcanized natural rubber, with the maximum number of cross-links is the hard rigid material known as 'Ebonite'.

The very hard and rigid thermosetting materials, such as Bakelite, possess three-dimensional network structures. From the foregoing it can be seen that the general form of a polymer will exert a major influence on mechanical properties. In addition the detailed shape of the polymer molecules, based on the structures of the manomers plays a great part in determining the final properties of the material. For example, the polymerisation of ethylene ($CH_2 = CH_2$) will give a simple chain molecule with only hydrogen atoms attached to the linear carbon chain. The molecules are flexible and so polyethylene is a soft and flexible material. Propylene ($CH_2 = CH - CH_3$) is a related monomer, but the polypropylene molecule has a structure with $—CH_3$ groups attached to alternate carbon atoms in the molecular chain. This gives a material that is harder, and has a greater rigidity, than polyethylene. In the case of polystyrene the material is even harder, and more rigid because of the bulky $—C_6H_5$ groups attached to the carbon chain (Fig. 12.11).

Some polymeric materials possess very great extensibility. Natural rubber and a number of synthetic polymers can be stretched in an elastic manner to a considerable extent. The amount of extension observed is far greater than can be accounted for by the straightening of bond angles in a linear chain. Materials that show this type of elastic behaviour are termed *elastomers.* Within elastomers the individual linear molecules tend to coil and each molecule becomes, in effect, a minute helical spring.

The terms *thermoplastic* and *thermosetting* are used in connection with polymeric materials. A thermoplastic material is one that becomes more plastic when heated and becomes rigid again on cooling. Polymers with linear structures and also those that are branched or lightly cross-linked are thermoplastic. A thermosetting material is one that on first heating becomes plastic and can be moulded, but almost immediately chemical reactions occur that either complete the polymerisation of a part polymerised material to give a full network structure, or cause extensive cross-linking to occur. In either case the result is a hard, rigid material. Once this setting or 'curing' has taken place the material cannot again be made plastic. Some materials are cold-setting. This usually involves mixing the polymer resin with a 'hardener'. Curing takes place over a short period of time at ordinary temperature to produce a hard and rigid network structure.

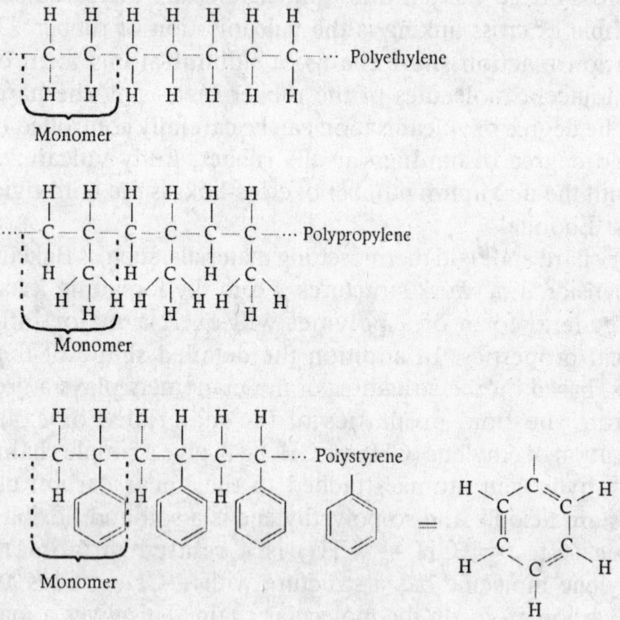

Fig. 12.11: Structures of three linear polymers. Slip between adjacent molecules of polyethylene could occur fairly easily. Slip would be more difficult with polypropylene and polystyrene owing to the presence of bulky side groups

12.5 NANOSTRUCTURED MATERIALS

The discovery of fullerenes and carbon nanotubes has opened a new research area in physics, chemistry and material science. Nanostructured materials may be defined as those materials whose structural element clusters, crystallites or molecules—have dimensions in the 1 to 100 nm range. Clusters of atoms consisting of typically hundreds to thousands on nanometer (nm) scale are commonly called as nanoclusters. These small groups of atoms, in general, go by different names such as nanoparticles, nanocrystals, quantum dots and quantum boxes. Substantial work is being carried out in the domain of nanostructured materials and nanotubes in the past one decade. The explosion in both academic and industrial interest in these nanomaterials over the past decade arises from the remarkable variations in fundamental electrical, optical and magnetic properties that occur as one progresses from an 'infinitely extended' solid to a particle of material consisting of a countable number of atoms. Nanostructured materials has led to new basic science phenomena and several applications for the short, medium and long term. Examples are: nanoelectronic devices, quantum wires, electron field emitters for ultra-thin TV screens, nanoprobes, high resolution tips for scanning and atomic force microscopes, sensors, ultrahigh strength composites, gas storage, nanodevices, and parts of nanomachines, among others. Carbon-based nanomaterials and nanostructures including fullerenes and nanotubes play an increasingly pervasive role in nanoscale science and technology. Carbon nanotubes are currently being studied in an effort to understand their novel structural, electronic and mechanical properties to explore their immense potential for many applications in nanoelectronics, and as actuators and sensors.

Fullerenes and carbon nanotubes can be seen as curves pieces of graphite. Graphite is another polymorph of carbon.

Graphite

Graphite has a crystal structure distinctly different from that of diamond and is also more stable than diamond at ambient temperature and pressure (Fig. 12.12). Graphite is formed by flat hexagonal layers of carbon atoms separated 3.35 Å. The C-C distance is 1.42 Å and the structure belongs to P 63 mc space groups with lattice constants a = 2.46 Å and c = 6.71 Å. Graphite is highly anisotropic solid, structurally its interplaner spacing (3.35 Å) is quite large compared to the in plane interatomic spacing (1.42 Å). Physically its stiffness along the plane is quite large because of strong σ bonds and in the perpendicular direction it is weak because of the Van der Waal's force. The graphite structure is composed of layers of hexagonally arranged carbon atoms; within the layers, each carbon atom is bonded to three coplanar neighbour atoms by strong covalent bonds. Carbon when sp^2 hybridized contains three strong π-bonds which lie in a triangular plane with bond angle of 120° between them and one weakly bonding π-bond perpendicular to the plane. Each carbon atom in the plane is covalently bonded to three others in plane with strong σ bonds of the sp^2 hybridization,

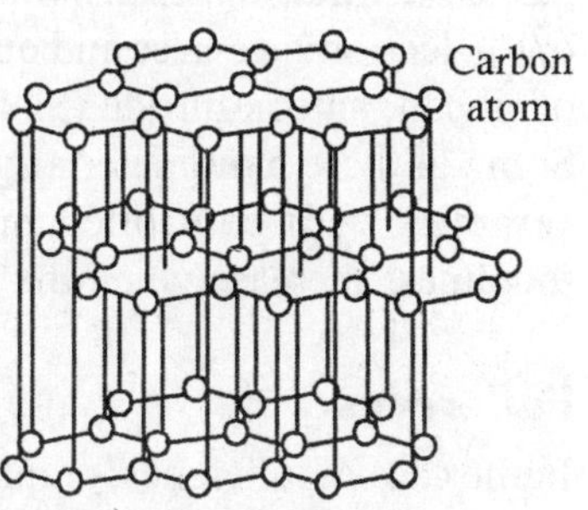

Fig. 12.12: The structure of Graphite

this network forms a continuous sheet of carbon. The fourth bonding electron participates in a weak Van der Waals type of bond between the layers. As a consequence of these weak interplanar bonds, interplanar cleavage is facile, which gives rise to the excellent lubricative properties of graphite. Also, the electrical conductivity is relatively high in crystallographic directions parallel to the hexagonal sheets. Graphite has metallic conductivity along the plane and semiconducting perpendicular to the plane. The anisotropy in conductivity is about 10^3.

Other desirable properties of graphite include: high strength and good chemical stability at elevated temperatures and in non oxidizing atmospheres, high thermal conductivity, low coefficient of thermal expansion and high resistance to thermal shock, high absorption of gases, and good machinability. Graphite is commonly used in heating elements for electric furnaces, as electrodes for AC welding, in metallurgical crucibles, in casting moulds for metal alloys and ceramics, for high temperature refractories and insulations, in rocket nozzles, in chemical reactor vessels, for electrical contacts, brushes and resistors, as electrodes in batteries, and in air purification devices.

There are other layered materials which also can acquire curvature to generate new nanomaterials with novel applications. The other types of nanostructured materials which has been considered for applications in opto-electronic devices and quantum-optic devices are nonosized powders of silicon, silicon-nitride (SiN), silicon carbide (SiC) and their thin films. Some of these nanomaterials, e.g. SiC and SiN are also used as advanced ceramics with controlled microstructures because their strength and toughness increase when the grain size diminishes.

Fullerenes

In the case of fullerenes, graphite acquires curvature due to the presence of pentagonal rings in the hexagonal graphite sheet. Fullerenes exists in discrete molecular form and consists of a hollow spherical cluster of sixty carbon atoms; a single molecule is denoted by C_{60}. Each molecule is composed of groups of carbon atoms that are bonded to one another to form both hexagon (six-carbon atom) and pentagon (five carbon atom) geometrical configurations. One such molecule (Fig. 12.13), is found to consist of 20 hexagons and 12 pentagons which are arrayed such that no two pentagons share a common side; the molecular surface thus exhibits the symmetry of soccer ball. The material composed of C_{60} molecules that contains sixty carbon atoms in a network of sp^2 bonding which forms a spherical structure, is known as *buckminsterfullerene*, from a number of perspectives, a molecule almost 'custom built for nanoscience. Discovered in 1985 by Kroto *et al.*, C_{60} is approximately 1 nm in diameter (the cage diameter is 0.71 nm; the intermolecular space in fullerite, the bull crystal is 1.004 nm readily synthesized in large (i.e. gram) quantities and of key significance—its physical and chemical properties may be tuned via the addition of a range of elemental and molecular species either to the fullerene lattice, to the fullerene cage or within the cage. The term *fullerene* is used to denote the class of materials that are composed of this type of molecule. Fullerenes are made up of hexagonal and pentagonal rings of carbon arranged exactly as of an Indian football.

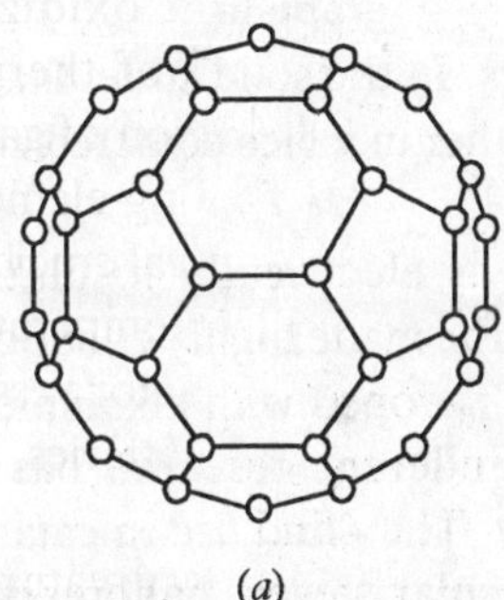

(*a*)

Face centered cubic crystal of C_{60} (fullerite)

(*b*)

Fig. 12.13: (*a*) The structure of C_{60} molecule (*b*) Face centered cubic crystal of C_{60} fullerite

Based on fullerenes (C_{60}) presently we find the existence of C_{36}, C_{70} (Fig. 12.14), C_{80}, C_{100} and also few members of geometrically allowed fullerene family. These fullerene molecules differ in the number of hexagonal rings present. The shape of this molecule changes as the number of carbon atom differs. The C_{70} looks similar to a rugby ball and C_{120} (Fig. 12.15) takes a dumbbell shape.

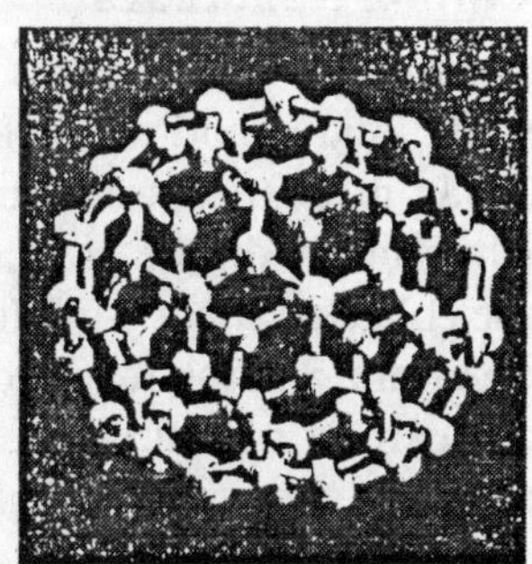

Fig. 12.14: Structure of fullerene C_{70}

Diamond and graphite are what may be termed *network solids*, in that all of the carbon atoms form primary bonds with adjacent atoms throughout the entirety of the solid. The common *synthetic graphite* is usually manufactured as composites. In this process petroleum coke is mixed into a paste with a coal tar pitch and then heated to around 120° C-1400° C. This step drives out all volatile material from the petroleum coke. Further heating this to 2500°

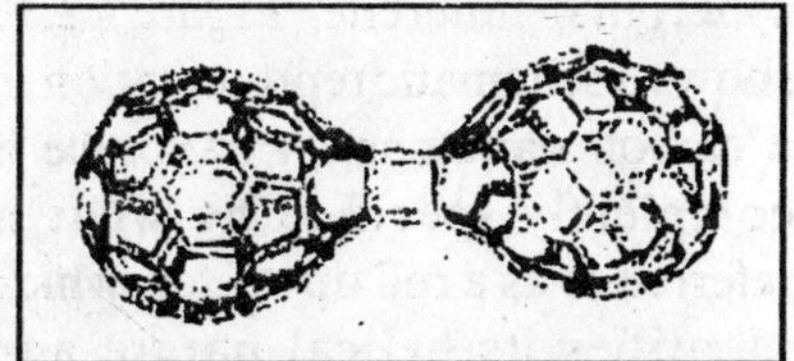

Fig. 12.15: Structure of C_{120} fullerene

C-3000° C causes an ordering of the carbon atoms to graphitize the mixture into a completely graphitic structure (Fig. 12.16).

Fig. 12.16: AFM image of (HOPG) graphite

Presently, there are methods to prepare C_{60} molecules in large quantities, so the study of this molecular form of carbon is advancing in all possible arenas. By way of contrast, the carbon atoms in the buckminsterfullerene bond together so as to form these spherical molecules. In the solid state, the C_{60} units form a crystalline structure and pack together in a face centered cubic array.

As a pure crystalline solid, this material is electrically insulating. However, with proper impurity additions, it can be made highly conductive and semiconductive. Presently we know that C_{60} doped with alkali metals become superconducting with $T_C \sim 50$ K. Fullerene research has got immense scope in nanoscience and technology. They find use in catalytic chemistry, biomolecular recognition, as molecular sieves, nanoreactors, and also as inhibitors to the activity of HIV virus, etc.

Carbon Nanotubes

Another molecular forms of carbon has recently been discovered that has some unique and technologically promising properties. Ijiyama of NEC Japan in 1991 first announced the synthesis of carbon nanotubes. In his experiments on the arc discharge between graphite electrodes, alongwith the collected shoot particles there were some fine tube like structures. This on careful analysis revealed that they are carbon nanotubes (CNTs). Its structure consists of a single sheet of graphite, rolled into a tube, both ends of which are capped with C_{60} fullerene hemispheres. The tubes can be either open at their ends or capped at one or both ends with half a spheroidal fullerene. Figure 12.17 shows a schematic representation of a carbon nanotube. A nanotube is completely specified by what is referred to as a roll up vector, which identifies its helical nature and diameter. Depending on its roll-up

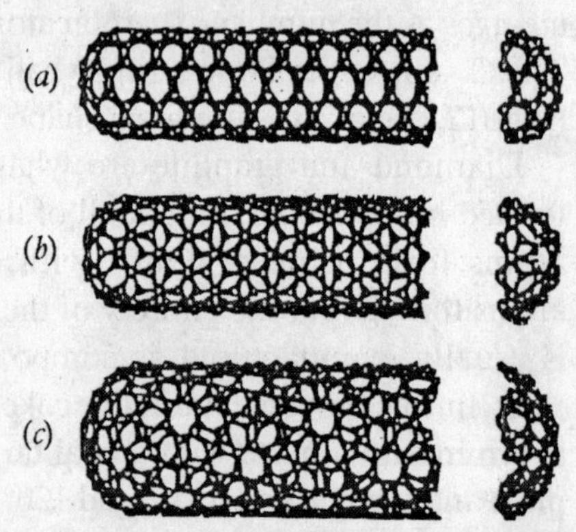

Fig. 12.17: The structure of carbon nanotubes (*a*) metallic (*b*) and (c) semiconducting

vector, a nanotube can have either metallic Fig, 12.17 (*a*) or semiconducting (Fig. 12.17 *b*, *c*) properties.

There are a vast number of ways in which a graphite sheet can be rolled up to form a seamless cylinder and hence a wide variety of nanotubes exist. Nanotubes are characterized by their *chiral* (or wrapping) vector, c such that $c = na_1 + ma_2$, where a_1 and a_2 are the basis vectors of the graphite (more correctly graphene) lattice and m, n are integers (Fig. 12.18). The chiral vector spans the circumference of the tube formed by joining the dotted lines shown in Fig. 12.18. Those tubes with chiral vectors of the form (n, o) are termed zig-zag tubes (Fig. 12.19 (*a*)) whereas $n = m$ a so called 'arm chair' tube results Fig. 12.19 (*b*)). All other values of n and m produce a chiral tube (Fig. 12.19 (*c*)).

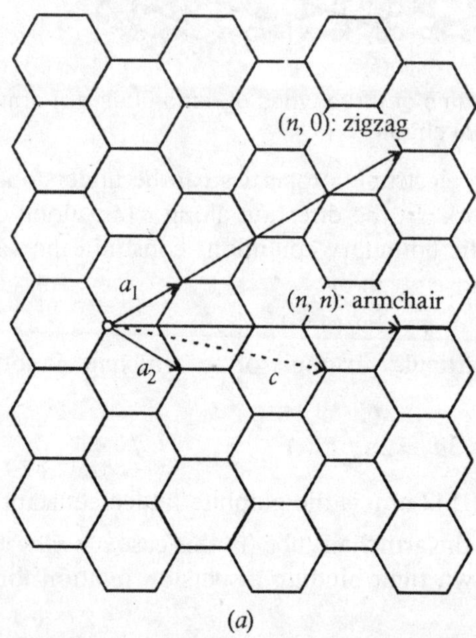

(*a*)

Fig. 12.18: The relationship between the graphite lattice basis vectors $\boldsymbol{a}_1$, $\boldsymbol{a}_2$ and the chiral vector $\boldsymbol{c} = n\boldsymbol{a}_1 + m\boldsymbol{a}_2$ used to characterize carbon nanotubes. Two limiting cases are shown: (n, 0) indices are associated with zig-zag tubes whereas (n, n) indices are associated with armchair tubes. All other tubes are chiral

Closely following the synthesis of nanotubes came the remarkable theoretical prediction that their electronic properties could be changed between metallic and semiconducting simply by varying the tube diameter or its helicity *i.e.* by changing the values of n and m. A key theoretical result is that armchair nanotubes are metallic, whereas for all other SWNTs, when n-m = 3 1 (1 an integer) the tubes are metallic; otherwise they are semiconducting.

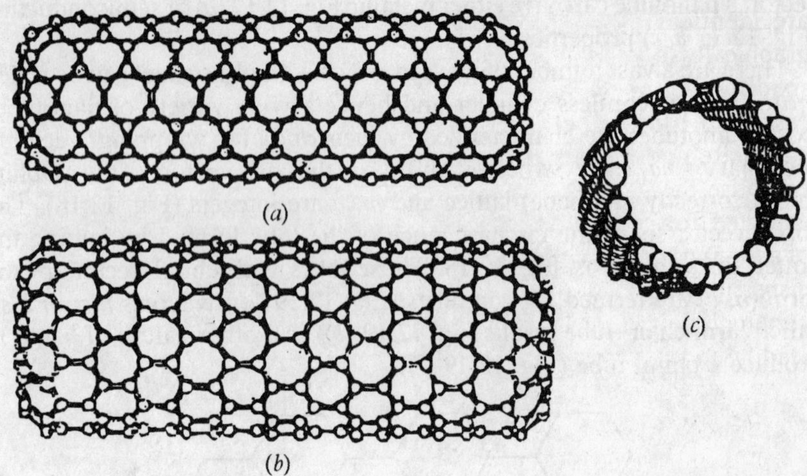

Fig. 12.19: Diagrams of three types of nanotubes: (*a*) armchair (*b*) zig-zag and (*c*) chiral

These striking electronic properties can be understood within a tight-binding framework. In the direction along c (*i.e.* along circumference of the tube), periodic boundary conditions constrain the values of electron wavevector to those given by

$$c \cdot \boldsymbol{k} = 2\pi q \, (q = 1, \ldots, n) \tag{8}$$

Taking the particular example of an armchair nanotube, this can be reduced to

$$k_x n \sqrt{3} a_0 = 2\pi q \tag{9}$$

where $a_0 = (\sqrt{3})0.142$ nm is the graphite lattice constant and defines the chiral vector for the armchair tube (in this case $c = n(a_1 + a_2)$.

The well-known tight binding dispersion relation for a 2 D graphite sheet is

$$E_{2D}(k_x, k_y) = +\gamma_0 \left\{1 + 4\cos\left(\frac{\sqrt{3}k_x a_0}{2}\right)\cos\left(\frac{k_y a_0}{2}\right) + 4\cos^2\left(\frac{k_y a_0}{2}\right)\right\}^{1/2} \tag{10}$$

γ_0 is the C-C overlap integral. Substituting into (10) the allowed values of k_x, given by (9) yields

$$E_{10}^{A}(k_y) = \pm\gamma_0 \left\{1 \pm 4\cos\left(\frac{q\pi}{n}\right)\cos\left(\frac{k_y a_0}{2}\right) + 4\cos^2\left(\frac{k_y a_0}{2}\right)\right\}^{1/2} \quad (-\pi < k_y a_0 < \pi) \tag{11}$$

where the supercript A denotes that (11) is valid only for armchair tubes. It is assumed that the value of the overlap integral for graphite and nanotube are identical. The 1 D dispersion relations given by (11) are shown graphically in Fig. 12.20 for the case of (5, 5) armchair SWNT. There are six dispersion curves for both the valence and conduction bands. In each case, four bands are doubly degenerate (bold continuous lines) and two bands are non-degenerate (thin, dashed lines). Thus ten electronic levels comprise the valence bands and ten comprise the conduction bands, as expected from the fact that there are ten hexagons around the circumference of a (5, 5) nanotube. There are two particularly noteworthy features of the dispersion relations shown in Fig. (12.19 (*a*), (*b*) and (*c*)). First there is a very high degree of degeneracy at the Brillouin Zone boundary. One can readily understood this from the fact that at $k_y a_0 = \pi$, Eq. (11) reduces to

$$E_{1D}^{A} = \pm \gamma_0$$

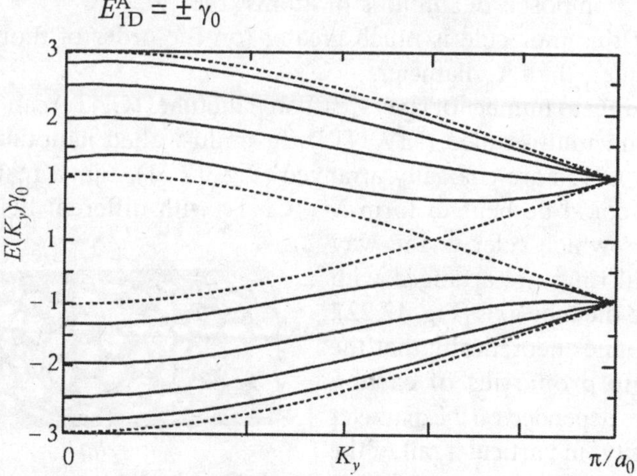

Fig. 12.20: One dimensional dispersion relation for an armchair (5, 5) nanotube

The second and more important feature of Fig. 12.19 (*a*), (*b*) and (*c*) in terms of electronic properties of armchair nanotubes is that the highest valence band and lowest conduction band are degenerate at $k_y = \pm 2\pi/(3a_0)$ and each cross the Fermi energy ($E = 0$) at that point. This is true of all armchair nanotubes-thus all armchair nanotubes be metallic.

We must note that Eqs. (9)-(11) represent taking a 1D slice-in a direction given by

$$E(\theta) = -KV\cos^2\theta$$

where K is the magnetic anisotropy constant, V the volume of the particle, θ the angle between the magnetization vector and an easy direction of magnetization and $E(\theta)$ is the magnetic anisotropy energy for a uniaxial particle, through the 2 D graphite band structure. The crossing at $E = 0$ in Fig. 12.20 occurs because the corresponding 2 D graphite bands cross at the K points (*i.e.* the corners) of the 2 D Brillouin Zone. For armchair tubes, there is always an allowed ***K***-vector that goes through K point of the 2 D lattice. However, this alignment of the tube wavevector and *K*-points of the graphite Brillouin zone does not automatically occur for zig-zag or chiral tubes and hence these can have semiconducting character. Only for $n - m = 31$ is there an alignment of an allowed **k**-vector and a *K*-point of the 2D Brillouin Zone, giving rise to metallic behaviour.

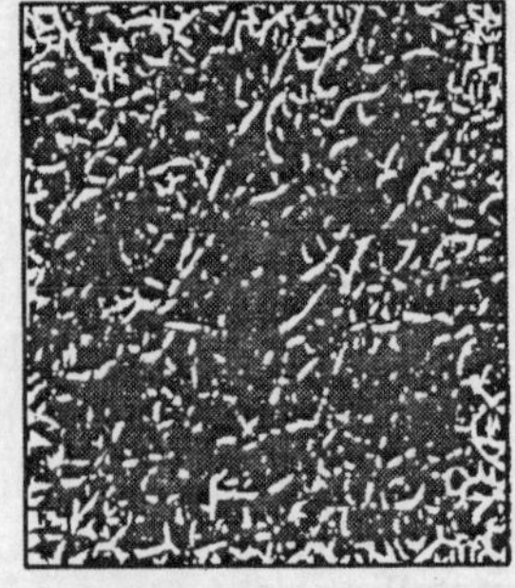

Fig. 12.21: SEM image of CNTs

The *nano* in the nanotube prefix denotes that tube diameters are of the order of a nanometer (*i.e.*, 100 nm or less). Each nanotube is a single molecule composed of millions of atoms; the length of this molecule is much greater (on the order of thousands of times greater) than its diameter.

According to number of layers, carbon nanotubes (CNTs) can be single-walled and multi-walled (MWNTs). In multi-walled nanotubes, more than one CNTs are coaxially arranged (Fig. 12.21). Iijima realized that graphite could be bent to form MWCNTs with different helicities or chiralities which refer to the way hexagonal rings are arranged with respect to the tube axis (Fig. 12.22).

It is found theoretically that the electronic properties of carbon nanotubes depended on the diameter and helicity: in particular, all of the so-called armchair type nanotubes are conductors (Fig. 12.22) and most zig-zag nanotubes are semiconductors (Fig. 12.22).

These nanotubes are extremely strong and stiff, and relatively ductile. According to indirect measurements, it has been found that carbon nanotubes are 100 times stronger than steel and six times lighter. For single walled nanotubes, tensile strength range between 50 and 200 GPa (approximately an order of magnitude greater than for carbon fibres); this is the strongest known material. Elastic modulus values are on the order of one

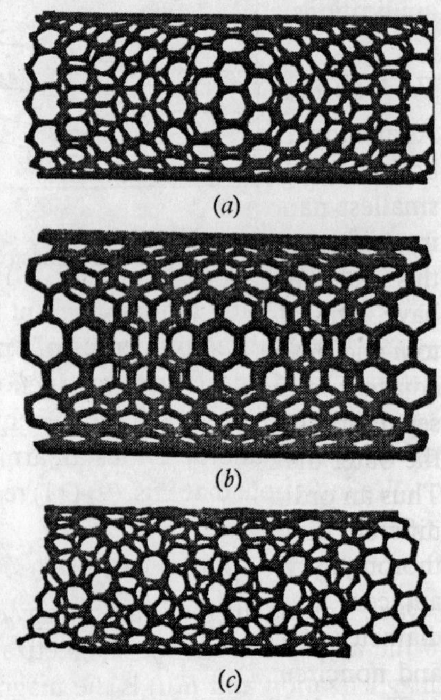

Fig. 12.22: Structures of carbon nanotubes (*a*) Zig-zag nanotube; (*b*) Ann-chair nanotube; and (*c*) Chiral or helical nanotube

tetrapascal [TPa: (1 TPa = 10^3 GPa)], with fracture strain between 5% and 20%. Furthermore, nanotubes have relatively low densities. On the basis of unique properties, the carbon nanotube has been termed the 'ultimate fibre' and is extremely promising as a reinforcement in composite materials

Carbon nanotubes also have unique and structure sensitive electrical characteristics. Depending on the orientation of the hexagonal units in the graphite plane, *i.e.*, tube well, with the tube axis, the nanotube may behave electrically as either a metal or a semiconductor. Semiconducting and metallic nanotubes recently are put into use in field effect transistors (FET) and single electron transistors. The flat-panel and full-colour displays, *i.e.* TV and computer monitors, have been fabricated using carbon nanotubes as field emitters. These displays should be cheaper to produce and will have lower power requirements than CRT and liquid crystal displays. The mechanical strength of CNT is amazingly 600 times tougher than steel, which finds applications in micro electro mechanical systems (MEMS) and in aerospace. Progress has been made in using CNTs as field emission devices e.g. diodes, transistors, etc. for high resolution display systems. Construction of nanoprobes and electrodes in biological and other applications are also being aimed.

Various Forms of Carbon of Nanotubes

As already mentioned, the carbon nanotube is a generic term that includes both single and multiwalled nanotubes (SWNTs and MWNTs). The smallest nanotube reported to date has a diameter of only 4 Å. Since increasing the curvature of the graphite wall increases the strain energy, this is also predicted to be the smallest possible diameter a nanotube can have. Typical diameters for SWNTs are between 1 and 2 nm, corresponding to n and m numbers of around 10. In MWNTs, the distance between the adjacent layers is about 0.34 nm, which is much more than the atomic separation (1.42 Å) within the individual layers. While typical values for the outer diameter are 10-20 nm, the inner diameter is typically 3-5 nm. Thus an ordinary MWNT has 10-30 layers, each of which are SWNTs of different diameters. The shell structure of MWNTs has not been studied thoroughly yet, but from geometrical considerations it is concluded that adjacent layers will generally be incommensurate. Also, the larger the diameter of a MWNT is, the more likely the outer layers are disordered and noncircular.

The ends of nanotubes are generally closed in the growth process. While the wall of a carbon nanotube is made up of an exclusively hexagonal pattern, pentagons are needed to cap the tubes. More generally, pentagonal and heptagonal (5- and 7-corpered polygons, respectively) defects enable the graphitic sheet to take up more complicated structures than a simple tube, as we will discuss below. Unless special setups are used for the growth of SWNTs, they arc usually assembled into ropes by their mutual Van der Waals attraction. For example, a rope of a typical diameter of about 10 nm contains ~100 SWNTs. Since multiwalled nanotubes have a much higher bending stiffness, they do not arrange similarly.

Nanotubes grown under suitable conditions have a very low concentration of defects over μm-distances that is, over hundreds or even thousands of interatomic spacings. However, if one were able to control the occurrence of defects, very useful and interesting nanotube structures would emerge. We will next show what kind of modified SWNTs and MWNTs have already been observed due to the occurrence of defects within the nanotubes, A single, stable defect structure in the hexagonal lattice of a graphite sheet is made up of a pentagon-heptagon pair, that is 5- and 7-sided polygons. One such defect can cause a sharp bend in a SWNT. Defective nanotubes are especially interesting for electronic applications where the defect site may act as a tunnelling barrier. In the case of MWNTs, highly defective tubes may be obtained with certain synthesis conditions, for example with relatively low growth temperatures. While high-quality MWNTs are very straight and stiff, very defective ones have a continuous and smooth curvature. On the other hand, the curvature can be highly regular and so result in helices. An SWNT, or a single shell of a MWNT, can have a second nanotube branching out. The beginning point of the branch has similarly to the end sections pentagons dispersed among the hexagons to accommodate the necessary curvature. Such branching has been observed in a few cases to date.

As mentioned earlier, the ends of carbon nanotubes are usually closed. It is, however, possible to open nanotubes by a suitable chemical treatment. This is due to the stronger chemical reactivity of the end section with its larger curvature and pentagon structures. The opened nanotubes can be utilized for creating yet another side branch in the science of nanotubes, namely filled tubes. Both SWNTs and MWNTs have been filled with various materials, such as fullerenes, simple metals, and molecular compounds. Nanotubes with fullerenes inside are called peapods, and are presently intensively investigated. The study of the electronic properties of filled nanotubes is still in its infancy.

Chemistry of Carbon Nanotubes

The chemistry of carbon nanotubes, which encompasses the aspects of purification and functionalization, is a very important field for tailoring their structural and electronic properties and thus for the development of future applications. Purification aims at removing the nontubular material in raw nanotube samples. Several techniques, more or less complex, have been proposed, including solvent treatment followed by ultrafiltration, flocculalion using aqueous surfactants, oxidation and acid washing coupled with centrifugation, resuspension in surfactant solution, and cross-flow filtration steps. These methods however lead to a mixture of nanotubes of differing lengths and still containing potential contamination with nontubular material. Polymer suspensions and chromatographic purification are useful to remove nontubular material. Field-flow fractionation on purified, shortened nanotubes, and size-exclusion chromatography on raw nanotubes have demonstrated some success at producing fractions separated by length. Capillary electrophoresis was reported to allow separation based on tube length.

Several methods for attachment of oxygen-containing groups to the nanotube surfaces have been developed. Carboxylic, carbonyl, and hydroxyl groups were found covering the nanotube walls. The concentration of acid groups on the surfaces of HNO_3-treated nanotubes was higher than that found for graphite treated under the same conditions. Sonication of nanotubes prior to the functionalization in acids increases the concentration of acid groups. Oxidation of nanotubes with a H_2SO_4-HNO_3 mixture also leads to a higher concentration of functional groups on the surface. Long-term acid treatment of nanotubes also induces modifications. Functional groups can be removed from the surface of nanotubes by healing. It should be noted that SWNTs obtained by different methods exhibit different behaviour in the course of acid treatment and subsequent oxidation. The sensitivity of SWNTs to chemical processing was investigated by TEM.

To enhance the possibilities of atomic force microscopes with nanotube probe tips, the tip ends were functionalized using both solutions and gases. Modification of tips using a discharge in N_2 results in the formation of nitrogen-containing heterocycles at the tip ends. High degree of saturation of nanotube tips with functional groups is achieved by treatment with a low-pressure ammonia plasma followed by oxidation with a $NaClO_3$ solution.

The reactions with silicon of individual nanotubes or nanotube bundles at 970° C in ultrahigh vacuum resulted in the formation of Si-β-SiC (nanorod) nanotube heterojunctions.

Oriented or Localized Growth

The term "oriented" is used here to describe the formation of large quantities of CNTs aligned perpendicularly to the substrate surface, usually MWNTs of controlled dimensions (diameter 20-400 nm, length 0.1-60 μm). The key steps arc substrate patterning, catalyst deposition, and the appropriate use of different CVD routes.

The term localized" describes the self-assembled nanointerconnections made by CNTs between islands on prepatterned substrate. The directed growth of SWNTs parallel to the plane of a silicon substrate was reported. The SWNTs are suspended bridges grown from catalyst material placed on top of regularly patterned silicon tower structures. The key synthetic step in this approach involves developing a series of liquid-phase catalyst precursor materials, allowing for uniform film formation and large-scale catalyst patterning. Contact printing techniques were used to selectively deposit catalyst precursors on top of tac silicon tower arrays. Calcination led to the formation of catalyst particles confined on the tower tops and subsequent CVD (CH_3. 900° C, 20 min) growth yielded SWNTs emanating from the towers. Directed free-standing SWNT networks were formed by nanotubes growing to adjacent towers and becoming suspended above the surface. It is noted that a rational design of tower arrangements could lead to a variety directed SWNT architectures. The liquid-phase catalyst precursors consist of three general components: inorganic chloride precursors, a removable triblock copolymer serving as the structure-directing agent for the chlorides, and an appropriate alcohol for dissolution of the inorganic and polymer compounds. The increase in yield was achieved through a so-called conditioning step that uses bulk amounts of catalysts placed near (upstream) the catalyst-patterned substrate in the CVD growth environment.

Electromechanical Properties

Along with their electronic properties, carbon nanotubes have been praised for their mechanical properties. The praise is based on high strength, flexibility, and large aspect ratio. A carbon nanotube is a freestanding conductor and its mechanical properties are highly relevant for the electronic properties and the potential technological applications of the tube. For example, nanotube-based mechanical oscillators may have applications in high-frequency electronic circuits, and carbon nanotubes are already used in scanning probe microscopy both as STM and AFM tips. Thin, strong, and durable nanotubes come extremely close to an ideal scanning tip.

Mechanical properties of carbon nanotubes are closely related to the properties of a graphite sheet, but the tubular anisotropic form affects the mechanical behaviour. The basis is the graphite sp^2 bond, which is the strongest of chemical bonds. In nanotubes the overall density of defects can be extremely low depending on the synthesizing method and prevailing synthesizing parametres. This has led to predictions of a very high axial strength.

Basic Mechanical Properties: Theoretical calculations of Young's modulus for individual SWNTs center around 1 TPa or slightly higher, but values as high as 5.5 TPa have been presented. The spread is due to different interaction models and, also to differing values of nanotube wall thickness—the wall thickness (if an isolated nanotube is not a well-defined quantity. Most of the theoretical attention has been on SWNTs because modelling the interlayer interaction in MWNTs is a complicated matter. Young's modulus values for multiwalled tubes as well as SWNTs and obtains values from 0.97 TPa to 1.11 TPa with the value increasing slightly with the number of layers.

The small size of carbon nanotubes presents challenges also for experimental characterization. Nevertheless measurements have been performed. The first Young's modulus measurement related thermal vibration amplitudes of MWNTs to their Young's modulus and obtained an average value of 1.8 TPa with a large spread. After that, with AFM techniques, values such as 1.28 ±0.59 TPa for arc-discharge-produced tubes have been obtained. The latest measuring technique by Poncharal *et al.*, who induced vibrations on MWNTs by alternating potential and have measured the vibration frequencies . Young's modulus values between 0.7 TPa and 1.3 TPa for tubes with a diameter less than 12 nm and between 0.1 TPa and. 0.3 TPa for thicker tubes were reported. The drop is explained by an onset of a wave-like bending mode of the nanotube wall at the inner arc of the bending. Measuring SWNTs is more complicated than measuring MWNTs due to there small diameters and the tendency to bundle. Measurement of individual SWNTs using the thermal vibrations method to provide an average value of 1.25 TPa. Although the current measurements suffer from inaccuracies due to vibration amplitude measurement and assumptions made on AFM tip characteristics, the current agreement is that defect-free nanotubes, both SWNTs and MWNTs, have a Young's modulus value around slightly above 1 TPa, which is extremely high and sets nanotubes as the strongest known material albeit challenged by other nanotubular structures such as boron nitride tubes.

Theoretically, nanotube tensile strength is high, and this is supported by calculations in which SWNTs support as high as 30% of axial strain before brittle failure and by more recent kinetic activation based calculations that give a yield strain of 17% with a chirality and temperature-dependent defect formation activation energy barriers. Tensile strength values ranging from 11 GPa to 63 GPa are reported. The role of the inner layers is not, however, clear, and the failure process may be more complicated. For individual SWNTs, the experimental value of tensile strength is still an open question, but for bundles of SWNTs tensile strength values ranging between a few GPa and several tens of GPa depending on the bundle and measurement characteristic.

As a result of the high tensile strength, nanotubes can bear large strain before plastic deformation or brittle failure. The tubes buckle, flatten, form ripples, and generally deform under strain but plastic deformation and relaxation occur only at elevated strain levels and temperatures. An example of simulated nanotube behaviour under strain is presented in Fig. 12.23.

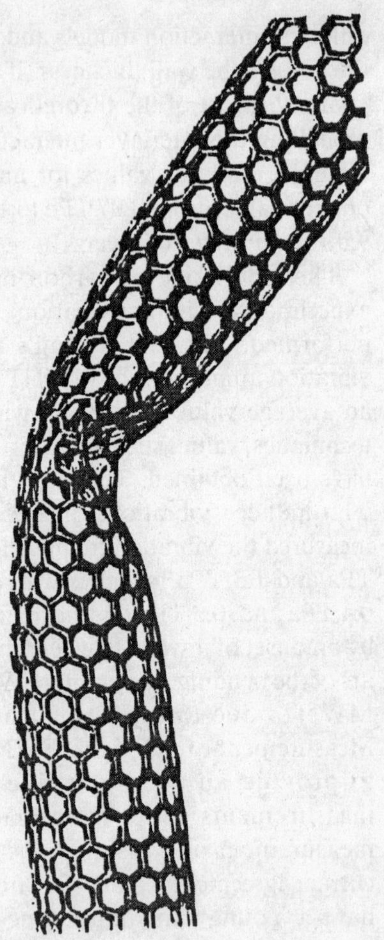

Fig. 12.23: A molecular dynamics simulations based example of nanotube behaviour under strain. The image shows portion of a (14,0) nanotube of 2632 atoms (tube length is 20 nm) bent 36 degrees with constant bending rate in 300 K temperature and then, to enhance structural reconstruction, annealed in an elevated 3300 K temperature for a sequence of three 130 ps annealing and cooling simulation runs. In the bending process the tube buckled but remained hexagonally bonded. The defects visible in the image relax the strain and are a result of the third annealing.

Defects in Carbon Nanotubes: As in any material, defects play an important role in nanotube properties. Structurally, defects make the tube less strong and thus in general defects are not desirable from the purely mechanical point of view. However, they alter the electronic properties locally, which can be utilized in the creation of single-tube devices. Defects are generated in the synthesizing process, and they can also be caused by mechanical manipulation, or, for example, by ion or electron beam bombardment of the tube. The most typical structural defects are fivefold (pentagon) and sevenfold (heptagon) rings in the sixfold (hexagonal) lattice. Other types of typical defects are vacancies and miscellaneous bonding configurations such as amorphous diamond. Noncarbon based defects include substitutional atoms or atom groups. In addition to these, MWNTs exhibit diverse defects based on discontinuous inner layers. Defects may alter the tube form from a straight tube to a bulging, kinked, spiral, or even more miscellaneous form. Defects and deformations induce scattering and localization, which result in conductivity drops, symmetry breaking based gap opening, and conductivity barriers. The scattering effects of isolated point defects are averaged over the circumference, and therefore their influence decreases with an increasing diameter and is small for all but the very thinnest tubes. However, if the defects or deformations accumulate locally, a tunnelling harrier may form. AFM manipulation has been used to make sharp bends in SWNTs, which has resulted in the formation of tunnelling barriers. The created changes are not, however, necessarily permanent. Nanotube defect manipulation offers interesting prospects both for devices will permanent characteristics and for devices will tunable properties, but a breakthrough requires more accurate and flexible mass manipulation of tube properties than is currently available.

Carbon Nanotubes as Mechanical Oscillators: Due to their small size, combined with a low mass density and high Young's modulus, nanotubes arc predicted to have characteristic vibration frequencies in the gigahertz regime, which is extremely high for a mechanical vibrator. Such frequencies are not attainable with the current silicon-based micro-mechanical oscillator production techniques with which eigenfrequencies in the 10 MHz to 100 MHz regime can be attained. A first approximation for the fundamental modes of a nanotube vibrator can be obtained from elasticity theory and is for a thin cylindrical transversally vibrating rod that is clamped from one end

$$f_1 = 0.28\frac{d_1}{L^2}\sqrt{\frac{Y}{\rho}} \tag{12}$$

If values for tube diameter d_1 = 1 nm, Young's modulus Y = 1 TPa, and density ρ = 750 kg/m^3 arc used (The mass density of a SWNT is not a well-defined quantity; the value presented here is a rough estimate based on the nanotube atomic spacial density and the volume encompassed by a tubular cylinder depicting the tube of 1 nm radius, to compare with graphite mass density is 1900-2300 kg/m^3, a frequency of 10 MHz is obtained for a tube length of 1 μm and a frequency of 1 GHz for a length of 100 nm. For a rod clamped from both ends, the constant in front of the equation is altered according to the boundary conditions, but the overall dependency on d_1, L, Y, and ρ remains. Based on the discussion above, nanotubes, and especially SWNTs, arc capable of significantly smaller dimensions and higher frequencies than conventional silicon-based vibrators.

Currently, only a few studies of excited mechanical vibrations in nanotubes exist. Pancharal *et al.* apply alternating potentials to MWNTs and measure Young's modulus and Roulet *et al.* have performed a basic study of acoustoelectric coupling in ropes of SWNTs, and they report that depending on the excitation power, either electron heating or phase coherence breaking is observed.

There are many things in the mechanical properties that have not been measured, and the measurements that exist suffer from a degree of inaccuracy. Most theory concentrates on SWNTs, and although the SWNTs often show more clearly the nanotube characteristic and desirable electronic properties, the theoretical aspects due to multiple layers still have many questions to answer, as do defectous-structure related matters. However, in nanoscale science and technology, electronic and mechanical will become more intimately connected, and therefore in the future the electromechanical properties of carbon nanotubes will play an important role.

Electronic Structure and Properties

Tight Binding Model: The carbon atoms in nanotubes are sp^2-hybridized, thus each atom contributes one unpaired π-electron to the tube. Carbon nanotubes are electronically active materials due to these π-electrons. The basic electronic properties of the carbon nanotube are calculated by extending the light-binding approach of 2D graphite to the geometrical configuration of the 1D nanotube system. The crystal momentum vector perpendicular to the nanotube axis is quantized due to the periodic boundary condition imposed by the finite circumference of the nanotube. This relatively simple model has been proven, mainly by low temperature STM spectroscopy, to explain well the basic electronic structure of the carbon nanotube. According to this model, the chirality of the carbon nanotube determines its division into metallic and semiconducting tubes. In the model, the curvature of graphitic wall of the nanotube is ignored. This has lately been shown to bring about important modifications to the model. Still the model survives well enough to continue to serve as a basis for classification of the nanotubes into metallic and semiconducting ones.

In the following brief presentation of the basic tightbinding model of the single shell carbon nanotube. Due to the tubular 1D structure, the nanotube has N separate 1D energy bands. In a general form these are derived from the graphite energy levels by the general formula

$$E_{\mu}(k) = E_{g2D}\left(k\frac{K_2}{|K_2|} + \mu K_1\right)\left(\mu = 0,\ldots N-1, -\frac{\pi}{T} < k < \frac{\pi}{T}\right) \quad (13)$$

E_{g2D} denotes the energy eigen values of a 2D graphite sheet, K_1 and K_2 are reciprocal lattice vectors corresponding to C and T, respectively. Figure 12.24 shows how the carbon nanotube k-vectors are projected to the graphite momentum space lattice. In graphite the conduction and valence bands meet at the so called K points in the momentum space, which is responsible for its semimetallic behaviour. In the tight-binding picture, the metallicity of a carbon nanotube is determined by whether the 1D lines of allowed k-vectors run through these points.

In an energy diagram, whether the energy bands will cross the Fermi energy or leave an energy gap depends on the (n, m) that is the chirality, as follows: The armchair nanotubes, with indices (n, n), have two bands that cross each other and the Fermi energy at the points $k = \pm 2\pi/3a$ they are thus always metallic, while only some of the zig-zag tubes are so. The general condition for the metallicity of a nanotube is that the factor $2n + m$ is a multiple of 3, a condition that indeed is fulfilled for the armchair tubes. As a further result, the theory predicts that one-third of all nanotubes are metallic while two-thirds are semiconducting. These rates of occurrence have not been confirmed by the presently performed experiments, which however is not surprising considering doping effects and the fact that almost nothing is known about the possible preference in the growth processes for certain nanotube symmetries over others.

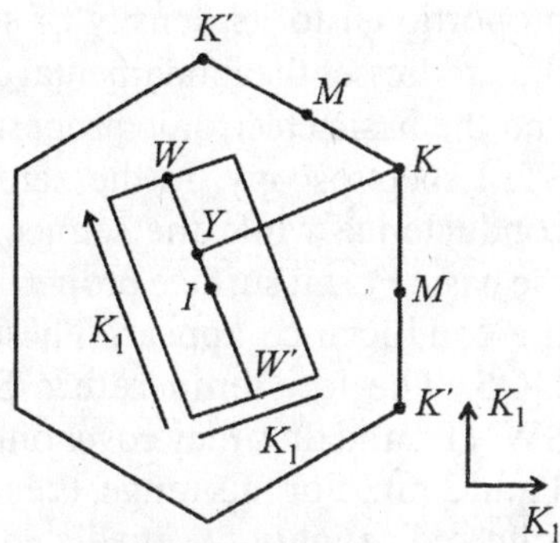

Fig. 12.24: Allowed K-vector or carbon nanotubes are formed by the vectors $\boldsymbol{K}_1$ and $\boldsymbol{K}_2$ in the reciprocal lattice space of graphite. The nanotube is metallic if the ratio of $\boldsymbol{YK}$ to $\boldsymbol{K}_1$ is an integer

As a general result, in all metallic 1D systems with half-filled bands, the so-called Peierls distortion will, by an atomic displacement, change the unit cell so that the band becomes filled and a gap opens up at the Fermi energy. The phenomenon is well known among conducting polymers, charge transfer salts, etc. In carbon nanotubes, however, it has been shown theoretically that the effect very quickly diminishes as a function of nanotube radius.

Experimental Studies: STM Imaging and Spectroscopy: The electronic structure of the SWNT that is predicted by the tight-binding calculations described above was put to test at first primarily by scanning tunneling spectroscopy, done with low temperature STM. In such experiments, it is possible to image the nanotube with atomic resolution along the length of the tube and with a width of a few unit cells. The width of the imaging is of course limited by the curvature of the nanotube. This imaging reveals (in the best cases) the structure and diameter of the nanotube, enabling one to determine the (n, m) indices of the nanotube with some confidence. Furthermore, the STM tip be positioned at any point along the nanotube, separated by a tunneling distance from the nanotube surface, and spectroscopic measurements be performed by taking the current-voltage characteristics. The I-V characteristics reveal the metallic or semiconducting character of the tube since the measured current is proportional to its density of states (DOS). This experimental technique thus addresses the fundamental question of the relation between the structure and the basic electronic properties of the individual nanotube. As seen by STM spectroscopy, in the zero-bias region the metallic tube has a finite conductance while the semiconducting tube has zero, corresponding to the basic Fermi surface properties. At higher voltages asymmetric peaks in the conductance appear. These are the Van Hove peaks due to the ID DOS. The low-temperature STM experiments were carried out with SWNTs, and thus they cover only a very limited range of diameters, between 1 and 2 nm. For this range, the measured E_g for semiconducting nanotubes-followed roughly the predicted $1/d_1$ dependence, being in the range 0.3-0.7 eV.

In more recent scanning tunneling spectroscopy measurements on SWNTs, a curvature-induced "extra" gap was found in zig-zag type metallic nanotubes, but not in tubes of armchair type. This gap is also inversely dependent on the nanotube radius. The measured results were theoretically anticipated in refined calculations The curvature-induced gap has a size of a few 10s of meV for subnanometer tube diameters. From this energy scale we see that for all but the very smallest of possible tubes, the effect is washed out at room temperature.

Other scanning tunneling spectroscopy experiments have been performed that explore other topics such as individual quantum slates in short SWNTs and MWNTs doped nanotubes crossing nanotubes, defects and junctions in SWNTs and peapods (SWNTs filled with C_{60})

Experimental Studies: Raman Spectroscopy: Another important tool for nanotube characterization is Raman spectroscopy. Particularly, the strong, so-called radial breathing mode is derived from the tubular structure of the nanotube and is absent from, all other 2D graphitic forms of carbon. It is widely utilized to give a measure on the nanotube diameter. Due to the singularities in the DOS of the 1D nanotubes and the electron-phonon coupling, very strong resonant Raman effects are observable in studies on carbon nanotubes. This enables the use of Raman spectroscopy to investigate the electronic structure of carbon nanotubes, particularly the occurrence of semiconducting and metallic properties. A possibly very important recent development is the ability to measure the Raman signal from individual nanotubes with the so-called micro-Raman technique. It was claimed to be possible to determine the (n, m) indices of the isolated nanotube with this technique. This of course opens up the possibility of transport measurements on a nanotube of known symmetry.

Optical Properties: The optical physics of nanotubes has not moved forward in pace with the other developments in the field. This is due to unability to make samples consisting of nanotubes of a single symmetry.

Thus all optical signals are a mixture of signals from a very large variety of nanotubes. Furthermore, since nanotubes tend to aggregate together, especially SWNTs, any signal from semiconducting tubes is easily hidden by interaction with neighbouring metallic tubes. Unlike transport measurements, optical measurements on single nanotubes are experimentally less feasible. On the other hand, however, the 1D singularities in the optical response of nanotubes give strong signals, just as in the case of Raman spectroscopy. Also, if aligned assemblies of nanotubes can be produced, the optical anisotropy of such a system is potentially huge. In a recent development, SWNTs were processed by sonication and surface-active chemicals to produce solutions of truly individually solvated tubes, coated with the surface-active hydrocarbon molecules. These SWNTs were rather short (due to the strong sonication), with an average length of only 130 nm. For perhaps the first time, photoexcited luminescence from nanotube solutions were observed in this case. The luminescence spectrum was closely associated with the measured absorption spectrum, with only a minor red-shift. At the moment of writing, this work is intensively being pursued and is likely significantly widen research in carbon nanotube systems. Photoconductivity measurements on single nanotubes have so far not been successful.

The research on the optical properties of carbon nanotube may in the near future bring forward a number of' potential applications within optoelectroates, thus adding to the list that already exists within other fields of nanotube-based nanotechnology.

Electronic Transport Properties

Within the realm of electronic transport physics, the carbon nanotube offers exciting possibilities both as a testing ground in basic research as well as in nanotechnology. In the case of the former, the near perfection in structure for lengths exceeding 1 μm makes the nanotube the best concrete example of a 1D quantum wire. On the other hand, in nanotechnology the prospect of devices made from a single nanotube, that is, intrananotube devices, offer the possibility of truly molecular-level electronics. Even if molecular electronics is eventually pursued with smaller molecules, carbon nanotubes may be used as connecting wires between metal structures of ~100 nm size and molecules of ~1 nm size.

Although the carbon nanotube is mainly "marketed" as a perfect 1D object, this is not always the case. In fact, depending on the circumstances one may have to treat the nanotube as anything from a 0D to a 3D object. Obviously, if a SWNT is short enough, it becomes a 0D object, or a quantum dot. If only the outer shell of a MWNT is considered, and the diameter of the tube is large enough with respect to some critical length scale of the particular situation, this outer shell becomes a 2D object. Finally, if we again consider a larger MWNT (in diameter), and the charge carrier coupling between the layers is strong enough, then we may have at hand an anisotropic 3D object. We can thus see that even without considering nanotubes of different charalities and more elaborate topologies than the simple tubular one, the subject of electron transport in carbon nanotubes can be rather complicated.

As has been discussed, the energy gap in semiconducting SWNTs is ~0.5 eV, which is high enough with a good margin for the nanotube to function at room temperature as the active component for example, in field effect transistors. Other effects, such as the energy level quantization in short nanotubes (L = 10–100 nm), require cryogenic temperatures for the effects to be observable. Roughly it can be said that in metallic nanotubes the interesting transport phenomena occur at low temperatures, while in semiconducting tubes much of the work is carried out at room temperature.

Bulk Transport: In this section we review work on the transport physics of single SWNTs, SWNT ropes, and MWNTs. This is clearly more significant than the subject of transport in macroscopic amounts of carbon nanotubes. However, the carbon nanotube may offer interesting applications as the conductive component in composites, when mixed together with an insulating host material. In order to have a composite conducting, the volume fraction of the conductive component has to exceed some critical value. Typically tile conductive material consists µm-sized particles of a more or less rounded shape. For conduction to occur, the particles have to touch each other frequently enough so that the conductive channels are formed over macroscopic distances. As prescribed by percolation theory, this occurs at a volume fraction of 16% whereby the conductivity of the composite rises very sharply (as a function of filling percentage) with several orders of magnitude. However, it has already been demonstrated with conductive material consisting of 1D objects, for example fibrils of polyaniline a conducting polymer, that there is no apparent sharp threshold for the onset of conduction, and more importantly, the composite becomes conducting at a much lower volume fraction of the conductive component. Obviously, this should also be achievable with nanotubes as the conductive components. In fact nanotube-based composite systems are already being considered for commercial applications.

Nanowires

Fullerenes and CNT are hollow, so it should be possible to fill them up with different elements and compounds. Introducing a metal inside CNT might form a nanowire. Using electric arc discharge with a metal powder inside the anode produces CNT filled with metal carbides. A difficulty from the method mentioned above is that the tubes are filled partially; this might affect the conductivity properties. It is found that nanotubes could be filled more if we use small amounts of selenium, sulfur and germanium with the electric arc discharge method. Moreover, there are chemical techniques in which by nitric acid treatment, the tubes can be opened and then materials such as palladium, silver, gold, cobalt, iron, uranium and nickel oxides, molybdenum, tin, nyodinium, europium, and cadmium, can be inserted in their interior. It is important to know that enzymes and proteins have been put inside nanotubes. Capillarity effects have also been used to introduce elements inside CNT such as tin, lead, bismuth, cesium, sulfur and selenium. Nanowires have been used to prepare ultimate interconnects in integrated circuits.

Carbon Nanocones

Theoretical studies predicted the formation of graphitic cones. Subsequently, isolated graphitic cones were produced by carbon condensation on a graphite substrate and by pyrolysis of heavy oils. More recently, single walled aggregates of conical graphitic structures have been prepared by laser ablation of graphite targets. In addition, conical structures consisting of other layered materials such as BN have also been prepared by reacting boron oxide vapours with multi-walled CNTs.

Recently, it is reported that pyrolysis of palladium precursors, always produce conical nanofibers. An important feature of these new nanostructures is that they are held together by Van der Waals forces since the fibre is composed of an arrangement of stacked cones which can be opened (lamp-shade), or closed. It is believed than nanocones may be good electron field emitters. Calculations of the electronic properties of nanocones reveal that there is a charge accumulation towards the tip and that there are localized states near the Fermi level. These features make them suitable as field emitters.

Haeckelites

The presence of defects such as pentagons and heptagons in fullerenes change the electronic properties. A new hypothetical type of graphene sheets has been proposed which admit pentagons, heptagons, and hexagons, observing that the number of heptagons and pentagons should be the same in order to compensate the negative curvature of the heptagons and the positive curvature of pentagons (Fig. 12.25). These arrangements have been named as haeckelites. They all exhibit a considerable signal at the Fermi level in the local density of states. Obviously, all these systems should possess a metallic behaviour. Thus, it is possible to roll up haeckelites to form nanotubes, which will be conductors independent of the diameter and the helical arrangement.

Another property of haeckelite tubes is that they keep the stiffness of classical graphite nanotubes composed with just hexagons; the Young's modulus of haeckelite tube is around 1.0 Tera Pa; of the same order as classical carbon nanotubes. Haeckelites also exhibit local rugosity due to the local bending produced by heptagons and pentagons. It might be possible, that electron irradiation experiments can generate heptagon-pentagon pairs (rugosity) in SWCNT.

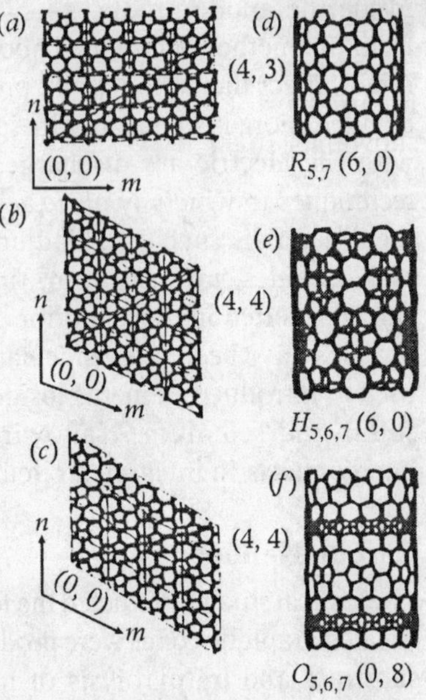

Fig. 12.25: Haeckelites (a) Rectangular haeckelite sheet (*R*); (*b*) Hexagonal haeckelite sheet (*H*); (*c*) Oblique haeckelite sheet (*O*); (*d*) to (*f*) are the corresponding haeckel tubes. All haeckelites are metallic (sheets and tubes)

Production Methods for CNTs

CNTs can be obtained by using four different methods: (*i*) arc-discharge (*ii*) pyrolysis (*iii*) laser vaporization and (*iv*) electrolysis.

The electric arc discharge set-up is shown in Fig. 12.26. This method is very similar to the one used for obtaining fullerenes with two main differences: first, the pressure are higher, around 500 torr (for fullerenes the pressure is around 100 torr), and second, the nanotubes are found deposited in the cathode and not in the soot. This method produces well ordered MWCNT with diameters from 2 nm upto 30 nm. The separation between graphite layers is around 3.4 Å, which is very close to that of graphite 3.35 Å and corresponds to turbostratric graphite. The length of these nanotubes can be upto 30 μ. The optimum conditions for generating nanotubes with this method consist of using currents of 150 amps, voltages of 25 volts and graphite electrodes between 6 mm and 8 mm in diameter with a separation between them of 1 mm under helium atmosphere. The electric arc reaction is too violent, so it is very difficult to control the formation reaction.

The pyrolysis method (Fig. 12.27) consists of heating a hydrocarbon or a precursor which contains carbon under the presence of a catalyst such as nickel, cobalt or iron. Prior to the discovery of C_{60}, it was well known that carbon fibres could be obtained by pyrolytic methods using hydrocarbons. We must note that carbon fibres are not nanotubes since the fibres are bigger in diameter and exhibit a great amount of defects and impurities. Nevertheless, the application of carbon fibres is widespread, showing a strength comparable to steel.

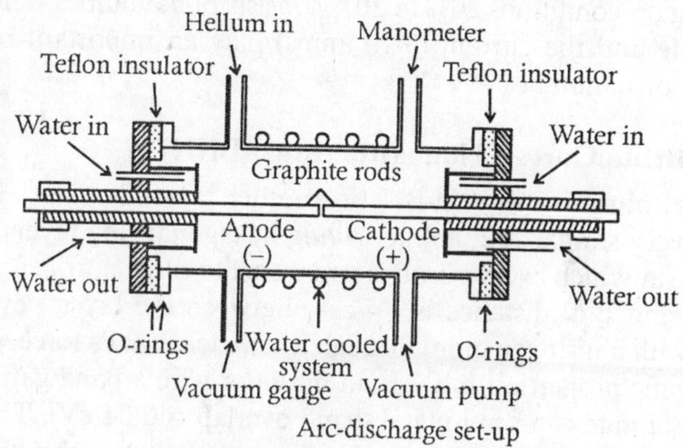

Fig. 12.26: Arc-discharge set-up

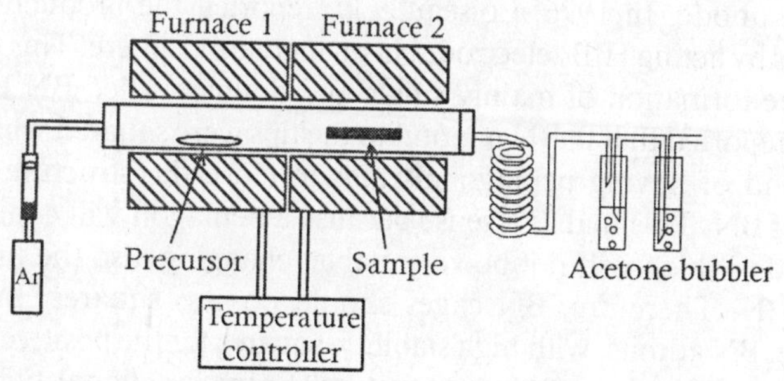

Fig. 12.27: Pyrolysis set-up

In 1994, Ajayan *et al.* found that carbon nanotubes could be aligned by embedding them in a polymer, thus forming a composite material. When the polymer was cut, they observed an aligned arrangement of nanotubes. In this context, a better alternative is to use laser ablation of cobalt thin films and then pyrolyse them with an organic precursor such as 2-amino 4, 6-dichloro-5-tryazine. The nanotubes obtained in this way are aligned in bundles which can grow upto 40 μ and diameters between 30 nm and 50 nm.

The laser vaporization method involves firing a high power laser towards a graphite target inside a furnace at 1200° C. The condensation of material generated by the laser is responsible for the nanotube formation. If we add nickel or cobalt to the graphite target, we obtain single-walled carbon nanotubes (SWCNT). The SWCNT show diameters around 14 Å forming a two-dimensional crystal with a lattice constant of 17 Å. Arrangements of SWCNT can also be obtained by electric arc discharge with mixtures of carbon, nickel and ytrium.

The electrolysis method is based on using graphite electrodes inside a molten salt such as lithium chloride (700° C) under an argon atmosphere. Depending on conditions 20% to 30% consists of nanotubes. The depth of the cathode and the current (3-10 amps) play an important role in the formation of nanotubes.

Boron Nitride Curved Nanostructures (BNNTs)

Hexagonal boron nitride (h-BN) is another layered material with the structure very similar to graphite having hexagonal ring layers separated by 3.33 Å, in which every boron is connected to three nitrogen atoms and *vice-versa*; and B-N distance is 1.44 Å. Between the layers, every boron interacts with a nitrogen atom through a Van der Waal's force. Regarding the electronic properties, h-BN is an insulator with a band gap of 5.8 eV, whereas graphite is a semi-metal (band overlap = 0.04 eV). The first BN nanotube were prepared using the electric arc discharge with a high melting point metal (tantalum and tungsten for e.g.) electrodes and BN powder inside the anode. In 1996, Loiseau *et al.* reported the production of BN nanotubes by acting HfB_2 electrodes in an inert atmosphere. This technique lead to the formation of mainly single-and double-walled BN tubes.

It was reported that in BN nanotubes the tips were squared which means that instead of having pentagonal rings to close the structure there are squares of BN. This makes a sense because a pentagon will exhibit B-B or N-N bonds, which will not be very stable, changing also the proportions of B and N. Therefore, BN cages should possess squares. In terms of curvature, BN admits, with high stability, squares for the positive curvature and octagons for the negative curvature. From functional tight binding calculations, it is found that nitrogen excess cages with B and N are stable. Theoretical studies revealed that BNNTs are semiconductors with a band gap of 5.5 eV, smaller than in bulk BN. This property of BNNTs is independent of tube diameter, chirality, and number of walls. These BNNTs may also be extremely resistant to oxidation and should exhibit high Young's modulii. Therefore, the BNNTs may find important uses in ceramics.

BNNTs bundles can also be produced when carbon nanotubes are exposed to mixture of B_2O_3, V_2O_5, and M_0O_3 in a nitrogen atmosphere at elevated temperatures. There are alternative approaches to generate BNNTs and other BN nanostructures: (a) thermolysis of borazine ($B_3N_6H_6$) over nickel boride catalyst particles at 1000-1100°C, (b) thermal annealing of powdered rhornbohedral boron and h-BN mixtures at 1200°C in lithium vapour, (c) plasma jets and (d) ball milling of h-BN, followed by thermal annealing at 1300°C.

At this stage, novel composites and nanoscale components need to be fabricated employing BNNTs, which appear to be extremely resistant to oxidation. Experimentally obtained values of Young's modulus of BNNTs (Y or E = 1.22 ± 0.24 TPa) agrees well with the values obtained using theoretical approaches, and reveals that BN tubes are highly crystalline and probably, they may be the strongest insulating nanofibres.

Graphite and h-BN have opened new avenues for novel nanomaterials with different combinations of the type $B_xC_yN_z$. Pyrolysis experiments with melamine have been found very helpful in synthesizing C_xN_y nanofibres.

Boron doped carbon nanotubes have been synthesized by electric arc discharge with h-BN inside the graphite anode. During this experiment, it is observed that carbon nanotubes are highly graphitic and also very long (~20μ). It was found that B was located at the tips of the nanotubes and that the zig-zag chirality was preferred. It is reported by ab-initio calculations that boron was acting as surfactant avoiding the zig-zag tubes closure. In addition, boron doped nanotubes exhibit conducting properties.

It is reported that nitrogen doped carbon nanotubes can also exhibit interesting features such as pyridine like bonding in the graphitic mesh, which causes peak close to the Fermi level in the LDOS. All these nanotubes, arm chair and zig-zag are found to be conductors. Obviously, nitrogen is donating electrons to the nanotube system to produce conductivity.

Key Issues in Nanomanufacturing

Similar to large-scale fabrication, nanomanufacturing research issues revolve around precursor materials; fabrication processes and characterization techniques; instrumentation and equipment; theoretical modelling and control; and design and integration of structures into devices and systems.

Nanostructures in Motion

Micromechanics, micromachines and moving nanostructures represent a new "micro-revolution". However, the concepts related to micromechanics and micromachining of silicon are not new. A number of commercial products including accelerometers, pressure transducers, thermal print heads and ink jet nozzle arrays have been fabricated for internal industrial use and for the open market,

Since 1982, the field of micromechanics has expanded considerably, adding micrometers, movable linkages, gears, micro-turbines, electrical drives and microweezers to an ever growing suite of micromechanical devices.

Nanomaterial Advantage

Nanocrystalline materials are three dimensional solids composed of nanometersized grains, or crystallites. Because of their unique structure, which is characterized by ultrafine grains and a rather high density of crystal lattice defects, these materials have extra ordinary fundamental properties that could be exploited to make 'next-generation' superstrong materials.

Strength and ductility are the central mechanical properties of any material. They are determined by the physical nature of plastic deformation, which in conventional, coarse-grained metals is mainly carried by dislocations-line defects of regular crystal lattice-within individual grains. Study suggest, however, that the mechanism of plastic deformation in nanocrystalline materials may be different and so may lead to novel mechanical properties.

High strength, or hardness has already been observed in many nanomaterials. But in most cases such nanomaterials have very low ductility—they fail when their shape is changed. Some even become brittle when a force or deforming load is applied. Strength and ductility are usually opposing characteristics: the higher the strength, the lower the ductility, and *vice versa.* This correlation is associated with the nature of plasticity: the more difficult it is for dislocations to appear and to move, the stronger but brittle and less ductile is any crystalline material.

Very recently, Wang *et al*. report their success in engineering an unusual nanocrystalline material that combines two useful properties that are often mutually exclusive-strength and ductility. They created a copper nanostructure by rolling the metal at temperature below 77 K and then heating it to around 450 K. The result was a 'bimodel' structure of micrometer-sized grains (at a volume fraction of around 25%) embedded in a matrix of nanocrystalline grains. The material showed extraordinary high ductility, but also retained its high strength. The reason for this behaviour seems to be that while the nanocrystalline grains provide strength, the embedded larger grains stabilize the tensile deformation of the material.

Processing nanomaterials to achieve both strength and ductility can enhance the material's resistance to fatigue. More than anything else, it is fatigue that at present limits the life-time, and hence the range of applications, of many advanced materials.

Transmission Electron Microscope (TEM)

Transmission electron microscope is very important tool in materials science for investigating the fine scale structure of materials. Fig. 12.28 shows a transmission electron microscope.

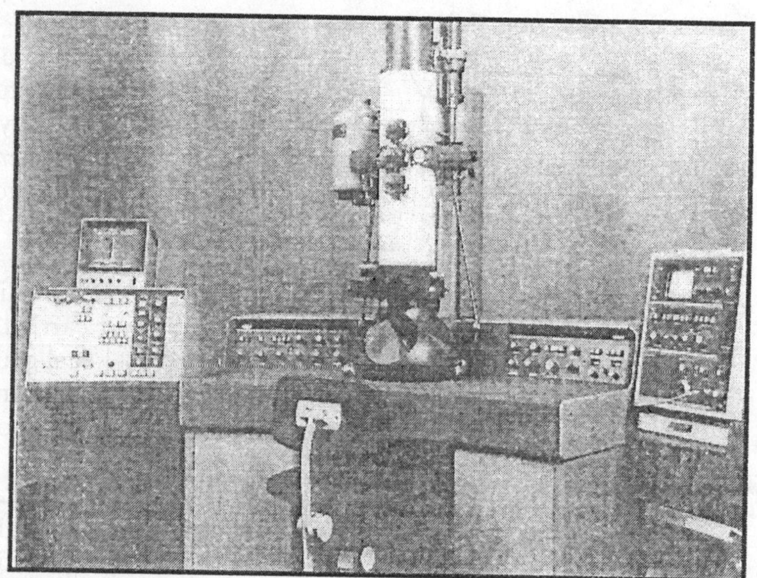

Fig. 12.28: Photograph of a modern transmission electron microscope

TEM is a microscopy technique whereby a beam of electrons is transmitted through an ultra thin specimen, interacting with the specimen magnified and focused by an objective lens and appears on an imaging screen, a fluorescent screen in most TEMs, plus a monitor, or on a layer of photographic film, or to be detected by a sensor such as a CCD camera. The first practical TEM was built by A Arebus and J. Hillier at the University of Tornto in 1938.

Figure 12.29 shows schematically the the basic principles of its operation. An electron beam is produced by a heated tungsten filament at the top of an evacuated column and is accelerated down the column by a high voltage (usually from 75 to 120 kV). Electromagnetic coils are used to condense the electron beam, and then it is passed through a very thin (i.e. about 100 nm thick or less) section of a thin metal specimen placed on the specimen stage (Fig. 12.29). The specimen area examined must be very thin so that some of the electrons entering are able to pass through it. As the electrons pass through the specimen, some are absorbed and some are scattered so that they change direction. Differences in crystal atomic arrangements will cause electron scattering. After the electron beam has passed through the specimen, it is focused with the objective coil and then enlarged and projected on a fluorescent screen (Fig. 12.29). A region in a metal specimen that tends to scatter electrons to a high degree will appear dark on the viewing screen. Thus dislocations that have an irregular linear atomic arrangement will appear as dark lines on the electron microscope screen, as observed in Fig. 12.30.

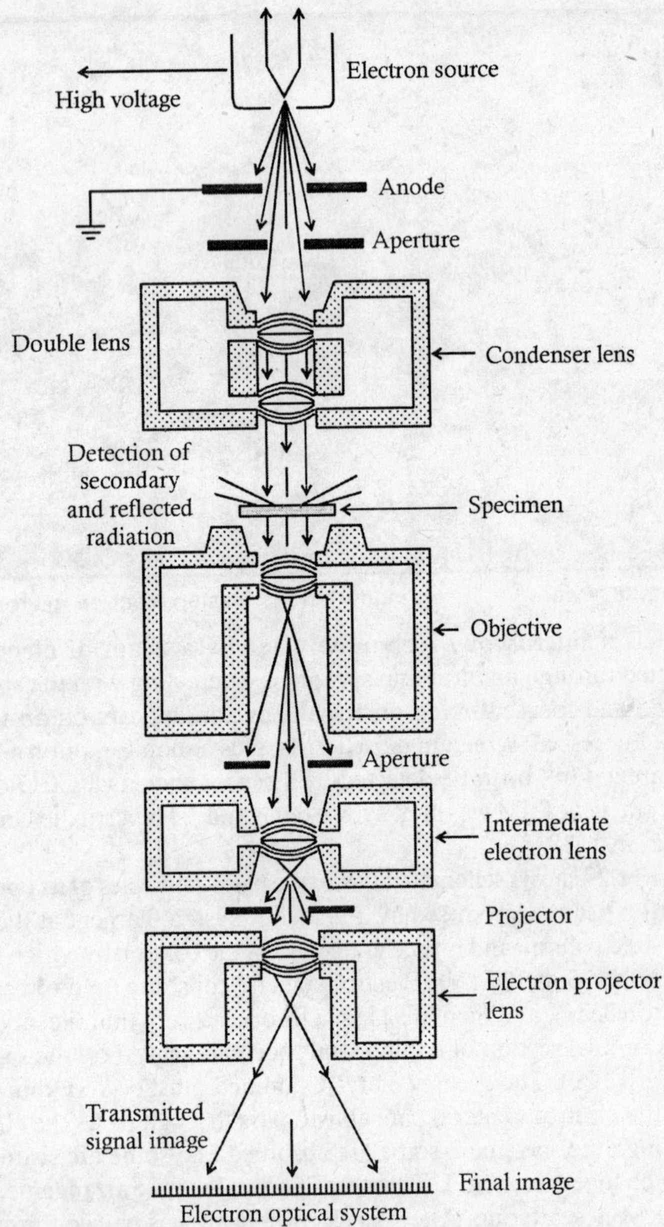

Fig. 12.29: Schematic arrangement of electron lens system in a transmission electron microscope. All the lenses are enclosed in a column that is evacuated during operation. The path of the electron beam from the electron source to the final projected transmitted image is indicated by arrows. A specimen thin enough to allow an electron beam to be transmitted through it is placed between the condenser and objective lenses as indicated

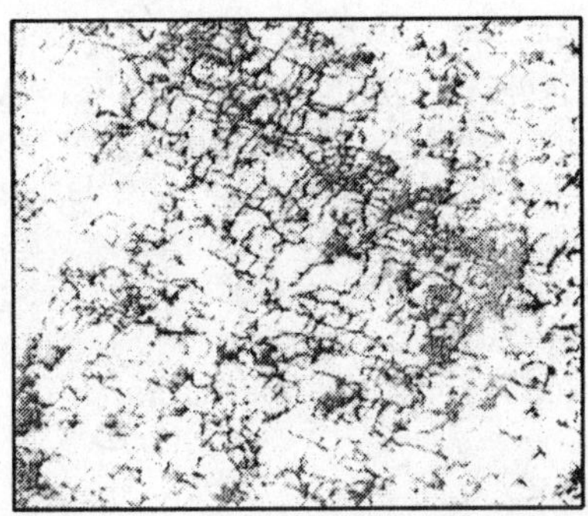

Fig. 12.30: Dislocation structure in iron deformed 14 per cent at –195° C. The dislocations appear as dark lines because electrons have been scattered along the linear irregular atomic arrangements of the dislocations (Thin foil specimen; magnification 40,000×)

A TEM works much like a slide projector. A projector shines a beam of light through (transmits) the slide, as the light passes through, it is affected by the structures and objects one the slide. These effects result in only certain parts of the light beam being transmitted through certain parts of the slide. This transmitted beam is then projected into the viewing screen, forming an enlarged image of the slide.

TEMs work the same way except that they shine a beam of electrons (like the light) through the specimen (like the slide). Whatever part is transmitted is projected into a phosphor screen for the user to see. The working of typical TEMs is as follows:

The 'source' at the top represents the *electron gun*, producing a stream of monochromatic electrons.

- This stream is focused to a small, thin, coherent beam by the use of condenser lenses. The first lens largely determines the 'spot' size, the general size range of the final spot that strikes the sample. The second lens usually (usually controlled by the "intensity of brightness knob") actually changes the size of the spot on the sample, changing it from a wide dispersed spot to a pinpoint beam.
- The beam is restricted by the condenser aperture (usually user selectable), knocking out high angle electrons (those for from the optic axis).
- The beam strikes the specimen and parts of it are transmited.
- This transmitted portion is focused by the objective lens into an image.
- Optional objective and selected area metal opertures can restrict the beam, the objective aperature enhancing contrast by blocking out high-angle diffracted electrons, the selected area aperture enabling the user to examine the periodic diffraction of electrons by ordered arrangements of atoms in the sample.

- The image is passed down the column through the intermediate and projector lenses, being enlarged all the way.
- The image strikes the phosphor image screen and light is generated allowing the user to see the image. The darker areas of the image represent those areas of the sample that fewer electrons were transmitted through (They are thicker or denser). The lighter areas of the image represent those area of the sample that more electrons were transmitted through (they are thinner or lens dense).

Applications of TEM

The TEM is used heavily in both material science/metallurgy and the biological sciences. In both cases the specimen must be very thin and able to withstand the high vacuum present inside the instrument.

For biological specimen, the maximum speciman thickness is roughly 1 micrometre. To withstand the instrument vacuum, biological specimen are typically held at liquid nitrogen temperatures after embedding in vitreous ice, or fixated using a negative staining material such as uranyl acetate or by plastic embedding. Typical biological applications include tomographic reconstructions of small cells or thin sections of larger cells and 3-D reconstructions of individual molecules via Single Particle Reconstruction.

Fig. 12.31: SEM image of a thin TEM sample milled by FIB. The thin membrane shown here is suitable for TEM examination; however at ~300 nm thick, it would not be suitable for high resolution TEM without further milling.

In material science/metallurgy the specimen tend to be naturally resistant to vacuum, but must be prepared as a thin foil, or etched so some portion of the speciman is thin enough for the beam to penetrate. Preparation techniques to obtain an electron transparent region include ion beam milling and wedge polishing. The focused ion beam (FIB) is a relatively new technique to prepare thin samples for TEM examination from larger specimen. Because the FIB can be used to micro-machine samples very precisely, it is possible to mill very thin membranes from a specific area of a sample, such as a semiconductor or metal. Materials that have dimensions small enough to be electron transparent, such as powders or nanotubes, can be quickly produced by the deposition of a dilute sample containing the specimen onto support grids. The suspension is normally a volatile solvent, such as ethanol, ensuring that the solvent rapidly evaporates allowing a sample that can be rapidly analysed.

The imaging techniques explained below are particularly important in materials science. Faults in crystals affect both the mechanical and the electronic properties of materials, so understanding how they behave gives a powerful insight. By carefully selecting the orientation of the sample, it is possible not just to determine the position of defects but also to determine the type of defect present. If the sample is orientated so that one particular plane is only slightly tilted away from the strongest diffracting angle (known as the Bragg Angle), any distortion of the crystal plane that locally tilts the plane to the Bragg angle will produce particularly strong contrast variations. However, defects that produce only displacement of atoms that do not tilt the crystal to the Bragg angle (i.e. displacements parallel to the crystal plane) will not produce strong contrast.

Furthermore, the HRTEM technique allows the direct observation of crystal structure and therefore has an advantage over other methods in that there is no displacement between the location of a defect and the contrast variation caused in the image. However, it is not always possible to interpret the lattice images directly in terms of sample structure or composition. This is because the image is sensitive to a number of factors (specimen thickness and orientation, objective lens defocus, spherical and chromatic aberration), and although quantitative interpretation of the contrast shown in lattice images is possible, it is inherently complicated and may require extensive simulation of the images. Computer modelling of these images has added a new layer of understanding to the study of crystalline materials.

Imaging in the TEM

The contrast in a TEM image is not like the contrast in a light microscope image. A crystalline material interacts with the electron beam mostly by diffraction rather than absorption, although the intensity of the transmitted beam is still affected by the volume and density of the material through which it passes. The intensity of the diffraction depends on the orientation of the planes of atoms in a crystal relative to the electron beam; at certain angles the electron beam is diffracted strongly from the axis of the incoming beam, while at other angles the beam is largely transmitted. Modem TEMs are often equipped with specimen holders that allow the user to tilt the specimen to a range of angles in order to obtain specific diffraction conditions, and apertures placed below the specimen allow the user to select electrons diffracted in a particular direction.

A high-contrast image can therefore be formed by blocking electrons deflected away from the optical axis of the microscope by placing the aperture to allow only unscattered electrons through. This produces a variation in the electron intensity that reveals information on the crystal structure, and can be viewed on a fluorescent screen, or recorded on photographic film or captured electronically.

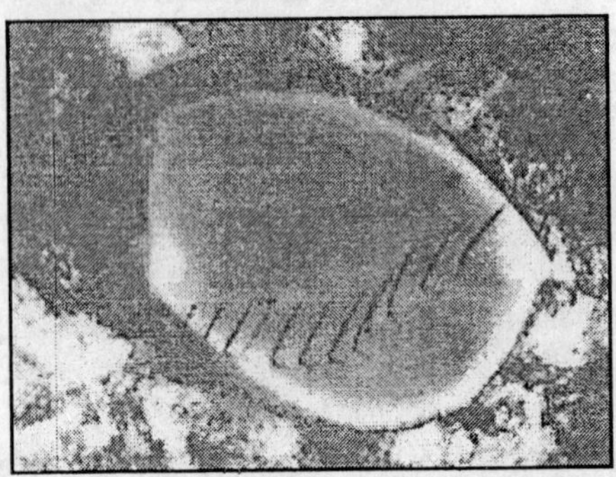

Fig. 12.32: Transmission Electron Micrograph of Dislocations

This technique (known as *Bright Field* or *Light Field*) is particularly sensitive to extended crystal lattice defects in an otherwise ordered crystal, such as dislocations. As the local distortion of the crystal around the defect changes the angle of the crystal plane, the intensity of the scattering will vary around the defect. As the image is formed by the distortion of the crystal planes around the defect, the contrast in these images does not normally coincide exactly with the defect, but is slightly to one side.

It is also possible to produce an image from electrons deflected by a particular crystal plane. By either moving the aperture to the position of the deflected electrons, or tilting the electron beam so that the deflected electrons pass through the centred aperture, an image can be formed of only deflected electrons, known as a *Dark Field* image.

In the most powerful diffraction contrast TEM instruments, crystal structure can also be investigated by High Resolution Transmission Electron Microscopy (HRTEM), also known as phase contrast imaging as the images are formed due to differences in phase of electron waves scattered through a thin speciman.

Resolution of the HRTEM is limited by spherical and chromatic aberration, but a new generation of aberration correctors has been able to overcome spherical aberration. Software correction of spherical aberration has allowed the production of images with sufficient resolution to show carbon atoms in diamond separated by only 0.89 Å (89 pm, one angstrom is 0.0000000001 of a meter or 100 picometers) and atoms in silicon at 0.78 Å (78 pm) at magnifications of 50 million times. Improved resolution has also allowed the imaging of lighter atoms that scatter electrons less efficiently—lithium atoms have been imaged in lithium battery materials. The ability to determine the positions of atoms within materials has made the HRTEM an indispensable tool for nanotechnology research and development in many fields, including heterogeneous catalysis and the development of semiconductor devices for electronics and photonics.

Limitations

There are a number of drawbacks to the TEM technique. Many materials require extensive sample preparation to produce a sample thin enough to be electron transparent, which makes TEM analysis a relatively time consuming process with a low throughput of samples. The structure of the sample may also be changed during the preparation process. Also the field of view is relatively small, raising the possibility that the region analysed may not be characteristic of the whole sample. There is potential that the sample may be damaged by the electron beam, particularly in the case of biological materials.

Semiconductor Quantum Dots (Q Dots)

These are nanometers-sized semiconductor crystals, or electrostatically confined electrons. These hold the promise of a true revolution in bio-imaging quality and sensitivity.

Q Dots takes advantage of the *quantum confinement effect* predicted by quantum mechanical theory to fluoresce extremely brightly when excited by a light source such as laser. Q Dots don't have any of the disadvantages of conventional technologies. By varying the size of crystals one can cause a rainbow of colours to fluoresce. In addition, Q Dots stay lit for much longer periods of time than conventional dyes often hours or days.

Q Dots enable the tagging of a variety of different biological components like proteins or different strands of DNA with specific colors. Q Dots are designed to bond with and illuminate individual biological targets of choice whether genes, nuliec acids, proteins, small molecules, cancer cells, or even entire blood vessels. Q Dots are applied in high content drug screening and initial drug screens using Q Dots as biological read-out have been completed.

Somicondutor quantum dots signify a class of materials in which quantum containment effects are investigated in greated detail. They are also refereed to as semiconductor nanocrystals. In fact these constitute a sub-class of a broad family of nanoparticles, which include semiconductor, metal, insulator, organic etc. particles. 'Quantum dot's is a term referred only to semiconductor particles, while nanocrystal can be an inorganic entity in which there is a crystalline arrangement of constituent atoms/ions. In a microscope semiconductor crystal, the energy levels form bands. The valence band is filled and the conducting band is completely empty at 0 K. The bands are separated with a specific energy gap E_g. When an electrons gas excited due to thermal excitations an electron-hole pair is created. The electron in the conducting band and the hole in the valence band can be bound when they approach each other at a finite distance. This bond pair is called an 'exciton', which is delocalized throughout the crystal. The Bohr radius of the exciton can be given as $a = h^2\varepsilon/4\pi^2e^2(1/m_e + 1/m_h)$, where ε is the dielectric constant of the material, m_e and m_h are the effective masses of electron and hole, respectively, and e is the elementary charge. Quantum size effects are manifested when the length of the nanocrystal made, d, is comparable to the exciton radius a. This is ~56 Å for CdSe. Note that the unit cell dimensions of the semiconductor are much smaller than the characteristic length. The de Broglie wavelength in materials $\lambda = h/mv$, is in the range of nanometers and strung confinement effects are manifested only when the particle dimension approaches this value. At this dimension, most materials have structures similar to those of their bulk counterparts, at least in the core of the particle.

The electronic structure of materials is strongly related to the nature of the material. In a three-dimensional object of large size, the electronic structure is not restricted by the dimension of the material. The wavelength of electrons is much smaller than the typical length of the material. When the electronic motion is confined in one dimension, and it is free in the other two dimensions, it results in the creation of 'quantum wells' or 'quantum films'. The quantum well notation implies that the electrons feel a potential well as they are trapped in the film. The quantum film notation is self-explanatory. Here the density of states shows a step-like behaviour. In tile case of a one-dimensional system, *i.e.* when the electrons are free to move only in one direction, we get a situation wherein the density of states shows a Lorenzian line shape. Such a situation can be seen in carbon nanotubes. If the electrons are confined to a point, we get a zero-dimensional system, wherein the electrons are not free to move at all. Here we get states which are molecular in nature. The situation is schematically depicted in Fig. 12.33. What is shown in Fig. 12.33 is that while the density of states is smoothly varying in bulk materials, it shows discontinuities in confined systems. This will lead to steps in two-dimensional confinement, singularities in one-dimensional confinement and discrete lines in zero-dimensional confinement.

Some of the properties which change drastically as a function of size arc the optical properties, including both the absorption and emission of liglit, which is evident from Fig. 12.33. Nanocrystals have discrete orbitals. The energy of the first level will be shifted from the position of the bulk value by $h^2/8ma^2$, where a is the diameter of the particle. Remember that the particle in a box model predicts the energies of the levels as $n^2h^2/8m_ea^2$ ($n = 1, 2, \ldots$). The simplest model of a quantum dot would be a particle in a sphere model, assuming that the nanocrystal is a sphere. This does not make a difference in the energy level description mentioned above. The energy gap increases with a decrease in *a*. As a consequence of the CdSe nanocrystals emit light anywhere from 4500 to 6500 Å, so that any color from blue to red is achievable, depending on the size of the particle.

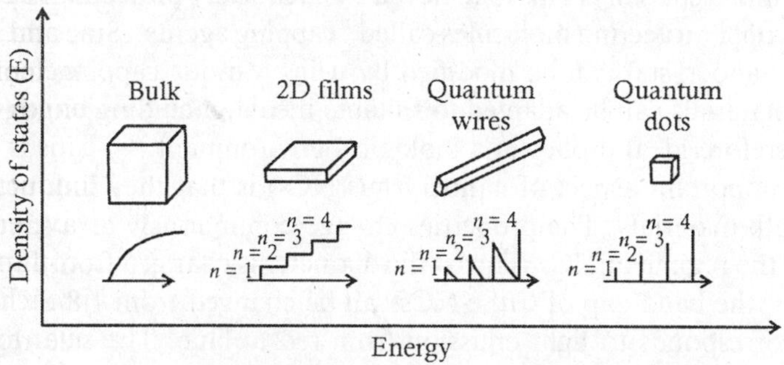

Fig. 12.33: Quantization of electronic density of states as a result of variation in the dimensionality materials

It is important to describe the issue of confinement again. These particles are called quantum dots as their electrons are confined to a point in space. They have no freedom in any dimension and electrons are said to be localized at a point, implying that a change in all directions changes the properties (in reality, a dot is a three-dimensional object comprising several hundreds or thousands of atoms, with finite shape). Compare this tiny unit of matter with semiconductor structures which are grown by thin film evaporation methods, which facilitate the creation of ultra thin films wlierrin the thickness is comparable to *a*, the diameter mentioned above. Such a material is called a 'quantum well", implying that the electron confined within a two-dimensional area which is said to be a 21 confinement. It is important to recall that the well itself is made by evaporation and the material of interest in confined inside another material of larger dielectric constant. Confinement along one direction results in a quantum wire, leaving the electron free to move only along one direction. This is known as a 1D confinement. In the quantum dot, there is no freedom along any direction.

An ideal quantum dot is realizable only when the electronic states within the dot face a discontinuity at the edge of the material. Due to this, the electron within the dot feels an insurmountable barrier at the edge. When a material is truncated at the surface, the surface atoms have unsatisfied valencies. In order to reduce the surface energy, the surface reconstructs, which leads to energy levels in the forbidden gap of the semiconductor. The electrical and optical properties of the material are degraded by these traps in an ideal semiconductor nanocrystal, the surface atoms are bonded to other materials in such a way as to remove the defects. This is what is done when a dot is covered with a material of larger band gap. In an ideal quantum dot, when there are no defect sites for charges to get trapped, the quantum yield of luminescence will be very high, nearly one. The emission will also be sharp. Better light emission occurs as the electrons and holes arc confined spatially. This is achieved by chemically protecting the surface with proper protecting molecules called "capping agents'. Since the surface of the nanocrystal can be modified by using various capping molecules, these materials can be adapted to suitable media, including biology. They arc therefore ideal probes in a biological environment.

An important aspect of nanocrystals(NCs) is that they link molecules and bulk materials. The properties change continuously as a function of size in the regime of NCs. Thus it the diameter is changed from 11.5 nm to 1.2 nm, the band gap of CdSe NCs can be changed from 1.8 eV to 3 eV. This corresponds to light emission from red to blue. The size regime of nanocrystals is difficult to pinpoint and depends on the material. In the case of most materials, it lies in the range of 100 to 1,00,000 atoms. The lowest size regime arises because at this point the structure changes to that of molecules (and consequently the properties). In the larger size regime, the system is nearly bulk-like as its energy level spacing is comparable to that of thermal energy.

Synthesis of Quantum Dots (QDs)

The kind of method used for the synthesis of quantum dots depends on the properties. Ideally nanocrystals should have the following properties:

1. Monodispersity
2. Possibility of future chemical derivatization.
3. High degree of crystallinity and specificity (avoiding polymorphic phase)
4. Chemical integrity
5. Lack of defects

There are several methods for the chemical synthesis of nanocrystals. All of them can be grouped under certain broad classes. It may be noted that there are also other approaches such as 'biological synthesis' which has been receiving some attention recently.

General Strategies: In the synthetic protocol, two kinds of general approaches are used, the *top-down approach* and the *bottom-up approach*. In the top-down method, the bulk material is brought into smaller dimension by various tools. Numerous methods are included in this category such as the various tools of patterning used to make structures in semiconductor electronics. These include chemical etching, optical lithography, use of particle beams (such as electron, ion and atom), etc. Apart from these, various methods are used to create ultra thin films of the material under investigation. These include thin film evaporation, molecular beam epitaxy, etc. However, there are other methods of creating powders starting from bulk materials. These techniques utilize various methods of milling and grinding, of which ball milling is the most commonly used one.

In the bottom-up approach, one can use gas phase or liquid state approaching. In the gas phase approach, the material to be synthesized is mixed in the atomic state in the gas phase itself. For this, the atoms of the materials are produced *in-situ* in an evaporation apparatus. This may also involve reacting atoms of one element with a gas phase species (such as oxygen). The prepared material in the gas phase is condensed to get the bulk material. At the stage of condensation, stabilizers may be added. Monodisperse metal particles are made in this way. Bulk powders of several oxides can be conveniently made by this route. This methodology may also be referred to as top-down as the synthesis starts from the bulk materials, which are subsequently evaporated.

The principal synthetic strategy used to make nanoparticles in a solution can be classified as "arrested precipitation'. Here at some stage of the growth of the particles, the surface is stabilized and further growth is arrested. This is commonly done by surfactants which bind to the surface of the growing nanocrystal. This synthetic approach, which is used for various kinds of materials is similar to that used for metal nanoparticlcs. Various kinds of surfactants, both ionic and covalent, have been used for this purpose. Arresting of the growth can be achieved in cavities such as those of micelles, zeolites, membranes, etc. Arresting can also be brought about by sudden variation in temperature and pressure conditions leadingto quenching.

All the nanoparticles formed are stabilized in the solution by the presence ot an electrical double layer when the protecting agents present are ionic in nature.This is typically seen in the case of metal nanoparticles such as gold when the reduction is done with sodium citrate. The surface of the nanoparticle is protected with anions and cations. In the specific case mentioned, the anions are citrate and chloride, while the cations are sodium and protons (as gold is completely reduced). The electrical potential created the double laver is large so that Coulomb repulsion prevents aggregation. On closer contact between the particles, the interaction potential rises sharply due to charge repulsion. However, there is Van der Waal attraction at larger distances and the net result is that there is a weak potential minimum at a moderate distance. This stabilizes the nanoparticle dispersion. These dispersions cannot be concentrated beyond point as the particles agglomerate. However, in dilute solutions, many colloidal solutions are stable for extended periods, such as in the case ot gold. If they are precipitated, metallic mirrors are obtained. In the case of semiconductor materials, the bulk material is formed. However, it is possible to change the surface cover of a covalent one and achieve subsequent purification. In the case of covalently bound ligands, there is no net charge and the particle behaves like a molecule. These materials can be precipitated, re-dispersed and stored tor extended periods.

Synthesis of Confined Media: In this approach, nanoparticlcs are synthesized in a space that is already available. The chemical reaction occurs inside a reactor, which is prepared by one of the several ways. Among the various confined media, reverse micelles, Langmuir-Blodgett tilnis, zeolites, porous membranes, clays, etc. are worth mentioning. They all have spaces of the order of nanometers, in which ions can be put. A proper stoicheometric mixture can be provided and the reaction conditions can be altered to produce the required nanocrystal. The chemistry in the solution phase can be conditioned to allow the usage of proper moleccules for surface passivation.

For example, one can conduct the reaction in reverie micelles (water in oil). Here oil is the majority phase while water is the minority phase, and nanoscopic containers are formed by micelles. One can add metal ions and organometallic reagents into the nanoreactors leading to the following reaction:

$$CdCl_2 + Se(SiMe_3)_2 \rightarrow CdSe + 2Me_3SiCl$$

The medium can have capping agents as a result of which the surface prepared is passivatcd. The oil/water ratio and the temperature conditions can be adjusted to obtain suitable particle dimensions. The materials can be separated from the solvent system and further annealed by heating in a higher boiling solvent to improve the crystallinity.

In confined cages such as those of zeolites, the maximum dimension of the particle formed is fixed. In a typical approach, the exchangeable ions such as Na^+ in the zeolite are exchanged with Cd^+ by washing the zeolite with a Cd^{2+} solution. The sample is then exposed to H_2S. Depending on the extent of Cd^{2+} present in the sample, various nanoparticle sizes are obtained. The same method can be used in membranes such as Nation® which have empty spaces of the order of nanometers.

Molecular Precursors: In the precursor route, nanocrystal seeds are parepared in a medium which can control the growth of the particles by co-ordinating with it. The solvent universally chosen for this appoach is trioctylphosphine oxide (TOPO), which has high thermal stability and can co-ordinate with inorganic surfaces. The cordinating TOPO can be exchanged with other ligands after isolation. The nanocrystals prepared this way can undergo Ostwald ripening (growth of larger particles at the expense of smaller ones) as a result of which monodisperse particles can be formed. A number of materials have been prepared by using the TOPO route. In a standard method, the metal ion precursors are added into hot TOPO while stirring continusouly (an injection of the ion precursors is made into a hot solution, as seen in Fig. 12.34). The difficulty of using pyrophoric compounds at elevated temperatures constitutes a risk and to avoid this, a single compound delivering both the constituent elements in tile semiconductor has been developed. Cadmium dithiocarbomates (e.g. $Cd(S_2CNEt_2)_2$) can produce CdS nanocrystals. A variation of chemicals can get other semiconductors. The synthetic protocol can be modified by changing the solvent (using a mixture of surfactants instead of TOPO), synthesis parameters such as volume and the number of injections, etc. These modifications have resulted in the formation of various morphologies, apart from the common spherical particles. It may be noted that even the spherical particles are faceted.

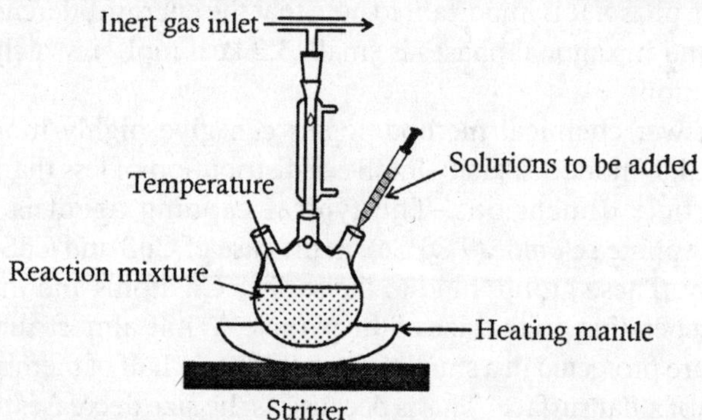

Fig. 12.34: Typical chemical synthesis approach for making nanoparticles, especially QDs. These are made at high temperatures, produced by a heating mantle. The process is carried out in an inert atmosphere. Solutions may be added simultaneously

Chemical Synthesis Using Clusters: Nanocrystals are large molecules. Therefore it is natural to think of building them starting from atoms. This atomic building approach has many advantages, one of which is the removal of all possible defects while the other is the possibility to have truly monodisperse materials. A larger cluster can be constructed from a smaller one. Using a starting cluster compound, $[(NMe_4)_4Cd_{10}S4(SC_6H_5)_{16}]$ the cluster $Cd_{32}S_{14}(SC_6H_5)_{24}\cdot 4DMF$ has been prepared. This is a cluster with a well-defined structure similar to the structure of bulk CdS. The absorption spectrum of the cluster is blue shifted in comparison to the bulk CdS (358 nm). More complex structures have been built by using this approach.

It is important to consider the crystalline state of the nanocrystal while deciding its properties. The effective masses of electrons and holes depend on the crystal structure of the material as a result of which all the optical, photophysical and photocaralytic properties will be affected by the structure. In the case of CdS and ZnS, there are two distinct crystalline forms, cubic (zincblende) and hexagonal (wurtzite) structures. CdS exists in the bulk in the wurtzite form while ZnS is find in the cubic form. When synthesized CdS mostly adopts the cubic form, which is metastable. However, hexagonal CdS can be synthesized in the nano form. By controlling the surface functionalization, ZnS has been made in different forms. When protected with compounds such as 1-hexanethiol, 1-decanethiol benzoic acid, etc. ZnS nanocrystals exist in the cubic form, while in the unprotected case, they exist in the hexagonal form. When the surface of the nanocrystal is not bound to the interacting molecules, one gets the metastable hexagonal phase. Thus it appears that surface stabilization leads to the formation of the cubic phase. Obviously, surface energy plays a significant role, as the reduction in surface energy by surface stabilization leads to the formation of the stable phase. It is important to note that the energy difference between the cubic and hexagonal phases is small (3.2 kcal mol^{-1}), which facilitates transformation.

Current wet chemical methodologies can give highly monodisperse particles. High monodispersity implies a distribution of less than 5 per cent in the particle dimensions. The typical capping agent is TOPO or trioctylphosphine selenide (TOPSe), in the case of CdS and CdSe particles, respectively. These groups bind to the surface Cd atoms and the coverage changes, depending on the size of the particle. While almost all the surface Cd atoms are protected in a smaller crystallite, only half of them are covered in the case of a flat surface. This is because as the size decreases, the surface area increases and molecules get adequate space to arrange themselves.

Modification of the Surface Nanocrystals: The surface modification of nanocrystals is important for the following reasons:

1. It removes surface states, thus making near band gap emission possible. Due to the removal of defects, the emission becomes narrow.
2. It adds chemical versatility to the system allowing it to become part of a larger structure. This is important in the use of such materials in polymers, inorganic material, etc. It is also important in making the system biocompatible.
3. Surface passivation and suitable functionalization make the system chemically inert and thermally stable. In fact various attributes can be added to the nanocrystal by appropriate modification. This includes chemical and biological compatibility, hydrophohicity, hydrophylicity, etc.

The most coniinon approach used is that of functionalizing the surface of the nanocrystal with various molecules, which can be thiols, amines, alkyl or silyl groups, etc. In all these cases, the chemical functionality of the molecule interacts with the metal atom at the surface of the nanocrystal. The Other approach involves the functionalization of the molecule further. For example, aminopropyltriethoxysilane coupling agent can be used to functionalize the surface of the nanocrystal. The surface will have ethoxy groups which can be hydrolyzed. Uy using a silica forming precursor, one can form a silica shell around the nanocrystal. The coating of CdS with a thin layer of HgS and further with CdS-yielding CdS/HgS/CdS structures, has also been reported. This structure would be called a 'quantum dot-quantum well' structure in the sense that a quantum dot is covered with a two-dimensional structure with a large dielectric barrier. This construction becomes possible as cubic CdS has a lattice constant similar to that of cubic HgS.

One of the important advantages of such a shell structure is the retardation of charge recombination in semiconductors upon photoexcitation. The electron that is created in the large band gap material upon photoexcitation can be stabilized in the lower lying conduction band of the other material. The direction of electron flow can be controlled in this fashion and a rectifying action has been observed.

InP Nanoparticles: As an example of the preparation of semiconductor nanocrystals, we will discuss the protocol used for InI quantum dots. This is called a group III-V semiconductor as In belongs to group III and phosphorus to group V. The indium precursor used in indium oxalate, indium chloride or indium fluoride. Trimethylsilylphosphine is used as the phosphorus precursor. A mixture of TOPO and trioctyl phosphine (TOP) is used as the colloidal stabilizer. The precursor species are mixed in an atomic ratio of 1 : 1 to make an InP precursor in the presence of the colloidal stabilizer. TOP is used in a ratio that is ten times larger than that of TOPO. The mixed solution which forms the transparent precursor is heated at 250-300° C for three days to obtain the colloidal solution. Depending on the size range desired and the extent of crystallinity, the dispersion can be heated to varying time/temperature. The decomposition temperature of the precursor is >200° C. The particle size can be modified by controlling the rate of decomposition. The material synthesized has a monolayer cover of TOPO/TOP. This makes the material disperse in a hexanebutabol mixture (0.9 : 0.1 by volume) containing 1 per cent TOPO. Precipitation of the nanocrystals can be achieved by adding methanol to the dispersion. This process is repeated. Dispersion and precipitation are carried out to remove unreacted materials or other products. TOPO is added as repeated washing can remove it from the surface of the nanocrystal and thereby affect colloidal stability. The protective cover can be replaced with thiols, amines, fatty acids, sulfonic acids, polymers, etc. It is also possible to bind the surface with proteins. Almost any functionality can be added by suitable chemistry. The material thus prepared can be precipitated, purified and redispersed as in the case of metal nanoparticles. Depending upon the functionality they will be dispersable in appropriate solvents: polar or non-polar. The polarity of the solution can be changed gradually. This approach helps precipitate particles gradually. The narrower the distribution of the parent material, the narrower are the precipitated particles as well.

The preparation method can get *n*-doped and *p*-doped particles. *N*-doping is achieved the addition of S. While *p*-doping is done by adding Zn. These are done by mixing suitable precursors to the InP precursor-forming solution. Sucb InP and InP : Zn nanoparticles are also stable.

Electronic Structure of Nanocrystals

The electronic structure of complex systems can be understood on the basis of simpler systems. For example, graphite can be understood starting trom an assembly of benzene fragments or diamond can be thought of as an assembly of tetrahedrally connected sp^3 carbon atoms. Benzene has discrete π and π^* states. As the number of rings increases, the highest occupied molecular orbital moves up in energy while the lowest unoccupied orbital moves down in energy. As a result, the energy of the transition decreases. This can be seen from ethylene to polyenes and from benzene to pentacene. The energy difference between π and π^* levels decreases and eventually it becomes smaller than thermal energy. At this stage, the levels are considered to be merged to form bands. In the case of an infinite ring containing solid, as in the case of a single sheet of graphite, the gap reduces to zero and the top of the valence band (the HOMO) touches the bottom of the conduction band (LUMO). This makes it a semimetal. Other similar situations may be envisaged in the kind of infinite chains such as polyenes and polyynes. Wlien the gap cannot reduce to zero, we obtain situations categorized as semiconductors or insulators. The reduction in the energy gap with an increase in the number of electrons is illustrated in Fig. 12.35.

This discussion obviously implies that a material will behave differently when the size regime is smaller than that required for bulk properties. Although this discussion was in the context of electronic properties, every other property will change as most of them are electronic in origin. This variation makes the question, 'At what size do properties become bulk-like', valid. This depends on the material of choice. In the case of elements such as carbon, with the smaller size regime containing a few tens to hundreds of carbon atoms, we have molecules called fullerenes. Their electronic structure is similar to that of the molecules which we are familiar with. Even in the regime of hundreds of atoms, their bonding and several other properties bear close similarity with those of bulk materials such as graphite. This is true of many elements. In the case of silicon, smaller clusters are entirely different and a bulk-like structure is exhibited only in the larger size regime of 10 atoms. Although the lattice constants may be similar to bulk, the electronic properties may not reach the bulk limit. In a lew cases, this bulk-co-molecular changeover has been investigated.

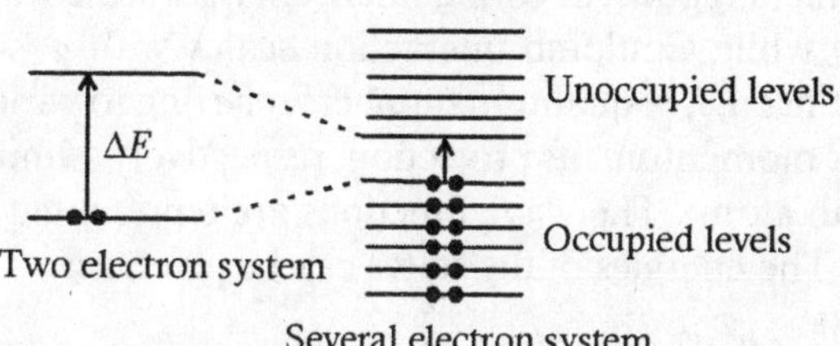

Fig. 12.35: The change in the electronic energy levels of the system when the number of structural units increases, as in the case of a change from one double bond to many double bonds. The energy gap between the levels, corresponding to the first excitation energy, ΔE decreases. When the number of double bonds increases significantly, the energy levels merge and gap becomes comparable to that of thermal energy.

The above discussion suggests that the bulk band gap increases with a decrease in the size of the material. Electronic transitions across the band gap result in the spectroscopic properties of the system. The variation in absorption edge can be seen as a function of the particle size. This aspect has been investigated in detail. In addition to the states which are created outside the band gap. In a real nanocrystal, there are also states within the band gap, which are created as a result ot defects. It is possible to calculate the diameter of the nanoparticle from the position of the absorption edge. The nanocrystal can be considered as a three-dimensional box and the energy gap can be related to the diameter of the particle or the width of the well.

While dealing with nanocrystals, one has to distinguish between three different kinds of size regimes. Various effects in these size regimes have been the subject of investigation, but we will not discuss this in great detail. The size regimes depend on the nanocrystal radius, r and the bulk exciton radius a. In the strong confinement regime, $r << a$, the Coulombic interaction between the electron and the hole is much smaller than the confinement energies, and therefore, electron and hole can be considered as separate particles. A weak confinement regime, $r >> a$ occurs when the electron and hole motions are strongly correlated, as electron-hole interactions are significant in comparison to the confinement energies. In the intermediate regime, $r \sim a$, the electronic structure depends strongly on both quantum confinement and Coulomb interaction.

The simplest model used to represent the energy plates of a nanocrystal is a spherical quantum well with an infinite potential barrier. Although the model is simple, if we include the Coulombic interaction between the electron and the hole, analytical solutions for the Schrodinger equation will not be possible. Disregarding the *e-h* interaction is possible in the strong confinement regime, as confinement energies scale with d^2 (as energy goes as n^2/d^2) while Coulomb interaction scales with d^{-1}. This result in states with distinct *n, l, m* quantum numbers referring to various symmetry, orbital angular momentum, its projection, respectively (similar to electrons in orbitals of an atom). The wave functions are represented as products of several terms. The energies of the states can be given as:

$$E_{l,n}^{e,h} = h^2 n^2 / 8\pi^2 m_{e,h} d^2$$

wher *n* is a quantum number. The exact nature of the wave function and the quantum number are not introduced here. The wave functions correspond to *S, P, D*, ..., etc. states depending on the orbital angular momentum *l*. There is one more quantum number, *m* which decides the degeneracy of the states. A pictorial representation of the energy states is given in Fig. 12.36. The energies are measured from the bottom of the conduction (valence) band for electrons (holes). The energy increases as one goes higher in the quantum number. Since the electron mass is much smaller than that of the hole ($m_h/m_e \sim 6$ in CdSe), the electron levels are separated more widely than the hole levels.

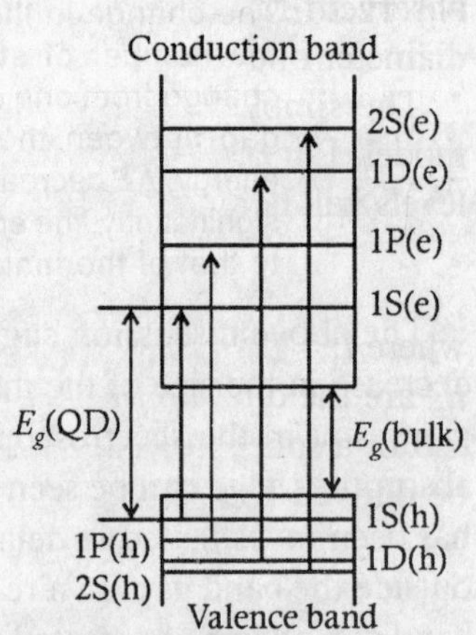

Fig. 12.36: The electronic states of a nanocrystal. The allowed optical transitions are shown

Electronic transitions are possible between various energy levels. However, the wave functions corresponding to different *n* and/or *l* are orthogonal and therefore it is not possible to observe al these transitions. Optical transitions between states of the same symmetry can, however, be observed. The intensity of the transition will be related to the degeneracy of the states in question. The transitions observed are far more complex than those described by the spherical quantum well model. The description given here is inadequate to describe the hole states. Spin-orbit and Coulomb *e-h* interactions have to be considered to improve the energy level picture.

One of the important aspects to be considered while interpreting experimental spectra is the size range of particles prepared in a typical synthesis. Spcctroscopic size selection can be done by using techniques such as fluorescence line narrowing (FLN), spectral hole burning and photoluminescence excitation (PLE). In these techniques, a narrow energy window is used for excitation (first two) or detection (last). This makes the technique sensitive only to a specific particle size. Size selection with the red region of the spectrum is preferable as it helps select the particles of the largest size in the ensemble.

Correlation of Properties with Size

The absorption band edge shifts to the blue as a result of a decrease in the size of the particle. The effective band gap measured, which explains this shift. The bulk bad gap is 2.42 eV (512 nm) and for all the particles, the measured band gap is higher than this. For several other materials, the band gap has been determined. In all the cases, systematic shifts have been observed. Models have been developed to correlate this shift with the diameter of the particles.

The simplest of these approaches is to consider the particle in a box model. Por one electron confined in a three-dimensional box, the energy levels will be

$$E_n = h^2 / 8m_e L^2 (n_x^2 + n_y^2 + n_z^2)$$

where $L = Lx = Ly = Lz$, the length of the box in three dimensions, n_x, n_y, n_z, are the quantum numbers and m_e is electron mass.

For a spherical semiconductor quantum dot, the energy expression can be derived by using the confined Wannier exciton Hamiltonian and we get

$$E_n = E_g + n^2 h^2 / 8m^* R^2 - e^2 / 4\pi\varepsilon_0 \varepsilon R$$

where

E_g = band gap of the bulk semiconductor.

n = quantum number (1, 2, 3, ...).

h = Planck's constant.

$m^* = (m_e \times m_h) / (m_e + m_h)$

m_e = effective mass of electron, m_h = effective mass of hole

R = radius of the quantum dots.

ε = dielectric constant of the semiconductor.

For the electronic transition from the valence band to the conduction band (n = 1), we can write (note that we have only one electron in the system)

$$E_g(R) = E_g + h^2 / 8R^2 (1/m_e + 1/m_h)$$

This neglects the third term.

The increase in the band gap can be given as

$$\Delta E_g(R) = E_g(R) - E_g = h^2 / 8R^2 (1/m_e + 1/m_h)$$

As shown here, there will be a $1/R^2$ dependence of the shift in the band gap.

Just like the absorption spectral shift, corresponding shifts are observed in emission too. In the case of CdS, the red emission seen is a result of the sulphur vacancies. The peak shifts to blue as a result of decrease in the particle size. The peak intensity increases and a blue shift occurs as the temperature is decreased. The emission occurs as the electrons are de-excited from the traps. This results in delayed emission.

Uses

Semiconductor nanocrystals find applications in a number of areas. A summary of the various applications is presented in Fig. 12.37. As can be seen, there are applications almost every area. Light can be absorbed by using nanoparticles, especially those with band gaps in the visible region. The absorbed light makes an electronic excitation possible in the material, leaving the electron free in the conduction band. These free electrons can move throughout the material and can be collected at an electrode. The electron may be put back into the material by a separate event, which creates a category of solar cells. By using efficient dye molecules which have LUMO levels placed near the conduction band of the semiconductor, it is possible to inject the cliarge from the excited dye molecule into the semiconductor. This has been an active area for some time and the device is referred to as dye sensitized solar cell.

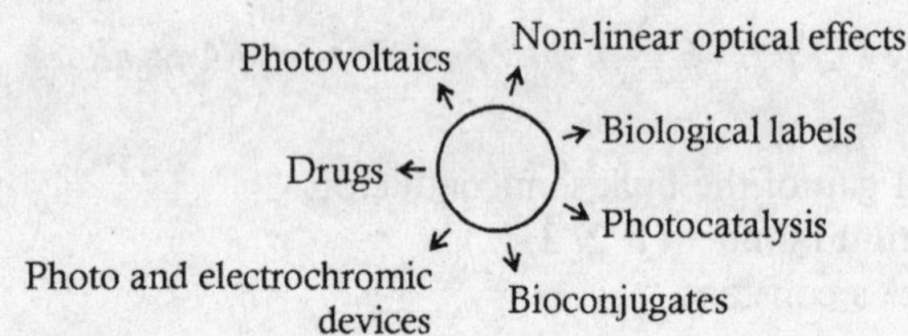

Fig. 12.37: Diverse applications of quantum dots

Chemical Properties: Electron injection into the conduction bund can lead to numerous possibilities. The surface of the particles will be covered with functional groups such as hydroxyls in the case of oxides. The hydroxyl groups take up the holes and form hydroxyl radicals, which can result in radical mediated oxidation. Pollutants in water and soil can be efficiently degraded in this way. Photocatalysis can also lead to reduction, using the electron in the conduction band which generates hydroxyl ions or oxide species. All these can make both oxidative and reductive processes feasible. Organic synthesis has utilized the power of semiconductor particles. To quote an example, aromatic ketones and olefins have been converted into alcohols and corresponding saturated compounds. The available electron can be used to fix gas phase CO_2 into organic compounds. In an attempt, CdS nanocrystals have been used to fix CO_2 into benzophenone, acetophenone and benzyl halides producing various compounds. Various other means of photocatalysis such as the reduction of nitrogen to ammonia, nitrogen oxides to nitrogen, decomposition of pollutants, etc. have also been attempted. In almost all these cases, TiO_2 is used as the photocatalyst. Tlie applications in this field are expanding enormously.

Single Electron Devices: Quantum dots can be used in single electron devices. In order to understand the elementary details of such a system, we consider the example of an electroneutral dot, so-called because the number of electrons that it has is the same as the number of holes. In order to put an electron inside, one has to apply a weak force. Tunneling is the most common means of putting a charge into a device of this kind as the device most often has an insulating barrier around it. The charge that the dot now possesses, $Q = -e$ produces an electric field, ε which repels any incoming negative charge. The fundamental charge is only 1.6×10^{-19} C, but the field it produces on the surface of a dot can be large, of the order of 140 kV on a 10 nm dimeter particle. A more important parameter in discussing single electron phenomenon has been recognized as the charging energy, $E_c = e^2/C$. At smaller dimensions, the electronic energy states in the material arc quantized. The energy required for the addition (of charge) is a sum of the charging energy and the electron kinetic energy in the material. In the larger size regime, of the order of 100 nm, the addition energy is dominated by the charging energy and the energy required for addition is of the order of *μcV*. Thus thermal energy is enough to do the job. Therefore, single electron effects will be manifested only at lower temperatures when thermal energy is small. However, in the lower size regime of the order of 10 nm, addition energy will be of the order of *meV* and will be dominated by electron kinetic energy or will become comparable to tile charging energy.

Single electron transition and corresponding effects are manifested dramatically in current-voltage measurements of nanocrystals. Imagine that a device structure is constructed in such a way that a single nanoparticle is trapped between an electron source and an electrode. As the electrode is separated at a large distance such that appreciable tunneling does not occur, a potential U is applied. If one measures the charge of the particle, Q as a function of the 'external charge', $Q = CU$, we get a step-like function. This is called the "Coulomb staircase'. Q is a step function of U, with the distance between the neighbouring steps = e. $\Delta Q = e$ or $\Delta U = e/C$. However, when thermal energy becomes comparable to the charging energy, the thermal fluctuations smear out the staircase.

Such a staircase implies that electrons can be transferred one at a time. However, the device by itself cannot be used for properties such as rectification or memory. These properties can be attained only by the use of more complicated devices such as single electron transistors, single electron traps, etc. The fundamental aspect of such device structures is the capability to charge or discharge nanosized regions selectively.

REVIEW QUESTIONS

1. What do you understand by metallic glasses ? Describe the melt spinning technique to prepare metallic glasses.
2. Mention some important properties and application of metglasses.
3. Explain how metallic glasses are obtained ? What is glass transition temperature (T_g)
4. What are different types of metglasses. Give examples of each.
5. What are the different techniques used for the preparation of metallic glasses ?
6. What are metallic glasses ? What is the advantages of metallic glasses as core in transformers ?
7. What are shape memory alloys ? Explain the temperature induced and stress induced transformations.
8. Explain the crystal structure of shape memory alloys.
9. Mention important applications of shape memory alloys in different fields.
10. Classify shape memory alloys.
11. What are shape memory alloys ? What causes the shape memory alloys to remember their shape ?
12. How does phase transformation takes place in shape memory alloys? Explain superelasticity or pseudoelasticity in SMAs.
13. What are ferromagnetic shape memory alloys ?
14. What are the advantages and disadvantages of shape memory alloys.
15. What are nonlinear materials? Mention their properties and applications.
16. What do you understand by passive and active nonlinear materials?
17. What are nanomaterials? Mention their properties.
18. What do you understand by metallic nanoclusters?
19. What are carbon nanotubes. Explain their electrical properties.
20. Write a note on semiconductor quantum dots.
21. Mention important properties of nanoparticles.
22. Can a quantum dot be made with a metallic element?
23. How is quantum confinement manifested in various measurements?
24. How would one make and stabilize a quantum dot?
25. What are the different types of quantum dots investigated?
26. What makes quantum dot luminescence attractive?
27. How do we correlate absorption spectra with size of the quantum dot?
28. What are the unique chemical properties of quantum dots? Give specific examples and illustrate how these are possible.

29. Why functionalize these particles ?

30. How do make biocompatible quantum dots?

SHORT ANSWERS QUESTION

1. How metallic glasses are obtained ?

Ans. These are obtained by cooling a molten metal so rapidly at a rate of about 2 × 10^{60} C/s. During this process of solidification, the atoms do not find enough time or energy to get themselves rearranged for crystal nucleation. This is why, the liquid upon attaining the glass transition temperature (T_g) solidifies as a metallic glass.

2. Define glass transition temperature (T_g) ?

Ans. This is the temperature at which, the liquid like atomic arrangement frozen into a solid.

3. How many types of metallic glasses are there ?

Ans. Two: (*i*) Metal glasses: Ni-Nb, Mg-Zn and Cu-Zr (ii) Metal-Metalloid glasses: $Fe_{67}\,Co_{18}\,B_{14}\,Si_1$; $Fe_{40}\,Ni_{38}\,Mo_4\,B_{18}$

4. How metallic glasses are advantageous as core in transfomers ?

Ans. We know that met glasses are ferromagnetic materials with low hysteresis loss, high permeability and high electrical resistivity. These properties helps for core loss to be less then in comparision to conventional transformers. In comparison to conventional transformer, the size and weight is also less.

5. What are metallic glasses ?

Ans. Metallic glasses are a new class of metallic materials which exhibit a disordered atomic structure and closely resemble that of glass rather than the conventional crystalline structure of metals. Unlike the covalent and/or ionic bonds of ceramic glasses, the bonding of metallic glasses is metallic in nature which confers on the material ductility and cohesiveness similar to crystalline metals. However, the glassy structure of metallic glass is highly metastable. These are produced when molten metals or alloys are cooled rapidly (at a cooling rate of 10^6-10^8 Ks^{-1}). This technique is known as *rapid solidification*.

6. What is the uniqueness about metallic glasses ?

Ans. Metallic glasses show very high strength levels (approaching that of dislocation free single crystals or hard drawn piano wire, for example, $Fe_{80}B_{20}$ metallic glass has a tensile yield strength of 3.63 GPa), good ductility in bending or compression (but not in tension), excellent corrosion resistance (due to absence of grain boundaries and formation of a resistant surface film), high electrical resistivity with a small variation with temperature, superconductivity and radiation stability. However, these attributes are not necessarily present simultaneously in a given metallic glass. Metallic glasses suffer from the major disadvantage of poor fracture and fatigue behaviour.

7. Based on the mechanical properties, suggest some potential uses of metallic glasses.

Ans. Based on mechanical properties the potential applications of metallic glasses include high strength control cables, pressure vessels, inertial energy storages devices such as flywheels, cutting tools, spring components, extrusion dies, casting dies, abrasive wheels, mechanical transmission belts, torque transmission tubes and reinforcing belts in rubber tyre. Metallic glass also finds use as reinforcement in concrete and refractories. Metallic glasses as brazing foil for diffusion bonding is another application with some commercial success.

8. Name some metallic glasses used for good mechanical properties.

Ans. $Fe_{80}B_{20}$, $Fe_{80}P_{15}C_5$, $Fe_{72}Cr_8P_{12}C_8$, $Fe_{75}B_{25}$, $Fe_{80}P_{16}C_3B_1$, $Fe_{76}P_{16}C_4Si_2A_{12}$, $Fe_{54}Cr_{16}Mo_{12}C_{18}$, $Cu_{57}Zr_{43}$, $Ni_{49}Fe_{29}P_{14}B_6Si_2$, $Ti_{60}Be_{35}Si_5$, and $Cu_{60}Ti_{40}$ are some examples of metallic glasses used for their mechanical properties.

9. Write some electrical/electronic properties of metallic glasses.

Ans. The electrical resistance of metallic glasses is high, consistent with structural disorder. In addition, they have has a much lower temperature coefficient of resistance, about an order of magnitude lower than the crystalline counterpart.

10. Based on electrical/electronic properties suggest some potential applications of metallic glasses.

Ans. Glassy alloys with a very small and often zero value of temperature coefficient of resistivity can be used as resistance standards and also as magnetoresistance sensors for detecting the presence of magnetic fields, notably in bubble memory devices. Other potential uses include heating elements (low temperature heating element instead of high temperature Nichrome) required for products such as TVs, radios, refrigerators, inductance thermometers, toasters; resistance thermometers, electron sources, electrical fuses, temperature alarms, batteries, sensors, superconductors, timing devices, delay lines, flash bulbs, etc.

11. Name some metallic glasses used for electrical applications.

Ans. The examples of metallic glasses used for electrical applications include $Fe_{78}B_{20}Mo_2$, $Ni_{36}Fe_{32}Cr_{14}P_{12}B_6$, $Pd_{40}Cu_{40}P_{20}$, $Pd_{80}Si_{20}$, $Pd_{40}Ni_{40}P_{20}$ and $Cu_{60}al_{40}$.

12. Give some examples of superconducting metallic glasses.

Ans. Some examples of superconducting metallic glasses are, $La_{80}Al_{20}$, $Ti_{45}Nb_{40}Si_{12}B_3$, $Nb_{58}Rh_{42}$, $Nb_{83}Si_{17}$, $Zr_{70}Be_{30}$, $Zr_{70}Co_{30}$, $Mo_{80}P_{10}B_{10}$, $Ti_{45}Nb_{40}Si_{15}$ and $Zr_{85}Si_{15}$.

13. Discuss the electrochemical properties of metallic glasses.

Ans. Many of the metallic glasses have good corrosion resistance since they are free of grain boundaries, precipitates, dislocations and chemical heterogeneities which act as preferred sites of attack by corrosives in crystalline materials. In fact it has been found that of all metallic glasses only some have good corrosion resistance. The best results have been shown by iron base glasses containing more than 8% chromium and 13% phosphorous (by atom %). It is believed that phosphorous increases the initial dissolution rate of the glassy alloy resulting in rapid enrichment of the chromium which forms a uniform layer of highly protective film.

14. Based on chemical properties suggest some potential uses of metallic glasses.

Ans. The major applications of metallic glasses as corrosion resistant materials include chemical filters for liquid sodium used in fast breeder reactors and for separation of iron oxide from kaolin clay, oil and water. The glasses used for these applications are Fe-Cr-P-C alloy and Ni-P alloy. The other potential uses include naval aircraft cables, torpedo tubes, suture clips, reaction vessels, long-life razor blades, electric razors, watch cases, oil filters, electrodes, and surgical scalpels.

15. Name a typical iron base metallic glass used for corrosion resistance applications and what is its popular name?

Ans. The typical iron base metallic glass used for corrosion resistance applications is $Fe_{70}Cr_{10}P_{13}C_7$. It is popularly known as *super stainless steel.*

16. Describe the magnetic properties of metallic glasses.

Ans. Among the ferromagnetic materials, ferrous magnetic glasses are the most easily magnetized materials. This is because of the absence of grain boundaries which otherwise act as obstacles for movement of domain walls. Also magnetocrystalline anisotropy is absent in metallic glasses. The high resistivity of metallic glass damps eddy currents. These glasses are characterized by very high permeability and low coercive force. Their magnetization curves are characterized by square shaped hysteresis loops.

17. Based on magnetic properties give some potential uses of metallic glasses.

Ans. On the basis of their magnetic properties, three major areas of potential applications emerge for metallic glasses. The foremost and by far the major application, is perhaps in power devices, particularly distribution transformers. The second area is in information handling equipment such as tape recorder heads, where the initial permeability is the most important characteristic requirement. The third area of application includes transducers and magnetic delay lines. The other applications include electronic devices, material for magnetic shielding, transformers, leakage current, overcurrent alarm, small power wattmeter, phase meters, speaker diaphragms, magnetic amplifiers, chokes, and electrical motors.

18. Give some of the examples of magnetic metallic glasses.

Ans. Examples of magnetic metallic glasses are $Fe_{80}B_{20}$, $Fe_{40}Ni_{40}P_{14}B_6$, $Fe_{67}Co_{18}B_{14}Si_1$, $Co_{70}Fe_5Si_{15}B_{10}$ and $Fe_{81}B_{13.5}Si_{3.5}C_2$

19. Name the famous industry that produced metallic glasses first on commercial scale and the trade name given to such commercial glass.

Ans. Allied Corporation of USA was the first to produce metallic glasses commercially. The trade name given to these glasses was METGLAS.

20. What were the dimensions of Metglas produced by Allied Corporation?

Ans. The Metglas produced at Allied Corporation was in the form of continuous ribbons with width 10 cm and thickness up to about 40 μm.

21. What are sharp memory alloys (SMAs) ?

Ans. These are metals which exhibit two very unique properties: Shape memory effect (SME) and Superelasticity (SE). Interestingly, SMAs are unique class of materials, which remember their shape even after severe deformation.

22. How many types of shape memory alloys are there?

Ans. Two types: (i) One way shape memory alloys and (ii) Two way shape memory alloys.

The materials which exhibit shape memory effect only upon heating are termed as one way shape memory alloys. However, those materials which exhibit both during heating and cooling are termed as two way shape memory alloys.

23. What do you understand by smart (intelligent) materials?

Ans. These materials are group of new and state of the art materials now being developed that will have a significant influence on many of our technologies. The adjective "Smart" implies that these materials are able to sense changes in their environments and then respond to these changes in predetermined manners—traits that are also found in living organism. In addition, this "Smart" concept is being extended to rather sophisticated system that consist of both smart and traditional materials.

Components of a smart material (or system) include some types of sensor (that detects an input signal), and an actuator that performs a responsive and adaptive function. Actuators may be called upon to change shape, position, natural frequency or mechanical characterstic in response to changes in temperature, electric fields, and/or magnetic fields.

Four types of materials are commonly used for actuators: Shape memory alloys, piezoelectric ceramic, magnetostrictive materials, and electrorheological/ magnetorheological fluids.

Materials/devices employed as sensors include optical fibers piezoelectric materials (including some polymers) and micro electromechanical devices (MEMS).

24. What are different classes of SMAs?

Ans. Broadly speaking there are three classes: (i) Thermal induced SMAs (ii) Stres induced SMAs and (iii) Ferromagnetic SMAs.

25. Mention the phases or crystal structure exhibited by SMAs at low and high temperatures.

Ans. At low temperature: Martensite phase at high temperature: Austenite phase.

26. Why SMAs remember their shape?

Ans. At the atomic level, the structural changes causes SMAs to remember their shape, *i.e.* the easy to and fro martensite transformation is found to be responsible for this.

27. How active materials are prepared?

Ans. A material might be filled with nano-scale sensors, computers, and actuators so the material can probe its environment, compute a response, and act.

28. What are nonlinear optical materials?

Ans. The integration of photonic and electronic devices is called optoelectronics. The materials that exhibit and that are used in phonic and optoelectronic devices are generally referred to as nonlinear optical materials (NLO)

29. What are nanomaterials?

Ans. These are the materials containing nanocrystals, *i.e.* their grain size is in the range 1 to 100 nm range. Nano materials may be metals, alloys, intermetallics and ceramics.

30. What is nano technology?

Ans. Study of the properties of nano particles is termed 'nanotechnology' the 'nano' prefix denotes that the dimensions of these structural entities are on the order of a nanometer (10^{-9}m). Example is carbon nanotube.

31. What is a carbon nanotube?

Ans. Structure of carbon nanotube consists of single sheet of graphite, rolled into a tube, both ends of which are capped with C_{60} fullerene hemispheres. Tube diameters are of the order of 100 nm or less. Each nanotube is a single molecule composed of millions of atoms; the length of this molecule is much greater (or the order of thousands times greater) than its diameter. Multi-walled carbon nanotubes consists of concentric cylinders.

32. What are the properties of carbon nanotubes?

Ans. Carbon nanotubes are extremely strong and stiff, and relatively ductile. For single-walled nanotubes tensile strength range between 50 and 200 GPa (approximately an order of magnitude greater than for carbon fibers); this is the strongest known material. Elastic modulus values are on the order of one tetrapascal [TPa (1 TPa = 10^3 GPa)], with fracture strains between about 5% and 20%. These tubes have relatively low densities. On the basis of these properties, the carbon nanotubes has been termed as the "ultimate fiber" and is extremely promising as a reinforcement in composite materials. Carbon nanotubes also have unique and structure-sensitive electrical properties.

33. What is 'Semiconductor quantum dot'?

Ans. Q Dots are nano meter-sized semiconductor crystals, or electrostatically confined electrons. These hold the promise of a true revolution in bio-imaging quality and sensitivity.

34. What is a nano sensor?

Ans. A chemical or physical sensor constructed using nano scale components, usually microscope or submicroscopic in size.

35. Why nano material are also called as superplastic material?

Ans. Due to smaller grain size the superplastic behaviour occurs at a lower temperature. This causes in an increase in the strain rate. Such an extensive or fracture is termed as superplastic state. Since the nano material micro structural features of superplastic materials and hence they are also called as super plastic materials.

36. Why nano materials exhibit altered physical and mechanical properties?

Ans. The grain size in nano materials is in the range of 1 to 100 nm. As the grain size decreases. There is a significant increase in the volume fraction (10-50%) at grain boundaries or interfaces. This characterstic of nanomaterials strongly alters the mechanical and physical properties.

OBJECTIVE QUESTIONS

1. The temperature at which the liquid like atomic arrangement can be frozen into a solid is called

(a) superconducting transition temperature
(b) Neel temperature
(c) Curie temperature
(d) Glass transition temperature.

2. The grain size of nanocrystals is

(a) 1 to 100 mm (b) 1 to 100 m
(c) 1 to 100 cm (d) 1 to 100 nm

3. Nano materials are also called as

(a) superplastic materials
(b) amorphous materials
(c) superconducting materials
(d) none of the above.

4. Shape memory effect is exhibited by

(a) shape memory alloys
(b) normal metals

(c) plasma
(d) none of the above.

5. Ferromagnetic shape memory alloys are............driven.

ANSWERS

1. d 2. d 3. a 4. a 5. Magnetically

Engineering Physics University Examination (December 2006)

Q. 1 Aluminium has thermal conductivity of 270 $Wm^{-1}K^{-1}$ at 293 K. Calculate its electrical resistivity at the same temperature.

Sol. Given, $K = 270\ Wm^{-1}K^{-1}$, $T = 293$ K, $\rho = ?$

we have $$\frac{K}{\sigma} = LT$$

or $$\sigma = \frac{K}{LT} \text{ where } L = 2.44 \times 10^{-8}\ W\Omega K^{-2}.$$

$$\therefore \quad \sigma = \frac{270}{2.44 \times 10^{-8} \times 293} = \mathbf{37.766 \times 10^6\ \Omega^{-1}m^{-1}}.$$

But, we have $$\rho = \frac{1}{\sigma} = \frac{1}{37.766 \times 10^6}$$
$$= \mathbf{2.64785 \times 10^{-8}\ \Omega m.}$$

Q. 2 Show that the probability that a state corresponding to $E_1 = (E_F + \delta E)$ occupied by an electron is equal to the probability that a state corresponding to $E_2 = (E_F - \delta E)$ is vacant.

or

Show that the probability that a state ΔE above the Fermi level E_F is filled equals the Probability that a state ΔE below is empty.

Sol. Given, $$E_1 = (E_F + \delta E)$$
$$E_2 = (E_F - \delta E)$$

Using the Fermi function, the probability that the energy level E_1 is occupied is given by,

$$f(E_1) = f(E_F + \delta E) = \frac{1}{1 + e^{\left(\frac{E_F + \delta E - E_F}{k_B T}\right)}}$$

or
$$f(E_1) = \frac{1}{1 + e^{\left(\frac{\delta E}{k_B T}\right)}} \qquad (1)$$

Similarly, the probability that the energy level E_2 is vacant (i.e. not occupied) is given by, $1 - f(E_2)$.

$$\therefore \quad 1 - f(E_2) = 1 - f(E_F - \delta E) = 1 - \frac{1}{1 + e^{\left(\frac{E_F - \delta E - E_F}{k_B T}\right)}}$$

or
$$1 - f(E_2) = 1 - \frac{1}{1 + e^{\left(\frac{-\delta E}{k_B T}\right)}} = \frac{1 + e^{\left(\frac{-\delta E}{k_B T}\right)} - 1}{1 + e^{\left(\frac{-\delta E}{k_B T}\right)}}$$

On rearranging the above equation, one obtains

$$1 - f(E_2) = 1 - \frac{1}{1 + e^{\left(\frac{-\delta E}{k_B T}\right)} \Big/ e^{\left(\frac{-\delta E}{k_B T}\right)}} = \frac{1}{\frac{1}{e^{(-\delta E / k_B T)}} + 1}$$

$$\therefore \quad 1 - f(E_2) = \frac{1}{e^{(\delta E / k_B T)} + 1} \qquad (2)$$

Therefore, from equation (1) and (2), one finds

$f(E_F + \delta E) = 1 - f(E_F - \delta E)$.

The above equation implies that the probability function is symmetrical about E_F at all temperatures other than $0\,K$.

Q. 3 The Fermi temperature of a metal is 24600 K. Calculate the Fermi velocity.

Sol. Given, $T_F = 24600$ K and $V_F = ?$

Formula $V_F = \sqrt{\frac{2E_F}{m}}$

But, $E_F = k_B T_F$

Therefore, $V_F = \sqrt{\frac{2k_B T_F}{m}} = \sqrt{\frac{2 \times 1.38 \times 10^{-23} \times 24600}{9.11 \times 10^{-31}}}$

$$= \sqrt{\frac{6.789 \times 10^{-19}}{9.11 \times 10^{-31}}} = \sqrt{7.452908 \times 10^{11}}$$

or $\quad$ V_F = **863302.3162 ms^{-1}** or **0.86330 × 10^{6} ms^{-1}**.

Q. 4 A silicon plate of thickness 1 mm, breadth 10 mm and length 100 mm is placed in a magnetic field of 0.5 Wb/m^2 acting perpendicular to its thickness. If 10^{-2} A current flow along its length, calculate the hall voltage developed if the hall coefficient is 3.66×10^{-4} m^3/C.

Sol. Given, $t = 1$ mm $= 1 \times 10^{-3}$ m, $b = 10$ mm $= 10 \times 10^{-3}$ m

$l = 100$ mm $= 100 \times 10^{-3}$ m, $B = 0.5$ Wb/m^2

$I = 10^{-2}$ A, $R_H = 3.66 \times 10^{-4}$ m^3/C, $V_H = ?$

we have $$V_H = \frac{R_H IB}{t} = \frac{3.66 \times 10^{-4} \times 0.5}{1 \times 10^{-3}}$$

$$= \mathbf{1.83 \times 10^{-3}\ V} \quad \text{or} \quad \mathbf{1.83\ mV.}$$

Q. 5 A glass of a dielectric constant 3.9 is subjected to a dc potential of 150 V. What is the polarization within the glass? (Thickness of glass is 1 mm).

Sol. Given, $\varepsilon_r = 3.9$, $V = 150$ V, Thickness (t) = 1 mm = 1×10^{-3} m, $P = ?$

we have $P = \varepsilon_0(\varepsilon_r - 1)E$, where $E = V/t$.

$\therefore$ $$E = \frac{150}{1 \times 10^{-3}} = \mathbf{150\ kV/m}$$

Hence, $P = 8.85 \times 10^{-12} \times (3.9 - 1) \times 150 \times 10^3$

$= 8.85 \times 10^{-12} \times 2.9 \times 150 \times 10^3$

$= \mathbf{3.84975 \times 10^{-6}\ Cm^{-2}.}$

B.E./Tech Degree Examination, May/June 2007
Engineering Physics
Annual Pattern

1. Define (*a*) Magnetostriction effect and (*b*) Piezo-electric effect.
2. Write note on SONAR.
3. What is known as Numerical Aperture?
4. Differentiate between active and passive fiber optic sensors.

   Hint:

   Active Sensor: It is an intrinsic sensor. The quantity to be measured acts directly on the fiber itself. Fiber itself acts as a transducer element and modifies the light. **Example**, liquid level sensor and Faraday sensor.

   Passive Sensor: It is an extrinsic sensor. The quantity to be measured acts directly on the transducing material which modifies the light reaching it through a fiber, i.e. in this sensor the fiber mere acts as a wave guide. **Example**, displacement, temperature sensors etc.
5. Define Wiedemann-Franz Law.
6. Outline the drawbacks of classical free electron theory.
7. Define Fermi energy.
8. The resistivity of an intrinsic semiconductor is 4.5 ohm-m at 20° C and 2.0 ohm-m at 32° C. Find the energy band gap in eV. (Baltzmann constant = 8.617×10^{-5} eV/K).

   Solution: $\rho_1 = 4.5\ \Omega\text{-m},\ T_1 = 20°\ C = 293$ k

   Given, $\rho_2 = 2.0\ \Omega\text{-m},\ T_2 = 32°\ C = 305$ k and $k_B = 8.617 \times 10^{-5}$ eV/K

   $E_g = ?$ (in eV)

   we have $\sigma = Ce^{\left(\frac{-E_g}{2k_BT}\right)}$

   Therefore, $\rho = \frac{1}{C}e^{\left(\frac{E_g}{2k_BT}\right)}$

   Hence, $\rho_1 = \frac{1}{C}e^{\left(\frac{E_g}{2k_BT_1}\right)}$ and $\rho_2 = \frac{1}{C}e^{\left(\frac{E_g}{2k_BT_2}\right)}$

   $$\therefore \frac{\rho_1}{\rho_2} = \frac{\frac{1}{C}e^{\left(\frac{E_g}{2k_BT_1}\right)}}{\frac{1}{C}e^{\left(\frac{E_g}{2k_BT_2}\right)}}$$

or $$\frac{\rho_1}{\rho_2} = e^{\left(\frac{E}{2k_B}\left(\frac{1}{T_1} - \frac{1}{T_2}\right)\right)} = e^{\left(\frac{E_g}{2k_B}\left(\frac{T_2 - T_1}{T_1 T_2}\right)\right)}$$

(Taking log on both sides and on rearranging, one obtains)

$$E_g = 2k_B \ln\left(\frac{\rho_1}{\rho_2}\right)\left(\frac{T_1 T_2}{T_2 - T_1}\right)$$

$$= 2 \times 8.617 \times 10^{-5} \times \ln\left(\frac{4.5}{2}\right) \times \frac{293 \times 305}{305 - 293}$$

$$= 2 \times 8.617 \times 10^{-5} \times \ln(2.25) \times 7447.083$$

Therefore, $\mathbf{E_g = 1.029}$ **eV.**

9. Write a note on nanomaterials.

10. Write a note on biomaterials.

11. (*a*) (*i*) Define reverberation time.

(*ii*) Derive Sabine's formula for the reverberation time.

(*b*) (*i*) Define Packing factor.

(*ii*) Deduce the c/a ratio and packing factor for HCP structure.

(*iii*) Calculate the inter-planar spacing for (101) and (221) planes in a simple cubic lattice whose lattice constant is 0.42 nm.

Solution: Given, $a = 0.42$ nm $= 0.42 \times 10^{-9}$ m, $(hkl) = (101)$ and (221)

$$d_{hkl} = ?$$

we have $$d_{hkl} = \frac{a}{\sqrt{h^2 + k^2 + l^2}}$$

Therefore, $$d_{(101)} = \frac{0.42 \times 10^{-9}}{\sqrt{1^2 + 0^2 + l^2}} = \frac{0.42 \times 10^{-9}}{\sqrt{2}} = 2.9698 \times 10^{-10} \text{ m.}$$

or $\mathbf{d_{(101)} = 2.9698}$ **Å.**

Similarly, $$d_{(221)} = \frac{0.42 \times 10^{-9}}{\sqrt{2^2 + 2^2 + l^2}} = \frac{0.42 \times 10^{-9}}{\sqrt{5}} = 1.8782 \times 10^{-10} \text{ m.}$$

or $\mathbf{d_{(221)} = 1.8782}$ **Å.**

12. (*a*)

(*i*) Explain how the wavelength of monochromatic source of light can be experimentally determined using Michelson's Interferometer.

(*ii*) Distinguish between (1) circularly polarised light and unpolarised and (2) elliptically polarised and partially polarised light.

Solution: (1) To distinguish between circularly polarised light and unpolarised light the given beam of light is first passed through a Q.W.P. and then examined through a Nicol prism and observed for the following variation in the intensity of the light.

(*a*) If the intensity of light varies with two maxima intensity the given light is circularly polarised.

(*b*) On the other hand, if there is no variation in intensity, the given beam of light is unpolarised.

(2) To distinguish between elliptically polarised and partially polarised light, the given beam of light is passed through a Q.W.P. and then through a rotating Nicol prism and observed for the following variation in the intensity of light.

(*a*) If the intensity varies with two maxima and two complete zero minima the given light is elliptically polarised.

(*b*) On the other hand, if the incident light is partially polarised, the intensity will vary between a maximum and a non-zero minimum during one complete rotation.

or

(*b*) (*i*) Explain the lasing schemes and working of Nd: YAG laser.

(*ii*) Explain the construction and working of a CO_2 laser.

13. (*a*)

(*i*) Define Compton effect.

(*ii*) Derive an expression for the wavelength of an *X*-ray photon. (Compton Shift.)

(*iii*) Find the change in wavelength of an *X*-ray photon when it is scattered through an angle of 135° by a free electron.

Solution: Given $\theta = 135°$, $h = 6.63 \times 10^{-34}$ J–s, $m_o = 9.1 \times 10^{-31}$ kg and $c = 3 \times 10^8$ m/s, $\Delta\lambda = ?$

we have $$\Delta\lambda = \frac{h}{m_o c}(1 - \cos\theta)$$

$$= \frac{6.63 \times 10^{-34}}{9.1 \times 10^{-31} \times 3 \times 10^8}(1 - \cos(135°))$$

$$= \frac{6.63 \times 10^{-34}}{9.1 \times 10^{-31} \times 3 \times 10^8}(1 - (0.70710))$$

$$= \frac{6.63 \times 10^{-34} \times 1.70710}{9.1 \times 10^{-31} \times 3 \times 10^8} = \frac{1.131811 \times 10^{-33}}{2.73 \times 10^{-22}}$$

Therefore, $\Delta\lambda = 4.14582 \times 10^{-12}$ m
or **0.041458 Å.**

or

(*b*) Derive an expression for the density of energy states and carrier concentration in a solid material (Metal) by using the Fermi distribution function.

14. (*a*)

(*i*) Define Hall Effect.

(*ii*) Explain in detail how Hall coefficient can be determined experimentally.

(*iii*) The donor density of an *n*-type germanium sample is $10^{21}/\text{m}^3$. The sample is arranged in a Hall experiment having magnetic field of 0.5 tesla and the current density is 500 Ampere/m^2. Find the Hall voltage if the sample is 3 mm wide.

Solution: Given, $n_e = 10^{21}/\text{m}^3$, $B = 0.5$ tesla, $J = 500$ A/m^2, $b = 3$ mm $= 3 \times 10^{-3}$ m, $V_H = ?$

Formula $$V_H = \frac{R_H IB}{t} \quad \left(\text{But, } R_H = \frac{1}{n_e e} \text{ and } I = \text{Jbt}\right)$$

Therefore, $$V_H = \frac{\text{JbtB}}{n_e \text{et}}$$

$$= \frac{\text{JbB}}{n_e e} = \frac{500 \times 3 \times 10^{-3} \times 0.5}{10^{21} \times 1.6 \times 10^{-19}} = 4.6875 \times 10^{-3} \text{ V}$$

or $V_H = 4.6875$ mV.

(*b*) Write a note on:

(*i*) Meissner Effect.

(*ii*) Type I and Type II Superconductors.

(*iii*) High T_c Superconductors.

(*iv*) Applications of Superconductors.

15. (*a*) Derive an expression for Internal field and deduce the Clausius Mosotti equation.

(*b*)

(*i*) Write a note on Liquid Penetrant method of NDT.

(*ii*) Explain in detail the Ultrasonic method of flaw detection with a suitable experimental system. (Block diagram.)

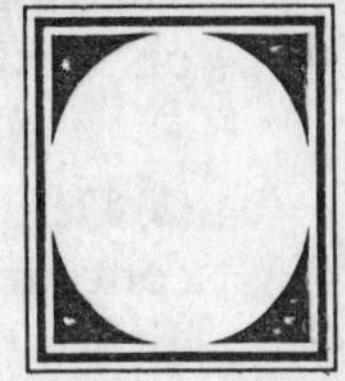

Suggested Readings

- Kakani, S. L., '*Mechanics*', Viva Book, New Delhi-2 (2005).
- Beiser, A., '*Modern Physics*', 6th ed., McGraw Hill (2003).
- Alonso and Finn, *University Physices*, 2nd ed.
- Kakani, S. L. and Amit Kakani, '*Material Science*,' New Age Int. Pubs., New Delhi-2 (2004).
- Kakani, S. L. and K. C. Bhandari, '*Electronics Theory and Applications*', 3rd ed., New Age (2004).
- Kakani, S. L. and H. M. Chandalia, '*Quantum Mechanics*', 4th ed., Sultan Chand, New Delhi-2 (2004).
- Kakani, S. L. and C. Hemrajani, '*Solid State Physics*', 4th ed., Sultan Chand (2005).
- Kakani, S. L. and K. C. Bhandari, '*Optics*', 2nd ed., Sultan Chand (2003).
- Kakani, S. L. and Shubhra Kakani, '*Modern Physics*', Viva Books (2006)
- Kakani, S. L. and C. Hemrajani, '*Electromagnetics*,' CBS Pubs., New Delhi-2 (2002).
- Kakani, S. L. and C. Hemrajani, '*Waves and Oscillations*', CBS Pubs. (2002).
- Krane, K. S., '*Modern Physics*,' John Wiley 2nd ed. (1998).
- Beiser, A., '*Perspectives of Modern Physics*,' McGraw Hill 5th ed. (1969).
- Kakani, S. L. and C. Hemrajani, '*Mathematical Physics*,' CBS Pubs. New Delhi-2 (2003).
- Bernstein *et al.*, '*Modern Physics*,' Pearson Education (2000).
- Kakani, S. L. and Shubhra Kakani, '*Superconductivity*', New Age Int. (2007).
- Kakani, S. L., '*Thermodynamics*', Sultan Chand, 2nd edn. (2007).
- Kakani, S. L. and Shubhra Kakani, '*Nuclear Physics*', Viva Books (In Press)

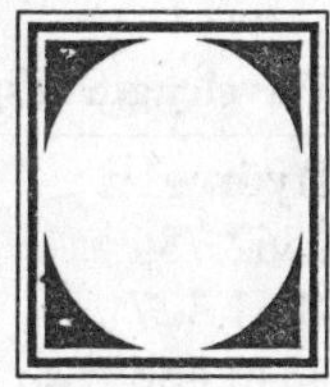

Appendices

APPENDIX 1

Physical Constants

Gravitational constant G	$= 6.67 \times 1000^{-11}$ newton-m^2/kg
Boltzmann constant (k_B)	$= 1.38 \times 10^{-23}$ Joule/K
Mass of H_2 atom (m_H)	$= 1.67399 \times 10^{-27}$ kg
Mass of proton (m_H)	$= 1.67399 \times 10^{-27}$ kg
Mass of electron (m_e)	$= 9.1083 \times 10^{-31}$ kg
Charge on electron (e)	$= 1.6 \times 10^{-31}$ Coulomb
Velocity of light in vacuum (c)	$= 3 \times 10^{8}$ m/s
Planck's constant (G)	$= 6.63 \times 10^{-34}$ J-s

APPENDIX 2

Referactive index of substances at 15 °C for D-line of sofium relative of air

Substance	μ	*Substance*	μ
Solids		**Liquid**	
Glass crown	1.50	Aniline	1.590
Glass flint	1.55	Benzene	1.504
Diamond	2.417	Chloroform	1.53
Mica	1.56-1.69	Glycerine	1.449
Sugar	1.56	Sulphuric acid	1.47
Quartz	1.544	Turpentine	1.43
		Water	1.333

APPENEDIX 3

Wavelength of spectral lines (Å)

Hydrogen	Mercury	Neon
6562.784	4047 V	5765 y
4861.327	4078 V	5853 y
4340.466	4358 V	5882 o
4101.736	4916 bg	6507
Sodium	4960 g	**Cadmium**
5890 D_1	5461 g	6438
5896 D_2	5770 y	5085
Helium	5791 y	4799
3889		4678
4026		4662
4471		
5876		

APPENDIX 4

Specific rotation

Optically Active Substance	*Solvent*	*Specific rotation*
Cane sugar	Water	+66.5° C
Glucose	Water	+52°
Fructose	Water	–91°
Camphor	Alcohol	+41°
Turpentine	Pure	–37°
Nicotine	Pure	–122°

APPENDIX 5

Specific resistance and temperature coefficient of resistance

Substance	*Composition*	*Sp. resistance ohm× cm × 10^{-6}*	*Temperature co-efficient per°C × 10^{-4}*
Constantan	60% Cu, 40% Zn	49.0	–0.4 to + 0.1
Silver	......	1.64	36.0
German Silver	62% Cu, 15% Ni, 22% Zn	26.6	2.3 to 6.0
Copper	......	1.78	42.8
Nichrome	80% Ni, 20% Cr	110.0	1.7
Mercury	......	99.8	9.0
Brass	70% Cu, 30% Zn	6.6	10.0
Platinum	......	11.0	37.0
Manganin	84% Cu, 4% Ni, 12% Zn	43.3	0.02 to 0.5

APPENDIX 6

Thermo E.M.F.

Thermocouple	*Thermo E.M.F.*
Copper-constantan	41.8 Microvolt/°C
Copper-Iron	8.6 Microvolt/°C
Antimony-Bismuth	113 Microvolt/°C

APPENDIX 7

Internal resistance and E.M.F. of cells

Cell	*Internal resistance (ohms)*	*E.M.F.(volts)*
Cadmium	900 Ω (very high)	1.0183 volt at 20° C
Daniel	3-4 Ω (fairly constant)	1.08
Leclanche	High, increases with usage	
Alkali accumulator	Low	1.46
Lead accumulator	Very low	2.1

Notes